Birth of Our Solar System

Our solar system was born from the gravitational collapse of an interstellar cloud of gas about $4\frac{1}{2}$ billion years ago, or about September 3 on the cosmic calendar. The Sun formed at the center of the cloud while the planets, including Earth, formed in a disk surrounding it.

Life on Earth

We do not know exactly when life arose on Earth, but fossil evidence indicates that it was within a few hundred million years after Earth's formation. Nearly three billion more years passed before complex plant and animal life evolved.

SEPTEMBER	OCTOBER	NOVEMBER	DECEMBER

SEP 3   **SEP 22**   **DEC 17**

This rock formation in West Greenland holds the oldest known evidence of life on Earth, dating to more than 3.85 billion years ago, or September 22 on the cosmic calendar.

Fossil evidence shows a remarkable increase in animal diversity beginning about 540 million years ago — December 17 on the cosmic calendar. We call this the Cambrian explosion.

Dinosaurs arose about 225 million years ago — December 26 on the cosmic calendar. Mammals arose around the same time.

...s what the solar system may have looked like shortly before ...finished forming.

Dinosaurs went extinct, probably due to an asteroid or comet impact, about 65 million years ago, which was only yesterday (December 30) on the cosmic calendar.

Human History

On the cosmic calendar, our hominid ancestors arose only a few hours ago, and all of recorded human history has occurred in just the last 15 seconds before midnight.

You

The average human life span is only about two-tenths of a second on the cosmic calendar.

DECEMBER 31

a.m.

12

p.m.

12

DEC
26

DEC
30

On the cosmic calendar, our ancestors began to master agriculture only 25 seconds ago ...

11:59:35 PM

Our early ancestors had smaller brains, but probably were walking upright by about 5 million years ago—December 31, 9 PM on the cosmic calendar.

Modern humans arose about 40,000 years ago, which is only about two minutes ago (December 31, 11:58 PM) on the cosmic calendar.

11:59:49 PM

...the Egyptians built the pyramids only 11 seconds ago ...

11:59:59 PM

...we learned that Earth is a planet orbiting the Sun only 1 second ago ...

11:59:59.95 PM

...and a typical college student was born only 0.05 second ago.

THE
COSMIC
PERSPECTIVE
THE SOLAR SYSTEM

Astronauts get a unique opportunity to experience a cosmic perspective. Here, astronaut John Grunsfeld has a CD of *The Cosmic Perspective* floating in front of him while orbiting Earth during the Space Shuttle's final servicing mission to the Hubble Space Telescope (May 2009).

THE

COSMIC
PERSPECTIVE
THE SOLAR SYSTEM

SEVENTH EDITION

JEFFREY BENNETT
University of Colorado at Boulder

MEGAN DONAHUE
Michigan State University

NICHOLAS SCHNEIDER
University of Colorado at Boulder

MARK VOIT
Michigan State University

PEARSON

Boston   Columbus   Indianapolis   New York   San Francisco   Upper Saddle River
Amsterdam   Cape Town   Dubai   London   Madrid   Milan   Munich   Paris   Montréal   Toronto
Delhi   Mexico City   São Paulo   Sydney   Hong Kong   Seoul   Singapore   Taipei   Tokyo

Publisher: James Smith
Executive Editor: Nancy Whilton
Director of Development: Laura Kenney
Development Editor: Jody Larson
Media Producer: Kate Brayton
Director of Marketing: Christy Lesko
Marketing Manager: Will Moore
Associate Director of Production: Erin Gregg
Managing Editor: Corinne Benson
Project Manager: Mary O'Connell
Production Management: Lifland et al., Bookmakers

Composition: Cenveo Publisher Services
Illustrations: Dartmouth Publishing, Inc.
Interior and Cover Design: Gary Hespenheide, Hespenheide
    Design
Design Manager: Mark Ong
Manufacturing Buyer: Jeffrey Sargent
Photo Research and Management: Maya Melenchuk
Permissions Specialist: Joseph Croscup
Printer and Binder: RR Donnelley, Willard
Cover Printer: Lehigh-Phoenix Color

Cover Images:
    Main Edition
        Petroglyphs/Comet Hale-Bopp, Photographer: Frank Zullo/Getty Images
    The Solar System
        Lunar Eclipse, Photographer: Sean Bagshaw/Getty Images
    Stars, Galaxies, and Cosmology
        Gemini Observatory and the Milky Way, Photographer: Philip Rosenberg/Getty Images

Library of Congress Cataloging-in-Publication Data

The cosmic perspective / Jeffrey Bennett ... [et al.]. — 7th ed.
    p. cm.
  Summary: "The story of modern astronomy and the new perspective that astronomy gives us of ourselves and our planet: the cosmic perspective"—Provided by publisher.
  Includes bibliographical references and index.
  ISBN-13: 978-0-321-83955-8 (student edition)
  ISBN-10: 0-321-83955-2 (student edition)
  ISBN-13: 978-0-321-84099-8 (professional copy)
  ISBN-10: 0-321-84099-2 (professional copy)
  [etc.]
 1. Astronomy—Textbooks. I. Bennett, Jeffrey O.
  QB43.3.C68 2014
  520—dc23
                        2012038541

ISBN-10-digit 0-321-83955-2; 13-digit 978-0-321-83955-8 (Student edition)
ISBN-10-digit 0-321-84099-2; 13-digit 978-0-321-84099-8 (Professional copy)
ISBN-10-digit 0-321-84106-9; 13-digit 978-0-321-84106-3 (The Solar System)
ISBN-10-digit 0-321-84107-7; 13-digit 978-0-321-84107-0 (Stars, Galaxies and Cosmology)

DEDICATION

To all who have ever wondered about the mysteries
of the universe. We hope this book will answer some
of your questions—and that it will also raise new
questions in your mind that will keep you curious
and interested in the ongoing human adventure
of astronomy. And, especially, to Michaela, Emily,
Sebastian, Grant, Nathan, Brooke, and Angela.
The study of the universe begins at birth, and we
hope that you will grow up in a world with far less
poverty, hatred, and war so that all people will have
the opportunity to contemplate the mysteries of the
universe into which they are born.

BRIEF CONTENTS

The chapters included in this volume are printed in bold type.

DETAILED CONTENTS

PART I
DEVELOPING PERSPECTIVE

PREFACE

We humans have gazed into the sky for countless generations. We have wondered how our lives are connected to the Sun, Moon, planets, and stars that adorn the heavens. Today, through the science of astronomy, we know that these connections go far deeper than our ancestors ever imagined. This book tells the story of modern astronomy and the new perspective, *The Cosmic Perspective*, that astronomy gives us of ourselves and our planet.

This book grew out of our experience teaching astronomy to both college students and the general public over the past 30 years. During this time, a flood of new discoveries fueled a revolution in our understanding of the cosmos but had little impact on the basic organization and approach of most astronomy textbooks. We felt the time had come to rethink how to organize and teach the major concepts in astronomy to reflect this revolution in scientific understanding. This book is the result.

Who Is This Book For?

The Cosmic Perspective is designed as a textbook for college courses in introductory astronomy, but is suitable for anyone who is curious about the universe. We assume no prior knowledge of astronomy or physics, and the book is especially suited to students who do not intend to major in mathematics or science.

The Cosmic Perspective provides a comprehensive survey of modern astronomy, and it contains enough material for a two-semester introductory astronomy sequence. It may also be used for one-semester survey courses if professors choose their areas of emphasis. However, instructors of one-term courses may also wish to consider our two shorter versions of this book: *The Essential Cosmic Perspective*, which covers a smaller set of topics and is tailored to meet the needs of comprehensive one-semester survey courses in astronomy, and *The Cosmic Perspective Fundamentals*, which covers only the most fundamental topics in astronomy and is designed for courses that address a more limited set of topics.

New to This Edition

The underlying philosophy, goals, and structure of *The Cosmic Perspective* remain the same as in past editions, but we have thoroughly updated the text and made a number of other improvements. Here, briefly, is a list of the significant changes you'll find in the seventh edition:

- **Major Chapter-Level Changes:** We have made numerous significant changes to both update science and improve the pedagogical flow in this edition. The full list is too long to put here, but major changes include the following:

- **Chapter 1** has been reorganized so that Section 1.1 now focuses on our cosmic address and the scale of space, while Section 1.2 focuses on our cosmic origins and the scale of time. We believe these changes will make it easier for students to learn key ideas that they'll need for the rest of their course in astronomy.

- In **Chapter 3**, we have significantly reorganized and rewritten Sections 3.1 and 3.2, with the goal of making it easier for students to focus on the key take-away ideas in each of these sections.

- In **Chapter 5**, we have consolidated the last two sections from the prior edition into a single Section 5.4, thereby making it easier for students to understand all the key ideas that we learn from spectra.

- **Chapter 8** has been significantly reorganized and rewritten so that the four major features of the solar system introduced in Chapter 7 are now all explained in a single section, making it much easier for students to draw connections between the features and how they are natural consequences of the nebular theory.

- **Chapter 13** has been completely rewritten to incorporate new developments in the study of extrasolar planets and planetary systems, including results from the *Kepler* mission. In the time since the last edition was published, the number of known extrasolar planets has more than quintupled, the number of known Earth-size planets has gone from zero to many, and our general picture of planetary systems has evolved to the point where we can now make strong statistical statements about planets throughout the universe.

- In **Chapter 20**, we have reorganized so that Hubble's law is introduced along with other distance measurement techniques in Section 20.2, and Section 20.3 now focuses on the age of the universe.

- **Chapters 22 and 23** have been thoroughly reorganized so that we now cover the Big Bang theory before we cover dark matter, dark energy, and the fate of the universe. This change was motivated by the spectacular progress in cosmic microwave background observations during the past decade and allows light-element abundances and microwave background observations to be presented as primary evidence that dark matter consists of exotic particles.

- **Mathematical Insight** boxes have been substantially reworked throughout the book. Although we have kept the same basic mathematical content and level, we have revised them to make the explanations shorter and simpler, and therefore more accessible to students.

- **Fully Updated Science:** Astronomy is a fast-moving field, and numerous new developments have occurred since the prior edition was published. In addition to the major chapter-level changes above, other scientific updates in this edition include

  - New results and images from spacecraft exploring our solar system, including *Phoenix*, the *Mars Reconnaissance Orbiter*, and *Curiosity* at Mars; *Venus Express* at Venus; *Cassini* at Saturn; *MESSENGER* at Mercury; *Stardust* and *Dawn* at Vesta; *SOHO* and *TRACE* at the Sun; and more
  - Recent results from major space observatories, including Hubble, Kepler, Spitzer, Chandra, and Fermi
  - Updated data and models on topics including global warming, cycling of gas in the Milky Way, and galaxy formation and evolution
  - New research on the timing and possible origin of life on Earth

- **New Group Work Questions:** The end-of-chapter exercise sets now include a subsection of problems that can be easily done in class to foster peer learning and in-class participation.

- **New Content in MasteringAstronomy®:** We have reached the point where *The Cosmic Perspective* is no longer just a textbook; rather it is a "learning package" that combines a printed book with deeply integrated, interactive media developed to support every chapter of our book. For students, the MasteringAstronomy Study Area provides a wealth of tutorials and activities to build understanding, while quizzes and exercises allow them to test what they've learned. For instructors, the MasteringAstronomy Item Library provides the unprecedented ability to quickly build, post, and automatically grade pre- and post-lecture diagnostic tests, weekly homework assignments, and exams of appropriate difficulty, duration, and content coverage. It also provides the ability to record detailed information on the step-by-step work of every student directly into a powerful and easy-to-use gradebook, and to evaluate results with a sophisticated suite of diagnostics. Among the changes you'll find to the MasteringAstronomy site for this edition are numerous new and revised interactive figures, including many narrated video tours; numerous new tutorials in the Item Library; and a fully updated set of reading, concept, and visual quizzes in both the Study Area and the Item Library.

Themes of *The Cosmic Perspective*

The Cosmic Perspective offers a broad survey of modern understanding of the cosmos and of how we have built that understanding. Such a survey can be presented in a number of different ways. We have chosen to interweave a few key themes throughout the book, each selected to help make the subject more appealing to students who may never have taken any formal science courses and who may begin the course with little understanding of how science works. We built our book around the following five key themes:

- *Theme 1: We are a part of the universe and thus can learn about our origins by studying the universe.* This is the overarching theme of *The Cosmic Perspective*, as we continually emphasize that learning about the universe helps us understand ourselves. Studying the intimate connections between human life and the cosmos gives students a reason to care about astronomy and also deepens their appreciation of the unique and fragile nature of our planet and its life.
- *Theme 2: The universe is comprehensible through scientific principles that anyone can understand.* The universe is comprehensible because the same physical laws appear to be at work in every aspect, on every scale, and in every age of the universe. Moreover, while professional scientists generally have discovered the laws, anyone can understand their fundamental features. Students can learn enough in one or two terms of astronomy to comprehend the basic reasons for many phenomena that they see around them—phenomena ranging from seasonal changes and phases of the Moon to the most esoteric astronomical images that appear in the news.
- *Theme 3: Science is not a body of facts but rather a process through which we seek to understand the world around us.* Many students assume that science is just a laundry list of facts. The long history of astronomy can show them that science is a process through which we learn about our universe—a process that is not always a straight line to the truth. That is why our ideas about the cosmos sometimes change as we learn more, as they did dramatically when we first recognized that Earth is a planet going around the Sun rather than the center of the universe. In this book, we continually emphasize the nature of science so that students can understand how and why modern theories have gained acceptance and why these theories may still change in the future.
- *Theme 4: A course in astronomy is the beginning of a lifelong learning experience.* Building upon the prior themes, we emphasize that what students learn in their astronomy course is not an end but a beginning. By remembering a few key physical principles and understanding the nature of science, students can follow astronomical developments for the rest of their lives. We therefore seek to motivate students enough that they will continue to participate in the ongoing human adventure of astronomical discovery.
- *Theme 5: Astronomy affects each of us personally with the new perspectives it offers.* We all conduct the daily business of our lives with reference to some "world view"—a set of personal beliefs about our place and purpose in the universe that we have developed through a combination of schooling, religious training, and personal thought. This world view shapes our beliefs and many of our actions. Although astronomy does not mandate a particular set of beliefs, it does provide perspectives on the architecture of the universe that can influence how we view ourselves and our world, and these perspectives can potentially affect our behavior. For example, someone who believes Earth to be at the center

of the universe might treat our planet quite differently from someone who views it as a tiny and fragile world in the vast cosmos. In many respects, the role of astronomy in shaping world views may be to represent the deepest connection between the universe and the everyday lives of humans.

Pedagogical Principles of *The Cosmic Perspective*

No matter how an astronomy course is taught, it is very important to present material according to a clear set of pedagogical principles. The following list briefly summarizes the major pedagogical principles that we apply throughout this book. (The *Instructor Guide* describes these principles in more detail.)

- *Stay focused on the big picture.* Astronomy is filled with interesting facts and details, but they are meaningless unless they fit into a big picture view of the universe. We therefore take care to stay focused on the big picture (essentially the themes discussed above) at all times. A major benefit of this approach is that although students may forget individual facts and details after the course is over, the big picture framework should stay with them for life.
- *Always provide context first.* We all learn new material more easily when we understand why we are learning it. In essence, this is simply the idea that it is easier to get somewhere when you know where you are going. We therefore begin the book (Chapter 1) with a broad overview of modern understanding of the cosmos, so that students can know what they will be studying in the rest of the book. We maintain this "context first" approach throughout the book by always telling students what they will be learning, and why, before diving into the details.
- *Make the material relevant.* It's human nature to be more interested in subjects that seem relevant to our lives. Fortunately, astronomy is filled with ideas that touch each of us personally. For example, the study of our solar system helps us better understand and appreciate our planet Earth, and the study of stars and galaxies helps us learn how we have come to exist. By emphasizing our personal connections to the cosmos, we make the material more meaningful, inspiring students to put in the effort necessary to learn it.
- *Emphasize conceptual understanding over "stamp collecting" of facts.* If we are not careful, astronomy can appear to be an overwhelming collection of facts that are easily forgotten when the course ends. We therefore emphasize a few key conceptual ideas that we use over and over again. For example, the laws of conservation of energy and conservation of angular momentum (introduced in Section 4.3) reappear throughout the book, and we find that the wide variety of features found on the terrestrial planets can be understood through just a few basic geological processes. Research shows that, long after the course is over, students are far more likely to retain such conceptual learning than individual facts or details.

- *Proceed from the more familiar and concrete to the less familiar and abstract.* It's well known that children learn best by starting with concrete ideas and then generalizing to abstractions later. The same is true for many adults. We therefore always try to "build bridges to the familiar"— that is, to begin with concrete or familiar ideas and then gradually draw more general principles from them.
- *Use plain language.* Surveys have found that the number of new terms in many introductory astronomy books is larger than the number of words taught in many first-year courses on a foreign language. In essence, this means the books are teaching astronomy in what looks to students like a foreign language! Clearly, it is much easier for students to understand key astronomical concepts if they are explained in plain English without resorting to unnecessary jargon. We have gone to great lengths to eliminate jargon as much as possible or, at minimum, to replace standard jargon with terms that are easier to remember in the context of the subject matter.
- *Recognize and address student misconceptions.* Students do not arrive as blank slates. Most students enter our courses not only lacking the knowledge we hope to teach but often holding misconceptions about astronomical ideas. Therefore, to teach correct ideas, we must also help students recognize the paradoxes in their prior misconceptions. We address this issue in a number of ways, the most obvious being the presence of many Common Misconceptions boxes. These summarize commonly held misconceptions and explain why they cannot be correct.

The Topical (Part) Structure of *The Cosmic Perspective*

The Cosmic Perspective is organized into seven broad topical areas (the seven Parts in the table of contents), each approached in a distinctive way designed to help maintain the focus on the themes discussed earlier. Here, we summarize the guiding philosophy through which we have approached each topic. Every Part concludes with one of our two-page Cosmic Context spreads, which tie together into a coherent whole the diverse ideas covered in the individual chapters.

Part I: Developing Perspective (Chapters 1–3, S1)

Guiding Philosophy: Introduce the big picture, the process of science, and the historical context of astronomy.

The basic goal of these chapters is to give students a big picture overview and context for the rest of the book, and to help them develop an appreciation for the process of science and how science has developed through history. Chapter 1 outlines our modern understanding of the cosmos, so that students gain perspective on the entire universe before diving into its details. Chapter 2 introduces basic sky phenomena, including seasons and phases of the Moon, and provides perspective on how phenomena we experience every day are tied to the broader cosmos. Chapter 3 discusses the nature of science, offering a historical perspective on the development of science and giving students perspective on how science works and how it differs from nonscience. The supplementary (optional) Chapter S1 goes into more detail about the sky, including celestial timekeeping and navigation.

The *Cosmic Context* figure for Part I appears on **pp. 108–109.**

Part II: Key Concepts for Astronomy (Chapters 4–6)

Guiding Philosophy: Connect the physics of the cosmos to everyday experiences.

These chapters lay the groundwork for understanding astronomy through what is sometimes called the "universality of physics"— the idea that a few key principles governing matter, energy, light, and motion explain both the phenomena of our daily lives and the mysteries of the cosmos. Each chapter begins with a section on science in everyday life in which we remind students how much they already know about scientific phenomena from their everyday experiences. We then build on this everyday knowledge to help students learn the formal principles of physics needed for the rest of their study of astronomy. Chapter 4 covers the laws of motion, the crucial conservation laws of angular momentum and energy, and the universal law of gravitation. Chapter 5 covers the nature of light and matter, the formation of spectra, and the Doppler effect. Chapter 6 covers telescopes and astronomical observing techniques.

The *Cosmic Context* figure for Part II appears on **pp. 188–189.**

Part III: Learning from Other Worlds (Chapters 7–13)

Guiding Philosophy: We learn about our own world and existence by studying about other planets in our solar system and beyond.

This set of chapters begins in Chapter 7 with a broad overview of the solar system, including a 10-page tour that highlights some of the most important and interesting features of the Sun and each of the planets in our solar system. In the remaining chapters of this Part, we seek to explain these features through a true *comparative planetology* approach, in which the discussion emphasizes the *processes* that shape the planets rather than the "stamp collecting" of facts about them. Chapter 8 uses the concrete features of the solar system presented in Chapter 7 to build student understanding of the current theory of solar system formation. Chapters 9 and 10 focus on the terrestrial planets, covering key ideas of geology and atmospheres, respectively. In both chapters, we start with examples from our own planet Earth to help students understand the types of features that are found throughout the terrestrial worlds and the fundamental processes that explain how these features came to be. We then complete each of these chapters by summarizing how the various processes have played out on each individual world. Chapter 11 covers the jovian planets and their moons and rings. Chapter 12 covers small bodies in the solar system, including asteroids, comets, and dwarf planets. It also covers cosmic collisions, including the impact linked to the extinction of the dinosaurs and a discussion of how seriously we should take the ongoing impact threat. Finally, Chapter 13 turns to the exciting topic of other planetary systems that have been discovered in recent years. Note that Part III is essentially independent of Parts IV through VII, and thus can be covered either before or after them.

The *Cosmic Context* figure for Part III appears on **pp. 396–397.**

Part IV: A Deeper Look at Nature (Chapters S2–S4)

Guiding Philosophy: Ideas of relativity and quantum mechanics are accessible to anyone.

Nearly all students have at least heard of things like the prohibition on faster-than-light travel, curvature of spacetime, and the uncertainty principle. But few (if any) students enter an introductory astronomy course with any idea of what these things mean, and they are naturally curious about them. Moreover, a basic understanding of the ideas of relativity and quantum mechanics makes it possible to gain a much deeper appreciation of many of the most important and interesting topics in modern astronomy including black holes, gravitational lensing, and the overall geometry of the universe. The three chapters of Part IV cover special relativity (Chapter S2), general relativity (Chapter S3), and key astronomical ideas of quantum mechanics (Chapter S4). The main thrust throughout is to demystify relativity and quantum mechanics by convincing students that they are capable of understanding the key ideas despite the reputation of these subjects for being hard or counterintuitive. ***These chapters are labeled "supplementary" because coverage of them is optional.*** Covering them will give your students a deeper understanding of the topics that follow on stars, galaxies, and cosmology, but the later chapters are self-contained so that they may be covered without having covered Part IV at all.

The *Cosmic Context* figure for Part IV appears on **pp. 462–463.**

Part V: Stars (Chapters 14–18)

Guiding Philosophy: We are intimately connected to the stars.

These are our chapters on stars and stellar life cycles. Chapter 14 covers the Sun in depth so that it can serve as a concrete model for building an understanding of other stars. Chapter 15 describes the general properties of other stars, how we measure these properties, and how we classify stars with the H-R diagram. Chapter 16 covers star birth, and the rest of stellar evolution is discussed in Chapter 17. Chapter 18 covers the end points of stellar evolution: white dwarfs, neutron stars, and black holes, along with a discussion of gamma-ray bursts.

The *Cosmic Context* figure for Part V appears on **pp. 576–577.**

Part VI: Galaxies and Beyond (Chapters 19–23)

Guiding Philosophy: Present galaxy evolution and cosmology together as intimately related topics.

These chapters cover galaxies and cosmology. Chapter 19 presents the Milky Way as a paradigm for galaxies in much the same way that Chapter 14 uses the Sun as a paradigm for stars.

Chapter 20 presents the properties of galaxies and shows how the quest to measure galactic distances led to Hubble's law and laid the foundation for modern cosmology. Chapter 21 discusses how the current state of knowledge regarding galaxy evolution has emerged from our ability to look back through time. Chapter 22 then presents the Big Bang theory and the evidence supporting it, setting the stage for Chapter 23, which explores dark matter and its role in galaxy formation, as well as dark energy and its implications for the fate of the universe.

The *Cosmic Context* figure for Part VI appears on pp. 694–695.

Part VII: Life on Earth and Beyond (Chapter 24)

Guiding Philosophy: The study of life on Earth helps us understand the search for life in the universe.

This Part consists of a single chapter. It may be considered optional, to be used as time allows. Those who wish to teach a more detailed course on astrobiology may wish to consider the text *Life in the Universe*, by Bennett and Shostak.

The *Cosmic Context* figure for Part VII appears on pp. 722–723.

Pedagogical Features of *The Cosmic Perspective*

Alongside the main narrative, *The Cosmic Perspective* includes a number of pedagogical devices designed to enhance student learning:

- **Learning Goals** Presented as key questions, motivational learning goals begin every chapter, and every section of every chapter is carefully written to address the specific learning goal in the title. This helps students stay focused on the big picture and stay motivated by the understanding they will gain.
- **Chapter Summary** The end-of-chapter summary offers a concise review of the learning goal questions, helping reinforce student understanding of key concepts from the chapter. Thumbnail figures are included to remind students of key illustrations and photos in the chapter.
- **Annotated Figures** Key figures in each chapter now include the research-proven technique of "annotation"— carefully crafted text placed on the figure (in blue) to guide students through interpreting graphs, following process figures, and translating between different representations.
- **Cosmic Context Two-Page Figures** These two-page spreads provide visual summaries of key processes and concepts.
- **Wavelength/Observatory Icons** For astronomical images, simple icons indicate whether the image is a photo, artist's impression, or computer simulation; whether a photo came from ground-based or space-based observations; and the wavelength band used to take the photo.
- **MasteringAstronomy® Resources** Specific resources from the MasteringAstronomy site, such as Interactive Figures or Photos and Self-Guided Tutorials, are referenced above specific section titles to direct students to help when they need it.

- **Think About It** This feature, which appears throughout the book as short questions integrated into the narrative, gives students the opportunity to reflect on important new concepts. It also serves as an excellent starting point for classroom discussions.
- **See It for Yourself** This feature also occurs throughout the book, integrated into the narrative, and gives students the opportunity to conduct simple observations or experiments that will help them understand key concepts.
- **Common Misconceptions** These boxes address popularly held but incorrect ideas related to the chapter material.
- **Special Topic Boxes** These boxes contain supplementary discussion topics related to the chapter material but not prerequisite to the continuing discussion.
- **Mathematical Insight Boxes** These boxes contain most of the mathematics used in the book and can be covered or skipped depending on the level of mathematics that you wish to include in your course. The Mathematical Insights use a three-step problem-solving strategy—Understand, Solve, and Explain—that gives students a consistent and explicit structure for solving quantitative homework problems.
- **The Big Picture** Every chapter narrative ends with this feature. It helps students put what they've learned in the chapter into the context of the overall goal of gaining a broader perspective on ourselves and our planet.
- **Visual Skills Check** This set of questions is designed to help students build their skills at interpreting the many types of visual information used in astronomy. Answers to these questions appear in Appendix J.
- **End-of-Chapter Exercises** Each chapter includes an extensive set of exercises that can be used for study, discussion, or assignment. All of the end-of-chapter exercises are organized into the following subsets:

  - Review Questions: Questions that students should be able to answer from the reading alone.
  - Does It Make Sense? (or similar title): A set of short statements for which students are expected to determine whether each statement makes sense, and to explain why or why not. These exercises are generally easy once students understand a particular concept, but very difficult otherwise; thus, they are an excellent probe of comprehension.
  - Quick Quiz: A short multiple-choice quiz that allows students to check their progress.
  - Process of Science Questions: Essay or discussion questions that help students focus on how science progresses over time.
  - NEW! Group Work Exercises: Questions designed for collaborative learning in class.
  - Short-Answer/Essay Questions: Questions that go beyond the Review Questions in asking for conceptual interpretation.
  - Quantitative Problems: Problems that require some mathematics, usually based on topics covered in the Mathematical Insight boxes.
  - Discussion Questions: Open-ended questions for class discussions.

- **Cross-References** When a concept is covered in greater detail elsewhere in the book, we include a boldface cross-reference in brackets to the relevant section (e.g., [**Section 5.2**]).
- **Glossary** A detailed glossary makes it easy for students to look up important terms.
- **Appendixes** The appendixes include a number of useful references and tables including key constants (Appendix A), key formulas (Appendix B), key mathematical skills (Appendix C), numerous data tables and star charts (Appendixes D–I), and answers to the Visual Skills Check questions (Appendix J).

MasteringAstronomy®—A New Paradigm in Astronomy Teaching

What is the single most important factor in student success in astronomy? Both research and common sense reveal the same answer: *study time*. No matter how good the teacher or how good the textbook, students learn only when they spend adequate time studying. Unfortunately, limitations on resources for grading have prevented most instructors from assigning much homework despite its obvious benefits to student learning. And limitations on help and office hours have made it difficult for students to make sure they use self-study time effectively. That, in a nutshell, is why we have created MasteringAstronomy. For students, it provides adaptive learning designed to coach them *individually*—responding to their errors with specific, targeted feedback and providing optional hints for those who need additional guidance. For professors, MasteringAstronomy provides unprecedented ability to automatically monitor and record students' step-by-step work and evaluate the effectiveness of assignments and exams. As a result, we believe that MasteringAstronomy can change the way astronomy courses are taught: It is now possible, even in large classes, to ensure that each student spends his or her study time on optimal learning activities outside of class.

MasteringAstronomy provides students with a wealth of self-study resources including interactive tutorials targeting the most difficult concepts of the course, interactive versions of key figures and photos, self-study quizzes, and other activities for self-assessment covering every chapter. For professors, MasteringAstronomy provides a library of tutoring activities that is periodically updated based on the performance of students nationwide. You can create assignments tailored to your specific class goals from among hundreds of activities and problems including pre- and post-lecture diagnostic quizzes, tutoring activities, end-of-chapter problems from this textbook, and test bank questions. Visit the MasteringAstronomy website to learn more.

Finally, in a world where everyone claims to have the best website, we'd like to point out three reasons why you'll discover that MasteringAstronomy really does stand out from the crowd:

- MasteringAstronomy specifically supports the structure and pedagogy of *The Cosmic Perspective.* You'll find the same concepts emphasized in the book and on the website, using the same terminology and the same pedagogical approaches. This type of consistency will ensure that students focus on the concepts, without the risk of becoming confused by different presentations.
- Nearly all MasteringAstronomy content has been developed either directly by *The Cosmic Perspective* author team or by this author team in close collaboration with outstanding educators including Jim Dove, Jim Cooney, Jonathan Williams, Richard Gelderman, Lauren Jones, Ed Prather, Tim Slater, and Daniel Loranz. The direct involvement of *The Cosmic Perspective* authors ensures that you can expect the same high level of quality in our website that you have come to expect in our textbook.
- The MasteringAstronomy platform uses the same unique student-driven engine as the highly successful MasteringPhysics® product (the most widely adopted physics homework and tutorial system), developed by a group led by MIT physicist David Pritchard. This robust platform gives instructors unprecedented power not only to tailor content to their own courses but also to evaluate the effectiveness of assignments and exams.

Additional Supplements for The Cosmic Perspective

The Cosmic Perspective is much more than just a textbook. It is a complete package of teaching, learning, and assessment resources designed to help both teachers and students. In addition to MasteringAstronomy (described above), the following supplements are available with this book:

- *SkyGazer 5.0* (Access code card ISBN 0-321-76518-4, CD ISBN 0-321-89843-5) Based on Voyager IV, one of the world's most popular planetarium programs, SkyGazer 5.0 makes it easy for students to learn constellations and explore the wonders of the sky through interactive exercises and demonstrations. Accompanying activities are available in *LoPresto's Astronomy Media Workbook,* seventh edition. Both SkyGazer and LoPresto's workbook are available for download. Ask your Pearson sales representative for details.
- *Starry Night™ College* (ISBN 0-321-71295-1) Now available as an additional option with *The Cosmic Perspective,* Starry Night™ College has been acclaimed as the world's most realistic desktop planetarium software. This special version has an easy-to-use point-and-click interface and is available as an additional bundle. The *Starry Night Activity Workbook,* consisting of thirty-five worksheets for homework or lab, based on Starry Night Planetarium software, is available for download in the MasteringAstronomy Study Area or with a Starry Night College access code. Ask your Pearson sales representative for details.
- *Astronomy Active Learning In-Class Tutorials* (ISBN 0-805-38296-8) by Marvin L. De Jong. This workbook provides fifty 20-minute in-class tutorial activities to choose from. Designed for use in large lecture classes, these activities are also suitable for labs. These short, structured activities are designed for students to complete on their own or in peer-learning groups. Each activity

targets specific learning objectives such as understanding Newton's laws, understanding Mars's retrograde motion, tracking stars on the H-R diagram, or comparing the properties of planets.

- **Lecture Tutorials for Introductory Astronomy** (0-321-82046-0), by Prather, Slater, Adams and Brissenden, includes forty-four lecture tutorials to engage students in critical reasoning and spark classroom discussion.
- **Sky and Telescope: Special Student Supplement** (0-321-70620-X), which includes nine articles with assessment insert covering general review, Process of Science, Scale of the Universe, and Our Place in the Universe, is available for bundling. Ask your Pearson representative for details.
- **Observation Exercises in Astronomy** (ISBN 0-321-63812-3) This workbook by Lauren Jones includes fifteen observation activities that can be used with a number of different planetarium software packages.

Instructor-Only Supplements

Several additional supplements are available for instructors only. Contact your local Pearson sales representative to find out more about the following supplements:

- **Instructor Resources** This comprehensive collection of instructor resources includes high-resolution JPEGs of all images from the book; Interactive Figures and Photos™ based on figures in the text; additional applets and animations to illustrate key concepts; PowerPoint® Lecture Outlines that incorporate figures, photos, checkpoint questions, and multimedia; and PRS-enabled clicker quizzes based on the book and book-specific interactive media, to make preparing for lectures quick and easy. These resources are located in MasteringAstronomy for easy use.
- **Clickers in the Astronomy Classroom** (ISBN 0-805-39616-0) This 100-page handbook by Douglas Duncan provides everything you need to know to successfully introduce or enhance your use of CRS (clicker) quizzing in your astronomy class—the research-proven benefits, common pitfalls to avoid, and a wealth of thought-provoking astronomy questions for every week of your course.
- **Instructor Guide** (ISBN 0-321-89659-9) The *Instructor Guide* contains a detailed overview of the text, sample syllabi for courses of different emphasis and duration, suggested teaching strategies, answers or discussion points for all Think About It and See It For Yourself questions in the text, solutions to all end-of-chapter problems, and a detailed reference guide summarizing media resources available for every chapter and section of the book.
- **Test Bank** (ISBN 0-321-84100-X) Available in both Word and TestGen formats on the Instructor Resource Center and MasteringAstronomy, the Test Bank contains a broad set of multiple-choice, true/false, and free-response questions for each chapter. The Test Bank is also assignable through MasteringAstronomy.

Acknowledgments

Our textbook carries only four author names, but in fact it is the result of hard work by a long list of committed individuals. We could not possibly list everyone who has helped, but we would like to call attention to a few people who have played particularly important roles. First, we thank our editors and friends at Pearson, who have stuck with us through thick and thin, including Adam Black, Nancy Whilton, Jim Smith, Michael Gillespie, Will Moore, Christy Lesko, Mary O'Connell, and Corinne Benson. Special thanks to our production teams, especially Sally Lifland; our art and design team, led by Mark Ong; and our Web team, led by Deb Perry, Kate Brayton, Wendy Romaniecki, Amir Said, and Caroline Power.

We've also been fortunate to have an outstanding group of reviewers, whose extensive comments and suggestions helped us shape the book. We thank all those who have reviewed drafts of the book in various stages, including

Marilyn Akins, *Broome Community College*
Christopher M. Anderson, *University of Wisconsin*
John Anderson, *University of North Florida*
Peter S. Anderson, *Oakland Community College*
Keith Ashman
Simon P. Balm, *Santa Monica College*
Nadine Barlow, *Northern Arizona University*
John Beaver, *University of Wisconsin at Fox Valley*
Peter A. Becker, *George Mason University*
Timothy C. Beers, *National Optical Astronomy Observatory*
Jim Bell, *Cornell University*
Priscilla J. Benson, *Wellesley College*
Bernard W. Bopp, *University of Toledo*
Sukanta Bose, *Washington State University*
David Brain, *University of Colorado*
David Branch, *University of Oklahoma*
John C. Brandt, *University of New Mexico*
James E. Brau, *University of Oregon*
Jean P. Brodie, *UCO/Lick Observatory, University of California, Santa Cruz*
Erik Brogt, *University of Canterbury*
James Brooks, *Florida State University*
Daniel Bruton, *Stephen F. Austin State University*
Amy Campbell, *Louisiana State University*
Eugene R. Capriotti, *Michigan State University*
Eric Carlson, *Wake Forest University*
David A. Cebula, *Pacific University*
Supriya Chakrabarti, *University of Massachusetts, Lowell*
Kwang-Ping Cheng, *California State University Fullerton*
Dipak Chowdhury, *Indiana University—Purdue University Fort Wayne*
Chris Churchill, *New Mexico State University*
Josh Colwell, *University of Central Florida*
Anita B. Corn, *Colorado School of Mines*
Philip E. Corn, *Red Rocks Community College*
Kelli Corrado, *Montgomery County Community College*
Peter Cottrell, *University of Canterbury*
John Cowan, *University of Oklahoma*
Kevin Crosby, *Carthage College*

Christopher Crow, *Indiana University—Purdue University Fort Wayne*
Manfred Cuntz, *University of Texas at Arlington*
Christopher De Vries, *California State University Stanislaus*
John M. Dickey, *University of Minnesota*
Bryan Dunne, *University of Illinois, Urbana-Champaign*
Suzan Edwards, *Smith College*
Robert Egler, *North Carolina State University at Raleigh*
Paul Eskridge, *Minnesota State University*
David Falk, *Los Angeles Valley College*
Timothy Farris, *Vanderbilt University*
Robert A. Fesen, *Dartmouth College*
Tom Fleming, *University of Arizona*
Douglas Franklin, *Western Illinois University*
Sidney Freudenstein, *Metropolitan State College of Denver*
Martin Gaskell, *University of Nebraska*
Richard Gelderman, *Western Kentucky University*
Harold A. Geller, *George Mason University*
Donna Gifford, *Pima Community College*
Mitch Gillam, *Marion L. Steele High School*
Bernard Gilroy, *The Hun School of Princeton*
Owen Gingerich, *Harvard–Smithsonian* (Historical Accuracy Reviewer)
David Graff, *U.S. Merchant Marine Academy*
Richard Gray, *Appalachian State University*
Kevin Grazier, *Jet Propulsion Laboratory*
Robert Greeney, *Holyoke Community College*
Henry Greenside, *Duke University*
Alan Greer, *Gonzaga University*
John Griffith, *Lin-Benton Community College*
David Griffiths, *Oregon State University*
David Grinspoon, *Denver Museum of Nature and Science*
John Gris, *University of Delaware*
Bruce Gronich, *University of Texas at El Paso*
Thomasana Hail, *Parkland University*
Jim Hamm, *Big Bend Community College*
Charles Hartley, *Hartwick College*
J. Hasbun, *University of West Georgia*
Joe Heafner, *Catawba Valley Community College*
Scott Hildreth, *Chabot College*
Mark Hollabaugh, *Normandale Community College*
Richard Holland, *Southern Illinois University, Carbondale*
Joseph Howard, *Salisbury University*
James Christopher Hunt, *Prince George's Community College*
Richard Ignace, *University of Wisconsin*
James Imamura, *University of Oregon*
Douglas R. Ingram, *Texas Christian University*
Assad Istephan, *Madonna University*
Bruce Jakosky, *University of Colorado*
Adam G. Jensen, *University of Colorado*
Adam Johnston, *Weber State University*
Lauren Jones, *Gettysburg College*
William Keel, *University of Alabama*
Julia Kennefick, *University of Arkansas*
Steve Kipp, *University of Minnesota, Mankato*
Kurtis Koll, *Cameron University*
Ichishiro Konno, *University of Texas at San Antonio*
John Kormendy, *University of Texas at Austin*

Eric Korpela, *University of California, Berkeley*
Kevin Krisciunas, *Texas A&M*
Ted La Rosa, *Kennesaw State University*
Kristine Larsen, *Central Connecticut State University*
Ana Marie Larson, *University of Washington*
Stephen Lattanzio, *Orange Coast College*
Larry Lebofsky, *University of Arizona*
Patrick Lestrade, *Mississippi State University*
Nancy Levenson, *University of Kentucky*
David M. Lind, *Florida State University*
Abraham Loeb, *Harvard University*
Michael LoPresto, *Henry Ford Community College*
William R. Luebke, *Modesto Junior College*
Darrell Jack MacConnell, *Community College of Baltimore City*
Marie Machacek, *Massachusetts Institute of Technology*
Loris Magnani, *University of Georgia*
Steven Majewski, *University of Virginia*
Phil Matheson, *Salt Lake Community College*
John Mattox, *Fayetteville State University*
Marles McCurdy, *Tarrant County College*
Stacy McGaugh, *Case Western University*
Barry Metz, *Delaware County Community College*
William Millar, *Grand Rapids Community College*
Dinah Moche, *Queensborough Community College of City University, New York*
Stephen Murray, *University of California, Santa Cruz*
Zdzislaw E. Musielak, *University of Texas at Arlington*
Charles Nelson, *Drake University*
Gerald H. Newsom, *Ohio State University*
Brian Oetiker, *Sam Houston State University*
John P. Oliver, *University of Florida*
Stacy Palen, *Weber State University*
Russell L. Palma, *Sam Houston State University*
Bryan Penprase, *Pomona College*
Eric S. Perlman, *Florida Institute of Technology*
Peggy Perozzo, *Mary Baldwin College*
Charles Peterson, *University of Missouri, Columbia*
Cynthia W. Peterson, *University of Connecticut*
Jorge Piekarewicz, *Florida State University*
Lawrence Pinsky, *University of Houston*
Stephanie Plante, *Grossmont College*
Jascha Polet, *California State Polytechnic University, Pomona*
Matthew Price, *Oregon State University*
Harrison B. Prosper, *Florida State University*
Monica Ramirez, *Aims College, Colorado*
Christina Reeves-Shull, *Richland College*
Todd M. Rigg, *City College of San Francisco*
Elizabeth Roettger, *DePaul University*
Roy Rubins, *University of Texas at Arlington*
Carl Rutledge, *East Central University*
Bob Sackett, *Saddleback College*
Rex Saffer, *Villanova University*
John Safko, *University of South Carolina*
James A. Scarborough, *Delta State University*
Britt Scharringhausen, *Ithaca College*
Ann Schmiedekamp, *Pennsylvania State University, Abington*

Joslyn Schoemer, *Denver Museum of Nature and Science*
James Schombert, *University of Oregon*
Gregory Seab, *University of New Orleans*
Larry Sessions, *Metropolitan State College of Denver*
Anwar Shiekh, *Colorado Mesa University*
Ralph Siegel, *Montgomery College, Germantown Campus*
Philip I. Siemens, *Oregon State University*
Caroline Simpson, *Florida International University*
Paul Sipiera, *William Harper Rainey College*
Earl F. Skelton, *George Washington University*
Evan Skillman, *University of Minnesota*
Michael Skrutskie, *University of Virginia*
Mark H. Slovak, *Louisiana State University*
Norma Small-Warren, *Howard University*
Dale Smith, *Bowling Green State University*
James R. Sowell, *Georgia Technical University*
Kelli Spangler, *Montgomery County Community College*
John Spencer, *Lowell Observatory*
Darryl Stanford, *City College of San Francisco*
George R. Stanley, *San Antonio College*
John Stolar, *West Chester University*
Jack Sulentic, *University of Alabama*
C. Sean Sutton, *Mount Holyoke College*
Beverley A. P. Taylor, *Miami University*
Brett Taylor, *Radford University*
Donald M. Terndrup, *Ohio State University*
Frank Timmes, *Arizona State University*
David Trott, *Metro State College*
David Vakil, *El Camino College*
Trina Van Ausdal, *Salt Lake Community College*
Licia Verde, *Institute of Cosmological Studies, Barcelona*
Nicole Vogt, *New Mexico State University*
Darryl Walke, *Rariton Valley Community College*
Fred Walter, *State University of New York, Stony Brook*
James Webb, *Florida International University*
Mark Whittle, *University of Virginia*
Paul J. Wiita, *The College of New Jersey*
Lisa M. Will, *Mesa Community College*
Jonathan Williams, *University of Hawaii*
J. Wayne Wooten, *Pensacola Junior College*
Scott Yager, *Brevard College*
Andrew Young, *Casper College*
Arthur Young, *San Diego State University*
Min S. Yun, *University of Massachusetts, Amherst*
Dennis Zaritsky, *University of Arizona*
Robert L. Zimmerman, *University of Oregon*

In addition, we thank the following colleagues who helped us clarify technical points or checked the accuracy of technical discussions in the book:

Nahum Arav, *Virginia Technical University*
Phil Armitage, *University of Colorado*
Thomas Ayres, *University of Colorado*
Cecilia Barnbaum, *Valdosta State University*
Rick Binzel, *Massachusetts Institute of Technology*
Howard Bond, *Space Telescope Science Institute*
David Brain, *University of Colorado*
Humberto Campins, *University of Central Florida*
Robin Canup, *Southwest Research Institute*

Clark Chapman, *Southwest Research Institute*
Kelly Cline, *Carroll College*
Josh Colwell, *University of Central Florida*
Mark Dickinson, *National Optical Astronomy Observatory*
Jim Dove, *Metropolitan State College of Denver*
Dan Fabrycky, *University of Chicago*
Harry Ferguson, *Space Telescope Science Institute*
Andrew Hamilton, *University of Colorado*
Todd Henry, *Georgia State University*
Dennis Hibbert, *Everett Community College*
Dave Jewitt, *University of California, Los Angeles*
Julia Kregenow, *Penn State University*
Emily Lakdawalla, *The Planetary Society*
Hal Levison, *Southwest Research Institute*
Mario Livio, *Space Telescope Science Institute*
J. McKim Malville, *University of Colorado*
Geoff Marcy, *University of California Berkeley and San Francisco State University*
Mark Marley, *Ames Research Center*
Linda Martel, *University of Hawaii*
Kevin McLin, *University of Colorado*
Michael Mendillo, *Boston University*
Steve Mojzsis, *University of Colorado*
Francis Nimmo, *University of California, Santa Cruz*
Tyler Nordgren, *University of Redlands*
Rachel Osten, *Space Telescope Science Institute*
Bob Pappalardo, *Jet Propulsion Laboratory*
Bennett Seidenstein, *Arundel High School*
Michael Shara, *American Museum of Natural History*
Brad Snowder, *Western Washington University*
Bob Stein, *Michigan State University*
Glen Stewart, *University of Colorado*
John Stolar, *West Chester University*
Jeff Taylor, *University of Hawaii*
Dave Tholen, *University of Hawaii*
Nick Thomas, *University of Bern*
Dimitri Veras, *Cambridge University*
John Weiss, *Carleton College*
Francis Wilkin, *Union College*
Jeremy Wood, *Hazard Community College*
Iason Wright, *Penn State University*
Don Yeomans, *Jet Propulsion Laboratory*

Finally, we thank the many people who have greatly influenced our outlook on education and our perspective on the universe over the years, including Tom Ayres, Fran Bagenal, Forrest Boley, Robert A. Brown, George Dulk, Erica Ellingson, Katy Garmany, Jeff Goldstein, David Grinspoon, Robin Heyden, Don Hunten, Geoffrey Marcy, Joan Marsh, Catherine McCord, Dick McCray, Dee Mook, Cherilynn Morrow, Charlie Pellerin, Carl Sagan, Mike Shull, John Spencer, and John Stocke.

Jeff Bennett
Megan Donahue
Nick Schneider
Mark Voit

ABOUT THE AUTHORS

Jeffrey Bennett

Jeffrey Bennett holds a B.A. (1981) in biophysics from the University of California, San Diego, and an M.S. and Ph.D. (1987) in astrophysics from the University of Colorado, Boulder. He has taught at every level from preschool through graduate school, including more than 50 college classes in astronomy, physics, mathematics, and education. He served 2 years as a visiting senior scientist at NASA headquarters, where he created NASA's "IDEAS" program, started a program to fly teachers aboard NASA's airborne observatories (including *SOFIA*), and worked on numerous educational programs for the Hubble Space Telescope and other space science missions. He also proposed the idea for and helped develop both the Colorado Scale Model Solar System on the CU-Boulder campus and the *Voyage* Scale Model Solar System on the National Mall in Washington, D.C. (He is pictured here with the model Sun.) In addition to this astronomy textbook, he is also lead author of college-level textbooks in astrobiology, mathematics, and statistics (all from Pearson); of critically acclaimed books for the general public including *Beyond UFOs* (Princeton University Press, 2008/2011) and *Math for Life* (Roberts & Co., 2012); and of the award-winning children's books *Max Goes to the Moon, Max Goes to Mars, Max Goes to Jupiter*, and *The Wizard Who Saved the World*. When not working, he enjoys participating in masters swimming and in the daily adventures of life with his wife, Lisa; his children, Grant and Brooke; and his dog, Cosmo. His personal website is www.jeffreybennett.com.

Megan Donahue

Megan Donahue is a professor in the Department of Physics and Astronomy at Michigan State University and a Fellow of the American Association for the Advancement of Science. Her current research is mainly about using X-ray, UV, infrared, and visible light to study clusters of galaxies: their contents—dark matter, hot gas, galaxies, active galactic nuclei—and what they reveal about the contents of the universe and how galaxies form and evolve. She grew up on a farm in Nebraska and received an S.B. in physics from MIT, where she began her research career as an X-ray astronomer. She has a Ph.D. in astrophysics from the University of Colorado. Her Ph.D. thesis on theory and optical observations of intergalactic and intracluster gas won the 1993 Trumpler Award from the Astronomical Society for the Pacific for an outstanding astrophysics doctoral dissertation in North America. She continued postdoctoral research as a Carnegie Fellow at Carnegie Observatories in Pasadena, California, and later as an STScI Institute Fellow at Space Telescope. Megan was a staff astronomer at the Space Telescope Science Institute until 2003, when she joined the MSU faculty. Megan is married to Mark Voit, and they collaborate on many projects, including this textbook and the raising of their children, Michaela, Sebastian, and Angela. Between the births of Sebastian and Angela, Megan qualified for and ran the Boston Marathon. These days, Megan runs trails, orienteers, and plays piano and bass guitar whenever her children allow it.

Nicholas Schneider

Nicholas Schneider is an associate professor in the Department of Astrophysical and Planetary Sciences at the University of Colorado and a researcher in the Laboratory for Atmospheric and Space Physics. He received his B.A. in physics and astronomy from Dartmouth College in 1979 and his Ph.D. in planetary science from the University of Arizona in 1988. In 1991, he received the National Science Foundation's Presidential Young Investigator Award. His research interests include planetary atmospheres and planetary astronomy. One research focus is the odd case of Jupiter's moon Io. Another is the mystery of Mars's lost atmosphere, which he hopes to answer by serving as science lead on the Imaging UV Spectrograph on NASA's *MAVEN* mission. Nick enjoys teaching at all levels and is active in efforts to improve undergraduate astronomy education. In 2010 he received the Boulder Faculty Assembly's Teaching Excellence Award. Off the job, Nick enjoys exploring the outdoors with his family and figuring out how things work.

Mark Voit

Mark Voit is a professor in the Department of Physics and Astronomy at Michigan State University and a Fellow of the American Association for the Advancement of Science. He earned his A.B. in astrophysical sciences at Princeton University and his Ph.D. in astrophysics at the University of Colorado in 1990. He continued his studies at the California Institute of Technology, where he was a research fellow in theoretical astrophysics, and then moved on to Johns Hopkins University as a Hubble Fellow. Before going to Michigan State, Mark worked in the Office of Public Outreach at the Space Telescope, where he developed museum exhibitions about the Hubble Space Telescope and helped design NASA's award-winning HubbleSite. His research interests range from interstellar processes in our own galaxy to the clustering of galaxies in the early universe. He is married to coauthor Megan Donahue, and cooks terrific meals for her and their three children. Mark likes getting outdoors whenever possible and particularly enjoys running, mountain biking, canoeing, orienteering, and adventure racing. He is also author of the popular book *Hubble Space Telescope: New Views of the Universe*.

HOW TO SUCCEED
IN YOUR ASTRONOMY COURSE

If Your Course Is	Time for Reading the Assigned Text (per week)	Time for Homework or Self-Study (per week)	Time for Review and Test Preparation (average per week)	Total Study Time (per week)
3 credits	2 to 4 hours	2 to 3 hours	2 hours	6 to 9 hours
4 credits	3 to 5 hours	2 to 4 hours	3 hours	8 to 12 hours
5 credits	3 to 5 hours	3 to 6 hours	4 hours	10 to 15 hours

The Key to Success: Study Time

The single most important key to success in any college course is to spend enough time studying. A general rule of thumb for college classes is that you should expect to study about 2 to 3 hours per week *outside* of class for each unit of credit. For example, based on this rule of thumb, a student taking 15 credit hours should expect to spend 30 to 45 hours each week studying outside of class. Combined with time in class, this works out to a total of 45 to 60 hours spent on academic work—not much more than the time a typical job requires, and you get to choose your own hours. Of course, if you are working while you attend school, you will need to budget your time carefully.

As a rough guideline, your studying time in astronomy might be divided as shown in the table above. If you find that you are spending fewer hours than these guidelines suggest, you can probably improve your grade by studying longer. If you are spending more hours than these guidelines suggest, you may be studying inefficiently; in that case, you should talk to your instructor about how to study more effectively.

Using This Book

Each chapter in this book is designed to make it easy for you to study effectively and efficiently. To get the most out of each chapter, you might wish to use the following study plan.

■ A textbook is not a novel, and you'll learn best by reading the elements of this text in the following order:

1. Start by reading the Learning Goals and the introductory paragraphs at the beginning of the chapter so that you'll know what you are trying to learn.
2. Get an overview of key concepts by studying the illustrations and reading their captions and annotations. The illustrations highlight almost all of the major concepts, so this "illustrations first" strategy gives you an opportunity to survey the concepts before you read about them in depth. You will find the two-page Cosmic Context figures especially useful. Also watch for the Interactive Figure icons—when you see one, go to the MasteringAstronomy® website to try the interactive version.

3. Read the chapter narrative, trying the Think About It questions and the See It for Yourself activities as you go along, but save the boxed features (Common Misconceptions, Special Topics, Mathematical Insights) to read later. As you read, make notes on the pages to remind yourself of ideas you'll want to review later. Avoid using a highlight pen (or a highlighting tool if you are using an e-book), which makes it too easy to highlight mindlessly. For a printed book, underlining with pen or pencil is far more effective, because it forces you to take greater care and therefore helps keep you alert as you study; be careful to underline selectively—it won't help you later if you've underlined everything. For an e-book, write notes to remind yourself why you marked a block of text as particularly important.
4. After reading the chapter once, go back through and read the boxed material. You should read all of the Common Misconceptions and Special Topics boxes; whether you choose to read the Mathematical Insights is up to you and your instructor. Also watch for the MasteringAstronomy tutorial icons throughout the chapter; if a concept is giving you trouble, go to the MasteringAstronomy site to try the relevant tutorial.
5. Finally, turn your attention to the Chapter Summary. The best way to use the summary is to try to answer the Learning Goal questions for yourself before reading the short answers given in the summary.

■ After completing the reading as outlined above, test your understanding with the end-of-chapter exercises. A good way to begin is to make sure you can answer all of the Review Questions; if you don't know an answer, look back through the chapter until you figure it out. Then try the Does It Make Sense? and Quick Quiz questions.

■ You can further check your understanding and get feedback on difficulties by trying the online quizzes in the Study Area at the MasteringAstronomy site. Each chapter has three quizzes: a Reading Quiz, a Concept Quiz, and a Visual Quiz. Try the Reading Quiz first. Once you clear up any difficulties you have with it, try the Concept and Visual quizzes.

- If your course has a quantitative emphasis, work through all of the examples in the Mathematical Insights before trying the quantitative problems for yourself. Remember that you should always try to answer questions qualitatively before you begin plugging numbers into a calculator. For example, make an order-of-magnitude estimate of what your answer should be so that you'll know your calculation is on the right track, and be sure that your answer makes sense and has the appropriate units.
- If you have done all the above, you will have already made use of numerous resources on the MasteringAstronomy site. Don't stop there; visit the site again and make use of other resources that will help you further build your understanding. These resources have been developed specifically to help you learn the most important ideas in your astronomy course, and they have been extensively tested to make sure they are effective. They really do work, and the only way you'll gain their benefits is by going to the website and using them.

General Strategies for Studying

- Budget your time effectively. Studying 1 or 2 hours each day is more effective, and far less painful, than studying all night before homework is due or before exams.
- Engage your brain. Learning is an active process, not a passive experience. Whether you are reading, listening to a lecture, or working on assignments, always make sure that your mind is actively engaged. If you find your mind drifting or find yourself falling asleep, make a conscious effort to revive yourself, or take a break if necessary.
- Don't miss class. Listening to lectures and participating in discussions is much more effective than reading someone else's notes. Active participation will help you retain what you are learning. Also, be sure to complete any assigned reading *before* the class in which it will be discussed. This is crucial, since class lectures and discussions are designed to help reinforce key ideas from the reading.
- Take advantage of resources offered by your professor, whether it be e-mail, office hours, review sessions, online chats, or other opportunities to talk to and get to know your professor. Most professors will go out of their way to help you learn in any way that they can.
- Start your homework early. The more time you allow yourself, the easier it is to get help if you need it. If a concept gives you trouble, do additional reading or studying beyond what has been assigned. And if you still have trouble, ask for help: You surely can find friends, peers, or teachers who will be glad to help you learn.
- Working together with friends can be valuable in helping you understand difficult concepts. However, be sure that you learn *with* your friends and do not become dependent on them.
- Don't try to multitask. A large body of research shows that human beings simply are not good at multitasking: When we attempt it, we do more poorly at all of the individual tasks. And in case you think you are an exception, the same research found that those people who believed they were best at multitasking were actually the worst! So when it is time to study, turn off your electronic devices, find a quiet spot, and give your work a focused effort at concentration.

Preparing for Exams

- Study the Review Questions, and rework problems and other assignments; try additional questions to be sure you understand the concepts. Study your performance on assignments, quizzes, or exams from earlier in the term.
- Work through the relevant online tutorials and chapter quizzes available at the MasteringAstronomy site.
- Study your notes from lectures and discussions. Pay attention to what your instructor expects you to know for an exam.
- Reread the relevant sections in the textbook, paying special attention to notes you have made on the pages.
- Study individually *before* joining a study group with friends. Study groups are effective only if every individual comes prepared to contribute.
- Don't stay up too late before an exam. Don't eat a big meal within an hour of the exam (thinking is more difficult when blood is being diverted to the digestive system).
- Try to relax before and during the exam. If you have studied effectively, you are capable of doing well. Staying relaxed will help you think clearly.

Presenting Homework and Writing Assignments

All work that you turn in should be of *collegiate quality*: neat and easy to read, well organized, and demonstrating mastery of the subject matter. Future employers and teachers will expect this quality of work. Moreover, although submitting homework of collegiate quality requires "extra" effort, it serves two important purposes directly related to learning:

1. The effort you expend in clearly explaining your work solidifies your learning. In particular, research has shown that writing and speaking trigger different areas of your brain. Writing something down—even when you think you already understand it—reinforces your learning by involving other areas of your brain.
2. If you make your work clear and self-contained (that is, make it a document that you can read without referring to the questions in the text), you will have a much more useful study guide when you review for a quiz or exam.

The following guidelines will help ensure that your assignments meet the standards of collegiate quality:

- Always use proper grammar, proper sentence and paragraph structure, and proper spelling. Do not use texting shorthand.
- Make all answers and other writing fully self-contained. A good test is to imagine that a friend will be reading

your work and to ask yourself whether the friend will understand exactly what you are trying to say. It is also helpful to read your work out loud to yourself, making sure that it sounds clear and coherent.

- In problems that require calculation:
  - Be sure to *show your work* clearly so that both you and your instructor can follow the process you used to obtain an answer. Also, use standard mathematical symbols, rather than "calculator-ese." For example, show multiplication with the × symbol (not with an asterisk), and write 10^5, not 10^5 or 10E5.
  - *Check that word problems have word answers.* That is, after you have completed any necessary calculations, make sure that any problem stated in words is answered with one or more *complete sentences* that describe the point of the problem and the meaning of your solution.

- Express your word answers in a way that would be *meaningful* to most people. For example, most people would find it more meaningful if you expressed a result of 720 hours as 1 month. Similarly, if a precise calculation yields an answer of 9,745,600 years, it may be more meaningfully expressed in words as "nearly 10 million years."

- Include illustrations whenever they help explain your answer, and make sure your illustrations are neat and clear. For example, if you graph by hand, use a ruler to make straight lines. If you use software to make illustrations, be careful not to make them overly cluttered with unnecessary features.

- If you study with friends, be sure that you turn in your own work stated in your own words—you should avoid anything that might give even the *appearance* of possible academic dishonesty.

FOREWORD
THE MEANING OF
THE COSMIC PERSPECTIVE

by Neil deGrasse Tyson

*Astrophysicist Neil deGrasse Tyson is the Frederick P. Rose Director of New York City's Hayden Planetarium at the American Museum of Natural History. He has written numerous books and articles, hosts the PBS series NOVA scienceNOW, and was named one of the "Time 100"—*Time *Magazine's list of the 100 most influential people in the world. He contributed this essay*

© Neil deGrasse Tyson

about the meaning of "The Cosmic Perspective," abridged from his 100th essay written for Natural History *magazine.*

Of all the sciences cultivated by mankind, Astronomy is acknowledged to be, and undoubtedly is, the most sublime, the most interesting, and the most useful. For, by knowledge derived from this science, not only the bulk of the Earth is discovered . . .; but our very faculties are enlarged with the grandeur of the ideas it conveys, our minds exalted above [their] low contracted prejudices.

—*James Ferguson,* Astronomy Explained Upon Sir Isaac Newton's Principles, and Made Easy To Those Who Have Not Studied Mathematics (1757)

Long before anyone knew that the universe had a beginning, before we knew that the nearest large galaxy lies two and a half million light-years from Earth, before we knew how stars work or whether atoms exist, James Ferguson's enthusiastic introduction to his favorite science rang true.

But who gets to think that way? Who gets to celebrate this cosmic view of life? Not the migrant farm worker. Not the sweatshop worker. Certainly not the homeless person rummaging through the trash for food. You need the luxury of time not spent on mere survival. You need to live in a nation whose government values the search to understand humanity's place in the universe. You need a society in which intellectual pursuit can take you to the frontiers of discovery, and in which news of your discoveries can be routinely disseminated.

When I pause and reflect on our expanding universe, with its galaxies hurtling away from one another, embedded with the ever-stretching, four-dimensional fabric of space and time, sometimes I forget that uncounted people walk this Earth without food or shelter, and that children are disproportionately represented among them.

When I pore over the data that establish the mysterious presence of dark matter and dark energy throughout the universe, sometimes I forget that every day—every twenty-four-hour rotation of Earth—people are killing and being killed. In the name of someone's ideology.

When I track the orbits of asteroids, comets, and planets, each one a pirouetting dancer in a cosmic ballet choreographed by the forces of gravity, sometimes I forget that too many people act in wanton disregard for the delicate interplay of Earth's atmosphere, oceans, and land, with consequences that our children and our children's children will witness and pay for with their health and well-being.

And sometimes I forget that powerful people rarely do all they can to help those who cannot help themselves.

I occasionally forget those things because, however big the world is—in our hearts, our minds, and our outsize atlases—the universe is even bigger. A depressing thought to some, but a liberating thought to me.

Consider an adult who tends to the traumas of a child: a broken toy, a scraped knee, a schoolyard bully. Adults know that kids have no clue what constitutes a genuine problem, because inexperience greatly limits their childhood perspective.

As grown-ups, dare we admit to ourselves that we, too, have a collective immaturity of view? Dare we admit that our thoughts and behaviors spring from a belief that the world revolves around us? Part the curtains of society's racial, ethnic, religious, national, and cultural conflicts, and you find the human ego turning the knobs and pulling the levers.

Now imagine a world in which everyone, but especially people with power and influence, holds an expanded view of our place in the cosmos. With that perspective, our problems would shrink—or never arise at all—and we could celebrate our earthly differences while shunning the behavior of our predecessors who slaughtered each other because of them.

■■■

Back in February 2000, the newly rebuilt Hayden Planetarium featured a space show called "Passport to the Universe," which took visitors on a virtual zoom from New York City to the edge of the cosmos. En route the audience saw Earth, then the solar system, then the 100 billion stars of the Milky Way galaxy shrink to barely visible dots on the planetarium dome.

I soon received a letter from an Ivy League professor of psychology who wanted to administer a questionnaire to visitors, assessing the depth of their depression after viewing the show. Our show, he wrote, elicited the most dramatic feelings of smallness he had ever experienced.

How could that be? Every time I see the show, I feel alive and spirited and connected. I also feel large, knowing that the goings-on within the three-pound human brain are what enabled us to figure out our place in the universe.

Allow me to suggest that it's the professor, not I, who has misread nature. His ego was too big to begin with, inflated by delusions of significance and fed by cultural assumptions that human beings are more important than everything else in the universe.

In all fairness to the fellow, powerful forces in society leave most of us susceptible. As was I . . . until the day I learned in biology class that more bacteria live and work in one centimeter of my colon than the number of people who have ever existed in the world. That kind of information makes you think twice about who—or what—is actually in charge.

From that day on, I began to think of people not as the masters of space and time but as participants in a great cosmic chain of being, with a direct genetic link across species both living and extinct, extending back nearly 4 billion years to the earliest single-celled organisms on Earth.

■■■

Need more ego softeners? Simple comparisons of quantity, size, and scale do the job well.

Take water. It's simple, common, and vital. There are more molecules of water in an eight-ounce cup of the stuff than there are cups of water in all the world's oceans. Every cup that passes through a single person and eventually rejoins the world's water supply holds enough molecules to mix 1,500 of them into every other cup of water in the world. No way around it: some of the water you just drank passed through the kidneys of Socrates, Genghis Khan, and Joan of Arc.

How about air? Also vital. A single breathful draws in more air molecules than there are breathfuls of air in Earth's entire atmosphere. That means some of the air you just breathed passed through the lungs of Napoleon, Beethoven, Lincoln, and Billy the Kid.

Time to get cosmic. There are more stars in the universe than grains of sand on any beach, more stars than seconds have passed since Earth formed, more stars than words and sounds ever uttered by all the humans who ever lived.

Want a sweeping view of the past? Our unfolding cosmic perspective takes you there. Light takes time to reach Earth's observatories from the depths of space, and so you see objects and phenomena not as they are but as they once were. That means the universe acts like a giant time machine: the farther away you look, the further back in time you see—back almost to the beginning of time itself. Within that horizon of reckoning, cosmic evolution unfolds continuously, in full view.

Want to know what we're made of? Again, the cosmic perspective offers a bigger answer than you might expect. The chemical elements of the universe are forged in the fires of high-mass stars that end their lives in stupendous explosions, enriching their host galaxies with the chemical arsenal of life as we know it. We are not simply in the universe. The universe is in us. Yes, we are stardust.

■■■

Again and again across the centuries, cosmic discoveries have demoted our self-image. Earth was once assumed to be astronomically unique, until astronomers learned that Earth is just another planet orbiting the Sun. Then we presumed the Sun was unique, until we learned that the countless stars of the night sky are suns themselves. Then we presumed our galaxy, the Milky Way, was the entire known universe, until we established that the countless fuzzy things in the sky are other galaxies, dotting the landscape of our known universe.

The cosmic perspective flows from fundamental knowledge. But it's more than just what you know. It's also about having the wisdom and insight to apply that knowledge to assessing our place in the universe. And its attributes are clear:

- The cosmic perspective comes from the frontiers of science, yet is not solely the provenance of the scientist. It belongs to everyone.
- The cosmic perspective is humble.
- The cosmic perspective is spiritual—even redemptive— but is not religious.
- The cosmic perspective enables us to grasp, in the same thought, the large and the small.
- The cosmic perspective opens our minds to extraordinary ideas but does not leave them so open that our brains spill out, making us susceptible to believing anything we're told.
- The cosmic perspective opens our eyes to the universe, not as a benevolent cradle designed to nurture life but as a cold, lonely, hazardous place.
- The cosmic perspective shows Earth to be a mote, but a precious mote and, for the moment, the only home we have.
- The cosmic perspective finds beauty in the images of planets, moons, stars, and nebulae but also celebrates the laws of physics that shape them.
- The cosmic perspective enables us to see beyond our circumstances, allowing us to transcend the primal search for food, shelter, and sex.
- The cosmic perspective reminds us that in space, where there is no air, a flag will not wave—an indication that perhaps flag waving and space exploration do not mix.

■ The cosmic perspective not only embraces our genetic kinship with all life on Earth but also values our chemical kinship with any yet-to-be discovered life in the universe, as well as our atomic kinship with the universe itself.

■ ■ ■

At least once a week, if not once a day, we might each ponder what cosmic truths lie undiscovered before us, perhaps awaiting the arrival of a clever thinker, an ingenious experiment, or an innovative space mission to reveal them. We might further ponder how those discoveries may one day transform life on Earth.

Absent such curiosity, we are no different from the provincial farmer who expresses no need to venture beyond the county line, because his forty acres meet all his needs. Yet if all our predecessors had felt that way, the farmer would instead be a cave dweller, chasing down his dinner with a stick and a rock.

During our brief stay on planet Earth, we owe ourselves and our descendants the opportunity to explore—in part because it's fun to do. But there's a far nobler reason. The day our knowledge of the cosmos ceases to expand, we risk regressing to the childish view that the universe figuratively and literally revolves around us. In that bleak world, arms-bearing, resource-hungry people and nations would be prone to act on their "low contracted prejudices." And that would be the last gasp of human enlightenment—until the rise of a visionary new culture that could once again embrace the cosmic perspective.

KEY TO WAVELENGTH ICONS ON FIGURES

You'll see the following icons on figures throughout the book. They are used to indicate the wavelength of light shown in each image, and to identify photo-realistic artworks and images made by computer simulations.

Indicates an artist's representation

Indicates a graphic generated using computer simulations

Indicates an image based on data from an observatory on Earth observing radio waves

Indicates an image based on data from a spacecraft observing radio waves

Indicates an image based on data from an observatory on Earth observing infrared light

Indicates an image based on data from a spacecraft observing infrared light

Indicates an image based on data from an observatory on Earth observing visible light

Indicates an image based on data from a spacecraft observing visible light

Indicates an image based on data from a rover observing visible light

Indicates an image based on data from a spacecraft observing ultraviolet light

Indicates an image based on data from a spacecraft observing X rays

Indicates an image based on data from a spacecraft observing gamma rays

THE
COSMIC
PERSPECTIVE
THE SOLAR SYSTEM

1

A MODERN VIEW OF THE UNIVERSE

LEARNING GOALS

1.1 THE SCALE OF THE UNIVERSE

- What is our place in the universe?
- How big is the universe?

1.2 THE HISTORY OF THE UNIVERSE

- How did we come to be?
- How do our lifetimes compare to the age of the universe?

1.3 SPACESHIP EARTH

- How is Earth moving through space?
- How do galaxies move within the universe?

1.4 THE HUMAN ADVENTURE OF ASTRONOMY

- How has the study of astronomy affected human history?

We shall not cease from exploration
And the end of all our exploring
Will be to arrive where we started
And know the place for the first time.

—T. S. Eliot

Far from city lights on a clear night, you can gaze upward at a sky filled with stars. Lie back and watch for a few hours, and you will observe the stars marching steadily across the sky. Confronted by the seemingly infinite heavens, you might wonder how Earth and the universe came to be. If you do, you will be sharing an experience common to humans around the world and in thousands of generations past.

Modern science offers answers to many of our fundamental questions about the universe and our place within it. We now know the basic content and scale of the universe. We know the age of Earth and the approximate age of the universe. And, although much remains to be discovered, we are rapidly learning how the simple ingredients of the early universe developed into the incredible diversity of life on Earth.

In this first chapter, we will survey the scale, history, and motion of the universe. This "big picture" perspective on our universe will provide a base on which you'll be able to build a deeper understanding in the rest of the book.

1.1 THE SCALE OF THE UNIVERSE

For most of human history, our ancestors imagined Earth to be stationary and located at the center of a relatively small universe. These ideas made sense at a time when understanding was built upon everyday experience. After all, we cannot feel the constant motion of Earth as it rotates on its axis and orbits the Sun, and if you observe the sky you'll see that the Sun, Moon, planets, and stars all appear to revolve around us each day. Nevertheless, we now know that Earth is a planet orbiting a rather average star in a vast universe.

The historical path to this knowledge was long and complex. In later chapters, we'll see that the ancient belief in an Earth-centered (or *geocentric*) universe changed only when people were confronted by strong evidence to the contrary, and we'll explore how the method of learning that we call *science* enabled us to acquire this evidence. First, however, it's useful to have a general picture of the universe as we know it today.

What is our place in the universe?

Take a look at the remarkable photo that opens this chapter (on page 1). This photo, taken by the Hubble Space Telescope, shows a piece of the sky so small that you could block your view of it with a grain of sand held at arm's length. Yet it encompasses an almost unimaginable expanse of both space and time: Nearly every object within it is a *galaxy* filled with billions of stars, and some of the smaller smudges are galaxies so far away that their light has taken billions of years to reach us. Let's begin our study of astronomy by exploring what a photo like this one tells us about our own place in the universe.

Our Cosmic Address The galaxies that we see in the Hubble Space Telescope photo are just one of several levels of structure in our universe. A good way to build context on these levels is to consider what we might call our "cosmic address," illustrated in **FIGURE 1.1**.

Earth is a planet in our **solar system**, which consists of the Sun, the planets and their moons, and countless smaller objects that include rocky *asteroids* and icy *comets*. Keep in mind that our Sun is a *star*, just like the stars we see in our night sky.

Our solar system belongs to the huge, disk-shaped collection of stars called the **Milky Way Galaxy**. A **galaxy** is a great island of stars in space, containing between a few hundred million and a trillion or more stars. The Milky Way is a relatively large galaxy, containing more than 100 billion stars. Our solar system is located a little over halfway from the galactic center to the edge of the galactic disk.

Billions of other galaxies are scattered throughout space. Some galaxies are fairly isolated, but many others are found in groups. Our Milky Way, for example, is one of the two largest among about 40 galaxies in the **Local Group.** Groups of galaxies with more than a few dozen members are often called **galaxy clusters.**

On a very large scale, galaxies and galaxy clusters appear to be arranged in giant chains and sheets with huge voids between them; the background of Figure 1.1 shows this large-scale structure. The regions in which galaxies and galaxy clusters are most tightly packed are called **superclusters,** which are essentially clusters of galaxy clusters. Our Local Group is located in the outskirts of the Local Supercluster.

Together, all these structures make up our **universe.** In other words, the universe is the sum total of all matter and energy, encompassing the superclusters and voids and everything within them.

> ### THINK ABOUT IT
> Some people think that our tiny physical size in the vast universe makes us insignificant. Others think that our ability to learn about the wonders of the universe gives us significance despite our small size. What do *you* think?

Astronomical Distance Measurements Notice that Figure 1.1 is labeled with an approximate size for each structure in kilometers. In astronomy, many of the distances are so large that kilometers are not the most convenient unit. Instead, we often use two other units:

- One **astronomical unit (AU)** is Earth's average distance from the Sun, which is about 150 million kilometers (93 million miles). We commonly describe distances within our solar system in astronomical units.

- One **light-year (ly)** is the distance that light can travel in 1 year, which is about 10 trillion kilometers (6 trillion miles). We generally use light-years to describe the distances of stars and galaxies.

Universe

approx. size: 10^{21} km ≈ 100 million ly

Local Supercluster

approx. size: 3 x 10^{19} km ≈ 3 million ly

Local Group

approx. size: 10^{18} km ≈ 100,000 ly

Milky Way Galaxy

Solar System
(*not to scale*)

Earth

approx. size: 10^{10} km ≈ 60 AU

approx. size: 10^4 km

FIGURE 1.1 Our cosmic address. These diagrams show key levels of structure in our universe. For a more detailed view, see the "You Are Here in Space" foldout diagram in the front of the book.

Be sure to note that a light-year is a unit of *distance*, not of time. Light travels at the speed of light, which is 300,000 kilometers per second. We therefore say that one *light-second* is about 300,000 kilometers, because that is the distance light travels in one second. Similarly, one light-minute is the distance that light travels in one minute, one light-hour is the distance that light travels in one hour, and so on. Mathematical Insight 1.1 shows that light travels about 10 trillion kilometers in one year, so that distance represents a light-year.

Looking Back in Time The speed of light is extremely fast by earthly standards. It is so fast that if you could make light go in circles, it could circle Earth nearly eight times in a single second. Nevertheless, even light takes time to travel the vast distances in space. Light takes a little more than 1 second to reach Earth from the Moon, and about 8 minutes to reach Earth from the Sun. Stars are so far away that their light takes years to reach us, which is why we measure their distances in light-years.

Consider Sirius, the brightest star in the night sky, which is located about 8 light-years away. Because it takes light 8 years to travel this distance, we see Sirius not as it is today, but rather as it was 8 years ago. The effect is more dramatic at greater distances. The Orion Nebula (**FIGURE 1.2**) is a giant cloud in which stars and planets are forming. It is located about 1500 light-years from Earth, which means we see it as it looked about 1500 years ago—about the time of the fall of the Roman Empire. If any major events have occurred in the Orion Nebula since that time, we cannot yet know about them because the light from these events has not yet reached us.

The general idea that light takes time to travel through space leads to a remarkable fact:

The farther away we look in distance, the further back we look in time.

The Andromeda Galaxy (**FIGURE 1.3**) is about 2.5 million light-years away, which means we see it as it looked about 2.5 million years ago. We see more distant galaxies as they were even further in the past. Some of the galaxies in the Hubble Space Telescope photo that opens the chapter are billions of light-years away, meaning we see them as they were billions of years ago.

SEE IT FOR YOURSELF

The glow from the central region of the Andromeda Galaxy is faintly visible to the naked eye and easy to see with binoculars. Use a star chart to find it in the night sky. Contemplate the fact that you are seeing light that spent 2.5 million years in space before reaching your eyes. If students on a planet in the Andromeda Galaxy were looking at the Milky Way, what would they see? Could they know that we exist here on Earth?

Basic Astronomical Definitions

ASTRONOMICAL OBJECTS

star A large, glowing ball of gas that generates heat and light through nuclear fusion in its core. Our Sun is a star.

planet A moderately large object that orbits a star and shines primarily by reflecting light from its star. According to a definition adopted in 2006, an object can be considered a planet only if it (1) orbits a star, (2) is large enough for its own gravity to make it round, and (3) has cleared most other objects from its orbital path. An object that meets the first two criteria but has not cleared its orbital path, like Pluto, is designated a **dwarf planet.**

moon (or **satellite**) An object that orbits a planet. The term *satellite* is also used more generally to refer to any object orbiting another object.

asteroid A relatively small and rocky object that orbits a star.

comet A relatively small and ice-rich object that orbits a star.

small solar system body An asteroid, comet, or other object that orbits a star but is too small to qualify as a planet or dwarf planet.

COLLECTIONS OF ASTRONOMICAL OBJECTS

solar system The Sun and all the material that orbits it, including planets, dwarf planets, and small solar system bodies. Although the term *solar system* technically refers only to our own star system (*solar* means "of the Sun"), it is often applied to other star systems as well.

star system A star (sometimes more than one star) and any planets and other materials that orbit it.

galaxy A great island of stars in space, containing from a few hundred million to a trillion or more stars, all held together by gravity and orbiting a common center.

cluster (or **group**) **of galaxies** A collection of galaxies bound together by gravity. Small collections (up to a few dozen galaxies) are generally called *groups,* while larger collections are called *clusters.*

supercluster A gigantic region of space in which many groups and clusters of galaxies are packed more closely together than elsewhere in the universe.

universe (or **cosmos**) The sum total of all matter and energy—that is, all galaxies and everything between them.

observable universe The portion of the entire universe that can be seen from Earth, at least in principle. The observable universe is probably only a tiny portion of the entire universe.

ASTRONOMICAL DISTANCE UNITS

astronomical unit (AU) The average distance between Earth and the Sun, which is about 150 million kilometers. More technically, 1 AU is the length of the semimajor axis of Earth's orbit.

light-year The distance that light can travel in 1 year, which is about 9.46 trillion kilometers.

TERMS RELATING TO MOTION

rotation The spinning of an object around its axis. For example, Earth rotates once each day around its axis, which is an imaginary line connecting the North and South Poles.

orbit (**revolution**) The orbital motion of one object around another due to gravity. For example, Earth orbits the Sun once each year.

expansion (**of the universe**) The increase in the average distance between galaxies as time progresses.

FIGURE 1.2 The Orion Nebula, located about 1500 light-years away. The inset shows its location in the constellation Orion.

FIGURE 1.3 The Andromeda Galaxy (M31). When we look at this galaxy, we see light that has been traveling through space for 2.5 million years.

It's also amazing to realize that any "snapshot" of a distant galaxy is a picture of both space and time. For example, because the Andromeda Galaxy is about 100,000 light-years in diameter, the light we currently see from the far side of the galaxy must have left on its journey to us some 100,000 years before the light we see from the near side. Figure 1.3 therefore shows different parts of the galaxy spread over a time period of 100,000 years. When we study the universe, it is impossible to separate space and time.

The Observable Universe As we'll discuss in Section 1.2, astronomers estimate that the universe is about 14 billion years old. This fact, combined with the fact that looking deep into space means looking far back in time, places a limit on the portion of the universe that we can see, even in principle. **FIGURE 1.4** shows the idea. If we look at a galaxy that is 7 billion light-years away, we see it as it looked 7 billion years

ago*—which means we see it as it was when the universe was half its current age. If we look at a galaxy that is 12 billion light-years away (like the most distant ones in the Hubble Space Telescope photo), we see it as it was 12 billion years ago, when the universe was only 2 billion years old. And if we tried to look beyond 14 billion light-years, we'd be looking to a time more than 14 billion years ago—which is before the universe existed and therefore means that there is nothing to see. This distance of 14 billion light-years therefore marks the boundary (or *horizon*) of our **observable universe**—the portion of the entire universe that we can potentially observe. Note that this fact does not put any limit on the size of the

---

*As we'll see in Chapter 20, distances to faraway galaxies must be defined carefully in an expanding universe; distances like those given here are based on the time it has taken a galaxy's light to reach us (called the *lookback time*).

Far: We see a galaxy 7 billion light-years away as it was 7 billion years ago—when the universe was about half its current age of 14 billion years.

Farther: We see a galaxy 12 billion light-years away as it was 12 billion years ago—when the universe was only about 2 billion years old.

The limit of our observable universe: Light from nearly 14 billion light-years away shows the universe as it looked shortly after the Big Bang, before galaxies existed.

Beyond the observable universe: We cannot see anything farther than 14 billion light-years away, because its light has not had enough time to reach us.

7 billion light-years

12 billion light-years

14 billion light-years

FIGURE 1.4 interactive figure The farther away we look in space, the further back we look in time. The age of the universe therefore puts a limit on the size of the *observable* universe—the portion of the entire universe that we can observe, at least in principle.

You've probably heard people say things like "It will take me light-years to finish this homework!" But a statement like this one doesn't make sense, because light-years are a unit of *distance*, not time. If you are unsure whether the term *light-year* is being used correctly, try testing the statement by using the fact that 1 light-year is about 10 trillion kilometers, or 6 trillion miles. The statement then reads "It will take me 6 trillion miles to finish this homework," which clearly does not make sense.

entire universe, which may be far larger than our observable universe. We simply have no hope of seeing or studying anything beyond the bounds of our observable universe.

 Scale of the Universe Tutorial, Lessons 1–3

How big is the universe?

Figure 1.1 put numbers on the sizes of different structures in the universe, but these numbers have little meaning for most people—after all, they are literally astronomical. Therefore, to help you develop a greater appreciation of our modern view of the universe, we'll discuss a few ways of putting these numbers into perspective.

The Scale of the Solar System One of the best ways to develop perspective on cosmic sizes and distances is to imagine our solar system shrunk down to a scale that would allow you to walk through it. The Voyage scale model solar system in Washington, D.C., makes such a walk possible (**FIGURE 1.5**). The Voyage model shows the Sun and the planets, and the distances between them, at *one ten-billionth* of their actual sizes and distances.

FIGURE 1.6a shows the Sun and planets at their correct sizes (but not distances) on the Voyage scale. The model Sun is about the size of a large grapefruit, Jupiter is about the size of a marble, and Earth is about the size of the ball point in a pen. You can immediately see some key facts about our solar system. For example, the Sun is far larger than any of the planets; in mass, the Sun outweighs all the planets combined by a factor of nearly 1000. The planets also vary considerably in size: The storm on Jupiter known as the Great Red Spot (visible near Jupiter's lower left in the painting) could swallow up the entire Earth.

The scale of the solar system is even more remarkable when you combine the sizes shown in Figure 1.6a with the distances illustrated by the map of the Voyage model in **FIGURE 1.6b**. For example, the ball-point-size Earth is located about 15 meters (16.5 yards) from the grapefruit-size Sun, which means you can picture Earth's orbit as a circle of radius 15 meters around a grapefruit.

MATHEMATICAL INSIGHT 1.1

How Far Is a Light-Year? An Introduction to Astronomical Problem Solving

 Math Review Video: Problem Solving Part 1

We can develop greater insight into astronomical ideas by applying mathematics. The key to using mathematics is to approach problems in a clear and organized way. One simple approach uses the following three steps:

Step 1 Understand the problem: Ask yourself what the solution will look like (for example, what units will it have? will it be big or small?) and what information you need to solve the problem. Draw a diagram or think of a simpler analogous problem to help you decide how to solve it.

Step 2 Solve the problem: Carry out the necessary calculations.

Step 3 Explain your result: Be sure that your answer makes sense, and consider what you've learned by solving the problem.

You can remember this process as "Understand, Solve, and Explain," or USE for short. You may not always need to write out the three steps explicitly, but they may help if you are stuck.

EXAMPLE: How far is a light-year?

SOLUTION: Let's use the three-step process.

Step 1 Understand the problem: The question asks how *far*, so we are looking for a *distance*. In this case, the definition of a light-year tells us that we are looking for the *distance that light can travel in 1 year*. We know that light travels at the speed of light, so we are looking for an equation that gives us distance from speed. If you don't remember this equation, just think of a simpler but analogous problem, such as

"If you drive at 50 kilometers per hour, how far will you travel in 2 hours?" You'll realize that you simply multiply the speed by the time: distance = speed × time. In this case, the speed is the speed of light, or 300,000 km/s, and the time is 1 year.

Step 2 Solve the problem: From Step 1, our equation is that 1 light-year is the speed of light times one year. To make the units consistent, we convert 1 year to seconds by remembering that there are 60 seconds in 1 minute, 60 minutes in 1 hour, 24 hours in 1 day, and 365 days in 1 year. (See Appendix C.3 to review unit conversions.) We now carry out the calculations:

$$1 \text{ light-year} = (\text{speed of light}) \times (1 \text{ yr})$$

$$= \left(300,000 \, \frac{\text{km}}{\text{s}}\right) \times \left(1 \, \text{yr} \times \frac{365 \, \text{days}}{1 \, \text{yr}}\right.$$

$$\left. \times \frac{24 \, \text{hr}}{1 \, \text{day}} \times \frac{60 \, \text{min}}{1 \, \text{hr}} \times \frac{60 \, \text{s}}{1 \, \text{min}}\right)$$

$$= 9,460,000,000,000 \text{ km} \ (9.46 \text{ trillion km})$$

Step 3 Explain your result: In sentence form, our answer is "One light-year is about 9.46 trillion kilometers." This answer makes sense: It has the expected units of distance (kilometers) and it is a long way, which we expect for the distance that light can travel in a year. We say "about" in the answer because we know it is not exact. For example, a year is not exactly 365 days long. In fact, for most purposes, we can approximate the answer further as "One light-year is about 10 trillion kilometers."

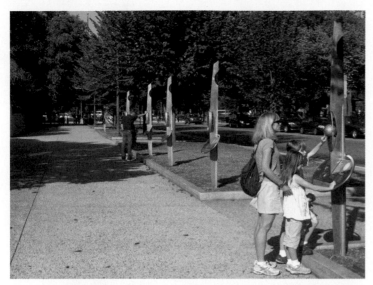

FIGURE 1.5 This photo shows the pedestals housing the Sun (the gold sphere on the nearest pedestal) and the inner planets in the Voyage scale model solar system (Washington, D.C.). The model planets are encased in the sidewalk-facing disks visible at about eye level on the planet pedestals. The building at the left is the National Air and Space Museum.

Perhaps the most striking feature of our solar system when we view it to scale is its emptiness. The Voyage model shows the planets along a straight path, so we'd need to draw each planet's orbit around the model Sun to show the full extent of our planetary system. Fitting all these orbits would require an area measuring more than a kilometer on a side—an area equivalent to more than 300 football fields arranged in a grid. Spread over this large area, only the grapefruit-size Sun, the planets, and a few moons would be big enough to see. The rest of it would look virtually empty (that's why we call it *space!*).

Seeing our solar system to scale also helps put space exploration into perspective. The Moon, the only other world on which humans have ever stepped (**FIGURE 1.7**), lies only about 4 centimeters ($1\frac{1}{2}$ inches) from Earth in the Voyage model. On this scale, the palm of your hand can cover the entire region of the universe in which humans have so far traveled. The trip to Mars is more than 150 times as far as the trip to the Moon, even when Mars is on the same side of its orbit as Earth. And while you can walk from the Sun to Pluto in a few minutes on the Voyage scale, the *New Horizons* spacecraft that is making the real journey will have been in space nearly a decade when it flies past Pluto in July 2015.

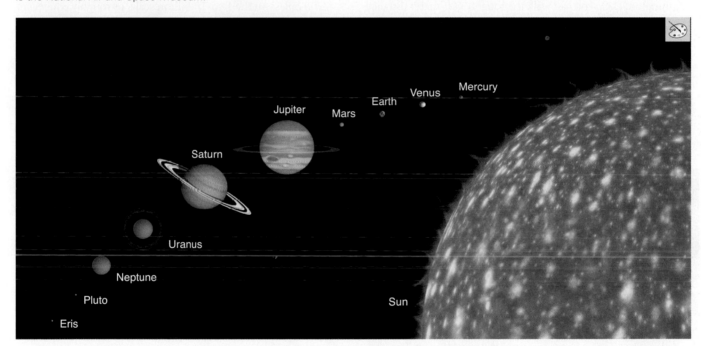

a The scaled sizes (but not distances) of the Sun, planets, and two largest known dwarf planets.

b Locations of the Sun and planets in the Voyage model, Washington, D.C.; the distance from the Sun to Pluto is about 600 meters (1/3 mile). Planets are lined up in the model, but in reality each planet orbits the Sun independently and a perfect alignment never occurs.

FIGURE 1.6 interactive figure The Voyage scale model represents the solar system at *one ten-billionth* of its actual size. Pluto is included in the Voyage model, which was built before the International Astronomical Union reclassified Pluto as a dwarf planet.

FIGURE 1.7 This famous photograph from the first Moon landing (*Apollo 11* in July 1969) shows astronaut Buzz Aldrin, with Neil Armstrong reflected in his visor. Armstrong was the first to step onto the Moon's surface, saying, "That's one small step for a man, one giant leap for mankind."

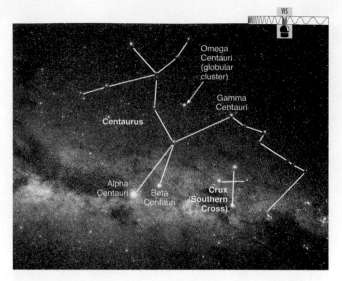

FIGURE 1.8 This photograph and diagram show the constellation Centaurus, which is visible from tropical and southern latitudes. Alpha Centauri's real distance of 4.4 light-years is 4400 kilometers on the 1-to-10-billion Voyage scale.

Distances to the Stars If you visit the Voyage model in Washington, D.C., you can walk the roughly 600-meter distance from the Sun to Pluto in just a few minutes. How much farther would you have to walk to reach the next star on this scale?

Amazingly, you would need to walk to California. If this answer seems hard to believe, you can check it for yourself. A light-year is about 10 trillion kilometers, which becomes 1000 kilometers on the 1-to-10-billion scale (because 10 trillion ÷ 10 billion = 1000). The nearest star system to our own, a three-star system called Alpha Centauri (**FIGURE 1.8**), is

about 4.4 light-years away. That distance is about 4400 kilometers (2700 miles) on the 1-to-10-billion scale, or roughly equivalent to the distance across the United States.

The tremendous distances to the stars give us some perspective on the technological challenge of astronomy. For example, because the largest star of the Alpha Centauri system is roughly the same size and brightness as our Sun, viewing it in the night sky is somewhat like being in Washington, D.C., and seeing a very bright grapefruit in San Francisco (neglecting the problems introduced by the curvature of Earth). It may seem

SPECIAL TOPIC

How Many Planets Are There in Our Solar System?

Until recently, children were taught that our solar system had nine planets. However, in 2006 astronomers voted to demote Pluto to a *dwarf planet*, leaving our solar system with only eight official planets (**FIGURE 1**). Why the change?

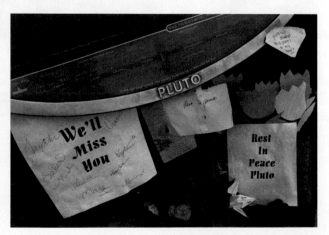

FIGURE 1 Notes left at the Voyage scale model solar system Pluto plaque upon Pluto's demotion to dwarf planet.

When Pluto was discovered in 1930, it was assumed to be similar to other planets. But as we'll discuss in Chapter 12, we now know that Pluto is much smaller than any of the first eight planets and that it shares the outer solar system with thousands of other icy objects. Still, as long as Pluto was the largest known of these objects, most astronomers were content to leave the planetary status quo. Change was forced by the 2005 discovery of an object called Eris. Because Eris is slightly larger than Pluto, astronomers could no longer avoid the question of what objects should count as planets.

Official decisions on astronomical names and definitions rest with the International Astronomical Union (IAU), an organization made up of professional astronomers from around the world. The question of Pluto's status was voted upon during the IAU's 2006 meeting. The result was the new definition of "planet" that you see in the Basic Astronomical Definitions box on page 4, and the addition of the "dwarf planet" category to accommodate objects like Pluto and Eris.

Not all astronomers are happy with the new definitions, but for now they seem likely to hold. Of course, some people are likely to keep thinking of Pluto as a planet regardless of what professional astronomers say, much as many people still talk of Europe and Asia as separate continents even though both belong to the same land mass (Eurasia). So if you're a Pluto fan, don't despair: It's good to know the official classifications, but it's better to understand the science behind them.

remarkable that we can see the star at all, but the blackness of the night sky allows the naked eye to see it as a faint dot of light. It looks much brighter through powerful telescopes, but we still cannot see features of the star's surface.

Now, consider the difficulty of detecting *planets* orbiting nearby stars, which is equivalent to looking from Washington, D.C., and trying to find ball points or marbles orbiting grape-fruits in California or beyond. When you consider this challenge, it is all the more remarkable to realize that we now have technology capable of finding such planets [**Section 13.1**].

The vast distances to the stars also offer a sobering lesson about interstellar travel. Although science fiction shows like *Star Trek* and *Star Wars* make such travel look easy, the reality is far different. Consider the *Voyager 2* spacecraft. Launched in 1977, *Voyager 2* flew by Jupiter in 1979, Saturn in 1981, Uranus in 1986, and Neptune in 1989. It is now bound for the stars at a speed of close to 50,000 kilometers per hour—about 100 times as fast as a speeding bullet. But even at this speed, *Voyager 2* would take about 100,000 years to reach Alpha Centauri if it were headed in that direction (which it's not). Convenient interstellar travel remains well beyond our present technology.

The Size of the Milky Way Galaxy The vast separation between our solar system and Alpha Centauri is typical of

the separations among star systems in our region of the Milky Way Galaxy. We therefore cannot use the 1-to-10-billion scale for thinking about distances beyond the nearest stars, because more distant stars would not fit on Earth with this scale. To visualize the galaxy, let's reduce our scale by another factor of 1 billion (making it a scale of 1 to 10^{19}).

On this new scale, each light-year becomes 1 millimeter, and the 100,000-light-year diameter of the Milky Way Galaxy becomes 100 meters, or about the length of a football field. Visualize a football field with a scale model of our galaxy centered over midfield. Our entire solar system is a microscopic dot located around the

---

MATHEMATICAL INSIGHT 1.2

The Scale of Space and Time

 Math Review Video: Scientific Notation, Parts 1 to 3

Making a scale model usually requires nothing more than division. For example, in a 1-to-20 architectural scale model, a building that is actually 6 meters tall will be only $6 \div 20 = 0.3$ meter tall. The idea is the same for astronomical scaling, except that we usually divide by such large numbers that it's easier to work in *scientific notation*—that is, with the aid of powers of 10. (See Appendixes C.1 and C.2 to review these concepts.)

EXAMPLE 1: How big is the Sun on a 1-to-10-billion scale?

SOLUTION:

Step 1 Understand: We are looking for the scaled *size* of the Sun, so we simply need to divide its actual radius by 10 billion, or 10^{10}. Appendix E.1 gives the Sun's radius as 695,000 km, or 6.95×10^5 km in scientific notation.

Step 2 Solve: We carry out the division:

$$\text{scaled radius} = \frac{\text{actual radius}}{10^{10}}$$
$$= \frac{6.95 \times 10^5 \, \text{km}}{10^{10}}$$
$$= 6.95 \times 10^{(5-10)} \, \text{km} = 6.95 \times 10^{-5} \, \text{km}$$

Notice that we used the rule that dividing powers of 10 means subtracting their exponents [**Appendix C.1**].

Step 3 Explain: We have found an answer, but because most of us don't have a good sense of what 10^{-5} kilometer looks like, the answer will be more meaningful if we convert it to units that will be easier to interpret. In this case, because there are 1000 (10^3) meters in a kilometer and 100 (10^2) centimeters in a meter, we convert to centimeters:

$$6.95 \times 10^{-5} \, \text{km} \times \frac{10^3 \, \text{m}}{1 \, \text{km}} \times \frac{10^2 \, \text{cm}}{1 \, \text{m}} = 6.95 \, \text{cm}$$

We've found that on the 1-to-10-billion scale the Sun's radius is about 7 centimeters, which is a diameter of about 14 centimeters—about the size of a large grapefruit.

EXAMPLE 2: What scale allows the 100,000-light-year diameter of the Milky Way Galaxy to fit on a 100-meter-long football field?

SOLUTION:

Step 1 Understand: We want to know *how many times larger* the actual diameter of the galaxy is than 100 meters, so we'll divide the actual diameter by 100 meters. To carry out the division, we'll need both numbers in the same units. We can put the galaxy's diameter in meters by using the fact that a light-year is about 10^{13} kilometers (see Mathematical Insight 1.1) and a kilometer is 10^3 meters; because we are working with powers of 10, we'll write the galaxy's 100,000-light-year diameter as 10^5 ly.

Step 2 Solve: We now convert the units and carry out the division:

$$\frac{\text{galaxy diameter}}{\text{football field diameter}} = \frac{10^5 \, \text{ly} \times \dfrac{10^{13} \, \text{km}}{1 \, \text{ly}} \times \dfrac{10^3 \, \text{m}}{1 \, \text{km}}}{10^2 \, \text{m}}$$
$$= 10^{(5+13+3-2)} = 10^{19}$$

Note that the answer has no units, because it simply tells us how many times larger one thing is than the other.

Step 3 Explain: We've found that we need a scale of 1 to 10^{19} to make the galaxy fit on a football field.

20-yard line. The 4.4-light-year separation between our solar system and Alpha Centauri becomes just 4.4 millimeters on this scale—smaller than the width of your little finger. If you stood at the position of our solar system in this model, millions of star systems would lie within reach of your arms.

Another way to put the galaxy into perspective is to consider its number of stars—more than 100 billion. Imagine that tonight you are having difficulty falling asleep (perhaps because you are contemplating the scale of the universe). Instead of counting sheep, you decide to count stars. If you are able to count about one star each second, how long would it take you to count 100 billion stars in the Milky Way? Clearly, the answer is 100 billion (10^{11}) seconds, but how long is that? Amazingly, 100 billion seconds is more than 3000 years. (You can confirm this by dividing 100 billion by the number of seconds in 1 year.) You would need thousands of years just to *count* the stars in the Milky Way Galaxy, and this assumes you never take a break— no sleeping, no eating, and absolutely no dying!

THINK ABOUT IT

Contemplate the fact that it would take more than 3000 years just to count out loud the stars in our galaxy, and that each star is a potential sun for a system of planets. How does this perspective affect your thoughts about the possibilities for finding life—or intelligent life—beyond Earth? Explain.

The Observable Universe As incredible as the scale of our galaxy may seem, the Milky Way is only one of roughly 100 billion galaxies in the observable universe. Just as it would take thousands of years to count the stars in the Milky Way, it would take thousands of years to count all the galaxies.

Think for a moment about the total number of stars in all these galaxies. If we assume 100 billion stars per galaxy, the total number of stars in the observable universe is roughly 100 billion × 100 billion, or 10,000,000,000,000,000,000,000 (10^{22}). How big is this number? Visit a beach. Run your hands through the fine-grained sand. Imagine counting each tiny grain of sand as it slips through your fingers. Then imagine counting every grain of sand on the beach and continuing to count *every* grain of dry sand on *every* beach on Earth (see Mathematical Insight 1.3). If you could actually complete this task, you would find that the number of grains of sand is comparable to the number of stars in the observable universe (**FIGURE 1.9**).

THINK ABOUT IT

Study the foldout in the front of this book, which illustrates the ideas covered in this section in greater detail. Overall, how does visualizing Earth to scale affect your perspective on our planet and on human existence? Explain.

MATHEMATICAL INSIGHT 1.3

Order of Magnitude Estimation

In astronomy, numbers are often so large that an estimate can be useful even if it's good only to about the nearest power of 10. For example, when we multiplied 100 billion stars per galaxy by 100 billion galaxies to estimate that there are about 10^{22} stars in the observable universe, we knew that the "ballpark" nature of these numbers means the actual number of stars could easily be anywhere from about 10^{21} to 10^{23}. Estimates good to about the nearest power of 10 are called **order of magnitude estimates**.

EXAMPLE: Verify the claim that the number of grains of (dry) sand on all the beaches on Earth is comparable to the number of stars in the observable universe.

SOLUTION:

Step 1 Understand: To verify the claim, we need to estimate the number of grains of sand and see if it is close to our estimate of 10^{22} stars. We can estimate the total number of sand grains by dividing the *total volume* of sand on Earth's beaches by the *average volume* of an individual sand grain. Volume is equal to length times width times depth, so the total volume is the total length of sandy beach on Earth multiplied by the typical width and depth of dry sand. That is,

$$\text{total sand grains} = \frac{\text{total volume of beach sand}}{\text{average volume of 1 sand grain}}$$
$$= \frac{\text{beach length} \times \text{beach width} \times \text{beach depth}}{\text{average volume of 1 sand grain}}$$

We now need numbers to put into the equation. We can estimate the average volume of an individual sand grain by measuring out a small

volume of sand, counting the number of grains in this volume, and then dividing the volume by the number of grains. If you do this, you'll find that a reasonable order of magnitude estimate is one-tenth of a cubic millimeter, or 10^{-10} m³, per sand grain. We can estimate beach width and depth from experience or photos of beaches. Typical widths are about 20 to 50 meters and typical sand depth is about 2 to 5 meters, so we can make the numbers easy by assuming that the product of beach width times depth is about 100 square meters, or 10^2 m². The total length of sandy beach on Earth is more difficult to estimate, but you can look online and find that it is less than about 1 million kilometers, or 10^9 m.

Step 2 Solve: We already have our equation and all the numbers we need, so we just put them in; note that we group beach width and depth together, since we estimated them together in Step 1:

$$\text{total sand grains} = \frac{\text{beach length} \times (\text{beach width} \times \text{beach depth})}{\text{average volume of 1 sand grain}}$$
$$= \frac{10^9 \text{ m} \times 10^2 \text{ m}^2}{10^{-10} \text{ m}^3}$$
$$= 10^{[9+2-(-10)]} = 10^{21}$$

Step 3 Explain: Our order of magnitude estimate for the total number of grains of dry sand on all the beaches on Earth is 10^{21}, which is within a factor of 10 of the estimated 10^{22} stars in the observable universe. Because both numbers could easily be off by a factor of 10 or more, we cannot say with certainty that one is larger than the other, but the numbers are clearly comparable.

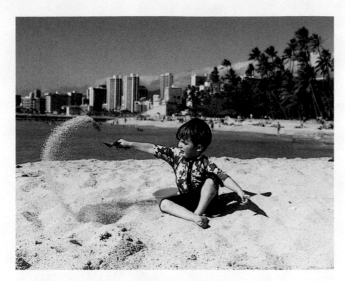

FIGURE 1.9 The number of stars in the observable universe is comparable to the number of grains of dry sand on all the beaches on Earth.

1.2 THE HISTORY OF THE UNIVERSE

Our universe is vast not only in space, but also in time. In this section, we will briefly discuss the history of the universe as we understand it today.

Before we begin, you may wonder how we can claim to know anything about what the universe was like in the distant past. We'll devote much of the rest of this textbook to understanding how science enables us to do this, but you already know part of the answer: Because looking farther into space means looking further back in time, we can actually *see* parts of the universe as they were long ago, simply by looking far enough away. In other words, our telescopes act somewhat like time machines, enabling us to observe the history of the universe. At great distances, we see the universe as it was long ago, when it was much younger than it is today.

How did we come to be?

FIGURE 1.10 (pages 12–13) summarizes the history of the universe according to modern science. Let's start at the upper left of the figure, and discuss the key events and what they mean.

The Big Bang, Expansion, and the Age of the Universe
Telescopic observations of distant galaxies show that the entire universe is *expanding,* meaning that average distances between galaxies are increasing with time. This fact implies that galaxies must have been closer together in the past, and if we go back far enough, we must reach the point at which the expansion began. We call this beginning the **Big Bang,** and scientists use the observed rate of expansion to calculate that it occurred about 14 billion years ago. The three cubes in the upper left portion of Figure 1.10 represent the expansion of a small piece of the universe through time.

The universe as a whole has continued to expand ever since the Big Bang, but on smaller size scales the force of gravity has drawn matter together. Structures such as galaxies and galaxy clusters occupy regions where gravity has won out against the overall expansion. That is, while the universe as a whole continues to expand, individual galaxies and galaxy clusters (and objects within them such as stars and planets) do *not* expand. This idea is also illustrated by the three cubes in Figure 1.10. Notice that as the cube as a whole grew larger, the matter within it clumped into galaxies and galaxy clusters. Most galaxies, including our own Milky Way, formed within a few billion years after the Big Bang.

Stellar Lives and Galactic Recycling
Within galaxies like the Milky Way, gravity drives the collapse of clouds of gas and dust to form stars and planets. Stars are not living organisms, but they nonetheless go through "life cycles." A star is born when gravity compresses the material in a cloud to the point at which the center becomes dense enough and hot enough to generate energy by **nuclear fusion,** the process in which lightweight atomic nuclei smash together and stick (or fuse) to make heavier nuclei. The star "lives" as long as it can shine with energy from fusion, and "dies" when it exhausts its usable fuel.

In its final death throes, a star blows much of its content back out into space. The most massive stars die in titanic explosions called *supernovae.* The returned matter mixes with other matter floating between the stars in the galaxy, eventually becoming part of new clouds of gas and dust from which new generations of stars can be born. Galaxies therefore function as cosmic recycling plants, recycling material expelled from dying stars into new generations of stars and planets. This cycle is illustrated in the lower right of Figure 1.10. Our own solar system is a product of many generations of such recycling.

Star Stuff
The recycling of stellar material is connected to our existence in an even deeper way. By studying stars of different ages, we have learned that the early universe contained only the simplest chemical elements: hydrogen and helium (and a trace of lithium). We and Earth are made primarily of other elements, such as carbon, nitrogen, oxygen, and iron. Where did these other elements come from? Evidence shows that they were manufactured by stars, some through the nuclear fusion that makes stars shine, and others through nuclear reactions accompanying the explosions that end stellar lives.

By the time our solar system formed, about $4\frac{1}{2}$ billion years ago, earlier generations of stars had already converted about 2% of our galaxy's original hydrogen and helium into heavier elements. Therefore, the cloud that gave birth to our solar system was made of about 98% hydrogen and helium and 2% other elements. This 2% may sound small, but it was more than enough to make the small rocky planets of our solar system, including Earth. On Earth, some of these elements became the raw ingredients of life, which ultimately blossomed into the great diversity of life on Earth today.

In summary, most of the material from which we and our planet are made was created inside stars that lived and died before the birth of our Sun. As astronomer Carl Sagan (1934–1996) said, we are "star stuff."

Throughout this book we will see that human life is intimately connected with the development of the universe as a whole. This illustration presents an overview of our cosmic origins, showing some of the crucial steps that made our existence possible.

(1) **Birth of the Universe:** The expansion of the universe began with the hot and dense Big Bang. The cubes show how one region of the universe has expanded with time. The universe continues to expand, but on smaller scales gravity has pulled matter together to make galaxies.

(4) **Earth and Life:** By the time our solar system was born, 4¹/₂ billion years ago, about 2% of the original hydrogen and helium had been converted into heavier elements. We are therefore "star stuff," because we and our planet are made from elements manufactured in stars that lived and died long ago.

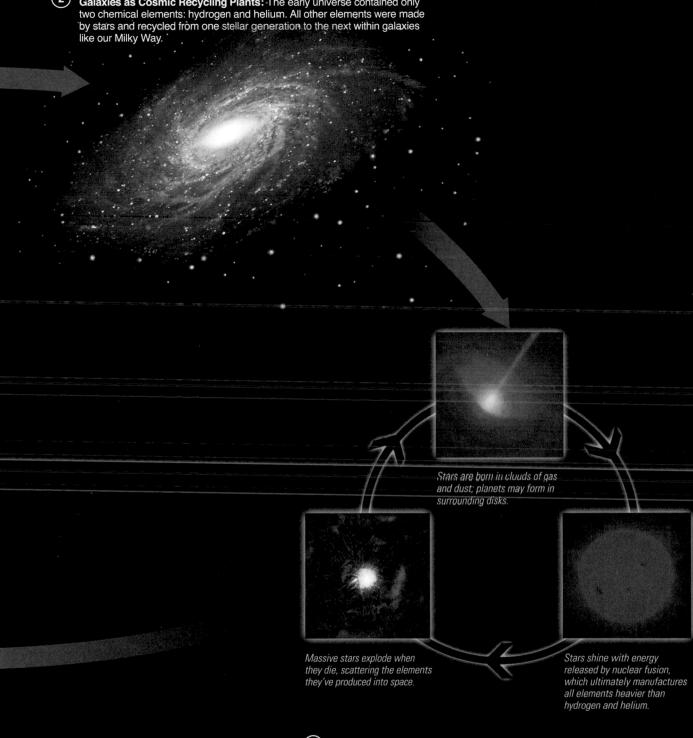

Galaxies as Cosmic Recycling Plants: The early universe contained only two chemical elements: hydrogen and helium. All other elements were made by stars and recycled from one stellar generation to the next within galaxies like our Milky Way.

Stars are born in clouds of gas and dust; planets may form in surrounding disks.

Massive stars explode when they die, scattering the elements they've produced into space.

Stars shine with energy released by nuclear fusion, which ultimately manufactures all elements heavier than hydrogen and helium.

③ **Life Cycles of Stars:** Many generations of stars have lived and died in the Milky Way.

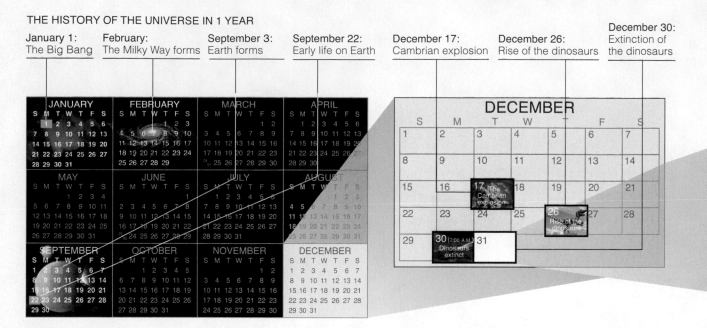

January 1:
The Big Bang

February:
The Milky Way forms

September 3:
Earth forms

September 22:
Early life on Earth

December 17:
Cambrian explosion

December 26:
Rise of the dinosaurs

December 30:
Extinction of
the dinosaurs

FIGURE 1.11 The cosmic calendar compresses the 14-billion-year history of the universe into 1 year, so each month represents a little more than 1 billion years. This cosmic calendar is adapted from a version created by Carl Sagan. (For a more detailed version, see the "You Are Here in Time" foldout diagram in the front of the book.)

How do our lifetimes compare to the age of the universe?

We can put the 14-billion-year age of the universe into perspective by imagining this time compressed into a single year, so each month represents a little more than 1 billion years. On this *cosmic calendar*, the Big Bang occurred at the first instant of January 1 and the present is the stroke of midnight on December 31 (**FIGURE 1.11**).

On this time scale, the Milky Way Galaxy probably formed in February. Many generations of stars lived and died in the subsequent cosmic months, enriching the galaxy with the "star stuff" from which we and our planet are made.

Our solar system and our planet did not form until early September on this scale ($4\frac{1}{2}$ billion years ago in real time). By late September, life on Earth was flourishing. However, for most of Earth's history, living organisms remained relatively primitive and microscopic. On the scale of the cosmic calendar, recognizable animals became prominent only in mid-December. Early dinosaurs appeared on the day after Christmas. Then, in a cosmic instant, the dinosaurs disappeared forever—probably because of the impact of an asteroid or a comet [**Section 12.4**]. In real time the death of the dinosaurs occurred some 65 million years ago, but on the cosmic calendar it was only yesterday. With the dinosaurs gone, small furry mammals inherited Earth. Some 60 million years later, or around 9 p.m. on December 31 of the cosmic calendar, early hominids (human ancestors) began to walk upright.

Perhaps the most astonishing fact about the cosmic calendar is that the entire history of human civilization falls into just the last half-minute. The ancient Egyptians built the pyramids only about 11 seconds ago on this scale. About 1 second ago, Kepler and Galileo proved that Earth orbits the Sun rather than vice versa. The average college student

was born about 0.05 second ago, around 11:59:59.95 p.m. on the cosmic calendar. On the scale of cosmic time, the human species is the youngest of infants, and a human lifetime is a mere blink of an eye.

THINK ABOUT IT

Study the backside of the foldout in the front of this book, which shows a more detailed version of the cosmic calendar. How does an understanding of the scale of time affect your view of human civilization? Explain.

1.3 SPACESHIP EARTH

Wherever you are as you read this book, you probably have the feeling that you're "just sitting here." Nothing could be further from the truth. As we'll discuss in this section, all of us are moving through space in so many ways that noted inventor and philosopher R. Buckminster Fuller (1895–1983) described us as travelers on *spaceship Earth.*

How is Earth moving through space?

As you "sit" on spaceship Earth, you are in fact being spun in circles as Earth rotates, you are racing around the Sun in Earth's orbit, you are circling the galactic center with our Sun, and you are careening through the cosmos in the Milky Way Galaxy. Let's explore each of these motions in a little more detail.

Rotation and Orbit The most basic motions of Earth are its daily **rotation** (spin) and its yearly **orbit** (or *revolution*) around the Sun.

Earth rotates once each day around its axis, which is the imaginary line connecting the North Pole to the South Pole. Earth rotates from west to east—counterclockwise as

December 31:

9:00 pm:
Early hominids evolve

11:58 pm:
Modern humans evolve

25 seconds ago:
Agriculture arises

11 seconds ago:
Pyramids built

1 second ago:
Kepler and Galileo
show that Earth
orbits the Sun

Now

DECEMBER 31

Morning...
12:00 noon
1:00 pm
2:00 pm
3:00 pm
4:00 pm
5:00 pm
6:00 pm
7:00 pm
8:00 pm
9:00 pm
10:00 pm
11:00 pm
11:58 pm
11:59 pm
12:00 midnight

viewed from above the North Pole—which is why the Sun and stars appear to rise in the east and set in the west each day. Although the physical effects of rotation are so subtle that our ancestors assumed the heavens revolved around us, the rotation speed is substantial (**FIGURE 1.12**): Unless you live quite far north or south, you are whirling around Earth's axis at a speed of more than 1000 kilometers per hour (600 miles per hour)—faster than most airplanes travel.

At the same time as it is rotating, Earth also orbits the Sun, completing one orbit each year (**FIGURE 1.13**). Earth's orbital distance varies slightly over the course of each year, but as we discussed earlier, the average distance is one astronomical unit (AU), which is about 150 million kilometers. Again, even though we don't feel this motion, the speed is impressive: We are racing around the Sun at a speed in excess of

100,000 kilometers per hour (60,000 miles per hour), which is faster than any spacecraft yet launched.

As you study Figure 1.13, notice that Earth's orbital path defines a flat plane that we call the **ecliptic plane**. Earth's axis is tilted by $23\frac{1}{2}°$ from a line *perpendicular* to the ecliptic plane. This **axis tilt** happens to be oriented so that the axis points almost directly at a star called *Polaris*, or the *North Star*. Keep in mind that the idea of axis tilt makes sense only in relation to the ecliptic plane. That is, the idea of "tilt" by itself has no meaning in space, where there is no absolute up or down. In space, "up" and "down" mean only "away from the center of Earth" (or another planet) and "toward the center of Earth," respectively.

THINK ABOUT IT

If there is no up or down in space, why do you think that most globes and maps have the North Pole on top? Would it be equally correct to have the South Pole on top or to turn a globe sideways? Explain.

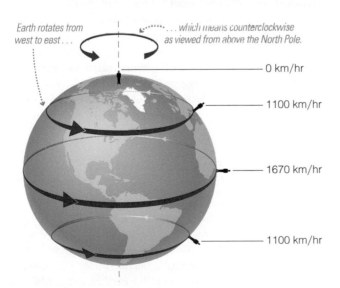

Earth rotates from west to east . . .

. . . which means *counterclockwise* as viewed from above the North Pole.

0 km/hr

1100 km/hr

1670 km/hr

1100 km/hr

FIGURE 1.12 As Earth rotates, your speed around Earth's axis depends on your location: The closer you are to the equator, the faster you travel with rotation.

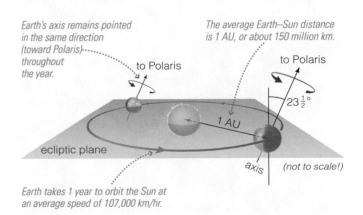

Earth's axis remains pointed in the same direction (toward Polaris) throughout the year.

to Polaris

The average Earth–Sun distance is 1 AU, or about 150 million km.

to Polaris

$23\frac{1}{2}°$

1 AU

ecliptic plane

axis

(not to scale!)

Earth takes 1 year to orbit the Sun at an average speed of 107,000 km/hr.

FIGURE 1.13 Earth orbits the Sun at a surprisingly high speed. Notice that Earth both rotates and orbits counterclockwise as viewed from above the North Pole.

Notice also that Earth orbits the Sun in the same direction that it rotates on its axis: counterclockwise as viewed from above the North Pole. This is not a coincidence but a consequence of the way our planet was born. As we'll discuss in Chapter 8, strong evidence indicates that Earth and the other planets were born in a spinning disk of gas that surrounded our Sun when it was young, and Earth rotates and orbits in the same direction that the disk was spinning.

Motion Within the Local Solar Neighborhood

Rotation and orbit are only a small part of the travels of spaceship Earth. Our entire solar system is on a great journey within the Milky Way Galaxy. There are two major components to this motion, both shown in **FIGURE 1.14**. Let's begin with our motion relative to other stars in our *local solar neighborhood*, by which we mean the region of the Sun and nearby stars.

To get a sense of the size of our local solar neighborhood relative to the galaxy, imagine drawing a tiny dot on the painting of the galaxy. Because the galaxy contains at least 100 billion stars, even a dot that is 10,000 times smaller than the whole painting will cover a region representing more than 10 million stars (because 100 billion ÷ 10,000 = 10 million). We usually think of our local solar neighborhood as a region containing just a few thousand to a few million of the nearest stars.

The arrows in the box in Figure 1.14 indicate that stars in our local solar neighborhood move essentially at random relative to one another. The speeds are quite fast: On average, our Sun is moving relative to nearby stars at a speed of about 70,000 kilometers per hour (40,000 miles per hour), almost three times as fast as the Space Station orbits Earth.

Given these high speeds, you might wonder why we don't see stars racing around our sky. The answer lies in their vast distances from us. You've probably noticed that a distant airplane appears to move through your sky more slowly than one flying close overhead. Stars are so far away that even at speeds of 70,000 kilometers per hour, their motions would be noticeable to the naked eye only if we watched them for thousands of years. That is why the patterns in the constellations seem to remain fixed. Nevertheless, in 10,000 years the constellations will be noticeably different from those we see today.

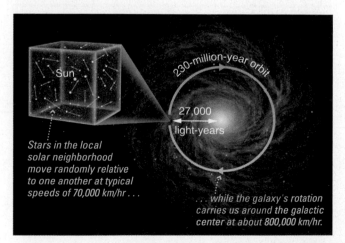

Stars in the local solar neighborhood move randomly relative to one another at typical speeds of 70,000 km/hr . . .

. . . while the galaxy's rotation carries us around the galactic center at about 800,000 km/hr.

230-million-year orbit

27,000 light-years

Sun

FIGURE 1.14 This painting illustrates the motion of the Sun both within the local solar neighborhood and around the center of the galaxy.

In 500,000 years they will be unrecognizable. If you could watch a time-lapse movie made over millions of years, you *would* see stars racing across our sky.

Galactic Rotation If you look closely at leaves floating in a stream, their motions relative to one another might appear random, just like the motions of stars in the local solar neighborhood. As you widen your view, you see that all the leaves are being carried in the same general direction by the downstream current. In the same way, as we widen our view beyond the local solar neighborhood, the seemingly random motions of its stars give way to a simpler and even faster motion: rotation of the Milky Way Galaxy. Our solar system, located about 27,000 light-years from the galactic center, completes one orbit of the galaxy in about 230 million years. Even if you could watch from outside our galaxy, this motion would be unnoticeable to your naked eye. However, if you calculate the speed of our solar system as we orbit the center of the galaxy, you will find that it is close to 800,000 kilometers (500,000 miles) per hour.

Careful study of the galaxy's rotation reveals one of the greatest mysteries in science. Stars at different distances from the galactic center orbit at different speeds, and we can learn how mass is distributed in the galaxy by measuring these speeds. Such studies indicate that the stars in the disk of the galaxy represent only the "tip of the iceberg" compared to the mass of the entire galaxy (**FIGURE 1.15**). Most of the mass of the galaxy seems to be located outside the visible disk (occupying the galactic *halo* that surrounds and encompasses the disk), but the matter that makes up this mass is completely invisible to our telescopes. We therefore know very little about the nature of this matter, which we refer to as *dark matter* (because of the lack of light from it). Studies of other galaxies suggest that they also are made mostly of dark matter, which means this mysterious matter must significantly outweigh the ordinary matter that makes up planets and stars. We know even less about the mysterious *dark energy* that seems to make up much of the total energy content of the universe. We'll discuss the mysteries of dark matter and dark energy in Chapter 23.

How do galaxies move within the universe?

The billions of galaxies in the universe also move relative to one another. Within the Local Group (see Figure 1.1), some of the galaxies move toward us, some move away from us, and at least two small galaxies (known as the Large and Small Magellanic Clouds) apparently orbit our Milky Way Galaxy. Again, the speeds are enormous by earthly standards. For example, the Milky Way is moving toward the Andromeda Galaxy at about 300,000 kilometers per hour (180,000 miles per hour). Despite this high speed, we needn't worry about a

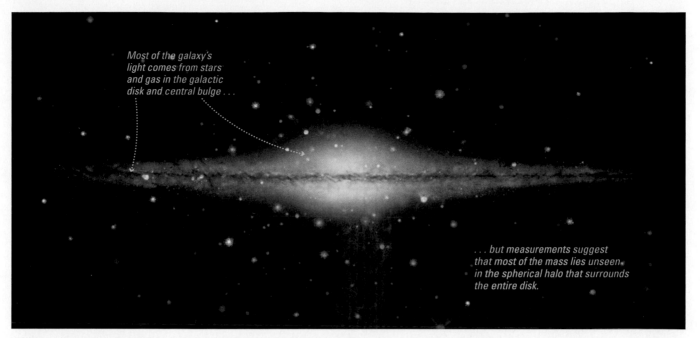

Most of the galaxy's light comes from stars and gas in the galactic disk and central bulge . . .

. . . but measurements suggest that most of the mass lies unseen in the spherical halo that surrounds the entire disk.

FIGURE 1.15 This painting shows an edge-on view of the Milky Way Galaxy. Study of galactic rotation shows that although most visible stars lie in the central bulge or thin disk, most of the mass lies in the halo that surrounds and encompasses the disk. Because this mass emits no light that we have detected, we call it *dark matter*.

MATHEMATICAL INSIGHT 1.4

Speeds of Rotation and Orbit

 MA **Math Review Video: Problem Solving, Part 3**

Building upon prior Mathematical Insights, we will now see how simple formulas—such as the formula for the circumference of a circle—expand the range of astronomical problems we can solve.

EXAMPLE 1: How fast is a person on Earth's equator moving with Earth's rotation?

SOLUTION:

Step 1 Understand: The question *how fast* tells us we are looking for a *speed*. If you remember that highway speeds are posted in miles (or kilometers) per hour, you'll realize that speed is a distance (such as miles) divided by a time (such as hours). In this case, the distance is Earth's equatorial circumference, because that is how far a person at the equator travels with each rotation (see Figure 1.12); we'll therefore use the formula for the circumference of a circle, $C = 2 \times \pi \times \text{radius}$. The time is 24 hours, because that is how long each rotation takes.

Step 2 Solve: From Appendix E.1, Earth's equatorial radius is 6378 km, so its circumference is $2 \times \pi \times 6378 \text{ km} = 40{,}074 \text{ km}$. We divide this distance by the time of 24 hours:

$$\text{rotation speed at equator} = \frac{\text{equatorial circumference}}{\text{length of day}}$$

$$= \frac{40{,}074 \text{ km}}{24 \text{ hr}} = 1670 \frac{\text{km}}{\text{hr}}$$

Step 3 Explain: A person at the equator is moving with Earth's rotation at a speed of about 1670 kilometers per hour, which is a little over 1000 miles per hour, or about twice the flying speed of a commercial jet.

EXAMPLE 2: How fast is Earth orbiting the Sun?

SOLUTION:

Step 1 Understand: We are again asked *how fast* and therefore need to divide a distance by a time. In this case, the distance is the circumference of Earth's orbit, and the time is the 1 year that Earth takes to complete each orbit.

Step 2 Solve: Earth's average distance from the Sun is 1 AU, or about 150 million (1.5×10^8) km, so the orbit circumference is about $2 \times \pi \times 1.5 \times 10^8 \text{ km} \approx 9.40 \times 10^8 \text{ km}$. The orbital speed is this distance divided by the time of 1 year, which we convert to hours so that we end up with units of km/hr:

$$\text{orbital speed} = \frac{\text{orbital circumference}}{1 \text{ yr}}$$

$$= \frac{9.40 \times 10^8 \text{ km}}{1 \text{ yr} \times \frac{365 \text{ day}}{\text{yr}} \times \frac{24 \text{ hr}}{\text{day}}} \approx 107{,}000 \frac{\text{km}}{\text{hr}}$$

Step 3 Explain: Earth orbits the Sun at an average speed of about 107,000 km/hr (66,000 mi/hr). Most "speeding bullets" travel between about 500 and 1000 km/hr, so Earth's orbital speed is more than 100 times as fast as a speeding bullet.

collision anytime soon. Even if the Milky Way and Andromeda Galaxies are approaching each other head-on, it will be billions of years before any collision begins.

When we look outside the Local Group, however, we find two astonishing facts recognized in the 1920s by Edwin Hubble, for whom the Hubble Space Telescope was named:

1. Virtually every galaxy outside the Local Group is moving *away* from us.

2. The more distant the galaxy, the faster it appears to be racing away.

These facts might make it sound as if we suffered from a cosmic case of chicken pox, but there is a much more natural explanation: *The entire universe is expanding.* We'll save the details for later in the book, but you can understand the basic idea by thinking about a raisin cake baking in an oven.

The Raisin Cake Analogy Imagine that you make a raisin cake in which the distance between adjacent raisins is 1 centimeter. You place the cake into the oven, where it expands as it bakes. After 1 hour, you remove the cake, which has expanded so that the distance between adjacent raisins has increased to 3 centimeters (**FIGURE 1.16**). The expansion of the cake seems fairly obvious. But what would you see if you lived *in* the cake, as we live in the universe?

Pick any raisin (it doesn't matter which one) and call it the Local Raisin. Figure 1.16 shows one possible choice, with three nearby raisins also labeled. The accompanying table summarizes what you would see if you lived within the Local Raisin. Notice, for example, that Raisin 1 starts out at a distance of 1 centimeter before baking and ends up

at a distance of 3 centimeters after baking, which means it moves a distance of 2 centimeters away from the Local Raisin during the hour of baking. Hence, its speed as seen from the Local Raisin is 2 centimeters per hour. Raisin 2 moves from a distance of 2 centimeters before baking to a distance of 6 centimeters after baking, which means it moves a distance of 4 centimeters away from the Local Raisin during the hour. Hence, its speed is 4 centimeters per hour, or twice the speed of Raisin 1. Generalizing, the fact that the cake is expanding means that all the raisins are moving away from the Local Raisin, with more distant raisins moving away faster.

Hubble's discovery that galaxies are moving in much the same way as the raisins in the cake, with most moving away from us and more distant ones moving away faster, implies that the universe is expanding much like the raisin cake. If you now imagine the Local Raisin as representing our Local Group of galaxies and the other raisins as representing more distant galaxies or clusters of galaxies, you have a basic picture of the expansion of the universe. Like the expanding dough between the raisins in the cake, *space* itself is growing between galaxies. More distant galaxies move away from us faster because they are carried along with this expansion like the raisins in the expanding cake. Many billions of light-years away, we see galaxies moving away from us at speeds approaching the speed of light.

The Real Universe There's at least one important distinction between the raisin cake and the universe: A cake has a center and edges, but we do not think the same is true

From an outside perspective, the cake expands uniformly as it bakes . . .

. . . but from the point of view of the Local Raisin, all other raisins move farther away during baking, with more distant raisins moving faster.

Distances and Speeds as Seen from the Local Raisin

Raisin Number	Distance Before Baking	Distance After Baking (1 hour later)	Speed
1	1 cm	3 cm	2 cm/hr
2	2 cm	6 cm	4 cm/hr
3	3 cm	9 cm	6 cm/hr
⋮	⋮	⋮	⋮

FIGURE 1.16 interactive figure An expanding raisin cake offers an analogy to the expanding universe. Someone living in one of the raisins inside the cake could figure out that the cake is expanding by noticing that all other raisins are moving away, with more distant raisins moving away faster. In the same way, we know that we live in an expanding universe because all galaxies outside our Local Group are moving away from us, with more distant ones moving faster.

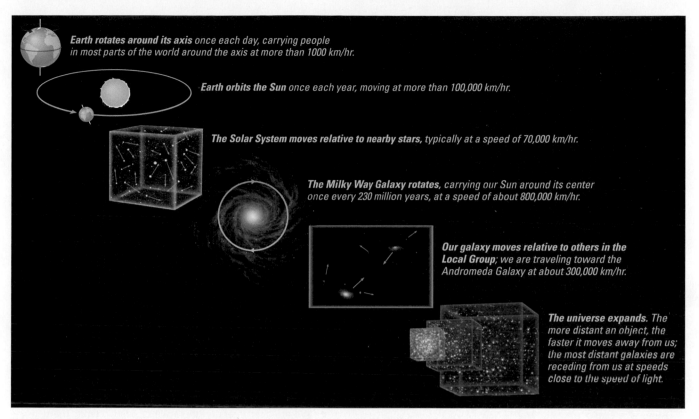

FIGURE 1.17 This figure summarizes the basic motions of Earth in the universe, along with their associated speeds.

of the entire universe. Anyone living in any galaxy in an expanding universe sees just what we see—other galaxies moving away, with more distant ones moving away faster. Because the view from each point in the universe is about the same, no place can claim to be more "central" than any other place.

It's also important to realize that, unlike the case with a raisin cake, we can't actually *see* galaxies moving apart with time—the distances are too vast for any motion to be noticeable on the time scale of a human life. Instead, we measure the speeds of galaxies by spreading their light into spectra and observing what we call *Doppler shifts* [**Section 5.4**]. This illustrates how modern astronomy depends both on careful observations and on using current understanding of the laws of nature to explain what we see.

Motion Summary FIGURE 1.17 summarizes the motions we have discussed. As we have seen, we are never truly sitting still. We spin around Earth's axis at more than 1000 kilometers per hour, while our planet orbits the Sun at more than 100,000 kilometers per hour. Our solar system moves among the stars of the local solar neighborhood at typical speeds of 70,000 kilometers per hour, while also orbiting the center of the Milky Way Galaxy at a speed of about 800,000 kilometers per hour. Our galaxy moves among the other galaxies of the Local Group, while all other galaxies move away from us at speeds that grow greater with distance in our expanding universe. Spaceship Earth is carrying us on a remarkable journey.

1.4 THE HUMAN ADVENTURE OF ASTRONOMY

In relatively few pages, we've laid out a fairly complete overview of modern scientific ideas about the universe. But our goal in this book is not simply for you to be able to recite these ideas. Rather, it is to help you understand the evidence that supports them and the extraordinary story of how they developed.

How has the study of astronomy affected human history?

Astronomy is a human adventure in the sense that it affects everyone—even those who have never looked at the sky—because the history of astronomy has been so deeply intertwined with the development of civilization. Revolutions in astronomy have gone hand in hand with the revolutions in science and technology that have shaped modern life.

Witness the repercussions of the *Copernican revolution*, which showed us that Earth is not the center of the universe but rather just one planet orbiting the Sun. This revolution, which we will discuss further in Chapter 3, began when Copernicus published his idea of a Sun-centered solar system in 1543. Three later figures—Tycho Brahe, Johannes Kepler, and Galileo—provided the key evidence that eventually led to wide acceptance of the Copernican idea. The revolution culminated with Isaac Newton's uncovering of the laws of motion and gravity. Newton's work, in turn, became the foundation of physics that helped fuel the industrial revolution.

More recently, the development of space travel and the computer revolution have helped fuel tremendous progress in astronomy. We've sent probes to all the planets in our solar system, and many of our most powerful observatories, including the Hubble Space Telescope, reside in space. On the ground, computer design and control have led to tremendous growth in the size and power of telescopes.

Many of these efforts, and the achievements they spawned, led to profound social change. The most famous example is the fate of Galileo, whom the Vatican put under house arrest in 1633 for his claims that Earth orbits the Sun. Although the Church soon recognized that Galileo was right, he was formally vindicated only in 1992 with a statement by Pope John Paul II. In the meantime, his case spurred great debate in religious circles and profoundly influenced both theological and scientific thinking.

As you progress through this book and learn about astronomical discovery, try to keep in mind the context of the human adventure. You will then be learning not just about a science, but also about one of the great forces that has shaped our modern world.

These forces will continue to play a role in our future. What will it mean to us when we learn the nature of dark matter and dark energy? How will our view of Earth change when we learn whether life is common or rare in the universe? Only time may answer these questions, but the chapters ahead will give you the foundation you need to understand how we changed from a primitive people looking at patterns in the night sky to a civilization capable of asking deep questions about our existence.

The Big Picture

Putting Chapter 1 into Context

In this first chapter, we developed a broad overview of our place in the universe. As we consider the universe in more depth in the rest of the book, remember the following "big picture" ideas:

- Earth is not the center of the universe but instead is a planet orbiting a rather ordinary star in the Milky Way Galaxy. The Milky Way Galaxy, in turn, is one of billions of galaxies in our observable universe.

- Cosmic distances are literally astronomical, but we can put them in perspective with the aid of scale models and other scaling techniques. When you think about these enormous scales, don't forget that every star is a sun and every planet is a unique world.

- We are "star stuff." The atoms from which we are made began as hydrogen and helium in the Big Bang and were later fused into heavier elements by massive stars. Stellar deaths released these atoms into space, where our galaxy recycled them into new stars and planets. Our solar system formed from such recycled matter some $4\frac{1}{2}$ billion years ago.

- We are latecomers on the scale of cosmic time. The universe was already more than half its current age when our solar system formed, and it took billions of years more before humans arrived on the scene.

- All of us are being carried through the cosmos on spaceship Earth. Although we cannot feel this motion in our everyday lives, the associated speeds are surprisingly high. Learning about the motions of spaceship Earth gives us a new perspective on the cosmos and helps us understand its nature and history.

SUMMARY OF KEY CONCEPTS

1.1 THE SCALE OF THE UNIVERSE

- **What is our place in the universe?** Earth is a planet orbiting the Sun. Our Sun is one of more than 100 billion stars in the **Milky Way Galaxy**. Our galaxy is one of about 40 galaxies in the **Local Group**. The Local Group is one small part of the **Local Supercluster**, which is one small part of the **universe**.

- **How big is the universe?** If we imagine our Sun as a large grapefruit, Earth is a ball point that orbits 15 meters away; the nearest stars are thousands of kilometers away on the same scale. Our galaxy contains more than 100 billion stars—so many that it would take thousands of years just to count them out loud.

The observable universe contains roughly 100 billion galaxies, and the total number of stars is comparable to the number of grains of dry sand on all the beaches on Earth.

1.2 THE HISTORY OF THE UNIVERSE

- **How did we come to be?** The universe began in the **Big Bang** and has been expanding ever since, except in localized regions where gravity has caused matter to collapse into galaxies and stars. The Big Bang essentially produced only two chemical elements: hydrogen and helium. The rest have been produced by stars and recycled within galaxies from one generation of stars to the next, which is why we are "star stuff."

- **How do our lifetimes compare to the age of the universe?** On a cosmic calendar that compresses the history of the universe into 1 year, human civilization is just a few seconds old, and a human lifetime lasts only a fraction of a second.

1.3 SPACESHIP EARTH

■ **How is Earth moving through space?** Earth **rotates** on its axis once each day and **orbits** the Sun once each year. At the same time, we move with our Sun

in random directions relative to other stars in our local solar neighborhood, while the galaxy's rotation carries us around the center of the galaxy every 230 million years.

■ **How do galaxies move within the universe?** Galaxies move essentially at random within the Local Group, but all galaxies beyond the Local Group are moving away from us. More distant galaxies are moving faster, which tells us that we live in an expanding universe.

1.4 THE HUMAN ADVENTURE OF ASTRONOMY

■ **How has the study of astronomy affected human history?** Throughout history, astronomy has developed hand in hand with social and technological development. Astronomy thereby touches all of us and is a human adventure that all can enjoy.

EXERCISES AND PROBLEMS

MasteringAstronomy®

For instructor-assigned homework go to MasteringAstronomy®.

REVIEW QUESTIONS

Short-Answer Questions Based on the Reading

1. Briefly describe the major levels of structure (such as planet, star, galaxy) in the universe.
2. Define *astronomical unit* and *light-year*.
3. Explain the statement *The farther away we look in distance, the further back we look in time.*
4. What do we mean by the *observable universe*? Is it the same thing as the entire universe?
5. Using techniques described in the chapter, put the following into perspective: the size of our solar system; the distance to nearby stars; the size and number of stars in the Milky Way Galaxy; the number of stars in the observable universe.
6. What do we mean when we say that the universe is *expanding*, and how does expansion lead to the idea of the *Big Bang* and our current estimate of the age of the universe?
7. In what sense are we "star stuff"?

8. Use the cosmic calendar to describe how the human race fits into the scale of time.

9. Briefly explain Earth's daily rotation and annual orbit, defining the terms *ecliptic plane* and *axis tilt*.

10. Briefly describe our solar system's location and motion within the Milky Way Galaxy.

11. Where does *dark matter* seem to reside in our galaxy? What makes dark matter and *dark energy* so mysterious?

12. What key observations lead us to conclude that the universe is expanding? Use the raisin cake model to explain how these observations imply expansion.

TEST YOUR UNDERSTANDING

Does It Make Sense?

Decide whether the statement makes sense (or is clearly true) or does not make sense (or is clearly false). Explain clearly; not all of these have definitive answers, so your explanation is more important than your chosen answer.

Example: I walked east from our base camp at the North Pole.

Solution: The statement does not make sense because east has no meaning at the North Pole—all directions are south from the North Pole.

13. Our solar system is bigger than some galaxies.

14. The universe is billions of light-years in age.

15. It will take me light-years to complete this homework assignment!

16. Someday we may build spaceships capable of traveling a light-year in only a decade.

17. Astronomers recently discovered a moon that does not orbit a planet.

18. NASA plans soon to launch a spaceship that will photograph our Milky Way Galaxy from beyond its halo.

19. The observable universe is the same size today as it was a few billion years ago.

20. Photographs of distant galaxies show them as they were when they were much younger than they are today.

21. At a nearby park, I built a scale model of our solar system in which I used a basketball to represent Earth.

22. Because nearly all galaxies are moving away from us, we must be located at the center of the universe.

Quick Quiz

Choose the best answer to each of the following. Explain your reasoning with one or more complete sentences.

23. Which of the following correctly lists our "cosmic address" from small to large? (a) Earth, solar system, Milky Way Galaxy, Local Group, Local Supercluster, universe (b) Earth, solar system, Local Group, Local Supercluster, Milky Way Galaxy, universe (c) Earth, Milky Way Galaxy, solar system, Local Group, Local Supercluster, universe.

24. An astronomical unit is (a) any planet's average distance from the Sun. (b) Earth's average distance from the Sun. (c) any large astronomical distance.

25. The star Betelgeuse is about 600 light-years away. If it explodes tonight, (a) we'll know because it will be brighter than the full Moon in the sky. (b) we'll know because debris from the explosion will rain down on us from space. (c) we won't know about it until about 600 years from now.

26. If we represent the solar system on a scale that allows us to walk from the Sun to Pluto in a few minutes, then (a) the planets are the size of basketballs and the nearest stars are a few miles away. (b) the planets are marble-size or smaller and the nearest stars are thousands of miles away. (c) the planets are microscopic and the stars are light-years away.

27. The total number of stars in the observable universe is roughly equivalent to (a) the number of grains of sand on all the beaches on Earth. (b) the number of grains of sand on Miami Beach. (c) infinity.

28. When we say the universe is *expanding,* we mean that (a) everything in the universe is growing in size. (b) the average distance between galaxies is growing with time. (c) the universe is getting older.

29. If stars existed but galaxies did not, (a) we would probably still exist anyway. (b) we would not exist because life on Earth depends on the light of galaxies. (c) we would not exist because we are made of material that was recycled in galaxies.

30. Could we see a galaxy that is 50 billion light-years away? (a) Yes, if we had a big enough telescope. (b) No, because it would be beyond the bounds of our observable universe. (c) No, because a galaxy could not possibly be that far away.

31. The age of our solar system is about (a) one-third of the age of the universe. (b) three-fourths of the age of the universe. (c) two billion years less than the age of the universe.

32. The fact that nearly all galaxies are moving away from us, with more distant ones moving faster, helped us to conclude that (a) the universe is expanding. (b) galaxies repel each other like magnets. (c) our galaxy lies near the center of the universe.

PROCESS OF SCIENCE

Examining How Science Works

33. *Earth as a Planet.* For most of human history, scholars assumed Earth was the center of the universe. Today, we know that our Sun is just one star in a vast universe. How did science make it possible for us to learn these facts about Earth?

34. *Thinking About Scale.* One key to success in science is finding simple ways to evaluate new ideas, and making a simple scale model is often helpful. Suppose someone tells you that the reason it is warmer during the day than at night is that the day side of Earth is closer to the Sun than the night side. Evaluate this idea by thinking about the size of Earth and its distance from the Sun in a scale model of the solar system.

35. *Looking for Evidence.* In this first chapter, we have discussed the scientific story of the universe but have not yet discussed most of the evidence that backs it up. Choose one idea presented in this chapter—such as the idea that there are billions of galaxies in the universe, or that the universe was born in the Big Bang, or that the galaxy contains more dark matter than ordinary matter—and briefly discuss the type of evidence you would want to see before accepting the idea. (*Hint:* It's okay to look ahead in the book to see the evidence presented in later chapters.)

GROUP WORK EXERCISE

36. *Counting the Milky Way's Stars.* In this exercise, you will first make an estimate of the number of stars in the Milky Way and then apply some scientific thinking to your estimation method. Before you begin, assign the following roles to the people in your group: *Scribe* (takes notes on the group's activities), *Proposer* (proposes explanations to the group), *Skeptic* (points out weaknesses in proposed explanations), and *Moderator* (leads group discussion and makes sure everyone contributes).
    a. Estimate the number of stars in the Milky Way as follows. First, count the number of stars within 12 light-years of the Sun, which are listed in Appendix F. Assuming that the Milky Way's disk is 100,000 light-years across and 1000 light-years thick, its volume is about 1 billion times the volume of the region of your star count. You should therefore multiply your count by 1 billion to get an estimate of the total number of stars in the Milky Way. **b.** Your estimate from part a is based on the number of stars near the Sun. Compare

it to the value given in this chapter and determine whether your estimate is an underestimate or an overestimate of the total number of stars in the Milky Way. Write down a list of possible reasons why your technique gave you an under/overestimate.

INVESTIGATE FURTHER

In-Depth Questions to Increase Your Understanding

Short-Answer/Essay Questions

37. *Alien Technology.* Some people believe that Earth is regularly visited by aliens who travel here from other star systems. For this to be true, how much more advanced than our own technology would the alien space travel technology have to be? Write one to two paragraphs to give a sense of the technological difference. (*Hint:* Use the scale model from this chapter to contrast the distance the aliens would have to travel easily with the distances we currently are capable of traveling.)

38. *Raisin Cake Universe.* Suppose that all the raisins in a cake are 1 centimeter apart before baking and 4 centimeters apart after baking.
    a. Draw diagrams to represent the cake before and after baking.
    b. Identify one raisin as the Local Raisin on your diagrams. Construct a table showing the distances and speeds of other raisins as seen from the Local Raisin. **c.** Briefly explain how your expanding cake is similar to the expansion of the universe.

39. *Scaling the Local Group of Galaxies.* Both the Milky Way Galaxy and the Andromeda Galaxy (M31) have a diameter of about 100,000 light-years. The distance between the two galaxies is about 2.5 million light-years.
    a. Using a scale on which 1 centimeter represents 100,000 light-years, draw a sketch showing both galaxies and the distance between them to scale. **b.** How does the separation between galaxies compare to the separation between stars? Based on your answer, discuss the likelihood of galactic collisions in comparison to the likelihood of stellar collisions.

40. *The Cosmic Perspective.* Write a short essay describing how the ideas presented in this chapter affect your perspectives on your own life and on human civilization.

Quantitative Problems

Be sure to show all calculations clearly and state your final answers in complete sentences.

41. *Distances by Light.* Just as a light year is the distance that light can travel in 1 year, we define a light-second as the distance that light can travel in 1 second, a light minute as the distance that light can travel in 1 minute, and so on. Calculate the distance in both kilometers and miles represented by each of the following:
    a. 1 light-second. **b.** 1 light-minute. **c.** 1 light-hour. **d.** 1 light-day.

42. *Spacecraft Communication.* We use radio waves, which travel at the speed of light, to communicate with robotic spacecraft. How long does it take a message to travel from Earth to a spacecraft at
    a. Mars at its closest to Earth (about 56 million km)? **b.** Mars at its farthest from Earth (about 400 million km)? **c.** Pluto at its average distance from Earth (about 5.9 billion km)?

43. *Saturn vs. the Milky Way.* Photos of Saturn and photos of galaxies can look so similar that children often think the photos show similar objects. In reality, a galaxy is far larger than any planet. About how many times larger is the diameter of the Milky Way Galaxy than the diameter of Saturn's rings? (Data: Saturn's rings are about 270,000 km in diameter; the Milky Way is 100,000 light-years in diameter.)

44. *Galaxy Scale.* Consider the 1-to-10^{19} scale on which the disk of the Milky Way Galaxy fits on a football field. On this scale, how far is it

from the Sun to Alpha Centauri (real distance: 4.4 light-years)? How big is the Sun itself on this scale? Compare the Sun's size on this scale to the actual size of a typical atom (about 10^{-10} m in diameter).

45. *Universal Scale.* Suppose we wanted to make a scale model of the Local Group of galaxies, in which the Milky Way Galaxy was the size of a marble (about 1 cm in diameter).
    a. How far from the Milky Way Galaxy would the Andromeda Galaxy be on this scale? **b.** How far would the Sun be from Alpha Centauri on this scale? **c.** How far would it be from the Milky Way Galaxy to the most distant galaxies in the observable universe on this scale?

46. *Driving Trips.* Imagine that you could drive your car at a constant speed of 100 km/hr (62 mi/hr), even across oceans and in space. (In reality, the law of gravity would make driving through space at a constant speed all but impossible.) How long would it take to drive
    a. around Earth's equator? **b.** from the Sun to Earth? **c.** from the Sun to Pluto? **d.** to Alpha Centauri?

47. *Faster Trip.* Suppose you wanted to reach Alpha Centauri in 100 years.
    a. How fast would you have to go, in km/hr? **b.** How many times faster is the speed you found in part a than the speeds of our fastest current spacecraft (around 50,000 km/hr)?

48. *Galactic Rotation Speed.* We are located about 27,000 light-years from the galactic center and we orbit the center about once every 230 million years. How fast are we traveling around the galaxy, in km/hr?

49. *Earth Rotation Speed.* Mathematical Insight 1.3 shows how to find Earth's equatorial rotation speed. To find the rotation speed at any other latitude, you need the following fact: The radial distance from Earth's axis at any latitude is equal to the equatorial radius times the *cosine* of the latitude. Use this fact to find the rotation speed at the following latitudes. (*Hint:* When using the cosine (cos) function, be sure your calculator is set to recognize angles in degree mode, not in radian or gradient mode.)
    a. 30°N **b.** 60°N **c.** your latitude.

Discussion Questions

50. *Eliot Quote.* Think carefully about the chapter opening quotation from T. S. Eliot. What do you think he means? Explain clearly.

51. *Infant Species.* In the last few tenths of a second before midnight on December 31 of the cosmic calendar, we have developed an incredible civilization and learned a great deal about the universe, but we also have developed technology with which we could destroy ourselves. The midnight bell is striking, and the choice for the future is ours. How far into the next cosmic year do you think our civilization will survive? Defend your opinion.

52. *A Human Adventure.* Astronomical discoveries clearly are important to science, but are they also important to our personal lives? Defend your opinion.

Web Projects

53. *Astronomy on the Web.* The Web contains a vast amount of astronomical information. Spend at least an hour exploring astronomy on the Web. Write two or three paragraphs summarizing what you learned from your research. What was your favorite astronomical website, and why?

54. *NASA Missions.* Visit the NASA website to learn about upcoming astronomy missions. Write a one-page summary of the mission you believe is most likely to give us new astronomical information before the end of your astronomy course.

55. *The Hubble Ultra Deep Field.* The photo that opens this chapter is called the Hubble Ultra Deep Field. Find this photo on the Hubble Space Telescope website. Learn how it was taken, what it shows, and what we've learned from it. Write a short summary of your findings.

2 DISCOVERING THE UNIVERSE FOR YOURSELF

We had the sky, up there, all speckled with stars, and we used to lay on our backs and look up at them, and discuss about whether they was made, or only just happened.
　　　—Mark Twain, *Huckleberry Finn*

This is an exciting time in the history of astronomy. A new generation of telescopes is scanning the depths of the universe. Increasingly sophisticated space probes are collecting new data about the planets and other objects in our solar system. Rapid advances in computing technology are allowing scientists to analyze the vast amount of new data and to model the processes that occur in planets, stars, galaxies, and the universe.

One goal of this book is to help *you* share in the ongoing adventure of astronomical discovery. One of the best ways to become a part of this adventure is to do what other humans have done for thousands of generations: Go outside, observe the sky around you, and contemplate the awe-inspiring universe of which you are a part. In this chapter, we'll discuss a few key ideas that will help you understand what you see in the sky.

2.1 PATTERNS IN THE NIGHT SKY

Today we take for granted that we live on a small planet orbiting an ordinary star in one of many galaxies in the universe. But this fact is not obvious from a casual glance at the night sky, and we've learned about our place in the cosmos only through a long history of careful observations. In this section, we'll discuss major features of the night sky and how we understand them in light of our current knowledge of the universe.

What does the universe look like from Earth?

Shortly after sunset, as daylight fades to darkness, the sky appears to slowly fill with stars. On clear, moonless nights far from city lights, more than 2000 stars may be visible to your naked eye, along with the whitish band of light that we call the Milky Way (**FIGURE 2.1**). As you look at the stars, your mind may group them into patterns that look like familiar shapes or objects. If you observe the sky night after night or year after year, you will recognize the same patterns of stars. These patterns have not changed noticeably in the past few thousand years.

Constellations People of nearly every culture gave names to patterns they saw in the sky. We usually refer to such patterns as constellations, but to astronomers the term has a more precise meaning: A **constellation** is a *region* of the sky with well-defined borders; the familiar patterns of stars merely help us locate the constellations. Just as every spot of land in the continental United States is part of some state, every point in the sky belongs to some constellation. **FIGURE 2.2** shows the borders of the constellation Orion and several of its neighbors.

FIGURE 2.1 This photo shows the Milky Way over Haleakala crater on the island of Maui, Hawaii. The bright spot just below (and slightly left of) the center of the band is the planet Jupiter.

The names and borders of the 88 official constellations (Appendix H) were chosen in 1928 by members of the International Astronomical Union (IAU). Most of the IAU members lived in Europe or the United States, so they chose names familiar in the western world. That is why the official names for constellations visible in the Northern Hemisphere can be traced back to civilizations of the ancient Middle East,

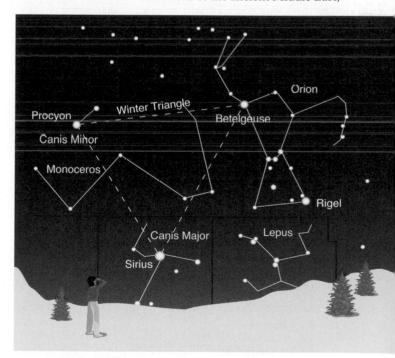

FIGURE 2.2 Red lines mark official borders of several constellations near Orion. Yellow lines connect recognizable patterns of stars within constellations. Sirius, Procyon, and Betelgeuse form a pattern that spans several constellations and is called the *Winter Triangle*. This view shows how it appears on winter evenings from the Northern Hemisphere.

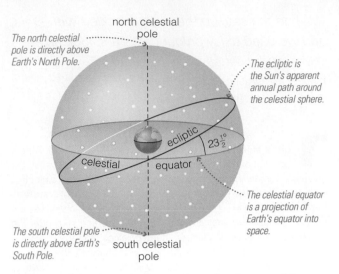

FIGURE 2.4 This schematic diagram shows key features of the celestial sphere.

FIGURE 2.3 The stars and constellations appear to lie on a celestial sphere that surrounds Earth. This is an illusion created by our lack of depth perception in space, but it is useful for mapping the sky.

while Southern Hemisphere constellations carry names that originated with 17th-century European explorers.

Recognizing the patterns of just 20 or so constellations is enough to make the sky seem as familiar as your own neighborhood. The best way to learn the constellations is to go out and view them, guided by a few visits to a planetarium, star charts (Appendix I), or sky-viewing apps for smart phones and tablets.

The Celestial Sphere The stars in a particular constellation appear to lie close to one another but may be quite far apart in reality, because they may lie at very different distances from Earth. This illusion occurs because we lack depth perception when we look into space, a consequence of the fact that the stars are so far away [**Section 1.1**]. The ancient Greeks mistook this illusion for reality, imagining the stars and constellations to lie on a great **celestial sphere** that surrounds Earth (**FIGURE 2.3**).

We now know that Earth seems to be in the center of the celestial sphere only because it is where we are located as we look into space. Nevertheless, the celestial sphere is a useful illusion, because it allows us to map the sky as seen from Earth. For reference, we identify four special points and circles on the celestial sphere (**FIGURE 2.4**).

- The **north celestial pole** is the point directly over Earth's North Pole.

- The **south celestial pole** is the point directly over Earth's South Pole.

- The **celestial equator,** which is a projection of Earth's equator into space, makes a complete circle around the celestial sphere.

- The **ecliptic** is the path the Sun follows as it appears to circle around the celestial sphere once each year. It crosses the celestial equator at a $23\frac{1}{2}°$ angle, because that is the tilt of Earth's axis.

The Milky Way The band of light that we call the Milky Way circles all the way around the celestial sphere, passing through more than a dozen constellations. The widest and brightest parts of the Milky Way are most easily seen from the Southern Hemisphere, which probably explains why the Aborigines of Australia gave names to patterns within the Milky Way in the same way other cultures named patterns of stars.

Our Milky Way Galaxy gets its name from this band of light, and the two "Milky Ways" are closely related: *The Milky Way in the night sky traces our galaxy's disk of stars—the galactic plane—as it appears from our location within the Milky Way Galaxy.* **FIGURE 2.5** shows the idea. Our galaxy is shaped like a thin pancake with a bulge in the middle. We view the universe from our location a little more than halfway out from the center of this "pancake." In all directions that we look within the pancake, we see the countless stars and vast interstellar clouds that make up the Milky Way in the night sky; that is why the band of light makes a full circle around our sky. The Milky Way appears somewhat wider in the direction of the constellation Sagittarius, because that is the direction in which we are looking toward the galaxy's central bulge. We have a clear view to the distant universe only when we look *away* from the galactic plane, along directions that have relatively few stars and clouds to block our view.

The dark lanes that run down the center of the Milky Way contain the densest clouds, obscuring our view of stars behind them. In fact, these clouds generally prevent us from seeing more than a few thousand light-years into our galaxy's disk. As a result, much of our own galaxy remained hidden from view until just a few decades ago, when new technologies allowed us to peer through the clouds by observing forms of light that are invisible to our eyes (such as radio waves and X rays [**Section 5.2**]).

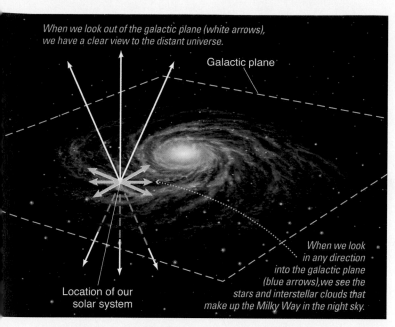

When we look out of the galactic plane (white arrows), we have a clear view to the distant universe.

Galactic plane

When we look in any direction into the galactic plane (blue arrows), we see the stars and interstellar clouds that make up the Milky Way in the night sky.

Location of our solar system

FIGURE 2.5 This painting shows how our galaxy's structure affects our view from Earth.

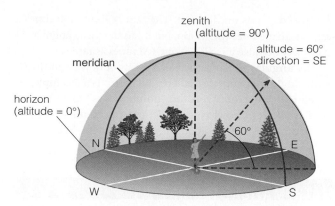

zenith (altitude = 90°)

altitude = 60° direction = SE

meridian

horizon (altitude = 0°)

60°

N   E

W   S

FIGURE 2.6 From any place on Earth, the local sky looks like a dome (hemisphere). This diagram shows key reference points in the local sky. It also shows how we can describe any position in the local sky by its altitude and direction.

THINK ABOUT IT

Consider a distant galaxy located in the same direction from Earth as the center of our own galaxy (but much farther away). Could we see it with our eyes? Explain.

The Local Sky The celestial sphere provides a useful way of thinking about the appearance of the universe from Earth. But it is not what we actually see when we go outside. Instead, your **local sky**—the sky as seen from wherever you happen to be standing—appears to take the shape of a hemisphere or dome, which explains why people of many ancient cultures imagined that we lived on a flat Earth under a great dome encompassing the world. The dome shape arises from the fact that we see only half of the celestial sphere at any particular moment from any particular location, while the ground blocks the other half from view.

FIGURE 2.6 shows key reference features of the local sky. The boundary between Earth and sky defines the **horizon.** The point directly overhead is the **zenith.** The **meridian** is an imaginary half circle stretching from the horizon due south, through the zenith, to the horizon due north.

We can pinpoint the position of any object in the local sky by stating its **direction** along the horizon (sometimes stated as *azimuth*, which is degrees clockwise from due north) and its **altitude** above the horizon. For example, Figure 2.6 shows a person pointing to a star located in the direction of southeast at an altitude of 60°. Note that the zenith has altitude 90° but no direction, because it is straight overhead.

Angular Sizes and Distances Our lack of depth perception on the celestial sphere means we have no way to judge the true sizes or separations of the objects we see in the sky. However, we can describe the *angular* sizes or separations of objects without knowing how far away they are.

The **angular size** of an object is the angle it appears to span in your field of view. For example, the angular sizes of the Sun and Moon are each about $\frac{1}{2}$° (**FIGURE 2.7a**). Note that angular size does not by itself tell us an object's true size, because angular

Moon

$\frac{1}{2}$

a The angular sizes of the Sun and the Moon are about 1/2°.

Big Dipper

to Polaris

Southern Cross

5°

6°

b The angular distance between the "pointer stars" of the Big Dipper is about 5°, and the angular length of the Southern Cross is about 6°.

20°   1°

10°

Stretch out your arm as shown here.

c You can estimate angular sizes or distances with your outstretched hand.

FIGURE 2.7 We measure *angular sizes* or *angular distances,* rather than actual sizes or distances, when we look at objects in the sky.

size also depends on distance. The Sun is about 400 times as large in diameter as the Moon, but it has the same angular size in our sky because it is also about 400 times as far away.

The **angular distance** between a pair of objects in the sky is the angle that appears to separate them. For example, the angular distance between the "pointer stars" at the end of the Big Dipper's bowl is about 5° and the angular length of the Southern Cross is about 6° (**FIGURE 2.7b**). You can use your outstretched hand to make rough estimates of angles in the sky (**FIGURE 2.7c**).

For more precise astronomical measurements, we subdivide each degree into 60 **arcminutes** and subdivide each arcminute into 60 **arcseconds** (**FIGURE 2.8**). We abbreviate arcminutes with the symbol ′ and arcseconds with the symbol ″. For example, we read 35°27′15″ as "35 degrees, 27 arcminutes, 15 arcseconds."

MATHEMATICAL INSIGHT 2.1

Angular Size, Physical Size, and Distance

An object's angular size depends on its physical (actual) size and distance. **FIGURE 1a** shows the basic idea: An object's physical size does not change as you move it farther from your eye, but its *angular size* gets smaller, making it appear smaller against the background.

FIGURE 1b shows a simple approximation that we can use to find a formula relating angular size to physical size and distance. As long as an object's angular size is relatively small (less than a few degrees), its physical size (diameter) is similar to that of a small piece of a circle going all the way around your eye with a radius equal to the object's distance from your eye. The object's angular size (in degrees) is therefore the *same fraction* of the full 360° circle as its physical size is of the circle's full circumference (given by the formula $2\pi \times distance$). That is,

$$\frac{\text{angular size}}{360°} = \frac{\text{physical size}}{2\pi \times \text{distance}}$$

We solve for the angular size by multiplying both sides by 360°:

$$\text{angular size} = \text{physical size} \times \frac{360°}{2\pi \times \text{distance}}$$

This formula is often called the **small-angle formula**, because it is valid only when the angular size is small.

EXAMPLE 1: The two headlights on a car are separated by 1.5 meters. What is their angular separation when the car is 500 meters away?

SOLUTION:

Step 1 Understand: We can use the small-angle formula by thinking of the "separation" between the two lights as a "size." That is, if we set the physical size to the actual separation of 1.5 meters, the small-angle formula will tell us the angular separation.

Step 2 Solve: We simply plug in the given values and solve:

$$\begin{aligned}\text{angular} \atop \text{separation} &= \text{physical separation} \times \frac{360°}{2\pi \times \text{distance}} \\ &= 1.5 \text{ m} \times \frac{360°}{2\pi \times 500 \text{ m}} \approx 0.17°\end{aligned}$$

Step 3 Explain: We have found that the angular separation of the two headlights is 0.17°. This small angle will be easier to interpret if we convert it to arcminutes. There are 60 arcminutes in 1°, so 0.17° is equivalent to $0.17 \times 60 = 10.2$ arcminutes. In other words, the

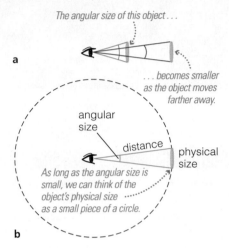

FIGURE 1 Angular size depends on physical size and distance.

angular separation of the headlights is about 10 arcminutes, or about a third of the 30 arcminute (0.5°) angular diameter of the Moon.

EXAMPLE 2: Estimate the Moon's actual diameter from its angular diameter of about 0.5° and its distance of about 380,000 km.

SOLUTION:

Step 1 Understand: We are seeking to find a physical size (diameter) from an angular size and distance. We therefore need to solve the small-angle formula for the *physical size*, which we do by switching its left and right sides and multiplying both sides by $(2\pi \times distance)/360°$:

$$\text{physical size} = \text{angular size} \times \frac{2\pi \times \text{distance}}{360°}$$

Step 2 Solve: We now plug in the given values of the Moon's angular size and distance:

$$\text{physical size} = 0.5° \times \frac{2\pi \times 380,000 \text{ km}}{360°} \approx 3300 \text{ km}$$

Step 3 Explain: We have used the Moon's approximate angular size and distance to find that its diameter is about 3300 kilometers. We could find a more exact value (3476 km) by using more precise values for the angular diameter and distance.

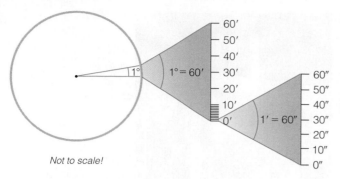

Not to scale!

FIGURE 2.8 We subdivide each degree into 60 arcminutes and each arcminute into 60 arcseconds.

Why do stars rise and set?

If you spend a few hours out under a starry sky, you'll notice that the universe seems to be circling around us, with stars moving gradually across the sky from east to west. Many ancient people took this appearance at face value, concluding that we lie at the center of a universe that rotates around us each day. Today we know that the ancients had it backward: It is Earth that rotates daily, not the rest of the universe.

We can picture the movement of the sky by imagining the celestial sphere rotating around Earth (**FIGURE 2.9**). From this perspective you can see how the universe seems to turn around us: Every object on the celestial sphere appears to make a simple daily circle around Earth. However, the motion can look a little more complex in the local sky, because the horizon cuts the celestial sphere in half. **FIGURE 2.10** shows the idea for a location in the United States. If you study the figure carefully,

you'll notice the following key facts about the paths of various stars through the local sky:

- Stars near the north celestial pole are **circumpolar**, meaning that they remain perpetually above the horizon, circling (counterclockwise) around the north celestial pole each day.

- Stars near the south celestial pole never rise above the horizon at all.

- All other stars have daily circles that are partly above the horizon and partly below it, which means they appear to rise in the east and set in the west.

The time-exposure photograph that opens this chapter (page 24), taken at Arches National Park in Utah, shows a part of the daily paths of stars. Paths of circumpolar stars are visible within the arch; notice that the complete daily circles for these stars are above the horizon, although the photo shows only a portion of each circle. The north celestial pole lies at the center of these circles. The circles grow larger for stars farther from

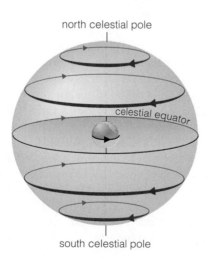

FIGURE 2.9 Earth rotates from west to east (black arrow), making the celestial sphere *appear* to rotate around us from east to west (red arrows).

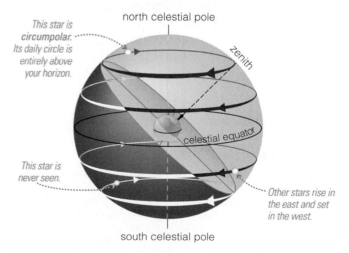

FIGURE 2.10 The local sky for a location in the United States (40°N). The horizon slices through the celestial sphere at an angle to the equator, causing the daily circles of stars to appear tilted in the local sky. Note: It may be easier to follow the star paths in the local sky if you rotate the page so that the zenith points up.

a We can locate any place on Earth's surface by its latitude and longitude.

b The entrance to the Old Royal Greenwich Observatory, near London. The line emerging from the door marks the prime meridian.

FIGURE 2.11 Definitions of latitude and longitude.

the north celestial pole. If they are large enough, the circles cross the horizon, so that the stars rise in the east and set in the west. The same ideas apply in the Southern Hemisphere, except that circumpolar stars are those near the south celestial pole and they circle clockwise rather than counterclockwise.

Why do the constellations we see depend on latitude and time of year?

If you stay in one place, the basic patterns of motion in the sky will stay the same from one night to the next. However, if you travel far north or south, you'll see a different set of

constellations than you see at home. And even if you stay in one place, you'll see different constellations at different times of year. Let's explore why.

Variation with Latitude **Latitude** measures north-south position on Earth, and **longitude** measures east-west position (**FIGURE 2.11**). Latitude is defined to be 0° at the equator, increasing to 90°N at the North Pole and 90°S at the South Pole. By international treaty, longitude is defined to be 0° along the **prime meridian**, which passes through Greenwich, England. Stating a latitude and a longitude pinpoints a location on Earth. For example, Miami lies at about 26°N latitude and 80°W longitude.

Latitude affects the constellations we see because it affects the locations of the horizon and zenith relative to the celestial sphere. **FIGURE 2.12** shows how this works for the latitudes of the North Pole (90°N) and Sydney, Australia (34°S). Note that although the sky varies with latitude, it does *not* vary with longitude. For example, Charleston (South Carolina) and San Diego (California) are at about the same latitude, so people in both cities see the same set of constellations at night.

You can learn more about how the sky varies with latitude by studying diagrams like those in Figures 2.10 and 2.12. For example, at the North Pole, you can see only objects that lie on the northern half of the celestial sphere, and they are all circumpolar. That is why the Sun remains above the horizon for 6 months at the North Pole: The Sun lies north of the celestial equator for half of each year (see Figure 2.3), so during these 6 months it circles the sky at the North Pole just like a circumpolar star.

The diagrams also show a fact that is very important to navigation:

The altitude of the celestial pole in your sky is equal to your latitude.

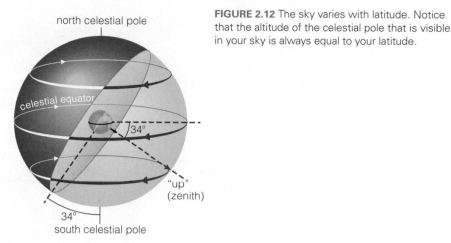

FIGURE 2.12 The sky varies with latitude. Notice that the altitude of the celestial pole that is visible in your sky is always equal to your latitude.

a The local sky at the North Pole (latitude 90°N).

b The local sky at latitude 34°S.

For example, if you see the north celestial pole at an altitude of 40° above your north horizon, your latitude is 40°N. Similarly, if you see the south celestial pole at an altitude of 34° above your south horizon, your latitude is 34°S. You can therefore determine your latitude simply by finding the celestial pole in your sky (**FIGURE 2.13**). Finding the north celestial pole is fairly easy, because it lies very close to the star Polaris, also known as the North Star (Figure 2.13a). In the Southern Hemisphere, you can find the south celestial pole with the

aid of the Southern Cross (Figure 2.13b). We'll discuss celestial navigation and how the sky varies with latitude in more detail in Chapter S1.

SEE IT FOR YOURSELF

What is *your* latitude? Use Figure 2.13 to find the celestial pole in your sky, and estimate its altitude with your hand as shown in Figure 2.7c. Is its altitude what you expect?

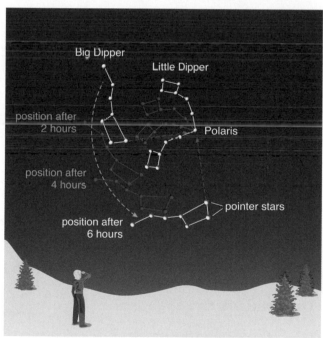

looking northward in the Northern Hemisphere

a The pointer stars of the Big Dipper point to the North Star, Polaris, which lies within 1° of the north celestial pole. The sky appears to turn *counterclockwise* around the north celestial pole.

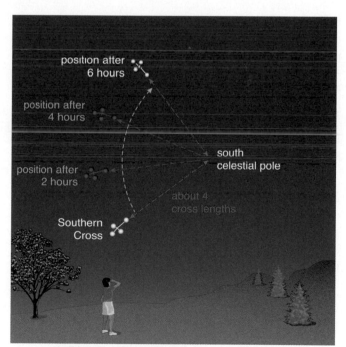

looking southward in the Southern Hemisphere

b The Southern Cross points to the south celestial pole, which is not marked by any bright star. The sky appears to turn *clockwise* around the south celestial pole.

FIGURE 2.13 interactive figure You can determine your latitude by measuring the altitude of the celestial pole in your sky.

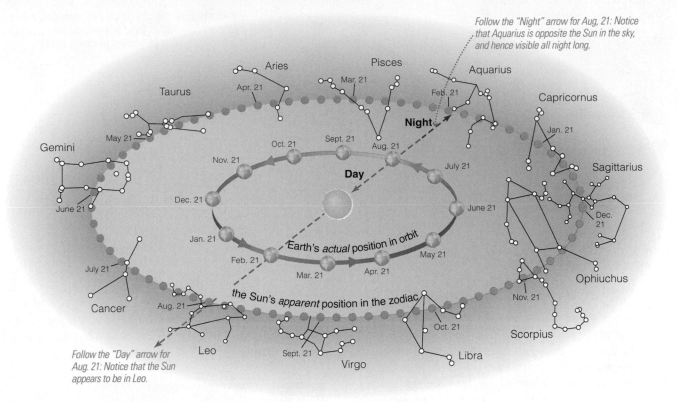

Follow the "Night" arrow for Aug. 21: Notice that Aquarius is opposite the Sun in the sky, and hence visible all night long.

Follow the "Day" arrow for Aug. 21: Notice that the Sun appears to be in Leo.

FIGURE 2.14 interactive figure The Sun appears to move steadily eastward along the ecliptic as Earth orbits the Sun, so we see the Sun against the background of different zodiac constellations at different times of year. For example, on August 21 the Sun appears to be in Leo, because it is between us and the much more distant stars that make up Leo.

Variation with Time of Year The night sky changes throughout the year because of Earth's changing position in its orbit around the Sun. **FIGURE 2.14** shows how this works. As Earth orbits, the Sun *appears* to move steadily eastward along the ecliptic, with the stars of different constellations in the background at different times of year. The constellations along the ecliptic make up what we call the **zodiac**; tradition places 12 constellations along the zodiac, but the official borders include a thirteenth constellation, Ophiuchus.

The Sun's apparent location along the ecliptic determines which constellations we see at night. For example, Figure 2.14 shows that the Sun appears to be in Leo in late August. We therefore cannot see Leo at this time (because it is in our daytime sky), but we can see Aquarius all night long because of its location opposite Leo on the celestial sphere. Six months later, in February, we see Leo at night while Aquarius is above the horizon only in the daytime.

COMMON MISCONCEPTIONS

What Makes the North Star Special?

Most people are aware that the North Star, Polaris, is a special star. Contrary to a relatively common belief, however, it is *not* the brightest star in the sky. More than 50 other stars are just as bright or brighter. Polaris is special not because of its brightness, but because it is so close to the north celestial pole and therefore very useful in navigation.

SEE IT FOR YOURSELF

Based on Figure 2.14 and today's date, in what constellation does the Sun currently appear? What constellation of the zodiac will be on your meridian at midnight? What constellation of the zodiac will you see in the west shortly after sunset? Go outside at night to confirm your answers to the last two questions.

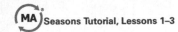 **MA** Seasons Tutorial, Lessons 1–3

2.2 THE REASON FOR SEASONS

We have seen how Earth's rotation makes the sky appear to circle us daily and how the night sky changes as Earth orbits the Sun each year. The combination of Earth's rotation and orbit also leads to the progression of the seasons.

What causes the seasons?

You know that we have seasonal changes, such as longer and warmer days in summer and shorter and cooler days in winter. But why do the seasons occur? The answer is that the tilt of Earth's axis causes sunlight to fall differently on Earth at different times of year.

FIGURE 2.15 (pages 34–35) illustrates the key ideas. Step 1 illustrates the tilt of Earth's axis, which remains pointed in the same direction in space (toward Polaris) throughout the year.

As a result, the orientation of the axis *relative to the Sun* changes over the course of each orbit: The Northern Hemisphere is tipped toward the Sun in June and away from the Sun in December, while the reverse is true for the Southern Hemisphere. That is why the two hemispheres experience opposite seasons. The rest of the figure shows how the changing angle of sunlight on the two hemispheres leads directly to seasons.

Step 2 shows Earth in June, when axis tilt causes sunlight to strike the Northern Hemisphere at a steeper angle and the Southern Hemisphere at a shallower angle. The steeper sunlight angle makes it summer in the Northern Hemisphere for two reasons. First, as shown in the zoom-out, the steeper angle means more concentrated sunlight, which tends to make it warmer. Second, if you visualize what happens as Earth rotates each day, you'll see that the steeper angle also means the Sun follows a longer and higher path through the sky, giving the Northern Hemisphere more hours of daylight during which it is warmed by the Sun. The opposite is true for the Southern Hemisphere at this time: The shallower sunlight angle makes it winter there because sunlight is less concentrated and the Sun follows a shorter, lower path through the sky.

The sunlight angle gradually changes as Earth orbits the Sun. At the opposite side of Earth's orbit, Step 4 shows that it has become winter for the Northern Hemisphere and summer for the Southern Hemisphere. In between these two extremes, Step 3 shows that both hemispheres are illuminated equally in March and September. It is therefore spring for the hemisphere that is on the way from winter to summer, and fall for the hemisphere on the way from summer to winter.

Notice that the seasons on Earth are caused only by the axis tilt and *not* by any change in Earth's distance from the Sun. Although Earth's orbital distance varies over the course of each year, the variation is fairly small: Earth is only about 3% farther from the Sun at its farthest point than at its nearest. The difference in the strength of sunlight due to this small change in distance is easily overwhelmed by the effects caused by the axis tilt. If Earth did not have an axis tilt, we would not have seasons.

THINK ABOUT IT

Jupiter has an axis tilt of about 3°, small enough to be insignificant. Saturn has an axis tilt of about 27°, slightly greater than that of Earth. Both planets have nearly circular orbits around the Sun. Do you expect Jupiter to have seasons? Do you expect Saturn to have seasons? Explain.

Solstices and Equinoxes To help us mark the changing seasons, we define four special moments in the year, each of which corresponds to one of the four special positions in Earth's orbit shown in Figure 2.15.

- The **summer (June) solstice**, which occurs around June 21, is the moment when the Northern Hemisphere is tipped most directly toward the Sun and receives the most direct sunlight.

- The **winter (December) solstice**, which occurs around December 21, is the moment when the Northern Hemisphere receives the least direct sunlight.

- The **spring (March) equinox**, which occurs around March 21, is the moment when the Northern Hemisphere goes from being tipped slightly away from the Sun to being tipped slightly toward the Sun.

- The **fall (September) equinox**, which occurs around September 22, is the moment when the Northern Hemisphere first starts to be tipped away from the Sun.

The exact dates and times of the solstices and equinoxes vary from year to year, but stay within a couple of days of the dates given above. In fact, our modern calendar includes leap years in a pattern specifically designed to keep the solstices and equinoxes around the same dates [**Section S1.1**].

We can mark the dates of the equinoxes and solstices by observing changes in the Sun's path through our sky (**FIGURE 2.16**). The equinoxes occur on the only two days of the year on which the Sun rises precisely due east and sets precisely due west. The June solstice occurs on the day on which the Sun follows its longest and highest path through the Northern Hemisphere sky (and its shortest and lowest path through the Southern Hemisphere sky). It is therefore the day on which the Sun rises and sets farther to the north than on any other day of the year, and on which the noon Sun reaches its highest point in the Northern Hemisphere sky. The opposite is true on the day of the December solstice, when the Sun rises and sets farthest to the south and the noon Sun is lower in the Northern Hemisphere sky than on any other day of the year. **FIGURE 2.17** shows how the Sun's position in the sky varies over the course of the year.

First Days of Seasons We usually say that each equinox and solstice marks the first day of a season. For example, the day of the summer solstice is usually called the "first day of summer." Notice, however, that the summer (June) solstice occurs when the Northern Hemisphere has its *maximum* tilt toward the Sun. You might then wonder why we consider the solstice to be the beginning rather than the midpoint of summer.

The choice is somewhat arbitrary, but it makes sense in at least two ways. First, it was much easier for ancient people to identify the days on which the Sun reached extreme positions in the sky—such as when it reached its highest point on the summer solstice—than other days in between.

Earth's seasons are caused by the tilt of its rotation axis, which is why the seasons are opposite in the two hemispheres. The seasons do *not* depend on Earth's distance from the Sun, which varies only slightly throughout the year.

① **Axis Tilt:** Earth's axis points in the same direction throughout the year, which causes changes in Earth's orientation *relative to the Sun*.

② **Northern Summer/Southern Winter:** In June, sunlight falls more directly on the Northern Hemisphere, which makes it summer there because solar energy is more concentrated and the Sun follows a longer and higher path through the sky. The Southern Hemisphere receives less direct sunlight, making it winter.

Summer (June) Solstice

The Northern Hemisphere is tipped most directly toward the Sun.

Noon rays of sunlight hit the ground at a steeper angle in the Northern Hemisphere, meaning more concentrated sunlight and shorter shadows.

Noon rays of sunlight hit the ground at a shallower angle in the Southern Hemisphere, meaning less concentrated sunlight and longer shadows.

Interpreting the Diagram

To interpret the seasons diagram properly, keep in mind:

1. Earth's size relative to its orbit would be microscopic on this scale, meaning that both hemispheres are at essentially the same distance from the Sun.

2. The diagram is a side view of Earth's orbit. A top-down view (below) shows that Earth orbits in a nearly perfect circle and comes closest to the Sun in January.

Spring Equinox

147.1 million km   *January 3*

152.1 million km

July 4

Fall Equinox

③ **Spring/Fall:** Spring and fall begin when sunlight falls equally on both hemispheres, which happens twice a year: In March, when spring begins in the Northern Hemisphere and fall in the Southern Hemisphere; and in September, when fall begins in the Northern Hemisphere and spring in the Southern Hemisphere.

④ **Northern Winter/Southern Summer:** In December, sunlight falls less directly on the Northern Hemisphere, which makes it winter because solar energy is less concentrated and the Sun follows a shorter and lower path through the sky. The Southern Hemisphere receives more direct sunlight, making it summer.

Spring (March) Equinox

The Sun shines equally on both hemispheres.

The variation in Earth's orientation relative to the Sun means that the seasons are linked to four special points in Earth's orbit:

Solstices *are the two points at which sunlight becomes most extreme for the two hemispheres.*

Equinoxes *are the two points at which the hemispheres are equally illuminated.*

Winter (December) Solstice

The Southern Hemisphere is tipped most directly toward the Sun.

Fall (September) Equinox

The Sun shines equally on both hemispheres.

Noon rays of sunlight hit the ground at a shallower angle in the Northern Hemisphere, meaning less concentrated sunlight and longer shadows.

Noon rays of sunlight hit the ground at a steeper angle in the Southern Hemisphere, meaning more concentrated sunlight and shorter shadows.

meridian   zenith

N — — — — — — — — — — — — — — S

E

W

Sun's path on summer (June) solstice

Sun's path on equinoxes

Sun's path on winter (December) solstice

FIGURE 2.16 interactive figure This diagram shows the Sun's path on the solstices and equinoxes for a Northern Hemisphere sky (latitude 40°N). The precise paths are different for other latitudes; for example, at latitude 40°S, the paths look similar except tilted to the north rather than to the south. Notice that the Sun rises exactly due east and sets exactly due west only on the equinoxes.

Second, we usually think of the seasons in terms of weather, and the solstices and equinoxes correspond well with the beginnings of seasonal weather patterns. For example, although the Sun's path through the Northern Hemisphere sky is longest and highest around the time of the summer solstice, the warmest days tend to come 1 to 2 months later. To understand why, think about what happens when you heat a pot of cold soup. Even though you may have the stove turned on high from the start, it takes a while for the soup to warm up. In the same way, it takes some time for sunlight to heat the ground and oceans from the cold of winter to the warmth of summer. "Midsummer" in terms of weather therefore comes in late July and early August, which makes the summer solstice a pretty good choice for the "first day of summer." For similar reasons, the winter solstice is a good choice for the first day of winter, and the spring and fall equinoxes are good choices for the first days of those seasons.

Seasons Around the World Notice that the names of the solstices and equinoxes reflect the northern seasons, and therefore sound backward to people who live in the Southern Hemisphere. For example, Southern Hemisphere

FIGURE 2.17 This composite photograph shows images of the Sun taken at the same time of morning (technically, at the same "mean solar time") and from the same spot (over a large sundial in Carefree, Arizona) at 7- to 11-day intervals over the course of a year; the photo looks eastward, so north is to the left and south is to the right. Because this location is in the Northern Hemisphere, the Sun images that are high and to the north represent times near the summer solstice and the images that are low and to the south represent times near the winter solstice. The "figure 8" shape (called an *analemma*) arises from a combination of Earth's axis tilt and Earth's varying speed as it orbits the Sun (see the Special Topic on page 91).

winter begins when Earth is at the orbital point usually called the *summer* solstice. This apparent injustice to people in the Southern Hemisphere arose because the solstices and equinoxes were named long ago by people living in the Northern Hemisphere. A similar injustice is inflicted on people living in equatorial regions. If you study Figure 2.15 carefully, you'll see that Earth's equator gets its most direct sunlight on the two equinoxes and its least direct sunlight on the solstices. People living near the equator therefore don't experience four seasons in the same way as people living at mid-latitudes. Instead, equatorial regions generally have rainy and dry seasons, with the rainy seasons coming when the Sun is higher in the sky.

In addition, seasonal variations around the times of the solstices are more extreme at high latitudes. For example, Vermont has much longer summer days and much longer winter nights than Florida. At the Arctic Circle (latitude $66\frac{1}{2}°$), the Sun remains above the horizon all day long on the summer solstice (**FIGURE 2.18**), and never rises on the winter solstice. The most extreme cases occur at the North and South Poles, where the Sun remains above the horizon for 6 months in summer and below the horizon for 6 months in winter.*

Why Orbital Distance Doesn't Affect Our Seasons

We've seen that the seasons are caused by Earth's axis tilt, not by Earth's slightly varying distance from the Sun. Still, we might expect the varying orbital distance to play at least some role. For example, the Northern Hemisphere has winter when

---

*These statements are true for the Sun's *real* position, but the bending of light by Earth's atmosphere makes the Sun *appear* to be about 0.6° higher than it really is when it is near the horizon.

| Approximate time: | Midnight | 6:00 A.M. | Noon | 6:00 P.M. |
| Direction: | due north | due east | due south | due west |

FIGURE 2.18 This sequence of photos shows the progression of the Sun around the horizon on the summer solstice at the Arctic Circle. Notice that the Sun skims the northern horizon at midnight, then gradually rises higher, reaching its highest point when it is due South at noon.

Earth is closer to the Sun and summer when Earth is farther away (see the lower left diagram in Figure 2.15), so we might expect the Northern Hemisphere to have more moderate seasons than the Southern Hemisphere. In fact, weather records show that the opposite is true: Northern Hemisphere seasons are slightly more extreme than those of the Southern Hemisphere.

The main reason for this surprising fact becomes clear when you look at a map of Earth (**FIGURE 2.19**). Most of Earth's land lies in the Northern Hemisphere, with far more ocean in the Southern Hemisphere. As you'll notice at any beach, lake, or pool, water takes longer to heat or cool than soil or rock (largely because sunlight heats bodies of water to a depth of many meters while heating only the very top layer of land). The water temperature therefore remains fairly steady both day and night, while the ground can heat up and cool down dramatically. The Southern Hemisphere's larger amount of ocean moderates its climate. The Northern Hemisphere, with more land and less ocean, heats up and cools down more easily, which is why it has the more extreme seasons.

FIGURE 2.19 Most land lies in the Northern Hemisphere while most ocean lies in the Southern Hemisphere. The climate-moderating effects of water make Southern Hemisphere seasons less extreme than Northern Hemisphere seasons.

Although distance from the Sun plays no role in Earth's seasons, the same is not true for planets that have much greater distance variations. For example, Mars has about the same axis tilt as Earth and therefore has similar seasonal patterns. However, because Mars is more than 20% closer to the Sun during its Southern Hemisphere summer than during its Northern Hemisphere summer, its Southern Hemisphere experiences much more extreme seasonal changes.

How does the orientation of Earth's axis change with time?

We have now discussed both daily and seasonal changes in the sky, but there are other changes that occur over longer periods of time. One of the most important of these slow changes is **precession**, a gradual wobble that alters the orientation of Earth's axis in space.

Precession occurs with many rotating objects. You can see it easily by spinning a top (**FIGURE 2.20a**). As the top spins rapidly, you'll notice that its axis also sweeps out a circle at a slower rate. We say that the top's axis *precesses*. Earth's axis precesses in much the same way, but far more slowly (**FIGURE 2.20b**). Each cycle of Earth's precession takes about 26,000 years, gradually changing where the axis points in space. Today, the axis points toward Polaris, making it our North Star. Some 13,000 years from now, Vega will be the bright star closest to true north. At most other times, the axis does not point near any bright star.

Notice that precession does not change the *amount* of the axis tilt (which stays close to $23\frac{1}{2}°$) and therefore does not affect the pattern of the seasons. However, because the solstices and equinoxes correspond to points in Earth's orbit that depend on the direction the axis points in space, their positions in the orbit gradually shift with the cycle of precession. As a result, the constellations associated with the solstices and equinoxes

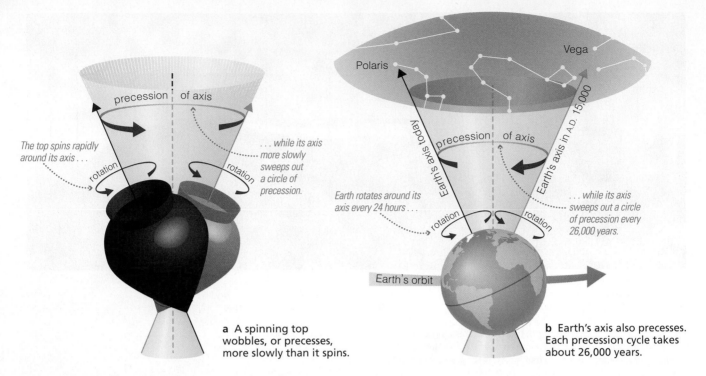

precession of axis

The top spins rapidly around its axis . . .

rotation

rotation

. . . while its axis more slowly sweeps out a circle of precession.

a A spinning top wobbles, or precesses, more slowly than it spins.

Polaris

Vega

precession of axis

Earth's axis today

Earth's axis in A.D. 15,000

Earth rotates around its axis every 24 hours . . .

rotation

rotation

. . . while its axis sweeps out a circle of precession every 26,000 years.

Earth's orbit

b Earth's axis also precesses. Each precession cycle takes about 26,000 years.

FIGURE 2.20 interactive figure Precession affects the orientation of a spinning object's axis but not the amount of its tilt.

change over time. For example, a couple thousand years ago the Sun appeared in the constellation Cancer on the day of the summer solstice, but now it appears in Gemini on that day. This explains something you can see on any world map: The latitude at which the Sun is directly overhead on the summer solstice ($23\frac{1}{2}°N$) is called the *Tropic of Cancer,* telling us that it got its name back when the Sun used to appear in Cancer on the summer solstice.

THINK ABOUT IT

What constellation will the Sun be in on the summer solstice about 2000 years from now? (*Hint:* Figure 2.14 shows the names and order of the zodiac constellations.)

Precession is caused by gravity's effect on a tilted, rotating object. You have probably seen how gravity affects a top. If you try to balance a nonspinning top on its point, it will fall over almost immediately. This happens because a top will inevitably lean a little to one side. No matter how slight this lean, gravity will quickly tip the nonspinning top over. However, if you spin the top rapidly, it does not fall over so easily. The spinning top stays upright because rotating objects tend to keep spinning around the same rotation axis (a consequence of the *law of conservation of angular momentum* [**Section 4.3**]). This tendency prevents gravity from immediately pulling the spinning top over, since falling over would mean a change in the spin axis from near-vertical to horizontal. Instead, gravity succeeds only in making the axis trace circles of precession. As friction slows the top's spin, the circles of

precession get wider and wider, and ultimately the top falls over. If there were no friction to slow its spin, the top would spin and precess forever.

The spinning (rotating) Earth precesses because of gravitational tugs from the Sun and Moon. Earth is not quite a perfect sphere, because it bulges at its equator. Because the equator is tilted $23\frac{1}{2}°$ to the ecliptic plane, the gravitational attractions of the Sun and Moon try to pull the equatorial bulge into the ecliptic plane, effectively trying to "straighten out" Earth's axis tilt. However, like the spinning top, Earth tends to keep rotating around the same axis. Gravity therefore does not succeed in straightening out Earth's axis tilt and instead makes the axis precess. To gain a better understanding of precession and how it works, you might wish to experiment with a simple toy gyroscope. *Gyroscopes* are essentially

COMMON MISCONCEPTIONS

Sun Signs

You probably know your astrological "Sun sign." When astrology began a few thousand years ago, your Sun sign was supposed to represent the constellation in which the Sun appeared on your birth date. However, because of precession, this is no longer the case for most people. For example, if your birthday is March 21, your Sun sign is Aries even though the Sun now appears in Pisces on that date. The problem is that astrological Sun signs are based on the positions of the Sun among the stars as they were almost 2000 years ago. Because Earth's axis has moved about 1/13 of the way through its 26,000-year precession cycle since that time, the Sun signs are off by nearly a month from the actual positions of the Sun among the constellations today.

rotating wheels mounted in a way that allows them to move freely, which makes it easy to see how their spin rate affects their motion. (The fact that gyroscopes tend to keep the same rotation axis makes them very useful in aircraft and space-craft navigation.)

Phases of the Moon Tutorial, Lessons 1–3

2.3 THE MOON, OUR CONSTANT COMPANION

Aside from the seasons and the daily circling of the sky, the most familiar pattern of change in the sky is that of the changing phases of the Moon. We will explore these changes in this section—along with the rarer changes that occur with eclipses—and see that they are consequences of the Moon's orbit around Earth.

Why do we see phases of the Moon?

As the Moon orbits Earth, it returns to the same position relative to the Sun in our sky (such as along the Earth-Sun line) about every $29\frac{1}{2}$ days. This time period marks the cycle of **lunar phases**, in which the Moon's appearance in our sky changes as its position relative to the Sun changes. This $29\frac{1}{2}$-day period is also the origin of the word *month* (think "moonth").

The Moon's Orbit to Scale FIGURE 2.21 shows the Moon's orbit on the 1-to-10-billion scale model of the solar system that we discussed in Chapter 1. Recall that on this scale, the Sun is about the size of a large grapefruit, which means the entire orbit of the Moon could fit inside it. When you then consider the fact that the Sun is located 15 meters away on this scale, you'll realize that for practical purposes the Sun's rays all hit Earth and the Moon from the same direction. This fact is helpful to understanding the Moon's phases, because it means we can think of sunlight coming from a single direction at both Earth and the Moon, an effect shown clearly in the inset photo. The figure also shows the elliptical shape of the Moon's orbit, which causes the Earth-Moon distance to vary between about 356,000 and 407,000 kilometers.

Understanding Phases The easiest way to understand the lunar phases is with the simple demonstration illustrated in FIGURE 2.22. Take a ball outside on a sunny day. (If it's dark

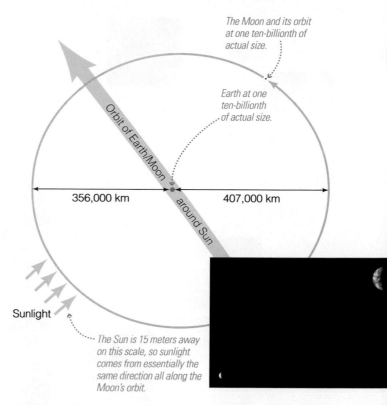

FIGURE 2.21 The Moon's orbit on the 1-to-10-billion scale introduced in Chapter 1 (see Figure 1.6); black labels indicate the Moon's actual distance at its nearest to and farthest from Earth. The orbit is so small compared to the distance to the Sun that sunlight strikes the entire orbit from the same direction. You can see this in the inset photo, which shows the Moon and Earth photographed from Mars by the Mars Reconnaissance Orbiter.

or cloudy, you can use a flashlight instead of the Sun; put the flashlight on a table a few meters away and shine it toward you.) Hold the ball at arm's length to represent the Moon while your head represents Earth. Slowly spin counterclockwise so that the ball goes around you the way the Moon orbits Earth. (If you live in the Southern Hemisphere, spin clockwise because you view the sky "upside down" compared to the Northern Hemisphere.) As you turn, you'll see the ball go through phases just like the Moon's. If you think about what's happening, you'll realize that the phases of the ball result from just two basic facts:

1. Half the ball always faces the Sun (or flashlight) and therefore is bright, while the other half faces away from the Sun and therefore is dark.

2. As you look at the ball at different positions in its "orbit" around your head, you see different combinations of its bright and dark faces.

For example, when you hold the ball directly opposite the Sun, you see only the bright portion of the ball, which represents the "full" phase. When you hold the ball at its "first-quarter" position, half the face you see is dark and the other half is bright.

We see lunar phases for the same reason. Half the Moon is always illuminated by the Sun, but the amount of this illuminated half that we see from Earth depends on the Moon's position in its orbit. The photographs in Figure 2.22 show how the phases look.

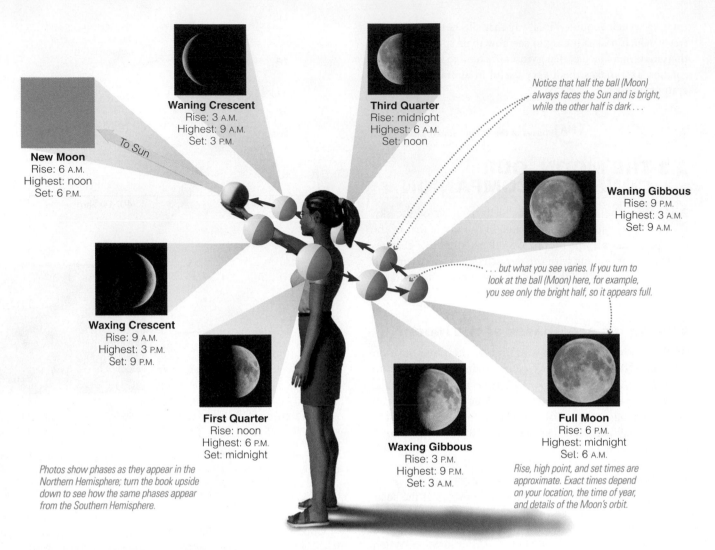

Waning Crescent
Rise: 3 A.M.
Highest: 9 A.M.
Set: 3 P.M.

Third Quarter
Rise: midnight
Highest: 6 A.M.
Set: noon

Notice that half the ball (Moon) always faces the Sun and is bright, while the other half is dark . . .

New Moon
Rise: 6 A.M.
Highest: noon
Set: 6 P.M.

To Sun

Waning Gibbous
Rise: 9 P.M.
Highest: 3 A.M.
Set: 9 A.M.

Waxing Crescent
Rise: 9 A.M.
Highest: 3 P.M.
Set: 9 P.M.

. . . but what you see varies. If you turn to look at the ball (Moon) here, for example, you see only the bright half, so it appears full.

First Quarter
Rise: noon
Highest: 6 P.M.
Set: midnight

Waxing Gibbous
Rise: 3 P.M.
Highest: 9 P.M.
Set: 3 A.M.

Full Moon
Rise: 6 P.M.
Highest: midnight
Set: 6 A.M.

Photos show phases as they appear in the Northern Hemisphere; turn the book upside down to see how the same phases appear from the Southern Hemisphere.

Rise, high point, and set times are approximate. Exact times depend on your location, the time of year, and details of the Moon's orbit.

FIGURE 2.22 interactive figure A simple demonstration illustrates the phases of the Moon. The half of the ball (Moon) facing the Sun is always illuminated while the half facing away is always dark, but you see the ball go through phases as it orbits around your head (Earth). (The new moon photo shows blue sky, because a new moon is always close to the Sun in the sky and hence hidden from view by the bright light of the Sun.)

The Moon's phase is directly related to the time it rises, reaches its highest point in the sky, and sets. For example, the full moon must rise around sunset, because it occurs when the Moon is opposite the Sun in the sky. It therefore reaches its highest point in the sky at midnight and sets around sunrise. Similarly, a first-quarter moon must rise around noon, reach its highest point around 6 p.m., and set around midnight, because it occurs when the Moon is about 90° east of the Sun in our sky. Figure 2.22 lists the approximate rise, highest point, and set times for each phase.

Notice that the phases from new to full are said to be *waxing*, which means "increasing." Phases from full to new

are *waning*, or "decreasing." Also notice that no phase is called a "half moon." Instead, we see half the Moon's face at first-quarter and third-quarter phases; these phases mark the times when the Moon is one quarter or three quarters of the way through its monthly cycle (which begins at new moon). The phases just before and after new moon are called *crescent*, while those just before and after full moon are called *gibbous* (pronounced with a hard *g* as in "gift"). A gibbous moon is essentially the opposite of a crescent moon—a crescent moon has a small sliver of light while a gibbous moon has a small sliver of dark. The term *gibbous* literally means "humpbacked," so you can see how the gibbous moon got its name.

The Moon's Synchronous Rotation Although we see many *phases* of the Moon, we do not see many *faces*. From Earth we always see (nearly*) the same face of the Moon. This happens because the Moon rotates on its axis in the same amount of time it takes to orbit Earth, a trait called **synchronous rotation**. A simple demonstration shows the idea. Place a ball on a table to represent Earth while you represent the Moon. Start by facing the ball. If you do not rotate as you walk around the ball, you'll be looking away from it by the time you are halfway around your orbit (**FIGURE 2.23a**). The only way you can face the ball at all times is by completing exactly one rotation while you complete one orbit (**FIGURE 2.23b**). Note that the Moon's synchronous rotation is *not* a coincidence; rather, it is a consequence of Earth's gravity affecting the Moon in much the same way the Moon's gravity causes tides on Earth [**Section 4.5**].

The View from the Moon A good way to solidify your understanding of the lunar phases is to imagine that you live on the side of the Moon that faces Earth. For example, what would you see if you looked at Earth when people on Earth saw a new moon? By remembering that a new moon occurs when the Moon is between the Sun and Earth, you'll realize that from the Moon you'd be looking at Earth's daytime side and hence would see a *full earth*. Similarly, at full moon you would be facing the night side of Earth and would see a *new earth*. In general, you'd always see Earth in a phase opposite the phase of the Moon seen by people on

---

*Because the Moon's orbital speed varies (in accord with Kepler's second law [**Section 3.3**]) while its rotation rate is steady, the visible face appears to wobble slightly back and forth as the Moon orbits Earth. This effect, called *libration*, allows us to see a total of about 59% of the Moon's surface over the course of a month, even though we see only 50% of the Moon at any single time.

Earth at the same time. Moreover, because the Moon always shows nearly the same face to Earth, Earth would appear to hang nearly stationary in your sky as it went through its cycle of phases.

Thinking about the view from the Moon clarifies another interesting feature of the lunar phases: The dark portion of the lunar face is not *totally* dark. Just as we can see at night by the light of the Moon, if you were in the dark area of the Moon during crescent phase your moonscape would be illuminated by a nearly full (gibbous) Earth. In fact, because Earth is much larger than the Moon, the illumination would be much greater than what the full moon provides on Earth. In other words, sunlight reflected by Earth faintly illuminates the "dark" portion of the Moon's face. We call this illumination the *ashen light*, or *earthshine*, and it enables us to see the outline of the full face of the Moon even when the Moon is not full.

a If you do not rotate while walking around the model, you will not always face it.

b You will face the model at all times only if you rotate exactly once during each orbit.

FIGURE 2.23 The fact that we always see the same face of the Moon means that the Moon must rotate once in the same amount of time it takes to orbit Earth once. You can see why by walking around a model of Earth while imagining that you are the Moon.

What causes eclipses?

Occasionally, the Moon's orbit around Earth causes events much more dramatic than lunar phases. The Moon and Earth both cast shadows in sunlight, and these shadows can create **eclipses** when the Sun, Earth, and Moon fall into a straight line. Eclipses come in two basic types:

- A **lunar eclipse** occurs when Earth lies directly between the Sun and Moon, so Earth's shadow falls on the Moon.

- A **solar eclipse** occurs when the Moon lies directly between the Sun and Earth, so the Moon's shadow falls on Earth.

Note that, because Earth is much larger than the Moon, Earth's shadow can cover the entire Moon during a lunar eclipse. Therefore, a lunar eclipse can be seen by anyone on the night side of Earth when it occurs. In contrast, the Moon's shadow can cover only a small portion of Earth at any one moment, so you must be living within the relatively small pathway through which the shadow moves to see a solar eclipse. That is why you see lunar eclipses more often than solar eclipses, even though both types occur about equally often.

Conditions for Eclipses Look once more at Figure 2.22. The figure makes it look as if the Sun, Earth, and Moon line up with every new and full moon. If this figure told the whole story of the Moon's orbit, we would have both a lunar and a solar eclipse every month—but we don't.

The missing piece of the story in Figure 2.22 is that the Moon's orbit is slightly inclined (by about 5°) to the ecliptic

COMMON MISCONCEPTIONS

Moon in the Daytime and Stars on the Moon

Night is so closely associated with the Moon in traditions and stories that many people mistakenly believe that the Moon is visible only in the nighttime sky. In fact, the Moon is above the horizon as often in the daytime as at night, though it is easily visible only when its light is not drowned out by sunlight. For example, a first-quarter moon is easy to spot in the late afternoon as it rises through the eastern sky, and a third-quarter moon is visible in the morning as it heads toward the western horizon.

Another misconception appears in illustrations that show a star in the dark portion of the crescent moon. The star in the dark portion appears to be in front of the Moon, which is impossible because the Moon is much closer to us than is any star.

plane (the plane of Earth's orbit around the Sun). To visualize this inclination, imagine the ecliptic plane as the surface of a pond, as shown in **FIGURE 2.24**. Because of the inclination of its orbit, the Moon spends most of its time either above or below this surface. It crosses *through* this surface only twice during each orbit: once coming out and once going back in. The two points in each orbit at which the Moon crosses the surface are called the **nodes** of the Moon's orbit.

Notice that the nodes are aligned approximately the same way (diagonally on the page in Figure 2.24) throughout the year, which means they lie along a nearly straight line with

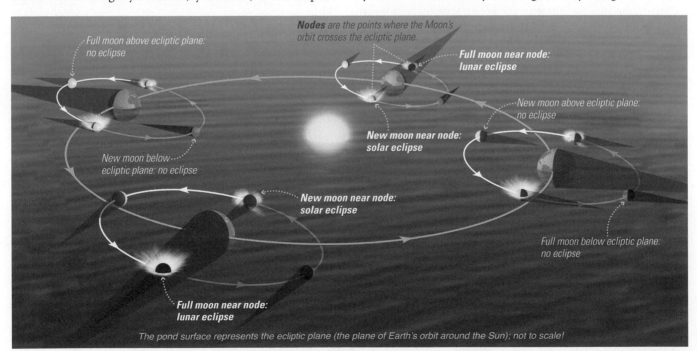

Nodes are the points where the Moon's orbit crosses the ecliptic plane.

Full moon above ecliptic plane: no eclipse

Full moon near node: lunar eclipse

New moon above ecliptic plane: no eclipse

New moon near node: solar eclipse

New moon below ecliptic plane: no eclipse

New moon near node: solar eclipse

Full moon below ecliptic plane: no eclipse

Full moon near node: lunar eclipse

The pond surface represents the ecliptic plane (the plane of Earth's orbit around the Sun); not to scale!

FIGURE 2.24 This illustration represents the ecliptic plane as the surface of a pond. The Moon's orbit is tilted by about 5° to the ecliptic plane, so the Moon spends half of each orbit above the plane (the pond surface) and half below it. Eclipses occur only when the Moon is at both a node (passing through the pond surface) and a phase of either new moon (for a solar eclipse) or full moon (for a lunar eclipse)—as is the case with the lower left and top right orbits shown.

the Sun and Earth about twice each year. We therefore find the following conditions for an eclipse to occur:

1. The phase of the Moon must be full (for a lunar eclipse) or new (for a solar eclipse).

2. The new or full moon must occur during one of the periods when the nodes of the Moon's orbit are aligned with the Sun and Earth.

Inner and Outer Shadows Figure 2.24 shows the Moon and Earth each casting only a simple "shadow cone" (extending away from the Sun) at each position shown. However, a closer look at the geometry shows that the shadow of the Moon or Earth actually consists of two distinct regions: a central **umbra,** where sunlight is completely blocked, and a surrounding **penumbra,** where sunlight is only partially blocked (**FIGURE 2.25**). The shadow cones in Figure 2.24 represent only the umbra, but both shadow regions are important to understanding eclipses.

Lunar Eclipses A lunar eclipse begins at the moment when the Moon's orbit first carries it into Earth's penumbra. After that, we will see one of three types of lunar eclipse (**FIGURE 2.26**). If the Sun, Earth, and Moon are nearly perfectly aligned, the

FIGURE 2.27 This multiple-exposure photograph shows the progression (left to right) of a total lunar eclipse observed from Tenerife, Canary Islands (Spain). Totality began (far right) just before the Moon set in the west. Notice Earth's curved shadow advancing across the Moon during the partial phases, and the redness of the full moon during totality.

Moon passes through Earth's umbra and we see a **total lunar eclipse.** If the alignment is somewhat less perfect, only part of the full moon passes through the umbra (with the rest in the penumbra) and we see a **partial lunar eclipse.** If the Moon passes through *only* Earth's penumbra, we see a **penumbral lunar eclipse.** Penumbral eclipses are the most common type of lunar eclipse, but they are the least visually impressive because the full moon darkens only slightly.

Total lunar eclipses are the most spectacular. The Moon becomes dark and eerily red during **totality,** when the Moon is entirely engulfed in the umbra. Totality usually lasts about an hour, with partial phases both before and after. The curvature of Earth's shadow during partial phases shows that Earth is round (**FIGURE 2.27**). To understand the redness during totality, consider the view of an observer on the eclipsed Moon, who would see Earth's night side surrounded by the reddish glow of all the sunrises and sunsets occurring on the Earth at that moment. It is this reddish light that illuminates the Moon during total eclipse.

FIGURE 2.25 The shadow cast by an object in sunlight. Sunlight is fully blocked in the umbra and partially blocked in the penumbra.

Solar Eclipses We can also see three types of solar eclipse (**FIGURE 2.28**). If a solar eclipse occurs when the Moon is in a part of its orbit where it is relatively close to Earth (see Figure 2.21), the Moon's umbra can cover a small area of Earth's surface (up to about 270 kilometers in diameter). Within this area you will see a **total solar eclipse.** If the eclipse occurs when the Moon is in a part of its orbit that puts it farther from Earth, the umbra may not reach Earth's surface at all. In that case, you will see an **annular eclipse**—a ring of sunlight surrounding the Moon—in the small region of Earth directly behind the umbra. In either case, the region of totality or annularity will be surrounded by a much larger region (typically about 7000 kilometers in diameter) that falls within the Moon's penumbral shadow. Here you will see a **partial solar eclipse,** in which only part of the Sun is blocked from view. The combination of Earth's rotation and the Moon's orbital motion causes the Moon's shadows to race across the

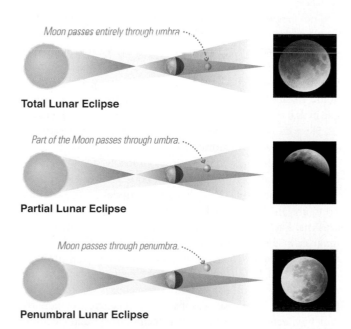

FIGURE 2.26 interactive figure The three types of lunar eclipse.

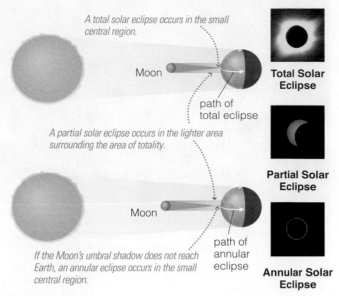

A total solar eclipse occurs in the small central region.

Moon

path of total eclipse

A partial solar eclipse occurs in the lighter area surrounding the area of totality.

Moon

If the Moon's umbral shadow does not reach Earth, an annular eclipse occurs in the small central region.

path of annular eclipse

Total Solar Eclipse

Partial Solar Eclipse

Annular Solar Eclipse

a The three types of solar eclipse. The diagrams show the Moon's shadow falling on Earth; note the dark central umbra surrounded by the much lighter penumbra.

b This photo from Earth orbit shows the Moon's shadow (umbra) on Earth during a total solar elipse. Notice that only a small region of Earth experiences totality at any one time.

FIGURE 2.28 interactive figure During a solar eclipse, the Moon's small shadow moves rapidly across the face of Earth.

face of Earth at a typical speed of about 1700 kilometers per hour. As a result, the umbral shadow traces a narrow path across Earth, and totality never lasts more than a few minutes in any particular place.

A total solar eclipse is a spectacular sight. It begins when the disk of the Moon first appears to touch the Sun. Over the next couple of hours, the Moon appears to take a larger and larger "bite" out of the Sun. As totality approaches, the sky darkens and temperatures fall. Birds head back to their nests, and crickets begin their nighttime chirping. During the few minutes of totality, the Moon completely blocks the normally visible disk of the Sun, allowing the faint *corona* to be seen (**FIGURE 2.29**). The surrounding sky takes on a twilight glow,

FIGURE 2.29 This multiple-exposure photograph shows the progression of a total solar eclipse over La Paz, Mexico. Totality (central image) lasts only a few minutes, during which time we can see the faint corona around the outline of the Sun. The foreground church was photographed at a different time of day.

and planets and bright stars become visible in the daytime. As totality ends, the Sun slowly emerges from behind the Moon over the next couple of hours. However, because your eyes have adapted to the darkness, totality appears to end far more abruptly than it began.

Predicting Eclipses Few phenomena have so inspired and humbled humans throughout the ages as eclipses. For many cultures, eclipses were mystical events associated with fate or the gods, and countless stories and legends surround them. One legend holds that the Greek philosopher Thales (c. 624–546 B.C.) successfully predicted the year (but presumably not the precise time) that a total eclipse of the Sun would be visible in the area where he lived, which is now part of Turkey. The eclipse occurred as two opposing armies (the Medes and the Lydians) were massing for battle, and it so frightened them that they put down their weapons, signed a treaty, and returned home. Because modern research shows that the only eclipse visible in that part of the world at about that time occurred on May 28, 585 B.C., we know the precise date on which the treaty was signed—the earliest historical event that can be dated precisely.

Much of the mystery of eclipses probably stems from the relative difficulty of predicting them. Look again at Figure 2.24. The two periods each year when the nodes of the Moon's orbit are nearly aligned with the Sun are called **eclipse seasons.** Each eclipse season lasts a few weeks. Some type of lunar eclipse occurs during each eclipse season's full moon, and some type of solar eclipse occurs during its new moon.

If Figure 2.24 told the whole story, eclipse seasons would occur every 6 months and predicting eclipses would be easy. For example, if eclipse seasons always occurred in January and July, eclipses would always occur on the dates of new and full moons in those months. Actual eclipse prediction is

TABLE 2.1   Lunar Eclipses 2013–2016*

Date	Type	Where You Can See It
April 25, 2013	partial	Europe, Africa, Asia, Australia
May 25, 2013	penumbral	Americas, Africa
Oct. 18, 2013	penumbral	Americas, Europe, Africa, Asia
April 15, 2014	total	Australia, Pacific, Americas
Oct. 8, 2014	total	Asia, Australia, Pacific, Americas
April 4, 2015	total	Asia, Australia, Pacific, Americas
Sept. 28, 2015	total	Americas, Europe, Africa
March 23, 2016	penumbral	Asia, Australia, Pacific, western Americas
Sept. 16, 2016	penumbral	Europe, Africa, Asia, Australia

*Dates are based on Universal Time and hence are those in Greenwich, England, at the time of the eclipse; check a news source for the local time and date. Eclipse predictions by Fred Espenak, NASA GSFC.

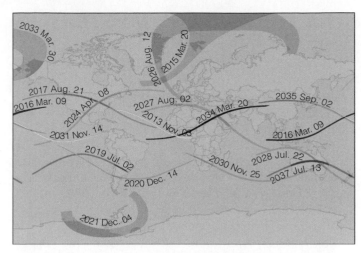

FIGURE 2.30 This map shows the paths of totality for solar eclipses from 2013 through 2037. Paths of the same color represent eclipses occurring in successive saros cycles, separated by 18 years 11 days. For example, the 2034 eclipse occurs 18 years 11 days after the 2016 eclipse (both shown in red). Eclipse predictions by Fred Espenak, NASA GSFC.

more difficult than this because of something the figure does not show: The nodes slowly move around the Moon's orbit, so the eclipse seasons occur slightly less than 6 months apart (about 173 days apart).

The combination of the changing dates of eclipse seasons and the $29\frac{1}{2}$ day cycle of lunar phases makes eclipses recur in a cycle of about 18 years, $11\frac{1}{3}$ days, called the **saros cycle.** Astronomers in many ancient cultures identified the saros cycle and used it to make eclipse predictions. For example, in the Middle East the Babylonians achieved remarkable success at predicting eclipses more than 2500 years ago, and the Mayans achieved similar success in Central America; in fact, the Mayan calendar includes a cycle (the *sacred round*) of 260 days—almost exactly $1\frac{1}{2}$ times the 173.32 days between successive eclipse seasons.

However, while the saros cycle allows you to predict when an eclipse will occur, it does not allow you to predict exactly where or the precise type of eclipse. For example, if a total solar eclipse occurred today, another would occur 18 years $11\frac{1}{3}$ days from now, but it would not be visible from the same places on Earth and might be annular or partial rather than total. No ancient culture achieved the ability to predict eclipses in every detail.

Today, we can predict eclipses because we know the precise details of the orbits of Earth and the Moon. **TABLE 2.1** lists upcoming lunar eclipses; notice that, as we expect, eclipses generally come a little less than 6 months apart. **FIGURE 2.30** shows paths of totality for upcoming total solar eclipses (but not for partial or annular eclipses), using color coding to show eclipses that repeat with the saros cycle.

THINK ABOUT IT

In Table 2.1, notice that there's one exception to the "rule" of eclipses coming a little less than 6 months apart: the 2013 lunar eclipses of April 25 and May 25. How can eclipses occur just a month apart? Explain.

SPECIAL TOPIC

Does the Moon Influence Human Behavior?

From myths of werewolves to stories of romance under the full moon, human culture is filled with claims that the Moon influences our behavior. Can we say anything scientific about such claims?

The Moon clearly has important influences on Earth, perhaps most notably through its role in creating tides [**Section 4.5**]. Although the Moon's tidal force cannot directly affect objects as small as people, the ocean tides have indirect effects. For example, fishermen, boaters, and surfers all adjust at least some of their activities to the cycle of the tides.

Another potential influence might come from the lunar phases. Physiological patterns in many species appear to follow the lunar phases; for example, some crabs and turtles lay eggs only at full moon.

No human trait is so closely linked to lunar phases, but the average human menstrual cycle is so close in length to a lunar month that it is difficult to believe the similarity is mere coincidence.

Nevertheless, aside from the physiological cycles and the influence of tides on people who live near the oceans, claims that the lunar phase affects human behavior are difficult to verify scientifically. For example, although it is possible that the full moon brings out certain behaviors, it may also simply be that some behaviors are easier to engage in when the sky is bright. A beautiful full moon may bring out your desire to walk on the beach under the moonlight, but there is no scientific evidence to suggest that the full moon would affect you the same way if you were confined to a deep cave.

2.4 THE ANCIENT MYSTERY OF THE PLANETS

We've now covered the appearance and motion of the stars, Sun, and Moon in the sky. That leaves us with the planets yet to discuss. As you'll soon see, planetary motion posed an ancient mystery that played a critical role in the development of modern civilization.

Five planets are easy to find with the naked eye: Mercury, Venus, Mars, Jupiter, and Saturn. Mercury is visible infrequently, and only just after sunset or just before sunrise because it is so close to the Sun. Venus often shines brightly in the early evening in the west or before dawn in the east. If you see a very bright "star" in the early evening or early morning, it is probably Venus. Jupiter, when it is visible at night, is the brightest object in the sky besides the Moon and Venus. Mars is often recognizable by its reddish color, though you should check a star chart to make sure you aren't looking at a bright red star. Saturn is also easy to see with the naked eye, but because many stars are just as bright as Saturn, it helps to know where to look. (It also helps to know that planets tend not to twinkle as much as stars.) Sometimes several planets may appear close together in the sky, offering a particularly beautiful sight (**FIGURE 2.31**).

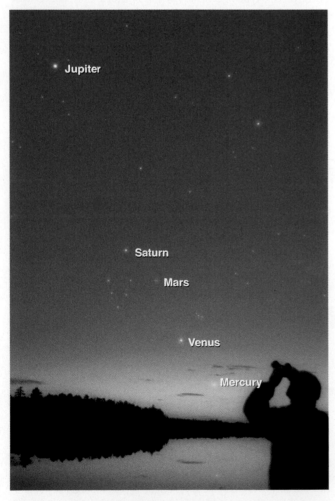

FIGURE 2.31 This photograph shows a grouping in our sky of all five planets that are easily visible to the naked eye. It was taken near Chatsworth, New Jersey, on April 23, 2002. The next such close grouping of these five planets in our sky will occur in September 2040.

SEE IT FOR YOURSELF

Using astronomical software or the Web, find out what planets are visible tonight and where to look for them, then go out and try to find them. Are they easy or difficult to identify?

Why was planetary motion so hard to explain?

Over the course of a single night, planets behave like all other objects in the sky: Earth's rotation makes them appear to rise in the east and set in the west. But if you continue to watch the planets night after night, you will notice that their movements among the constellations are quite complex. Instead of moving steadily eastward relative to the stars, like the Sun and Moon, the planets vary substantially in both speed and brightness; in fact, the word *planet* comes from the Greek for "wandering star." Moreover, while the planets *usually* move eastward through the constellations, they occasionally reverse course, moving westward through the zodiac (**FIGURE 2.32**). These periods of **apparent retrograde motion** (*retrograde* means "backward") last from a few weeks to a few months, depending on the planet.

For ancient people who believed in an Earth-centered universe, apparent retrograde motion was very difficult to explain; after all, what could make planets sometimes turn around and go backward if everything moves in circles around Earth? The ancient Greeks came up with some very clever ways to explain it, but their explanations (which we'll study in Chapter 3) were quite complex.

In contrast, apparent retrograde motion has a simple explanation in a Sun-centered solar system. You can demonstrate it for yourself with the help of a friend (**FIGURE 2.33a**). Pick a spot in an open area to represent the Sun. You can represent Earth by walking counterclockwise around the Sun, while your friend represents a more distant planet (such as Mars or Jupiter) by walking in the same direction around the Sun at a greater distance. Your friend should walk more slowly than you, because more distant planets orbit the Sun more slowly. As you walk, watch how your friend appears to

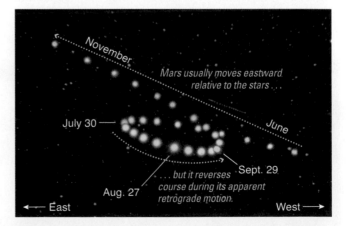

FIGURE 2.32 This composite of 29 individual photos (taken at 5- to 8-day intervals in 2003) shows a retrograde loop of Mars. Note that Mars is biggest and brightest in the middle of the retrograde loop, because that is where it is closest to Earth in its orbit. (The faint dots just right of center are images of the planet Uranus, which by coincidence was in the same part of the sky.)

Apparent retrograde motion occurs between positions 3 and 5, as the inner person (planet) passes the outer person (planet).

Follow the lines of sight from inner person (planet) to outer person (planet) to see where the outer one appears against the background.

a The retrograde motion demonstration: Watch how your friend (in red) usually appears to move forward against the background of the building in the distance but appears to move backward as you (in blue) catch up to and pass her in your "orbit."

b This diagram shows how the same idea applies to a planet. Follow the lines of sight from Earth to Mars in numerical order. Notice that Mars appears to move westward relative to the distant stars as Earth passes it by in its orbit (roughly from points 3 to 5).

FIGURE 2.33 interactive figure Apparent retrograde motion—the occasional "backward" motion of the planets relative to the stars—has a simple explanation in a Sun-centered solar system.

move relative to buildings or trees in the distance. Although both of you always walk the same way around the Sun, your friend will appear to move backward against the background during the part of your "orbit" in which you catch up to and pass him or her. **FIGURE 2.33b** shows how the same idea applies to Mars. Note that Mars never actually changes direction; it only *appears* to go backward as Earth passes Mars in its orbit. (To understand the apparent retrograde motions of Mercury and Venus, which are closer to the Sun than is Earth, simply switch places with your friend and repeat the demonstration.)

Why did the ancient Greeks reject the real explanation for planetary motion?

If the apparent retrograde motion of the planets is so readily explained by recognizing that Earth orbits the Sun, why wasn't this idea accepted in ancient times? In fact, the idea that Earth goes around the Sun was suggested as early as 260 B.C. by the Greek astronomer Aristarchus (see Special Topic, page 48). Nevertheless, Aristarchus's contemporaries rejected his idea, and the Sun-centered solar system did not gain wide acceptance until almost 2000 years later.

Although there were many reasons the Greeks were reluctant to abandon the idea of an Earth-centered universe, one of the most important was their inability to detect what we call **stellar parallax**. Extend your arm and hold up one finger. If you keep your finger still and alternately close your left eye and right eye, your finger will appear to jump back and forth against the background. This apparent shifting, called *parallax,* occurs because your two eyes view your finger from opposite sides of your nose. If you move your finger closer to your face, the parallax increases. If you look at a distant tree or

flagpole instead of your finger, you may not notice any parallax at all. In other words, parallax depends on distance, with nearer objects exhibiting greater parallax than more distant objects.

If you now imagine that your two eyes represent Earth at opposite sides of its orbit around the Sun and that the tip of your finger represents a relatively nearby star, you have the idea of stellar parallax. That is, because we view the stars from different places in our orbit at different times of year, nearby stars should *appear* to shift back and forth against the background of more distant stars (**FIGURE 2.34**).

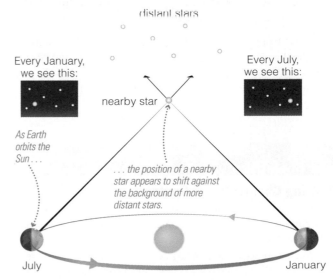

FIGURE 2.34 Stellar parallax is an apparent shift in the position of a nearby star as we look at it from different places in Earth's orbit. This figure is greatly exaggerated; in reality, the amount of shift is far too small to detect with the naked eye.

Who First Proposed a Sun-Centered Solar System?

You've probably heard of Copernicus, whose work in the 16th century started the revolution that ultimately overturned the ancient belief in an Earth-centered universe [**Section 3.3**]. However, the idea that Earth goes around the Sun was proposed much earlier by the Greek scientist Aristarchus (c. 310–230 B.C.).

Little of Aristarchus's work survives to the present day, so we cannot know what motivated him to suggest an idea so contrary to the prevailing view of an Earth-centered universe. However, it's likely that he was motivated by the fact that a Sun-centered system offers a much more natural explanation for the apparent retrograde motion of the planets. To account for the lack of detectable stellar parallax, Aristarchus suggested that the stars were extremely far away.

Aristarchus further strengthened his argument by estimating the sizes of the Moon and the Sun. By observing the shadow of Earth on the Moon during a lunar eclipse, he estimated the Moon's diameter to be about one-third of Earth's diameter—only slightly more than the actual value. He then used a geometric argument, based on measuring the angle between the Moon and the Sun at first- and third-quarter phases, to conclude that the Sun must be larger than

Earth. (Aristarchus's measurements were imprecise, so he estimated the Sun's diameter to be about 7 times Earth's rather than the correct value of about 100 times.) His conclusion that the Sun is larger than Earth may have been another reason he believed that Earth should orbit the Sun, rather than vice versa.

Although Aristarchus was probably the first to suggest that Earth orbits the Sun, his ideas built on the work of earlier scholars. For example, Heracleides (c. 388–315 B.C.) had previously suggested that Earth rotates, which offered Aristarchus a way to explain the daily circling of the sky in a Sun-centered system. Heracleides also suggested that not all heavenly bodies circle Earth: Based on the fact that Mercury and Venus always stay fairly close to the Sun in the sky, he argued that these two planets must orbit the Sun. In suggesting that *all* the planets orbit the Sun, Aristarchus was extending the ideas of Heracleides and others before him.

Aristarchus gained little support among his contemporaries, but his ideas never died, and Copernicus was aware of them when he proposed his own version of the Sun-centered system. Thus, our modern understanding of the universe owes at least some debt to the remarkable vision of a man born more than 2300 years ago.

Because the Greeks believed that all stars lie on the same celestial sphere, they expected to see stellar parallax in a slightly different way. If Earth orbited the Sun, they reasoned, at different times of year we would be closer to different parts of the celestial sphere and would notice changes in the angular separation of stars. However, no matter how hard they searched, they could find no sign of stellar parallax. They concluded that one of the following must be true:

1. Earth orbits the Sun, but the stars are so far away that stellar parallax is undetectable to the naked eye.

2. There is no stellar parallax because Earth remains stationary at the center of the universe.

Aside from a few notable exceptions such as Aristarchus, the Greeks rejected the correct answer (the first one) because they could not imagine that the stars could be *that* far away. Today, we can detect stellar parallax with the aid of telescopes,

providing direct proof that Earth really does orbit the Sun. Careful measurements of stellar parallax also provide the most reliable means of measuring distances to nearby stars [**Section 15.1**].

THINK ABOUT IT

How far apart are opposite sides of Earth's orbit? How far away are the nearest stars? Using the 1-to-10-billion scale from Chapter 1, describe the challenge of detecting stellar parallax.

The ancient mystery of the planets drove much of the historical debate over Earth's place in the universe. In many ways, the modern technological society we take for granted today can be traced directly to the scientific revolution that began in the quest to explain the strange wanderings of the planets among the stars in our sky. We will turn our attention to this revolution in the next chapter.

The Big Picture

Putting Chapter 2 into Context

In this chapter, we surveyed the phenomena of our sky. Keep the following "big picture" ideas in mind as you continue your study of astronomy:

■ You can enhance your enjoyment of astronomy by observing the sky. The more you learn about the appearance and apparent motions of the sky, the more you will appreciate what you can see in the universe.

■ From our vantage point on Earth, it is convenient to imagine that we are at the center of a great celestial sphere—even though

we really are on a planet orbiting a star in a vast universe. We can then understand what we see in the local sky by thinking about how the celestial sphere appears from our latitude.

■ Most of the phenomena of the sky are relatively easy to observe and understand. The more complex phenomena—particularly eclipses and apparent retrograde motion of the planets—challenged our ancestors for thousands of years. The desire to understand these phenomena helped drive the development of science and technology.

SUMMARY OF KEY CONCEPTS

2.1 PATTERNS IN THE NIGHT SKY

- **What does the universe look like from Earth?** Stars and other celestial objects appear to lie on a great **celestial** **sphere** surrounding Earth. We divide the celestial sphere into **constellations** with well-defined borders. From any location on Earth, we see half the celestial sphere at any one time as the dome of our **local sky,** in which the **horizon** is the boundary between Earth and sky, the **zenith** is the point directly overhead, and the **meridian** runs from due south to due north through the zenith.

- **Why do stars rise and set?** Earth's rotation makes stars appear to circle around Earth each day. A star whose complete circle lies above our horizon is said to be **circumpolar.** Other stars have circles that cross the horizon, making them rise in the east and set in the west each day.

- **Why do the constellations we see depend on latitude and time of year?** The visible constellations vary with time of year because our night sky lies in different directions in space as we orbit the Sun. The constellations vary with **latitude** because your latitude determines the orientation of your horizon relative to the celestial sphere. The sky does not vary with **longitude.**

2.2 THE REASON FOR SEASONS

- **What causes the seasons?** The tilt of Earth's axis causes the seasons. The axis points in the same direction throughout the year, so as Earth orbits the Sun, sunlight hits different parts of Earth more directly at different times of year.

- **How does the orientation of Earth's axis change with time?** Earth's 26,000-year cycle of **precession** changes the orientation of the axis in space, although the tilt remains about $23\frac{1}{2}°$. The changing orientation of the axis does not affect the pattern of seasons, but it changes the identity of the North Star and shifts the locations of the solstices and equinoxes in Earth's orbit.

2.3 THE MOON, OUR CONSTANT COMPANION

- **Why do we see phases of the Moon?** The **phase** of the Moon depends on its position relative to the Sun as it orbits Earth. The half of the Moon facing the Sun is always illuminated while the other half is dark, but from Earth we see varying combinations of the illuminated and dark faces.

- **What causes eclipses?** We see a **lunar eclipse** when Earth's shadow falls on the Moon and a **solar eclipse** when the Moon blocks our view of the Sun. We do not see an eclipse at every new and full moon because the Moon's orbit is slightly inclined to the ecliptic plane.

2.4 THE ANCIENT MYSTERY OF THE PLANETS

- **Why was planetary motion so hard to explain?** Planets generally move eastward relative to the stars over the course of the year, but for weeks or months they reverse course during periods of **apparent retrograde motion.** This motion occurs when Earth passes by (or is passed by) another planet in its orbit, but it posed a major mystery to ancient people who assumed Earth to be at the center of the universe.

- **Why did the ancient Greeks reject the real explanation for planetary motion?** The Greeks rejected the idea that Earth goes around the Sun in part because they could not detect **stellar parallax**—slight apparent shifts in stellar positions over the course of the year. To most Greeks, it seemed unlikely that the stars could be so far away as to make parallax undetectable to the naked eye, even though that is, in fact, the case.

Use the following questions to check your understanding of some of the many types of visual information used in astronomy. Answers are provided in Appendix J. For additional practice, try the Chapter 2 Visual Quiz at MasteringAstronomy®.

The figure above is a typical diagram used to describe Earth's seasons.

1. Which of the four labeled points (A through D) represents the beginning of summer for the Northern Hemisphere?
2. Which of the four labeled points represents the beginning of summer for the Southern Hemisphere?
3. Which of the four labeled points represents the beginning of spring for the Southern Hemisphere?
4. The diagram exaggerates the sizes of Earth and the Sun relative to the orbit. If Earth were correctly scaled relative to the orbit in the figure, how big would it be?
   a. about half the size shown   b. about 2 millimeters across
   c. about 0.1 millimeter across   d. microscopic
5. Given that Earth's actual distance from the Sun varies by less than 3% over the course of a year, why does the diagram look so elliptical?
   a. It correctly shows that Earth is closest to the Sun at points A and C and farthest at points B and D.
   b. The elliptical shape is an effect of perspective, since the diagram shows an almost edge-on view of a nearly circular orbit.
   c. The shape of the diagram is meaningless and is done only for artistic effect.

The figure above (based on Figure 2.14) shows the Sun's path through the constellations of the zodiac.

6. As viewed from Earth, in which zodiac constellation does the Sun appear to be located on April 21?
   a. Leo   b. Aquarius   c. Libra   d. Aries
7. If the date is April 21, what zodiac constellation will be visible on your meridian at midnight?
   a. Leo   b. Aquarius   c. Libra   d. Aries
8. If the date is April 21, what zodiac constellation will you see setting in the west shortly after sunset?
   a. Scorpius   b. Pisces   c. Taurus   d. Virgo

MasteringAstronomy®

For instructor-assigned homework go to MasteringAstronomy®.

REVIEW QUESTIONS

Short-Answer Questions Based on the Reading

1. What are *constellations*? How did they get their names?
2. Suppose you were making a model of the celestial sphere with a ball. Briefly describe all the things you would need to mark on your celestial sphere.
3. On a clear, dark night, the sky may appear to be "full" of stars. Does this appearance accurately reflect the way stars are distributed in space? Explain.
4. Why does the *local sky* look like a dome? Define *horizon, zenith,* and *meridian.* How do we describe the location of an object in the local sky?
5. Explain why we can measure only *angular sizes* and *angular distances* for objects in the sky. What are *arcminutes* and *arcseconds*?
6. What are *circumpolar stars*? Are more stars circumpolar at the North Pole or in the United States? Explain.

7. What are *latitude* and *longitude*? Does the sky vary with latitude? Does it vary with longitude? Explain.
8. What is the *zodiac,* and why do we see different parts of it at different times of year?
9. Suppose Earth's axis had no tilt. Would we still have seasons? Why or why not?
10. Briefly describe key facts about the solstices and equinoxes.
11. What is *precession,* and how does it affect what we see in our sky?
12. Briefly describe the Moon's cycle of *phases.* Can you ever see a full moon at noon? Explain.
13. Why do we always see the same face of the Moon?
14. Why don't we see an *eclipse* at every new and full moon? Describe the conditions needed for a *solar* or *lunar eclipse.*
15. What do we mean by the *apparent retrograde motion* of the planets? Why was this motion difficult for ancient astronomers to explain? How do we explain it today?

16. What is *stellar parallax*? How did an inability to detect it support the ancient belief in an Earth-centered universe?

TEST YOUR UNDERSTANDING

Does It Make Sense?

Decide whether the statement makes sense (or is clearly true) or does not make sense (or is clearly false). Explain clearly; not all of these have definitive answers, so your explanation is more important than your chosen answer.

17. The constellation Orion didn't exist when my grandfather was a child.
18. When I looked into the dark lanes of the Milky Way with my binoculars, I saw what must have been a cluster of distant galaxies.
19. Last night the Moon was so big that it stretched for a mile across the sky.
20. I live in the United States, and during my first trip to Argentina I saw many constellations that I'd never seen before.
21. Last night I saw Jupiter right in the middle of the Big Dipper. (*Hint:* Is the Big Dipper part of the zodiac?)
22. Last night I saw Mars move westward through the sky in its apparent retrograde motion.
23. Although all the known stars rise in the east and set in the west, we might someday discover a star that will rise in the west and set in the east.
24. If Earth's orbit were a perfect circle, we would not have seasons.
25. Because of precession, someday it will be summer everywhere on Earth at the same time.
26. This morning I saw the full moon setting at about the same time the Sun was rising.

Quick Quiz

Choose the best answer to each of the following. Explain your reasoning with one or more complete sentences.

27. Two stars that are in the same constellation (a) must both be part of the same cluster of stars in space. (b) must both have been discovered at about the same time. (c) may actually be very far away from each other.
28. The north celestial pole is 35° above your northern horizon. This tells you that (a) you are at latitude 35°N. (b) you are at longitude 35°E. (c) you are at latitude 35°S.
29. Beijing and Philadelphia have about the same latitude but very different longitudes. Therefore, tonight's night sky in these two places (a) will look about the same. (b) will have completely different sets of constellations. (c) will have partially different sets of constellations.
30. In winter, Earth's axis points toward the star Polaris. In spring, (a) the axis also points toward Polaris. (b) the axis points toward Vega. (c) the axis points toward the Sun.
31. When it is summer in Australia, the season in the United States is (a) winter. (b) summer. (c) spring.
32. If the Sun rises precisely due east, (a) you must be located at Earth's equator. (b) it must be the day of either the spring or the fall equinox. (c) it must be the day of the summer solstice.
33. A week after full moon, the Moon's phase is (a) first quarter. (b) third quarter. (c) new.
34. The fact that we always see the same face of the Moon tells us that (a) the Moon does not rotate. (b) the Moon's rotation period is the same as its orbital period. (c) the Moon looks the same on both sides.
35. If there is going to be a total lunar eclipse tonight, then you know that (a) the Moon's phase is full. (b) the Moon's phase is new. (c) the Moon is unusually close to Earth.

36. When we see Saturn going through a period of apparent retrograde motion, it means (a) Saturn is temporarily moving backward in its orbit of the Sun. (b) Earth is passing Saturn in its orbit, with both planets on the same side of the Sun. (c) Saturn and Earth must be on opposite sides of the Sun.

PROCESS OF SCIENCE

Examining How Science Works

37. *Earth-Centered or Sun-Centered?* Decide whether each of the following phenomena is consistent or inconsistent with a belief in an Earth-centered system. If consistent, describe how. If inconsistent, explain why, and also explain why the inconsistency did not immediately lead people to abandon the Earth-centered model.
    a. The daily paths of stars through the sky **b.** Seasons **c.** Phases of the Moon **d.** Eclipses **e.** Apparent retrograde motion of the planets
38. *Shadow Phases.* Many people incorrectly guess that the phases of the Moon are caused by Earth's shadow falling on the Moon. How would you convince a friend that the phases of the Moon have nothing to do with Earth's shadow? Describe the observations you would use to show that Earth's shadow can't be the cause of phases.

GROUP WORK EXERCISE

39. *Lunar Phases and Time of Day.* Before you begin, assign the following roles to the people in your group: *Scribe* (takes notes on the group's activities), *Proposer* (proposes explanations to the group), *Skeptic* (points out weaknesses in proposed explanations), and *Moderator* (leads group discussion and makes sure everyone contributes). Then each member of the group should draw a copy of the following diagram, which represents the Moon's orbit as seen from above Earth's North Pole (not to scale):

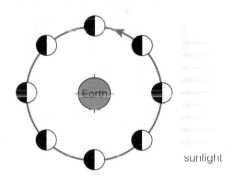

Discuss and answer the following questions as a group:
a. How would the Moon appear from Earth at each of the eight Moon positions? Label each one with the corresponding phase. **b.** What time of day corresponds to each of the four tick marks on Earth? Label each tick mark accordingly. **c.** Why doesn't the Moon's phase change during the course of one night? Explain your reasoning. **d.** At what times of day would a full moon be visible to someone standing on Earth? Write down when a full moon rises and explain why it appears to rise at that time. **e.** At what times of day would a third-quarter moon be visible to someone standing on Earth? Write down when a third-quarter moon sets and explain why it appears to set at that time. **f.** At what times of day would a waxing crescent moon be visible to someone standing on Earth? Write down when a waxing crescent moon rises and explain why it appears to rise at that time.

INVESTIGATE FURTHER

In-Depth Questions to Increase Your Understanding

Short-Answer/Essay Questions

40. *New Planet.* A planet in another solar system has a circular orbit and an axis tilt of 35°. Would you expect this planet to have seasons? If so, would you expect them to be more extreme than the seasons on Earth? If not, why not?

41. *Your View of the Sky.*
    a. What are your latitude and longitude? **b.** Where does the north (or south) celestial pole appear in your sky? **c.** Is Polaris a circumpolar star in your sky? Explain.

42. *View from the Moon.* Assume you live on the Moon, near the center of the face that looks toward Earth.
    a. Suppose you see a full earth in your sky. What phase of the Moon would people on Earth see? Explain. **b.** Suppose people on Earth see a full moon. What phase would you see for Earth? Explain. **c.** Suppose people on Earth see a waxing gibbous moon. What phase would you see for Earth? Explain. **d.** Suppose people on Earth are viewing a total lunar eclipse. What would you see from your home on the Moon? Explain.

43. *View from the Sun.* Suppose you lived on the Sun (and could ignore the heat). Would you still see the Moon go through phases as it orbits Earth? Why or why not?

44. *A Farther Moon.* Suppose the distance to the Moon were twice its actual value. Would it still be possible to have a total solar eclipse? Why or why not?

45. *A Smaller Earth.* Suppose Earth were smaller. Would solar eclipses be any different? If so, how? What about lunar eclipses?

46. *Observing Planetary Motion.* Find out which planets are currently visible in your evening sky. At least once a week, observe the planets and draw a diagram showing the position of each visible planet relative to stars in a zodiac constellation. From week to week, note how the planets are moving relative to the stars. Can you see any of the apparently wandering features of planetary motion? Explain.

47. *A Connecticut Yankee.* Find the book *A Connecticut Yankee in King Arthur's Court* by Mark Twain. Read the portion that deals with the Connecticut Yankee's prediction of an eclipse. In a one- to two-page essay, summarize the episode and explain how it helped the Connecticut Yankee gain power.

Quantitative Problems

Be sure to show all calculations clearly and state your final answers in complete sentences.

48. *Arcminutes and Arcseconds.* There are 360° in a full circle.
    a. How many arcminutes are in a full circle? **b.** How many arcseconds are in a full circle? **c.** The Moon's angular size is about $\frac{1}{2}°$. What is this in arcminutes? In arcseconds?

49. *Latitude Distance.* Earth's radius is approximately 6370 km.
    a. What is Earth's circumference? **b.** What distance is represented by each degree of latitude? **c.** What distance is represented by each arcminute of latitude? **d.** Can you give similar answers for the distances represented by a degree or arcminute of longitude? Why or why not?

50. *Angular Conversions I.* The following angles are given in degrees and fractions of degrees. Rewrite them in degrees, arcminutes, and arcseconds.
    a. 24.3° **b.** 1.59° **c.** 0.1° **d.** 0.01° **e.** 0.001°

51. *Angular Conversions II.* The following angles are given in degrees, arcminutes, and arcseconds. Rewrite them in degrees and fractions of degrees.
    a. 7°38′42″ **b.** 12°54′ **c.** 1°59′59″ **d.** 1′ **e.** 1″

52. *Angular Size of Your Finger.* Measure the width of your index finger and the length of your arm. Based on your measurements, calculate the angular width of your index finger at arm's length. Does your result agree with the approximations shown in Figure 2.7c? Explain.

53. *Find the Sun's Diameter.* The Sun has an angular diameter of about 0.5° and an average distance of about 150 million km. What is the Sun's approximate physical diameter? Compare your answer to the actual value of 1,390,000 km.

54. *Find a Star's Diameter.* Estimate the diameter of the supergiant star Betelgeuse, using its angular diameter of about 0.05 arcsecond and distance of about 600 light-years. Compare your answer to the size of our Sun and the Earth-Sun distance.

55. *Eclipse Conditions.* The Moon's precise equatorial diameter is 3476 km, and its orbital distance from Earth varies between 356,400 and 406,700 km. The Sun's diameter is 1,390,000 km, and its distance from Earth ranges between 147.5 and 152.6 million km.
    a. Find the Moon's angular size at its minimum and maximum distances from Earth. **b.** Find the Sun's angular size at its minimum and maximum distances from Earth. **c.** Based on your answers to parts a and b, is it possible to have a total solar eclipse when the Moon and Sun are both at their maximum distance? Explain.

Discussion Questions

56. *Earth-Centered Language.* Many common phrases reflect the ancient Earth-centered view of our universe. For example, the phrase "the Sun rises each day" implies that the Sun is really moving over Earth. We know that the Sun only *appears* to rise as the rotation of Earth carries us to a place where we can see the Sun in our sky. Identify other common phrases that imply an Earth-centered viewpoint.

57. *Flat Earth Society.* Believe it or not, there is an organization called the Flat Earth Society. Its members hold that Earth is flat and that all indications to the contrary (such as pictures of Earth from space) are fabrications made as part of a conspiracy to hide the truth from the public. Discuss the evidence for a round Earth and how you can check it for yourself. In light of the evidence, is it possible that the Flat Earth Society is correct? Defend your opinion.

Web Projects

58. *Sky Information.* Search the Web for sources of daily information about sky phenomena (such as lunar phases, times of sunrise and sunset, or dates of equinoxes and solstices). Identify and briefly describe your favorite source.

59. *Constellations.* Search the Web for information about the constellations and their mythology. Write a short report about one or more constellations.

60. *Upcoming Eclipse.* Find information about an upcoming solar or lunar eclipse. Write a short report about how you could best observe the eclipse, including any necessary travel to a viewing site, and what you could expect to see. Bonus: Describe how you could photograph the eclipse.

THE SCIENCE OF ASTRONOMY

LEARNING GOALS

3.1 THE ANCIENT ROOTS OF SCIENCE

- In what ways do all humans use scientific thinking?
- How is modern science rooted in ancient astronomy?

3.2 ANCIENT GREEK SCIENCE

- Why does modern science trace its roots to the Greeks?
- How did the Greeks explain planetary motion?

3.3 THE COPERNICAN REVOLUTION

- How did Copernicus, Tycho, and Kepler challenge the Earth-centered model?

- What are Kepler's three laws of planetary motion?
- How did Galileo solidify the Copernican revolution?

3.4 THE NATURE OF SCIENCE

- How can we distinguish science from nonscience?
- What is a scientific theory?

3.5 ASTROLOGY

- How is astrology different from astronomy?
- Does astrology have any scientific validity?

We especially need imagination in science. It is not all mathematics, nor all logic, but is somewhat beauty and poetry.

—Maria Mitchell (1818–1889), astronomer and the first woman elected to the American Academy of Arts and Sciences

Today we know that Earth is a planet orbiting a rather ordinary star, in a galaxy of more than a hundred billion stars, in an incredibly vast universe. We know that Earth, along with the entire cosmos, is in constant motion. We know that, on the scale of cosmic time, human civilization has existed for only the briefest moment. How did we manage to learn these things?

It wasn't easy. In this chapter, we will trace how modern astronomy grew from its roots in ancient observations, including those of the Greeks. We'll pay special attention to the unfolding of the Copernican revolution, which overturned the ancient belief in an Earth-centered universe and laid the foundation for the rise of our technological civilization. Finally, we'll explore the nature of modern science and how science can be distinguished from nonscience.

3.1 THE ANCIENT ROOTS OF SCIENCE

The rigorous methods of modern science have proven to be among the most valuable inventions in human history. These methods have enabled us to discover almost everything we now know about nature and the universe, and they also have made our modern technology possible. In this section, we will explore the ancient roots of science, which grew out of experiences common to nearly all people and all cultures.

In what ways do all humans use scientific thinking?

Scientific thinking comes naturally to us. By about a year of age, a baby notices that objects fall to the ground when she drops them. She lets go of a ball—it falls. She pushes a plate of food from her high chair—it falls, too. She continues to drop all kinds of objects, and they all plummet to Earth. Through her powers of observation, the baby learns about the physical world, finding that things fall when they are unsupported. Eventually, she becomes so certain of this fact that, to her parents' delight, she no longer needs to test it continually.

One day someone gives the baby a helium balloon. She releases it, and to her surprise it rises to the ceiling! Her understanding of nature must be revised. She now knows that the principle "all things fall" does not represent the whole truth, although it still serves her quite well in most situations. It will be years before she learns enough about the atmosphere, the force of gravity, and the concept of density to understand *why* the balloon rises when most other objects fall. For now, she is delighted to observe something new and unexpected.

The baby's experience with falling objects and balloons exemplifies scientific thinking. In essence, science is a way of learning about nature through careful observation and trial-and-error experiments. Rather than thinking differently than other people, modern scientists simply are trained to organize everyday thinking in a way that makes it easier for them to share their discoveries and use their collective wisdom.

Just as learning to communicate through language, art, or music is a gradual process for a child, the development of science has been a gradual process for humanity. Science in its modern form requires painstaking attention to detail, relentless testing of each piece of information to ensure its reliability, and a willingness to give up old beliefs that are not consistent with observed facts about the physical world. For professional scientists, these demands are the "hard work" part of the job. At heart, professional scientists are like the baby with the balloon, delighted by the unexpected and motivated by those rare moments when they—and all of us—learn something new about the universe.

How is modern science rooted in ancient astronomy?

Astronomy has been called the oldest of the sciences, because its roots stretch deepest into antiquity. Ancient civilizations did not always practice astronomy in the same ways or for the same reasons that we study it today, but they nonetheless had some amazing achievements. Understanding this ancient astronomy can give us a greater appreciation of how and why science developed through time.

Practical Benefits of Astronomy No one knows exactly how or when humans first began making careful observations of the sky, but we know observation has been going on for many thousands of years. This interest in astronomy probably comes in part from our inherent curiosity as humans, but ancient cultures also discovered that astronomy had practical benefits for timekeeping, keeping track of seasonal changes, and navigation.

One amazing example comes from people of central Africa. Although we do not know exactly when they developed the skill, people in some regions learned to predict rainfall patterns by making careful observations of the Moon. **FIGURE 3.1** shows how the method works. The orientation of the "horns" of a waxing crescent moon (relative to the horizon) varies over the course of the year, primarily because the angle at which the ecliptic intersects the horizon changes during the year. (The orientation also depends on latitude.) In tropical regions in which there are distinct rainy and dry seasons—rather than the four seasons familiar at temperate latitudes—the orientation of the crescent moon can be used to predict how much rainfall should be expected over coming days and weeks.

FIGURE 3.1 In central Nigeria, the orientation of the "horns" of a waxing crescent moon (shown along the top) correlates with the average amount of rainfall at different times of year. Local people could use this fact to predict the weather with reasonable accuracy. (Adapted from *Ancient Astronomers* by Anthony F. Aveni.)

Astronomy and Measures of Time The impact of ancient astronomical observations is still with us in our modern measures of time. The length of our day is the time it takes the Sun to make one full circuit of the sky. The length of a month comes from the Moon's cycle of phases [**Section 2.3**], and our year is based on the cycle of the seasons [**Section 2.2**]. The seven days of the week were named after the seven "planets" of ancient times (**TABLE 3.1**), which were the Sun, the Moon, and the five planets that are easily visible to the naked eye: Mercury, Venus, Mars, Jupiter, Saturn. Note that the ancient definition of *planet*, which comes from a Greek word meaning "wanderer," applied to any object that appeared to wander among the fixed stars. That is why the Sun and Moon were on the list while Earth was not, because we don't see our own planet moving in the sky.

Because timekeeping was so important and required precise observations, many ancient cultures built structures or created special devices to help with it. Let's briefly investigate a few of the ways in which ancient cultures kept track of time.

Determining the Time of Day In the daytime, ancient peoples could tell time by observing the Sun's path through the sky. Many cultures probably used sticks and the shadows they cast as simple sundials [**Section S1.3**]. The ancient Egyptians built huge obelisks, often decorated in homage to the Sun, which probably also served as simple clocks (**FIGURE 3.2**).

At night, ancient people could estimate the time from the position and phase of the Moon (see Figure 2.22) or by observing the constellations visible at a particular time (see Figure 2.14). For example, ancient Egyptian star clocks, often

TABLE 3.1 The Seven Days of the Week and the Astronomical Objects They Honor

The seven days were originally linked directly to the seven objects. The correspondence is no longer perfect, but the overall pattern is clear in many languages; some English names come from Germanic gods.

Object	Germanic God	English	French	Spanish
Sun	—	Sunday	dimanche	domingo
Moon	—	Monday	lundi	lunes
Mars	Tiw	Tuesday	mardi	martes
Mercury	Woden	Wednesday	mercredi	miércoles
Jupiter	Thor	Thursday	jeudi	jueves
Venus	Fria	Friday	vendredi	viernes
Saturn	—	Saturday	samedi	sábado

FIGURE 3.2 This ancient Egyptian obelisk, which stands 83 feet tall and weighs 331 tons, resides in St. Peter's Square at the Vatican in Rome. It is one of 21 surviving obelisks from ancient Egypt, most of which are now scattered around the world. Shadows cast by the obelisks may have been used to tell time.

a The remains of Stonehenge today.

b This sketch shows how archaeologists believe Stonehenge looked upon its completion in about 1550 B.C. Several astronomical alignments are shown as they appear from the center. For example, the Sun rises directly over the Heel Stone on the summer solstice.

FIGURE 3.3 Stonehenge, in southern England, was built in stages from about 2750 B.C. to about 1550 B.C.

found painted on the coffin lids of Egyptian pharaohs, cataloged where particular stars appeared in the sky at various times of night throughout the year. By knowing the date from their calendar and observing the positions of the cataloged stars in the sky, the Egyptians could use the star clocks to estimate the time of night.

In fact, we can trace the origins of our modern clock to ancient Egypt, some 4000 years ago. The Egyptians divided daytime and nighttime into 12 equal parts each, which is how we got our 12 hours each of a.m. and p.m. The abbreviations *a.m.* and *p.m.* stand for the Latin terms *ante meridiem* and *post meridiem,* respectively, which mean "before the middle of the day" and "after the middle of the day."

By about 1500 B.C., Egyptians had abandoned star clocks in favor of clocks that measure time by the flow of water through an opening of a particular size, just as hourglasses measure time by the flow of sand through a narrow neck.* These *water clocks* had the advantage of working even when the sky was cloudy. They eventually became the primary timekeeping instruments for many cultures, including the Greeks, Romans, and Chinese. Water clocks, in turn, were replaced by mechanical clocks in the 17th century and by electronic clocks in the 20th century. Despite the availability of other types of clocks, sundials were common throughout ancient times and remain popular today both for their decorative value and as reminders that the Sun and stars once were our only guides to time.

Marking the Seasons Many ancient cultures built structures to help them mark the seasons. Stonehenge (**FIGURE 3.3**) is a well-known example that served both as an astronomical device and as a social and religious gathering place. In the

Americas, one of the most spectacular structures was the Templo Mayor (**FIGURE 3.4**) in the Aztec city of Tenochtitlán (in modern-day Mexico City), which featured twin temples on a flat-topped pyramid. From the vantage point of a royal observer watching from the opposite side of the plaza, the Sun rose through the notch between the temples on the equinoxes. Before the Conquistadors destroyed it, Spanish visitors reported elaborate rituals at the Templo Mayor, sometimes including human sacrifice, that were held at times determined by astronomical observations. After its destruction, stones from the Templo Mayor were used to build a cathedral in the great plaza of Mexico City.

Many cultures aligned their buildings and streets with the cardinal directions (north, south, east, and west), which made it easier to keep track of the changing rise and set positions of the Sun over the course of the year. This type of alignment is

FIGURE 3.4 This scale model shows the Templo Mayor and the surrounding plaza as they are thought to have looked before the Spanish Conquistadors destroyed them. The structure was used to help mark the seasons.

---

*Hourglasses using sand were not invented until about the 8th century A.D., long after the advent of water clocks. Natural sand grains vary in size, so making accurate hourglasses required technology for making uniform grains of sand.

FIGURE 3.5 This large structure, more than 20 meters in diameter, is a kiva in Chaco Canyon, New Mexico. It was built by Ancestral Pueblo People approximately 1000 years ago. Its main axis is aligned almost precisely north-south.

found at such diverse sites as the Egyptian pyramids and the Forbidden City in China and among ceremonial kivas built by the Ancestral Pueblo People of the American southwest (**FIGURE 3.5**). Many modern cities retain this layout, which is why you'll find so many streets that run directly north-south or east-west.

Other structures were used to mark the Sun's position on special dates such as the winter or summer solstice. Many such structures can be found around the world, but one of the most amazing is the *Sun Dagger*, made by the Ancestral Pueblo People in Chaco Canyon, New Mexico (**FIGURE 3.6**). Three large slabs of rock lie in front of a carved spiral in such a way that they produced special patterns of light and shadow at different times of year. For example, a single dagger of sunlight pierced the center of the spiral only at noon on the summer solstice, while two daggers of light bracketed the spiral at the winter solstice.

FIGURE 3.6 The Sun Dagger. Three large slabs of rock in front of the carved spiral produced patterns of light and shadow that varied throughout the year. Here, we see the single dagger of sunlight that pierced the center of the spiral only at noon on the summer solstice. (Unfortunately, within just 12 years of the site's 1977 discovery, the rocks shifted so that the effect no longer occurs; the shifts probably were due to erosion of the trail below the rocks caused by large numbers of visitors.)

The Sun Dagger may also have been used to mark a special cycle of the Moon that had ritual significance to the Ancestral Pueblo People. The rise and set positions of the full moon vary in an 18.6-year cycle (the cycle of precession of the Moon's orbit), so the full moon rises at its most southerly point along the eastern horizon only once every 18.6 years. At this time, known as a "major lunar standstill," the shadow of the full moon passes through the slabs of rock to lie tangent to the edge of the spiral in the Sun Dagger; then, 9.3 years later, the lunar shadow cuts through the center of the spiral. The major lunar standstill can also be observed with structures at nearby Chimney Rock and in cliff dwellings at Colorado's Mesa Verde National Park.

Solar and Lunar Calendars The tracking of the seasons eventually led to the advent of written calendars. Today, we use a *solar calendar*, meaning a calendar that is synchronized with the seasons so that seasonal events such as the solstices and equinoxes occur on approximately the same dates each year. The origins of our modern solar calendar go back to ancient Egypt, though many details (such as the timing of leap years) have been refined throughout history to keep the calendar well synchronized to the seasons [**Section S1.1**].

Solar calendars are not the only option. Many cultures created *lunar calendars* that aimed to stay synchronized with the Moon's $29\frac{1}{2}$-day cycle of phases, so that the Moon's phase was always the same on the first day of each month. A basic lunar calendar has 12 months, with some months lasting 29 days and others lasting 30 days; the lengths are chosen to make the average agree with the approximately $29\frac{1}{2}$-day lunar cycle. A 12-month lunar calendar therefore has only 354 or 355 days, or about 11 days fewer than a calendar based on the Sun. Such a calendar is still used in the Muslim religion. That is why the month long fast of Ramadan (the ninth month) begins about 11 days earlier with each subsequent year.

Some cultures that used lunar calendars apparently did not like the idea of having their months cycle through the seasons over time, so they modified their calendars to take advantage of an interesting coincidence: 19 years on a solar calendar is almost precisely 235 months on a lunar calendar. As a result, the lunar phases repeat on the same dates about every 19 years (a pattern known as the *Metonic cycle*, so named because it was recognized by the Greek astronomer Meton in 432 B.C.). For example, there was a full moon on December 28, 2012, and there will be a full moon 19 years later, on December 28, 2031. Because an ordinary lunar calendar has only $19 \times 12 = 228$ months in a 19-year period, adding 7 extra months (to make 235) can keep the lunar calendar roughly synchronized to the seasons. One way of adding the 7 months is used in the Jewish calendar, which adds a thirteenth month in the third, sixth, eighth, eleventh, fourteenth, seventeenth, and nineteenth years of each 19-year cycle. This scheme keeps the dates of Jewish holidays within about a 1-month range on a solar calendar, with precise dates repeating every 19 years. It also explains why the date of Easter changes from year to year: The New Testament ties the date of Easter to the Jewish festival of Passover. In a slight modification of the original scheme, most Western Christians now celebrate Easter on the

FIGURE 3.7 This photo shows a model of the celestial sphere and other instruments on the roof of the ancient astronomical observatory in Beijing. The observatory was built in the 15th century; the instruments shown here were built later and show a European influence brought by Jesuit missionaries.

first Sunday after the first full moon after March 21. If the full moon falls on Sunday, Easter is the following Sunday. (Eastern Orthodox churches calculate the date of Easter differently, because they base the date on the Julian rather than the Gregorian calendar [**Section S1.1**].)

Learning About Ancient Achievements The study of ancient astronomical achievements is a rich field of research. Many ancient cultures made careful observations of planets and stars, and some left remarkably detailed records. The Chinese, for example, began recording astronomical observations at least 5000 years ago, allowing ancient Chinese astronomers to make many important discoveries. By the 15th century, the Chinese had built a great observatory in Beijing, which still stands today (**FIGURE 3.7**). We can also study written records from ancient Middle Eastern civilizations such as those of Egypt and Babylonia.

Other cultures either did not leave clear written records or had records that were lost or destroyed, so we must piece together their astronomical achievements by studying the physical evidence they left behind. This type of study is usually called *archaeoastronomy,* a word that combines archaeology and astronomy. The astronomical uses of most of the structures we've discussed so far were discovered by researchers working in archaeoastronomy.

The cases we've discussed to this point have been fairly straightforward for archaeoastronomers to interpret, but many other cases are more ambiguous. For example, ancient people in what is now Peru etched hundreds of lines and patterns in the sand of the Nazca desert. Many of the lines point to places where the Sun or bright stars rise at particular times of year, but that doesn't prove anything: With hundreds of lines, random chance ensures that many will have astronomical alignments no matter how or why they were made. The patterns, many of which are large figures of animals (**FIGURE 3.8**), have evoked even more debate. Some people think they may be representations of constellations recognized by the people who lived in the region, but we do not know for sure.

FIGURE 3.8 Hundreds of lines and patterns are etched in the sand of the Nazca desert in Peru. This aerial photo shows a large figure of a hummingbird.

In some cases, scientists studying archaeoastronomy can use other clues to establish the intentions of ancient builders. For example, lodges built by the Pawnee people in Kansas feature strategically placed holes for observing the passage of constellations that figure prominently in Pawnee folklore. The correspondence between the folklore and the structural features provides a strong case for deliberate intent rather than coincidence. Similarly, traditions of the Inca Empire of South America held that its rulers were descendents of the Sun and therefore demanded close watch of the movements of the Sun and stars. This fact supports the idea that astronomical alignments in Inca cities and ceremonial centers, such as the World Heritage Site of Machu Picchu (**FIGURE 3.9**), were deliberate rather than accidental.

FIGURE 3.9 The World Heritage Site of Machu Picchu has structures aligned with sunrise at the winter and summer solstices.

FIGURE 3.10 A Micronesian stick chart, an instrument used by Polynesian Navigators to represent swell patterns around islands.

Why does modern science trace its roots to the Greeks?

Greek philosophers developed at least three major innovations that helped pave the way for modern science. First, they developed a tradition of trying to understand nature without relying on supernatural explanations and of working communally to debate and challenge each other's ideas. Second, the Greeks used mathematics to give precision to their ideas, which allowed them to explore the implications of new ideas in much greater depth than would have otherwise been possible. Third, while much of their philosophical activity consisted of subtle debates grounded only in thought and was not scientific in the modern sense, the Greeks also saw the power of reasoning from observations. They understood that an explanation could not be right if it disagreed with observed facts.

Models of Nature Perhaps the greatest Greek contribution to science came from the way they synthesized all three innovations into the idea of creating models of nature, a practice that is central to modern science. Scientific models differ somewhat from the models you may be familiar with in everyday life. In our daily lives, we tend to think of models as miniature physical representations, such as model cars or airplanes. In contrast, a scientific **model** is a conceptual representation created to explain and predict observed phenomena. For example, a model of Earth's climate uses logic and mathematics to represent what we know about how the climate works. Its purpose is to explain and predict climate changes, such as the changes that may occur with global warming. Just as a model airplane does not faithfully represent every aspect of a real airplane, a scientific model may not fully explain all our observations of nature. Nevertheless, even the failings of a scientific model can be useful, because they often point the way toward building a better model.

From Greece to the Renaissance The Greeks created models that sought to explain many aspects of nature, including the properties of matter and the principles of motion. For our purposes, the most important of the Greek models was their Earth-centered model of the universe. Before we turn to its details, however, it's worth briefly discussing how ancient Greek philosophy was passed to Europe, where it ultimately grew into the principles of modern science.

Greek philosophy first began to spread widely with the conquests of Alexander the Great (356–323 B.C.). Alexander had a deep interest in science, perhaps in part because Aristotle (see Special Topic, page 61) had been his personal tutor. Alexander founded the city of Alexandria in Egypt, and shortly after his death the city commenced work on a great research center and library. The Library of Alexandria (**FIGURE 3.11**) opened in about 300 B.C. and remained the world's preeminent center of research for some 700 years. At its peak, it may have held as many as a half million books, handwritten on papyrus scrolls. Most of these scrolls were ultimately burned when the library was destroyed, their contents lost forever.

A different type of evidence makes a convincing case for the astronomical sophistication of ancient Polynesians, who lived and traveled among the islands of the mid- and South Pacific. Navigation was crucial to their survival because the next island in a journey usually was too distant to be seen. The most esteemed position in Polynesian culture was that of the Navigator, a person who had acquired the knowledge necessary to navigate great distances among the islands. Navigators used a combination of detailed knowledge of astronomy and an understanding of the patterns of waves and swells around different islands (**FIGURE 3.10**). The stars provided the broad navigational sense, while wave and swell patterns guided them to a precise landing point. A Navigator memorized all his knowledge and passed it to the next generation through a well-developed training program. Unfortunately, with the advent of modern navigational technology, many of the skills of the Navigators have been lost.

3.2 ANCIENT GREEK SCIENCE

Before a structure such as Stonehenge or the Templo Mayor could be built, careful observations had to be made and repeated over and over to ensure their accuracy. Careful, repeatable observations also underlie modern science. Elements of modern science were therefore present in many early human cultures. If the circumstances of history had been different, almost any culture might have been the first to develop what we consider to be modern science. In the end, however, history takes only one of countless possible paths. The path that led to modern science emerged from the ancient civilizations of the Mediterranean and the Middle East—especially from ancient Greece.

Greece gradually rose as a power in the Middle East beginning around 800 B.C. and was well established by about 500 B.C. Its geographical location placed it at a crossroads for travelers, merchants, and armies from northern Africa, Asia, and Europe. Building on the diverse ideas brought forth by the meeting of these many cultures, ancient Greek philosophers soon began their efforts to move human understanding of nature from the mythological to the rational.

a This rendering shows an artist's reconstruction of the Great Hall of the ancient Library of Alexandria.

b A rendering similar to part a, showing a scroll room in the ancient library.

c The New Library of Alexandria in Egypt, which opened in 2003.

FIGURE 3.11 The ancient Library of Alexandria thrived for some 700 years, starting in about 300 B.C.

Much of the Library of Alexandria's long history remains unknown today, in part because the books that recorded its history were destroyed along with the library. Nevertheless, historians are confident that the library's demise was intertwined with the life and death of a woman named Hypatia (A.D. 370–415), one of the few prominent female scholars of the ancient world. Hypatia was one of the last resident scholars of the library, as well as the director of the observatory in Alexandria and one of the leading mathematicians and astronomers of her time. Tragically, she became a scapegoat during a time of rising sentiment against free inquiry and was murdered in A.D. 415. The final destruction of the library took place not long after her death. In commemoration of the ancient library, Egypt built a New Library of Alexandria (the *Bibliotheca Alexandrina*, which opened in 2003), with hopes that it will once again make Alexandria a global center for scientific research.

The relatively few books that survived the destruction of the Library of Alexandria were preserved primarily thanks to the rise of a new center of intellectual inquiry in Baghdad (in present-day Iraq). While European civilization fell into the period of intellectual decline known as the Dark Ages, scholars of the new religion of Islam sought knowledge of mathematics and astronomy in hopes of better understanding the wisdom of Allah. During the 8th and 9th centuries A.D., scholars working in the Muslim Empire translated and thereby saved many ancient Greek works.

Around A.D. 800, the Islamic leader Al-Mamun (A.D. 786–833) established a "House of Wisdom" in Baghdad with a mission

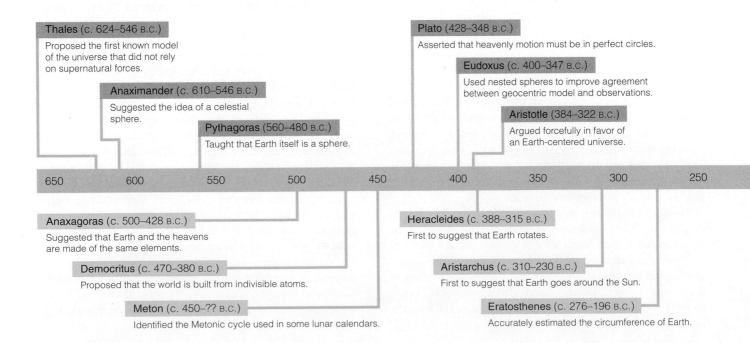

Thales (c. 624–546 B.C.)
Proposed the first known model of the universe that did not rely on supernatural forces.

Plato (428–348 B.C.)
Asserted that heavenly motion must be in perfect circles.

Anaximander (c. 610–546 B.C.)
Suggested the idea of a celestial sphere.

Eudoxus (c. 400–347 B.C.)
Used nested spheres to improve agreement between geocentric model and observations.

Pythagoras (560–480 B.C.)
Taught that Earth itself is a sphere.

Aristotle (384–322 B.C.)
Argued forcefully in favor of an Earth-centered universe.

650   600   550   500   450   400   350   300   250

Anaxagoras (c. 500–428 B.C.)
Suggested that Earth and the heavens are made of the same elements.

Heracleides (c. 388–315 B.C.)
First to suggest that Earth rotates.

Democritus (c. 470–380 B.C.)
Proposed that the world is built from indivisible atoms.

Aristarchus (c. 310–230 B.C.)
First to suggest that Earth goes around the Sun.

Meton (c. 450–?? B.C.)
Identified the Metonic cycle used in some lunar calendars.

Eratosthenes (c. 276–196 B.C.)
Accurately estimated the circumference of Earth.

Aristotle

Aristotle (384–322 B.C.) is among the best-known philosophers of the ancient world. Both his parents died when he was a child, and he was raised by a family friend. In his 20s and 30s, he studied under Plato (428–348 B.C.) at Plato's Academy. He later founded his own school, called the Lyceum, where he studied and lectured on virtually every subject. Historical records tell us that his lectures were collected and published in 150 volumes. About 50 of these volumes survive to the present day.

Many of Aristotle's scientific discoveries were about the nature of plants and animals. He studied more than 500 animal species in detail, dissecting specimens of nearly 50 species, and came up with a strikingly modern classification system. For example, he was the first person to recognize that dolphins should be classified with land mammals rather than with fish. In mathematics, he is known for laying the foundations of mathematical logic. Unfortunately, he was far less successful in physics and astronomy, areas in which many of his claims turned out to be wrong.

Despite his wide-ranging discoveries and writings, Aristotle's philosophies were not particularly influential until many centuries after his death. His books were preserved and valued by Islamic scholars but were unknown in Europe until they were translated into Latin in the 12th and 13th centuries. Aristotle's work gained great influence only after his philosophy was integrated into Christian theology by St. Thomas Aquinas (1225–1274). In the ancient world, Aristotle's greatest influence came indirectly, through his role as the tutor of Alexander the Great.

much like that of the destroyed Library of Alexandria. Founded in a spirit of openness and tolerance, the House of Wisdom employed Jews, Christians, and Muslims, all working together in scholarly pursuits. Using the translated Greek scientific manuscripts as building blocks, these scholars developed the mathematics of algebra and many new instruments and techniques for astronomical observation. Most of the official names of constellations and stars come from Arabic because of the work of the scholars at Baghdad. If you look at a star chart, you will see that the names of many bright stars begin with *al* (e.g., Aldebaran, Algol), which means "the" in Arabic.

The Islamic world of the Middle Ages was in frequent contact with Hindu scholars from India, who in turn brought knowledge of ideas and discoveries from China. Hence, the intellectual center in Baghdad achieved a synthesis of the surviving work of the ancient Greeks and that of the Indians and the Chinese. The accumulated knowledge of the Baghdad scholars spread throughout the Byzantine empire (part of the former Roman Empire). When the Byzantine capital of Constantinople (modern-day Istanbul) fell to the Turks in 1453, many Eastern scholars headed west to Europe, carrying with them the knowledge that helped ignite the European Renaissance.

How did the Greeks explain planetary motion?

The Greek **geocentric model** of the cosmos—so named because it placed a spherical Earth at the center of the universe—developed gradually over a period of several centuries. Because this model was so important in the history of science, let's briefly trace its development. FIGURE 3.12 will help you keep track of some of the personalities we will encounter.

Early Development of the Geocentric Model

We generally trace the origin of Greek science to the philosopher Thales (c. 624–546 B.C.; pronounced *thay-lees*). We encountered Thales earlier because of his legendary prediction

FIGURE 3.12 Timeline for major Greek figures in the development of astronomy. (All these individuals are discussed in this book, but not necessarily in this chapter.)

of a solar eclipse [**Section 2.3**]. Thales was the first person known to have addressed the question "What is the universe made of?" without resorting to supernatural explanations. His own guess—that the universe fundamentally consists of water and that Earth is a flat disk floating in an infinite ocean—was not widely accepted even in his own time. Nevertheless, just by asking the question he suggested that the world is inherently understandable and thereby inspired others to come up with better models for the structure of the universe.

A more sophisticated idea followed soon after, proposed by a student of Thales named Anaximander (c. 610–546 B.C.). Anaximander suggested that Earth floats in empty space surrounded by a sphere of stars and two separate rings along which the Sun and Moon travel. We therefore credit him with inventing the idea of a celestial sphere [**Section 2.1**]. Interestingly, Anaximander imagined Earth itself to be cylindrical rather than spherical in shape. He probably chose this shape because he knew Earth had to be curved in a north-south direction to explain changes in the constellations with latitude. Because the visible constellations do not change with longitude, he saw no need for curvature in the east-west direction.

We do not know precisely when the Greeks first began to think that Earth is round, but this idea was taught as early as about 500 B.C. by the famous mathematician Pythagoras (c. 560–480 B.C.). He and his followers envisioned Earth as a sphere floating at the center of the celestial sphere. Much of their motivation for adopting a spherical Earth probably was philosophical: The Pythagoreans had a mystical interest in mathematical perfection, and they considered a sphere to be geometrically perfect. More than a century later, Aristotle cited observations of Earth's curved shadow on the Moon during lunar eclipses as evidence for a spherical Earth.

The Pythagorean interest in "heavenly perfection" became deeply ingrained in most Greek philosophers. It took on even more significance after Plato (428–348 B.C.) asserted that all heavenly objects move in perfect circles at constant speeds and therefore must reside on huge spheres encircling Earth (**FIGURE 3.13**). The Platonic belief in perfection influenced astronomical models for the next 2000 years. Of course,

FIGURE 3.13 This model represents the Greek idea of the heavenly spheres (c. 400 B.C.). Earth is a sphere that rests in the center. The Moon, the Sun, and the planets all have their own spheres. The outermost sphere holds the stars.

those Greeks who made observations found Plato's model problematic: The apparent retrograde motion of the planets [**Section 2.4**], already well known by that time, clearly showed that planets do *not* move at constant speeds around Earth.

An ingenious solution came from Plato's colleague Eudoxus (c. 400–347 B.C.), who created a model in which the Sun, the Moon, and the planets each had their own spheres nested within several other spheres. Individually, the nested spheres turned in perfect circles. By carefully choosing the sizes, rotation axes, and rotation speeds for the invisible spheres, Eudoxus was able to make them work together in a way that reproduced many of the observed motions of the Sun, Moon, and planets in our sky. Other Greeks refined the model by comparing its predictions to observations and adding more spheres to improve the agreement.

This is how things stood when Aristotle (384–322 B.C.) arrived on the scene. Whether Eudoxus and his followers thought of the nested spheres as real physical objects is not clear, but Aristotle certainly did. In Aristotle's model, all the spheres responsible for celestial motion were transparent and interconnected like the gears of a giant machine. Earth's position at the center was explained as a natural consequence of gravity. Aristotle argued that gravity pulled heavy things toward the center of the universe (and allowed lighter things to float toward the heavens), thereby causing all the dirt, rock, and water of the universe to collect at the center and form the spherical Earth. We now know that Aristotle was wrong about both gravity and Earth's location. However, largely because of his persuasive arguments for an Earth-centered universe, the geocentric view dominated Western thought for almost 2000 years.

COMMON MISCONCEPTIONS

Columbus and a Flat Earth

A widespread myth gives credit to Columbus for learning that Earth is round, but knowledge of Earth's shape predated Columbus by nearly 2000 years. Not only were scholars of Columbus's time well aware that Earth is round, but they even knew its approximate size: Earth's circumference was first measured in about 240 B.C. by the Greek scientist Eratosthenes. In fact, a likely reason Columbus had so much difficulty finding a sponsor for his voyages was that he tried to argue a point on which he was wrong: He claimed the distance by sea from western Europe to eastern Asia to be much less than scholars knew it to be. When he finally found a patron in Spain and left on his journey, he was so woefully underprepared that the voyage would almost certainly have ended in disaster if the Americas hadn't stood in his way.

Ptolemy's Synthesis Greek modeling of the cosmos culminated in the work of Claudius Ptolemy (c. A.D. 100–170; pronounced *TOL-e-mee*). Ptolemy's model still placed Earth at the center of the universe, but it differed in significant ways from the nested spheres of Eudoxus and Aristotle. We refer to Ptolemy's geocentric model as the **Ptolemaic model** to distinguish it from earlier geocentric models.

To explain the apparent retrograde motion of the planets, the Ptolemaic model applied an idea first suggested by Apollonius (c. 240–190 B.C.). This idea held that each planet moved around Earth on a small circle that turned upon a larger circle (**FIGURE 3.14**). (The small circle is sometimes called an *epicycle,* and the larger circle is called a *deferent.*) A planet following this circle-upon-circle motion would trace a loop as seen from Earth, with the backward portion of the loop mimicking apparent retrograde motion.

Ptolemy also relied heavily on the work of Hipparchus (c. 190–120 B.C.), considered one of the greatest of the Greek astronomers. Among his many accomplishments, Hipparchus developed the circle-upon-circle idea of Apollonius into a model that could predict planetary positions. To do this, Hipparchus had to add several features to the basic idea; for example, he included even smaller circles that moved upon the original set of small circles, and he positioned the large circles slightly off-center from Earth.

Ptolemy's great accomplishment was to adapt and synthesize earlier ideas into a single system that agreed quite well with the astronomical observations available at the time. In the end, he created and published a model that could correctly forecast future planetary positions to within a few degrees of arc, which is about the angular size of your hand held at arm's length against the sky. This was sufficiently accurate to keep

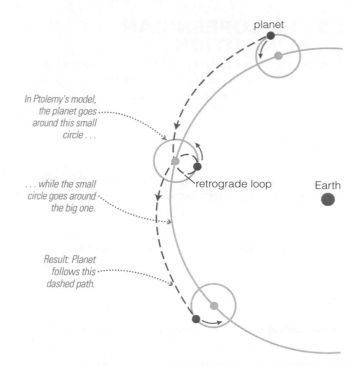

In Ptolemy's model, the planet goes around this small circle . . .

. . . while the small circle goes around the big one.

retrograde loop

Earth

Result: Planet follows this dashed path.

planet

FIGURE 3.14 interactive figure This diagram shows how the Ptolemaic model accounted for apparent retrograde motion. Each planet is assumed to move around a small circle that turns upon a larger circle. The resulting path (dashed) includes a loop in which the planet goes backward as seen from Earth.

the model in use for the next 1500 years. When Ptolemy's book describing the model was translated by Arabic scholars around A.D. 800, they gave it the title *Almagest,* derived from words meaning "the greatest compilation."

SPECIAL TOPIC

Eratosthenes Measures Earth

In a remarkable feat, the Greek scientist Eratosthenes accurately estimated the size of Earth in about 240 B.C. He did it by comparing the altitude of the Sun on the summer solstice in the Egyptian cities of Syene (modern-day Aswan) and Alexandria

Eratosthenes knew that the Sun passed directly overhead in Syene on the summer solstice. He also knew that in Alexandria, to the north, the Sun came within only 7° of the zenith on the summer solstice. He therefore reasoned that Alexandria must be 7° of latitude to the north of Syene (**FIGURE 1**). Because 7° is $\frac{7}{360}$ of a circle, he concluded that the north-south distance between Alexandria and Syene must be $\frac{7}{360}$ of the circumference of Earth.

Eratosthenes estimated the north-south distance between Syene and Alexandria to be 5000 stadia (the *stadium* was a Greek unit of distance). Thus, he concluded that

$$\frac{7}{360} \times \text{circumference of Earth} = 5000 \text{ stadia}$$

From this he found Earth's circumference to be about 250,000 stadia.

We don't know exactly what distance a stadium meant to Eratosthenes, but from sizes of actual Greek stadiums, it must have been about $\frac{1}{6}$ kilometer. Thus, Eratosthenes estimated the circumference of Earth to be about $\frac{250,000}{6} = 42,000$ kilometers—impressively close to the real value of just over 40,000 kilometers.

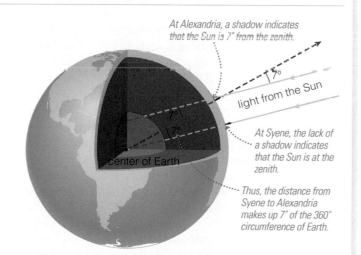

At Alexandria, a shadow indicates that the Sun is 7° from the zenith.

light from the Sun

At Syene, the lack of a shadow indicates that the Sun is at the zenith.

center of Earth

Thus, the distance from Syene to Alexandria makes up 7° of the 360° circumference of Earth.

FIGURE 1 At noon on the summer solstice, the Sun appears at the zenith in Syene but 7° shy of the zenith in Alexandria. Thus, 7° of latitude, which corresponds to a distance of $\frac{7}{360}$ of Earth's circumference, must separate the two cities.

3.3 THE COPERNICAN REVOLUTION

The Greeks and other ancient peoples developed many important scientific ideas, but what we now think of as science arose during the European Renaissance. Within a half century after the fall of Constantinople, Polish scientist Nicholas Copernicus began the work that ultimately overturned the Earth-centered Ptolemaic model.

How did Copernicus, Tycho, and Kepler challenge the Earth-centered model?

The ideas introduced by Copernicus fundamentally changed the way we perceive our place in the universe. The story of this dramatic change, known as the **Copernican revolution**, is in many ways the story of the origin of modern science. It is also the story of several key personalities, beginning with Copernicus himself.

Copernicus Nicholas Copernicus was born in Toruń, Poland, on February 19, 1473. His family was wealthy and he received an education in mathematics, medicine, and law. He began studying astronomy in his late teens.

By that time, tables of planetary motion based on the Ptolemaic model had become noticeably inaccurate. But few people were willing to undertake the difficult calculations required to revise the tables. The best tables available had been compiled some two centuries earlier under the guidance of Spanish monarch Alphonso X (1221–1284). Commenting on the tedious nature of the work, the monarch is said to have complained, "If I had been present at the creation, I would have recommended a simpler design for the universe."

Copernicus (1473–1543)

In his quest for a better way to predict planetary positions, Copernicus decided to try Aristarchus's Sun-centered idea, first proposed more than 1700 years earlier [**Section 2.4**]. He had read of Aristarchus's work, and recognized the much simpler explanation for apparent retrograde motion offered by a Sun-centered system (see Figure 2.33). But he went far beyond Aristarchus in working out mathematical details of the model. Through this process, Copernicus discovered simple geometric relationships that allowed him to calculate each planet's orbital period around the Sun and its relative distance from the Sun in terms of the Earth-Sun distance (see Mathematical Insight S1.1). The model's success in providing a geometric layout for the solar system convinced him that the Sun-centered idea must be correct.

Despite his own confidence in the model, Copernicus was hesitant to publish his work, fearing that his suggestion that Earth moved would be considered absurd. However, he discussed his system with other scholars, including high-ranking officials of the Catholic Church, who urged him to publish a book. Copernicus saw the first printed copy of his book, *De Revolutionibus Orbium Coelestium* ("Concerning the Revolutions of the Heavenly Spheres"), on the day he died—May 24, 1543.

Publication of the book spread the Sun-centered idea widely, and many scholars were drawn to its aesthetic advantages. Nevertheless, the Copernican model gained relatively few converts over the next 50 years, for a good reason: It didn't work all that well. The primary problem was that while Copernicus had been willing to overturn Earth's central place in the cosmos, he had held fast to the ancient belief that heavenly motion must occur in perfect circles. This incorrect assumption forced him to add numerous complexities to his system (including circles on circles much like those used by Ptolemy) to get it to make decent predictions. In the end, his complete model was no more accurate and no less complex than the Ptolemaic model, and few people were willing to throw out thousands of years of tradition for a new model that worked just as poorly as the old one.

Tycho Part of the difficulty faced by astronomers who sought to improve either the Ptolemaic or the Copernican system was a lack of quality data. The telescope had not yet been invented, and existing naked-eye observations were not very accurate. Better data were needed, and they were provided by the Danish nobleman Tycho Brahe (1546–1601), usually known simply as Tycho (pronounced *tie-koe*).

Tycho became interested in astronomy as a young boy, but his family discouraged this interest. He therefore kept his passion secret, learning the constellations from a miniature model of a celestial sphere that he kept hidden. As he grew older, Tycho was often arrogant about both his noble birth and his intellectual abilities. At age 20, he fought a duel with another student over which of them was the better mathematician. Part of Tycho's nose was cut off, and he designed a replacement piece made of silver and gold.

Tycho Brahe (1546–1601)

In 1563, Tycho decided to observe a widely anticipated alignment of Jupiter and Saturn. To his surprise, the alignment occurred nearly 2 days later than the date Copernicus had predicted. Resolving to improve the state of astronomical prediction, he set about compiling careful observations of stellar and planetary positions in the sky.

Tycho's fame grew after he observed what he called a *nova*, meaning "new star," in 1572. By measuring its parallax and comparing it to the parallax of the Moon, he proved that the nova was much farther away than the Moon. (Today, we know that Tycho saw a *supernova*—the explosion of a distant star [**Section 17.3**].) In 1577, Tycho made similar observations of a comet and proved that it too lay in the realm of the heavens. Others, including Aristotle, had argued that comets were phenomena of Earth's atmosphere. King Frederick II of Denmark decided to sponsor Tycho's ongoing work, providing him with money to build an unparalleled observatory for naked-eye observations (**FIGURE 3.15**). After Frederick II died in 1588, Tycho moved to Prague, where his work was supported by German emperor Rudolf II.

Over a period of three decades, Tycho and his assistants compiled naked-eye observations accurate to within less than 1 arcminute—less than the thickness of a fingernail viewed at arm's length. Because the telescope was invented shortly after his death, Tycho's data remain the best set of naked-eye observations ever made. Despite the quality of his observations, Tycho never succeeded in coming up with a satisfying explanation for planetary motion. He was convinced that the *planets* must orbit the Sun, but his inability to detect stellar parallax [**Section 2.4**] led him to conclude that Earth must remain stationary. He therefore advocated a model in which the Sun orbits Earth while all other planets orbit the Sun. Few people took this model seriously.

Kepler Tycho failed to explain the motions of the planets satisfactorily, but he succeeded in finding someone who could: In 1600, he hired the young German astronomer Johannes Kepler (1571–1630). Kepler and Tycho had a strained relationship, but Tycho recognized the talent of his young apprentice. In 1601, as he lay on his deathbed, Tycho begged Kepler to find a system that would make sense of his observations so "that it may not appear I have lived in vain."*

Johannes Kepler (1571–1630)

Kepler was deeply religious and believed that understanding the geometry of the heavens would bring him closer to God. Like Copernicus, he believed that planetary orbits should be perfect circles, so he worked diligently to match circular motions to Tycho's data.

Kepler labored with particular intensity to find an orbit for Mars, which posed the greatest difficulties in matching the data to a circular orbit. After years of calculation, Kepler found a circular orbit that matched all of Tycho's observations of Mars's position along the ecliptic (east-west) to within 2 arcminutes. However, the model did not correctly predict Mars's positions north or south of the ecliptic. Because Kepler sought a physically realistic orbit for Mars, he could not (as Ptolemy and Copernicus had done) tolerate one model for the east-west positions and another for the north-south positions. He attempted to find a unified model with a circular orbit. In doing so, he found that some of his predictions differed from Tycho's observations by as much as 8 arcminutes.

Kepler surely was tempted to ignore these discrepancies and attribute them to errors by Tycho. After all, 8 arcminutes is barely one-fourth the angular diameter of the full moon. But Kepler trusted Tycho's careful work. The small discrepancies finally led Kepler to abandon the idea of circular orbits—and to find the correct solution to the ancient riddle of planetary motion. About this event, Kepler wrote:

> *If I had believed that we could ignore these eight minutes [of arc], I would have patched up my hypothesis accordingly. But, since it was not permissible to ignore, those eight minutes pointed the road to a complete reformation in astronomy.*

FIGURE 3.15 Tycho Brahe in his naked-eye observatory, which worked much like a giant protractor. He could sit and observe a planet through the rectangular hole in the wall as an assistant used a sliding marker to measure the angle on the protractor.

---

*For a particularly moving version of the story of Tycho and Kepler, see Episode 3 of Carl Sagan's *Cosmos* video series.

FIGURE 3.16

interactive figure An ellipse is a special type of oval. These diagrams show how an ellipse differs from a circle and how different ellipses vary in their eccentricity.

a Drawing a circle with a string of fixed length.

b Drawing an ellipse with a string of fixed length.

c Eccentricity describes how much an ellipse deviates from a perfect circle.

Kepler's key discovery was that planetary orbits are not circles but instead are a special type of oval called an **ellipse**. You can draw a circle by putting a pencil on the end of a string, tacking the string to a board, and pulling the pencil around (**FIGURE 3.16a**). Drawing an ellipse is similar, except that you must stretch the string around *two* tacks (**FIGURE 3.16b**). The locations of the two tacks are called the **foci** (singular, **focus**) of the ellipse. The long axis of the ellipse is called its *major axis,* each half of which is called a **semimajor axis**; as you'll see shortly, the length of the semimajor axis is particularly important in astronomy. The short axis is called the *minor axis.* By altering the distance between the two foci while keeping the length of string the same, you can draw ellipses of varying **eccentricity**, a quantity that describes how much an ellipse is stretched out compared to a perfect circle (**FIGURE 3.16c**). A circle is an ellipse with zero eccentricity, and greater eccentricity means a more elongated ellipse.

Kepler's decision to trust the data over his preconceived beliefs marked an important transition point in the history of science. Once he abandoned perfect circles in favor of ellipses, Kepler soon came up with a model that could predict planetary positions with far greater accuracy than Ptolemy's Earth-centered model. Kepler's model withstood the test of time and became accepted not only as a model of nature but also as a deep, underlying truth about planetary motion.

(**MA**) Orbits and Kepler's Laws Tutorial, Lessons 2–4

What are Kepler's three laws of planetary motion?

Kepler summarized his discoveries with three simple laws that we now call **Kepler's laws of planetary motion**. He published the first two laws in 1609 and the third in 1619.

Kepler's First Law Kepler's first law tells us that *the orbit of each planet around the Sun is an ellipse with the Sun at one focus* (**FIGURE 3.17**). (Nothing is at the other focus.) In essence,

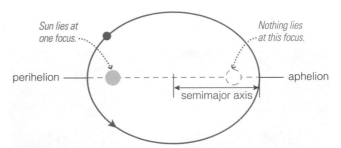

FIGURE 3.17 interactive figure Kepler's first law: The orbit of each planet about the Sun is an ellipse with the Sun at one focus. (The eccentricity shown here is exaggerated compared to the actual eccentricities of the planets.)

this law tells us that a planet's distance from the Sun varies during its orbit. It is closest at the point called **perihelion** (from the Greek for "near the Sun") and farthest at the point called **aphelion** (from the Greek for "away from the Sun"). The *average* of a planet's perihelion and aphelion distances is the length of its semimajor axis. We will refer to this simply as the planet's average distance from the Sun.

Kepler's Second Law Kepler's second law states that *as a planet moves around its orbit, it sweeps out equal areas in equal times.* As shown in **FIGURE 3.18**, this means the planet moves a greater distance when it is near perihelion than it does in the same amount of time near aphelion. That is, the planet travels faster when it is nearer to the Sun and slower when it is farther from the Sun.

Kepler's Third Law Kepler's third law tells us that *more distant planets orbit the Sun at slower average speeds, obeying a precise mathematical relationship.* The relationship is written

$$p^2 = a^3$$

where *p* is the planet's orbital period in years and *a* is its average distance from the Sun in astronomical units. **FIGURE 3.19a** shows the $p^2 = a^3$ law graphically. Notice that

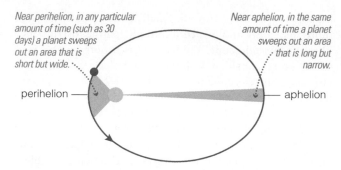

FIGURE 3.18 **interactive figure** Kepler's second law: As a planet moves around its orbit, an imaginary line connecting it to the Sun sweeps out equal areas (the shaded regions) in equal times.

THINK ABOUT IT

Suppose a comet has an orbit that brings it quite close to the Sun at its perihelion and beyond Mars at its aphelion, but with an average distance (semimajor axis) of 1 AU. According to Kepler's laws, how long does the comet take to complete each orbit? Does it spend most of its time close to the Sun, far from the Sun, or somewhere in between? Explain.

the square of each planet's orbital period (p^2) is indeed equal to the cube of its average distance from the Sun (a^3). Because Kepler's third law relates a planet's orbital distance to its orbital time (period), we can use the law to calculate a planet's average orbital speed.* **FIGURE 3.19b** shows the result, confirming that more distant planets orbit the Sun more slowly.

The fact that more distant planets move more slowly led Kepler to suggest that planetary motion might be the result of a force from the Sun. He even speculated about the nature of this force, guessing that it might be related to magnetism. (This idea, shared by Galileo, was first suggested by William Gilbert [1544–1603], an early believer in the Copernican system.) Kepler was right about the existence of a force but wrong in his guess of magnetism. A half century later, Isaac Newton finally explained planetary motion as a consequence of gravity [**Section 4.4**].

---

*To calculate orbital speed from Kepler's third law, remember that speed = distance/time. For a planetary orbit, the distance is the orbital circumference, or $2\pi a$ (where a is the semimajor axis, roughly the "radius" of the orbit), and the time is the orbital period p, so the orbital speed is $(2\pi a)/p$. From Kepler's third law, $p = a^{3/2}$. Plugging this value for p into the orbital speed equation, we find that a planet's orbital speed is $2\pi/\sqrt{a}$; the graph of this equation is the curve in Figure 3.19b.

How did Galileo solidify the Copernican revolution?

The success of Kepler's laws in matching Tycho's data provided strong evidence in favor of Copernicus's placement of the Sun, rather than Earth, at the center of the solar system. Nevertheless, many scientists still voiced reasonable objections to the Copernican view. There were three basic objections, all rooted in the 2000-year-old beliefs of Aristotle and other ancient Greeks.

- First, Aristotle had held that Earth could not be moving because, if it were, objects such as birds, falling stones, and clouds would be left behind as Earth moved along its way.

- Second, the idea of noncircular orbits contradicted Aristotle's claim that the heavens—the realm of the Sun, Moon, planets, and stars—must be perfect and unchanging.

- Third, no one had detected the stellar parallax that should occur if Earth orbits the Sun [**Section 2.4**].

Galileo Galilei (1564–1642), usually known by only his first name, answered all three objections.

Galileo's Evidence Galileo defused the first objection with experiments that almost single-handedly overturned the Aristotelian view of physics. In particular, he used experiments with rolling balls to demonstrate that a moving object remains in motion *unless* a force acts to stop it (an idea now codified in Newton's first law of motion [**Section 4.2**]). This

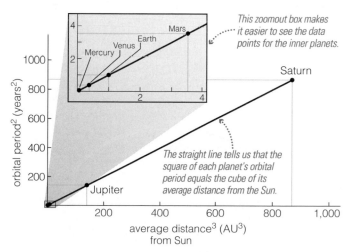

a This graph shows that Kepler's third law ($p^2 = a^3$) holds true; the graph shows only the planets known in Kepler's time.

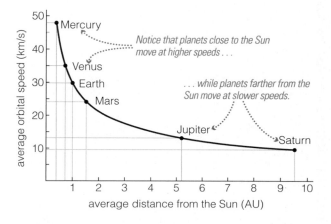

b This graph, based on Kepler's third law and modern values of planetary distances, shows that more distant planets orbit the Sun more slowly.

FIGURE 3.19 Graphs based on Kepler's third law.

Galileo (1564–1642)

insight explained why objects that share Earth's motion through space—such as birds, falling stones, and clouds—should *stay* with Earth rather than falling behind as Aristotle had argued. This same idea explains why passengers stay with a moving airplane even when they leave their seats.

The second objection had already been challenged by Tycho's supernova and comet observations, which proved that the heavens could change. Galileo then shattered the idea of heavenly perfection after he built a telescope in late 1609. (Galileo did not invent the telescope; it was patented by Hans Lippershey in 1608. However, Galileo took what was little more than a toy and turned it into a scientific instrument.) Through his telescope, Galileo saw sunspots on the Sun, which were considered "imperfections" at the time. He also used his telescope to prove that the Moon has mountains and valleys like the "imperfect" Earth by noticing the shadows cast near the dividing line between the light and dark portions of the lunar face (**FIGURE 3.20**). If the heavens were in fact not perfect, then the idea of elliptical orbits (as opposed to "perfect" circles) was not so objectionable.

The third objection—the absence of observable stellar parallax—had been of particular concern to Tycho. Based on his estimates of the distances of stars, Tycho believed that his naked-eye observations were sufficiently precise to detect stellar parallax if Earth did in fact orbit the Sun. Refuting Tycho's argument required showing that the stars were more distant than Tycho had thought and therefore too distant for him to have observed stellar parallax. Although Galileo didn't actually prove this fact, he provided strong evidence in its favor. For example, he saw with his telescope that the Milky Way

---

MATHEMATICAL INSIGHT 3.1

Eccentricity and Planetary Orbits

We describe how much a planet's orbit differs from a perfect circle by stating its orbital eccentricity. There are several equivalent ways to define the eccentricity of an ellipse, but the simplest is shown in **FIGURE 1**. We define c to be the distance from each focus to the center of the ellipse and a to be the length of the semimajor axis. The eccentricity, e, is then defined to be

$$e = \frac{c}{a}$$

Notice that $c = 0$ for a perfect circle, because a circle is an ellipse with both foci *in* the center, so this formula gives an eccentricity of 0 for a perfect circle, just as we expect.

You can find the orbital eccentricities for the planets in tables such as Table E.2 in Appendix E of this book. Once you know the eccentricity, the following formulas allow you to calculate the planet's perihelion and aphelion distances (**FIGURE 2**):

$$\text{perihelion distance} = a(1 - e)$$

$$\text{aphelion distance} = a(1 + e)$$

EXAMPLE: What are Earth's perihelion and aphelion distances?

SOLUTION:

Step 1 Understand: To use the given formulas, we need to know Earth's orbital eccentricity and semimajor axis length. From Table E.2, Earth's orbital eccentricity is $e = 0.017$ and its semimajor axis (average distance from the Sun) is 1 AU, or $a = 149.6$ million km.

Step 2 Solve: We plug these values into the equations:

$$\begin{aligned}
\text{Earth's perihelion distance} &= a(1 - e) \\
&= (149.6 \times 10^6 \text{ km})(1 - 0.017) \\
&= 147.1 \times 10^6 \text{ km}
\end{aligned}$$

$$\begin{aligned}
\text{Earth's aphelion distance} &= a(1 + e) \\
&= (149.6 \times 10^6 \text{ km})(1 + 0.017) \\
&= 152.1 \times 10^6 \text{ km}
\end{aligned}$$

Step 3 Explain: Earth's perihelion (nearest to the Sun) distance is 147.1 million kilometers and its aphelion (farthest from the Sun) distance is 152.1 million kilometers. In other words, Earth's distance from the Sun varies between 147.1 and 152.1 million kilometers.

FIGURE 1

FIGURE 2

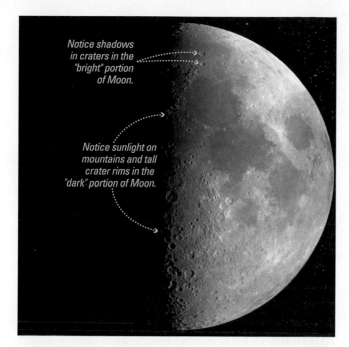

Notice shadows in craters in the "bright" portion of Moon.

Notice sunlight on mountains and tall crater rims in the "dark" portion of Moon.

FIGURE 3.20 The shadows cast by mountains and crater rims near the dividing line between the light and dark portions of the lunar face prove that the Moon's surface is not perfectly smooth.

FIGURE 3.21 A page from Galileo's notebook written in 1610. His sketches show four "stars" near Jupiter (the circle) but in different positions at different times (with one or more sometimes hidden from view). Galileo soon realized that the "stars" were actually moons orbiting Jupiter.

resolved into countless individual stars. This discovery helped him argue that the stars were far more numerous and more distant than Tycho had believed.

Sealing the Case In hindsight, the final nails in the coffin of the Earth-centered model came with two of Galileo's earliest discoveries through the telescope. First, he observed four moons clearly orbiting Jupiter, *not* Earth (**FIGURE 3.21**). By itself, this observation still did not rule out a stationary, central Earth. However, it showed that moons can orbit a moving planet like Jupiter, which overcame some critics' complaints that the Moon could not stay with a moving Earth. Soon thereafter, he observed that Venus goes through phases in a

way that makes sense only if it orbits the Sun and not Earth (**FIGURE 3.22**).

With Earth clearly removed from its position at the center of the universe, the scientific debate turned to the question of whether Kepler's laws were the correct model for our solar system. The most convincing evidence came in 1631, when astronomers observed a transit of Mercury across the Sun's

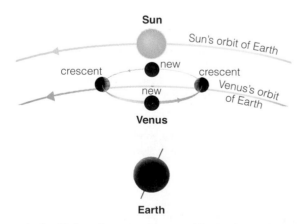

Ptolemaic View of Venus

a In the Ptolemaic system, Venus orbits Earth, moving around a smaller circle on its larger orbital circle; the center of the smaller circle lies on the Earth-Sun line. If this view were correct, Venus's phases would range only from new to crescent.

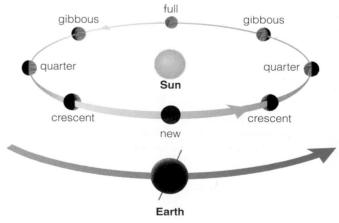

Copernican View of Venus

b In reality, Venus orbits the Sun, so from Earth we can see it in many different phases. This is just what Galileo observed, allowing him to prove that Venus orbits the Sun.

FIGURE 3.22 Galileo's telescopic observations of Venus proved that it orbits the Sun rather than Earth.

face. Kepler's laws had predicted the transit with overwhelmingly better success than any competing model.

Galileo and the Church Although we now recognize that Galileo won the day, the story was more complex in his own time, when Catholic Church doctrine still held Earth to be the center of the universe. On June 22, 1633, Galileo was brought before a Church inquisition in Rome and ordered to recant his claim that Earth orbits the Sun. Nearly 70 years old and fearing for his life, Galileo did as ordered and his life was spared. However, legend has it that as he rose from his knees he whispered under his breath, *Eppur si muove*—Italian for "And yet it moves." (Given the likely consequences if Church officials had heard him say this, most historians doubt the legend; see Special Topic, page 74.)

The Church did not formally vindicate Galileo until 1992, but Church officials gave up the argument long before that: In 1757, all works backing the idea of a Sun-centered solar system were removed from the Church's index of banned books. Today, Catholic scientists are at the forefront of much astronomical research, and official Church teachings are compatible not only with Earth's planetary status but also with the theories of the Big Bang and the subsequent evolution of the cosmos and of life.

3.4 THE NATURE OF SCIENCE

The story of how our ancestors gradually figured out the basic architecture of the cosmos exhibits many features of what we now consider "good science." For example, we have seen how models were formulated and tested against observations and modified or replaced when they failed those tests. The story also illustrates some classic mistakes, such as the apparent failure of anyone before Kepler to question the belief that orbits must be circles. The ultimate success of the Copernican revolution led scientists, philosophers, and theologians to reassess the various modes of thinking that played a role in the 2000-year process of discovering Earth's place in the universe. Let's examine how the principles of modern science emerged from the lessons learned in the Copernican revolution.

How can we distinguish science from nonscience?

It's surprisingly difficult to define the term *science* precisely. The word comes from the Latin *scientia*, meaning "knowledge," but not all knowledge is science. For example, you may know what music you like best, but your musical taste is not a result of scientific study.

Approaches to Science One reason science is difficult to define is that not all science works in the same way. For example, you've probably heard that science is supposed to proceed according to something called the "scientific method." As an idealized illustration of this method, consider what you would do if your flashlight suddenly stopped working. In hopes of fixing the flashlight, you might *hypothesize* that its batteries have died. This type of tentative explanation,

---

MATHEMATICAL INSIGHT 3.2

Kepler's Third Law

When Kepler discovered his third law, $p^2 = a^3$, he did so only by looking at planet orbits. In fact, it applies much more generally. We'll see its most general form in Mathematical Insight 4.3, but even in its original form we can apply it to any object *if*

1. the object is *orbiting the Sun* or another star of the same mass as the Sun and
2. we measure orbital *periods in years* and *distances in AU*.

EXAMPLE 1: What is the orbital period of the dwarf planet (and largest asteroid) Ceres, which orbits the Sun at an average distance (semimajor axis) of 2.77 AU?

SOLUTION:

Step 1 Understand: We can apply Kepler's third law because both conditions above are met. The first is met because Ceres orbits the Sun. The second is met because we are given the orbital distance in AU, which means Kepler's third law will tell us the orbital period in years.

Step 2 Solve: We want the period p, so we solve Kepler's third law for p by taking the square root of both sides; we then substitute the given value $a = 2.77$ AU:

$$p^2 = a^3 \implies p = \sqrt{a^3} = \sqrt{2.77^3} = 4.6$$

Note that because of the special conditions attached to the use of Kepler's third law in its original form, we do *not* include units when working with it; we know we'll get a period in years as long as we start with a distance in AU.

Step 3 Explain: Ceres has an orbital period of 4.6 years, meaning it takes 4.6 years to complete each orbit around the Sun.

EXAMPLE 2: A new planet is discovered to be orbiting a star with the same mass as our Sun. The planet orbits the star every 3 months. What is its average distance from its star?

SOLUTION:

Step 1 Understand: We can use Kepler's third law in its original form if the problem meets the two conditions above. The first condition is met because the planet is orbiting a star with the same mass as our Sun. To meet the second condition, we must convert the orbital period from 3 months to $p = 0.25$ year.

Step 2 Solve: We want the distance a, so we solve Kepler's third law for a by taking the cube root of both sides; we then substitute the orbital period $p = 0.25$ year:

$$p^2 = a^3 \implies a = \sqrt[3]{p^2} = \sqrt[3]{0.25^2} = 0.40$$

Step 3 Explain: The planet orbits its star at an average distance of 0.4 AU. By comparing this result to the distances of planets in our own solar system given in Table E.2, we find that this planet's average orbital distance is just slightly larger than that of the planet Mercury in our own solar system.

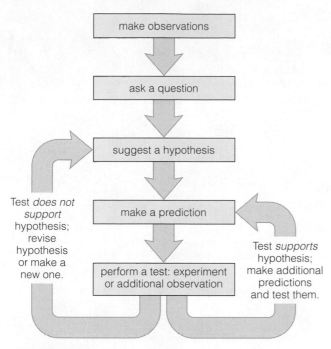

FIGURE 3.23 This diagram illustrates what we often call the scientific method.

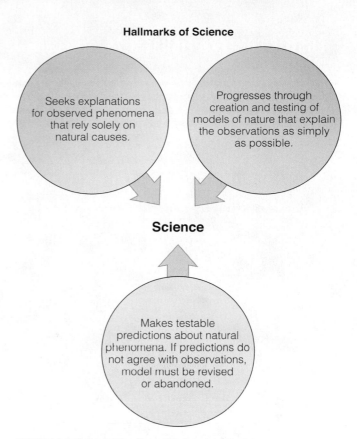

FIGURE 3.24 **interactive figure** Hallmarks of science.

or **hypothesis**, is sometimes called an *educated guess*—in this case, it is "educated" because you already know that flashlights need batteries. Your hypothesis allows you to make a simple prediction: If you replace the batteries with new ones, the flashlight should work. You can test this prediction by replacing the batteries. If the flashlight now works, you've confirmed your hypothesis. If it doesn't, you must revise or discard your hypothesis, perhaps in favor of some other one that you can also test (such as that the bulb is burned out). **FIGURE 3.23** illustrates the basic flow of this process.

The scientific method can be a useful idealization, but real science rarely progresses in such an orderly way. Scientific progress often begins with someone going out and looking at nature in a general way, rather than conducting a careful set of experiments. For example, Galileo wasn't looking for anything in particular when he pointed his telescope at the sky and made his first startling discoveries. Furthermore, scientists are human beings, and their intuition and personal beliefs inevitably influence their work. Copernicus, for example, adopted the idea that Earth orbits the Sun not because he had carefully tested it but because he believed it made more sense than the prevailing view of an Earth-centered universe. While his intuition guided him to the right general idea, he erred in the specifics because he still held Plato's ancient belief that heavenly motion must be in perfect circles.

Given that the idealized scientific method is an overly simplistic characterization of science, how can we tell what is science and what is not? To answer this question, we must look a little deeper into the distinguishing characteristics of scientific thinking.

Hallmarks of Science One way to define scientific thinking is to list the criteria that scientists use when they judge competing models of nature. Historians and philosophers of

science have examined (and continue to examine) this issue in great depth, and different experts express different viewpoints on the details. Nevertheless, everything we now consider to be science shares the following three basic characteristics, which we will refer to as the "hallmarks" of science (**FIGURE 3.24**):

- Modern science seeks explanations for observed phenomena that rely solely on natural causes.

- Science progresses through the creation and testing of models of nature that explain the observations as simply as possible.

- A scientific model must make testable predictions about natural phenomena that will force us to revise or abandon the model if the predictions do not agree with observations.

Each of these hallmarks is evident in the story of the Copernican revolution. The first shows up in the way Tycho's careful measurements of planetary motion motivated Kepler to come up with a better explanation for those motions. The second is evident in the way several competing models were compared and tested, most notably those of Ptolemy, Copernicus, and Kepler. We see the third in the fact that each model could make precise predictions about the future motions of the Sun, Moon, planets, and stars in our sky. Kepler's model gained acceptance because it worked, while the competing models lost favor because their predictions failed to match the observations. The Cosmic Context spread in **FIGURE 3.25** summarizes the key scientific changes that occurred with the Copernican revolution and explains how they illustrate the hallmarks of science.

Ancient Earth-centered models of the universe easily explained the simple motions of the Sun and Moon through our sky, but had difficulty explaining the more complicated motions of the planets. The quest to understand planetary motions ultimately led to a revolution in our thinking about Earth's place in the universe that illustrates the process of science. This figure summarizes the major steps in that process.

1 Night by night, planets usually move from west to east relative to the stars. However, during periods of *apparent retrograde motion,* they reverse direction for a few weeks to months [Section 2.4]. The ancient Greeks knew that any credible model of the solar system had to explain these observations.

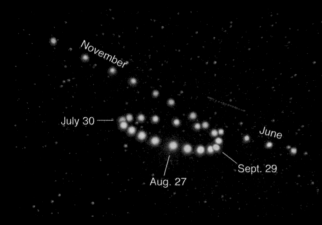

This composite photo shows the apparent retrograde motion of Mars.

2 Most ancient Greek thinkers assumed that Earth remained fixed at the center of the solar system. To explain retrograde motion, they therefore added a complicated scheme of circles moving upon circles to their Earth-centered model. However, at least some Greeks, such as Aristarchus, preferred a Sun-centered model, which offered a simpler explanation for retrograde motion.

The Greek geocentric model explained apparent retrograde motion by having planets move around Earth on small circles that turned on larger circles.

HALLMARK OF SCIENCE   **A scientific model must seek explanations for observed phenomena that rely solely on natural causes.** The ancient Greeks used geometry to explain their observations of planetary motion.

(Left page)
A schematic map of the universe from 1539 with Earth at the center and the Sun (Solis) orbiting it between Venus (Veneris) and Mars (Martis).

(Right page)
A page from Copernicus's De Revolutionibus, *published in 1543, showing the Sun (Sol) at the center and Earth (Terra) orbiting between Venus and Mars.*

LIBRI COSMO.   Fo.V.

Schema huius præmiſſæ diuiſionis

Sphærarum.

3 By the time of Copernicus (1473–1543), predictions based on the Earth-centered model had become noticeably inaccurate. Hoping for improvement, Copernicus revived the Sun-centered idea. He did not succeed in making substantially better predictions because he retained the ancient belief that planets must move in perfect circles, but he inspired a revolution continued over the next century by Tycho, Kepler, and Galileo.

Apparent retrograde motion is simply explained in a Sun-centered system. Notice how Mars appears to change direction as Earth moves past it.

HALLMARK OF SCIENCE **Science progresses through creation and testing of models of nature that explain the observations as simply as possible.** Copernicus developed a Sun-centered model in hopes of explaining observations better than the more complicated Earth-centered model.

4 Tycho exposed flaws in both the ancient Greek and Copernican models by observing planetary motions with unprecedented accuracy. His observations led to Kepler's breakthrough insight that planetary orbits are elliptical, not circular, and enabled Kepler to develop his three laws of planetary motion.

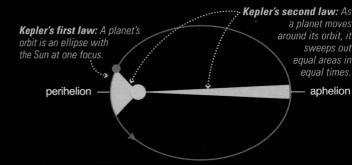

Kepler's first law: *A planet's orbit is an ellipse with the Sun at one focus.*

Kepler's second law: *As a planet moves around its orbit, it sweeps out equal areas in equal times.*

perihelion — — aphelion

Kepler's third law: *More distant planets orbit at slower average speeds, obeying* $p^2 = a^3$.

HALLMARK OF SCIENCE **A scientific model makes testable predictions about natural phenomena. If predictions do not agree with observations, the model must be revised or abandoned.** Kepler could not make his model agree with observations until he abandoned the belief that planets move in perfect circles.

5 Galileo's experiments and telescopic observations overcame remaining scientific objections to the Sun-centered model. Together, Galileo's discoveries and the success of Kepler's laws in predicting planetary motion overthrew the Earth-centered model once and for all.

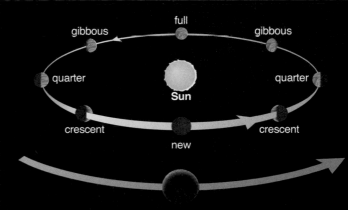

With his telescope, Galileo saw phases of Venus that are consistent only with the idea that Venus orbits the Sun rather than Earth.

Occam's Razor The criterion of simplicity in the second hallmark deserves additional explanation. Remember that the original model of Copernicus did *not* match the data noticeably better than Ptolemy's model. If scientists had judged Copernicus's model solely on the accuracy of its predictions, they might have rejected it immediately. However, many scientists found elements of the Copernican model appealing, such as its simple explanation for apparent retrograde motion. They therefore kept the model alive until Kepler found a way to make it work.

If agreement with data were the sole criterion for judgment, we could imagine a modern-day Ptolemy adding millions or billions of additional circles to the geocentric model in an effort to improve its agreement with observations. A sufficiently complex geocentric model could in principle reproduce the observations with almost perfect accuracy—

but it still would not convince us that Earth is the center of the universe. We would still choose the Copernican view over the geocentric view because its predictions would be just as accurate but follow a much simpler model of nature. The idea that scientists should prefer the simpler of two models that agree equally well with observations is called *Occam's razor,* after the medieval scholar William of Occam (1285–1349).

Verifiable Observations The third hallmark of science forces us to face the question of what counts as an "observation" against which a prediction can be tested. Consider the claim that aliens are visiting Earth in UFOs. Proponents of this claim say that thousands of eyewitness observations of UFO encounters provide evidence that it is true. But do these personal testimonials count as *scientific* evidence? On the

SPECIAL TOPIC

And Yet It Moves

The case of Galileo is often portrayed as a simple example of conflict between science and religion, but the reality was much more complex, with deep divisions inside the Church hierarchy. Perhaps the clearest evidence for a more open-minded Church comes from the case of Copernicus, whose revolutionary work was strongly supported by many Church officials. A less-well-known and even earlier example concerns Nicholas of Cusa (1401–1464), who published a book arguing for a Sun-centered solar system in 1440, more than a century before Copernicus. Nicholas was ordained a priest in the same year that his book was published, and he was later elevated to Cardinal. Clearly, his views caused no problems for Church officials of the time. (Copernicus probably was not aware of this work by Nicholas of Cusa.)

Many other scientists received similar support from within the Church. In fact, for most of his life, Galileo counted Cardinals (and even the Pope who later tried him) among his friends. Some historians suspect that Galileo got into trouble less for his views than for the way in which he portrayed them. In 1632—just a year before his famous trial—he published a book in which two fictional characters debated the geocentric and Sun-centered views. He named the character taking the geocentric position Simplicio—essentially "simple-minded"—and someone apparently convinced the Pope that the character was meant to represent him.

If it was personality rather than belief that got Galileo into trouble, he was not the only one. Another early supporter of Copernicus, Giordano Bruno (1548–1600), drew the wrath of the Church after essentially writing that no rational person could disagree with him (not just on the Copernican system but on other matters as well). Bruno was branded a heretic and burned at the stake.

The evidence supporting the idea that Earth rotates and orbits the Sun was quite strong by the time of Galileo's trial in 1633, but it was still indirect. Today, we have much more direct proof that Galileo was correct when he supposedly whispered of Earth, *Eppur si muove*—"And yet it moves."

French physicist Jean Foucault provided the first direct proof of rotation in 1851. Foucault built a large pendulum that he carefully started swinging. Any pendulum tends to swing always in the same plane, but Earth's rotation made Foucault's pendulum appear to twist

slowly in a circle. Today, *Foucault pendulums* are a popular attraction at many science centers and museums (**FIGURE 1**). A second direct proof that Earth rotates is provided by the *Coriolis effect,* first described by French physicist Gustave Coriolis (1792–1843). The Coriolis effect [**Section 10.2**], which would not occur if Earth were not rotating, is responsible for things such as the swirling of hurricanes and the fact that missiles that travel great distances on Earth deviate from straight-line paths.

The first direct proof that Earth orbits the Sun came from English astronomer James Bradley (1693–1762). To understand Bradley's proof, imagine that starlight is like rain, falling straight down. If you are standing still, you should hold your umbrella straight over your head, but if you are walking through the rain, you should tilt your umbrella forward, because your motion makes the rain appear to be coming down at an angle. Bradley discovered that observing light from stars requires that telescopes be tilted slightly in the direction of Earth's motion—just like the umbrella. This effect is called the *aberration of starlight.* Stellar parallax also provides direct proof that Earth orbits the Sun, and it was first measured in 1838 by German astronomer Friedrich Bessel.

FIGURE 1 A Foucault pendulum at the Science Museum of Virginia.

Eggs on the Equinox

One of the hallmarks of science holds that you needn't take scientific claims on faith. In principle, at least, you can always test them for yourself. Consider the claim, repeated in news reports every year, that the spring equinox is the only day on which you can balance an egg on its end. Many people believe this claim, but you'll be immediately skeptical if you think about the nature of the spring equinox. The equinox is merely a point in time at which sunlight strikes both hemispheres equally (see Figure 2.15). It's difficult to see how sunlight could affect an attempt to balance eggs (especially if the eggs are indoors), and the strength of either Earth's gravity or the Sun's gravity is no different on that day than on any other day.

More important, you can test this claim directly. It's not easy to balance an egg on its end, but with practice you can do it on any day of the year, not just on the spring equinox. Not all scientific claims are so easy to test for yourself, but the basic lesson should be clear: Before you accept any scientific claim, you should demand at least a reasonable explanation of the evidence that backs it up.

surface, the answer isn't obvious, because all scientific studies involve eyewitness accounts on some level. For example, only a handful of scientists have personally made detailed tests of Einstein's theory of relativity, and it is their personal reports of the results that have convinced other scientists of the theory's validity. However, there's an important difference between personal testimony about a scientific test and an observation of a UFO: The first can be verified by anyone, at least in principle, while the second cannot.

Understanding this difference is crucial to understanding what counts as science and what does not. Even though you may never have conducted a test of Einstein's theory of relativity yourself, there's nothing stopping you from doing so. It might require several years of study before you had the necessary background to conduct the test, but you could then confirm the results reported by other scientists. In other words, while you may currently be trusting the eyewitness testimony of scientists, you always have the option of verifying their testimony for yourself.

In contrast, there is no way for you to verify someone's eyewitness account of a UFO. Without hard evidence such as photographs or pieces of the UFO, there is nothing that you could evaluate for yourself, even in principle. (And in those cases where "hard evidence" for UFO sightings has been presented, scientific study has never yet found the evidence to be strong enough to support the claim of alien spacecraft [**Section 24.5**].) Moreover, scientific studies of eyewitness testimony show it to be notoriously unreliable, because different eyewitnesses often disagree on what they saw even immediately after an event has occurred. As time passes, memories of the event may change further. In some cases in which memory has been checked against reality, people have reported vivid memories of events that never happened at all. This explains something that virtually all of us have experienced: disagreements with a friend about who did what and when. Since both people cannot be right in such cases, at least one person must have a memory that differs from reality.

The demonstrated unreliability of eyewitness testimony explains why it is generally considered insufficient for a

Logic and Science

In science, we attempt to acquire knowledge through logical reasoning. A logical argument begins with a set of premises and leads to one or more conclusions. There are two basic types of logical argument: deductive and inductive.

In a *deductive argument*, the conclusion follows automatically from the premises, as in this example:

> PREMISE: All planets orbit the Sun in ellipses with the Sun at one focus.
>
> PREMISE: Earth is a planet.
>
> CONCLUSION: Earth orbits the Sun in an ellipse with the Sun at one focus.

Note that the first premise is a general statement that applies to all planets, and the conclusion is a specific statement that applies only to Earth. In other words, we use a deductive argument to *deduce* a specific prediction from a more general theory. If the specific prediction proves to be false, then something must be wrong with the premises from which it was deduced. If it proves true, then we've acquired a piece of evidence in support of the premises.

Now consider the following example of an *inductive argument*:

> PREMISE: Birds fly up into the air but eventually come back down.

> PREMISE: People who jump into the air fall back down.
>
> PREMISE: Rocks thrown into the air come back down.
>
> PREMISE: Balls thrown into the air come back down.
>
> CONCLUSION: What goes up must come down.

Notice that the inductive argument begins with specific facts that are used to *generalize* to a broader conclusion. In this case, each premise supports the conclusion, which may explain why the conclusion was thought to be true for thousands of years. However, no amount of additional examples could ever prove the conclusion to be true, and we need only a single counterexample—such as a rocket leaving Earth—to prove the conclusion to be false.

Both types of argument are important in science. We use inductive arguments to build scientific theories, because we use them to *infer* general principles from observations and experiments. We use deductive arguments to make specific predictions from hypotheses and theories, which we can then test. This explains why theories can never be proved true beyond all doubt—they can only be shown to be consistent with ever larger bodies of evidence. Theories *can* be proved false, however, if they fail to account for observed or experimental facts.

conviction in criminal court; at least some other evidence, such as motive, is required. For the same reason, we cannot accept eyewitness testimony by itself as evidence in science, no matter who reports it or how many people offer similar testimony.

Science and Pseudoscience It's important to realize that science is not the only valid way of seeking knowledge. For example, suppose you are shopping for a car, learning to play drums, or pondering the meaning of life. In each case, you might make observations, exercise logic, and test hypotheses. Yet these pursuits clearly are not science, because they are not directed at developing testable explanations for observed natural phenomena. As long as nonscientific searches for knowledge make no claims about how the natural world works, they do not conflict with science.

However, you will often hear claims about the natural world that seem to be based on observational evidence but do not treat evidence in a truly scientific way. Such claims are often called **pseudoscience**, which literally means "false science." To distinguish real science from pseudoscience, a good first step is to check whether a particular claim exhibits all three hallmarks of science. Consider the example of people who claim a psychic ability to "see" the future and use it to make specific, testable predictions. In this sense, "seeing" the future sounds scientific, since we can test it. However, numerous studies have tested the predictions of "seers" and have found that their predictions come true no more often than would be expected by pure chance. If the "seers" were scientific, they would admit that this evidence undercuts their claim of psychic abilities. Instead, they generally make excuses, such as saying that the predictions didn't come true because of "psychic interference." Making testable claims but then ignoring the results of the tests marks the claimed ability to see the future as pseudoscience.

Objectivity in Science The idea that science is objective, meaning that all people should be able to find the same results, is very important to the validity of science as a means of seeking knowledge. However, there is a difference between the overall objectivity of science and the objectivity of individual scientists.

Science is practiced by human beings, and individual scientists may bring their personal biases and beliefs to their scientific work. For example, most scientists choose their research projects based on personal interests rather than on some objective formula. In extreme cases, scientists have even been known to cheat—either deliberately or subconsciously—to obtain a result they desire. For example, in the late 19th century, astronomer Percival Lowell claimed to see a network of artificial canals in blurry telescopic images of Mars, leading him to conclude that there was a great Martian civilization [**Section 9.4**]. But no such canals actually exist, so Lowell must have allowed his beliefs about extraterrestrial life to influence the way he interpreted what he saw—in essence, a form of cheating, though probably not intentional.

Bias can sometimes show up even in the thinking of the scientific community as a whole. Some valid ideas may not be considered by any scientist because the ideas fall too far outside the general patterns of thought, or **paradigm**, of the time. Einstein's theory of relativity provides an example. Many scientists in the decades before Einstein had gleaned hints of the theory but did not investigate them, at least in part because they seemed too outlandish.

The beauty of science is that it encourages continued testing by many people. Even if personal biases affect some results, tests by others should eventually uncover the mistakes. Similarly, if a new idea is correct but falls outside the accepted paradigm, sufficient testing and verification of the idea will eventually force a shift in the paradigm. In that sense, *science ultimately provides a means of bringing people to agreement*, at least on topics that can be subjected to scientific study.

What is a scientific theory?

The most successful scientific models explain a wide variety of observations in terms of just a few general principles. When a powerful yet simple model makes predictions that survive repeated and varied testing, scientists elevate its status and call it a **theory**. Some famous examples are Isaac Newton's theory of gravity, Charles Darwin's theory of evolution, and Albert Einstein's theory of relativity.

Note that the scientific meaning of the word *theory* is quite different from its everyday meaning, in which we equate a theory more closely with speculation or a hypothesis. For example, someone might get a new idea and say "I have a new theory about why people enjoy the beach." Without the support of a broad range of evidence that others have tested and confirmed, this "theory" is really only a guess. In contrast, Newton's theory of gravity qualifies as a scientific theory because it uses simple physical principles to explain many observations and experiments.

Despite its success in explaining observed phenomena, a scientific theory can never be proved true beyond all doubt, because future observations may disagree with its predictions. However, anything that qualifies as a scientific theory must be supported by a large, compelling body of evidence.

In this sense, a scientific theory is not at all like a hypothesis or any other type of guess. We are free to change a hypothesis at any time, because it has not yet been carefully tested. In contrast, we can discard or replace a scientific theory only if we have an alternative way of explaining the evidence that supports it.

Again, the theories of Newton and Einstein offer good examples. A vast body of evidence supports Newton's theory of gravity, but in the late 19th century scientists began to discover cases where its predictions did not perfectly match observations. These discrepancies were explained only when Einstein developed his general theory of relativity in the early 20th century. Still, the many successes of Newton's theory could not be ignored, and Einstein's theory would not have gained acceptance if it had not been able to explain these successes equally well. It did, and that is why we now view Einstein's theory as a broader theory of gravity than Newton's theory. Some scientists today are seeking a theory of gravity that will go beyond Einstein's. If any new theory ever gains acceptance, it will have to match all the successes of Einstein's theory as well as work in new realms where Einstein's theory does not.

3.5 ASTROLOGY

We have discussed the development of astronomy and the nature of science in some depth. Now let's talk a little about a subject often confused with the science of astronomy: *astrology*. Although the terms *astrology* and *astronomy* sound very similar, today they describe very different practices. In ancient times, however, astrology and astronomy often went hand in hand, and astrology played an important role in the historical development of astronomy.

How is astrology different from astronomy?

The basic tenet of astrology is that the apparent positions of the Sun, Moon, and planets among the stars in our sky influence human events. The origins of this idea are easy to understand. After all, the position of the Sun in the sky certainly influences our lives, since it determines the seasons and the times of daylight and darkness, and the Moon's position determines the tides. Because planets also move among the stars, it probably seemed natural to imagine that they might also influence our lives, even if the influences were more subtle.

Ancient astrologers hoped to learn how the positions of the Sun, Moon, and planets influence our lives by charting the skies and seeking correlations with events on Earth. For example, if an earthquake occurred when Saturn was entering the constellation Leo, might Saturn's position have been the cause of the earthquake? If the king became ill when Mars appeared in the constellation Gemini and the first-quarter moon appeared in Scorpio, might another tragedy be in store for the king when this particular alignment of the Moon and Mars next recurred? Surely, the ancient astrologers thought, the patterns of influence would eventually become clear, and they would then be able to forecast human events with the same reliability with which astronomical observations of the Sun could be used to forecast the coming of spring.

Because forecasts of the seasons and forecasts of human events were imagined to be closely related, astrologers and astronomers usually were one and the same in the ancient world. For example, in addition to his books on astronomy, Ptolemy published a treatise on astrology called *Tetrabiblios,* which remains the foundation for much of astrology today. Interestingly, Ptolemy himself recognized that astrology stood upon a far shakier foundation than astronomy. In the introduction to *Tetrabiblios,* he wrote:

[Astronomy], which is first both in order and effectiveness, is that whereby we apprehend the aspects of the movements of sun, moon, and stars in relation to each other and to the earth . . . . I shall now give an account of the second and less sufficient method [of prediction (astrology)] in a proper philosophical way, so that one whose aim is the truth might never compare its perceptions with the sureness of the first, unvarying science . . . .

Other ancient scientists also recognized that their astrological predictions were far less reliable than their astronomical ones. Nevertheless, confronted with even a slight possibility that astrologers could forecast the future, no king or political leader would dare to be without one. Astrologers held esteemed positions as political advisers in the ancient world and were provided with the resources they needed to continue charting the heavens and human history. Wealthy political leaders' support of astrology made possible much of the development of ancient astronomy.

Throughout the Middle Ages and into the Renaissance, many astronomers continued to practice astrology. For example, Kepler cast numerous *horoscopes*—the predictive charts of astrology (**FIGURE 3.26**)—even as he was discovering the laws of planetary motion. However, given Kepler's later descriptions of astrology as "the foolish stepdaughter of astronomy" and "a dreadful superstition," he may have cast the horoscopes solely as a source of much-needed income. Modern-day astrologers also claim Galileo as one of their own, in part for his having cast a horoscope for the Grand Duke of Tuscany. However, while Galileo's astronomical discoveries changed human history, the horoscope was just plain wrong: The Duke died a few weeks after Galileo predicted that he would have a long and fruitful life.

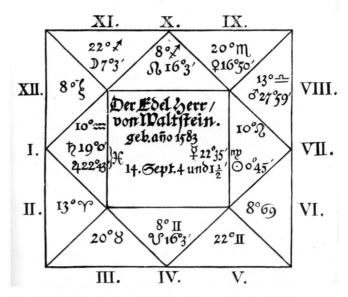

FIGURE 3.26 This chart, cast by Kepler, is an example of a horoscope.

The scientific triumph of Kepler and Galileo in showing Earth to be a planet orbiting the Sun heralded the end of the linkage between astronomy and astrology. Astronomy has since gained status as a successful science that helps us understand our universe, while astrology no longer has any connection to the modern science of astronomy.

Does astrology have any scientific validity?

Although astronomers gave up on it centuries ago, astrology remains popular with the public. Many people read their daily horoscopes, and some pay significant fees to have personal horoscopes cast by professional astrologers. With so many people giving credence to astrology, is it possible that it has some scientific validity after all?

Testing Astrology The validity of astrology can be difficult to assess, because there's no general agreement among astrologers even on such basic things as what astrology is or what it can predict. For example, "Western astrology" is quite different in nature from the astrology practiced in India and China. Some astrologers do not make testable predictions at all; rather, they give vague guidance about how to live one's life. Most daily horoscopes fall into this category. Although your horoscope may seem to ring true at first, a careful read will usually show it to be so vague as to be untestable. For example, a horoscope that says "It is a good day to spend time with your friends" may be good advice but doesn't offer much to test.

SEE IT FOR YOURSELF

Look online for today's local weather forecast and today's horoscope. Contrast the nature of their predictions. At the end of the day, you will know if the weather forecast was accurate. Will you also be able to say whether the horoscope was accurate? Explain.

Nevertheless, most professional astrologers still earn their livings by casting horoscopes that either predict future events in an individual's life or describe characteristics of the person's personality and life. If the horoscope predicts future events, we can check to see whether the predictions come true. If it describes a person's personality and life, the description can be checked for accuracy. A scientific test of astrology requires evaluating many horoscopes and comparing their accuracy to what would be expected by pure chance. For example, suppose a horoscope states that a person's best friend is female. Because roughly half the population of the United States is female, an astrologer who casts 100 such horoscopes would be expected by pure chance to be right about 50 times. We would be impressed with the predictive ability of the astrologer only if he or she were right much more often than 50 times out of 100.

In hundreds of scientific tests, astrological predictions have never proved to be substantially more accurate than expected from pure chance. Similarly, in tests in which astrologers are asked to cast horoscopes for people they have never met, the horoscopes fail to match actual personality profiles more often than expected by chance. The verdict is clear: The methods of astrology are useless for predicting the past, the present, or the future.

Examining the Underpinnings of Astrology In science, observations and experiments are the ultimate judge of any idea. No matter how outlandish an idea might appear, it cannot be dismissed if it successfully meets observational or experimental tests. The idea that Earth rotates and orbits the Sun seemed outlandish for most of human history, yet today it is so strongly supported by the evidence that we consider it a fact. The idea that the positions of the Sun, Moon, and planets among the stars influence our lives might sound outlandish today, but if astrology were to make predictions that came true, adherence to the principles of science would force us to take astrology seriously. However, given that scientific tests of astrology have never found any evidence that its predictive methods work, it is worth looking at its premises to see whether they make sense. Might there be a few kernels of wisdom buried within the lore of astrology?

Let's begin with one of the key premises of astrology: that there is special meaning in the patterns of the stars in the constellations. This idea may have seemed quite reasonable in ancient times, when the stars were assumed to be fixed on an unchanging celestial sphere, but today we know that the patterns of the stars in the constellations are accidents of the moment. Long ago the constellations did not look the same, and they will also look different in the future [**Section 1.3**]. Moreover, the stars in a constellation don't necessarily have any *physical* association, because two stars that are close together in the sky might lie at vastly different distances (see Figure 2.3). Constellations are only *apparent* associations of stars, with no more physical reality than the water in a desert mirage.

Astrology also places great importance on the positions of the planets among the constellations. Again, this idea might have seemed reasonable in ancient times, when it was thought that the planets truly wandered among the stars. Today we know that the planets only *appear* to wander among the stars, much as your hand might appear to move among distant mountains when you wave it. It is difficult to see how mere appearances could have profound effects on our lives.

Many other ideas at the heart of astrology are equally suspect. For example, most astrologers claim that a proper horoscope must account for the positions of *all* the planets. Does this mean that all horoscopes cast before the discovery of Neptune in 1846 were invalid? If so, why didn't astrologers notice that something was wrong with their horoscopes and predict the existence of Neptune? (In contrast, astronomers *did* predict its existence; see the Special Topic on page 312.) Most astrologers have included Pluto since its discovery in 1930; does this mean that they should now stop including it, since it has been demoted to dwarf planet, or that they need to include Eris and other dwarf planets, including some that may not yet have been discovered? And why stop with our own solar system; shouldn't horoscopes also depend on the positions of planets orbiting other stars? Given seemingly unanswerable questions like these, there seems little hope that astrology will ever meet its ancient goal of forecasting human events.

Putting Chapter 3 into Context

In this chapter, we focused on the scientific principles through which we have learned so much about the universe. Key "big picture" concepts from this chapter include the following:

- The basic ingredients of scientific thinking—careful observation and trial-and-error testing—are a part of everyone's experience. Modern science simply provides a way of organizing this thinking to facilitate the learning and sharing of new knowledge.

- Although our understanding of the universe is growing rapidly today, each new piece of knowledge builds on ideas that came before.

- The Copernican revolution, which overthrew the ancient Greek belief in an Earth-centered universe, unfolded over a period of more than a century. Many of the characteristics of modern science first appeared during this time.

- Science exhibits several key features that distinguish it from nonscience and that in principle allow anyone to come to the same conclusions when studying a scientific question.

- Astronomy and astrology once developed hand in hand, but today they represent very different things.

SUMMARY OF KEY CONCEPTS

3.1 THE ANCIENT ROOTS OF SCIENCE

- **In what ways do all humans use scientific thinking?**
Scientific thinking relies on the same type of trial-and-error thinking that we use in our everyday lives, but in a carefully organized way.

- **How is modern science rooted in ancient astronomy?**

Ancient astronomers were accomplished observers who learned to tell the time of day and the time of year, to track cycles of the Moon, and to observe planets and stars. The care and effort that went into these observations helped set the stage for modern science.

3.2 ANCIENT GREEK SCIENCE

- **Why does modern science trace its roots to the Greeks?**
The Greeks developed **models** of nature and emphasized the importance of agreement between the predictions of those models and observations of nature.

- **How did the Greeks explain planetary motion?** The Greek **geocentric model** reached its culmination with the **Ptolemaic model**, which explained apparent retrograde motion by having each planet move on a small circle whose center moves around Earth on a larger circle.

retrograde loop

3.3 THE COPERNICAN REVOLUTION

- **How did Copernicus, Tycho, and Kepler challenge the Earth-centered model?** Copernicus created a Sun-centered model of the solar system designed to replace the Ptolemaic model, but it was no more accurate than Ptolemy's because Copernicus still used perfect circles. Tycho's accurate, naked-eye observations provided the data needed to improve on Copernicus's model. Kepler developed a model of planetary motion that fit Tycho's data.

- **What are Kepler's three laws of planetary motion?**
(1) The orbit of each planet is an ellipse with the Sun at one focus.
(2) As a planet moves around its orbit, it sweeps out equal areas in

equal times. (3) More distant planets orbit the Sun at slower average speeds, obeying the mathematical relationship $p^2 = a^3$.

- **How did Galileo solidify the Copernican revolution?**
Galileo's experiments and telescopic observations overcame remaining objections to the Copernican idea of Earth as a planet

orbiting the Sun. Although not everyone accepted his results immediately, in hindsight we see that Galileo sealed the case for the Sun-centered solar system.

3.4 THE NATURE OF SCIENCE

- **How can we distinguish science from nonscience?**
Science generally exhibits three hallmarks: (1) Modern science seeks explanations for observed phenomena that rely solely on natural causes. (2) Science progresses through the creation and testing of models of nature that explain the observations as simply as possible. (3) A scientific model must make testable predictions about natural phenomena that would force us to revise or abandon the model if the predictions did not agree with observations.

- **What is a scientific theory?** A scientific **theory** is a simple yet powerful model that explains a wide variety of observations using just a few general principles and has been verified by repeated and varied testing.

3.5 ASTROLOGY

- **How is astrology different from astronomy?**
Astronomy is a modern science that has taught us much about the universe. Astrology is a search for hidden influences on human lives based on the apparent positions of planets and stars in the sky; it does not follow the tenets of science.

- **Does astrology have any scientific validity?** Scientific tests have shown that astrological predictions do not prove to be accurate more than we can expect by pure chance, showing that the predictions have no scientific validity.

Use the following questions to check your understanding of some of the many types of visual information used in astronomy. Answers are provided in Appendix J. For additional practice, try the Chapter 3 Visual Quiz at MasteringAstronomy®.

a

b

Study the two graphs above, based on Figure 3.19. Use the information in the graphs to answer the following questions.

1. Approximately how fast is Jupiter orbiting the Sun?
   a. This cannot be determined from the information provided.
   b. 20 km/s
   c. 10 km/s
   d. a little less than 15 km/s
2. An asteroid with an average orbital distance of 2 AU will orbit the Sun at an average speed that is
   a. a little slower than the orbital speed of Mars.
   b. a little faster than the orbital speed of Mars.
   c. the same as the orbital speed of Mars.
3. Uranus, not shown on graph **b**, orbits about 19 AU from the Sun. Based on the graph, its approximate orbital speed is between about
   a. 20 and 25 km/s.
   b. 15 and 20 km/s.
   c. 10 and 15 km/s.
   d. 5 and 10 km/s.
4. Kepler's third law is often stated as $p^2 = a^3$. The value a^3 for a planet is shown on
   a. the horizontal axis of graph **a**.   b. the vertical axis of graph **a**.
   c. the horizontal axis of graph **b**.   d. the vertical axis of graph **b**.

5. On graph **a**, you can see Kepler's third law ($p^2 = a^3$) from the fact that
   a. the data fall on a straight line.
   b. the axes are labeled with values for p^2 and a^3.
   c. the planet names are labeled on the graph.
6. Suppose graph **a** showed a planet on the red line directly above a value of 1000 AU³ along the horizontal axis. On the vertical axis, this planet would be at
   a. 1000 years².
   b. 1000^2 years².
   c. $\sqrt{1000}$ years².
   d. 100 years.
7. How far does the planet in question 6 orbit from the Sun?
   a. 10 AU
   b. 100 AU
   c. 1000 AU
   d. $\sqrt{1000}$ AU

MasteringAstronomy®

REVIEW QUESTIONS

Short-Answer Questions Based on the Reading

1. In what way is scientific thinking natural to all of us, and how does modern science build upon this everyday type of thinking?
2. Why did ancient peoples study astronomy? Describe an astronomical achievement of at least three ancient cultures.
3. Describe the astronomical origins of our day, week, month, and year.
4. What is a lunar calendar? How can it be kept roughly synchronized with a solar calendar?
5. What do we mean by a *model* in science?

6. Summarize the development of the Greek *geocentric model* to its culmination with Ptolemy. How did this model account for the apparent retrograde motion of planets in our sky?
7. What was the *Copernican revolution,* and how did it change the human view of the universe?
8. What is an *ellipse?* Define its *foci, semimajor axis,* and *eccentricity.*
9. State and explain the meaning of each of *Kepler's laws of planetary motion.*
10. Describe the three hallmarks of science and how we can see them in the Copernican revolution. What is *Occam's razor?* Why doesn't science accept personal testimony as evidence?

11. What is the difference between a *hypothesis* and a *theory* in science?
12. What is the basic idea behind *astrology*? Explain why this idea seemed reasonable in ancient times but is no longer accepted by scientists.

TEST YOUR UNDERSTANDING

Science or Nonscience?

Each of the following statements makes some type of claim. Decide in each case whether the claim could be evaluated scientifically or falls into the realm of nonscience. Explain clearly; not all of these have definitive answers, so your explanation is more important than your chosen answer.

13. The Yankees are the best baseball team of all time.
14. Several kilometers below its surface, Jupiter's moon Europa has an ocean of liquid water.
15. My house is haunted by ghosts who make the creaking noises I hear each night.
16. There are no lakes or seas on the surface of Mars today.
17. Dogs are smarter than cats.
18. Children born when Jupiter is in the constellation Taurus are more likely to be musicians than other children.
19. Aliens can manipulate time and memory so that they can abduct and perform experiments on people who never realize they were taken.
20. Newton's law of gravity works as well for explaining orbits of planets around other stars as it does for explaining orbits of the planets in our own solar system.
21. God created the laws of motion that were discovered by Newton.
22. A huge fleet of alien spacecraft will land on Earth and introduce an era of peace and prosperity on January 1, 2025.

Quick Quiz

Choose the best answer to each of the following. Explain your reasoning with one or more complete sentences.

23. In the Greek geocentric model, the retrograde motion of a planet occurs when (a) Earth is about to pass the planet in its orbit around the Sun. (b) the planet actually goes backward in its orbit around Earth. (c) the planet is aligned with the Moon in our sky.
24. Which of the following was *not* a major advantage of Copernicus's Sun-centered model over the Ptolemaic model? (a) It made significantly better predictions of planetary positions in our sky. (b) It offered a more natural explanation for the apparent retrograde motion of planets in our sky. (c) It allowed calculation of the orbital periods and distances of the planets.
25. When we say that a planet has a highly eccentric orbit, we mean that (a) it is spiraling in toward the Sun. (b) its orbit is an ellipse with the Sun at one focus. (c) in some parts of its orbit it is much closer to the Sun than in other parts.
26. Earth is closer to the Sun in January than in July. Therefore, in accord with Kepler's second law, (a) Earth travels faster in its orbit around the Sun in July than in January. (b) Earth travels faster in its orbit around the Sun in January than in July. (c) it is summer in January and winter in July.
27. According to Kepler's third law, (a) Mercury travels fastest in the part of its orbit in which it is closest to the Sun. (b) Jupiter orbits the Sun at a faster speed than Saturn. (c) all the planets have nearly circular orbits.
28. Tycho Brahe's contribution to astronomy included (a) inventing the telescope. (b) proving that Earth orbits the Sun. (c) collecting data that enabled Kepler to discover the laws of planetary motion.
29. Galileo's contribution to astronomy included (a) discovering the laws of planetary motion. (b) discovering the law of gravity. (c) making observations and conducting experiments that dispelled scientific objections to the Sun-centered model.
30. Which of the following is *not* true about scientific progress? (a) Science progresses through the creation and testing of models of nature. (b) Science advances only through the scientific method. (c) Science avoids explanations that invoke the supernatural.
31. Which of the following is *not* true about a scientific theory? (a) A theory must explain a wide range of observations or experiments. (b) Even the strongest theories can never be proved true beyond all doubt. (c) A theory is essentially an educated guess.
32. When Einstein's theory of gravity (general relativity) gained acceptance, it demonstrated that Newton's theory had been (a) wrong. (b) incomplete. (c) really only a guess.

PROCESS OF SCIENCE

Examining How Science Works

33. *What Makes It Science?* Choose a single idea in the modern view of the cosmos as discussed in Chapter 1, such as "The universe is expanding," "The universe began with a Big Bang," "We are made from elements manufactured by stars," or "The Sun orbits the center of the Milky Way Galaxy once every 230 million years." **a.** Describe how this idea reflects each of the three hallmarks of science, discussing how it is based on observations, how our understanding of it depends on a model, and how that model is testable. **b.** Describe a hypothetical observation that, if it were actually made, might cause us to call the idea into question. Then briefly discuss whether you think that, overall, the idea is likely or unlikely to hold up to future observations.
34. *Earth's Shape.* It took thousands of years for humans to deduce that Earth is spherical. For each of the following alternative models of Earth's shape, identify one or more observations that you could make for yourself that would invalidate the model. **a.** A flat Earth **b.** A cylindrical Earth, like that proposed by Anaximander **c.** A football-shaped Earth
35. *Scientific Test of Astrology.* Find out about at least one scientific test of the validity of astrology. Write a short summary of the methods and results of the test.
36. *Your Own Astrological Test.* Devise your own scientific test of astrology. Clearly define your methods and how you will evaluate the results. Carry out the test and write a short report about it.

GROUP WORK EXERCISE

37. *Galileo on Trial.* In this exercise, you will debate the evidence presented by Galileo in favor of the idea that Earth orbits the Sun. Before you begin, assign the following roles to the people in your group: *Scribe* (takes notes on the group's activities), *Galileo* (argues in favor of the idea that Earth orbits the Sun), *Prosecutor* (argues against the idea that Earth orbits the Sun), and *Moderator* (leads group discussion and makes sure the debate remains civil). Then consider each of the following three pieces of evidence: **a.** observations of mountains and valleys on the Moon. **b.** observations of moons orbiting Jupiter. **c.** observations of the phases of Venus. *Galileo* should explain why the evidence indicates that Earth orbits the Sun, and the *Prosecutor* should present a rebuttal. After the discussion, the *Scribe* and *Moderator* should decide whether the evidence is convincing beyond a reasonable doubt, somewhat convincing, or not convincing, and write down their verdict, along with an explanation of their reasoning.

In-Depth Questions to Increase Your Understanding

Short-Answer/Essay Questions

38. *Lunar Calendars.*
    a. Find the dates of the Jewish festival of Chanukah for this year and the next three years. Based on what you have learned in this chapter, explain why the dates change as they do. **b.** Find the dates of the Muslim fast for Ramadan for this year and the next three years. Based on what you have learned in this chapter, explain why the dates change as they do.

39. *Copernican Players.* Using a bulleted-list format, make a one-page "executive summary" of the major roles that Copernicus, Tycho, Kepler, and Galileo played in overturning the ancient belief in an Earth-centered universe.

40. *Influence on History.* Based on what you have learned about the Copernican revolution, write a one- to two-page essay about how you believe it altered the course of human history.

41. *Cultural Astronomy.* Choose a particular culture of interest to you, and research the astronomical knowledge and accomplishments of that culture. Write a two- to three-page summary of your findings.

42. *Astronomical Structures.* Choose an ancient astronomical structure of interest to you (e.g., Stonehenge, Templo Mayor, Pawnee lodges), and research its history. Write a two- to three-page summary of your findings. If possible, also build a scale model of the structure or create detailed diagrams to illustrate how the structure was used.

Quantitative Problems

Be sure to show all calculations clearly and state your final answers in complete sentences.

43. *The Metonic Cycle.* The length of our calendar year is 365.2422 days, and the Moon's monthly cycle of phases averages 29.5306 days in length. By calculating the number of days in each, confirm that 19 solar years is almost precisely equal to 235 cycles of the lunar phases. Show your work clearly; then write a few sentences explaining how this fact can be used to keep a lunar calendar roughly synchronized with a solar calendar.

44. *Chinese Calendar.* The traditional Chinese lunar calendar has 12 months in most years but adds a thirteenth month to 22 of every 60 years. How many days does this give the Chinese calendar in each 60-year period? How does this compare to the number of days in 60 years on a solar calendar? Based on your answers, explain how this scheme is similar to the scheme used by lunar calendars that follow the Metonic cycle. (*Hint:* You'll need the data given in Problem 43.)

45. *Method of Eratosthenes I.* You are an astronomer on planet Nearth, which orbits a distant star. It has recently been accepted that Nearth is spherical in shape, though no one knows its size. One day, while studying in the library of Alectown, you learn that on the equinox your sun is directly overhead in the city of Nyene, located 1000 kilometers due north of you. On the equinox, you go outside and observe that the altitude of your sun is 80°. What is the circumference of Nearth? (*Hint:* Apply the technique used by Eratosthenes to measure Earth's circumference.)

46. *Method of Eratosthenes II.* You are an astronomer on planet Tirth, which orbits a distant star. It has recently been accepted that Tirth is spherical in shape, though no one knows its size. One day, you learn that on the equinox your sun is directly overhead in the city of Tyene, located 400 kilometers due north of you. On the equinox, you go outside and observe that the altitude of your sun is 86°. What is the circumference of Tirth? (*Hint:* Apply the technique used by Eratosthenes to measure Earth's circumference.)

47. *Mars Orbit.* Find the perihelion and aphelion distances of Mars. (*Hint:* You'll need data from Appendix E.)

48. *Eris Orbit.* The recently discovered Eris, which is slightly larger than Pluto, orbits the Sun every 557 years. What is its average distance (semimajor axis) from the Sun? How does its average distance compare to that of Pluto?

49. *New Planet Orbit.* A newly discovered planet orbits a distant star with the same mass as the Sun at an average distance of 112 million kilometers. Its orbital eccentricity is 0.3. Find the planet's orbital period and its nearest and farthest orbital distances from its star.

50. *Halley Orbit.* Halley's Comet orbits the Sun every 76.0 years and has an orbital eccentricity of 0.97.
    a. Find its average distance from the Sun (semimajor axis). **b.** Find its perihelion and aphelion distances.

Discussion Questions

51. *The Impact of Science.* The modern world is filled with ideas, knowledge, and technology that developed through science and application of the scientific method. Discuss some of these things and how they affect our lives. Which of these impacts do you think are positive? Which are negative? Overall, do you think science has benefited the human race? Defend your opinion.

52. *The Importance of Ancient Astronomy.* Why was astronomy important to people in ancient times? Discuss both the practical importance of astronomy and the importance it may have had for religious or other traditions. Which do you think was more important in the development of ancient astronomy: its practical or its philosophical role? Defend your opinion.

53. *Astronomy and Astrology.* Why do you think astrology remains so popular around the world even though it has failed all scientific tests of its validity? Do you think the popularity of astrology has any positive or negative social consequences? Defend your opinions.

Web Projects

54. *Easter.* Research when different denominations of Christianity celebrate Easter and why they use different dates. Summarize your findings in a one- to two-page report.

55. *Greek Astronomers.* Many ancient Greek scientists had ideas that, in retrospect, seem well ahead of their time. Choose one ancient Greek scientist to study, and write a one- to two-page "scientific biography" of your chosen person.

56. *The Ptolemaic Model.* This chapter gives only a very brief description of Ptolemy's model of the universe. Investigate this model in greater depth. Using diagrams and text as needed, give a two- to three-page description of the model.

57. *The Galileo Affair.* In recent years, the Roman Catholic Church has devoted a lot of resources to learning more about the trial of Galileo and to understanding past actions of the Church in the Galilean case. Learn more about these studies and write a short report about the Vatican's current view of the case.

58. *Science or Pseudoscience.* Choose a pseudoscientific claim related to astronomy, and learn more about how scientists have "debunked" it. (A good starting point is the Bad Astronomy website.) Write a short summary of your findings.

S1 CELESTIAL TIMEKEEPING AND NAVIGATION
SUPPLEMENTARY CHAPTER

> **Socrates:** *Shall we make astronomy the next study? What do you say?*
>
> **Glaucon:** *Certainly. A working knowledge of the seasons, months, and years is beneficial to everyone, to commanders as well as to farmers and sailors.*
>
> **Socrates:** *You make me smile, Glaucon. You are so afraid that the public will accuse you of recommending unprofitable studies.*
>
> —*Plato,* Republic

As the opening quote from Plato shows, ancient astronomy served practical needs for timekeeping and navigation. These ancient uses may no longer seem so important in an age when we tell time with digital watches and navigate with the global positioning system (GPS). But knowing the celestial basis of timekeeping and navigation can help us understand the rich history of astronomical discovery, and occasionally still proves useful in its own right. In this chapter, we will explore the apparent motions of the Sun, Moon, and planets in enough detail to learn the basic principles of keeping time and navigating by the stars.

S1.1 ASTRONOMICAL TIME PERIODS

Although many people do not realize it, modern clocks and calendars are beautifully synchronized to the rhythms of the heavens. Precision measurements allow us to ensure that our clocks keep pace with the Sun's daily trek across our sky, while our calendar holds the dates of the equinoxes and solstices as steady as possible. In earlier chapters, we saw how this synchronicity took root in ancient observations of the sky. In this section, we will look more closely at basic measures of time and our modern, international system of timekeeping.

How do we define the day, month, year, and planetary periods?

By now you know that the length of the day corresponds to Earth's rotation, the length of the month to the cycle of lunar phases, and the length of the year to our orbit around the Sun. However, these correspondences are not quite as simple as we might at first guess, because there is more than one way to define the day, month, year, and planetary periods.

The Length of the Day We usually think of a day as the time it takes for Earth to rotate once. But if you measure Earth's rotation period, you'll find that it is actually about 4 minutes short of 24 hours. What's going on?

Remember that the daily circling of the stars in our sky is an illusion created by Earth's rotation (see Figure 2.9). You can therefore measure Earth's rotation period by measuring how long it takes for any star to go from its highest point in the sky one day to its highest point the next day (**FIGURE S1.1a**). This time period, which we call a **sidereal** (pronounced *sy-DEAR-ee-al*) **day**, is about 23 hours 56 minutes (more precisely, $23^h\ 56^m\ 4.09^s$). *Sidereal* means "related to the stars"; note that you'll measure the same time no matter what star you choose. For practical purposes, the sidereal day is Earth's precise rotation period.

Our 24-hour day, which we call a **solar day**, is based on the time it takes for the *Sun* to make one circuit around the local sky. You can measure this time period by measuring how long it takes the Sun to go from its highest point in the sky one day to its highest point the next day (**FIGURE S1.1b**). The solar day is indeed 24 hours on average, although it varies slightly (up to about 25 seconds longer or shorter than 24 hours) over the course of a year.

A simple demonstration shows why the solar day is about 4 minutes longer than the sidereal day. Set an object to represent the Sun on a table, and stand a few steps away to represent Earth. Point at the Sun and imagine that you also happen to be pointing toward a distant star that lies in the same direction. If you rotate (counterclockwise) while standing in place,

a A sidereal day is the time it takes any star to make a circuit of the local sky. It is about 23 hours 56 minutes.

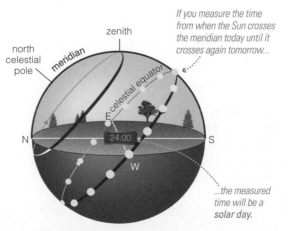

b A solar day is the time it takes the Sun to make a circuit of the local sky. Its precise length varies over the course of the year, but the average is 24 hours.

FIGURE S1.1 Using the sky to measure the length of a day.

to distant star

One full rotation means you are again pointing in the original direction . . .

. . . but you need a bit of extra rotation to point again at the Sun.

a One full rotation represents a sidereal day and returns you to pointing in your original direction, but you need to rotate a little extra to return to pointing at the Sun.

Earth travels 360° around orbit in 365 days, about 1° per day . . .

1°   1°

. . . so Earth must spin about 1° more than 360° from noon one day to noon the next.

Not to scale!

b Earth travels about 1° per day around its orbit, so a solar day requires about 361° of rotation.

FIGURE S1.2
A demonstration showing why a solar day is slightly longer than a sidereal day.

you'll again be pointing at both the Sun and the star after one full rotation. However, to show that Earth also orbits the Sun, you should take a couple of steps around the Sun (counterclockwise) as you rotate (**FIGURE S1.2a**). After one full rotation, you will again be pointing in the direction of the distant star, so this represents a sidereal day. But notice that you need to rotate a little extra to point back at the Sun. In fact, because Earth travels about 1° per day around its orbit, a solar day requires about 1° of extra rotation compared to a sidereal day (**FIGURE S1.2b**). This extra 1° of rotation takes about $\frac{1}{360}$ of Earth's rotation period, which is about 4 minutes.

The Length of the Month As we discussed in Chapter 2, our month comes from the Moon's $29\frac{1}{2}$-day cycle of phases (think "moonth"). More technically, this $29\frac{1}{2}$-day period is called a **synodic month**. The word *synodic* comes from the Latin *synod*, which means "meeting." A synodic month gets its name from the idea that the Sun and the Moon "meet" in the sky with every new moon.

Just as a solar day is not Earth's true rotation period, a synodic month is not the Moon's true orbital period. Earth's motion around the Sun means that the Moon must complete more than one full orbit of Earth from one new moon to the next (**FIGURE S1.3**). The Moon's true orbital period, or a **sidereal month**, is only about $27\frac{1}{3}$ days. Like the sidereal day, the sidereal month gets its name from the fact that it describes how long it takes the Moon to complete an orbit relative to the positions of distant stars.

The Length of the Year We can also define a year in two slightly different ways. The time it takes Earth to complete one orbit relative to the stars is called a **sidereal year**. However, our calendar is based on the cycle of the seasons, which we measure as the time from the spring equinox one year to the spring equinox the next year. This time period,

called a **tropical year**, is about 20 minutes shorter than the sidereal year. A 20-minute difference might not seem like much, but it would make a calendar based on the sidereal year get out of sync with the seasons by 1 day every 72 years—a difference that would add up over centuries.

The difference between the sidereal year and the tropical year arises from Earth's 26,000-year cycle of axis precession [**Section 2.2**]. Precession not only changes the orientation of the axis in space but also changes the locations in Earth's orbit at which the seasons occur. Each year, the location of the equinoxes and solstices among the stars shifts about $\frac{1}{26,000}$ of the way around the orbit. If you do the math, you'll find that $\frac{1}{26,000}$ of a year is about 20 minutes, which explains the 20-minute difference between the tropical year and the sidereal year.

Planetary Periods Although planetary periods are not used in our modern timekeeping, they were important to many ancient cultures. For example, the Mayan calendar was based in part on the apparent motions of Venus. In addition,

new moon

Earth travels about 30° per month around its orbit . . .

30°

new moon

. . . so from one new moon to the next, the Moon must complete a full 360° orbit . . .

360°

30°

. . . AND go an extra 30°.

FIGURE S1.3 interactive figure The Moon completes one 360° orbit in about $27\frac{1}{3}$ days (a sidereal month), but the time from new moon to new moon is about $29\frac{1}{2}$ days (a synodic month).

Copernicus's ability to determine orbital periods of planets with his Sun-centered model played an important role in keeping the model alive long enough for its ultimate acceptance (see Mathematical Insight S1.1).

A planet's **sidereal period** is the time the planet takes to orbit the Sun; again, the name comes from the fact that it is measured relative to distant stars. For example, Jupiter's sidereal period is 11.86 years, so it takes about 12 years for Jupiter to make a complete circuit through the constellations of the zodiac. Jupiter therefore appears to move through roughly one zodiac constellation each year. If Jupiter is currently in Gemini (as it is from about July 2013 to July 2014), it will be in Cancer at this time next year and Leo the following year, returning to Gemini in about 12 years.

A planet's **synodic period** is the time from when it is lined up with the Sun in our sky once to the next similar alignment. (As with the Moon, the term *synodic* refers to the planet's "meeting" the Sun in the sky.) **FIGURE S1.4** shows that the situation is somewhat different for planets nearer the Sun than Earth (that is, Mercury and Venus) than for planets farther away.

Look first at the situation for the more distant planet in Figure S1.4. As seen from Earth, this planet will sometimes line up with the Sun in what we call a **conjunction**. At other times, it will appear exactly opposite the Sun in our sky, or at **opposition**. We cannot see the planet during conjunction with the Sun because it is hidden by the Sun's glare and rises and sets with the Sun in our sky. At opposition, the planet moves through the sky like the full moon, rising at sunset, reaching the meridian at midnight, and setting at dawn. Note that the planet is closest to Earth at opposition and hence appears brightest in our sky at this time.

Now look at the planet that is *nearer* than Earth to the Sun in Figure S1.4. This planet never has an opposition but instead has two conjunctions—an "inferior conjunction" between Earth and the Sun and a "superior conjunction" when the planet appears behind the Sun as seen from Earth. Two other points are important for the inner planets: their points of **greatest elongation**, when they appear farthest from the Sun in our sky. At its greatest eastern elongation, Venus appears about 46° east of the Sun in our sky, which means it shines brightly in the evening. Similarly, at its greatest

western elongation, Venus appears about 46° west of the Sun and shines brightly before dawn. In between the times when Venus appears in the morning sky and the times when it appears in the evening sky, Venus disappears from view for a few weeks with each conjunction. Mercury's pattern is similar, but because it is closer to the Sun, it never appears more than about 28° from the Sun in our sky. That makes Mercury difficult to see, because it is almost always obscured by the glare of the Sun.

As you study Figure S1.4, you might wonder whether Mercury and Venus ever fall directly in front of the Sun at inferior conjunction, creating a mini-eclipse as they block a little of the Sun's light. They do, but only rarely, because their orbital planes are slightly tilted compared to Earth's orbital plane (the ecliptic plane). As a result, Mercury and Venus usually appear slightly above or below the Sun at inferior conjunction. But on rare occasions, we do indeed see Mercury or Venus appear to pass directly across the face of the Sun during inferior conjunction. Such events are called **transits** (**FIGURE S1.5**). Mercury transits occur an average of a dozen times per century; the first one in this century occurred on November 8, 2006, and the next will occur on May 9, 2016. Venus transits come in pairs 8 years apart, with more than a century between the second of one pair and the first of the next. If you missed both of the recent transits of Venus, which occurred on June 8, 2004 and June 6, 2012, you have a long wait for the next pair, which will occur in 2117 and 2125.

How do we tell the time of day?

Telling time seems simple, but in fact there are several different ways to define the time of day, even after we agree that time should be based on the 24-hour solar day. Let's explore some of the ways of telling time and see how they ultimately led to our modern system in which we can synchronize clocks anywhere in the world.

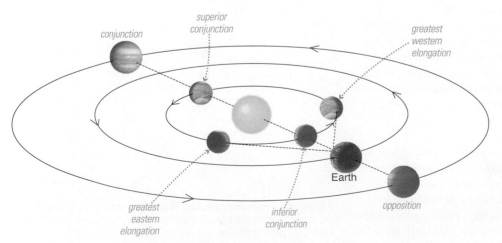

FIGURE S1.4 This diagram shows important positions of planets relative to Earth and the Sun. For a planet farther from the Sun than Earth (such as Jupiter), conjunction occurs when the planet appears aligned with the Sun in the sky, and opposition occurs when the planet appears on our meridian at midnight. Planets nearer the Sun (such as Venus) have two conjunctions and never get farther from the Sun in our sky than at their greatest elongations. (Adapted from *Advanced Skywatching*, by Burnham et al.)

FIGURE S1.5 This photo shows the transit of views that occurred on June 6, 2012—the last Venus transit of this century. It was taken in Germany at dawn. Venus is the small black dot visible near the upper center of the Sun's face.

Apparent Solar Time If we base time on the Sun's *actual* position in the local sky, as is the case when we use a sundial (**FIGURE S1.6**), we are measuring **apparent solar time**. Noon is the precise moment when the Sun is highest in the sky (on the meridian) and the sundial casts its shortest shadow. Before noon, when the Sun is rising upward through the sky, the apparent solar time is *ante meridiem* ("before the middle of the day"), or a.m. For example, if the Sun will reach the meridian 2 hours from now, the apparent solar time is 10 a.m. After noon, the apparent solar time is *post meridiem* ("after the middle of the day"), or p.m. If the Sun crossed the meridian 3 hours ago, the apparent solar time is 3 p.m. Note that, technically, noon and midnight are *neither* a.m. nor p.m. However, by convention we usually say that noon is 12 p.m. and midnight is 12 a.m.

THINK ABOUT IT
Is it daytime or nighttime at 12:01 a.m.? 12:01 p.m.? Explain.

Mean Solar Time Suppose you set a clock to precisely 12:00 when a sundial shows noon today. If every solar day were precisely 24 hours, your clock would always remain synchronized with the sundial. However, while 24 hours is the *average* length of the solar day, the actual length of the solar day varies throughout the year, so your clock is likely to read a few seconds before or after 12:00 when the sundial reads noon tomorrow, and within a few weeks your clock time may differ from the apparent solar time by several minutes.

If we average the differences between the time your clock would read and the time a sundial would read, we can define **mean solar time** (*mean* is another word for *average*). A clock set to mean solar time reads 12:00 each day at the time the

FIGURE S1.6 A basic sundial consists of a dial marked by numerals, and a stick, or *gnomon,* that casts a shadow. Here, the shadow is on the Roman numeral I, indicating that the apparent solar time is 1:00 p.m. (The portion of the dial without numerals represents nighttime hours.) Because the Sun's path across the local sky depends on latitude, a particular sundial will be accurate only for a particular latitude.

Sun crosses the meridian *on average*. The actual mean solar time at which the Sun crosses the meridian varies over the course of the year in a fairly complex way (see "Solar Days and the Analemma," page 91). The result is that, on any given day, a clock set to mean solar time may read anywhere from about 17 minutes before noon to 15 minutes after noon (that is, from 11:43 a.m. to 12:15 p.m.) when a sundial indicates noon.

Although the lack of perfect synchronization with the Sun might at first sound like a drawback, mean solar time is actually more convenient than apparent solar time (the sundial time)—as long as you have access to a mechanical or electronic clock. Once set, a reliable mechanical or electronic clock can always tell you the mean solar time. In contrast, precisely measuring apparent solar time requires a sundial, which is useless at night or when it is cloudy.

Like apparent solar time, mean solar time is a *local* measure of time. That is, it varies with longitude because of Earth's west-to-east rotation. For example, clocks in New York are set 3 hours ahead of clocks in Los Angeles. If clocks were set precisely to local mean solar time, they would vary even over relatively short east-west distances. For example, mean solar clocks in central Los Angeles would be about 2 minutes behind mean solar clocks in Pasadena, because Pasadena is slightly to the east.

Standard, Daylight, and Universal Time

Clocks displaying mean solar time were once common. But by the late 19th century, particularly in the United States, the growth of railroad travel made mean solar time increasingly problematic. Some states had dozens of different "official" times, usually corresponding to mean solar time in dozens of different cities, and each railroad company made schedules according to its own "railroad time." The many time systems made it difficult for passengers to follow train schedules.

On November 18, 1883, the railroad companies agreed to a new system that divided the United States into four time zones, setting all clocks within each zone to the same time. That was the birth of **standard time**, which today divides the entire world into time zones (**FIGURE S1.7**, page 90). Depending on where you live within a time zone, your standard time may vary somewhat from your mean solar time. In principle, the

---

MATHEMATICAL INSIGHT S1.1

The Copernican Layout of the Solar System

As discussed in Chapter 3, Copernicus favored the Sun-centered model partly because it allowed him to calculate orbital periods and distances for the planets. Let's see how.

We cannot directly measure a planet's orbital period, because we look at the planet from different points in our orbit at different times. However, we can measure its synodic period simply by seeing how much time passes between one particular alignment (such as opposition or inferior conjunction) and the next. **FIGURE 1** shows the geometry for a planet *farther from the Sun than Earth* (such as Jupiter), under the assumption of circular orbits (which Copernicus assumed). Note the following key facts:

- The dashed brown curve shows the planet's orbit, which takes a time of one orbital period, P_{orb}, to complete.

- The solid brown arrow shows how far the planet travels along its orbit from one opposition to the next. The time between oppositions is defined as its synodic period, P_{syn}.

- The dashed blue curve shows Earth's orbit; Earth takes $P_{Earth} = 1$ yr to complete an orbit.

- The solid red curve and extra red arrow show how far Earth goes during the planet's synodic period; it is *more* than one complete orbit because Earth must travel a little "extra" to catch back up with the planet, and the time required for this "extra" distance (the thick red arrow) is the planet's synodic period minus 1 year, or $P_{syn} - 1$ yr.

Now, notice that the angle that the planet sweeps out during its synodic period is equal to the angle that Earth sweeps out as it travels the "extra" distance. Therefore, the *ratio* of the planet's complete orbital period (P_{orb}) to its synodic period (P_{syn}) must equal the *ratio* of Earth's orbital period (1 yr) to the time required for the "extra" distance (see Appendix C.5 to review ratios). We already

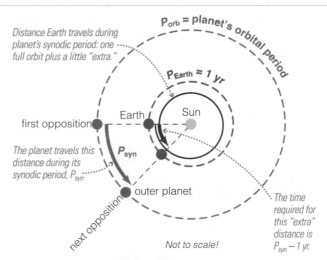

FIGURE 1

found that the time required for this extra distance is $P_{syn} - 1$ yr, so we write

$$\frac{P_{orb}}{P_{syn}} = \frac{1 \text{ yr}}{(P_{syn} - 1 \text{ yr})}$$

Multiplying both sides by P_{syn} gives us the final equation for a planet farther from the Sun than Earth:

$$\text{outer planets: } P_{orb} = P_{syn} \times \frac{1 \text{ yr}}{(P_{syn} - 1 \text{ yr})}$$

The geometry is slightly different for a planet closer to the Sun (Mercury or Venus). **FIGURE 2** shows that in this case the equal ratios are $1 \text{ yr}/P_{syn} = P_{orb}/(P_{syn} - P_{orb})$, leading to this equation for a planet closer to the Sun than Earth:

$$\text{inner planets: } P_{orb} = P_{syn} \times \frac{1 \text{ yr}}{(P_{syn} + 1 \text{ yr})}$$

standard time in a particular time zone was to be the mean solar time in the *center* of the time zone, so that local mean solar time within a 1-hour-wide time zone would never differ by more than a half-hour from standard time. However, time zones often have unusual shapes to conform to social, economic, and political realities, so larger variations between standard time and mean solar time sometimes occur.

In most parts of the United States, clocks are set to standard time for only part of the year. Between the second Sunday in March and the first Sunday in November, most of the United States changes to **daylight saving time**, which is 1 hour ahead of standard time. Because of the 1-hour advance with daylight saving time, clocks read around 1 p.m. (rather than around noon) when the Sun is on the meridian.

For purposes of navigation and astronomy, it is useful to have a single time for the entire Earth. For historical reasons, this "world" time was chosen to be the mean solar time in Greenwich, England—the place that also defines longitude 0° (see Figure 2.11). Today, this *Greenwich mean time (GMT)* is often called **universal time (UT)**. (Outside astronomy, it is more commonly called *universal coordinated time* [UTC]. Many airlines and weather services call it "Zulu time," because

Greenwich's time zone is designated "Z" and "zulu" is a common way of phonetically identifying the letter *Z*.)

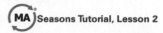 Seasons Tutorial, Lesson 2

When and why do we have leap years?

Our modern calendar is designed to stay synchronized with the seasons and is therefore based on the tropical year (the time from one spring equinox to the next). Getting this synchronization just right was a long process in human history.

The origins of our modern calendar go back to ancient Egypt. By 4200 B.C., the Egyptians were using a calendar that counted 365 days in a year. However, because the length of a year is about $365\frac{1}{4}$ days (rather than exactly 365 days), the Egyptian calendar drifted out of phase with the seasons by about 1 day every 4 years. For example, if the spring equinox occurred on March 21 one year, 4 years later it occurred on March 22, 4 years after that on March 23, and so on. Over many centuries, the spring equinox moved through many different months. To keep the seasons and the calendar synchronized, Julius Caesar decreed the adoption of a new

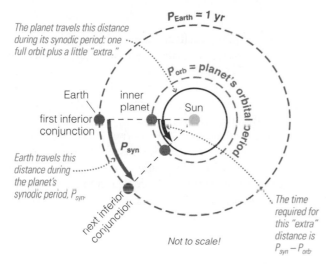

FIGURE 2

Copernicus knew the synodic periods of the planets and therefore could use the above equations (in a slightly different form) to calculate their true orbital periods. He then used the geometry of planetary alignments to compute distances in terms of the Earth-Sun distance. (That is, he calculated distances in AU.) His results were quite close to modern values.

EXAMPLE 1: Jupiter's synodic period is 398.9 days, or 1.092 years. What is its actual orbital period?

SOLUTION:

Step 1 Understand: We are given Jupiter's synodic period (P_{syn}), which is the only value we need to find its orbital period (P_{orb}).

Step 2 Solve: We use the equation for a planet farther from the Sun than Earth, with $P_{syn} = 1.092$ yr:

$$P_{orb} = P_{syn} \times \frac{1 \text{ yr}}{(P_{syn} - 1 \text{ yr})}$$

$$= 1.092 \text{ yr} \times \frac{1 \text{ yr}}{(1.092 \text{ yr} - 1 \text{ yr})} = 11.87 \text{ yr}$$

Step 3 Explain: We have found that Jupiter's orbital period is 11.87 years, or a little less than 12 years. Notice that, as we expect for a planet farther from the Sun, Jupiter's orbital period is longer than Earth's.

EXAMPLE 2: Venus's synodic period is 583.9 days, or 1.599 years. What is its actual orbital period?

SOLUTION:

Step 1 Understand: As in the first example, we can calculate Venus's orbital period from its given synodic period; the only difference is that we'll need the equation for a planet closer to the Sun than Earth.

Step 2 Solve: We use the equation for a planet closer to the Sun, with $P_{syn} = 1.599$ yr:

$$P_{orb} = P_{syn} \times \frac{1 \text{ yr}}{(P_{syn} + 1 \text{ yr})}$$

$$= 1.599 \text{ yr} \times \frac{1 \text{ yr}}{(1.599 \text{ yr} + 1 \text{ yr})} = 0.6152 \text{ yr}$$

Step 3 Explain: Venus takes 0.6152 year to orbit the Sun. This number is easier to interpret if we convert it to days or months; you should confirm that it is equivalent to 224.7 days, or about $7\frac{1}{2}$ months. As we expect, it is shorter than Earth's orbital period of 1 year.

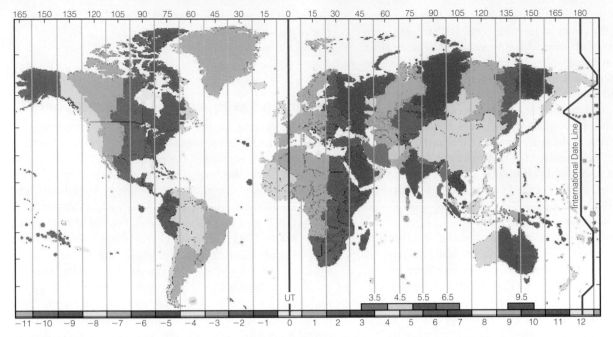

FIGURE S1.7 Time zones around the world. The numerical scale at the bottom shows hours ahead of (positive numbers) or behind (negative numbers) the time in Greenwich, England; the scale at the top is longitude. The vertical lines show standard time zones as they would be in the absence of political considerations. The color-coded regions show the actual time zones. Note, for example, that all of China uses the same standard time, even though the country is wide enough to span several time zones. Note also that a few countries use time zones centered on a half-hour (the upper set of colored bars), rather than an hour, relative to Greenwich time.

calendar in 46 B.C. This **Julian calendar** introduced the **leap year**: Every fourth year has 366 days, rather than 365, so that the average length of the calendar year is $365\frac{1}{4}$ days.

The Julian calendar originally had the spring equinox falling around March 24. If it had been perfectly synchronized with the tropical year, this calendar would have ensured that the spring equinox occurred on the same date every 4 years (that is, every leap-year cycle). It didn't work perfectly, however, because a tropical year is actually about 11 minutes short of $365\frac{1}{4}$ days. As a result, the moment of the spring equinox slowly advanced by an average of 11 minutes per year. By the late 16th century, the spring equinox was occurring on March 11.

Concerned by this drift in the date of the spring equinox, Pope Gregory XIII introduced a new calendar in 1582. This **Gregorian calendar** was much like the Julian calendar, with two important adjustments. First, Pope Gregory decreed that the day in 1582 following October 4 would be October 15. By eliminating the 10 dates from October 5 through October 14, 1582, he pushed the date of the spring equinox in 1583 from March 11 to March 21. (He chose March 21 because it was the date of the spring equinox in A.D. 325, which was the time of the Council of Nicaea, the first ecumenical council of the Christian church.) Second, the Gregorian calendar added an exception to the rule of having leap year every 4 years: Leap year is skipped when a century changes (for example, in years 1700, 1800, 1900) *unless* the century year is divisible by 400. Thus, 2000 was a leap year because it is divisible by 400 (2000 ÷ 400 × 5), but 2100 will *not* be a leap year. These adjustments make the average length of the Gregorian

calendar year almost exactly the same as the actual length of a tropical year, which ensures that the spring equinox will occur on March 21 every fourth year for thousands of years to come.

Today, the Gregorian calendar is used worldwide for international communication and commerce. (Many countries still use traditional calendars, such as the Chinese, Islamic, and Jewish calendars, for cultural purposes.) However, as you might guess, the pope's decree was not immediately accepted in regions not bound to the Catholic Church. For example, the Gregorian calendar was not adopted in England or in the American colonies until 1752, and it was not adopted in China until 1912 or in Russia until 1919.

S1.2 CELESTIAL COORDINATES AND MOTION IN THE SKY

We are now ready to turn our attention from timekeeping to navigation. The goal of celestial navigation is to use the Sun and the stars to find our position on Earth. Before we can do that, we first need to explore the apparent motions of the sky in more detail than we covered in Chapter 2.

How do we locate objects on the celestial sphere?

Recall from Chapter 2 that the celestial sphere is an illusion, but one that is quite useful when looking at the sky. We can make the celestial sphere even more useful by giving it a set of **celestial coordinates** that function much like the coordinates of latitude and longitude on Earth. Just as we can locate a city on Earth by its latitude and longitude, we can use an object's celestial coordinates to describe its precise location on the celestial sphere.

We have already discussed the key starting points for our coordinate system [**Section 2.1**]: the north and south celestial poles, the celestial equator, and the ecliptic (**FIGURE S1.8**). It is much easier to visualize the celestial sphere if you make your own model with a simple plastic ball. Use a felt-tip pen to mark the north and south celestial poles on your ball, and then add the celestial equator and the ecliptic. Note that the ecliptic crosses the celestial equator on opposite sides of

the celestial sphere at an angle of $23\frac{1}{2}°$ (because of the tilt of Earth's axis).

Equinoxes and Solstices Recall that the equinoxes and solstices are special moments that occur each year when Earth is at particular positions in its orbit (see Figure 2.15). These positions correspond to the apparent locations of the Sun along the ecliptic shown in Figure S1.8. For example, the spring equinox occurs at the moment when the Sun's path along the ecliptic crosses the celestial equator going from south to north, so we also use the term spring equinox to refer to this point on

SPECIAL TOPIC

Solar Days and the Analemma

The precise length of a solar day varies from its average of 24 hours for two reasons. The first is Earth's varying orbital speed. Recall that, in accord with Kepler's second law, Earth moves slightly faster—and therefore moves slightly farther along its orbit each day—when it is closer to the Sun in its orbit. The solar day therefore requires more than the average amount of "extra" rotation (see Figure S1.2) during these periods, making these solar days longer than average. Similarly, the solar day requires less than the average amount of "extra" rotation when it is in the portion of its orbit farther from the Sun, making these solar days shorter than average.

The second reason is the tilt of Earth's axis, which makes the ecliptic inclined $23\frac{1}{2}°$ to the celestial equator. Because the length of a solar day depends on the Sun's apparent *eastward* motion along the ecliptic, the inclination would cause solar days to vary in length even if Earth's orbit were perfectly circular. To see why, suppose the Sun appeared to move exactly 1° per day along the ecliptic. Around the times of the solstices, this motion would be entirely eastward,

making the solar day slightly longer than average. Around the times of the equinoxes, when the motion along the ecliptic has a significant northward or southward component, the solar day would be slightly shorter than average.

Together, the effects of varying orbital speed and tilt mean the actual length of a solar day can be up to about 25 seconds longer or shorter than the 24-hour average. Because the effects accumulate at particular times of year, the apparent solar time can differ by as much as 17 minutes from the mean solar time. The net result is often depicted visually by an **analemma** (**FIGURE 1**), which looks much like a figure 8. You'll find an analemma printed on many globes, and Figure 2.17 shows a photographic version.

You can use the horizontal scale on the analemma to convert between mean and apparent solar time for any date. (The vertical scale shows the Sun's declination.) For example, the dashed line shows that on November 10, a mean solar clock is about 17 minutes "behind the Sun," or behind apparent solar time; if the apparent solar time is 6:00 p.m. on November 10, the mean solar time is only 5:43 p.m. The discrepancy between mean and apparent solar times is called the **equation of time**. It is often plotted as a graph (**FIGURE 2**), which gives the same results as reading from the analemma.

The discrepancy between mean and apparent solar time also explains why the times of sunrise and sunset don't follow seasonal patterns perfectly. For example, the winter solstice around December 21 has the shortest daylight hours in the Northern Hemisphere, but the earliest sunset occurs around December 7, when the Sun is still well "behind" mean solar time.

FIGURE 1 The analemma shows the annual pattern of discrepancies between apparent and mean solar time. For example, the dashed red line shows that on November 10, a mean solar clock reads 17 minutes behind (earlier than) apparent solar time.

FIGURE 2 The discrepancies can also be plotted on a graph as the equation of time.

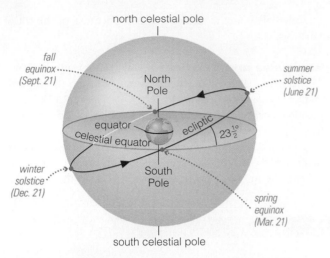

FIGURE S1.8 Schematic diagram of the celestial sphere without stars. The arrow along the ecliptic indicates the direction in which the Sun appears to move over the course of each year.

the celestial sphere. That is, the term *spring equinox* has a dual meaning: It is the *moment* when spring begins and the *point* on the ecliptic at which the Sun appears to be located at that moment. Figure S1.8 shows both the points on the celestial sphere and the approximate dates for each equinox and solstice.

SEE IT FOR YOURSELF

Using your plastic ball as a model of the celestial sphere (which you have already marked with the celestial poles, equator, and ecliptic), mark the locations and approximate dates of the equinoxes and solstices. Based on the dates for these points, approximately where along the ecliptic is the Sun on April 21? On November 21? How do you know?

You can find the locations of the equinoxes and solstices among the constellations with the aid of nearby bright stars (**FIGURE S1.9**). For example, the point marking the spring equinox is located in the constellation Pisces and can be found with the aid of the four bright stars that make up the "Great Square of Pegasus." Keep in mind that you can find this point any time it is above the horizon on a clear night, even though the Sun is located at this point only once each year (around March 21).

Celestial Coordinates We can now add a system of celestial coordinates to the celestial sphere. Let's begin by reviewing the two other coordinate systems we've used in this book: **FIGURE S1.10a** shows the coordinates of *altitude* and *direction* (or *azimuth**) that we use in the local sky, and **FIGURE S1.10b** shows the coordinates of *latitude* and *longitude* that we use on Earth's surface. Our system of celestial coordinates, called **declination (dec)** and **right ascension (RA)**, is shown in **FIGURE S1.10c**. Notice that declination on the celestial sphere is similar to latitude on Earth, and right ascension is similar to longitude.

To understand the correspondence better, notice the following key points about declination:

- Just as lines of latitude are parallel to Earth's equator, lines of declination are parallel to the celestial equator.

- Just as Earth's equator has lat $= 0°$, the celestial equator has dec $= 0°$.

---

*Azimuth is usually measured clockwise around the horizon from due north. By this definition, the azimuth of due north is $0°$, of due east is $90°$, of due south is $180°$, and of due west is $270°$.

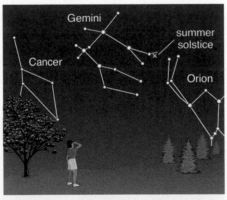

FIGURE S1.9 These diagrams show the locations among the constellations of the equinoxes and solstices. No bright stars mark any of these points, so you must find them by studying their positions relative to recognizable patterns. The time of day or night at which each point is above the horizon depends on the time of year.

a We use altitude and direction to pinpoint locations in the local sky.

b We use latitude and longitude to pinpoint locations on Earth.

c We use declination and right ascension to pinpoint locations on the celestial sphere.

FIGURE S1.10 Celestial coordinate systems.

- Latitude is labeled *north* or *south* relative to the equator, while declination is labeled *positive* or *negative*. For example, the North Pole has lat = 90°N, while the north celestial pole has dec = +90°; the South Pole has lat = 90°S, while the south celestial pole has dec = −90°.

We find a similar correspondence between right ascension and longitude:

- Just as lines of longitude extend from the North Pole to the South Pole, lines of right ascension extend from the north celestial pole to the south celestial pole.

- Just as there is no natural starting point for longitude, there is no natural starting point for right ascension. By international treaty, longitude zero (the prime meridian) is the line of longitude that runs through Greenwich, England. By convention, right ascension zero is the line of right ascension that runs through the spring equinox.

- Longitude is measured in *degrees* east or west of Greenwich, while right ascension is measured in *hours* (and minutes and seconds) east of the spring equinox. A full 360° circle around the celestial equator goes through 24 hours of right ascension, so each hour of right ascension represents an angle of 360° ÷ 24 = 15°.

As an example of how we use celestial coordinates to locate objects on the celestial sphere, consider the bright star Vega. Its coordinates are dec = +38°44′ and RA = 18^{h}35^m (**FIGURE S1.11**). The positive declination tells us that Vega is 38°44′ *north* of the celestial equator. The right ascension tells us that Vega is 18 hours 35 minutes east of the spring equinox. Translating the right ascension from hours to angular degrees, we find that Vega is about 279° east of the spring equinox (because 18 hours represents 18 × 15° = 270° and 35 minutes represents $\frac{35}{60}$ × 15° ≈ 9°).

We can also use the Vega example to see the benefit of measuring right ascension in units of time. All objects with a particular right ascension cross the meridian at the same time. For example, all stars with RA = 0^h cross the meridian at the same time the spring equinox crosses the meridian, all objects with RA = 1^h cross the meridian one hour after the spring equinox, and so on. Vega's right ascension of 18^{h}35^m tells us that it always crosses the meridian 18 hours 35 minutes after the spring equinox crosses the meridian. (This is 18 hours 35 minutes of *sidereal time* later, which is not exactly the same as 18 hours 35 minutes of solar time; see Mathematical Insight S1.2.) Generalizing, an object's right ascension tells us *how long after* the spring equinox the object crosses the meridian.

Note that while we generally think of declination and right ascension as fixed coordinates like latitude and longitude, they are not perfectly constant. Instead, they move slowly relative to distant stars because they are tied to the celestial equator, which moves gradually relative to the constellations with Earth's 26,000-year cycle of axis precession [**Section 2.2**]. (Axis precession does not affect Earth's orbit, so it does not affect the location of the ecliptic among the constellations.)

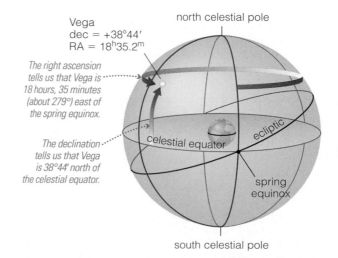

FIGURE S1.11 This diagram shows how we interpret the celestial coordinates of Vega.

Even over just a few decades, the resulting coordinate changes can be significant enough to make a difference in precise astronomical work, such as aiming a telescope at a particular object. As a result, careful observations require almost continual updating of celestial coordinates. Star catalogs therefore always state the year for which coordinates are given (for example, "epoch 2000"). Astronomical software can automatically calculate day-to-day celestial coordinates for any object in our sky.

Celestial Coordinates of the Sun Unlike stars, which remain fixed in the patterns of the constellations on the celestial sphere, the Sun moves gradually along the ecliptic. It takes a year for the Sun to make a full circuit of the ecliptic, which means it moves through all 24 hours of right ascension over the course of the year. The Sun therefore moves approximately one twelfth of the way around the ecliptic each month, meaning that its right ascension changes by about $24 \div 12 = 2$ hours per month. **FIGURE S1.12** shows the ecliptic marked with the Sun's monthly position and a scale of celestial coordinates. From this figure, we can create a table of the Sun's month-by-month celestial coordinates.

TABLE S1.1 starts from the spring equinox, when the Sun has declination 0° and right ascension 0^h. You can see in the shaded areas of the table that while RA advances steadily through the year, the Sun's declination changes much more slowly around the solstices than around the equinoxes. For

---

MATHEMATICAL INSIGHT S1.2

Time by the Stars

Our everyday clocks are set to solar time, ticking through 24 hours for each day of mean solar time. For astronomical observations, it is also useful to have clocks that tell time by the stars, or **sidereal time**. Just as *solar time* is defined according to the Sun's position relative to the meridian, *sidereal time* is based on the positions of stars relative to the meridian. We define the **hour angle (HA)** of any object on the celestial sphere to be the time *since* it last crossed the meridian (or the higher of its two meridian crossing points for a circumpolar star). For example:

- If a star is crossing the meridian *now*, its hour angle is 0^h.

- If a star crossed the meridian 3 hours ago, its hour angle is 3^h.

- If a star will cross the meridian 1 hour from now, its hour angle is -1^h or, equivalently, 23^h.

By convention, time by the stars is based on the hour angle of the spring equinox. That is, the **local sidereal time (LST)** is

$$LST = HA_{\text{spring equinox}}$$

For example, the local sidereal time is 00:00 when the spring equinox is *on* the meridian. Three hours later, when the spring equinox is 3 hours west of the meridian, the local sidereal time is 03:00.

Note that, because right ascension tells us how long after the spring equinox an object reaches the meridian, the local sidereal time is also equal to the right ascension (RA) of objects currently crossing your meridian. For example, if your local sidereal time is 04:30, stars with RA $= 4^h30^m$ are currently crossing your meridian. This idea leads to an important relationship among any object's current hour angle, the current local sidereal time, and the object's right ascension:

$$HA_{\text{object}} = LST - RA_{\text{object}}$$

This formula should make sense: The local sidereal time tells us how long it has been since the *spring equinox* was on the meridian and an object's right ascension tells us how long after the spring equinox it crosses the meridian. Therefore, the difference LST $-$ RA$_{\text{object}}$ must tell us how long it has been *since* the object crossed the meridian, which is the object's hour angle.

Sidereal time has one important subtlety: Sidereal clocks tick through 24 hours of sidereal time in one sidereal day, which is only about 23 hours 56 minutes of solar time. As a result, a sidereal hour is slightly shorter than a "normal" solar hour, and sidereal clocks gain about 4 minutes per day over solar clocks. Therefore, you cannot easily determine sidereal time from a solar clock. That is why astronomical observatories always have special sidereal clocks in addition to clocks that tell solar time.

EXAMPLE 1: Suppose the local solar time is 9:00 p.m. on the spring equinox (March 21). What is the local sidereal time?

SOLUTION:

Step 1 Understand: We are asked to find the local sidereal time, which is the hour angle of the spring equinox. We therefore need to know the current location of the spring equinox in the local sky. The key clue is that it is the day of the spring equinox, which is the one day on which the Sun is located in the same position as the spring equinox in the sky.

Step 2 Solve: We are told that the local solar time is 9:00 p.m., which means that the Sun is 9 hours past the meridian and therefore has an hour angle of 9 hours. Because the spring equinox and the Sun are located in the same place on this date, the spring equinox also has an hour angle of 9 hours.

Step 3 Explain: The hour angle of the spring equinox is 9 hours, which means the local sidereal time is LST $= 09:00$.

EXAMPLE 2: Suppose the local sidereal time is LST $= 04:00$. When will Vega (RA $= 18^h35^m$) cross the meridian?

SOLUTION:

Step 1 Understand: We are given the local sidereal time and Vega's right ascension, so we can use our formula to determine Vega's hour angle, which tells us its current position relative to the meridian.

Step 2 Solve: We put the given values into the formula to find Vega's hour angle:

$$HA_{\text{Vega}} = LST - RA_{\text{Vega}} = 4:00 - 18:35 = -14:35$$

Step 3 Explain: Vega's hour angle is -14 hours 35 minutes, which means Vega will cross your meridian 14 hours and 35 minutes of sidereal time from now. This also means that Vega crossed your meridian 9 hours and 25 minutes ago (because $14^h35^m + 9^h25^m = 24^h$).

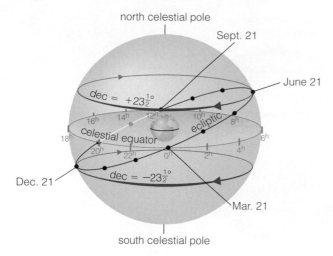

FIGURE S1.12 We can use this diagram of the celestial sphere to determine the Sun's right ascension and declination at monthly intervals.

TABLE S1.1 The Sun's Approximate Celestial Coordinates at 1-Month Intervals

Approximate Date	RA	Dec
Mar. 21 (spring equinox)	0^h	$0°$
Apr. 21	2^h	$+12°$
May 21	4^h	$+20°$
June 21 (summer solstice)	6^h	$+23\frac{1}{2}°$
July 21	8^h	$+20°$
Aug. 21	10^h	$+12°$
Sept. 21 (fall equinox)	12^h	$0°$
Oct. 21	14^h	$-12°$
Nov. 21	16^h	$-20°$
Dec. 21 (winter solstice)	18^h	$-23\frac{1}{2}°$
Jan. 21	20^h	$-20°$
Feb. 21	22^h	$-12°$

example, during the 2 months around the summer solstice (that is, between May 21 and July 21), the Sun's declination varies only between $+20°$ and its maximum of $+23\frac{1}{2}°$. In contrast, in the two months around the spring equinox, the Sun's declination changes by about 24°, from $-12°$ on February 21 to $+12°$ on April 21. These facts explain why the number of daylight hours increases rapidly in spring and decreases rapidly in fall, while remaining long and nearly constant for a couple of months around the summer solstice and short and nearly constant for a couple of months around the winter solstice.

SEE IT FOR YOURSELF

On your plastic ball model of the celestial sphere, add dots along the ecliptic to show the Sun's monthly positions. Based on your model, what are the Sun's approximate celestial coordinates on your birthday?

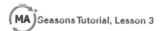
Seasons Tutorial, Lesson 3

How do stars move through the local sky?

Recall that Earth's rotation makes all celestial objects appear to circle around Earth each day (see Figure 2.9), but what we actually see in the local sky is more complex because we see only half the celestial sphere at one time (the ground blocks our view of the other half). We can now use our understanding of celestial coordinates to gain a deeper understanding of the local sky. As we'll see, the path of any star through your local sky depends only on (1) your latitude and (2) the declination of the star.

The Sky at the North Pole Let's begin by exploring the local sky at the North Pole, where the daily paths of stars are easiest to understand. **FIGURE S1.13a** shows your orientation relative to the celestial sphere when you are standing at the North Pole. Your "up" points toward the north celestial pole,

which therefore marks your zenith. Earth blocks your view of anything south of the celestial equator, which therefore runs along your horizon. To make it easier for you to visualize the local sky, **FIGURE S1.13b** shows your horizon extending to the celestial sphere. The horizon is marked with directions; note that all directions are south from the North Pole, which means we cannot define a meridian for the North Pole.

Notice that the daily circles of the stars keep them at constant altitudes above or below the North Polar horizon. Moreover, the altitude of any star is equal to its declination. For example, a star with declination $+60°$ circles the sky at an altitude of 60°, and a star with declination $-30°$ remains 30° below your horizon at all times. As a result, all stars north of the celestial equator are circumpolar at the North Pole, meaning that they never fall below the horizon. Stars south of the celestial equator can never be seen at the North Pole. (If you are having difficulty visualizing the star paths, it may help you to watch them as you rotate your plastic ball model of the celestial sphere.)

You should also notice that right ascension does not affect a star's path at all: The path depends only on declination. As we'll see shortly, this rule holds for all latitudes. Right ascension affects only the *time* of day and year at which a star is found in a particular position in your sky.

The Sky at the Equator Imagine that you are standing somewhere on Earth's equator (lat = 0°), such as in Ecuador, in Kenya, or on the island of Borneo. **FIGURE S1.14a** shows that "up" points directly away from (perpendicular to) Earth's rotation axis. **FIGURE S1.14b** shows the local sky more clearly by extending the horizon to the celestial sphere and rotating the diagram so the zenith is up. As it does everywhere except at the poles, the meridian extends from the horizon due south, through the zenith, to the horizon due north.

Look carefully at how the celestial sphere appears to rotate in the local sky. The north celestial pole remains stationary on your horizon due north, with its altitude equal to the equator's latitude of 0° [**Section 2.1**], and the south celestial

dec = +60°

dec = +30°

celestial equator

dec = 0°

dec = –30°

dec = –60°

north celestial pole

south celestial pole

North celestial pole
appears at zenith.

Any star circles the sky
daily at an altitude equal
to its declination.

dec = +60°

dec = +30°

S   S

dec = 0°

S   S

Celestial equator
circles the horizon.

dec = –30°

dec = –60°

south celestial pole

a The orientation of the local sky, relative to the celestial sphere, for an observer at the North Pole.

b Extending the horizon to the celestial sphere makes it easier to visualize the local sky at the North Pole.

FIGURE S1.13 The sky at the North Pole.

pole remains stationary on your horizon due south. Exactly half the celestial equator is visible, extending from the horizon due east, through the zenith, to the horizon due west. (The other half lies below the horizon.) As the equatorial sky appears to turn, all star paths rise straight out of the eastern horizon and set straight into the western horizon, with the following features:

- **Stars with dec = 0°** lie *on* the celestial equator and therefore rise due east, cross the meridian at the zenith, and set due west.

- **Stars with dec > 0°** rise north of due east, reach their highest point on the meridian in the north, and set north of due west. Their rise, set, and highest point depend on their declination. For example, a star with dec = +30° rises

30° north of due east, crosses the meridian 30° to the north of the zenith—that is, at an *altitude* of 90° − 30° = 60° in the north—and sets 30° north of due west.

- **Stars with dec < 0°** rise south of due east, reach their highest point on the meridian in the south, and set south of due west. For example, a star with dec = −50° rises 50° south of due east, crosses the meridian 50° to the south of the zenith—that is, at an *altitude* of 90° − 50° = 40° in the south—and sets 50° south of due west.

Because exactly half of any star's daily circle lies above the horizon, every star at the equator is above the horizon for exactly half of each sidereal day, or just under 12 hours (and below the horizon for the other half of the sidereal day).

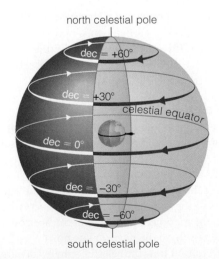

north celestial pole

dec = +60°

dec = +30°

celestial equator

dec = 0°

dec = –30°

dec = –60°

south celestial pole

a The orientation of the local sky, relative to the celestial sphere, for an observer at Earth's equator.

zenith

meridian

north celestial pole N

E

dec = +60°

dec = +30°

dec = 0°

celestial equator

W

dec = –30°

dec = –60°

south S celestial pole

b Extending the horizon and rotating the diagram make it easier to visualize the local sky at the equator.

FIGURE S1.14 The sky at the equator.

Skies at Other Latitudes Star tracks may at first seem more complex at other latitudes, with their mixtures of circumpolar stars and stars that rise and set. However, they are easy to understand if we apply the same basic strategy we've used for the North Pole and equator. Let's consider latitude 40°N, such as in Denver, Indianapolis, Philadelphia, or Beijing. First, as shown in **FIGURE S1.15a**, imagine standing at this latitude on a basic diagram of the rotating celestial sphere. Note that "up" points to a location on the celestial sphere with declination +40°. To make it easier to visualize the local sky, we next extend the horizon and rotate the diagram so that the zenith is up (**FIGURE S1.15b**).

As we expect, the north celestial pole appears 40° above the horizon due north, since its altitude in the local sky is always equal to your latitude. Half the celestial equator is visible, extending from the horizon due east, to the meridian at an altitude of 50° in the south, to the horizon due west. By comparing this diagram to that of the local sky for the equator, you'll notice the following general rule that applies to all latitudes except the poles:

The celestial equator always extends from due east on your horizon to due west on your horizon, crossing the meridian at an altitude of 90° minus your latitude.

The celestial equator crosses the meridian south of the zenith for locations in the Northern Hemisphere and north of the zenith for locations in the Southern Hemisphere.

If you study Figure S1.15b carefully, you'll notice the following features of the sky for latitude 40°N:

- **Stars with dec = 0°** lie *on* the celestial equator and therefore follow the celestial equator's path through the local sky. For latitude 40°N, these stars rise due east, cross the meridian at altitude 90° − 40° = 50° in the south, and set due west.

- **Stars with dec > (90° − lat)** are circumpolar. For latitude 40°N, stars with declination greater than 90° − 40° = 50° are circumpolar, because they lie *within* 40° of the north celestial pole.

- **Stars with dec > 0° that are not circumpolar** follow paths parallel to but north of the celestial equator: They rise north of due east and set north of due west, and cross the meridian to the north of the place where the celestial equator crosses it by an amount equal to their declination. For example, because the celestial equator at latitude 40° crosses the meridian at altitude 50° in the south, a star with dec = +30° crosses the meridian at altitude 50° + 30° = 80° in the south. Similarly, a star with dec = +60° crosses the meridian 60° farther north than the celestial equator, which means at altitude 70° in the north (because 50° + 60° = 110°, which means 20° past the zenith, which is 90° − 20° = 70°).

- **Stars with dec < (−90° + lat)** never rise above the horizon. For latitude 40°N, stars with declination less than −90° + 40° = −50° never rise above the horizon, because they lie within 40° of the south celestial pole.

- **Stars with dec < 0° that are sometimes visible** follow paths parallel to but south of the celestial equator: They rise south of due east and set south of due west, and cross the meridian south of the place where the celestial equator crosses it by an amount equal to their declination. For example, a star with dec = −30° crosses the meridian at altitude 50° − 30° = 20° in the south.

Note also that the fraction of any star's daily circle that is above the horizon—and hence the amount of time it is above the horizon each day—depends on its declination. Because

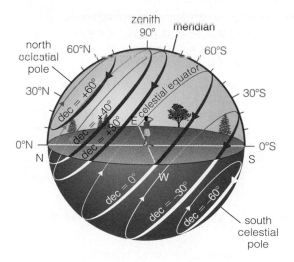

a The orientation of the local sky, relative to the celestial sphere, for an observer at latitude 40°N. Because latitude is the angle to Earth's equator, "up" points to the circle on the celestial sphere with declination +40°.

b Extending the horizon and rotating the diagram so that the zenith is up make it easier to visualize the local sky. The blue scale along the meridian shows altitudes and directions in the local sky.

FIGURE S1.15 The sky at 40°N latitude.

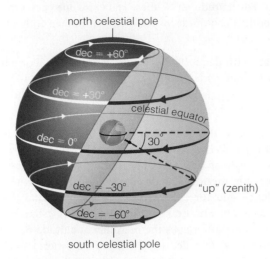

a The orientation of the local sky for an observer at latitude 30°S, relative to the celestial sphere. "Up" points to the circle on the celestial sphere with dec = –30°.

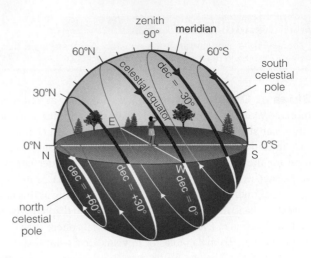

b Extending the horizon and rotating the diagram so that the zenith is up make it easier to visualize the local sky. Note that the south celestial pole is visible at altitude 30° in the south, while the celestial equator stretches across the northern half of the sky.

FIGURE S1.16 The sky at 30°S latitude.

exactly half the celestial equator is above the horizon, stars on the celestial equator (dec = 0°) are above the horizon for exactly half of each sidereal day, or about 12 hours. For northern latitudes like 40°N, stars with positive declinations have more than half their daily circles above the horizon and hence are above the horizon for more than 12 hours each day (with the range extending to 24 hours a day for the circumpolar stars). Stars with negative declinations have less than half their daily circles above the horizon and hence are above the horizon for less than 12 hours each day (with the range going to zero for stars that are never above the horizon).

We can apply the same strategy we used in Figure S1.15 to find star paths for other latitudes. **FIGURE S1.16** shows the local sky for latitude 30°S. Note that the south celestial pole is visible to the south and that the celestial equator passes through the northern half of the sky. If you study the diagram carefully, you can see how star tracks depend on declination.

How does the Sun move through the local sky?

Just as we've discussed for stars, the Sun's path on any particular day depends only on its declination and your latitude. However, because the Sun's declination changes over the course of the year, the Sun's path also changes.

FIGURE S1.17 shows the Sun's path on the equinoxes and solstices for latitude 40°N. On the equinoxes, the Sun is on

the celestial equator (dec = 0°) and therefore follows the celestial equator's path: It rises due east, crosses the meridian at altitude 50° in the south, and sets due west. Like other objects on the celestial equator, it is above the horizon for 12 hours. On the summer solstice, the Sun has dec = $+23\frac{1}{2}°$ (see Table S1.1) and therefore rises well north of due east,[*] reaches an altitude of $50° + 23\frac{1}{2}° = 73\frac{1}{2}°$ when it crosses the meridian in the south, and sets well north of due west. The daylight hours are long because much more than half the Sun's path is above the horizon. On the winter solstice, when the Sun has dec = $-23\frac{1}{2}°$, the Sun rises well south of due east, reaches an altitude of only $50° - 23\frac{1}{2}° = 26\frac{1}{2}°$ when it crosses the meridian in the south, and sets well south of due west.

---

[*]Calculating exactly how far north of due east the Sun rises is beyond the scope of this book, but astronomical software and websites can do these calculations for different latitudes.

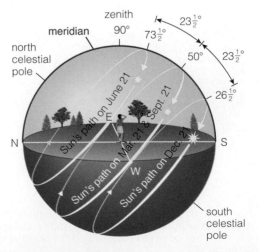

FIGURE S1.17 interactive figure The Sun's daily path on the equinoxes and solstices at latitude 40°N.

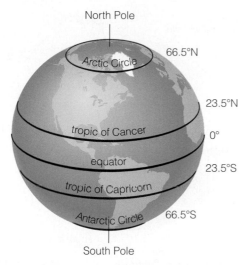

FIGURE S1.18 Special latitudes defined by the Sun's path through the sky.

The daylight hours are short because much less than half the Sun's path is above the horizon.

We could make a similar diagram to show the Sun's path on various dates for any latitude. However, the $23\frac{1}{2}°$ tilt of Earth's axis makes the Sun's path particularly interesting at the special latitudes shown in **FIGURE S1.18**. Let's investigate.

The Sun at the North and South Poles Recall that the celestial equator circles the horizon at the North Pole. **FIGURE S1.19** shows how we use this fact to find the Sun's path in the North Polar sky. Because the Sun appears *on* the celestial equator on the day of the spring equinox, the Sun circles the North Polar sky *on the horizon* on March 21, completing a full circle in 24 hours (1 solar day). Over the next 3 months, the Sun continues to circle the horizon each day, circling at gradually higher altitudes as its declination increases. It reaches its highest point on the summer solstice, when its declination of $+23\frac{1}{2}°$ means that it circles the North Polar sky at an altitude of $23\frac{1}{2}°$. After the summer solstice, the daily circles gradually fall lower over the next 3 months, reaching the horizon on the fall equinox.

Then, because the Sun's declination is negative for the next 6 months (until the following spring equinox), the Sun remains below the North Polar horizon. That is why the North Pole essentially has 6 months of daylight and 6 months of darkness, with an extended twilight that lasts a few weeks beyond the fall equinox and an extended dawn that begins a few weeks before the spring equinox.

The situation is the opposite at the South Pole. Here, the Sun's daily circle first reaches the horizon on the fall equinox. The daily circles then rise gradually higher, reaching a maximum altitude of $23\frac{1}{2}°$ on the *winter* solstice (when it is summer in the Antarctic), and then slowly fall back to the horizon on the spring equinox. That is, the South Pole has the Sun above the horizon during the 6 months it is below the North Polar horizon.

Two important caveats make the actual view from the Poles slightly different than we've described. First, the atmosphere bends light enough so that when the Sun is near the horizon, it *appears* to be about 1° higher than it really is, which means we can see the Sun even when it is slightly below the horizon. Second, the Sun's angular size of about $\frac{1}{2}°$ means that it does not fall below the horizon at a single moment but instead sets gradually. Together, these effects mean that the Sun *appears* above each polar horizon for several days longer than 6 months each year.

The Sun at the Equator At the equator, the celestial equator extends from the horizon due east, through the zenith, to the horizon due west. The Sun therefore follows this path on each equinox, reaching the zenith at local noon (**FIGURE S1.20**). Following the spring equinox, the Sun's increasing declination means that it follows a daily track that takes it gradually northward in the sky. It is farthest north on the summer solstice, when it rises $23\frac{1}{2}°$ north of due east, crosses the meridian at altitude $90° - 23\frac{1}{2}° = 66\frac{1}{2}°$ in the north, and sets $23\frac{1}{2}°$ north of due west. Over the next 6 months, it gradually tracks southward until the winter solstice, when its path is the mirror image (across the celestial equator) of its summer solstice path.

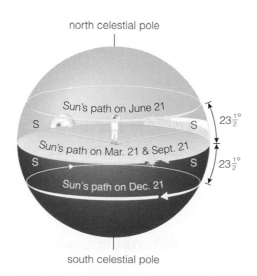

FIGURE S1.19 interactive figure Daily path of the Sun on the equinoxes and solstices at the North Pole.

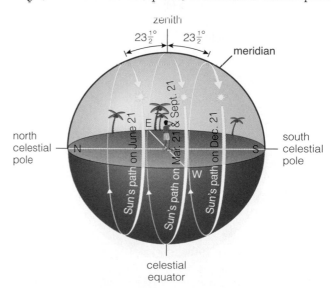

FIGURE S1.20 interactive figure Daily path of the Sun on the equinoxes and solstices at the equator.

Like all objects in the equatorial sky, the Sun is always above the horizon for half a day and below it for half a day. Moreover, the Sun's track is highest in the sky on the equinoxes and lowest on the summer and winter solstices. That is why equatorial regions do not have four seasons like temperate regions [**Section 2.2**]. The Sun's path in the equatorial sky also makes it rise and set perpendicular to the horizon every day of the year, making for a more rapid dawn and a briefer twilight than at other latitudes.

The Sun at the Tropics The circles of latitude 23.5°N and 23.5°S are called the **tropic of Cancer** and the **tropic of Capricorn**, respectively (see Figure S1.18). The region between these two circles, generally called the *tropics*, represents the parts of Earth where the Sun can sometimes reach the zenith at noon.

FIGURE S1.21 shows why the tropic of Cancer is special. The celestial equator extends from due east on the horizon to due west on the horizon, crossing the meridian in the south at an altitude of $90° - 23\frac{1}{2}°$ (the latitude) $= 66\frac{1}{2}°$, or $23\frac{1}{2}°$ short of the zenith. Therefore, the Sun reaches the zenith at local noon on the summer solstice, when it crosses the meridian $23\frac{1}{2}°$ northward of the celestial equator. The tropic of Cancer marks the northernmost latitude at which the Sun ever reaches the zenith. Similarly, at the tropic of Capricorn, the Sun reaches the zenith at local noon on the winter solstice (when it is summer for the Southern Hemisphere), making this the southernmost latitude at which the Sun ever reaches the zenith. Between the two tropic circles, the Sun passes through the zenith twice a year; the precise dates vary with latitude.

The Sun at the Arctic and Antarctic Circles At the equator, the Sun is above the horizon for 12 hours each day year-round. At latitudes progressively farther from the equator, the daily time that the Sun is above the horizon varies progressively more with the seasons. The special latitudes at which the Sun remains continuously above the horizon for a full day each year are the polar circles: the **Arctic Circle** at

FIGURE S1.21 interactive figure Daily path of the Sun on the equinoxes and solstices at the tropic of Cancer.

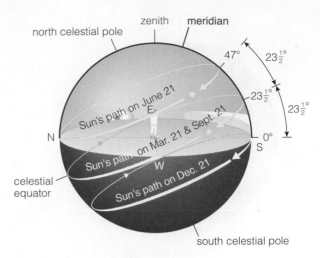

FIGURE S1.22 interactive figure Daily path of the Sun on the equinoxes and solstices at the Arctic Circle.

latitude 66.5°N and the **Antarctic Circle** at latitude 66.5°S (see Figure S1.18). Poleward of these circles, the length of continuous daylight (or darkness) increases beyond 24 hours, reaching the extreme of 6 months at the North and South Poles.

FIGURE S1.22 shows why the Arctic Circle is special. The celestial equator extends from due east on the horizon to due west on the horizon, crossing the meridian in the south at an altitude of $90° - 66\frac{1}{2}°$ (the latitude) $= 23\frac{1}{2}°$. As a result, the Sun's path is circumpolar on the summer solstice: The Sun skims the northern horizon at midnight, rises through the eastern sky to a noon maximum altitude of 47° in the south (which is the celestial equator's maximum altitude of $23\frac{1}{2}°$ plus the Sun's summer solstice declination of $23\frac{1}{2}°$), and then gradually falls through the western sky until it is back on the horizon at midnight (see Figure 2.18). At the Antarctic Circle, the Sun follows the same basic pattern on the winter solstice, except that it skims the horizon in the south and rises to a noon maximum altitude of 47° in the north.

Of course, what we *see* is subject to the same caveats we discussed for the North and South Poles: The bending of light by Earth's atmosphere and the Sun's angular size make the Sun *appear* to be slightly above the horizon even when it is slightly below it. As a result, at the Arctic Circle, the Sun seems not to set for several days around the summer solstice (rather than for a single day) and appears to peek above the horizon momentarily (rather than not at all) around the winter solstice. The same ideas hold for the opposite solstices at the Antarctic Circle.

S1.3 PRINCIPLES OF CELESTIAL NAVIGATION

Imagine that you're on a ship at sea, far from any landmarks. How can you figure out where you are? We now have all the background we need to answer this question.

How can you determine your latitude?

It's easy to determine your latitude if you can find the north or south celestial pole in your sky, because it is equal to the

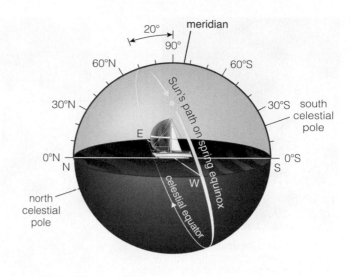

a Because Vega has dec = +38°44′, it crosses the meridian 38°44′ north of the celestial equator. Because Vega crosses the meridian at altitude 78°44′ in the south, the celestial equator must cross the meridian at altitude 40° in the south. Thus, the latitude must be 50°N.

b To determine latitude from the Sun's meridian crossing, you must know the Sun's declination, which you can determine from the date. The case shown is for the spring equinox, when the Sun's declination is 0° and hence the Sun follows the path of the celestial equator through the local sky. Because the celestial equator crosses the meridian at 70° in the north, the latitude must be 20°S.

FIGURE S1.23 Determining latitude from a star and from the Sun.

altitude of the celestial pole. In the Northern Hemisphere at night, you can determine your approximate latitude by measuring the altitude of Polaris, which lies within 1° of the north celestial pole. For example, if Polaris has altitude 17°, your latitude is between 16°N and 18°N.

If you want to be more precise, you can determine your latitude from the altitude of *any* star as it crosses your meridian. For example, suppose Vega happens to be crossing your meridian right now and it appears in your southern sky at altitude 78°44′. Because Vega has dec = +38°44′ (see Figure S1.11), it crosses your meridian 38°44′ north of the celestial equator. As shown in **FIGURE S1.23a**, you can conclude that the celestial equator crosses your meridian at an altitude of precisely 40° in the south. Your latitude must therefore be 50°N, because the celestial equator always crosses the meridian at an altitude of 90° minus the latitude. You know you are in the Northern Hemisphere because the celestial equator crosses the meridian in the south.

In the daytime, you can find your latitude from the Sun's altitude on your meridian if you know the date and the Sun's declination on that date. For example, suppose the date is March 21 and the Sun crosses your meridian at altitude 70° in the north (**FIGURE S1.23b**). Because the Sun has dec = 0° on March 21, you can conclude that the celestial equator also crosses your meridian in the north at altitude 70°. You must be in the Southern Hemisphere because the celestial equator crosses the meridian in the north. From the rule that the celestial equator crosses the meridian at an altitude of 90° minus the latitude, you can conclude that you are at latitude 20°S.

How can you determine your longitude?

You can determine your longitude by comparing the current position of an object in your sky with its position as seen

from some known longitude. As a simple example, suppose you use a sundial to determine that the apparent solar time is 1:00 p.m., which means the Sun passed the meridian 1 hour ago. You immediately call a friend in England and learn that it is 3:00 p.m. in Greenwich (or you carry a clock that keeps Greenwich time). You now know that your local time is 2 hours earlier than the local time in Greenwich, which means you are 2 hours west of Greenwich. (An earlier time means that you are *west* of Greenwich, because Earth rotates from west to east.) Each hour corresponds to 15° of longitude, so "2 hours west of Greenwich" means longitude 30°W.

At night, you can find your longitude by comparing the positions of stars in your local sky and at some known longitude. For example, suppose Vega is on your meridian and a call to your friend reveals that it won't cross the meridian in Greenwich until 6 hours from now. In this case, your local time is 6 hours later than the local time in Greenwich, which means you are 6 hours east of Greenwich, or at longitude 90°E (because 6 × 15° = 90°).

Celestial Navigation in Practice Although celestial navigation is easy in principle, at least three considerations make it more difficult in practice. First, finding either latitude or longitude requires a tool for measuring angles in the sky. One such device, called an *astrolabe,* was invented by the ancient Greeks and significantly improved by Islamic scholars during the Middle Ages. The astrolabe's faceplate (**FIGURE S1.24a**) could be used to tell time, because it consisted of a rotating star map and horizon plates for specific latitudes. Today you can buy similar rotatable star maps, called *planispheres.* Most astrolabes contained a sighting stick on the back that allowed users to measure the altitudes of bright stars in the sky. These measurements could then be correlated a The faceplate of an astrolabe. Many astrolabes had sighting sticks on the back for measuring positions of bright stars.

b A copper engraving of Italian explorer Amerigo Vespucci (for whom America was named) using an astrolabe to sight the Southern Cross. The engraving by Philip Galle, from the book *Nova Reperta*, was based on an original by Joannes Stradanus in the early 1580s.

c A woodcutting of Ptolemy holding a cross-staff (artist unknown).

d A sextant.

FIGURE S1.24 Navigational instruments.

against special markings under the faceplate (**FIGURE S1.24b**). Astrolabes were effective but difficult and expensive to make. As a result, medieval sailors often measured angles with a simple pair of calibrated perpendicular sticks, called a *cross-staff* or *Jacob's staff* (**FIGURE S1.24c**). A more modern device called a *sextant* allows much more precise angle determinations by incorporating a small telescope for sightings (**FIGURE S1.24d**). Sextants are still used for celestial navigation on many ships. If you want to practice celestial navigation yourself, you can buy an inexpensive plastic sextant at many science-oriented stores.

A second practical consideration is knowing the celestial coordinates of stars and the Sun so that you can determine their paths through the local sky. At night, you can use a table listing the celestial coordinates of bright stars. In addition to knowing the celestial coordinates, you must either know the constellations and bright stars extremely well or carry star charts to help you identify them. For navigating by the Sun in the daytime, you'll need a table listing the Sun's celestial coordinates on each day of the year.

The third practical consideration applies to determining longitude: You need to know the current position of the Sun (or a particular star) in a known location, such as Greenwich,

England. Although you could determine this by calling a friend who lives there, it's more practical to carry a clock set to universal time (the time in Greenwich). In the daytime, the clock makes it easy to determine your longitude. If apparent solar time is 1:00 p.m. in your location and the clock tells you that it is 3:00 p.m. in Greenwich, then you are 2 hours

COMMON MISCONCEPTIONS

Compass Directions

Most people determine direction with the aid of a compass rather than the stars. However, a compass needle doesn't actually point to true geographic north. Instead, the compass needle responds to Earth's magnetic field and points to *magnetic* north, which can be substantially different from true north. If you want to navigate precisely with a compass, you need a special map that takes into account local variations in Earth's magnetic field. Such maps are available at most camping stores. They are not perfectly reliable, however, because the magnetic field also varies with time. In general, celestial navigation is much more reliable for determining direction than using a compass.

west of Greenwich, or at longitude 30°W. The task is more difficult at night, because you must compare the position of a *star* in your sky to its current position in Greenwich. You can do this with the aid of detailed astronomical tables that allow you to determine the current position of any star in the Greenwich sky from the date and the universal time.

Historically, this third consideration created enormous problems for navigation. Before the invention of accurate clocks, sailors could easily determine their latitude but not their longitude. Indeed, most of the European voyages of discovery in the 15th century through the 17th century relied on little more than guesswork about longitude, although some sailors learned complex mathematical techniques for estimating longitude through observations of the lunar phases. More accurate longitude determination, upon which the development of extensive ocean commerce and travel depended, required the invention of a clock that would remain accurate on a ship rocking in the ocean swells. By the early 18th century, solving this problem was considered so important that the British government offered a substantial monetary prize for the solution. John

Harrison claimed the prize in 1761, with a clock that lost only 5 seconds during a 9-week voyage to Jamaica.*

The Global Positioning System Today, a new type of celestial navigation has supplanted traditional methods. It finds positions relative to satellites of the **global positioning system (GPS)**. In essence, these Earth-orbiting satellites function like artificial stars. The satellites' positions at any moment are known precisely from their orbital characteristics, and they transmit radio signals that can be picked up by GPS receivers in cars, smart phones, and other devices. Your GPS receiver locates three or more of the satellites and then does computations to calculate your position on Earth.

Navigation by GPS is so precise that the ancient practice of celestial navigation is in danger of becoming a lost art. Fortunately, many amateur clubs and societies are keeping the skills of celestial navigation alive.

---

*The story of the difficulties surrounding the measurement of longitude at sea and how Harrison finally solved the problem is chronicled in *Longitude*, by Dava Sobel (Walker and Company, 1995).

The Big Picture

Putting Chapter S1 into Context

In this chapter, we built upon concepts from the first three chapters to form a more detailed understanding of celestial timekeeping and navigation. You also learned how to determine paths for the Sun and the stars in the local sky. As you look back at what you've learned, keep in mind the following "big picture" ideas:

- Our modern systems of timekeeping are rooted in the apparent motions of the Sun through the sky. Although it's easy to forget these roots when you look at a clock or a calendar, the sky was the only guide to time for most of human history.

- The term *celestial navigation* sounds a bit mysterious, but it refers to simple principles that allow you to determine your location on Earth. Even if you're never lost at sea, you may find the basic techniques of celestial navigation useful to orient yourself at night (for example, on your next camping trip).

- If you understand the apparent motions of the sky discussed in this chapter and also learn the constellations and bright stars, you'll feel very much "at home" under the stars at night.

SUMMARY OF KEY CONCEPTS

S1.1 ASTRONOMICAL TIME PERIODS

- **How do we define the day, month, year, and planetary periods?** Each of these is defined in two ways. A **sidereal day** is Earth's rotation period, which is about 4 minutes shorter than the 24-hour **solar day** from noon one day to noon the next day. A **sidereal month** is the Moon's orbital period of about $27\frac{1}{3}$ days; a **synodic month** is the $29\frac{1}{2}$ days required for the Moon's cycle of phases. A **sidereal year** is Earth's orbital period, which is about 20 minutes longer than the **tropical year** from one spring

  equinox to the next. A planet's **sidereal period** is its orbital period, and its **synodic period** is the time from one opposition or conjunction to the next.

- **How do we tell the time of day?** There are several time measurement systems. **Apparent solar time** is based on the Sun's position in the local sky. **Mean solar time** is also local, but it averages the changes in the Sun's rate of motion over the year. **Standard time** and **daylight saving time** divide the world into

time zones. **Universal time** is the mean solar time in Greenwich, England.

- **When and why do we have leap years?** We usually have a **leap year** every 4 years because the length of the year is about $365\frac{1}{4}$ days. However, it is not exactly $365\frac{1}{4}$ days, so our calendar skips a leap year in century years not divisible by 400.

S1.2 CELESTIAL COORDINATES AND MOTION IN THE SKY

- **How do we locate objects on the celestial sphere?**

  Declination is given as an angle describing an object's position north or south of the celestial equator. **Right ascension**, usually measured in hours (and minutes and seconds), tells us how far east an object is located relative to the spring equinox.

■ **How do stars move through the local sky?** A star's path

through the local sky depends on its declination and your latitude. Latitude tells you the orientation of your sky relative to the celestial sphere, while declination tells you how a particular star's path compares to the path of the celestial equator through your sky.

■ **How does the Sun move through the local sky?** The

Sun's path also depends on its declination and your latitude, but it varies throughout the year because of the Sun's changing declination. The Sun's varying path helps define special latitudes, including the **tropics of Cancer** and **Capricorn** and the **Arctic** and **Antarctic Circles**.

S1.3 PRINCIPLES OF CELESTIAL NAVIGATION

■ **How can you determine your latitude?** You can deter-

mine your latitude from the altitude of the celestial pole in your sky or by measuring the altitude and knowing the declination of a star (or the Sun) as it crosses your meridian.

■ **How can you determine your longitude?** To determine longitude you must know the position of the Sun or a star in your sky and its position at the same time in the sky of Greenwich, England (or some other specific location). This is most easily done if you have a clock that tells universal time.

VISUAL SKILLS CHECK

Use the following questions to check your understanding of some of the many types of visual information used in astronomy. Answers are provided Appendix J. For additional practice, try the Chapter S1 Visual Quiz at MasteringAstronomy®.

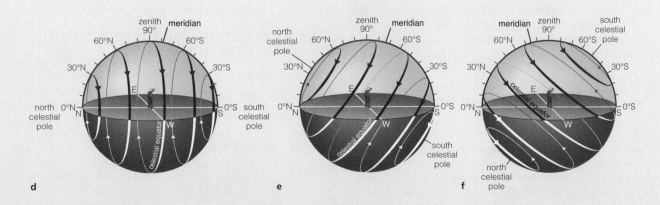

The six diagrams represent the sky at six different latitudes. Answer the following questions about them.

1. Which diagram represents the paths of stars at the North Pole?
2. Which diagram represents the paths of stars at the South Pole?
3. Which diagrams represent Southern Hemisphere skies?
4. What latitude is represented in diagram c?
5. Which diagram(s) represent(s) a latitude at which the Sun sometimes passes directly overhead?
6. Which diagram(s) represent(s) a latitude at which the Sun sometimes remains below the horizon during a full 24-hour period?
7. Each diagram shows five star circles. Look at the first circle to the north of the celestial equator on each diagram. Can you characterize the *declination* of stars on this circle? If so, what is it? Can you characterize the *right ascension* of stars on this circle? If so, what is it?

EXERCISES AND PROBLEMS

For instructor-assigned homework go to MasteringAstronomy®.

MasteringAstronomy®

REVIEW QUESTIONS

Short-Answer Questions Based on the Reading

1. Briefly explain the differences between (a) a *sidereal day* and a *solar day*, (b) a *sidereal month* and a *synodic month*, (c) a *sidereal year* and a *tropical year*, (d) a planet's *sidereal period* and its *synodic period*.
2. Define *opposition, conjunction*, and *greatest elongation* for planets. Explain both for planets closer than Earth to the Sun and for planets farther than Earth from the Sun.
3. Under what circumstances do we see a *transit* of a planet across the Sun?
4. Distinguish among *apparent solar time, mean solar time, standard time, daylight saving time*, and *universal time*.
5. Describe the origins of the Julian and Gregorian calendars. Which one do we use today?
6. What do we mean when we say the equinoxes and solstices are points on the celestial sphere? How are these points related to the times of year called the equinoxes and solstices?
7. What are *declination* and *right ascension*? How are these celestial coordinates similar to latitude and longitude on Earth? How are they different?
8. How and why do the Sun's celestial coordinates change over the course of each year?
9. Suppose you are at the North Pole. Where is the celestial equator? Where is the north celestial pole? Describe the daily motion of the sky. Do the same for the equator and for latitude 40°N.
10. Describe the Sun's path through the local sky on the equinoxes and on the solstices for latitude 40°N. Do the same for the North Pole, South Pole, and equator.
11. What is special about the tropics of Cancer and Capricorn? Describe the Sun's path on the solstices at these latitudes. Do the same for the Arctic and Antarctic Circles.
12. Briefly describe how you can use the Sun or stars to determine your latitude and longitude.

TEST YOUR UNDERSTANDING

Does It Make Sense?

Decide whether the statement makes sense (or is clearly true) or does not make sense (or is clearly false). Explain clearly; not all of these have definitive answers, so your explanation is more important than your chosen answer. (Hint: For statements that involve coordinates—such as altitude, longitude, or declination—check whether the correct coordinates are used for the situation. For example, it does not make sense to describe a location on Earth by an altitude, because altitude only describes positions of objects in the local sky.)

13. Last night I saw Venus shining brightly on the meridian at midnight.
14. The apparent solar time was noon, but the Sun was just setting.
15. My mean solar clock said it was 2:00 p.m., but a friend who lives east of here had a mean solar clock that said it was 2:11 p.m.
16. When the standard time is 3:00 p.m. in Baltimore, it is 3:15 p.m. in Washington, D.C.
17. Last night around 8:00 p.m., I saw Jupiter at an altitude of 45° in the south.
18. The latitude of the stars in Orion's belt is about 5°N.
19. Today the Sun is at an altitude of 10° on the celestial sphere.
20. Los Angeles is west of New York by about 3 hours of right ascension.
21. The summer solstice is east of the vernal equinox by 6 hours of right ascension.
22. Even though my UT clock had stopped, I was able to find my longitude by measuring the altitudes of 14 different stars in my local sky.

Quick Quiz

Choose the best answer to each of the following. Explain your reasoning with one or more complete sentences.

23. The time from one spring equinox to the next is the (a) sidereal day. (b) tropical year. (c) synodic month.
24. Jupiter is brightest when it is (a) at opposition. (b) at conjunction. (c) closest to the Sun in its orbit.
25. Venus is easiest to see in the evening when it is (a) at superior conjunction. (b) at inferior conjunction. (c) at greatest eastern elongation.
26. In the winter, your wristwatch tells (a) apparent solar time. (b) standard time. (c) universal time.
27. A star that is located 30° north of the celestial equator has (a) declination = 30°. (b) right ascension = 30°. (c) latitude = 30°.
28. A star's path through your sky depends on your latitude and the star's (a) declination. (b) right ascension. (c) both declination and right ascension.
29. At latitude 50°N, the celestial equator crosses the meridian at altitude (a) 50° in the south. (b) 50° in the north. (c) 40° in the south.
30. At the North Pole on the summer solstice, the Sun (a) remains stationary in the sky. (b) reaches the zenith at noon. (c) circles the horizon at altitude $23\frac{1}{2}°$.
31. If you know a star's declination, you can determine your latitude if you also (a) measure its altitude when it crosses the meridian. (b) measure its right ascension. (c) know the universal time.
32. If you measure the Sun's position in your local sky, you can determine your longitude if you also (a) measure its altitude when it crosses the meridian. (b) know its right ascension and declination. (c) know the universal time.

PROCESS OF SCIENCE

Examining How Science Works

33. *Transits and the Geocentric Universe.* Ancient people could not observe transits of Mercury or Venus across the Sun, because they lacked instruments for viewing a small dark spot against the Sun. But suppose they could have seen transits. How would transit observations have affected the debate over an Earth-centered versus Sun-centered solar system? Explain.

34. *Geometry and Science.* As discussed in Mathematical Insight S1.1, Copernicus found that a Sun-centered model led him to a simple geometric layout for the solar system, a fact that gave him confidence that his model was on the right track. Did the mathematics actually prove that the Sun-centered model was correct? Use your answer to briefly discuss the role of mathematics in science.

GROUP WORK EXERCISE

35. *Find Your Way Home.* Assign the following roles to the people in your group: *Scribe* (takes notes on the group's activities), *Proposer* (proposes methods for the group), *Skeptic* (points out weaknesses in proposed methods), and *Moderator* (leads group discussion and makes sure everyone contributes). You are an international spy who has been captured by a criminal mastermind and flown for many hours to a secret compound, which could be anywhere in the world. You escape … but all you have is a watch (set to your previous local time), a star chart, and a world map with longitude and latitude marked on it. How do you figure out where you are? How could you use celestial navigation to find your way back home?

INVESTIGATE FURTHER

In-Depth Questions to Increase Your Understanding

Short-Answer/Essay Questions

36. *Opposite Rotation.* Suppose Earth rotated in a direction opposite to its orbital direction; that is, suppose it rotated clockwise (as seen from above the North Pole) but orbited counterclockwise. Would the solar day still be longer than the sidereal day? Explain.

37. *No Precession.* Suppose Earth's axis did *not* precess. Would the sidereal year still be different from the tropical year? Explain.

38. *The Sun from Mars.* Mars has an axis tilt of 25.2°, only slightly larger than that of Earth. Compared to that on Earth, is the range of latitudes on Mars for which the Sun can reach the zenith larger or smaller? Is the range of latitudes for which the Sun is circumpolar larger or smaller? Make a sketch of Mars similar to the one for Earth in Figure S1.18.

39. *Fundamentals of Your Local Sky.* Answer each of the following for *your* latitude. **a.** Where is the north (or south) celestial pole in your sky? **b.** Describe the meridian in your sky, specifying at least three distinct points along it (such as the points at which it meets your horizon and its highest point). **c.** Describe the celestial equator in your sky, specifying at least three distinct points along it. **d.** Does the Sun ever appear at your zenith? If so, when? If not, why not? **e.** What range of declinations makes a star circumpolar in your sky? **f.** What is the range of declinations for stars that you can never see in your sky?

40. *Sydney Sky.* Repeat Problem 39 for the local sky in Sydney, Australia (latitude 34°S).

41. *Local Path of the Sun.* Describe the path of the Sun through your local sky for each of the following days: **a.** the spring and fall equinoxes. **b.** the summer (June) solstice. **c.** the winter (December) solstice. **d.** today. (*Hint:* You can estimate the Sun's RA and dec for today's date from data in Table S1.1.)

42. *Sydney Sun.* Repeat Problem 41 for the local sky in Sydney, Australia (latitude 34°S).

Quantitative Problems

Be sure to show all calculations clearly and state your final answers in complete sentences.

43. *Lost at Sea I.* During a vacation, you decide to take a solo boat trip. While contemplating the universe, you lose track of your location. Fortunately, you have some astronomical tables and instruments, as well as a UT clock. You thereby put together the following description of your situation:
    - It is the spring equinox.
    - The Sun is on your meridian at altitude 75° in the south.
    - The UT clock reads 22:00.
    a. What is your latitude? How do you know? **b.** What is your longitude? How do you know? **c.** Consult a map. Based on your position, where is the nearest land? Which way should you sail to reach it?

44. *Lost at Sea II.* Repeat Problem 43 for this situation:
    - It is the day of the summer solstice.
    - The Sun is on your meridian at altitude $67\frac{1}{2}°$ in the north.
    - The UT clock reads 06:00.

45. *Lost at Sea III.* Repeat Problem 43 for this situation:
    - Your local time is midnight.
    - Polaris appears at altitude 67° in the north.
    - The UT clock reads 01:00.

46. *Lost at Sea IV.* Repeat Problem 43 for this situation:
    - Your local time is 6 a.m.
    - From the position of the Southern Cross, you estimate that the south celestial pole is at altitude 33° in the south.
    - The UT clock reads 11:00.

47. *Orbital and Synodic Periods.* Use each object's given synodic period to find its actual orbital period. **a.** Saturn, synodic period = 378.1 days **b.** Mercury, synodic period = 115.9 days **c.** An asteroid with synodic period = 429 days

48. *Using the Analemma.* **a.** It's February 15 and your sundial tells you the apparent solar time is 18 minutes until noon. What is the mean solar time? **b.** It's July 1 and your sundial tells you that the apparent solar time is 3:30 p.m. What is the mean solar time?

49. $HA = LST - RA$. **a.** It is 4 p.m. on the spring equinox. What is the local sidereal time? **b.** The local sidereal time is 19:30. When will Vega cross your meridian? **c.** You observe a star that has an hour angle of 13 hours (13^h) when the local sidereal time is 8:15. What is the star's right ascension? **d.** The Orion Nebula has declination of about −5.5° and right ascension of 5^h25^m. If you are at latitude 40°N and the local sidereal time is 7:00, approximately where does the Orion Nebula appear in your sky?

50. *Meridian Crossings of the Moon and Phobos.* Estimate the time between meridian crossings of the Moon for a person standing on Earth. Repeat your calculation for meridian crossings of the Martian moon Phobos. Use the Appendixes in the back of the book if necessary.

51. *Mercury's Rotation Period.* Mercury's sidereal day is approximately $\frac{2}{3}$ of its orbital period, or about 58.6 days. Estimate the length of Mercury's solar day. Compare it to Mercury's orbital period of about 88 days.

Discussion Questions

52. *Northern Chauvinism.* Why is the solstice in June called the *summer solstice*, when it marks winter for places like Australia, New Zealand, and South Africa? Why is the writing on maps and globes usually oriented so that the Northern Hemisphere is at the top, even though there is no up or down in space? Discuss.

53. *Celestial Navigation.* Briefly discuss how you think the benefits and problems of celestial navigation might have affected ancient sailors. For example, how did they benefit from using the north celestial pole to tell directions, and what problems did they experience because of the difficulty in determining longitude? Can you explain why ancient sailors generally hugged coastlines as much as possible on their voyages? What dangers did this type of sailing pose? Why did the Polynesians become the best navigators of their time?

Web Projects

54. *Sundials.* Although they are no longer necessary for timekeeping, sundials remain popular for their cultural and artistic value. Search the Web for pictures and information about sundials around the world. Write a short report about three sundials that you find particularly interesting.

55. *Calendar History.* Investigate the history of the Julian or Gregorian calendar in greater detail. Write a short summary of an interesting aspect of the history you learn from your Web research. (For example, why did Julius Caesar allow one year to have 445 days? How did our months end up with 28, 30, or 31 days?)

56. *Global Positioning System.* Learn more about the global positioning system and its uses. Write a short report summarizing how new uses of GPS may affect our lives over the next 10 years.

Our perspective on the universe has changed dramatically throughout human history. This timeline summarizes some of the key discoveries that have shaped our modern perspective.

Stonehenge

Earth-centered model of the universe

Galileo's telescope

< 2500 B.C.	400 B.C. −170 A.D.	1543–1648 A.D.

(1) Ancient civilizations recognized patterns in the motion of the Sun, Moon, planets, and stars through our sky. They also noticed connections between what they saw in the sky and our lives on Earth, such as the cycles of seasons and of tides [Section 3.1].

(2) The ancient Greeks tried to explain observed motions of the Sun, Moon, and planets using a model with Earth at the center, surrounded by spheres in the heavens. The model explained many phenomena well, but could explain the apparent retrograde motion of the planets only with the addition of many complex features—and even then, its predictions were not especially accurate [Section 3.2].

(3) Copernicus suggested that Earth is a planet orbiting the Sun. The Sun-centered model explained apparent retrograde motion simply, though it made accurate predictions only after Kepler discovered his three laws of planetary motion. Galileo's telescopic observations confirmed the Sun-centered model, and revealed that the universe contains far more stars than had been previously imagined [Section 3.3].

Earth's rotation around its axis leads to the daily east-to-west motions of objects in the sky.

The tilt of Earth's rotation axis leads to seasons as Earth orbits the Sun.

Planets are much smaller than the Sun. At a scale of 1 to 10 billion, the Sun is the size of a grapefruit, Earth is the size of a ball point of a pen, and the distance between them is about 15 meters.

Yerkes Observatory

Edwin Hubble at the Mt. Wilson telescope

Hubble Space Telescope

1838–1920 A.D.   1924–1929 A.D.   1990 A.D.–present

(4) Larger telescopes and photography made it possible to measure the parallax of stars, offering direct proof that Earth really does orbit the Sun and showing that even the nearest stars are light-years away. We learned that our Sun is a fairly ordinary star in the Milky Way [Sections 2.4, 15.1].

(5) Edwin Hubble measured the distances of galaxies, showing that they lay far beyond the bounds of the Milky Way and proving that the universe is far larger than our own galaxy. He also discovered that more distant galaxies are moving away from us faster, telling us that the entire universe is expanding and suggesting that it began in an event we call the Big Bang [Sections 1.3, 20.2].

(6) Improved measurements of galactic distances and the rate of expansion have shown that the universe is about 14 billion years old. These measurements have also revealed still-unexplained surprises, including evidence for the existence of mysterious "dark matter" and "dark energy" [Sections 1.3, 22.1].

Distances between stars are enormous. At a scale of 1 to 10 billion, you can hold the Sun in your hand, but the nearest stars are thousands of kilometers away.

Our solar system is located about 27,000 light-years from the center of the Milky Way Galaxy.

The Milky Way Galaxy contains over 100 billion stars.

The observable universe contains over 100 billion galaxies.

4 MAKING SENSE OF THE UNIVERSE
UNDERSTANDING MOTION, ENERGY, AND GRAVITY

If I have seen farther than others, it is because I have stood on the shoulders of giants.

—Isaac Newton

The history of the universe is essentially a story about the interplay between matter and energy. This interplay began in the Big Bang and continues today in everything from the microscopic jiggling of atoms to gargantuan collisions of galaxies. Understanding the universe therefore depends on becoming familiar with how matter responds to the ebb and flow of energy.

You might guess that it would be difficult to understand the many interactions that shape the universe, but we now know that just a few physical laws govern the movements of everything from atoms to galaxies. The Copernican revolution spurred the discovery of these laws, and Galileo deduced some of them from his experiments. But it was Sir Isaac Newton who put all the pieces together into a simple system of laws describing both motion and gravity.

In this chapter, we'll discuss Newton's laws of motion, the laws of conservation of angular momentum and of energy, and the universal law of gravitation. By understanding these laws, you will be able to make sense of many of the wide-ranging phenomena you will encounter as you study astronomy.

4.1 DESCRIBING MOTION: EXAMPLES FROM DAILY LIFE

Think about what happens when you throw a ball to a dog. The ball leaves your hand, traveling in some particular direction at some particular speed. During its flight, the ball is pulled toward Earth by gravity, slowed by air resistance, and pushed by gusts of wind. Despite the complexity of the ball's motion, the dog still catches it.

We humans can perform an even better trick: We have learned how to figure out where the ball will land even before throwing it. In fact, we can use the same basic trick to predict the motions of objects throughout the universe, and we can perform it with such extraordinary precision that we can land a spaceship on target on Mars after sending it on a journey of hundreds of millions of kilometers.

Our primary goal in this chapter is to understand how humans have learned to make sense of motion in the universe. We all have experience with motion and a natural intuition as to what motion is, but in science we need to define our ideas and terms precisely. In this section, we'll use examples from everyday life to explore some of the fundamental ideas of motion.

How do we describe motion?

You are probably familiar with common terms used to describe motion in science, such as *velocity, acceleration,* and *momentum.* However, their scientific definitions may differ subtly from those you use in casual conversation. Let's investigate the precise meanings of these terms.

Speed, Velocity, and Acceleration A car provides a good illustration of the three basic terms that we use to describe motion:

- The **speed** of the car tells us how far it will go in a certain amount of time. For example, "100 kilometers per hour" (about 60 miles per hour) is a speed, and it tells us that the car will cover a distance of 100 kilometers if it is driven at this speed for an hour.

- The **velocity** of the car tells us both its speed and its direction. For example, "100 kilometers per hour going due north" describes a velocity.

- The car has an **acceleration** if its velocity is changing in any way, whether in speed or direction or both.

Note that while we normally think of *acceleration* as an increase in speed, in science we also say that you are accelerating when you slow down or turn (**FIGURE 4.1**). Slowing represents a negative acceleration, causing your velocity to decrease. Turning means a change in direction—which therefore means a change in velocity—so turning is a form of acceleration even if your speed remains constant.

You can often feel the effects of acceleration. For example, as you speed up in a car, you feel yourself being pushed back into your seat. As you slow down, you feel yourself being pulled forward. As you drive around a curve, you feel yourself being pushed away from the direction of your turn. In contrast, you don't feel such effects when moving at *constant velocity.* That is why you don't feel any sensation of motion when you're traveling in an airplane on a smooth flight.

The Acceleration of Gravity One of the most important types of acceleration is the acceleration caused by gravity. In a legendary experiment in which he supposedly dropped

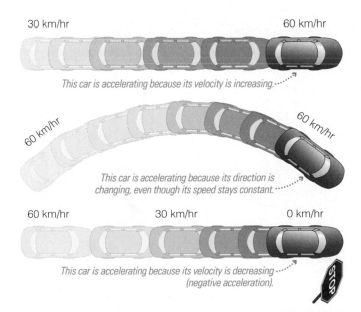

FIGURE 4.1 Speeding up, turning, and slowing down are all examples of acceleration.

weights from the Leaning Tower of Pisa, Galileo demonstrated that gravity accelerates all objects by the same amount, regardless of their mass. This fact may be surprising because it seems to contradict everyday experience: A feather floats gently to the ground, while a rock plummets. However, air resistance causes this difference in acceleration. If you dropped a feather and a rock on the Moon, where there is no air, both would fall at exactly the same rate.

The acceleration of a falling object is called the **acceleration of gravity**, abbreviated *g*. On Earth, the acceleration of gravity causes falling objects to fall faster by 9.8 meters per second (m/s), or about 10 m/s, with each passing second. For example, suppose you drop a rock from a tall building. At the moment you let it go, its speed is 0 m/s. After 1 second, the rock will be falling downward at about 10 m/s. After 2 seconds, it will be falling at about 20 m/s. In the absence of air resistance, its speed will continue to increase by about 10 m/s each second until it hits the ground (**FIGURE 4.2**). We therefore say that the acceleration of gravity is about 10 *meters per second per second*, or 10 *meters per second squared*, which we write as 10 m/s^2 (more precisely, $g = 9.8$ m/s^2).

Momentum and Force The concepts of speed, velocity, and acceleration describe how an individual object moves, but most of the interesting phenomena we see in the universe

Acceleration of gravity: Downward velocity increases by about 10 m/s with each passing second. (Gravity does not affect horizontal velocity.)

t = 0
v = 0

t = 1 s
v ≈ 10 m/s

t = 2 s
v ≈ 20 m/s

t = time
v = velocity (downward)

FIGURE 4.2 On Earth, gravity causes an unsupported object to accelerate downward at about 10 m/s^2, which means its downward velocity increases by about 10 m/s with each passing second. (Gravity does not affect horizontal velocity.)

result from interactions between objects. We need two additional concepts to describe these interactions:

- An object's **momentum** is the product of its mass and velocity; that is, momentum = mass × velocity.

- The only way to change an object's momentum is to apply a **force** to it.

We can understand these concepts by considering the effects of collisions. Imagine that you're stopped in your car at a red light when a bug flying at a velocity of 30 km/hr due south slams into your windshield. What will happen to your car? Not much, except perhaps a bit of a mess on your windshield. Next, imagine that a 2-ton truck runs the red light and hits you head-on with the same velocity as the bug. Clearly, the truck will cause far more damage. We can understand why by considering the momentum and force in each collision.

Before the collisions, the truck's much greater mass means it has far more momentum than the bug, even though both the truck and the bug are moving with the same velocity. During the collisions, the bug and the truck each transfer some of their momentum to your car. The bug has very little momentum to give to your car, so it does not exert much of a force. In contrast, the truck imparts enough of its momentum to cause a dramatic and sudden change in your car's momentum. You feel this sudden change in momentum as a force, and it can do great damage to you and your car.

The mere presence of a force does not always cause a change in momentum. For example, a moving car is always affected by forces of air resistance and friction with the road—forces that will slow your car if you take your foot off the gas pedal. However, you can maintain a constant velocity, and hence constant momentum, if you step on the gas pedal hard enough to overcome the slowing effects of these forces.

In fact, forces of some kind are always present, such as the force of gravity or the electromagnetic forces acting between atoms. The **net force** (or *overall force*) acting on an object represents the combined effect of all the individual forces put together. There is no net force on your car when you are driving at constant velocity, because the force generated by the engine to turn the wheels precisely offsets the forces of air resistance and road friction. A change in momentum occurs only when the net force is not zero.

Changing an object's momentum means changing its velocity, as long as its mass remains constant. A net force that is not zero therefore causes an object to accelerate. Conversely, whenever an object accelerates, a net force must be causing the acceleration. That is why you feel forces (pushing you forward, backward, or to the side) when you accelerate in your car. We can use the same ideas to understand many astronomical processes. For example, planets are always accelerating as they orbit the Sun, because their direction of travel constantly changes as they go around their orbits. We can therefore conclude that some force must be causing this acceleration. As we'll discuss shortly, Isaac Newton identified this force as gravity.

Moving in Circles Think about an ice skater spinning in place. She isn't going anywhere, so she has no overall velocity and hence no overall momentum. Nevertheless, every part of her body is moving in a circle as she spins, so these parts have momentum even though her overall momentum is zero. Is there a way to describe the total momentum from each part of her body as she spins? Yes—we say that her spin gives her **angular momentum**, which you can also think of as "circling momentum" or "turning momentum." (The term *angular* arises because a complete circle turns through an *angle* of 360°.)

Any object that is either spinning or moving along a curved path has angular momentum, which makes angular momentum very important in astronomy. For example, Earth has angular momentum due to its rotation (its *rotational angular momentum*) and to its orbit around the Sun (its *orbital angular momentum*).

Because angular momentum is a special type of momentum, an object's angular momentum can change only when a special type of force is applied to it. To see why, consider what happens when you try to open a swinging door. Opening the door means making it rotate on its hinges, which means giving the door some angular momentum. Pushing directly on the hinges will have no effect on the door, even if you push with a very strong force. However, even a light force can make the door rotate if you push on the part of the door that is farthest from the hinges. The type of force that can change an object's angular momentum is called a **torque**, which you can think of as a "twisting force." As the door example shows, the amount of torque depends not only on how much force is applied, but also on where it is applied.

Changing a tire offers another familiar example of torque. Turning the bolts on a tire means making them rotate, which requires giving them some angular momentum. A longer wrench means you can push from farther out than you can with a short wrench, so you can turn the bolts with less force. We will see many more applications of angular momentum in astronomy throughout the rest of the book.

How is mass different from weight?

In daily life, we usually think of *mass* as something you can measure with a bathroom scale, but technically the scale measures your *weight*, not your mass. The distinction between mass and weight rarely matters when we are talking about objects on Earth, but it is very important in astronomy:

- Your **mass** is the amount of matter in your body.

- Your **weight** (or *apparent weight**) is the *force* that a scale measures when you stand on it; that is, weight depends both on your mass and on the forces (including gravity) acting on your mass.

To understand the difference between mass and weight, imagine standing on a scale in an elevator (**FIGURE 4.3**). Your mass will be the same no matter how the elevator moves, but your weight can vary. When the elevator is stationary or moving at constant velocity, the scale reads your "normal" weight. When the elevator accelerates upward, the floor exerts

---

*Some physics texts distinguish between "true weight," due only to gravity, and "apparent weight," which also depends on other forces (as in an elevator). In this book the word *weight* means "apparent weight."

When the elevator moves at constant velocity (or is stationary)...
...your weight is normal.

*When the elevator **accelerates** upward...*
...you weigh more.

*When the elevator **accelerates** downward...*
...you weigh less.

*If the cable breaks so that you are in **free-fall**...*
...you are weightless.

FIGURE 4.3
interactive figure Mass is not the same as weight. The man's mass never changes, but his weight is different when the elevator accelerates.

a greater force than it does when you are at rest. You feel heavier, and the scale verifies your greater weight. When the elevator accelerates downward, the floor and the scale exert a weaker force on you, so the scale registers less weight. Note that the scale shows a weight different from your "normal" weight only when the elevator is *accelerating*, not when it is going up or down at constant speed.

Your mass therefore depends only on the amount of matter in your body and is the same anywhere, but your weight can vary because the forces acting on you can vary. For example, your mass would be the same on the Moon as on Earth, but you would weigh less on the Moon because of its weaker gravity.

Free-Fall and Weightlessness Now consider what happens if the elevator cable breaks (see the last frame in Figure 4.3). The elevator and you are suddenly in **free-fall**— falling without any resistance to slow you down. The floor drops away at the same rate that you fall, allowing you to "float" freely above it, and the scale reads zero because you are no longer held to it. In other words, your free-fall has made you **weightless**.

In fact, you are in free-fall whenever there's nothing to *prevent* you from falling. For example, you are in free-fall when you jump off a chair or spring from a diving board or trampoline. Surprising as it may seem, you have therefore experienced weightlessness many times in your life. You can experience it right now simply by jumping off your chair— though your weightlessness lasts for only the very short time until you hit the ground.

Weightlessness in Space You've probably seen videos of astronauts floating weightlessly in the Space Station. But why are they weightless? Many people guess that there's no gravity in space, but that's not true. After all, it is gravity that makes the Space Station orbit Earth. Astronauts feel weightless for the same reason you are weightless when you jump off a chair: They are in free-fall.

Astronauts are weightless the entire time they orbit Earth because they are in a *constant state of free-fall*. To understand this idea, imagine a tower that reaches all the way to the Space Station's orbit, about 350 kilometers above Earth (**FIGURE 4.4**). If you stepped off the tower, you would fall downward, remaining weightless until you hit the ground (or until air resistance had a noticeable effect on you). Now, imagine that instead of stepping off the tower, you ran and jumped out of the tower. You'd still fall to the ground, but because of your forward motion, you'd land a short distance away from the base of the tower.

The faster you ran out of the tower, the farther you'd go before landing. If you could somehow run fast enough—about

The faster you run from the tower, the farther you go before falling to Earth.

Using a rocket to gain enough speed, you could continually "fall" around Earth; that is, you'd be in orbit.

Not to scale!

FIGURE 4.4 interactive figure This figure explains why astronauts are weightless and float freely in space. If you could leap from a tall tower with enough speed, you could travel forward so fast that you'd orbit Earth. You'd then be in a constant state of free-fall, which means you'd be weightless. *Note:* On the scale shown here, the tower extends far higher than the Space Station's orbit; the rocket's orientation shows it rotating once with each orbit. (Adapted from *Space Station Science* by Marianne Dyson.)

28,000 km/hr (17,000 mi/hr) at the orbital altitude of the Space Station—a very interesting thing would happen: By the time gravity had pulled you downward as far as the length of the tower, you'd already have moved far enough around Earth that you'd no longer be going down at all. Instead, you'd be just as high above Earth as you'd been all along, but a good portion of the way around the world. In other words, you'd be orbiting Earth.

The Space Shuttle, the Space Station, and all other orbiting objects stay in orbit because they are constantly "falling around" Earth. Their constant state of free-fall makes these spacecraft and everything in them weightless.

MA Motion and Gravity Tutorial, Lesson 1

4.2 NEWTON'S LAWS OF MOTION

The complexity of motion in daily life might lead you to guess that the laws governing motion would also be complex. For example, if you watch a falling piece of paper waft lazily to the ground, you'll see it rock back and forth in a seemingly unpredictable pattern. However, the complexity of this motion arises because the paper is affected by a variety of forces, including gravity and the changing forces caused by air currents. If you could analyze the forces individually, you'd find that each force affects the paper's motion in a simple, predictable way. Sir Isaac Newton (1642–1727) discovered the remarkably simple laws that govern motion.

How did Newton change our view of the universe?

Newton was born in Lincolnshire, England, on Christmas Day in 1642. His father, a farmer who never learned to read or write, died 3 months before his birth. Newton had a difficult childhood and showed few signs of unusual talent. He attended Trinity College at Cambridge, where he earned his keep by performing menial labor, such as cleaning the boots and bathrooms of wealthier students and waiting on their tables.

The plague hit Cambridge shortly after Newton graduated, and he returned home. By his own account, he experienced a moment of inspiration in 1666 when he saw an apple fall to the ground. He suddenly realized that the gravity making the apple fall was the same force that held the Moon in orbit around Earth. In that moment, Newton shattered the remaining vestiges of the Aristotelian view of the world, which for centuries had been accepted as unquestioned truth.

Aristotle had made many claims about the physics of motion, using his ideas to support his belief in an Earth-centered cosmos. He had also maintained that the heavens were totally distinct from Earth, so physical laws on Earth did not apply to heavenly motion. By the time Newton saw the apple fall, the Copernican revolution had displaced Earth from a central position, and Galileo's experiments had shown that the laws of physics were not what Aristotle had believed [**Section 3.3**].

Newton's sudden insight delivered the final blow to Aristotle's view. By recognizing that gravity operates in the heavens as well as on Earth, Newton eliminated Aristotle's distinction between the two realms and brought the heavens and Earth together as one *universe*. This insight also heralded the birth of the modern science of *astrophysics* (although the term wasn't coined until much later), which applies physical laws discovered on Earth to phenomena throughout the cosmos.

Over the next 20 years, Newton's work completely revolutionized mathematics and science. He quantified the laws of motion and gravity, conducted crucial experiments regarding the nature of light, built the first reflecting telescopes, and invented the mathematics of calculus. The compendium of Newton's discoveries is so tremendous that it would take a complete book just to describe them, and many more books to describe their influence on civilization. When Newton died in 1727, at age 84, English poet Alexander Pope composed the following epitaph:

Sir Isaac Newton (1642–1727)

> *Nature, and Nature's laws lay hid in the Night.*
> *God said, Let Newton be! and all was Light.*

What are Newton's three laws of motion?

Newton published the laws of motion and gravity in 1687, in his book *Philosophiae Naturalis Principia Mathematica* ("Mathematical Principles of Natural Philosophy"), usually called *Principia*. He enumerated three laws that apply to all motion, which we now call **Newton's laws of motion**. These laws govern the motion of everything from our daily movements on Earth to the movements of planets, stars, and galaxies throughout the universe. **FIGURE 4.5** summarizes the three laws.

Newton's First Law Newton's first law of motion essentially restates Galileo's discovery that objects will remain in motion unless a force acts to stop them. It can be stated as follows:

> **Newton's first law:** *An object moves at constant velocity if there is no net force acting upon it.*

In other words, objects at rest (velocity = 0) tend to remain at rest, and objects in motion tend to remain in motion with no change in either their speed or their direction.

The idea that an object at rest should remain at rest is rather obvious: A car parked on a flat street won't suddenly start moving for no reason. But what if the car is traveling along a flat, straight road? Newton's first law says that the car should keep going at the same speed forever *unless* a force

Newton's first law of motion:
An object moves at constant velocity unless a net force acts to change its speed or direction.

Example: A spaceship needs no fuel to keep moving in space.

Newton's second law of motion:
Force = mass × acceleration

Example: A baseball accelerates as the pitcher applies a force by moving his arm. (Once the ball is released, the force from the pitcher's arm ceases, and the ball's path changes only because of the forces of gravity and air resistance.)

Newton's third law of motion:
For any force, there is always an equal and opposite reaction force.

Example: A rocket is propelled upward by a force equal and opposite to the force with which gas is expelled out its back.

FIGURE 4.5 Newton's three laws of motion.

acts to slow it down. You know that the car eventually will come to a stop if you take your foot off the gas pedal, so one or more forces must be stopping the car—in this case forces arising from friction and air resistance. If the car were in space, and therefore unaffected by friction or air, it would keep moving forever (though gravity would gradually alter its speed and direction). That is why interplanetary spacecraft need no fuel to keep going after they are launched into space, and why astronomical objects don't need fuel to travel through the universe.

Newton's first law also explains why you don't feel any sensation of motion when you're traveling in an airplane on a smooth flight. As long as the plane is traveling at constant velocity, no net force is acting on it or on you. Therefore, you feel no different from the way you would feel at rest. You can walk around the cabin, play catch with someone, or relax and go to sleep just as though you were "at rest" on the ground.

Newton's Second Law Newton's second law of motion tells us what happens to an object when a net force *is* present. We have already seen that a net force will change an object's momentum, accelerating it in the direction of the force.

Newton's second law, which quantifies this relationship, can be written in either of the following two forms:

Newton's second law:

$$force = mass \times acceleration\,(F = ma)$$

or $\quad force =$ rate of change in *momentum*

This law explains why you can throw a baseball farther than you can throw a shot in the shot put. The force your arm delivers to both the baseball and the shot equals the product of mass and acceleration. Because the mass of the shot is greater than that of the baseball, the same force from your arm gives the shot a smaller acceleration. Because of its smaller acceleration, the shot leaves your hand with less speed than the baseball and therefore travels a shorter distance before hitting the ground. Astronomically, Newton's second law explains why a large planet such as Jupiter has a greater effect on asteroids and comets than a small planet such as Earth [**Section 12.4**]. Because Jupiter is much more massive than Earth, it exerts a stronger gravitational force on passing asteroids and comets, and therefore sends them scattering with a greater acceleration.

MATHEMATICAL INSIGHT 4.1

Units of Force, Mass, and Weight

Newton's second law, $F = ma$, shows that the unit of force is equal to a unit of mass multiplied by a unit of acceleration. Consider a mass of 1 kilogram accelerating at 10 m/s²:

$$force = mass \times acceleration$$
$$= 1 \text{ kg} \times 10\,\frac{\text{m}}{\text{s}^2} = 10\,\frac{\text{kg} \times \text{m}}{\text{s}^2}$$
$$= 10 \text{ newtons}$$

This example shows that the standard unit of force is called the **newton**, which is equivalent to a *kilogram-meter per second squared*.

We can also use Newton's second law to clarify the difference between mass and weight. Imagine standing on a chair when it is suddenly pulled out from under you. You will immediately begin accelerating downward with the acceleration of gravity, which

means the force of gravity acting on you must be your mass times the acceleration of gravity. We define this force as your weight, and it is the same whether you are falling or standing still:

$$weight = mass \times acceleration\ of\ gravity$$

(Your apparent weight may differ if forces besides gravity are acting on you at the same time and is zero if you are in free-fall.)

Like any force, weight has units of mass times acceleration. Thus, although we commonly speak of weights in *kilograms*, this usage is not technically correct: Kilograms are a unit of mass, not of force. You may safely ignore this technicality as long as you are dealing with objects on Earth that are not accelerating. In space or on other planets, the distinction between mass and weight is important and cannot be ignored.

The inward force along the string keeps the ball moving in a circle.

$F = ma$

v

If the string breaks, the inward force is gone...

string breaks

v *...so the ball moves with constant velocity from the point of the break.*

FIGURE 4.6 Newton's second law of motion tells us that an object going around a curve has an acceleration pointing toward the inside of the curve.

a When you swing a ball on a string, the string exerts a force that pulls the ball inward.

b If the string breaks, the ball flies off in a straight line at constant velocity.

We can also use Newton's second law of motion to understand acceleration around curves. Suppose you swing a ball on a string around your head (**FIGURE 4.6a**). The ball is accelerating even if it has a steady speed, because it is constantly changing direction and therefore has a changing velocity. What makes it accelerate? According to Newton's second law, the taut string must be applying a force to the ball. We can understand this force by thinking about what happens when the string breaks (**FIGURE 4.6b**): With the force gone, the ball flies off in a straight line. Therefore, when the string is intact, the force must be pulling the ball *inward* to keep it from flying off. Because acceleration must be in the same direction as the force, we conclude that the ball has an inward acceleration as it moves around the circle.

The same idea helps us understand the force on a car moving around a curve or a planet orbiting the Sun. In the case of the car, the inward force comes from friction between the tires and the road. The tighter the curve (or the faster the car is going), the greater the force needed to keep the car moving around it. If the inward force due to friction is not great enough, the car skids outward. Similarly, a planet orbiting the Sun always has an acceleration in the direction of the Sun, and gravity is the inward force that causes this acceleration. Indeed, it was Newton's discovery of the precise nature of this acceleration that helped him deduce the law of gravity, which we'll discuss in Section 4.4.

COMMON MISCONCEPTIONS

What Makes a Rocket Launch?

If you've ever watched a rocket launch, it's easy to see why many people believe that the rocket "pushes off" the ground. However, the ground has nothing to do with the rocket launch, which is actually explained by Newton's third law of motion. To balance the force driving gas out the back of the rocket, an equal and opposite force must propel the rocket forward. Rockets can be launched horizontally as well as vertically, and a rocket can be "launched" in space (for example, from a space station) with no need for any solid ground.

Newton's Third Law Think for a moment about standing still on the ground. Your weight exerts a downward force, so if this force were acting alone, Newton's second law would demand that you accelerate downward. The fact that you are not falling means there must be no *net* force acting on you, which is possible only if the ground is exerting an upward force on you that precisely offsets the downward force you exert on the ground. The fact that the downward force you exert on the ground is offset by an equal and opposite force that pushes upward on you is one example of Newton's third law of motion, which tells us that any force is always paired with an equal and opposite reaction force.

> **Newton's third law:** *For any force, there is always an equal and opposite reaction force.*

This law is very important in astronomy, because it tells us that objects always attract *each other* through gravity. For example, your body always exerts a gravitational force on Earth identical to the force that Earth exerts on you, except that it acts in the opposite direction. Of course, the same force means a much greater acceleration for you than for Earth (because your mass is so much smaller than Earth's), which is why you fall toward Earth when you jump off a chair, rather than Earth falling toward you.

Newton's third law also explains how a rocket works: A rocket engine generates a force that drives hot gas out the back, which creates an equal and opposite force that propels the rocket forward.

4.3 CONSERVATION LAWS IN ASTRONOMY

Newton's laws of motion are easy to state, but they may seem a bit arbitrary. Why, for example, should every force be opposed by an equal and opposite reaction force? In the centuries since Newton first stated his laws, we have learned that they are not arbitrary at all, but instead reflect deeper aspects of nature known as *conservation laws*. In this section, we'll explore three of the most important conservation laws for astronomy: *conservation of momentum, conservation of angular momentum,* and *conservation of energy.* We'll see some immediate examples of how they apply to astronomy, and then use these laws over and over throughout the rest of the book.

Why do objects move at constant velocity if no force acts on them?

The first of our conservation laws, the law of **conservation of momentum**, states that as long as there are no external forces, the total momentum of interacting objects cannot change; that is, their total momentum is *conserved.* An individual object can gain or lose momentum only if some other object's momentum changes by a precisely opposite amount.

The law of conservation of momentum is implicit in Newton's laws. To see why, watch a game of pool. Newton's second law tells us that when one pool ball strikes another, it exerts a force that changes the momentum of the second ball. At the same time, Newton's third law tells us that the second ball exerts an equal and opposite force on the first one—which means that the first ball's momentum changes by precisely the same amount as the second ball's momentum, but in the opposite direction. The total combined momentum of the two balls remains the same both before and after the collision (**FIGURE 4.7**). Note that no *external* forces are accelerating the balls.

Rockets offer another good example of conservation of momentum in action. When you fire a rocket engine, the total momentum of the rocket and the hot gases it shoots out the back must stay the same. In other words, the amount of forward momentum the rocket gains is equal to the amount of backward momentum in the gas that shoots out the back. That is why forces between the rocket and the gases are always equal and opposite.

From the perspective of conservation of momentum, Newton's first law makes perfect sense. When no net force acts on an object, there is no way for the object to transfer any momentum to or from any other object. In the absence of a net force, an object's momentum must therefore remain unchanged—which means the object must continue to move exactly as it has been moving.

According to current understanding of the universe, conservation of momentum is an absolute law that always holds true. For example, it holds even when you jump up into the air. You may wonder, Where do I get the momentum that carries me upward? The answer is that as your legs propel you skyward, they are actually pushing Earth in the other direction, giving Earth's momentum an equal and opposite kick. However, Earth's huge mass renders its acceleration undetectable. During your brief flight, the gravitational force between you and Earth pulls you back down, transferring your momentum back to Earth. The total momentum of you and Earth remains the same at all times.

What keeps a planet rotating and orbiting the Sun?

Perhaps you've wondered how Earth manages to keep rotating and going around the Sun day after day and year after year. The answer comes from our second conservation law: the law of **conservation of angular momentum**. Recall that rotating or orbiting objects have angular momentum because they are moving in circles or going around curves, and that angular momentum can be changed only by a "twisting force," or *torque.* The law of conservation of angular momentum states that as long as there is no external torque, the total angular momentum of a set of interacting objects cannot change. An individual object can change its angular momentum only by transferring some angular momentum to or from another object. Because astronomical objects can have angular momentum due to both their rotation and their orbit, let's consider both cases.

Orbital Angular Momentum Consider Earth's orbit around the Sun. A simple formula tells us Earth's angular momentum at any point in its orbit:

$$\text{angular momentum} = m \times v \times r$$

where m is Earth's mass, v is its orbital velocity (or, more technically, the component of velocity perpendicular to r), and r is the "radius" of the orbit, by which we mean its distance from the Sun (**FIGURE 4.8**). Because there are no objects around to give or take angular momentum from Earth as it orbits the Sun, Earth's orbital angular momentum must always stay the same. This explains two key facts about Earth's orbit:

1. Earth needs no fuel or push of any kind to keep orbiting the Sun—it will keep orbiting as long as nothing comes along to take angular momentum away.

2. Because Earth's angular momentum at any point in its orbit depends on the product of its speed and orbital radius (distance from the Sun), Earth's orbital speed must be faster when it is nearer to the Sun (and the radius is smaller) and slower when it is farther from the Sun (and the radius is larger).

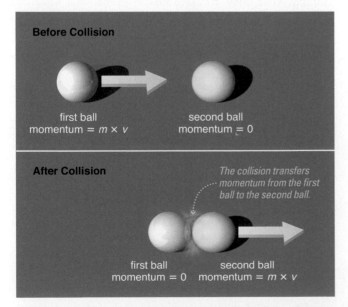

Before Collision

first ball
momentum = $m \times v$

second ball
momentum = 0

After Collision

The collision transfers momentum from the first ball to the second ball.

first ball
momentum = 0

second ball
momentum = $m \times v$

FIGURE 4.7 Conservation of momentum demonstrated with balls on a pool table.

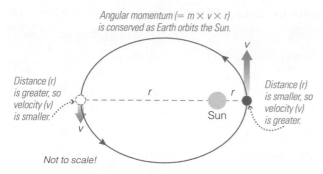

Angular momentum (= m × v × r) is conserved as Earth orbits the Sun.

Distance (r) is greater, so velocity (v) is smaller.

Distance (r) is smaller, so velocity (v) is greater.

Sun

Not to scale!

FIGURE 4.8 Earth's orbital angular momentum stays constant, so Earth moves faster when it is closer to the Sun and slower when it is farther from the Sun.

The second fact is just what Kepler's second law of planetary motion states [**Section 3.3**]. That is, the law of conservation of angular momentum tells us *why* Kepler's law is true.

Rotational Angular Momentum The same idea explains why Earth keeps rotating. As long as Earth isn't transferring any of the angular momentum of its rotation to another object, it keeps rotating at the same rate. (In fact, Earth is very gradually transferring some of its rotational angular momentum to the Moon, and as a result Earth's rotation is gradually slowing down; see Section 4.5.)

Conservation of angular momentum also explains why we see so many spinning disks in the universe, such as the disks of galaxies like the Milky Way and disks of material orbiting young stars. The idea is easy to illustrate with an ice skater spinning in place (**FIGURE 4.9**). Because there is so little friction on ice, the angular momentum of the ice skater remains essentially constant. When she pulls in her extended arms, she decreases her radius—which means her velocity of rotation

In the product m × v × r, extended arms mean larger radius and smaller velocity of rotation.

Bringing in her arms decreases her radius and therefore increases her rotational velocity.

FIGURE 4.9 A spinning skater conserves angular momentum.

must increase. Stars and galaxies are both born from clouds of gas that start out much larger in size. These clouds almost inevitably have some small net rotation, though it may be imperceptible. Like the spinning skater as she pulls in her arms, they must therefore spin faster as gravity makes them shrink in size. (We'll discuss why the clouds also flatten into disks in Chapter 8.)

 Energy Tutorial, Lesson 1

Where do objects get their energy?

Our third crucial conservation law for astronomy is the law of **conservation of energy**. This law tells us that, like momentum and angular momentum, energy cannot appear out of nowhere or disappear into nothingness. Objects can gain or lose energy only by exchanging energy with other objects. Because of this law, the story of the universe is a story of the interplay of energy and matter: All actions involve exchanges of energy or the conversion of energy from one form to another.

Throughout the rest of this book, we'll see numerous cases in which we can understand astronomical processes simply by studying how energy is transformed and exchanged. For example, we'll see that planetary interiors cool with time because they radiate energy into space, and that the Sun became hot because of energy released by the gas that formed it. By applying the laws of conservation of momentum, angular momentum, and energy, we can understand almost every major process that occurs in the universe.

Basic Types of Energy Before we can fully understand the law of conservation of energy, we need to know what energy is. In essence, energy is what makes matter move. Because this statement is so broad, we often distinguish between different types of energy. For example, we talk about the energy we get from the food we eat, the energy that makes our cars go, and the energy a light bulb emits. Fortunately, we can classify nearly all types of energy into just three major categories (**FIGURE 4.10**):

- Energy of motion, or **kinetic energy** (*kinetic* comes from a Greek word meaning "motion"). Falling rocks, orbiting planets, and the molecules moving in the air are all examples of objects with kinetic energy. Quantitatively, the kinetic energy of a moving object is $\frac{1}{2}mv^2$, where m is the object's mass and v is its speed.

- Energy carried by light, or **radiative energy** (the word *radiation* is often used as a synonym for *light*). All light carries energy, which is why light can cause changes in matter. For example, light can alter molecules in our eyes—thereby allowing us to see—or warm the surface of a planet.

Energy can be converted from one form to another.

kinetic energy
(energy of motion)

radiative energy
(energy of light)

potential energy
(stored energy)

FIGURE 4.10 The three basic categories of energy. Energy can be converted from one form to another, but it can never be created or destroyed, an idea embodied in the law of conservation of energy.

TABLE 4.1   Energy Comparisons

Item	Energy (joules)
Energy of sunlight at Earth (per m² per second)	1.3×10^3
Energy from metabolism of a candy bar	1×10^6
Energy needed to walk for 1 hour	1×10^6
Kinetic energy of a car going 60 mi/hr	1×10^6
Daily food energy need of average adult	1×10^7
Energy released by burning 1 liter of oil	1.2×10^7
Thermal energy of parked car	1×10^8
Energy released by fission of 1 kilogram of uranium-235	5.6×10^{13}
Energy released by fusion of hydrogen in 1 liter of water	7×10^{13}
Energy released by 1-megaton H-bomb	4×10^{15}
Energy released by magnitude 8 earthquake	2.5×10^{16}
Annual U.S. energy consumption	10^{20}
Annual energy generation of Sun	10^{34}
Energy released by a supernova	10^{44}–10^{46}

- Stored energy, or **potential energy**, which might later be converted into kinetic or radiative energy. For example, a rock perched on a ledge has *gravitational* potential energy because it will fall if it slips off the edge, and gasoline contains *chemical* potential energy that can be converted into the kinetic energy of a moving car.

Regardless of which type of energy we are dealing with, we can measure the amount of energy with the same standard units. For Americans, the most familiar units of energy are *Calories,* which are shown on food labels to tell us how much energy our bodies can draw from the food. A typical adult needs about 2500 Calories of energy from food each day. In science, the standard unit of energy is the **joule**. One food Calorie is equivalent to about 4184 joules, so the 2500 Calories used daily by a typical adult is equivalent to about 10 million joules. **TABLE 4.1** compares various energies in joules.

Thermal Energy—The Kinetic Energy of Many Particles Although there are only three major categories of energy, we sometimes divide them into various subcategories. In astronomy, the most important subcategory of kinetic energy is **thermal energy**, which represents the collective kinetic energy of the many individual particles (atoms and molecules) moving randomly within a substance like a rock or the air or the gas within a distant star. In such cases, it is much easier to talk about the thermal energy of the object than about the kinetic energies of its billions upon billions of individual particles. Note that all objects contain thermal energy even when they are sitting still, because the particles

within them are always jiggling about randomly. These random motions can contain substantial energy: The thermal energy of a parked car due to the random motion of its atoms is much greater than the kinetic energy of the car moving at highway speed.

Thermal energy gets its name because it is related to temperature, but temperature and thermal energy are not quite the same thing. Thermal energy measures the *total* kinetic energy of all the randomly moving particles in a substance, while **temperature** measures the *average* kinetic energy of the particles. For a particular object, a higher temperature simply means that the particles on average have more kinetic energy and hence are moving faster (**FIGURE 4.11**). You're probably familiar with temperatures measured in *Fahrenheit* or *Celsius,*

lower temperature   higher temperature

These particles are moving relatively slowly, which means low temperature . . .

. . . and now the same particles are moving faster, which means higher temperature.

FIGURE 4.11 Temperature is a measure of the average kinetic energy of the particles (atoms and molecules) in a substance. Longer arrows represent faster speeds.

FIGURE 4.12 Three common temperature scales: Kelvin, Celsius, and Fahrenheit. Scientists generally prefer the Kelvin scale. Note that the degree symbol (°) is not usually used with the Kelvin scale.

but in science we often use the **Kelvin** temperature scale (**FIGURE 4.12**). The Kelvin scale does not have negative temperatures, because it starts from the coldest possible temperature, known as *absolute zero* (0 K).

Thermal energy depends on temperature, because a higher average kinetic energy for the particles in a substance means a higher total energy. But thermal energy also depends on the number and density of the particles, as you can see by imagining that you quickly thrust your arm in and out of a hot oven and a pot of boiling water (don't try this!). The air in a hot oven is much higher in temperature than the water boiling in a pot (**FIGURE 4.13**). However, the boiling water would scald your arm almost instantly, while you can safely put your arm into the oven air for a few seconds. The reason for this difference is density. In both cases, because the air or water is hotter than your body, molecules striking your skin transfer thermal energy to molecules in your arm. The higher temperature in the oven means that the air molecules strike your skin harder, on average, than the molecules in

The air in a hot oven is hotter than the boiling water in the pot . . .

. . . but the water in the pot contains more thermal energy because of its much higher density.

212°F = 100°C   400°F ≈ 200°C

FIGURE 4.13 Thermal energy depends on both the temperature and the density of particles in a substance.

the boiling water. However, because the *density* of water is so much higher than the density of air (meaning water has far more molecules in the same amount of space), many more molecules strike your skin each second in the water. While each individual molecule that strikes your skin transfers a little less energy in the boiling water than in the oven, the sheer number of molecules hitting you in the water means that more thermal energy is transferred to your arm. That is why the boiling water causes a burn almost instantly.

The environment in space provides another example of the difference between temperature and heat. Surprisingly, the temperature in low Earth orbit can be several thousand degrees. However, astronauts working in Earth orbit are at much greater risk of getting cold than hot. The reason is the extremely low density: Although the particles striking an astronaut's space suit may be moving quite fast, there are not enough of them to transfer much thermal energy. (You may wonder how the astronauts become cold given that the low density also means the astronauts cannot transfer much of their own thermal energy to the particles in space. It turns out that they lose their body heat by emitting *thermal radiation*, which we will discuss in Section 5.3.)

Potential Energy in Astronomy Many types of potential energy are important in astronomy, but two are particularly important: *gravitational potential energy* and the potential energy of mass itself, or *mass-energy*.

An object's **gravitational potential energy** depends on its mass and how far it can fall as a result of gravity. An object has more gravitational potential energy when it is higher and less when it is lower. For example, if you throw a ball up into the air, it has more potential energy when it is high up than when it is near the ground. Because energy must be conserved during the ball's flight, the ball's kinetic energy increases when its gravitational potential energy decreases, and vice versa (**FIGURE 4.14a**). That is why the ball travels fastest (has the most kinetic energy) when it is closest to the ground, where it has the least gravitational potential energy. The higher it is, the more gravitational potential energy it has and the slower the ball travels (less kinetic energy). For an object near Earth's surface, its gravitational potential energy is *mgh*, where *m* is its mass, *g* is the acceleration of gravity, and *h* is its height above the ground.

The same general idea explains how stars become hot (**FIGURE 4.14b**). Before a star forms, its matter is spread out in a large, cold cloud of gas. Most of the individual gas particles are far from the center of this large cloud and therefore have a lot of gravitational potential energy. The particles lose gravitational potential energy as the cloud contracts under

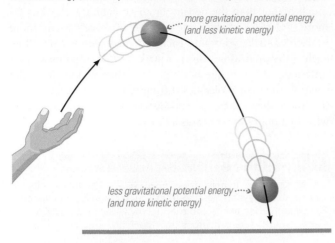

The total energy (kinetic + potential) is the same at all points in the ball's flight.

more gravitational potential energy
(and less kinetic energy)

less gravitational potential energy
(and more kinetic energy)

a The ball has more gravitational potential energy when it is high up than when it is near the ground.

Energy is conserved: As the cloud contracts, gravitational potential energy is converted to thermal energy and radiation.

less gravitational potential energy (and more thermal energy)

more gravitational potential energy (and less thermal energy)

b A cloud of interstellar gas contracting because of its own gravity has more gravitational potential energy when it is spread out than when it shrinks in size.

FIGURE 4.14 Two examples of gravitational potential energy.

its own gravity, and this "lost" potential energy ultimately gets converted into thermal energy, making the center of the cloud hot.

Einstein discovered that mass itself is a form of potential energy, often called **mass-energy**. The amount of potential energy contained in mass is described by Einstein's famous equation

$$E = mc^2$$

where E is the amount of potential energy, m is the mass of the object, and c is the speed of light. This equation tells us that a small amount of mass contains a huge amount of energy. For example, the energy released by a 1-megaton H-bomb comes from converting only about 0.1 kilogram of mass (about 3 ounces—a quarter of a can of soda) into energy (**FIGURE 4.15**). The Sun generates energy by converting a tiny fraction of its mass into energy through a similar process of nuclear fusion [**Section 14.2**].

Just as Einstein's formula tells us that mass can be converted into other forms of energy, it also tells us that energy can be transformed into mass. This process is especially important in understanding what we think happened during the early moments in the history of the universe, when some of the energy of the Big Bang turned into the mass from which all objects, including us, are made [**Section 22.1**]. Scientists also use this idea to search for undiscovered particles of matter,

---

MATHEMATICAL INSIGHT 4.2

Mass-Energy

It's easy to calculate mass-energies with Einstein's formula $E = mc^2$.

EXAMPLE: Suppose a 1-kilogram rock were completely converted to energy. How much energy would it release? Compare this to the energy released by burning 1 liter of oil.

SOLUTION:

Step 1 Understand: We can compute the total mass-energy of the rock from Einstein's formula and then compare it to the energy released by burning a liter of oil, from Table 4.1.

Step 2 Solve: The mass-energy of the rock is

$$E = mc^2 = 1 \text{ kg} \times \left(3 \times 10^8 \frac{\text{m}}{\text{s}}\right)^2$$

$$= 1 \text{ kg} \times \left(9 \times 10^{16} \frac{\text{m}^2}{\text{s}^2}\right)$$

$$= 9 \times 10^{16} \frac{\text{kg} \times \text{m}^2}{\text{s}^2} = 9 \times 10^{16} \text{ joules}$$

We divide to compare this mass-energy to the energy released by burning 1 liter of oil (12 million joules; see Table 4.1):

$$\frac{9 \times 10^{16} \text{ joules}}{1.2 \times 10^7 \text{ joules}} = 7.5 \times 10^9$$

Step 3 Explain: We have found that converting a 1-kilogram rock completely to energy would release 9×10^{16} joules of energy, which is about 7.5 billion times as much energy as we get from burning 1 liter of oil. In fact, the total amount of oil used by all cars in the United States is approximately 7.5 billion liters per week—which means that complete conversion of the mass of a 1-kilogram rock to energy could yield enough energy to power all the cars in the United States for a week. Unfortunately, no technology available now or in the foreseeable future can release all the mass-energy of a rock.

FIGURE 4.15 The energy released by this H-bomb comes from converting only about 0.1 kilogram of mass into energy in accordance with the formula $E = mc^2$.

using large machines called *particle accelerators* to create subatomic particles from energy.

Conservation of Energy We have seen that energy comes in three basic categories—kinetic, radiative, and potential—and explored several subcategories that are especially important in astronomy: thermal energy, gravitational potential energy, and mass-energy. Now we are ready to return to the question of where objects get their energy. Because energy cannot be created or destroyed, objects always get their energy from other objects. Ultimately, we can always trace an object's energy back to the Big Bang [**Section 1.2**], the beginning of the universe in which all matter and energy is thought to have come into existence.

For example, imagine that you've thrown a baseball. It is moving, so it has kinetic energy. Where did this kinetic energy come from? The baseball got its kinetic energy from the motion of your arm as you threw it. Your arm, in turn, got its kinetic energy from the release of chemical potential energy stored in your muscle tissues. Your muscles got this energy from the chemical potential energy stored in the foods you ate. The energy stored in the foods came from sunlight, which plants convert into chemical potential energy through photosynthesis. The radiative energy of the Sun was generated through the process of nuclear fusion, which releases some of the mass-energy stored in the Sun's supply of hydrogen. The mass-energy stored in the hydrogen came from the birth of the universe in the Big Bang. After you throw the ball, its kinetic energy will ultimately be transferred to molecules in the air or ground. It may be difficult to trace after this point, but it will never disappear.

4.4 THE UNIVERSAL LAW OF GRAVITATION

Newton's laws of motion describe how objects in the universe move in response to forces. The laws of conservation of momentum, angular momentum, and energy offer an alternative and often simpler way of thinking about what happens when a force causes some change in the motion of one or more objects. However, we cannot fully understand motion unless we also understand the forces that lead to changes in motion. In astronomy, the most important force is gravity, which governs virtually all large-scale motion in the universe.

Motion and Gravity Tutorial, Lesson 2

What determines the strength of gravity?

Isaac Newton discovered the basic law that describes how gravity works. Newton expressed the force of gravity mathematically with his **universal law of gravitation**. Three simple statements summarize this law:

- Every mass attracts every other mass through the force called *gravity*.

- The strength of the gravitational force attracting any two objects is *directly proportional* to the product of their masses. For example, doubling the mass of *one* object doubles the force of gravity between the two objects.

- The strength of gravity between two objects decreases with the *square* of the distance between their centers. We therefore say that the gravitational force follows an **inverse square law**. For example, doubling the distance between two objects weakens the force of gravity by a factor of 2^2, or 4.

These three statements tell us everything we need to know about Newton's universal law of gravitation. Mathematically, all three statements can be combined into a single equation, usually written like this:

$$F_g = G \frac{M_1 M_2}{d^2}$$

where F_g is the force of gravitational attraction, M_1 and M_2 are the masses of the two objects, and d is the distance between their centers (**FIGURE 4.16**). The symbol G is a constant called

FIGURE 4.16 The universal law of gravitation is an *inverse square law*, which means that the force of gravity declines with the *square* of the distance d between two objects.

the **gravitational constant**, and its numerical value has been measured to be $G = 6.67 \times 10^{-11}$ m^3/(kg $\times$ s^2).

THINK ABOUT IT
How does the gravitational force between two objects change if the distance between them triples? If the distance between them drops by half?

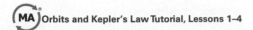

(MA) Orbits and Kepler's Law Tutorial, Lessons 1–4

How does Newton's law of gravity extend Kepler's laws?

By the time Newton published *Principia* in 1687, Kepler's laws of planetary motion [**Section 3.3**] had already been known and tested for some 70 years. Kepler's laws had proven so successful that there was little doubt about their validity. However, there was great debate among scientists about *why* Kepler's laws hold true.

Newton resolved the debate by showing that Kepler's laws are consequences of the laws of motion and the universal law of gravitation. In particular, with the aid of the mathematics of calculus that he invented, Newton showed that the inverse square law for gravity leads naturally to elliptical orbits for planets orbiting the Sun (with the Sun at one focus), which is Kepler's first law. As we've seen, Kepler's second law (a planet moves faster when it is closer to the Sun) then arises as a consequence of conservation of angular momentum. Kepler's third law (average orbital speed is slower for planets with larger average orbital distance) arises from the fact that gravity weakens with distance from the Sun. Newton also discovered that he could extend Kepler's laws into a more general set of rules about orbiting objects.

Newton's discoveries sealed the triumph of the Copernican revolution. Prior to Newton, it was still possible to see Kepler's model of planetary motion as "just" another model, though it fit the observational data far better than any previous model. By explaining Kepler's laws in terms of basic laws of physics, Newton removed virtually all remaining doubt about the legitimacy of the Sun-centered solar system. By extending the laws to other orbiting objects, he provided us with a way to explain the motions of objects throughout the universe. Let's explore four crucial ways in which Newton extended Kepler's laws.

Planets Are Not the Only Objects with Elliptical Orbits
Kepler wrote his first two laws for planets orbiting the Sun, but Newton showed that any object going around another object will obey these laws. For example, the orbits of a satellite around Earth, of a moon around a planet, and of an asteroid around the Sun are all ellipses in which the orbiting object moves faster at the nearer points in its orbit and slower at the farther points.

Ellipses Are Not the Only Possible Orbital Paths
Ellipses (which include circles) are the only possible shapes for **bound orbits**—orbits in which an object goes around

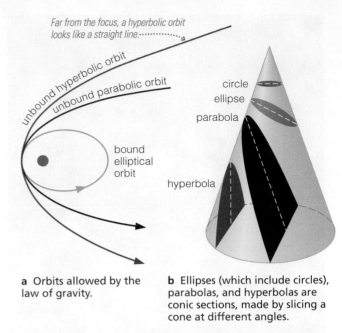

a Orbits allowed by the law of gravity.

b Ellipses (which include circles), parabolas, and hyperbolas are conic sections, made by slicing a cone at different angles.

FIGURE 4.17 Newton showed that ellipses are not the only possible orbital paths. Orbits can also be unbound, taking the mathematical shape of either parabolas or hyperbolas.

another object over and over again. (The term *bound orbit* comes from the idea that gravity creates a *bond* that holds the objects together.) However, Newton discovered that objects can also follow **unbound orbits**—paths that bring an object close to another object just once. For example, some comets that enter the inner solar system follow unbound orbits. They come in from afar just once, loop around the Sun, and never return.

More specifically, Newton showed that orbital paths can be ellipses, parabolas, or hyperbolas (**FIGURE 4.17a**). Bound orbits are ellipses, while unbound orbits can be either parabolas or hyperbolas. Together, these shapes are known in mathematics as the *conic sections*, because they can be made by slicing through a cone at different angles (**FIGURE 4.17b**). Note that objects on unbound orbits still obey the basic principle of Kepler's second law: They move faster when they are closer to the object they are orbiting, and slower when they are farther away.

Objects Orbit Their Common Center of Mass
We usually think of one object orbiting another object, like a planet orbiting the Sun or the Moon orbiting Earth. However, Newton showed that two objects attracted by gravity actually *both* orbit around their common **center of mass**—the point at which the two objects would balance if they were somehow connected (**FIGURE 4.18**). For example, in a binary star system in which both stars have the same mass, we would see both stars tracing ellipses around a point halfway between them. When one object is more massive than the other, the center of mass lies closer to the more massive object.

The idea that objects orbit their common center of mass holds even for the Sun and planets. However, the Sun is so much more massive than the planets that the center of mass between the Sun and any planet lies either inside or nearly

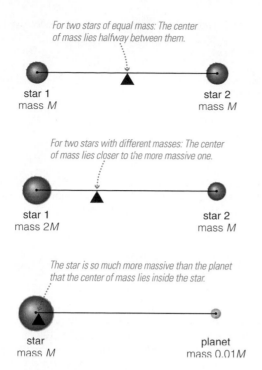

For two stars of equal mass: The center of mass lies halfway between them.

star 1
mass M

star 2
mass M

For two stars with different masses: The center of mass lies closer to the more massive one.

star 1
mass 2M

star 2
mass M

The star is so much more massive than the planet that the center of mass lies inside the star.

star
mass M

planet
mass 0.01M

FIGURE 4.18 **interactive figure** Two objects attracted by gravity orbit their common center of mass—the point at which they would balance if they were somehow connected.

FIGURE 4.19 Newton's version of Kepler's third law shows that when one object orbits a much more massive object, the orbital period depends only on its average orbital distance. The astronaut and the Space Shuttle share the same orbit and therefore stay together—even as both orbit Earth at a speed of 25,000 km/hr.

inside the Sun, making it difficult for us to notice the Sun's motion about this center. Nevertheless, with precise measurements we can detect the Sun's slight motion around this center of mass. As we will see in Chapter 13, astronomers have used this same idea to discover many planets around other stars.

Orbital Characteristics Tell Us the Masses of Distant Objects Recall that Kepler's third law is written $p^2 = a^3$, where p is a planet's orbital period in years and a is the planet's average distance from the Sun in AU. Newton found that this statement is actually a special case of a more general equation that we call **Newton's version of Kepler's third law** (see Mathematical Insight 4.3). This equation allows us to measure orbital period and distance in any units we wish (rather than only in years and AU), and also shows that the relationship between orbital period and average distance depends on the masses of the orbiting objects. When an object is much less massive than the object it orbits, we can calculate the mass of the central object from the orbital period and average distance of the orbiting object.

Newton's version of Kepler's third law is the primary means by which we determine masses throughout the universe. For example, it allows us to calculate the mass of the Sun from Earth's orbital period (1 year) and its average distance from the Sun (1 AU). Similarly, measuring the orbital period and average distance of one of Jupiter's moons allows us to calculate Jupiter's mass, and measuring the orbital periods and average distances of stars in binary star systems allows us to determine their masses.

Newton's version of Kepler's third law also explains another important characteristic of orbital motion. It shows that the orbital period of a *small* object orbiting a much more massive object depends only on its orbital distance, not on its mass. That is why an astronaut does not need a tether to stay close to the spacecraft during a space walk (**FIGURE 4.19**). The spacecraft and the astronaut are both much smaller in mass than Earth, so they stay together because they have the same orbital distance and hence the same orbital period.

4.5 ORBITS, TIDES, AND THE ACCELERATION OF GRAVITY

Newton's universal law of gravitation has applications that go far beyond explaining Kepler's laws. In this final section, we'll explore three important concepts that we can understand with the help of the universal law of gravitation: orbits, tides, and the acceleration of gravity.

How do gravity and energy allow us to understand orbits?

The law of gravitation explains Kepler's laws of planetary motion, which describe the simple and stable orbits of the planets, and Newton's extensions of Kepler's laws explain other stable orbits, such as the orbit of a satellite around Earth or of a moon around a planet. But orbits do not always stay the same. For example, you've probably heard of satellites crashing to Earth from orbit, proving that orbits can sometimes change dramatically. To understand how and why orbits sometimes change, we need to consider the role of energy in orbits.

Orbital Energy Consider the orbit of a planet around the Sun. The planet has both kinetic energy (because it is moving around the Sun) and gravitational potential energy (because it would fall toward the Sun if it stopped orbiting). Its kinetic energy depends on its orbital speed, and its gravitational potential energy depends on its distance from the Sun. Because the planet's distance and speed both vary as it orbits the Sun, its gravitational potential energy and kinetic energy also vary as it orbits (**FIGURE 4.20**). However, the planet's total **orbital energy**—the sum of its kinetic and gravitational potential energies—stays the same. This fact is a consequence of the law of conservation of energy. As long as no other object causes the planet to gain or lose orbital energy, its orbital energy cannot change and its orbit must remain the same.

Total orbital energy = gravitational potential energy + kinetic energy

FIGURE 4.20 The total orbital energy of a planet stays constant throughout its orbit, because its gravitational potential energy increases when its kinetic energy decreases, and vice versa.

MATHEMATICAL INSIGHT 4.3

Newton's Version of Kepler's Third Law

Newton's version of Kepler's third law relates the orbital periods, distances, and masses of any pair of orbiting objects, such as the Sun and a planet, a planet and a moon, or two stars in a binary star system. Mathematically, we write it as follows:

$$p^2 = \frac{4\pi^2}{G(M_1 + M_2)} a^3$$

where M_1 and M_2 are the object masses, p is their orbital period, and a is the average distance between their centers. The term $4\pi^2$ is simply a number ($4\pi^2 \approx 4 \times 3.14^2 = 39.44$); G is the gravitational constant, which is measured experimentally.

If we measure the orbital period and distance of one object orbiting another, we can use Newton's equation to calculate the sum $M_1 + M_2$ of the object masses. If one object is much more massive than the other, we essentially learn its mass. For example, when we apply the law to a planet orbiting the Sun, the sum $M_{Sun} + M_{planet}$ is pretty much just M_{Sun} because the Sun is so much more massive than any planet. We can therefore use any planet's orbital period and distance from the Sun to calculate the mass of the Sun.

EXAMPLE 1: Earth orbits the Sun in 1 year at an average distance of 150 million kilometers (1 AU). Calculate the Sun's mass.

SOLUTION:

Step 1 Understand: We will use Newton's version of Kepler's third law. For Earth's orbit around the Sun, this law takes the form

$$(p_{Earth})^2 = \frac{4\pi^2}{G(M_{Sun} + M_{Earth})} (a_{Earth})^3$$

The Sun is much more massive than Earth, so the sum $M_{Sun} + M_{Earth}$ is approximately the Sun's mass alone, M_{Sun}. We therefore rewrite the equation as

$$(p_{Earth})^2 \approx \frac{4\pi^2}{G \times M_{Sun}} (a_{Earth})^3$$

Step 2 Solve: We know Earth's orbital period (p_{Earth}) and average distance (a_{Earth}), so the above equation contains only one unknown: M_{Sun}. To solve for this unknown, we multiply both sides by M_{Sun} and divide both sides by $(p_{Earth})^2$:

$$M_{Sun} \approx \frac{4\pi^2 (a_{Earth})^3}{G(p_{Earth})^2}$$

We now plug in the values $p_{Earth} = 1$ yr, which is the same as 3.15×10^7 s; $a_{Earth} \approx 150$ million km, or 1.5×10^{11} m; and the experimentally measured value $G = 6.67 \times 10^{-11}$ m³/(kg × s²). The equation becomes

$$M_{Sun} \approx \frac{4\pi^2}{\left(6.67 \times 10^{-11} \dfrac{m^3}{kg \times s^2}\right)} \frac{(1.5 \times 10^{11} \text{ m})^3}{(3.15 \times 10^7 \text{ s})^2}$$

$$= 2 \times 10^{30} \text{ kg}$$

Step 3 Explain: Simply by knowing Earth's orbital period and average distance, along with the gravitational constant, G, we were able to use Newton's version of Kepler's third law to find that the Sun's mass is about 2×10^{30} kilograms. (Note: G was not measured until more than 100 years after Newton published *Principia*, so Newton was not able to calculate masses in absolute units.)

EXAMPLE 2: A *geosynchronous satellite* orbits Earth with the same period as that of Earth's rotation: 1 sidereal day, or about 23 hours, 56 minutes, 4 seconds [**Section S1.1**]. Calculate the orbital distance of a geosynchronous satellite.

SOLUTION:

Step 1 Understand: We need Newton's version of Kepler's third law in a form that we can apply to this problem. A satellite is much less massive than Earth ($M_{Earth} + M_{satellite} \approx M_{Earth}$), so we put the law in this form:

$$(p_{satellite})^2 \approx \frac{4\pi^2}{G \times M_{Earth}} (a_{satellite})^3$$

Step 2 Solve: To solve for the satellite's distance, $a_{satellite}$, we multiply both sides of the equation by $(G \times M_{Earth})/4\pi^2$ and then take the cube root of both sides:

$$a_{satellite} \approx \sqrt[3]{\frac{G \times M_{Earth}}{4\pi^2} (p_{satellite})^2}$$

If you now plug in the given value $p_{satellite} = 1$ sidereal day $\approx$ 86,164 s, along with Earth's mass and G, you will find that $a_{satellite} \approx$ 42,000 km. (You should try the calculation for yourself.)

Step 3 Explain: We have found that a geosynchronous satellite orbits at a distance of 42,000 kilometers above the *center* of Earth, which is about 35,600 kilometers above Earth's surface.

Generalizing from planets to other objects leads to an important idea about motion throughout the cosmos: *Orbits cannot change spontaneously.* Left undisturbed, planets would forever keep the same orbits around the Sun, moons would keep the same orbits around their planets, and stars would keep the same orbits in their galaxies.

Gravitational Encounters Although orbits cannot change spontaneously, they can change through exchanges of energy. One way that two objects can exchange orbital energy is through a **gravitational encounter**, in which they pass near enough so that each can feel the effects of the other's gravity. For example, in the rare cases in which a comet happens to pass near a planet, the comet's orbit can change dramatically. **FIGURE 4.21** shows a comet headed toward the Sun on an unbound orbit. The comet's close passage by Jupiter allows the comet and Jupiter to exchange energy. In this case, the comet loses so much orbital energy that its orbit changes from unbound to bound and elliptical. Jupiter gains exactly as much energy as the comet loses, but the effect on Jupiter is unnoticeable because of its much greater mass.

Spacecraft engineers can use the same basic idea in reverse. For example, on its way to Pluto, the *New Horizons* spacecraft was deliberately sent past Jupiter on a path that allowed it to gain orbital energy at Jupiter's expense. This extra orbital energy boosted the spacecraft's speed; without this boost, it would have needed four extra years to reach Pluto. The effect of the tiny spacecraft on Jupiter was negligible.

A similar dynamic sometimes occurs naturally and may explain why most comets orbit so far from the Sun. Astronomers think that most comets once orbited in the same region of the solar system as the large outer planets

[**Section 12.2**]. Gravitational encounters with Jupiter or the other large planets then caused some of these comets to be "kicked out" into much more distant orbits around the Sun; some may have been ejected from the solar system completely.

Atmospheric Drag Friction can cause objects to lose orbital energy. Consider a satellite orbiting Earth. If the orbit is fairly low—such as a few hundred kilometers above Earth's surface—the satellite experiences a bit of drag from Earth's thin upper atmosphere. This drag gradually causes the satellite to lose orbital energy until it finally plummets to Earth. The satellite's lost orbital energy is converted to thermal energy in the atmosphere, which is why a falling satellite usually burns up.

Friction may also have played a role in shaping the current orbits of some of the small moons of Jupiter and other planets. These moons may once have orbited the Sun independently, and their orbits could not have changed spontaneously. However, the outer planets probably once were surrounded by clouds of gas [**Section 8.2**], and friction would have slowed objects passing through this gas. Some of these small objects may have lost just enough energy to friction to allow them to be "captured" as moons. (Mars may have captured its two small moons in a similar way.)

Escape Velocity An object that gains orbital energy moves into an orbit with a higher average altitude. For example, if we want to boost the orbital altitude of a spacecraft, we can give it more orbital energy by firing a rocket. The chemical potential energy released by the rocket fuel is converted to orbital energy for the spacecraft.

If we give a spacecraft enough orbital energy, it may end up in an unbound orbit that allows it to *escape* Earth completely (**FIGURE 4.22**). For example, when we send a space probe to Mars, we must use a large rocket that gives the probe enough energy to leave Earth orbit. Although it would probably make more sense to say that the probe achieves "escape energy," we instead say that it achieves **escape velocity**. The escape velocity from Earth's surface is about 40,000 km/hr,

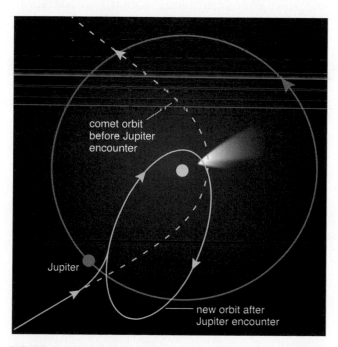

FIGURE 4.21 This diagram shows a comet in an unbound orbit of the Sun that happens to pass near Jupiter. The comet loses orbital energy to Jupiter, changing its unbound orbit to a bound orbit around the Sun.

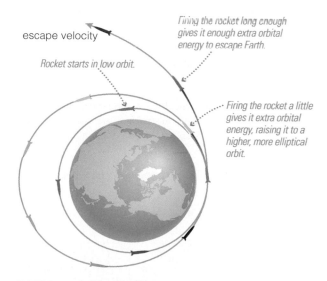

FIGURE 4.22 interactive figure An object with escape velocity has enough orbital energy to escape Earth completely.

or 11 km/s, meaning that this is the minimum velocity required to escape Earth's gravity for a spacecraft that starts near the surface.

Note that escape velocity does not depend on the mass of the escaping object—*any* object must travel at a velocity of 11 km/s to escape from Earth, whether it is an individual atom or molecule escaping from the atmosphere, a spacecraft being launched into deep space, or a rock blasted into the sky by a large impact. Escape velocity *does* depend on whether you start from the surface or from someplace high above the surface. Because gravity weakens with distance, it takes less energy—and hence a lower velocity—to escape from a point high above Earth than from Earth's surface.

How does gravity cause tides?

If you've spent time near an ocean, you've probably observed the rising and falling of the tides. In most places, tides rise and fall twice each day. We can understand the basic cause of tides by examining the gravitational attraction between Earth and the Moon. We'll then see how the same ideas explain many other phenomena that we can observe throughout the universe, including the synchronous rotation of our own Moon and many other worlds.

The Moon's Tidal Force Gravity attracts Earth and the Moon toward each other (with the Moon staying in orbit as

The Origin of Tides

Many people believe that tides arise because the Moon pulls Earth's oceans toward it. But if that were the whole story, there would be a bulge only on the side of Earth facing the Moon, and hence only one high tide each day. The correct explanation for tides must account for why Earth has *two* tidal bulges.

Only one explanation works: Earth must be stretching from its center in both directions (toward and away from the Moon). This stretching force, or tidal force, arises from the *difference* in the force of gravity attracting different parts of Earth to the Moon. In fact, stretching due to tides affects many objects, not just Earth. Many moons are stretched into slightly oblong shapes by tidal forces caused by their parent planets, and mutual tidal forces stretch close binary stars into teardrop shapes. In regions where gravity is extremely strong, such as near a black hole, tides can have even more dramatic effects.

it "falls around" Earth), but it affects different parts of Earth slightly differently: Because the strength of gravity declines with distance, the gravitational attraction of each part of Earth to the Moon becomes weaker as we go from the side of Earth facing the Moon to the side facing away from the Moon. This difference in attraction creates a "stretching force," or **tidal force**, that stretches the entire Earth to create two tidal bulges, one facing the Moon and one opposite the

MATHEMATICAL INSIGHT 4.4

Escape Velocity

A simple formula allows us to calculate the escape velocity from any planet, moon, or star:

$$v_{escape} = \sqrt{2 \times G \times \frac{M}{R}}$$

where M is the object's mass, R is the starting distance above the object's center, and G is the gravitational constant. If you use this formula to calculate the escape velocity from an object's *surface*, replace R with the object's radius.

EXAMPLE 1: Calculate the escape velocity from the Moon's surface. Compare it to the 11 km/s escape velocity from Earth.

SOLUTION:

Step 1 Understand: We use the above formula; because we seek the escape velocity from the Moon's *surface*, we use the Moon's radius as R. From Appendix E, the Moon's mass and radius are $M_{Moon} = 7.4 \times 10^{22}$ kg and $R_{Moon} = 1.7 \times 10^6$ m.

Step 2 Solve: We substitute the Moon's mass and radius into the escape velocity formula:

$$v_{escape} = \sqrt{2 \times G \times \frac{M_{Moon}}{R_{Moon}}}$$

$$= \sqrt{2 \times \left(6.67 \times 10^{-11} \frac{m^3}{kg \times s^2}\right) \times \frac{(7.4 \times 10^{22} \text{ kg})}{(1.7 \times 10^6 \text{ m})}}$$

$$\approx 2400 \text{ m/s} = 2.4 \text{ km/s}$$

Step 3 Explain: Escape velocity from the Moon's surface is 2.4 km/s, which is less than one-fourth the escape velocity (11 km/s) from Earth's surface.

EXAMPLE 2: Suppose a future space station orbits Earth in geosynchronous orbit, 42,000 kilometers above the center of Earth (see Mathematical Insight 4.3). At what velocity must a spacecraft be launched from the station to escape Earth?

SOLUTION:

Step 1 Understand: We seek the escape velocity from a satellite orbiting 42,000 kilometers above the center of Earth, so we use the escape velocity formula with the mass of Earth ($M_{Earth} = 6.0 \times 10^{24}$ kg) and R set to the satellite's distance ($R = 42,000$ km $= 4.2 \times 10^7$ m).

Step 2 Solve: With the above values, we find

$$v_{escape} = \sqrt{2 \times G \times \frac{M_{Earth}}{R_{orbit}}}$$

$$= \sqrt{2 \times \left(6.67 \times 10^{-11} \frac{m^3}{kg \times s^2}\right) \times \frac{(6.0 \times 10^{24} \text{ kg})}{(4.2 \times 10^7 \text{ m})}}$$

$$= 4400 \text{ m/s} = 4.4 \text{ km/s}$$

Step 3 Explain: The escape velocity from geosynchronous orbit is 4.4 km/s—considerably lower than the 11 km/s escape velocity from Earth's surface. It would therefore require substantially less fuel to launch a spacecraft from the space station than from Earth, which is why some people propose building future spacecraft at future space stations.

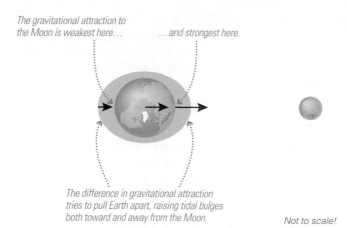

The gravitational attraction to the Moon is weakest here...   *...and strongest here.*

The difference in gravitational attraction tries to pull Earth apart, raising tidal bulges both toward and away from the Moon.   *Not to scale!*

FIGURE 4.23 Tides are created by the difference in the force of attraction between the Moon and different parts of Earth. The two daily high tides occur as a location on Earth rotates through the two tidal bulges. (The diagram greatly exaggerates the tidal bulges, which raise the oceans only about 2 meters and the land only about a centimeter.)

Moon (**FIGURE 4.23**). If you are still unclear about why there are *two* tidal bulges, think about a rubber band: If you pull on a rubber band, it will stretch in both directions relative to its center, even if you pull on only one side (while holding the other side still). In the same way, Earth stretches on both sides even though the Moon is tugging harder on only one side.

Tides affect both land and ocean, but we generally notice only the ocean tides because water flows much more readily than land. Earth's rotation carries any location through each of the two bulges each day, creating two high tides. Low tides occur when the location is at the points halfway between the two tidal bulges. The two "daily" high tides actually come slightly more than 12 hours apart. Because of its orbital motion around Earth, the Moon reaches its highest point in the sky at any location about every 24 hours 50 minutes, rather than every 24 hours. In other words, the tidal cycle of two high tides and two low tides takes about 24 hours 50 minutes, with each high tide occurring about 12 hours 25 minutes after the previous one.

The height and timing of ocean tides vary considerably from place to place on Earth, depending on factors such as latitude, the orientation of the coastline (such as whether it is north-facing or west-facing), and the depth and shape of any channel through which the rising tide must flow. For example, while the tide rises gradually in most locations, the incoming tide near the famous abbey on Mont-Saint-Michel, France, moves much faster than a person can swim (**FIGURE 4.24**). In centuries past, the Mont was an island twice a day at high tide but was connected to the mainland at low tide. Many pilgrims drowned when they were caught unprepared by the tide rushing in. (Today, a human-made causeway keeps the island connected to the mainland.) Another unusual tidal pattern occurs in coastal states along the northern shore of the Gulf of Mexico, where topography and other factors combine to make only one noticeable high tide and low tide each day.

The Tidal Effect of the Sun The Sun also exerts a tidal force on Earth, causing Earth to stretch along the Sun-Earth line. You might at first guess that the Sun's tidal force would be more than the Moon's, since the Sun's mass is more than a million times that of the Moon. Indeed, the *gravitational* force between Earth and the Sun is much greater than that between Earth and the Moon, which is why Earth orbits the Sun. However, the much greater distance to the Sun (than to the Moon) means that the *difference* in the Sun's pull on the near and far sides of Earth is relatively small.

The overall tidal force caused by the Sun is a little less than half that caused by the Moon (**FIGURE 4.25**). When the tidal forces of the Sun and the Moon work together, as is the case at both new moon and full moon, we get the especially pronounced *spring tides* (so named because the water tends to "spring up" from Earth). When the tidal forces of the Sun and the Moon counteract each other, as is the case at first- and third-quarter moon, we get the relatively small tides known as *neap tides*.

THINK ABOUT IT

Explain why any tidal effects on Earth caused by the other planets would be unnoticeably small.

FIGURE 4.24 Photographs of high and low tide at the abbey of Mont-Saint-Michel, France, one of the world's most popular tourist destinations. Here the tide rushes in much faster than a person can swim. Before a causeway was built (visible at the far left), the Mont was accessible by land only at low tide. At high tide, it became an island.

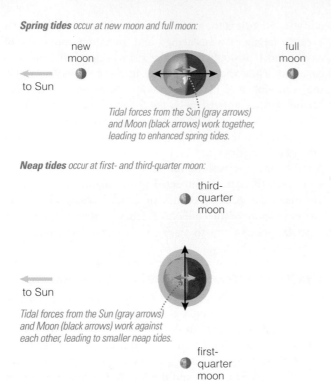

Spring tides occur at new moon and full moon:

new moon

full moon

to Sun

Tidal forces from the Sun (gray arrows) and Moon (black arrows) work together, leading to enhanced spring tides.

Neap tides occur at first- and third-quarter moon:

third-quarter moon

to Sun

Tidal forces from the Sun (gray arrows) and Moon (black arrows) work against each other, leading to smaller neap tides.

first-quarter moon

FIGURE 4.25 interactive figure The Sun exerts a tidal force on Earth less than half as strong as that from the Moon. When the tidal forces from the Sun and Moon work together at new moon and full moon, we get enhanced *spring tides*. When they work against each other at first- and third-quarter moons, we get smaller *neap tides*.

Tidal Friction So far, we have talked as if Earth rotated smoothly through the tidal bulges. But because tidal forces stretch Earth itself, the process creates friction, called **tidal friction**. FIGURE 4.26 shows the effects of this friction. In essence, the Moon's gravity tries to keep the tidal bulges on the Earth-Moon line, while Earth's rotation tries to pull the bulges around with it. The resulting "compromise" keeps the bulges just ahead of the Earth-Moon line at all times.

The slight misalignment of the tidal bulges with the Earth-Moon line causes two important effects. First, the Moon's gravity always pulls back on the bulges, slowing Earth's rotation. Second, the gravity of the bulges pulls the Moon slightly ahead in its orbit, causing the Moon to move farther from Earth. These effects are barely noticeable on human time scales—for example, tidal friction increases the length of a day

by only about 1 second every 50,000 years*—but they add up over billions of years. Early in Earth's history, a day may have been only 5 or 6 hours long and the Moon may have been one-tenth or less its current distance from Earth. These changes also provide a great example of conservation of angular momentum and energy: The Moon's growing orbit gains the angular momentum and energy that Earth loses as its rotation slows.

The Moon's Synchronous Rotation Recall that the Moon always shows (nearly) the same face to Earth, a trait called **synchronous rotation** (see Figure 2.23). Synchronous rotation may seem like an extraordinary coincidence, but it is a natural consequence of tidal friction.

Because Earth is more massive than the Moon, Earth's tidal force has a greater effect on the Moon than the Moon's tidal force has on Earth. This tidal force gives the Moon two tidal bulges along the Earth-Moon line, much like the two tidal bulges that the Moon creates on Earth. (The Moon does not have visible tidal bulges, but it does indeed have excess mass along the Earth-Moon line.) If the Moon rotated through its tidal bulges in the same way as Earth, the resulting friction would cause the Moon's rotation to slow down. This is exactly what we think happened long ago.

The Moon probably once rotated much faster than it does today. As a result, it *did* rotate through its tidal bulges, and its rotation gradually slowed. Once the Moon's rotation slowed to the point at which the Moon and its bulges rotated at the same rate—that is, synchronously with the orbital period—there was no further source for tidal friction. The Moon's synchronous rotation was therefore a natural outcome of Earth's tidal effects on the Moon. In fact, if the Moon and Earth stay together long enough (for another few hundred billion years), the gradual slowing of Earth's rotation will eventually make Earth keep the same face to the Moon as well.

Tidal Effects on Other Worlds Tidal forces and tidal friction affect many worlds. Synchronous rotation is especially common. For example, Jupiter's four large moons (Io, Europa, Ganymede, and Callisto) keep nearly the same face toward Jupiter at all times, as do many other moons. Pluto and its moon Charon *both* rotate synchronously: Like two dancers, they

---

*This effect is overwhelmed on short time scales by other effects due to slight changes in Earth's internal mass distribution; these changes can alter Earth's rotation period by a second or more per year, which is why "leap seconds" are occasionally added to or subtracted from the year.

If Earth didn't rotate, tidal bulges would be oriented along the Earth-Moon line.

Friction with the rotating Earth pulls the tidal bulges slightly ahead of the Earth-Moon line.

The Moon's gravity tries to pull the bulges back into line, slowing Earth's rotation.

The gravity of the bulges pulls the Moon ahead, increasing its orbital distance.

Moon

Not to scale!

FIGURE 4.26 Earth's rotation pulls its tidal bulges slightly ahead of the Earth-Moon line, leading to gravitational effects that gradually slow Earth's rotation and increase the Moon's orbital energy and distance.

always keep the same face toward each other. Many binary star systems also rotate in this way. Some moons and planets exhibit variations on synchronous rotation. For example, Mercury rotates exactly three times for every two orbits of the Sun. This pattern ensures that Mercury's tidal bulge always aligns with the Sun at perihelion, where the Sun exerts its strongest tidal force.

Tidal forces play other roles in the cosmos as well. They can alter the shapes of objects by stretching them along the line of tidal bulges. In Chapter 11, we'll see how tidal forces also lead to the astonishing volcanic activity of Jupiter's moon Io and the possibility of a subsurface ocean on its moon Europa. As you study astronomy, you'll encounter many more cases where tides and tidal friction play important roles.

Why do all objects fall at the same rate?

We will discuss many more applications of the universal law of gravitation in this book, but for now let's look at just one more: Galileo's discovery that the acceleration of a falling object is independent of its mass.

If you drop a rock, the force acting on the rock is the force of gravity. The two masses involved are the mass of Earth and the mass of the rock, which we'll denote M_{Earth} and M_{rock}, respectively. The distance is the distance from the *center of Earth* to the center of the rock. If the rock isn't too far above Earth's surface, this distance is approximately the radius of Earth, R_{Earth} (about 6400 kilometers), so the force of gravity acting on the rock is

$$F_g = G \frac{M_{Earth} M_{rock}}{d^2} \approx G \frac{M_{Earth} M_{rock}}{(R_{Earth})^2}$$

According to Newton's second law of motion ($F = ma$), this force is equal to the product of the rock's mass and acceleration. That is,

$$G \frac{M_{Earth} \cancel{M_{rock}}}{(R_{Earth})^2} = \cancel{M_{rock}} a_{rock}$$

Note that M_{rock} "cancels" because it appears on both sides of the equation (as a multiplier), giving Galileo's result that the acceleration of the rock—or of any falling object—does not depend on the object's mass.

MATHEMATICAL INSIGHT 4.5

The Acceleration of Gravity

The text shows that the acceleration of a falling rock near the surface of Earth is

$$a_{rock} = G \times \frac{M_{Earth}}{(R_{Earth})^2}$$

Because this formula applies to *any* falling object on Earth, it is the *acceleration of gravity, g*. Calculating g is easy. Simply plug in Earth's mass (6.0×10^{24} kg) and radius (6.4×10^6 m):

$$g_{Earth} = G \times \frac{M_{Earth}}{(R_{Earth})^2}$$

$$= \left(6.67 \times 10^{-11} \frac{m^3}{kg \times s^2}\right) \times \frac{6.0 \times 10^{24} \text{ kg}}{(6.4 \times 10^6 \text{ m})^2} = 9.8 \frac{m}{s^2}$$

EXAMPLE 1: What is the acceleration of gravity on the Moon?

SOLUTION:

Step 1 Understand: We want the acceleration of gravity on the Moon's surface, so we use the above formula with the Moon's mass (7.4×10^{22} kg) and radius (1.7×10^6 m).

Step 2 Solve: The formula becomes

$$g_{Moon} = G \times \frac{M_{Moon}}{(R_{Moon})^2}$$

$$= \left(6.67 \times 10^{-11} \frac{m^3}{kg \times s^2}\right) \times \frac{7.4 \times 10^{22} \text{ kg}}{(1.7 \times 10^6 \text{ m})^2} = 1.7 \frac{m}{s^2}$$

Step 3 Explain: The acceleration of gravity on the Moon is 1.7 m/s², or about one-sixth that on Earth, so objects on the Moon weigh about one-sixth of what they weigh on Earth.

EXAMPLE 2: The Space Station orbits at an altitude of roughly 350 kilometers above Earth's surface. What is the acceleration of gravity at this altitude?

SOLUTION:

Step 1 Understand: Because the Space Station is significantly above Earth's surface, we cannot use the approximation $d \approx R_{Earth}$ that we used in the text. Instead, we must go back to Newton's second law and set the gravitational force on the Space Station equal to its mass times acceleration. The acceleration in this equation is the acceleration of gravity at the Space Station's altitude.

Step 2 Solve: We write Newton's second law with the force being the force of gravity acting between Earth and the Space Station, which we set equal to the Space Station's mass times its acceleration:

$$G \times \frac{M_{Earth} M_{station}}{d^2} = M_{station} \times a_{station}$$

You should confirm that when we solve this equation for the acceleration of gravity, we find

$$a_{station} = G \times \frac{M_{Earth}}{d^2}$$

In this case, the distance d is the 6400-kilometer radius of Earth *plus* the 350-kilometer altitude of the Station, or $d = 6750$ km $= 6.75 \times 10^6$ m. The gravitational acceleration is

$$a_{station} = G \times \frac{M_{Earth}}{d^2}$$

$$= \left(6.67 \times 10^{-11} \frac{m^3}{kg \times s^2}\right) \times \frac{6.0 \times 10^{24} \text{ kg}}{(6.75 \times 10^6 \text{ m})^2} = 8.8 \frac{m}{s^2}$$

Step 3 Explain: The acceleration of gravity in low Earth orbit is 8.8 m/s², which is only about 10% less than the 9.8 m/s² acceleration of gravity at Earth's surface. We see again that lack of gravity cannot be the reason astronauts are weightless in orbit; rather, they are weightless because they are in free-fall.

The fact that objects of different masses fall with the same acceleration struck Newton as an astounding coincidence, even though his own equations showed it to be so. For the next 240 years, this seemingly odd coincidence remained just that—a coincidence—in the minds of most scientists. However, in 1915, Einstein showed that it is not a coincidence at all. Rather, it reveals something deeper about the nature of gravity and of the universe. Einstein described the new insights in his *general theory of relativity* (the topic of Chapter S3).

The Big Picture

Putting Chapter 4 into Context

We've covered a lot of ground in this chapter, from the scientific terminology of motion to the overarching principles that govern motion throughout the universe. Be sure you grasp the following "big picture" ideas:

- Understanding the universe requires understanding motion. Motion may seem complex, but it can be described simply using Newton's three laws of motion.

- Today, we know that Newton's laws of motion stem from deeper physical principles, including the laws of conservation of momentum, of angular momentum, and of energy. These principles enable us to understand a wide range of astronomical phenomena.

- Newton also discovered the universal law of gravitation, which explains how gravity holds planets in their orbits and much more—including how satellites can reach and stay in orbit, the nature of tides, and why the Moon rotates synchronously around Earth.

- Newton's discoveries showed that the same physical laws we observe on Earth apply throughout the universe. The universality of physics opens up the entire cosmos as a possible realm of human study.

SUMMARY OF KEY CONCEPTS

4.1 DESCRIBING MOTION: EXAMPLES FROM DAILY LIFE

- **How do we describe motion?** Speed is the rate at which an object is moving. **Velocity** is speed in a certain direction. **Acceleration** is a change in velocity, meaning a change in either speed or direction. **Momentum** is mass times velocity. A **force** can change an object's momentum, causing it to accelerate.

- **How is mass different from weight?** An object's mass is

the same no matter where it is located, but its weight varies with the strength of gravity or other forces acting on the object. An object becomes **weightless** when it is in **free-fall**, even though its mass is unchanged.

4.2 NEWTON'S LAWS OF MOTION

- **How did Newton change our view of the universe?** Newton showed that the same physical laws that operate on Earth also operate in the heavens, making it possible to learn about the universe by studying physical laws on Earth.

- **What are Newton's three laws of motion?** (1) An object moves at constant velocity if there is no net force acting upon it. (2) Force = mass × acceleration ($F = ma$). (3) For any force, there is always an equal and opposite reaction force.

4.3 CONSERVATION LAWS IN ASTRONOMY

- **Why do objects move at constant velocity if no force acts on them?** Conservation of momentum means that an object's momentum cannot change unless the object transfers momentum to or from other objects. When no force is present, no momentum can be transferred so an object must maintain its speed and direction.

- **What keeps a planet rotating and orbiting the Sun?**

Conservation of angular momentum means that a planet's rotation and orbit cannot change unless the planet transfers angular momentum to another object. The planets in our solar system do not exchange substantial angular momentum with each other or anything else, so their orbits and rotation rates remain steady.

- **Where do objects get their energy?** Energy is always

conserved—it can be neither created nor destroyed. Objects received whatever energy they now have from exchanges of energy with other objects. Energy comes in three basic categories—**kinetic, radiative,** and **potential**.

kinetic energy

radiative energy   potential energy

4.4 THE UNIVERSAL LAW OF GRAVITATION

- **What determines the strength of gravity?** The **universal**

law of gravitation states that every object attracts every other object with a gravitational force that is proportional to the product of the objects' masses and declines with the square of the distance between their centers:

$$F_g = G \frac{M_1 M_2}{d^2}$$

- **How does Newton's law of gravity extend Kepler's laws?** (1) Newton showed that any object going around another object will obey Kepler's first two laws. (2) He showed that ellipses (or circles), which define **bound orbits**, are not the only possible orbital shape—orbits can also be **unbound** and in the form of parabolas or hyperbolas. (3) He showed that two objects actually orbit their common **center of mass**. (4) **Newton's version of Kepler's third law** allows us to calculate the masses of orbiting objects from their orbital periods and distances.

4.5 ORBITS, TIDES, AND THE ACCELERATION OF GRAVITY

- **How do gravity and energy allow us to understand orbits?** Gravity determines orbits, and an object cannot change its orbit unless it gains or loses **orbital energy**—the sum of its kinetic and gravitational potential energies—through energy

transfer with other objects. If an object gains enough orbital energy, it may achieve **escape velocity** and leave the gravitational influence of the object it was orbiting.

- **How does gravity cause tides?** The Moon's gravity creates a **tidal force** that stretches Earth along the Earth-Moon line, causing Earth to bulge both toward and away from the Moon. Earth's rotation carries us through the two bulges each day, giving us two daily high tides and two daily low tides. Tidal forces also lead to **tidal friction**, which is gradually slowing Earth's rotation and explains the synchronous rotation of the Moon.

- **Why do all objects fall at the same rate?** Newton's equations show that the acceleration of gravity is independent of the mass of a falling object, so all objects fall at the same rate.

VISUAL SKILLS CHECK

Use the following questions to check your understanding of some of the many types of visual information used in astronomy. Answers are provided in Appendix J. For additional practice, try the Chapter 4 Visual Quiz at MasteringAstronomy®.

The figure above, based on Figure 4.23, shows how the Moon causes tides on Earth. Note that the North Pole is in the center of the diagram, so the numbers 1 through 4 label points along Earth's equator.

1. What do the three black arrows represent?
   a. the tidal force Earth exerts on the Moon
   b. the Moon's gravitational force at different points on Earth
   c. the direction in which Earth's water is flowing
   d. Earth's orbital motion
2. Where is it high tide?
   a. point 1 only
   b. point 2 only
   c. points 1 and 3
   d. points 2 and 4
3. Where is it low tide?
   a. point 1 only
   b. point 2 only
   c. points 1 and 3
   d. points 2 and 4

4. What time is it at point 1?
   a. noon
   b. midnight
   c. 6 a.m.
   d. cannot be determined from the information in the figure
5. The light blue ellipse represents tidal bulges. In what way are these bulges drawn inaccurately?
   a. There should be only one bulge rather than two.
   b. They should be aligned with the Sun rather than the Moon.
   c. They should be much smaller compared to Earth.
   d. They should be more pointy in shape.

REVIEW QUESTIONS

Short-Answer Questions Based on the Reading

1. Define *speed, velocity,* and *acceleration.* What are the units of acceleration? What is the *acceleration of gravity*?
2. Define *momentum* and *force.* What do we mean when we say that momentum can be changed only by a *net force*?
3. What is *free-fall,* and why does it make you *weightless*? Briefly describe why astronauts are weightless in the Space Station.
4. State *Newton's three laws of motion.* For each law, give an example of its application.
5. Describe the laws of conservation of momentum, angular momentum, and energy. Give an example of how each is important in astronomy.
6. Define *kinetic energy, radiative energy,* and *potential energy,* with at least two examples for each.
7. Define *temperature* and *thermal energy.* How are they related? How are they different?
8. What do we mean by *mass-energy*? Explain the formula $E = mc^2$.
9. Summarize the *universal law of gravitation* both in words and with an equation.
10. What is the difference between a *bound* and an *unbound orbit*? What orbital shapes are possible?
11. What do we need to know if we want to measure an object's mass with *Newton's version of Kepler's third law*? Explain.
12. Explain why orbits cannot change spontaneously, and how a *gravitational encounter* can cause a change. How can an object achieve *escape velocity*?
13. Explain how the Moon creates tides on Earth. Why do we have two high and low tides each day? How do the tides vary with the phase of the Moon?
14. What is *tidal friction*? What effects does it have on Earth? How does it explain the Moon's synchronous rotation?

TEST YOUR UNDERSTANDING

Does It Make Sense?

Decide whether the statement makes sense (or is clearly true) or does not make sense (or is clearly false). Explain clearly; not all of these have definitive answers, so your explanation is more important than your chosen answer.

15. If you could buy a pound of chocolate on the Moon, using a pound scale from Earth, you'd get more chocolate than if you bought a pound on Earth.
16. Suppose you could enter a vacuum chamber (a chamber with no air in it) on Earth. Inside this chamber, a feather would fall at the same rate as a rock.
17. When an astronaut goes on a space walk outside the Space Station, she will quickly float away from the station unless she has a tether holding her to the station.
18. I used Newton's version of Kepler's third law to calculate Saturn's mass from orbital characteristics of its moon Titan.
19. If the Sun were magically replaced with a giant rock that had precisely the same mass, Earth's orbit would not change.
20. The fact that the Moon rotates once in precisely the time it takes to orbit Earth once is such an astonishing coincidence that scientists probably never will be able to explain it.
21. Venus has no oceans, so it could not have tides even if it had a moon (which it doesn't).
22. If an asteroid passed by Earth at just the right distance, Earth's gravity would capture it and make it our second moon.
23. When I drive my car at 30 miles per hour, it has more kinetic energy than it does at 10 miles per hour.
24. Someday soon, scientists are likely to build an engine that produces more energy than it consumes.

Quick Quiz

Choose the best answer to each of the following. Explain your reasoning with one or more complete sentences.

25. Which one of the following describes an object that is *accelerating*? (a) A car traveling on a straight, flat road at 50 miles per hour. (b) A car traveling on a straight uphill road at 30 miles per hour. (c) A car going around a circular track at a steady 100 miles per hour.
26. Suppose you visited another planet. (a) Your mass and weight would be the same as they are on Earth. (b) Your mass would be the same as on Earth, but your weight would be different. (c) Your weight would be the same as on Earth, but your mass would be different.
27. Which person is weightless? (a) A child in the air as she plays on a trampoline. (b) A scuba diver exploring a deep-sea wreck. (c) An astronaut on the Moon.
28. Consider the statement "There's no gravity in space." This statement is (a) completely false. (b) false if you are close to a planet or moon, but true in between the planets. (c) completely true.
29. To make a rocket turn left, you need to (a) fire an engine that shoots out gas to the left. (b) fire an engine that shoots out gas to the right. (c) spin the rocket clockwise.
30. Compared to its angular momentum when it is farthest from the Sun, Earth's angular momentum when it is nearest to the Sun is (a) greater. (b) less. (c) the same.
31. The gravitational potential energy of a contracting interstellar cloud (a) stays the same at all times. (b) gradually transforms into other forms of energy. (c) gradually grows larger.
32. If Earth were twice as far from the Sun, the force of gravity attracting Earth to the Sun would be (a) twice as strong. (b) half as strong. (c) one quarter as strong.
33. According to the universal law of gravitation, what would happen to Earth if the Sun were somehow replaced by a black hole of the same mass? (a) Earth would be quickly sucked into the black hole. (b) Earth would slowly spiral into the black hole. (c) Earth's orbit would not change.
34. If the Moon were closer to Earth, high tides would (a) be higher than they are now. (b) be lower than they are now. (c) occur three or more times a day rather than twice a day.

PROCESS OF SCIENCE

Examining How Science Works

35. *Testing Gravity.* Scientists are continually trying to learn whether our current understanding of gravity is complete or must be modified. Describe how the observed motion of spacecraft headed out of our solar system (such as the *Voyager* spacecraft) can be used to test the accuracy of our current theory of gravity.
36. *How Does the Table Know?* Thinking deeply about seemingly simple observations sometimes reveals underlying truths that we might otherwise miss. For example, think about holding a golf ball in one hand and a bowling ball in the other. To keep them motionless you must actively adjust the tension in your arm muscles so that each arm exerts a different upward force that

exactly balances the weight of each ball. Now, think about what happens when you set the balls on a table. Somehow, the table also exerts exactly the right amount of upward force to keep the balls motionless, even though their weights are very different. How does a table "know" to make the same type of adjustment that you make when you hold the balls motionless in your hands? (*Hint:* Think about the origin of the force pushing upward on the objects.)

GROUP WORK EXERCISE

37. *Your Ultimate Energy Source.* According to the law of conservation of energy, the energy your body is using right now had to come from somewhere else. Your task in this exercise is to trace the flow of that energy as far back in time as you can. Before you begin, assign the following roles to the people in your group: *Scribe* (takes notes on the group's activities), *Proposer* (proposes explanations to the group), *Skeptic* (points out weaknesses in proposed explanations), and *Moderator* (leads group discussion and makes sure the group works as a team). After you have your roles, make a list going backwards in time describing how the energy you are using right now has proceeded through time. Then, for each item on the list, state whether that energy was in the form of kinetic energy, gravitational potential energy, chemical potential energy, electrical potential energy, mass-energy, or radiative energy.

INVESTIGATE FURTHER

In-Depth Questions to Increase Your Understanding

Short-Answer/Essay Questions

38. *Units of Acceleration.*
   a. If you drop a rock from a very tall building, how fast will it be going after 4 seconds? **b.** As you sled down a steep, slick street, you accelerate at a rate of 4 meters per second squared. How fast will you be going after 5 seconds? **c.** You are driving along the highway at a speed of 60 miles per hour when you slam on the brakes. If your acceleration is at an average rate of −20 miles per hour per second, how long will it take to come to a stop?

39. *Gravitational Potential Energy.* For each of the following, which object has more gravitational potential energy, and how do you know?
   a. A bowling ball perched on a cliff ledge or a baseball perched on the same ledge **b.** A diver on a 10-meter platform or a diver on a 3-meter diving board **c.** A 100-kilogram satellite orbiting Jupiter or a 100-kilogram satellite orbiting Earth (Assume both satellites orbit at the same distance from their planet's center.)

40. *Einstein's Famous Formula.*
   a. What is the meaning of the formula $E = mc^2$? Be sure to define each variable. **b.** How does this formula explain the generation of energy by the Sun? **c.** How does this formula explain the destructive power of nuclear bombs?

41. *The Gravitational Law.*
   a. How does quadrupling the distance between two objects affect the gravitational force between them? **b.** Suppose the Sun were somehow replaced by a star with twice as much mass. What would happen to the gravitational force between Earth and the Sun? **c.** Suppose Earth were moved to one-third of its current distance from the Sun. What would happen to the gravitational force between Earth and the Sun?

42. *Allowable Orbits?*
   a. Suppose the Sun were replaced by a star with twice as much mass. Could Earth's orbit stay the same? Why or why not? **b.** Suppose Earth doubled in mass (but the Sun stayed the same as it is now). Could Earth's orbit stay the same? Why or why not?

43. *Head-to-Foot Tides.* You and Earth attract each other gravitationally, so you should also be subject to a tidal force resulting from the difference between the gravitational attraction felt by your feet and that felt by your head (at least when you are standing). Explain why you can't feel this tidal force.

44. *Synchronous Rotation.* Suppose the Moon had rotated *more slowly* when it formed than it does now. Would it still have ended up in synchronous rotation? Why or why not?

45. *Geostationary Orbit.* A satellite in *geostationary* orbit appears to remain stationary in the sky as seen from any particular location on Earth.
   a. Briefly explain why a geostationary satellite must orbit Earth in 1 *sidereal* day, rather than 1 solar day. **b.** Explain why a geostationary satellite must be in orbit around Earth's equator, rather than in some other orbit (such as around the poles). **c.** Home satellite dishes (such as those used for television) receive signals from communications satellites. Explain why these satellites must be in geostationary orbit.

46. *Elevator to Orbit.* Some people have proposed building a giant elevator from Earth's surface to the altitude of geosynchronous orbit. The top of the elevator would then have the same orbital distance and period as any satellite in geosynchronous orbit.
   a. Suppose you were to let go of an object at the top of the elevator. Would the object fall? Would it orbit Earth? Explain. **b.** Briefly explain why (not counting the huge costs for construction) the elevator would make it much cheaper and easier to put satellites in orbit or to launch spacecraft into deep space.

Quantitative Problems

Be sure to show all calculations clearly and state your final answers in complete sentences.

47. *Energy Comparisons.* Use the data in Table 4.1 to answer each of the following questions.
   a. Compare the energy of a 1-megaton H-bomb to the energy released by a major earthquake. **b.** If the United States obtained all its energy from oil, how much oil would be needed each year? **c.** Compare the Sun's annual energy output to the energy released by a supernova.

48. *Moving Candy Bar.* We can calculate the kinetic energy of any moving object with a very simple formula: kinetic energy $= \frac{1}{2}mv^2$, where m is the object's mass and v is its velocity or speed. Table 4.1 shows that metabolizing a candy bar releases about 10^6 joules. How fast must the candy bar travel to have the same 10^6 joules in the form of kinetic energy? (Assume the candy bar's mass is 0.2 kilogram.) Is your answer faster or slower than you expected?

49. *Spontaneous Human Combustion.* Suppose that all the mass in your body were suddenly converted into energy according to the formula $E = mc^2$. How much energy would be released? Compare this to the energy released by a 1-megaton H-bomb (see Table 4.1). What effect would your disappearance have on your surroundings?

50. *Fusion Power.* No one has yet succeeded in creating a commercially viable way to produce energy through nuclear fusion. However, suppose we could build fusion power plants using the hydrogen in water as a fuel. Based on the data in Table 4.1, how much water would we need each minute to meet U.S. energy needs? Could such a reactor power the entire United States with the water flowing from your kitchen sink? Explain. (*Hint:* Use the annual U.S. energy consumption to find the energy consumption per minute, and then divide by the energy yield from fusing 1 liter of water to figure out how many liters would be needed each minute.)

51. *Understanding Newton's Version of Kepler's Third Law.* Find the orbital period for the planet in each case. (*Hint:* The calculations for this problem are so simple that you will not need a calculator.) **a.** A planet with twice Earth's mass orbiting at a distance of 1 AU from a star with the same mass as the Sun **b.** A planet with the same mass as Earth orbiting at a distance of 1 AU from a star with four times the Sun's mass

52. *Using Newton's Version of Kepler's Third Law.*
   a. Find Earth's approximate mass from the fact that the Moon orbits Earth in an average time of 27.3 days at an average distance of 384,000 kilometers. (*Hint:* The Moon's mass is only about $\frac{1}{80}$ of Earth's.) **b.** Find Jupiter's mass from the fact that its moon Io orbits every 42.5 hours at an average distance of 422,000 kilometers. **c.** You discover a planet orbiting a distant star that has about the same mass as the Sun, with an orbital period of 63 days. What is the planet's orbital distance? **d.** Pluto's moon Charon orbits Pluto every 6.4 days with a semimajor axis of 19,700 kilometers. Calculate the *combined* mass of Pluto and Charon. **e.** Calculate the orbital period of a spacecraft in an orbit 300 kilometers above Earth's surface. **f.** Estimate the mass of the Milky Way Galaxy from the fact that the Sun orbits the galactic center every 230 million years at a distance of 27,000 light-years. (As we'll discuss in Chapter 23, this calculation actually tells us only the mass of the galaxy *within* the Sun's orbit.)

53. *Escape Velocity.* Calculate the escape velocity from each of the following.
   a. The surface of Mars (mass $= 0.11 M_{Earth}$, radius $= 0.53 R_{Earth}$) **b.** The surface of Mars's moon Phobos (mass $= 1.1 \times 10^{16}$ kg, radius $= 12$ km) **c.** The cloud tops of Jupiter (mass $= 317.8 M_{Earth}$, radius $= 11.2 R_{Earth}$) **d.** Our solar system, starting from Earth's orbit (*Hint:* Most of the mass of our solar system is in the Sun; $M_{Sun} = 2.0 \times 10^{30}$ kg.) **e.** Our solar system, starting from Saturn's orbit

54. *Weights on Other Worlds.* Calculate the acceleration of gravity on the surface of each of the following worlds. How much would *you* weigh, in pounds, on each of these worlds?
   a. Mars (mass $= 0.11 M_{Earth}$, radius $= 0.53 R_{Earth}$) **b.** Venus (mass $= 0.82 M_{Earth}$, radius $= 0.95 R_{Earth}$) **c.** Jupiter (mass $= 317.8 M_{Earth}$, radius $= 11.2 R_{Earth}$) Bonus: Given that Jupiter has no solid surface, how could you weigh yourself on Jupiter? **d.** Jupiter's moon Europa (mass $= 0.008 M_{Earth}$, radius $= 0.25 R_{Earth}$) **e.** Mars's moon Phobos (mass $= 1.1 \times 10^{16}$ kg, radius $= 12$ km)

55. *Gees.* Acceleration is sometimes measured in *gees*, or multiples of the acceleration of gravity: 1 gee (1*g*) means $1 \times g$, or 9.8 m/s^2; 2 gees (2*g*) means $2 \times g$, or 2×9.8 m/s$^2 = 19.6$ m/s^2; and so on. Suppose you experience 6 gees of acceleration in a rocket.

a. What is your acceleration in meters per second squared? **b.** You will feel a compression force from the acceleration. How does this force compare to your normal weight? **c.** Do you think you could survive this acceleration for long? Explain.

56. *Extra Moon.* Suppose Earth had a second moon, called Swisscheese, with an average orbital distance double the Moon's and a mass about the same as the Moon's.
   a. Is Swisscheese's orbital period longer or shorter than the Moon's? Explain. **b.** The Moon's orbital period is about one month. Apply Kepler's third law to find the approximate orbital period of Swisscheese. (*Hint:* If you form the ratio of the orbital distances of Swisscheese and the Moon, you can solve this problem with Kepler's original version of his third law rather than looking up all the numbers you'd need to apply Newton's version of Kepler's third law.) **c.** In words, describe how tides would differ because of the presence of this second moon. Consider the cases when the two moons are on the same side of Earth, on opposite sides of Earth, and 90° apart in their orbits.

Discussion Questions

57. *Knowledge of Mass-Energy.* Einstein's discovery that energy and mass are equivalent has led to technological developments that are both beneficial and dangerous. Discuss some of these developments. Overall, do you think the human race would be better or worse off if we had never discovered that mass is a form of energy? Defend your opinion.

58. *Perpetual Motion Machines.* Every so often, someone claims to have built a machine that can generate energy perpetually from nothing. Why isn't this possible according to the known laws of nature? Why do you think claims of perpetual motion machines sometimes receive substantial media attention?

Web Projects

59. *Space Station.* Visit a NASA website with pictures from the Space Station. Choose two photos that illustrate some facet of Newton's laws of motion or gravity. Explain how what is going on is related to Newton's laws.

60. *Tide Tables.* Find a tide table or tide chart for a beach town that you'd like to visit. Explain how to read the table and discuss any differences between the actual tidal pattern and the idealized tidal pattern described in this chapter.

61. *Space Elevator.* Read more about space elevators (see Problem 46) and how they might make it easier and cheaper to get to Earth orbit or beyond. Write a short report about the feasibility of building a space elevator, and briefly discuss the pros and cons of such a project.

5

LIGHT AND MATTER
READING MESSAGES FROM THE COSMOS

LEARNING GOALS

5.1 LIGHT IN EVERYDAY LIFE
- How do we experience light?
- How do light and matter interact?

5.2 PROPERTIES OF LIGHT
- What is light?
- What is the electromagnetic spectrum?

5.3 PROPERTIES OF MATTER
- What is the structure of matter?
- What are the phases of matter?
- How is energy stored in atoms?

5.4 LEARNING FROM LIGHT
- What are the three basic types of spectra?
- How does light tell us what things are made of?
- How does light tell us the temperatures of planets and stars?
- How does light tell us the speed of a distant object?

May the warp be the white light of morning,
May the weft be the red light of evening,
May the fringes be the falling rain,
May the border be the standing rainbow.
Thus weave for us a garment of brightness.
　　　　　—*Song of the Sky Loom (Native American)*

Ancient observers could discern only the most basic features of the light that they saw, such as color and brightness. Over the past several hundred years, we have discovered that light carries far more information. Today, we can analyze the light of distant objects to learn what they are made of, how hot they are, how fast they are moving, and much more. Light is truly the cosmic messenger, bringing the stories of distant objects to Earth.

In this chapter, we will focus our attention on learning how to read the messages carried by light. We'll begin with a brief look at the basic interactions of light and matter that create those messages, and then study the properties of light and matter individually and in some detail. With that background, we'll be ready to explore how a spectrum forms, so that we can understand how light can encode so much information about distant objects.

(MA) **Light and Spectroscopy Tutorial, Lesson 1**

5.1 LIGHT IN EVERYDAY LIFE

What do you see as you look around you? You may be tempted to list nearby objects, but all you're really seeing is *light* that has interacted with those objects. Through intuition and experience, you're able to interpret the colors and patterns of the light and turn them into information about the objects and substances that surround you.

Astronomers study the universe in much the same way. Telescopes collect the light of distant objects, and we use the light to extract information about those objects. The more we understand about light and its interactions with matter, the more information we can extract. As a first step in developing this understanding, let's take a closer look at our everyday experience with light.

How do we experience light?

You can tell that light is a form of energy even without opening your eyes. Outside on a hot, sunny day you can feel your skin warm as it absorbs sunlight. Because greater warmth means more molecular motion, sunlight must be transferring its energy to the molecules in your skin. The energy that light carries is called *radiative energy*; recall that it is one of the three basic categories of energy, along with kinetic and potential energy (see Figure 4.10).

Energy and Power We measure energy in units of joules [**Section 4.3**], so we can use these energy units to measure the total amount of energy that light transfers to your skin. With light, however, we are usually more interested in the *rate* at which it carries energy toward or away from us than in the

total amount of energy it carries. After all, because light always travels through space at the speed of light, we cannot hold light in our hands in the same way that we can hold a hot potato, which has thermal energy, or a rock, which has gravitational potential energy. The rate of energy flow is called **power**, which we measure in units called **watts**. A power of 1 watt means an energy flow of 1 joule per second:

$$1 \text{ watt} = 1 \text{ joule/s}$$

For example, a 100-watt light bulb requires 100 joules of energy (which you buy from the electric company) for each second it is turned on. Interestingly, the power requirement of an average human—about 10 million joules per day—is about the same as that of a 100-watt light bulb.

Light and Color Everyday experience tells us that light comes in different forms that we call *colors*. You've probably seen a prism split light into the rainbow of light called a **spectrum** (**FIGURE 5.1**). The basic colors in a rainbowlike spectrum are red, orange, yellow, green, blue, and violet. We see *white* when these colors are mixed in roughly equal proportions. Light from the Sun or a light bulb is often called *white light*, because it contains all the colors of the rainbow. *Black* is what we perceive when there is no light and hence no color.

The wide variety of all possible colors comes from mixtures of just a few colors in varying proportions. Your television takes advantage of this fact to simulate a huge range of colors by combining only red, green, and blue light; these three colors are often called the *primary colors of vision*, because they are the colors directly detected by cells in your eyes. Colors tend to look different on paper, so artists generally work with an alternative set of primary colors: red, yellow, and blue. If you do any graphic design work, you may be familiar with the CMYK process, in which the four colors cyan, magenta, yellow, and black are mixed to produce a great variety of colors; the CMYK process was used to print this book.

FIGURE 5.1
When we pass white light through a prism, it disperses into a rainbow of color that we call a *spectrum*.

Reflection (mirror): angle of incidence = angle of reflection.

Scattering: The screen scatters light from the projector in many directions…

screen

projection booth

…so that every person in the audience sees light from all parts of the screen.

a A mirror *reflects* light along a simple path: The angle at which the light strikes the mirror is the same angle at which it is reflected.

b A movie screen *scatters* light in many different directions, so that each member of the audience can watch the movie. The pages in a book do the same thing, which is why you can read them from different angles and distances.

FIGURE 5.2 Reflection and scattering.

You can produce a spectrum with either a prism or a **diffraction grating**, which is a piece of plastic or glass etched with many closely spaced lines. If you have a DVD handy, you can make a spectrum for yourself. The bottom of a DVD is etched with many closely spaced circles and therefore acts like a diffraction grating. That is why you see rainbows of color on the bottom of the disc when you hold it up to light.

How do light and matter interact?

Light can interact with matter in four basic ways, all of which are familiar in everyday life:

- **Emission:** A light bulb *emits* visible light; the energy of the light comes from electrical potential energy supplied to the light bulb.

- **Absorption:** When you place your hand near an incandescent light bulb, your hand *absorbs* some of the light, and this absorbed energy warms your hand.

- **Transmission:** Some forms of matter, such as glass or air, *transmit* light, which means allowing it to pass through.

- **Reflection/scattering:** Light can bounce off matter, leading to what we call *reflection* when the bouncing is all in the same general direction or *scattering* when the bouncing is more random (**FIGURE 5.2**).

Materials that transmit light are said to be *transparent,* and materials that absorb light are called *opaque.* Many materials are neither perfectly transparent nor perfectly opaque. For example, dark sunglasses and clear eyeglasses are both partially transparent, but the dark glasses absorb more and transmit less light. Materials often interact differently with different colors of light. For example, red glass transmits red light but absorbs other colors, while a green lawn reflects (scatters) green light but absorbs all other colors.

Let's put these ideas together to understand what happens when you walk into a room and turn on the light switch (**FIGURE 5.3**). The light bulb begins to emit white light, which is a mix of all the colors in the spectrum. Some of this light exits the room, transmitted through the windows. The rest of the light strikes the surfaces of objects inside the room, and the material properties of each object determine the colors it absorbs or reflects. The light coming from each object therefore carries an enormous amount of information about the object's location, shape and structure, and composition. You acquire this information when light enters your eyes, where special cells in your retina absorb it and send signals to your brain. Your brain interprets the messages that light carries, recognizing materials and objects in the process we call *vision.*

All the information that light brings us from the cosmos was encoded by the same four basic interactions between light and matter common to our everyday experience. However, our eyes perceive only a tiny fraction of all the information contained in light. Modern instruments can break light into a much wider variety of colors and can analyze those colors in far greater detail. In order to understand how to decode that information, we need to examine the nature of light and matter more closely.

5.2 PROPERTIES OF LIGHT

Light is familiar to all of us, but its nature remained a mystery for most of human history. Experiments performed by Isaac Newton in the 1660s provided the first real insights into the nature of light. It was already known that passing white light through a prism produced a rainbow of color, but many people thought the colors came from the prism rather than

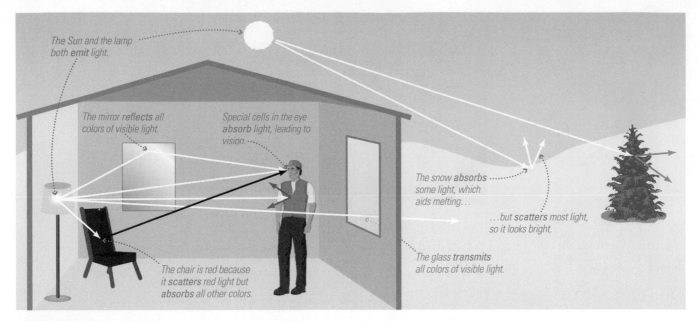

FIGURE 5.3 interactive figure This diagram shows examples of the four basic interactions between light and matter: emission, absorption, transmission, and reflection (or scattering).

The Sun and the lamp both emit light.

The mirror reflects all colors of visible light.

Special cells in the eye absorb light, leading to vision.

The snow absorbs some light, which aids melting…

…but scatters most light, so it looks bright.

The chair is red because it scatters red light but absorbs all other colors.

The glass transmits all colors of visible light.

from the light itself. Newton proved that the colors came from the light by placing a second prism in front of the light of just one color, such as red, from the first prism. If the rainbow of color had come from the prism itself, the second prism would have produced a rainbow just like the first. But it did not: When only red light entered the second prism, only red light emerged, proving that the color was a property of the light and not of the prism.

What is light?

Newton's work tells us something about the nature of color, but it still does not tell us exactly what light *is*. Newton himself guessed that light is made up of countless tiny particles. However, other scientists soon conducted experiments that demonstrated that light behaves like waves. Thus began one of the most important debates in scientific history: Is light a wave or a particle? To understand this question, and our modern answer to it, we must first understand the differences between particles and waves.

Particles and Waves in Everyday Life Marbles, baseballs, and individual atoms and molecules are all examples of *particles*. A particle of matter can sit still or it can move from one place to another. If you throw a baseball at a wall, it obviously travels from your hand to the wall.

In contrast, think about what happens when you toss a pebble into a pond, creating a set of outward moving ripples, or *waves* (**FIGURE 5.4**). These waves consist of *peaks*, where the water is higher than average, and *troughs*, where the water is lower than average. If you watch as the waves pass by a floating leaf, you'll see the leaf rise up with each peak and drop down with each trough, but the leaf itself will *not* travel across the pond's surface with the wave. We conclude that even though the waves are moving outward, the particles

(molecules) that make up the water are moving primarily up and down (along with a bit of sloshing back and forth). That is, the waves carry *energy* outward from the place where the pebble landed but do not carry matter along with them. In essence, a particle is a *thing*, while a wave is a *pattern* revealed by its interaction with particles.

Let's focus on three basic properties of waves: wavelength, frequency, and speed.* **Wavelength** is the distance from one

---

*There is also a fourth wave property, *amplitude,* defined as half the height from trough to peak. Amplitude is related to the brightness of light.

Wavelength is the distance from one peak to the next (or one trough to the next).

pebble thrown into pond

trough

peak

speed of wave moving outward

Leaf bobs up and down with the frequency of the waves.

FIGURE 5.4 interactive figure Tossing a pebble into a pond generates waves. The waves carry energy outward, but matter, such as a floating leaf and the molecules of the water, only bobs up and down (with a bit of sloshing back and forth) as the waves pass by.

peak to the next (or one trough to the next). **Frequency** is the number of peaks passing by any point each second. For example, if the leaf bobs up and down three times each second, then three peaks must be passing by it each second, which means the waves have a frequency of three **cycles per second**. "Cycles per second" are often called **hertz (Hz)**, so we can also describe this frequency as 3 Hz. The **speed** of the waves tells us how fast their peaks travel across the pond. Because the waves carry energy, the speed essentially tells us how fast the energy travels from one place to another.

A simple formula relates the wavelength, frequency, and speed of any wave. Suppose a wave has a wavelength of 1 centimeter and a frequency of 3 hertz. The wavelength tells us that each time a peak passes by, the wave peak has traveled 1 centimeter. The frequency tells us that three peaks pass by each second. The speed of the wave must therefore be 3 centimeters per second. If you try a few more similar examples, you'll find the general rule

$$\text{wavelength} \times \text{frequency} = \text{speed}$$

Light as an Electromagnetic Wave You've probably heard that light is a wave, but it isn't quite like the waves we see in everyday life. More familiar waves always move through some form of matter. For example, the waves on the pond move through the water, causing particles (molecules) of water to vibrate up and down and slosh back and forth, while sound waves move through air, causing air molecules to vibrate back and forth. The vibrations of matter allow the waves to transmit energy from one place to another, even though particles of matter do not travel along with the waves. In contrast to these everyday examples of waves, we do not see anything move up and down when light travels through space. So what, exactly, is "waving" when a light wave passes by?

The answer is what scientists call electric and magnetic fields. The concept of a **field** is a bit abstract, but it is used to describe the strength of force that a particle would experience at any point in space. For example, Earth creates a *gravitational field* that describes the strength of gravity at any distance from Earth, which means that the strength of the field declines with the square of the distance from Earth's center [**Section 4.4**]. Electricity and magnetism also create forces, so their strength in different places can be described in terms of *electric fields* and *magnetic fields*.

Light waves are traveling vibrations of both electric and magnetic fields, so we say that light is an **electromagnetic wave**. Just as the ripples on a pond will cause a leaf to bob up and down, the vibrations of the electric field in an electromagnetic wave will cause any charged particle, such as an electron, to bob up and down. If you could set up electrons in a row, they would wriggle like a snake as light passed by (**FIGURE 5.5a**). The distance between peaks in this row of electrons would tell us the wavelength of the light wave, while the number of times each electron bobbed up and down would tell us the frequency (**FIGURE 5.5b**).

All light travels through empty space at the same speed—the **speed of light** (represented by the letter *c*), which is about 300,000 kilometers per second. Because the speed of

If you could line up electrons, they would bob up and down with the vibrating electric field of a passing light wave.

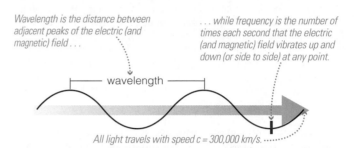

a Electrons move when light passes by, showing that light carries a vibrating electric field.

Wavelength is the distance between adjacent peaks of the electric (and magnetic) field . . .

. . . while frequency is the number of times each second that the electric (and magnetic) field vibrates up and down (or side to side) at any point.

wavelength

All light travels with speed c = 300,000 km/s.

b The vibrations of the electric field determine the wavelength and frequency of a light wave. Light also has a magnetic field (not shown) that vibrates perpendicular to the direction of the electric field vibrations.

FIGURE 5.5 interactive figure Light is an electromagnetic wave.

any wave is its wavelength times its frequency, we find a very important relationship between wavelength and frequency for light: *The longer the wavelength, the lower the frequency, and vice versa.* For example, light waves with a wavelength of 1 centimeter must have half the frequency of light waves with a wavelength of $\frac{1}{2}$ centimeter and one-fourth the frequency of light waves with a wavelength of $\frac{1}{4}$ centimeter (**FIGURE 5.6**).

Photons: "Particles" of Light Waves and particles appear distinctly different in everyday life. For example, no one would confuse the ripples on a pond with a baseball. However, experiments have shown that light behaves as *both* a wave and a particle. We say that light comes in individual "pieces," called **photons**, that have properties of both particles and waves. Like baseballs, photons of light can be counted

Longer wavelength means lower frequency

1 cm

wavelength = 1 cm, frequency = 30 GHz

0.5 cm

wavelength = $\frac{1}{2}$ cm, frequency = 2 × 30 GHz = 60 GHz

0.25 cm

wavelength = $\frac{1}{4}$ cm, frequency = 4 × 30 GHz = 120 GHz

Shorter wavelength means higher frequency.

FIGURE 5.6 Because all light travels through space at the same speed, light of longer wavelength must have lower frequency, and vice versa. (GHz stands for gigahertz, or 10^9 Hz.)

individually and can hit a wall one at a time. Like waves, each photon is characterized by a wavelength and a frequency. The idea that light can be both a wave and a particle may seem quite strange, but it is fundamental to our modern understanding of physics. (We will discuss some of the implications of this wave-particle duality in Chapter S4.)

Just as a moving baseball carries a specific amount of kinetic energy, each photon of light carries a specific amount of radiative energy. The shorter the wavelength of the light (or, equivalently, the higher its frequency), the higher the energy of the photons.

To sum up, our modern understanding maintains that (1) light is both a particle and a wave, an idea we describe by saying that light consists of individual photons characterized by wavelength, frequency, and energy, and (2) the wavelength, frequency, and energy of light are simply related because all photons travel through space at the same speed—the speed of light.

THINK ABOUT IT

If you assume that each of the three waves shown in Figure 5.6 represents a photon of light, which one has the most energy? Which one has the least energy? Explain.

What is the electromagnetic spectrum?

Newton's experiments proved that white light is a mix of all the colors in the rainbow. Later scientists found that there is light "beyond the rainbow" as well. Just as there are sounds that our ears cannot hear (such as the sound of a dog whistle), there is light that our eyes cannot see. In fact, light that we can see is only a tiny part of the complete spectrum of light, usually called the **electromagnetic spectrum**; light itself is often called **electromagnetic radiation**. FIGURE 5.7 shows the way the electromagnetic spectrum is commonly divided into regions according to wavelength (or, equivalently, frequency or energy). Keep in mind that despite the different names, everything in the electromagnetic spectrum represents a form of light and therefore consists of photons that travel through space at the speed of light.

The light that our eyes can see, which we call **visible light**, is found near the middle of the spectrum, with wavelengths ranging from about 400 nanometers at the blue or violet end of the rainbow to about 700 nanometers at the red end. (A nanometer [nm] is a billionth of a meter.) Light with wavelengths somewhat longer than red light is called **infrared**, because it lies beyond the red end of the rainbow. **Radio waves** are the longest-wavelength light. Wavelengths of light that fall near the border between infrared and radio waves, where wavelengths range from micrometers to centimeters, are sometimes given the name **microwaves**. In astronomy, you may occasionally hear portions of the microwave band described more specifically by wavelength. For example, the science conducted with telescopes optimized to detect microwaves with wavelengths of around 1 to a few millimeters is often called *millimeter astronomy*, and the science conducted with wavelengths of tenths of a millimeter is often called *submillimeter astronomy*.

On the other side of the spectrum, light with wavelengths somewhat shorter than those of blue light is called **ultraviolet**, because it lies beyond the blue (or violet) end of the rainbow. Light with even shorter wavelengths is called **X rays**, and the shortest-wavelength light is called **gamma rays**. Notice that visible light is an extremely small part of the entire electromagnetic spectrum: The reddest red that our eyes can see has only about twice the wavelength of the bluest blue, but the radio waves from your favorite radio station are a billion times longer than the X rays used in a doctor's office.

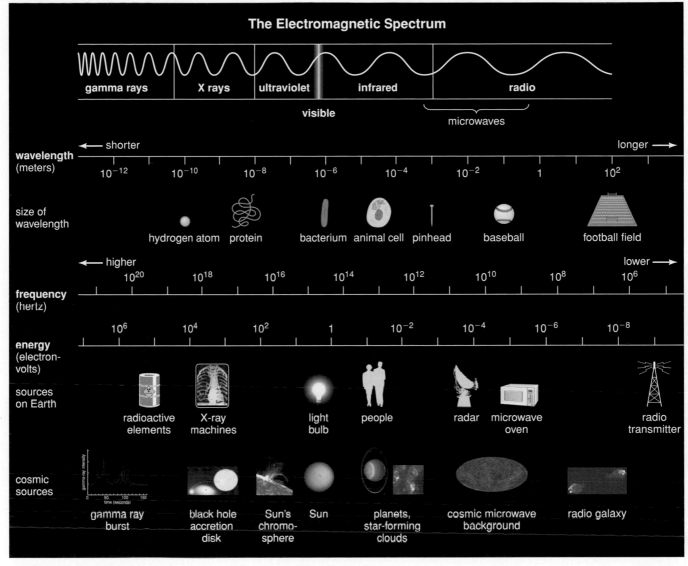

The Electromagnetic Spectrum

gamma rays	X rays	ultraviolet	infrared	radio	
		visible		microwaves	

wavelength (meters)
shorter ← → longer

10^{-12} 10^{-10} 10^{-8} 10^{-6} 10^{-4} 10^{-2} 1 10^2

size of wavelength

hydrogen atom protein bacterium animal cell pinhead baseball football field

higher ← → lower
10^{20} 10^{18} 10^{16} 10^{14} 10^{12} 10^{10} 10^8 10^6

frequency (hertz)

10^6 10^4 10^2 1 10^{-2} 10^{-4} 10^{-6} 10^{-8}

energy (electron-volts)

sources on Earth

radioactive elements   X-ray machines   light bulb   people   radar   microwave oven   radio transmitter

cosmic sources

gamma ray burst   black hole accretion disk   Sun's chromo-sphere   Sun   planets, star-forming clouds   cosmic microwave background   radio galaxy

FIGURE 5.7 interactive figure The electromagnetic spectrum. Notice that wavelength increases as we go from gamma rays to radio waves, while frequency and energy increase in the opposite direction. (Energy is given in units of *electron-volts*, eV: 1 eV = 1.60 × 10^{-19} joule.)

The different energies of different forms of light explain many familiar effects in everyday life. Radio waves carry so little energy that they have no noticeable effect on our bodies. However, radio waves can make electrons move up and down in an antenna, which is how your car radio receives the radio waves coming from a radio station. Molecules moving around in a warm object emit infrared light, which is why we sometimes associate infrared light with heat. Receptors in our eyes respond to visible-light photons, making vision possible. Ultraviolet photons carry enough energy to harm cells in our skin, causing sunburn or skin cancer. X-ray photons have enough energy to penetrate through skin and muscle but can be blocked by bones or teeth. That is why doctors and dentists can see our bone structures on photographs taken with X-ray light.

Just as different colors of visible light may be absorbed or reflected differently by the objects we see (see Figure 5.3), the various portions of the electromagnetic spectrum may interact in very different ways with matter. For example, a brick wall is opaque to visible light but transmits radio waves, which is why radios and cell phones work inside buildings. Similarly, glass that is transparent to visible light may be opaque to ultraviolet light. In general, certain types of matter tend to interact more strongly with certain types of light, so each type of light carries different information about distant objects in the universe. That is why astronomers seek to observe light of all wavelengths [**Section 6.4**].

5.3 PROPERTIES OF MATTER

Light carries information about matter across the universe, but we are usually more interested in the matter the light is coming from—such as planets, stars, and galaxies—than we are in the light itself. We must therefore explore the nature of matter if we are to decode the messages carried by light.

What is the structure of matter?

Like the nature of light, the nature of matter remained mysterious for most of human history. Nevertheless, ancient philosophers came up with some ideas that are still with us today.

The ancient Greek philosopher Democritus (c. 470–380 B.C.) wondered what would happen if we could break a piece of matter, such as a rock, into ever smaller pieces. He claimed that the rock would eventually break into particles so small that nothing smaller could be possible. He called these particles *atoms*, a Greek term meaning "indivisible." Building on the beliefs of earlier Greek philosophers, Democritus assumed that all materials were composed from four basic *elements:* fire, water, earth, and air. He proposed that the properties of different elements could be explained by the physical characteristics of their atoms. For example, Democritus suggested that atoms of water were smooth and round, so water flowed and had no fixed shape, while burns were painful because atoms of fire were thorny. He imagined atoms of earth to be rough and jagged, so they could fit together like pieces of a three-dimensional jigsaw puzzle, and he used this idea to suggest that the universe began as a chaotic mix of atoms that slowly clumped together to form our world.

Although Democritus was wrong in his specifics, he was on the right track. All ordinary matter is indeed composed of **atoms**, and the properties of ordinary matter depend on the physical characteristics of its atoms. However, by modern definition, atoms are *not* indivisible because they are composed of even smaller particles.

MATHEMATICAL INSIGHT 5.1

Wavelength, Frequency, and Energy

 Math Review Video: Problem Solving, Part 4

The relationship *wavelength* $\times$ *frequency* = *speed* holds for any wave. For light, which travels (in a vacuum) at speed $c = 3 \times 10^8$ m/s, this relationship becomes

$$\lambda \times f = c$$

where λ (the Greek letter *lambda*) stands for wavelength and f stands for frequency. Note that, because c is a constant, frequency must go up when wavelength goes down, and vice versa.

The radiative energy (E) carried by a photon of light is given by

$$E = h \times f$$

where h is *Planck's constant* ($h = 6.626 \times 10^{-34}$ joule $\times$ s). Energy therefore increases with frequency.

EXAMPLE 1: A radio station at 93.3 FM broadcasts radio waves with a frequency of 93.3 megahertz (MHz). What is their wavelength?

SOLUTION:

Step 1 Understand: Radio waves are a form of light, so they obey the relationship $\lambda \times f = c$. We are given the frequency (f) and know the speed of light (c), so we can simply solve for the wavelength (λ). There is one subtlety: The units of frequency, called hertz or "cycles per second," are really just "per second," or 1/s; the reason is that "cycles" is just a descriptive term, with no units itself. So we write 93.3 megahertz as 93.3×10^6 1/s.

Step 2 Solve: We solve for wavelength by dividing both sides of $\lambda \times f = c$ by f, which gives $\lambda = c/f$. We now plug in the speed of light and the frequency to find

$$\lambda = \frac{c}{f} = \frac{3 \times 10^8 \, \frac{m}{s}}{93.3 \times 10^6 \, \frac{1}{s}} = 3.2 \text{ m}$$

Step 3 Explain: Radio waves with a frequency of 93.3 MHz have a wavelength of 3.2 meters. That is why radio towers are so large; they must be taller than the waves they are transmitting.

EXAMPLE 2: The middle of the visible spectrum is green light with a wavelength of about 550 nanometers. What is its frequency?

SOLUTION:

Step 1 Understand: All light obeys the relation $\lambda \times f = c$. In this case we are given the wavelength, so we simply solve the equation for the frequency.

Step 2 Solve: Dividing both sides of the equation $\lambda \times f = c$ by λ gives $f = c/\lambda$. We plug in the speed of light and the wavelength ($\lambda = 550 \times 10^{-9}$ m) to find

$$f = \frac{c}{\lambda} = \frac{3 \times 10^8 \, \frac{m}{s}}{550 \times 10^{-9} \, m} = 5.45 \times 10^{14} \, \frac{1}{s}$$

Step 3 Explain: Green visible light has a frequency of about 5.5×10^{14} 1/s, which is 5.5×10^{14} Hz, or 550 trillion Hz. This high frequency is one reason the wave properties of light are not obvious in everyday life.

EXAMPLE 3: What is the energy of a visible-light photon with a wavelength of 550 nanometers?

SOLUTION:

Step 1 Understand: The energy of a photon is $E = h \times f$. We are given the photon's wavelength rather than frequency, so we use the fact that $f = c/\lambda$ to write

$$E = h \times f = h \times \frac{c}{\lambda}$$

Step 2 Solve: We plug in the wavelength and Planck's constant to find

$$E = h \times \frac{c}{\lambda}$$

$$= (6.626 \times 10^{-34} \text{ joule} \times s) \times \frac{3 \times 10^8 \, \frac{m}{s}}{550 \times 10^{-9} \, m}$$

$$= 3.6 \times 10^{-19} \text{ joule}$$

Step 3 Explain: The energy of a single visible-light photon is about 3.6×10^{-19} joule. Note that this is barely a *billion-trillionth* of the 100 joules of energy needed each second by a 100-watt light bulb.

Atoms come in different types, and each type corresponds to a different chemical **element**. Scientists have identified more than 100 chemical elements, and fire, water, earth, and air are *not* among them. Some of the most familiar chemical elements are hydrogen, helium, carbon, oxygen, silicon, iron, gold, silver, lead, and uranium. Appendix D gives the periodic table of all the elements.

Atomic Structure Each chemical element consists of a different type of atom, and atoms are in turn made of particles that we call **protons**, **neutrons**, and **electrons** (**FIGURE 5.8**). Protons and neutrons are found in the tiny **nucleus** at the center of the atom. The rest of the atom's volume contains electrons, which surround the nucleus. Although the nucleus is very small compared to the atom as a whole, it contains most of the atom's mass, because protons and neutrons are each about 2000 times as massive as an electron. Note that atoms are incredibly small: Millions could fit end to end across the period at the end of this sentence. The number of atoms in a single drop of water (typically, 10^{22} to 10^{23} atoms) may exceed the number of stars in the observable universe.

The properties of an atom depend mainly on the **electrical charge** in its nucleus. Electrical charge is a fundamental physical property that describes how strongly an object will interact with electromagnetic fields; total electrical charge is always

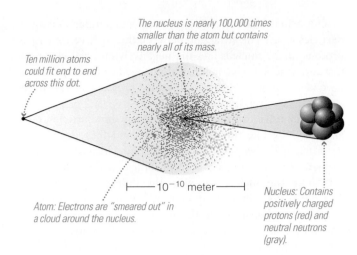

The nucleus is nearly 100,000 times smaller than the atom but contains nearly all of its mass.

Ten million atoms could fit end to end across this dot.

$\longmapsto 10^{-10}$ meter $\longmapsto$

Atom: Electrons are "smeared out" in a cloud around the nucleus.

Nucleus: Contains positively charged protons (red) and neutral neutrons (gray).

FIGURE 5.8 The structure of a typical atom. Note that atoms are extremely tiny: The atom shown in the middle is magnified to about 1 billion times its actual size, and the nucleus on the right is magnified to about 100 trillion times its actual size.

conserved, just as energy is always conserved. We define the electrical charge of a proton as the basic unit of positive charge, which we write as $+1$. An electron has an electrical charge that is precisely opposite that of a proton, so we say it has negative charge (-1). Neutrons are electrically neutral, meaning that they have no charge.

SPECIAL TOPIC

What Do Polarized Sunglasses Have to Do with Astronomy?

If you go to the store to buy a pair of sunglasses, you'll face a dizzying array of choices. Sunglasses come in different styles and different tints and with different efficiencies in blocking ultraviolet and infrared light. Most of these choices should make sense to you (well, perhaps not all of the styles), but one option may not be familiar: The labels on some sunglasses say that they are "polarized." What does this mean? The term comes from a property of light, called *polarization*, that has to do with the direction in which a light wave vibrates and how those vibrations change when light bounces off or passes through matter. Polarization is important not only to sunglasses but also to astronomy.

To explore this idea, think about how waves move on a string when you shake one end of it. The string vibrates either up and down or back and forth while the wave itself moves along it in a direction perpendicular to the direction of vibration. Light waves move in a similar way, with the electric and magnetic fields vibrating either up and down or side to side compared with the direction of travel. For example, the wave shown in Figure 5.5b is moving to the right on the page while its electric field vibrates up and down on the page.

The direction of vibration affects the way light interacts with matter. As Figure 5.5a indicates, an electric field that vibrates up and down will make electrons move up and down as the wave passes by. That is, the direction in which the electric field vibrates determines the direction in which charged particles vibrate as the wave passes by. Because the direction of wave vibration matters, we give it a name: the *polarization* of the wave. An individual wave moving toward you can be polarized with its vibrations either up and down or side to side or some combination of those two.

Each light wave (or, more technically, each individual photon) has a particular direction of polarization, although our eyes do not

detect it. If all the waves taken together have no preferential direction of vibration, we say that the light is *unpolarized*. However, some physical processes produce waves with a particular direction of polarization, which is where your sunglasses and astronomy come in.

When light reflects off a flat horizontal surface like the ground or a lake, all the reflected light tends to have its electric field vibrating horizontally. (Light with other directions of vibration is absorbed or transmitted.) In other words, the reflected light is horizontally polarized. Polarized sunglasses are designed to block light with horizontal polarization, which is often the cause of "glare." Of course, the polarized glasses work only if you are wearing them horizontally; if you turn a pair of polarized sunglasses so that the two lenses are no longer horizontal to the ground, they will not block glare effectively.

In astronomy, we aren't worried about glare from distant objects, but if we learn that a light source is producing polarized light, this tells us something about the nature of the source. For example, light that passes through clouds of interstellar dust tends to be polarized, telling us that the dust grains in the cloud must be preferentially absorbing light with electric fields vibrating in a particular direction. More detailed analysis has taught us that the polarization arises because the microscopic dust grains have an elongated shape, and all tend to be aligned in the same way as a result of magnetic fields within the clouds. Polarization arises in many other astronomical contexts as well, including the study of the leftover radiation from the Big Bang. Although polarization has provided important insights into many astronomical processes, its analysis can be fairly technical, and we will not discuss it further in this book.

Oppositely charged particles attract and similarly charged particles repel. The attraction between the positively charged protons in the nucleus and the negatively charged electrons that surround it is what holds an atom together. Ordinary atoms have identical numbers of electrons and protons, making them electrically neutral overall.*

Although we can think of electrons as tiny particles, they are not quite like tiny grains of sand and they don't orbit the nucleus the way planets orbit the Sun. Instead, the electrons in an atom form a kind of "smeared out" cloud that surrounds the nucleus and gives the atom its apparent size. The electrons aren't really cloudy, but it is impossible to pinpoint their positions in the atom. The electrons therefore give the atom a size far larger than that of its nucleus even though they represent only a tiny portion of the atom's mass. If you imagine an atom on a scale that makes its nucleus the size of your fist, its electron cloud would be many kilometers wide.

Atomic Terminology You've probably learned the basic terminology of atoms in past science classes, but let's review it just to be sure. **FIGURE 5.9** summarizes the key terminology we will use in this book.

Each different chemical element contains a different number of protons in its nucleus. This number is its **atomic number**. For example, a hydrogen nucleus contains just one proton, so its atomic number is 1. A helium nucleus contains two protons, so its atomic number is 2. The *combined* number

---

*You may wonder why electrical repulsion doesn't cause the positively charged protons in a nucleus to fly apart from one another. The answer is that an even stronger force, called the *strong force*, overcomes electrical repulsion and holds the nucleus together [**Section S4.2**].

of protons and neutrons in an atom is called its **atomic mass number**. The atomic mass number of ordinary hydrogen is 1 because its nucleus is just a single proton. Helium usually has two neutrons in addition to its two protons, giving it an atomic mass number of 4. Carbon usually has six protons and six neutrons, giving it an atomic mass number of 12.

Every atom of a given element contains exactly the same number of protons, but the number of neutrons can vary. For example, all carbon atoms have six protons, but they may have six, seven, or eight neutrons. Versions of an element with different numbers of neutrons are called **isotopes** of that element. Isotopes are named by listing their element name and atomic mass number. For example, the most common isotope of carbon has six protons and six neutrons, giving it atomic mass number $6 + 6 = 12$, so we call it *carbon-12*. The other isotopes of carbon are carbon-13 (six protons and seven neutrons) and carbon-14 (six protons and eight neutrons). We sometimes write the atomic mass number as a superscript to the left of the element symbol: ^{12}C, ^{13}C, ^{14}C. We read ^{12}C as "carbon-12."

Molecules The number of different material substances is far greater than the number of chemical elements because atoms can combine to form **molecules**. Some molecules consist of two or more atoms of the same element. For example, we breathe O_2, oxygen molecules made of two oxygen atoms. Other molecules, such as the water molecule, are made up of atoms of two or more different elements. (Molecules with two or more types of atom are often called *compounds*.) The symbol H_2O tells us that a water molecule contains two hydrogen atoms and one oxygen atom. The chemical properties of a molecule are different from those of its individual atoms. For example, molecular oxygen (O_2) behaves very differently from atomic oxygen (O), and water behaves very differently from pure hydrogen or pure oxygen.

---

atomic number = number of protons
atomic mass number = number of protons + neutrons
(A neutral atom has the same number of electrons as protons.)

Hydrogen (^{1}H)   **Helium (^{4}He)**   **Carbon (^{12}C)**

atomic number = 1   atomic number = 2   atomic number = 6
atomic mass   atomic mass   atomic mass
  number = 1   number = 4   number = 12
(1 electron)   (2 electrons)   (6 electrons)

---

*Different **isotopes** of a given element contain the same number of protons, but different numbers of neutrons.*

Isotopes of Carbon

carbon-12   carbon-13   carbon-14

^{12}C ^{13}C ^{14}C
(6 protons   (6 protons   (6 protons
+ 6 neutrons)   + 7 neutrons)   + 8 neutrons)

FIGURE 5.9 Terminology of atoms.

What are the phases of matter?

Interactions between light and matter depend on the physical state of the matter, which we usually describe by the matter's **phase**. For example, molecules of H_2O can exist in three familiar phases: as **solid** ice, as **liquid** water, and as the **gas** we call water vapor. But how can the *same* molecules (H_2O) look and act so different in different phases?

You are probably familiar with the idea of a **chemical bond**, the name we give to the interactions between electrons that hold the atoms in a molecule together. For example, we say that chemical bonds hold the hydrogen and oxygen atoms together in a molecule of H_2O. Similar but much weaker interactions among electrons hold together the many water molecules in a block of ice or a pool of water. We can think of the interactions that keep neighboring atoms or molecules close together as other types of bonds, with the phases of solid, liquid, and gas differing in the strength of the bonds between neighboring atoms and molecules. Phase changes occur when one type of bond is broken and replaced by another. Changes in either pressure or temperature (or both) can cause phase changes, but it's easier to think first about temperature: As a substance is heated, the average kinetic energy of its particles increases, enabling the particles to break the bonds holding them to their neighbors.

Phase Changes in Water Water is the only familiar substance that we see in all three phases (solid, liquid, gas) in everyday life, so let's consider what happens to water as an example of how phase changes occur as a substance heats up.

At low temperatures, water molecules have a relatively low average kinetic energy, allowing them to be tightly bound to their neighbors in the *solid* structure of ice. As long as the temperature remains below freezing, the water molecules in ice remain rigidly held together. However, the molecules within this rigid structure are always vibrating, and higher temperature means greater vibrations. If we start with ice at a very low temperature, the molecular vibrations grow gradually stronger as the temperature rises toward the melting point, which is 0°C at ordinary (sea level) atmospheric pressure.

The melting point is the temperature at which the molecules have enough energy to break the solid bonds of ice. The molecules can then move much more freely among one another, allowing the water to flow as a *liquid*. However, the molecules in liquid water are not completely free of one another, as we can tell from the fact that droplets of water can stay intact. Adjacent molecules in liquid water must therefore still be held together by a type of bond, though it is much looser than the bond that holds them together in solid ice.

If we continue to heat the water, the increasing kinetic energy of the molecules will ultimately break the bonds between neighboring molecules altogether. The molecules will then be able to move freely, and freely moving particles constitute a *gas*. Above the boiling point (100°C at sea level), all the bonds between adjacent molecules are broken so the water can exist only as a gas.

We see ice melting into liquid water and liquid water boiling into gas so often that it's tempting to think that's the end

of the story. However, a little thought should convince you that the reality has to be more complex. For example, you know that Earth's atmosphere contains water vapor that condenses to form clouds and rain. But Earth's surface temperature is well below the boiling point of water, so how is it that our atmosphere can contain water in the gas phase?

The answer lies in the fact that temperature is a measure of the *average* kinetic energy of the particles in a substance [**Section 4.3**]. Individual particles may have substantially lower or higher energies than the average. Even at the low temperatures at which most water molecules are bound together as ice or liquid, a few molecules will always have enough energy to break free of their neighbors and enter the gas phase. In other words, some gas (water vapor) is always present along with solid ice or liquid water. The process by which molecules escape from a solid is called **sublimation**, and the process by which molecules escape from a liquid is called **evaporation**. Higher temperatures lead to higher rates of sublimation or evaporation.

Molecular Dissociation and Ionization Above the boiling point, all the water will have entered the gas phase. What happens if we continue to raise the temperature?

The molecules in a gas move freely, but they often collide with one another. As the temperature rises, the molecules move faster and the collisions become more violent. At high enough temperatures, the collisions become so violent that they can break the chemical bonds holding individual water molecules together. The molecules then split into pieces, a process we call **molecular dissociation**. (In the case of water, molecular dissociation usually frees one hydrogen atom and leaves a negatively charged molecule that consists of one hydrogen atom and one oxygen atom [OH]; at even higher temperatures, the OH dissociates into individual atoms.)

At still higher temperatures, collisions can break the bonds holding electrons around the nuclei of individual atoms, allowing the electrons to go free. The loss of one or more negatively charged electrons leaves the remaining atom with a net positive charge. Charged atoms (whether positive or negative) are called **ions**, and the process of stripping electrons from atoms is called **ionization**. At temperatures of several thousand degrees, the process of ionization turns what once was water into a hot gas consisting of freely moving electrons and positively charged ions of hydrogen and oxygen. This type of hot gas, in which atoms have become ionized, is called a **plasma**. Because a plasma contains many charged particles, its interactions with light are different from those of a gas consisting of neutral atoms, which is one reason plasma is sometimes referred to as "the fourth phase of matter." However, because the electrons and ions are not bound to one another, it is also legitimate to call plasma a gas. That is why we sometimes say that the Sun is made of hot gas and sometimes say that it is made of plasma; both statements are correct.

The degree of ionization in a plasma depends on its temperature and composition. A neutral hydrogen atom contains only one electron, so hydrogen can be ionized only once; the remaining hydrogen ion, designated H^+, is simply a proton. Oxygen, with atomic number 8, has eight electrons when it is neutral, so it can be ionized multiple times. *Singly ionized* oxygen is missing one electron, so it has a charge of +1 and is designated O^+. *Doubly ionized* oxygen, or O^{+2}, is missing two electrons; *triply ionized* oxygen, or O^{+3}, is missing three electrons; and so on. At temperatures of several million degrees, oxygen can be *fully ionized,* in which case all eight electrons are stripped away and the remaining ion has a charge of +8.

FIGURE 5.10 summarizes the changes that occur as we heat water from ice to a fully ionized plasma. Other chemical substances go through similar phase changes, but the changes generally occur at different temperatures for different substances.

Phases and Pressure Temperature is the primary factor determining the phase of a substance and the ways in which light interacts with it, but pressure also plays a role. You're undoubtedly familiar with the idea of pressure in an everyday sense: For example, you can put more pressure on your arm by squeezing it. In science, we use a more precise definition: **Pressure** is the *force per unit area* pushing on an object's surface. You feel more pressure when you squeeze your arm because squeezing increases the force on each square centimeter of your arm's surface. Similarly, piling rocks on a table increases the weight (force) on the table, which therefore increases the pressure on the surface of the table; if the pressure becomes too great, the table breaks. The gas in an atmosphere also creates pressure, because the weight of the gas bears down on everything beneath it. For example, at sea level on Earth, the weight of the atmosphere creates a pressure of about 14.7 pounds per square inch. That is, the total weight of all the air above each square inch of Earth's surface is about 14.7 pounds [**Section 10.1**].

Pressure can affect phases in a variety of ways. For example, deep inside Earth, the pressure is so high that Earth's

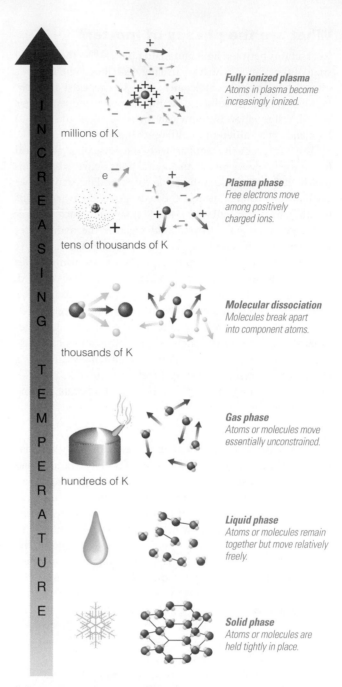

FIGURE 5.10 The general progression of phase changes in water.

inner metal core remains solid, even though the temperature is high enough that the metal would melt into liquid under less extreme pressure conditions [**Section 9.1**]. On a planetary surface, atmospheric pressure can determine whether water is stable in liquid form.

Remember that liquid water is always evaporating (or ice sublimating) at a low level, because a few molecules randomly get enough energy to break the bonds holding them to their neighbors. On Earth, enough liquid water has evaporated from the oceans to make water vapor an important ingredient of our atmosphere. Some of these atmospheric water vapor molecules collide with the ocean surface, where they can "stick" and rejoin the ocean—essentially the opposite of evaporation (**FIGURE 5.11**). The greater the pressure created

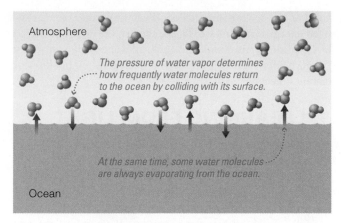

Atmosphere

The pressure of water vapor determines how frequently water molecules return to the ocean by colliding with its surface.

At the same time, some water molecules are always evaporating from the ocean.

Ocean

FIGURE 5.11 Evaporation of water molecules from the ocean is balanced in part by molecules of water vapor in Earth's atmosphere returning to the ocean. The rate at which these molecules return is directly related to the pressure created by water vapor in the atmosphere.

by water vapor molecules in our atmosphere,* the higher the rate at which water molecules return to the ocean. This direct return of water vapor molecules from the atmosphere helps keep the total amount of water in Earth's oceans fairly stable.† On the Moon, where the lack of atmosphere means no pressure from water vapor at all, liquid water would evaporate quite quickly (as long as the temperature were high enough that it did not freeze first). The same is true on Mars, because the atmosphere lacks enough water vapor to balance the rate of evaporation.

High pressure can also cause gases to dissolve in liquid water. For example, sodas are made by putting water in contact with high-pressure carbon dioxide gas. Because of the high pressure, many more carbon dioxide molecules enter the water than are released, so the water becomes "carbonated"—that is, it has a lot of dissolved carbon dioxide. When you open a bottle of carbonated water, exposing it to air with ordinary pressure, the dissolved carbon dioxide quickly bubbles up and escapes.

How is energy stored in atoms?

Now that we have reviewed the structure of matter and how its phase depends on temperature and pressure, it is time to return to the primary goal of this chapter: understanding how we learn about distant objects by studying their light. To produce light, these objects must somehow transform energy contained in matter into the vibrations of electric and magnetic fields that we call light. We therefore need to focus on the charged particles within atoms, particularly the electrons, because only particles that have charge can interact with light.

Atoms contain energy in three different ways. First, by virtue of their mass, they possess mass-energy in the amount mc^2.

*Technically, this is known as the *vapor pressure* of water in the atmosphere. We can also measure vapor pressure for other atmospheric constituents, and the total gas pressure is the sum of all the individual vapor pressures.

†Rain and snow also contribute, of course; however, even if Earth's temperature rose enough that raindrops and snowflakes could no longer form, only a small fraction of Earth's ocean water would evaporate before the return rate of water vapor molecules balanced the evaporation rate.

Second, they possess kinetic energy by virtue of their motion. Third, they contain *electrical potential energy* that depends on the arrangement of their electrons around their nuclei. To interpret the messages carried by light, we must understand how electrons store and release this electrical potential energy.

Energy Levels in Atoms The energy stored by electrons in atoms has a strange but important property: The electrons can have only particular amounts of energy, and not other energies in between. As an analogy, suppose you're washing windows on a building. If you use an adjustable platform to reach high windows, you can stop the platform at any height above the ground. But if you use a ladder, you can stand only at *particular* heights—the heights of the rungs of the ladder—and not at other heights in between. The possible energies of electrons in atoms are like the possible heights on a ladder. Only a few particular energies are possible; energies between these special few are not possible. The possible energies are known as the **energy levels** of an atom.

FIGURE 5.12 shows the energy levels of hydrogen, the simplest of all elements. The energy levels are labeled on the left in numerical order and on the right in units of *electron-volts*, or *eV* for short (1 eV = 1.60 × 10⁻¹⁹ joule). The lowest possible energy level—called level 1 or the *ground state*—is defined as an energy of 0 eV. Each of the higher energy levels (sometimes called *excited states*) is labeled with the extra energy of an electron in that level compared to an electron in the ground state.

Energy Level Transitions An electron can rise from a low energy level to a higher one or fall from a high level to a lower one. Such changes are called **energy level transitions**. Because energy must be conserved, energy level transition can occur only when an electron gains or loses the specific

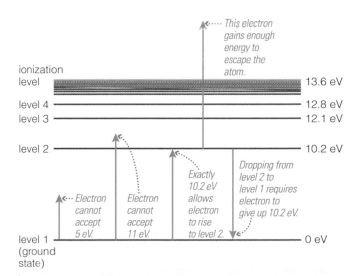

FIGURE 5.12 Energy levels for the electron in a hydrogen atom. The electron can change energy levels only if it gains or loses the amount of energy separating the levels. If the electron gains enough energy to reach the ionization level, it can escape from the atom, leaving behind a positively charged ion. (The many levels between level 4 and the ionization level are not labeled.)

amount of energy separating two levels. For example, an electron in level 1 can rise to level 2 only if it gains 10.2 eV of energy. If you try to give the electron 5 eV of energy, the electron won't accept it because it is not enough energy to reach level 2. Similarly, if you try to give the electron 11 eV, the electron won't accept it because it is too much for level 2 but not enough to reach level 3. Once in level 2, the electron can return to level 1 by giving up 10.2 eV of energy. Figure 5.12 shows several examples of allowed and disallowed energy level transitions.

Notice that the amount of energy separating the various levels gets smaller at higher levels. For example, it takes more energy to raise the electron from level 1 to level 2 than from level 2 to level 3, which in turn takes more energy than the transition from level 3 to level 4. If the electron gains enough energy to reach the *ionization level,* it escapes the atom completely, thereby ionizing the atom. Any excess energy beyond the amount needed for ionization becomes kinetic energy of the free-moving electron.

THINK ABOUT IT

Are there any circumstances under which an electron in a hydrogen atom can lose 2.6 eV of energy? Explain.

Quantum Physics If you think about it, the idea that electrons in atoms are restricted to particular energy levels is quite bizarre. It is as if you had a car that could go around a track only at particular speeds and not at speeds in between. How strange it would seem if your car suddenly changed its speed from 5 miles per hour to 20 miles per hour without first passing through a speed of 10 miles per hour! In scientific terminology, the electron's energy levels in an atom are said to be *quantized,* and the study of the energy levels of electrons (and other particles) is called *quantum physics* (or *quantum mechanics*). We will explore some of the astonishing implications of quantum physics in Chapter S4.

Electrons have quantized energy levels in all atoms, not just in hydrogen. Moreover, the allowed energy levels differ from element to element and from one ion of an element to another ion of the same element. Even molecules have quantized energy levels. As we will see shortly, the different energy levels of different atoms and molecules allow light to carry "fingerprints" that can tell us the chemical composition of distant objects.

5.4 LEARNING FROM LIGHT

Matter leaves its fingerprints whenever it interacts with light. Examining the color of an object is a crude way of studying the clues left by the matter it contains. For example, a red shirt absorbs all visible photons except those in the red part of the spectrum, so we know that it must contain a dye with these special light-absorbing characteristics. If we take light and disperse it into a spectrum, we can see the spectral fingerprints more clearly. The photograph that opens this chapter (page 137) shows the Sun's visible-light spectrum in great detail, with the rainbow of color stretching in horizontal rows from the upper left to the lower right of the photograph. We see similar dark or bright lines when we look at almost any spectrum, whether it is the spectrum of the flame from the gas grill in someone's backyard or the spectrum of a distant galaxy whose light we collect with a gigantic telescope. As long as we collect enough light to see details in the spectrum, we can learn many fundamental properties of the object we are viewing, no matter how far away the object is located.

The process of obtaining a spectrum and reading the information it contains is called **spectroscopy**. If you project a spectrum produced by a prism onto a wall, it looks like a rainbow (at least for visible light). However, it's often more useful to display spectra as graphs that show the amount, or **intensity**, of the light at each wavelength. For example, consider the spectrum in **FIGURE 5.13**, which plots the intensity of light from an astronomical object at wavelengths ranging from the ultraviolet on the left to the infrared on the right. At wavelengths where a lot of light is coming from the object, the intensity is high, while at wavelengths where there is little light, the intensity is low.*

---

*More technically, intensity is proportional to the total amount of *energy* transmitted by the light at each wavelength.

FIGURE 5.13 A schematic spectrum obtained from the light of a distant object. The "rainbow" at bottom shows how the light would appear if viewed with a prism or diffraction grating; of course, our eyes cannot see the ultraviolet or infrared light. The graph shows the corresponding intensity of the light at each wavelength. The intensity is high where the rainbow is bright and low where it is dim (such as in places where the rainbow shows dark lines).

Continuous Spectrum

The light bulb produces light of all visible wavelengths (colors). ······

hot light source

prism

The spectrum shows a smooth, continuous rainbow of light.

A graph of the spectrum is also continuous; notice that intensity varies slightly at different wavelengths.

intensity

wavelength

a

Emission Line Spectrum

The atoms in a warm gas cloud emit light only at specific wavelengths (colors) determined by the cloud's composition and temperature.

cloud of gas

prism

We see bright emission lines at specific wavelengths (colors), but no other light.

The graph shows an upward spike at the wavelength of each emission line.

intensity

wavelength

b

Absorption Line Spectrum

If light from a hot source passes through a cooler gas cloud, atoms in the cloud absorb light at wavelengths determined by the cloud's composition and temperature.

hot light source

cloud of gas

prism

We see dark absorption lines where the cloud has absorbed light of specific wavelengths (colors).

The graph shows a dip in intensity at the wavelength of each absorption line.

intensity

wavelength

c

FIGURE 5.14 interactive figure These diagrams show examples of the conditions under which we see the three basic types of spectra.

Our goal in this section is to learn how to interpret astronomical spectra like the one in Figure 5.13. The bumps and wiggles in that spectrum arise from several different processes, making it a good case study. We'll consider these processes one at a time, then return to interpret the full spectrum at the end of this section.

Light and Spectroscopy Tutorial, Lessons 2–4

What are the three basic types of spectra?

Laboratory studies show that spectra come in three basic types* (FIGURE 5.14):

1. The spectrum of a traditional, or incandescent, light bulb (which contains a heated wire filament) is a rainbow of color. Because the rainbow spans a broad range of wavelengths without interruption, we call it a **continuous spectrum**.

2. A thin or low density cloud of gas emits light only at specific wavelengths that depend on its composition and temperature. The spectrum therefore consists of bright **emission lines** against a black background and is called an **emission line spectrum**.

3. If the cloud of gas lies between us and a light bulb, we still see most of the continuous spectrum of the light bulb. However, the cloud absorbs light of specific wavelengths, so the spectrum shows dark **absorption lines** over the background rainbow,* making it what we call an **absorption line spectrum**.

Note that when the spectra are shown as graphs, absorption lines appear as dips on a background of relatively high-intensity light while emission lines look like spikes on a background with little or no intensity.

We can apply these ideas to the solar spectrum that opens this chapter, which shows numerous absorption lines over a

---

*The rules that specify the conditions producing each type are often called *Kirchhoff's laws*.

*More technically, we'll see an absorption line spectrum as long as the cloud is cooler in temperature than the source of background light, which in this case is the light bulb filament.

background rainbow of color. This tells us that we are essentially looking at a hot light source through gas that is absorbing some of the colors, much as when we look through the cloud of gas to the light bulb in Figure 5.14c. For the solar spectrum, the hot light source is the hot interior of the Sun, while the "cloud" is the relatively cool and low-density layer of gas at the top of the Sun's visible surface, or *photosphere* [**Section 14.1**].

How does light tell us what things are made of?

We have just seen *how* different viewing conditions lead to different types of spectra, so it is time to discuss *why*. Let's start with emission and absorption line spectra, in which the lines form as a direct consequence of the fact that each type of atom, ion, or molecule possesses a unique set of energy levels.

Emission Line Spectra The atoms in any cloud of gas are constantly colliding with one another, exchanging energy in each collision. Most of the collisions simply send the atoms flying off in new directions. However, a few of the collisions

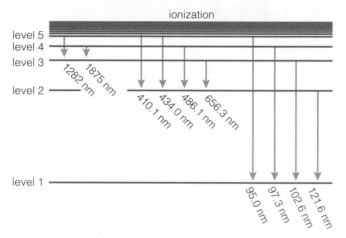

a Energy level transitions in hydrogen correspond to photons with specific wavelengths. Only a few of the many possible transitions are labeled.

b This spectrum shows emission lines produced by downward transitions between higher levels and level 2 in hydrogen.

c This spectrum shows absorption lines produced by upward transitions between level 2 and higher levels in hydrogen.

FIGURE 5.15 interactive figure An atom emits or absorbs light only at specific wavelengths that correspond to changes in the atom's energy as an electron undergoes transitions between its allowed energy levels.

transfer the right amount of energy to bump an electron from a low energy level to a higher energy level.

Electrons can't stay in higher energy levels for long. They always fall back down to the ground state, level 1, usually in a tiny fraction of a second. The energy the electron loses when it falls to a lower energy level must go somewhere, and often it goes into *emitting* a photon of light. The emitted photon must have the same amount of energy that the electron loses, which means that it has a specific wavelength and frequency. **FIGURE 5.15a** shows the energy levels in hydrogen that we saw in Figure 5.12, but it is also labeled with the wavelengths of the photons emitted by various downward transitions of an electron from a higher energy level to a lower one. For example, the transition from level 2 to level 1 emits an ultraviolet photon of wavelength 121.6 nm, and the transition from level 3 to level 2 emits a red visible-light photon of wavelength 656.3 nm.*

Although electrons that rise to higher energy levels in a gas quickly return to level 1, new collisions can raise other electrons to higher levels. As long as the gas remains moderately warm, collisions are always bumping some electrons to levels from which they fall back down and emit photons with some of the wavelengths shown in Figure 5.15a. The gas therefore emits light with these specific wavelengths. That is why a warm gas cloud produces an emission line spectrum, as shown in **FIGURE 5.15b**. The bright emission lines appear at the wavelengths that correspond to downward transitions of electrons, and the rest of the spectrum is dark (black). The specific set of lines that we see depends on the cloud's temperature as well as its composition: At higher temperatures, electrons are more likely to be bumped to higher energy levels.

THINK ABOUT IT

If nothing continues to heat the hydrogen gas, all the electrons eventually will end up at the lowest energy level (the ground state, or level 1). Use this fact to explain why we should *not* expect to see an emission line spectrum from a very cold cloud of hydrogen gas.

Absorption Line Spectra Now, suppose a light bulb illuminates the hydrogen gas from behind (as in Figure 5.14c). The light bulb emits light of all wavelengths, producing a spectrum that looks like a rainbow of color. However, the hydrogen atoms can absorb those photons that have the right amount of energy to raise an electron from a low energy level to a higher one. **FIGURE 5.15c** shows the result. It is an absorption line spectrum, because the light bulb produces a continuous rainbow of color while the hydrogen atoms absorb light at specific wavelengths.

Given that electrons at high energy levels quickly return to lower levels, you might wonder why photons emitted in downward transitions don't cancel out the effects of those

---

*Astronomers call transitions between level 1 and other levels the *Lyman* series of transitions. The transition between level 1 and level 2 is Lyman α, between level 1 and level 3 is Lyman β, and so on. Similarly, transitions between level 2 and higher levels are called *Balmer* transitions. Other sets of transitions also have names.

helium

sodium

neon

FIGURE 5.16 Visible-light emission line spectra for helium, sodium, and neon. The patterns and wavelengths of lines are different for each element, giving each a unique spectral fingerprint.

He He   He O⁺   He   Ne⁺⁺

hydrogen lines

FIGURE 5.17 The emission line spectrum of the Orion Nebula in a portion of the ultraviolet (about 350–400 nm). The lines are labeled with the chemical elements or ions that produce them (He = helium; O = oxygen; Ne = neon). The many hydrogen lines are all transitions from high levels to level 2.

absorbed in upward transitions. Finding the answer requires looking more deeply at what happens to the absorbed photons. Two things can happen after an electron absorbs a photon and rises to a higher energy level. The first is that the electron quickly returns to its original level, emitting a photon of the same energy as the one that it absorbed. However, the emitted photon can be going in any random direction, which means that we will still see an absorption line because photons that were coming toward us are redirected away from our line of sight. Alternatively, the electron can lose its energy in some other way, either by dropping back down to its original level in multiple steps (and therefore emitting photons with different energies than the originally absorbed photon) or by transferring its energy to another particle in a subsequent collision. Either way, we are left with an absorption line because photons of a specific wavelength have been removed from the spectrum of light that's coming toward us.

You can now see why the dark absorption lines in Figure 5.15c occur at the same wavelengths as the emission lines in Figure 5.15b: Both types of lines represent the same energy level transitions, except in opposite directions. For example, electrons moving downward from level 3 to level 2 in hydrogen can emit photons of wavelength 656.3 nm (producing an emission line at this wavelength), while electrons absorbing photons with this wavelength can rise up from level 2 to level 3 (producing an absorption line at this wavelength).

Chemical Fingerprints The fact that hydrogen emits and absorbs light at specific wavelengths makes it possible to detect its presence in distant objects. For example, imagine that you look through a telescope at an interstellar gas cloud, and its spectrum looks like that shown in Figure 5.15b. Because only hydrogen produces this particular set of lines, you can conclude that the cloud is made of hydrogen. In essence, the spectrum contains a "fingerprint" left by hydrogen atoms.

Real interstellar clouds are not made solely of hydrogen. However, the other chemical constituents in the cloud leave fingerprints on the spectrum in much the same way. Every type of atom has its own unique spectral fingerprint, because it has its own unique set of energy levels. For example, **FIGURE 5.16** shows emission line spectra for helium, sodium, and neon. Moreover, different ions (atoms with missing or extra electrons) also produce different fingerprints (**FIGURE 5.17**). For example, the wavelengths of lines produced by doubly ionized neon (Ne⁺²)

are different from those of singly ionized neon (Ne⁺), which in turn are different from those of neutral neon (Ne). These differences can help us determine the temperature of a hot gas or plasma, because more highly charged ions will be present at higher temperatures; this fact enables us to use spectra to measure the surface temperatures of stars [**Section 15.1**].

Molecules also produce spectral fingerprints. Like atoms, molecules can produce spectral lines when their electrons change energy levels. But molecules can also produce spectral lines in two other ways. Because they are made of two or more atoms bound together, molecules can vibrate and rotate (**FIGURE 5.18a**). Vibration and rotation also require energy, and the possible energies of rotation and vibration in molecules are quantized much like electron energy levels in atoms. A molecule can absorb or emit a photon when it changes its rate of vibration or rotation. The energy changes in molecules are usually smaller than those in atoms and therefore produce lower-energy photons, and the energy levels also tend to be bunched more closely together than in atoms. Molecules therefore produce spectra with many sets of tightly bunched lines, called **molecular bands** (**FIGURE 5.18b**), that are usually found in the infrared portion of the electromagnetic spectrum.

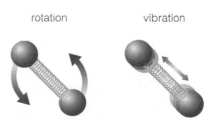

rotation   vibration

a We can think of a two-atom molecule as two balls connected by a spring. Although this model is simplistic, it illustrates how molecules can rotate and vibrate. The rotations and vibrations can have only particular amounts of energy and therefore produce unique spectral fingerprints.

b This spectrum of molecular hydrogen (H₂) consists of lines bunched into broad molecular bands.

FIGURE 5.18 Like atoms and ions, molecules emit or absorb light at specific wavelengths.

Over the past century, scientists have conducted laboratory experiments to identify the spectral lines of every chemical element and of many ions and molecules. As a result, when we see lines in the spectrum of a distant object, we can usually determine what produced them. For example, if we see spectral lines of hydrogen, helium, and carbon in the spectrum of a distant star, we know that all three elements are present in the star. More detailed analysis even allows us to determine the relative proportions of the various elements. That is how we have learned the chemical compositions of objects throughout the universe.

Reflected Light Spectra Some astronomical objects, such as planets and moons, reflect some of the light that falls on them. Reflected light also leaves a mark in spectra that can reveal information about the object, though not with the same level of detail as spectral lines. To understand why, consider the spectrum you would see from a red shirt on a sunny day. The red shirt absorbs blue light and reflects red light, so its visible spectrum will look like the spectrum of sunlight but with blue light missing. Because the shirt itself is too cool in temperature to emit visible light, the missing blue light must be telling you something about the dye in the shirt. In a similar way, the surface materials of a planet determine how much light of different colors is reflected or absorbed. The reflected light gives the planet its color, while the absorbed light heats the surface and helps determine its temperature. Careful study of which colors are absorbed and which are reflected can tell you at least something about the types of minerals on the surface.

How does light tell us the temperatures of planets and stars?

We have seen how emission and absorption line spectra form and how we can use them to determine the composition of a cloud of gas. Now we are ready to turn our attention to continuous spectra. Although continuous spectra can be produced in more than one way, light bulbs, planets, and stars produce a particular kind of continuous spectrum that can help us determine their temperatures.

Thermal Radiation: Every Body Does It In a cloud of gas that produces a simple emission or absorption line spectrum, the individual atoms or molecules are essentially independent of one another. Most photons pass easily through such a gas, except those that cause energy level transitions in the atoms or molecules of the gas. However, the atoms and molecules within most of the objects we encounter in everyday life—such as rocks, light bulb filaments, and people—cannot be considered independent and therefore have much more complex sets of energy levels. These objects tend to absorb light across a broad range of wavelengths, which means that light cannot easily pass through them and light emitted inside them cannot easily escape. The same is true of almost any large or dense object, including planets and stars.

In order to understand the spectra of such objects, let's consider an idealized case in which an object absorbs all photons that strike it and does not allow photons inside it to escape easily. Photons tend to bounce randomly around

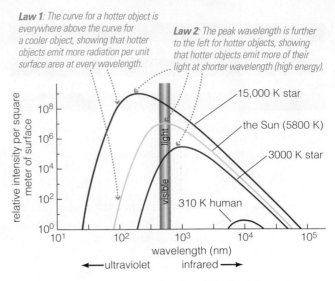

FIGURE 5.19 interactive figure Graphs of idealized thermal radiation spectra demonstrate the two laws of thermal radiation: (1) Each square meter of a hotter object's surface emits more light at all wavelengths; (2) hotter objects emit photons with a higher average energy. Notice that the graph uses power-of-10 scales on both axes, so that we can see all the curves even though the differences between them are quite large.

inside such an object, constantly exchanging energy with its atoms or molecules. By the time the photons finally escape the object, their radiative energies have become randomized so that they are spread over a wide range of wavelengths. The wide wavelength range of the photons explains why the spectrum of light from such an object is smooth, or *continuous,* like a pure rainbow without any absorption or emission lines.

Most important, the spectrum from such an object depends on only one thing: the object's *temperature.* To understand why, remember that temperature represents the average kinetic energy of the atoms or molecules in an object [**Section 4.3**]. Because the randomly bouncing photons interact so many times with those atoms or molecules, they end up with energies that match the kinetic energies of the object's atoms or molecules—which means the photon energies depend only on the object's temperature, regardless of what the object is made of. The temperature dependence of this light explains why we call it **thermal radiation** (sometimes known as *blackbody radiation*), and why its spectrum is called a **thermal radiation spectrum**.

No real object emits a perfect thermal radiation spectrum, but almost all familiar objects—including the Sun, the planets, rocks, and even you—emit light that approximates thermal radiation. **FIGURE 5.19** shows graphs of the idealized thermal radiation spectra of three stars and a human, each with its temperature given on the Kelvin scale (see Figure 4.12). Be sure to notice that these spectra show the intensity of light *per unit surface area*, not the total amount of light emitted by the object. For example, a very large 3000 K star can emit more total light than a small 15,000 K star, even though the hotter star emits much more light per unit area.

The Two Laws of Thermal Radiation If you compare the spectra in Figure 5.19, you'll see that they obey two laws of thermal radiation:

- **Law 1** (the Stefan-Boltzmann law): *Each square meter of a hotter object's surface emits more light at all wavelengths.* For example, each square meter on the surface of the 15,000 K star emits a lot more light at every wavelength than each square meter of the 3000 K star, and the hotter star emits light at some ultraviolet wavelengths that the cooler star does not emit at all.

- **Law 2** (Wien's [pronounced "veen's"] law): *Hotter objects emit photons with a higher average energy,* which means a shorter average wavelength. That is why the peaks of the spectra are at shorter wavelengths for hotter objects. For example, the peak for the 15,000 K star is in ultraviolet light, the peak for the 5800 K Sun is in visible light, and the peak for the 3000 K star is in the infrared.

You can see these laws in action with a fireplace poker (**FIGURE 5.20**). While the poker is still relatively cool, it emits only infrared light, which we cannot see. As it gets hot (above about 1500 K), it begins to glow with visible light, and it glows more brightly as it gets hotter, demonstrating the first law. Its color demonstrates the second law. At first it glows "red hot," because red light has the longest wavelengths of visible light. As it gets even hotter, the average wavelength of the emitted photons moves toward the blue (short-wavelength) end of the visible spectrum. The mix of colors emitted at this higher temperature makes the poker look white to your eyes, which is why "white hot" is hotter than "red hot."

At relatively low temperatures, the poker emits only infrared light that we cannot see.

As it gets hotter, it begins to glow.

It gets brighter as it heats up (demonstrating Law 1)…

…and changes from red to white in color (demonstrating Law 2).

FIGURE 5.20 interactive figure A fireplace poker shows the two laws of thermal radiation in action.

Because thermal radiation spectra depend only on temperature, we can use them to measure the temperatures of distant objects. In many cases we can estimate temperatures simply from the object's color. Notice that while hotter objects emit more light at *all* wavelengths, the biggest difference appears at the shortest wavelengths. At human body temperature of about 310 K, people emit mostly in the infrared and emit no visible light at all—which explains why we don't glow in the dark! A relatively cool star, with a 3000 K surface temperature, emits mostly red light. That is why some bright stars in our sky, such as Betelgeuse (in Orion) and Antares (in Scorpius), appear reddish in color. The Sun's 5800 K surface emits most strongly in green light (around 500 nm), but the Sun looks yellow or white to our eyes because it also emits other colors throughout the visible spectrum. Hotter stars emit mostly in the ultraviolet but appear blue-white in color because our eyes

---

MATHEMATICAL INSIGHT 5.2

Laws of Thermal Radiation

The two laws of thermal radiation have simple formulas.

Stefan-Boltzmann law (Law 1):

$$\text{emitted power (per square meter of surface)} = \sigma T^4$$

where σ (Greek letter *sigma*) is a constant with a measured value of $\sigma = 5.7 \times 10^{-8}$ watt/(m^2 × K^4) and T is on the Kelvin scale (K).

Wien's law (Law 2): $\quad \lambda_{max} \approx \dfrac{2,900,000}{T \text{ (Kelvin scale)}}$ nm

where λ_{max} (read as "lambda-max") is the wavelength (in nanometers) of maximum intensity, which is the peak of a thermal radiation spectrum.

EXAMPLE: Find the emitted power per square meter and the wavelength of peak intensity for a 15,000 K object that emits thermal radiation.

SOLUTION:

Step 1 Understand: We can calculate the emitted power per square meter from the Stefan-Boltzmann law and the wavelength of maximum intensity from Wien's law.

Step 2 Solve: We plug the object's temperature ($T = 15,000$ K) into the Stefan-Boltzmann law to find the emitted power per square meter:

$$\sigma T^4 = 5.7 \times 10^{-8} \frac{\text{watt}}{\text{m}^2 \times \text{K}^4} \times (15,000 \text{ K})^4$$
$$= 2.9 \times 10^9 \text{ watt/m}^2$$

We find the wavelength of maximum intensity with Wien's law:

$$\lambda_{max} \approx \frac{2,900,000}{15,000 \text{ (Kelvin scale)}} \text{ nm} \approx 190 \text{ nm}$$

Step 3 Explain: A 15,000 K object emits 2.9 billion watts per square meter of surface. Its wavelength of maximum intensity is 190 nm, which is in the ultraviolet. Note that we can learn about astronomical objects by using these facts in reverse. For example, if an object's thermal radiation spectrum peaks at a wavelength of 190 nm, its surface temperature must be about 15,000 K. We can then divide its total emitted power by the power it emits per square meter of surface to determine its surface area, from which we can calculate its radius.

The pitch this person hears . . .   *. . . is the same as the pitch this person hears.*

Behind the train, sound waves stretch to longer wavelength (lower frequency and pitch).   *In front of the train, sound waves bunch up to shorter wavelength (higher frequency and pitch).*

The light source is moving away from this person so the light appears redder (longer wavelength).   *The light source is moving toward this person so the light appears bluer (shorter wavelength).*

a The whistle sounds the same no matter where you stand near a stationary train.

b For a moving train, the sound you hear depends on whether the train is moving toward you or away from you.

c We get the same basic effect from a moving light source (although the shifts are usually too small to notice with our eyes).

FIGURE 5.21 The Doppler effect. Each circle represents the crests of sound (or light) waves going in all directions from the source. For example, the circles from the train might represent waves emitted 0.001 second apart.

cannot see their ultraviolet light. If an object were heated to a temperature of millions of degrees, it would radiate mostly X rays. Some astronomical objects are indeed hot enough to emit X rays, such as disks of gas encircling exotic objects like neutron stars and black holes (see Chapter 18).

(MA) The Doppler Effect Tutorial, Lessons 1, 2

How does light tell us the speed of a distant object?

There is still more that we can learn from light. In particular, we can learn about the motion of distant objects (relative to us) from changes in their spectra caused by the **Doppler effect**.

The Doppler Effect You've probably noticed the Doppler effect on the *sound* of a train whistle near train tracks. If the train is stationary, the pitch of its whistle sounds the same no matter where you stand (**FIGURE 5.21a**). But if the train is moving, the pitch sounds higher when the train is coming toward you and lower when it's moving away from you. Just as the train passes by, you can hear the dramatic change from high to low pitch—a sort of "weeeeeeee–ooooooooooh" sound. To understand why, we have to think about what happens to the sound waves coming from the train (**FIGURE 5.21b**). When the train is moving toward you, each pulse of a sound wave is emitted a little closer to you. The result is that waves are bunched up between you and the train, giving them a shorter wavelength and higher frequency (pitch). After the train passes you by, each pulse comes from farther away, stretching out the wavelengths and giving the sound a lower frequency.

The Doppler effect causes similar shifts in the wavelengths of light (**FIGURE 5.21c**). If an object is moving toward us, the light waves bunch up between us and the object, so its entire spectrum is shifted to shorter wavelengths. Because shorter wavelengths of visible light are bluer, the Doppler shift of an object coming toward us is called a **blueshift**. If an object

is moving away from us, its light is shifted to longer wavelengths. We call this Doppler shift a **redshift** because longer wavelengths of visible light are redder. For convenience, astronomers use the terms *blueshift* and *redshift* even when they aren't talking about visible light.

Spectral lines provide the reference points we use to identify and measure Doppler shifts (**FIGURE 5.22**). For example, suppose we recognize the pattern of hydrogen lines in the spectrum of a distant object. We know the **rest wavelengths** of the hydrogen lines—that is, their wavelengths in stationary clouds of hydrogen gas—from laboratory experiments in which a tube of hydrogen gas is heated so that the wavelengths of the spectral lines can be measured. If the hydrogen lines from the object appear at longer wavelengths, then we know they are redshifted and the object is moving away from us. The larger the shift, the faster the object is moving. If the lines appear at shorter wavelengths, then we know they are blueshifted and the object is moving toward us.

Laboratory spectrum
Lines at rest wavelengths.

Object 1 *Lines redshifted: Object moving away from us.*

Object 2 *Greater redshift: Object moving away faster than Object 1.*

Object 3 *Lines blueshifted: Object moving toward us.*

Object 4 *Greater blueshift: Object moving toward us faster than Object 3.*

FIGURE 5.22 interactive figure Spectral lines provide the crucial reference points for measuring Doppler shifts.

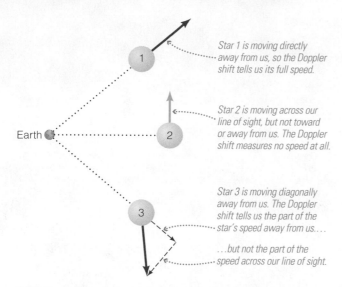

Star 1 is moving directly away from us, so the Doppler shift tells us its full speed.

Star 2 is moving across our line of sight, but not toward or away from us. The Doppler shift measures no speed at all.

Star 3 is moving diagonally away from us. The Doppler shift tells us the part of the star's speed away from us....

...but not the part of the speed across our line of sight.

FIGURE 5.23 interactive figure The Doppler shift tells us only the portion of an object's speed that is directed toward or away from us. It does not give us any information about how fast an object is moving across our line of sight.

Components of Motion It's important to note that a Doppler shift tells us only the part of an object's full motion that is directed toward or away from us (the object's *radial* component of motion). Doppler shifts do not give us any information

about how fast an object is moving across our line of sight (the object's *tangential* component of motion). For example, consider three stars all moving at the same speed, with one moving directly away from us, one moving across our line of sight, and one moving diagonally away from us (**FIGURE 5.23**). The Doppler shift will tell us the full speed of only the first star. It will not indicate any speed for the second star, because none of this star's motion is directed toward or away from us. For the third star, the Doppler shift will tell us only the part of the star's velocity that is directed away from us. To measure how fast an object is moving across our line of sight, we must observe it long enough to notice how its position gradually shifts across our sky.

Rotation Rates The Doppler effect not only tells us how fast a distant object is moving toward or away from us but also can reveal information about motion *within* the object. For example, suppose we look at spectral lines of a planet or star that happens to be rotating (**FIGURE 5.24**). As the object rotates, light from the part of the object rotating toward us will be blueshifted, light from the part rotating away from us will be redshifted, and light from the center of the object won't be shifted at all. The net effect, if we look at the whole object at once, is to make each spectral line appear *wider* than it would if the object were not rotating. The faster the object is rotating, the broader in wavelength the spectral lines become. We can therefore determine the rotation rate of a distant object by measuring the width of its spectral lines.

Putting It All Together FIGURE 5.25 shows the same spectrum we began with in Figure 5.13, but this time with labels indicating the processes responsible for its various features. The thermal emission peaks in the infrared, corresponding to a surface temperature of about 225 K, well below the 273 K freezing point of water. The absorption bands in the

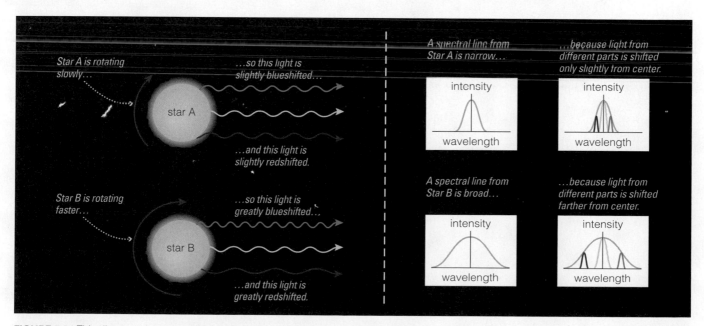

FIGURE 5.24 This diagram shows how the Doppler effect can tell us the rotation rate even of stars that appear as points of light to our telescopes. Rotation spreads the light of any spectral line over a range of wavelengths, so faster-rotating stars have broader spectral lines.

An astronomical spectrum contains an enormous amount of information. This figure shows a schematic spectrum of Mars. It is the same spectrum shown in Figure 5.13, but this time describing what we can learn from it.

① Continuous Spectrum: The visible light we see from Mars is actually reflected sunlight. The Sun produces a nearly continuous spectrum of light, which includes the full rainbow of color.

hot light source

prism

Like the Sun, a light bulb produces light of all visible wavelengths (colors).

② Scattered/Reflected Light: Mars is red because it absorbs most of the blue light from the Sun but reflects (scatters) most of the red light. This pattern of absorption and reflection helps us learn the chemical composition of the surface.

Like Mars, a red chair looks red because it absorbs blue light and scatters red light.

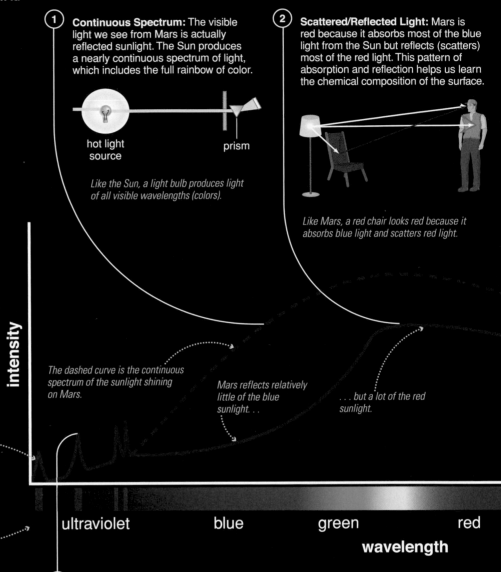

intensity

The dashed curve is the continuous spectrum of the sunlight shining on Mars.

Mars reflects relatively little of the blue sunlight. . .

. . . but a lot of the red sunlight.

The graph and the "rainbow" contain the same information. The graph makes it easier to read the intensity at each wavelength of light. . .

. . . while the "rainbow" shows how the spectrum appears to the eye (for visible light) or instruments (for non-visible light).

ultraviolet   blue   green   red

wavelength

④ Emission Lines: Ultraviolet emission lines in the spectrum of Mars tell us that the atmosphere of Mars contains hot gas at high altitudes.

cloud of gas

prism

We see bright emission lines from gases in which collisions raise electrons in atoms to higher energy levels. The atoms emit photons at specific wavelengths as the electrons drop to lower energy levels.

3 **Thermal Radiation:** Objects emit a continuous spectrum of thermal radiation that peaks at a wavelength determined by temperature. Thermal radiation from Mars produces a broad hump in the infrared, with a peak indicating a surface temperature of about 225 K.

All objects — whether a fireplace poker, planet, or star—emit thermal radiation. The hotter the object, (1) the more total light (per unit area), and (2) the higher the average energy (shorter average wavelength) of the emitted photons.

Mars's thermal radiation peaks in the infrared because it is much cooler than the Sun, which peaks in visible light.

infrared

5 **Absorption Lines:** These absorption lines reveal the presence of carbon dioxide in Mars's atmosphere.

hot light source

cloud of gas

prism

When light from a hot source passes through a cooler gas, the gas absorbs light at specific wavelengths that raise electrons to higher energy levels. Every different element, ion, and molecule has unique energy levels and hence its own spectral "fingerprint."

6 **Doppler Effect:** The wavelengths of the spectral lines from Mars are slightly shifted by an amount that depends on the velocity of Mars toward or away from us as it moves

A Doppler shift toward the red side of the spectrum tells us the object is moving away from

infrared come mainly from carbon dioxide, indicating a carbon dioxide atmosphere. The emission lines in the ultraviolet come from hot gas in a high, thin layer of the object's atmosphere. The reflected light looks like the Sun's 5800 K thermal radiation except that much of the blue light is missing, so the object must be reflecting sunlight and must look red in color. Perhaps by now you have guessed that this figure represents the spectrum of the planet Mars.

MATHEMATICAL INSIGHT 5.3

The Doppler Shift

We can calculate an object's radial (toward or away from us) velocity from its Doppler shift. For speeds small compared to the speed of light (less than a few percent of c), the formula is

$$\frac{v_{rad}}{c} = \frac{\lambda_{shift} - \lambda_{rest}}{\lambda_{rest}}$$

where v_{rad} is the radial velocity of the object, λ_{rest} is the rest wavelength of a particular spectral line, and λ_{shift} is the shifted wavelength of the same line. A positive answer means the object is redshifted and moving away from us; a negative answer means it is blueshifted and moving toward us.

EXAMPLE: One of the visible lines of hydrogen has a rest wavelength of 656.285 nm, but it appears in the spectrum of the star Vega at 656.255 nm. How is Vega moving relative to us?

SOLUTION:

Step 1 Understand: We can calculate the radial velocity from the given formula. Note that the line's wavelength in Vega's spectrum is slightly *shorter* than its rest wavelength, which means it is blueshifted and Vega's radial motion is *toward* us.

Step 2 Solve: We plug in the rest wavelength ($\lambda_{rest} = 656.285$ nm) and the wavelength in Vega's spectrum ($\lambda_{shift} = 656.255$ nm):

$$\frac{v_{rad}}{c} = \frac{\lambda_{shift} - \lambda_{rest}}{\lambda_{rest}}$$

$$= \frac{656.255 \text{ nm} - 656.285 \text{ nm}}{656.285 \text{ nm}}$$

$$= -4.5712 \times 10^{-5}$$

Step 3 Explain: We have found Vega's radial velocity as a fraction of the speed of light; it is negative because Vega is moving toward us. To convert to a velocity in km/s, we multiply by the speed of light:

$$v_{rad} = -4.5712 \times 10^{-5} \times c$$

$$= -4.5712 \times 10^{-5} \times (3 \times 10^5 \text{ km/s})$$

$$= -13.7 \text{ km/s}$$

Vega is moving *toward* us at 13.7 km/s. This speed is typical of stars in our neighborhood of the galaxy.

The Big Picture

Putting Chapter 5 into Context

This chapter was devoted to one essential purpose: understanding how we learn about the universe by observing the light of distant objects. "Big picture" ideas that will help you keep your understanding in perspective include the following:

- Light and matter interact in ways that allow matter to leave "fingerprints" on light. We can therefore learn a great deal about the objects we observe by carefully analyzing their light. Most of what we know about the universe comes from information that we receive from light.

- The visible light that our eyes can see is only a small portion of the complete electromagnetic spectrum. Different portions of the spectrum contain different pieces of the story of a distant object, so it is important to study all forms of light.

- There is far more to light than meets the eye. By dispersing the light of a distant object into a spectrum, we can determine the object's composition, surface temperature, motion toward or away from us, and more.

SUMMARY OF KEY CONCEPTS

5.1 LIGHT IN EVERYDAY LIFE

- **How do we experience light?** Light carries radiative energy that it can exchange with matter. **Power** is the *rate* of energy transfer, measured in **watts**: 1 watt = 1 joule/s. The colors of light contain a great deal of information about the matter with which it has interacted.

- **How do light and matter interact?** Matter can emit, absorb, transmit, or reflect (or scatter) light.

5.2 PROPERTIES OF LIGHT

- **What is light?** Light is an **electromagnetic wave**, but also comes in individual "pieces" called **photons**. Each photon has a precise wavelength, frequency, and energy: The shorter the wavelength, the higher the frequency and energy.

■ **What is the electromagnetic spectrum?** In order of decreasing wavelength (increasing frequency and energy), the forms of light are **radio waves, microwaves, infrared, visible light, ultraviolet, X rays**, and **gamma rays**.

5.3 PROPERTIES OF MATTER

■ **What is the structure of matter?** Ordinary matter is made of **atoms**, which are made of **protons, neutrons**, and **electrons**. Atoms of different chemical **elements** have different numbers of

protons. **Isotopes** of a particular chemical element all have the same number of protons but different numbers of neutrons. **Molecules** are made from two or more atoms.

■ **What are the phases of matter?** The appearance of matter depends on its **phase: solid, liquid**, or **gas**. Gas is always present along with solid or liquid phases; solids **sublimate** into gas and liquids **evaporate** into gas. At very high temperatures, **molecular dissociation** breaks up molecules and **ionization** strips electrons from atoms; an ionized gas is called a **plasma**.

■ **How is energy stored in atoms?** Electrons can exist at

particular **energy levels** within an atom. **Energy level transitions**, in which an electron moves from one energy level to another, can occur only when the electron gains or loses just the right amount of energy.

5.4 LEARNING FROM LIGHT

■ **What are the three basic types of spectra?** There are

three basic types of spectra: a **continuous spectrum**, which looks like a rainbow of light; an **absorption line spectrum**, in which specific colors are missing from the rainbow; and an **emission line spectrum**, in which we see light only with specific colors against a black background.

■ **How does light tell us what things are made of?** **Emission** or **absorption lines** occur only at specific wavelengths that correspond to particular energy level transitions in atoms or molecules. Every kind of atom, ion, and molecule produces a unique set of spectral lines, so we can determine composition by identifying these lines.

■ **How does light tell us the temperatures of planets and stars?** Objects such as planets and stars produce **thermal radiation** spectra, the most common type of continuous spectra.

We can determine temperature from these spectra because hotter objects emit more total radiation per unit area and emit photons with a higher average energy.

■ **How does light tell us the speed of a distant object?** The **Doppler effect** tells us how fast an object is moving toward or away from us. Spectral lines are shifted to shorter wave-

lengths (a **blueshift**) in objects moving toward us and to longer wavelengths (a **redshift**) in objects moving away from us.

VISUAL SKILLS CHECK

Use the following questions to check your understanding of some of the many types of visual information used in astronomy. Answers are provided in Appendix J. For additional practice, try the Chapter 5 Visual Quiz at MasteringAstronomy®.

The graph above is a schematic spectrum of the planet Mars; it is the same spectrum shown in Figure 5.13. Keeping in mind that Mars reflects visible sunlight and emits infrared light, refer to the numbered features of the graph and answer the following questions.

1. Which of the six numbered features represents emission lines?
2. Which of the six numbered features represents absorption lines?
3. Which portion(s) of the spectrum represent(s) reflected sunlight?
   a. 1 only   b. 2, 3, and 4   c. 3 and 6   d. the entire spectrum
4. What does the wavelength of the peak labeled 6 tell us about Mars?
   a. its color   b. its surface temperature
   c. its chemical composition   d. its orbital speed

5. What feature(s) of this spectrum indicate(s) that Mars appears red in color?
   a. the wavelength of the peak labeled 3
   b. the wavelength of the peak labeled 6
   c. the fact that the intensity of region 4 is higher than that of region 2
   d. the fact that the peak labeled 3 is higher than the peak labeled 6

REVIEW QUESTIONS

Short-Answer Questions Based on the Reading

1. What is the difference between *energy* and *power*? What units do we use to measure power?

2. What are the four major ways light and matter can interact? Give an example of each from everyday life.

3. What do we mean when we say that light is an *electromagnetic wave*? Describe the relationship among *wavelength, frequency,* and *speed* for light waves.

4. What is a *photon*? In what way is a photon like a particle? In what way is it like a wave?

5. List the different forms of light in order from lowest to highest energy. Would the list be different if you went in order from lowest to highest frequency? From shortest to longest wavelength? Explain.

6. Briefly describe the structure and size of an atom. How big is the *nucleus* in comparison to the entire atom?

7. What determines an atom's *atomic number*? What determines its *atomic mass number*? Under what conditions are two atoms different *isotopes* of the same element? What is a *molecule*?

8. What is *electrical charge*? Will an electron and a proton attract or repel each other? Will two electrons attract or repel each other? Explain.

9. Describe the phase changes of water as you heat it, starting from its solid phase, ice. What happens at very high temperatures? What is a *plasma*?

10. What do we mean when we say that *energy levels* are quantized in atoms? Under what circumstances can *energy level transitions* occur?

11. How do we convert a spectrum shown as a band of light (like a rainbow) into a graph of the spectrum?

12. Describe the conditions that would cause us to see each of the three basic types of spectra. What do we see in the Sun's spectrum shown on the opening page of this chapter?

13. How can we use *emission* or *absorption lines* to determine the chemical composition of a distant object?

14. Describe two ways in which the *thermal radiation spectrum* of an 8000 K star would differ from that of a 4000 K star.

15. Describe the *Doppler effect* for light and what we can learn from it. What does it mean to say that radio waves are *blueshifted*? Why does the Doppler effect widen the spectral lines of rotating objects?

16. Describe each of the key features of the spectrum in Figure 5.25 and explain what it tells us about the object.

TEST YOUR UNDERSTANDING

Does It Make Sense?

Decide whether the statement makes sense (or is clearly true) or does not make sense (or is clearly false). Explain clearly; not all of these have definitive answers, so your explanation is more important than your chosen answer.

17. The walls of my room are transparent to radio waves.

18. Because of their higher frequencies, X rays must travel through space faster than radio waves.

19. If you could see infrared light, you would see a glow from the backs of your eyelids when you closed your eyes.

20. If you had X-ray vision, you could read this entire book without turning any pages.

21. Two isotopes of the element rubidium differ in their number of protons.

22. A "white hot" object is hotter than a "red hot" object.

23. If the Sun's surface became much hotter (while the Sun's size remained the same), the Sun would emit more ultraviolet light but less visible light than it currently emits.

24. If you could view a spectrum of light reflecting off a blue sweatshirt, you'd find the entire rainbow of color (looking the same as a spectrum of white light).

25. Galaxies that show redshifts must be red in color.

26. If a distant galaxy has a substantial redshift (as viewed from our galaxy), then anyone living in that galaxy would see a substantial redshift in a spectrum of the Milky Way Galaxy.

Quick Quiz

Choose the best answer to each of the following. Explain your reasoning with one or more complete sentences.

27. Why is a sunflower yellow? (a) It emits yellow light. (b) It absorbs yellow light. (c) It reflects yellow light.

28. Blue light has higher frequency than red light. Thus, blue light has (a) higher energy and shorter wavelength than red light. (b) higher energy and longer wavelength than red light. (c) lower energy and shorter wavelength than red light.

29. Radio waves are (a) a form of sound. (b) a form of light. (c) a type of spectrum.

30. Compared to an atom as a whole, an atomic nucleus (a) is very tiny but has most of the mass. (b) is quite large and has most of the mass. (c) is very tiny and has very little mass.

31. Some nitrogen atoms have 7 neutrons and some have 8 neutrons, which makes these two forms of nitrogen (a) ions of each other. (b) phases of each other. (c) isotopes of each other.

32. Sublimation is the process by which (a) solid material enters the gas phase. (b) liquid material enters the gas phase. (c) solid material becomes a liquid.

33. If you heat a rock until it glows, its spectrum will be (a) a thermal radiation spectrum. (b) an absorption line spectrum. (c) an emission line spectrum.

34. The set of spectral lines that we see in a star's spectrum depends on the star's (a) interior temperature. (b) chemical composition. (c) rotation rate.

35. Compared to the Sun, a star whose spectrum peaks in the infrared is (a) cooler. (b) hotter. (c) larger.

36. A spectral line that appears at a wavelength of 321 nm in the laboratory appears at a wavelength of 328 nm in the spectrum of a distant object. We say that the object's spectrum is (a) redshifted. (b) blueshifted. (c) whiteshifted.

PROCESS OF SCIENCE

Examining How Science Works

37. *Elements in Space.* Astronomers claim that objects throughout the universe are made of the same chemical elements that exist here on Earth. Given that most of these objects are so far away that we can never hope to visit them, why are astronomers so confident that these objects are made from the same set of chemical elements, rather than some completely different types of materials?

38. *Newton's Prisms.* Look back at the brief discussion in this chapter of how Newton proved that the colors seen when light passed through a prism came from the light itself rather than from the prism. Suppose you wanted to test Newton's findings. Assuming you have two prisms and a white screen, describe how you would arrange the prisms to duplicate Newton's discovery.

GROUP WORK EXERCISE

39. *Light Around You.* Before you begin, assign the following roles to the people in your group: *Scribe* (takes notes on the group's activities), *Proposer* (proposes explanations to the group), *Skeptic* (points out weaknesses in proposed explanations), and *Moderator* (leads group discussion and makes sure everyone contributes). Look carefully at all the ways in which light and matter are interacting in the room around you to answer the following questions: **a.** What is emitting light? **b.** What is absorbing light? **c.** What is responsible for the colors you see? **d.** What would the room look like if you observed it with an infrared camera? With an ultraviolet camera? With an X-ray camera? **e.** Are there any radio waves in the room? Explain all your answers clearly.

INVESTIGATE FURTHER

In-Depth Questions to Increase Your Understanding

Short-Answer/Essay Questions

40. *Atomic Terminology Practice I.*
    a. The most common form of iron has 26 protons and 30 neutrons. State its atomic number, atomic mass number, and number of electrons (if it is neutral). **b.** Consider the following three atoms: Atom 1 has 7 protons and 8 neutrons; atom 2 has 8 protons and 7 neutrons; atom 3 has 8 protons and 8 neutrons. Which two are *isotopes* of the same element? **c.** Oxygen has atomic number 8. How many times must an oxygen atom be ionized to create an O^{+5} ion? How many electrons are in an O^{+5} ion?

41. *Atomic Terminology Practice II.*
    a. What are the atomic number and atomic mass number of a fluorine atom with 9 protons and 10 neutrons? If we could add a proton to this fluorine nucleus, would the result still be fluorine? What if we added a neutron to the fluorine nucleus? Explain. **b.** The most common isotope of gold has atomic number 79 and atomic mass number 197. How many protons and neutrons does the gold nucleus contain? If the isotope is electrically neutral, how many electrons does it have? If it is triply ionized, how many electrons does it have? **c.** Uranium has atomic number 92. Its most common isotope is ^{238}U, but the form used in nuclear bombs and nuclear power plants is ^{235}U. How many neutrons are in each of these two isotopes of uranium?

42. *The Fourth Phase of Matter.*
    a. Explain why nearly all the matter in the Sun is in the plasma phase. **b.** Based on your answer to part a, explain why plasma is the most common phase of matter in the universe. **c.** If plasma is the most common phase of matter in the universe, why is it so rare on Earth?

43. *Energy Level Transitions.* The following labeled transitions represent an electron moving between energy levels in hydrogen. Answer each of the following questions and explain your answers.

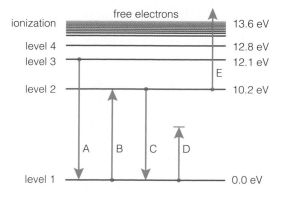

a. Which transition could represent an atom that *absorbs* a photon with 10.2 eV of energy? **b.** Which transition could represent an atom that *emits* a photon with 10.2 eV of energy? **c.** Which transition represents an electron that is breaking free of the atom? **d.** Which transition, as shown, is *not* possible? **e.** Would transition A represent emission or absorption of light? How would the wavelength of the emitted or absorbed photon compare to that of the photon involved in transition C? Explain.

44. *Spectral Summary.* Clearly explain how studying an object's spectrum can allow us to determine each of the following properties of the object.
    a. The object's surface chemical composition **b.** The object's surface temperature **c.** Whether the object is a low-density cloud of gas or something more substantial **d.** Whether the object has a hot upper atmosphere **e.** Whether the object is reflecting blue light from a star **f.** The speed at which the object is moving toward or away from us **g.** The object's rotation rate

45. *Orion Nebula.* Much of the Orion Nebula looks like a glowing cloud of gas. What type of spectrum would you expect to see from the glowing parts of the nebula? Why?

46. *The Doppler Effect.* In hydrogen, the transition from level 2 to level 1 has a rest wavelength of 121.6 nm. Suppose you see this line at a wavelength of 120.5 nm in Star A, 121.2 nm in Star B, 121.9 nm in Star C, and 122.9 nm in Star D. Which stars are coming toward us? Which are moving away? Which star is moving fastest relative to us? Explain your answers without doing any calculations.

Quantitative Problems

Be sure to show all calculations clearly and state your final answers in complete sentences.

47. *Human Wattage.* A typical adult uses about 2500 Calories of energy each day. Use this fact to calculate the typical adult's average *power* requirement, in watts. (*Hint:* 1 Calorie = 4184 joules.)

48. *Electric Bill.* Your electric utility bill probably shows your energy use for the month in units of *kilowatt-hours*. A kilowatt-hour is defined as the energy used in 1 hour at a rate of 1 kilowatt (1000 watts); that is, 1 kilowatt-hour = 1 kilowatt × 1 hour. Use this fact to convert 1 kilowatt-hour into joules. If your bill says you used 900 kilowatt-hours, how much energy did you use in joules?

49. *Radio Station.* What is the wavelength of a radio photon from an AM radio station that broadcasts at 1120 kilohertz? What is its energy?

50. *UV Photon.* What is the energy (in joules) of an ultraviolet photon with wavelength 120 nm? What is its frequency?

51. *X-Ray Photon.* What is the wavelength of an X-ray photon with energy 10 keV (10,000 eV)? What is its frequency? (1 eV = 1.60 × 10^{-19} joule.)

52. *How Many Photons?* Suppose that all the energy from a 100-watt light bulb came in the form of photons with wavelength 600 nm. (This is not quite realistic; see Problem 57.)
    a. Calculate the energy of a *single* photon with wavelength 600 nm. **b.** How many 600-nm photons must be emitted each second to account for all the light from this 100-watt light bulb? **c.** Based on your answer to part b, explain why we don't notice the particle nature of light in our everyday lives.

53. *Thermal Radiation Laws.*
    a. Find the emitted power per square meter and wavelength of peak intensity for a 3000 K object that emits thermal radiation. **b.** Find the emitted power per square meter and wavelength of peak intensity for a 50,000 K object that emits thermal radiation.

54. *Hotter Sun.* Suppose the surface temperature of the Sun were about 12,000 K, rather than 6000 K.

a. How much more thermal radiation would the Sun emit? **b.** What would happen to the Sun's wavelength of peak emission? **c.** Do you think it would still be possible to have life on Earth? Explain.

55. *Taking the Sun's Temperature.* The Sun radiates a total power of about 4×10^{26} watts into space. The Sun's radius is about 7×10^8 meters.

    a. Calculate the average power radiated by each square meter of the Sun's surface. (*Hint:* The formula for the surface area of a sphere is $A = 4\pi r^2$.) **b.** Using your answer from part a and the Stefan-Boltzmann law, calculate the average surface temperature of the Sun. (*Note:* The temperature calculated this way is called the Sun's *effective temperature.*)

56. *Doppler Calculations.* In hydrogen, the transition from level 2 to level 1 has a rest wavelength of 121.6 nm. Find the speed and direction (toward or away from us) for a star in which this line appears at wavelength

    a. 120.5 nm. **b.** 121.2 nm. **c.** 121.9 nm. **d.** 122.9 nm.

57. *Understanding Light Bulbs.* A traditional incandescent light bulb uses a hot tungsten coil to produce a thermal radiation spectrum. The temperature of this coil is typically about 3000 K.

    a. What is the wavelength of maximum intensity for this light bulb? Compare to the 500-nm wavelength of maximum intensity for the Sun. **b.** Overall, do you expect the light from this bulb to be the same as, redder than, or bluer than light from the Sun? Why? Use your answer to explain why professional photographers use a different type of film for indoor photography than for outdoor photography. **c.** Do incandescent light bulbs emit all their energy as visible light? Use your answer to explain why these light bulbs are usually hot to touch. **d.** *Fluorescent* light bulbs primarily produce emission line spectra rather than thermal radiation spectra. Explain why, if the emission lines are in the visible part of the spectrum, a fluorescent bulb can emit more visible light than a standard bulb of the same wattage. **e.** *Compact fluorescent* light bulbs are designed to produce so many emission lines in the visible part of the spectrum that their light looks very similar to the light of incandescent bulbs. However, they are much more energy efficient: A 15-watt compact fluorescent bulb typically emits as much visible light as a traditional incandescent 75-watt bulb. Although compact fluorescent bulbs generally cost more than incandescent bulbs, is it possible that they could save you money? Besides initial cost and energy efficiency, what other factors must be considered?

Discussion Questions

58. *The Changing Limitations of Science.* In 1835, French philosopher Auguste Comte stated that science would never allow us to learn the composition of stars. Although spectral lines had been seen in the Sun's spectrum by that time, not until the middle of the 19th century did scientists recognize that spectral lines give clear information about chemical composition (primarily through the work of Foucault and Kirchhoff). Why might our present knowledge have seemed unattainable in 1835? Discuss how new discoveries can change the apparent limitations of science. Today, other questions seem beyond the reach of science, such as the question of how life began on Earth. Do you think such questions will ever be answerable through science? Defend your opinion.

59. *Your Microwave Oven.* A microwave oven emits microwaves that have just the right wavelength to cause energy level changes in water molecules. Use this fact to explain how a microwave oven cooks your food. Why doesn't a microwave oven make a plastic dish get hot? Why do some clay dishes get hot in the microwave? Why do dishes that aren't themselves heated by the microwave oven sometimes still get hot when you heat food on them? (*Note:* It's not a good idea to put dishes without food or liquid in a microwave.)

60. *Democritus and the Path of History.* Besides his belief in atoms, Democritus held several other strikingly modern notions. For example, he maintained that the Moon was a world with mountains and valleys and that the Milky Way was composed of countless individual stars—ideas that weren't generally accepted until the time of Galileo, more than 2000 years later. Unfortunately, we know of Democritus's work only secondhand because none of the 72 books he is said to have written survived the destruction of the Library of Alexandria. Do you think history might have been different if the work of Democritus had not been lost? Defend your opinion.

Web Projects

61. *Kids and Light.* Visit one of the many websites designed to teach middle and high school students about light. Read the content and try the activities. If you were a teacher, would you find the site useful for your students? Why or why not? Write a one-page summary of your conclusions.

62. *Light Bulbs.* To save energy, in 2007 the U.S. Congress passed legislation designed to phase out the use of traditional incandescent light bulbs by 2014. Find out about the status of this phaseout; is it still on track? What types of alternative bulbs are available? Write a short report summarizing the advantages and disadvantages of each technology.

63. *Medical Imaging.* Learn about CT scans or other technologies for medical imaging of the human body. How do they work? How are such technologies similar to those used by astronomers to learn about the universe? Write a short report summarizing your findings.

TELESCOPES
PORTALS OF DISCOVERY

6

> *All of this has been discovered and observed these last days thanks to the telescope that I have [built], after having been enlightened by divine grace.*
>
> —Galileo

We are in the midst of a revolution in human understanding of the universe, sparked in large part by advances in telescope technology. New technologies are fueling the construction of more and larger telescopes while vastly improving the quality of data that we can obtain from older telescopes. Meanwhile, telescopes lofted into space are offering views of the heavens unobstructed by Earth's atmosphere while also allowing us to study light at wavelengths that do not penetrate to the ground.

Because telescopes are the portals through which we study the universe, understanding them can help us understand both the triumphs and the limitations of modern astronomy. In this chapter, we explore the basic principles by which telescopes work and some of the technological advances behind the current revolution in astronomy.

6.1 EYES AND CAMERAS: EVERYDAY LIGHT SENSORS

We learn about the world around us by observing with our five senses (touch, taste, smell, hearing, sight) and using our brains to analyze and interpret the data that our senses record. The science of astronomy progresses similarly. We collect data about the universe, and then we analyze and interpret the data. Within our solar system, we can analyze some matter directly, such as samples of Earth's surface, meteorites, and surfaces and atmospheres of worlds visited by spacecraft. Virtually all other data about the universe come to us in the form of light, which we collect with telescopes and record with cameras and other instruments. Because telescopes function much like giant eyes, we begin this chapter by examining the principles of eyes and cameras, our everyday light sensors.

How do eyes and cameras work?

Eyes and cameras work similarly, so let's begin with eyes. The eye is a remarkably complex organ, but its basic components are a *pupil*, a *lens*,* and a *retina* (**FIGURE 6.1**). The pupil controls how much light enters the eye; it dilates (opens wider) in low light and constricts in bright light. The lens bends light to form an image on the retina. The retina contains light-sensitive cells (called *cones* and *rods*) that, when triggered by light, send signals to the brain via the optic nerve.

SEE IT FOR YOURSELF

Use a mirror to compare the opening size of your pupils under normal lighting and right after looking at a bright light. What do you notice? Why do you think eye doctors dilate your pupils during eye exams?

---

*The lens actually works together with the cornea (the clear part of the eye in front of the pupil), but for simplicity we will consider their combined effects as the effects of the lens.

FIGURE 6.1 A simplified diagram of the human eye.

Bending Light The lens of the eye creates an image by bending light in much the same way as a simple glass lens. You can understand why light bends by imagining a light wave coming toward you from far away. The peaks and troughs of the electric and magnetic fields are perpendicular to the light wave's direction of travel, as shown for the approaching light wave in **FIGURE 6.2**. The wave slows down when it hits glass or your eye because light travels more slowly through denser matter than through air. For light coming in at an angle (as in Figure 6.2), this slowing affects the side of the wave nearest the surface first, allowing the far side to catch up. The result is bending (more technically known as *refraction*)—a change in the direction in which the light is traveling. **FIGURE 6.3** shows an example of how Earth's atmosphere bends light from space, distorting the Sun's image at sunset.

Image Formation We can visualize the bending of light by drawing simple *rays*, with each ray (drawn as an arrow) representing a series of light waves coming from a single direction. Light rays that enter the lens farther from the center are bent more, and rays that pass directly through the center are not bent at all. In this way, parallel rays of light, such as

FIGURE 6.2 Light that hits glass at an angle bends as it enters the glass, a phenomenon called *refraction*. The wide yellow ribbons in this figure represent light waves. The darker bands on those ribbons (perpendicular to the direction in which the light travels) represent the positions of wave peaks.

FIGURE 6.3 Earth's atmosphere also bends light. The Sun looks squashed at sunset because light from the lower portion of the Sun passes through more atmosphere and therefore bends slightly more than light from the upper portion.

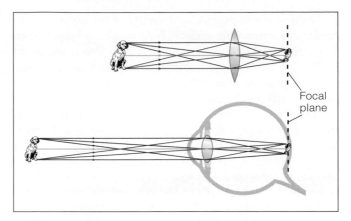

FIGURE 6.5 Light from different parts of an object focuses at different points to make an (upside-down) image of the object.

those from a distant star, converge to a point called the **focus** (or **focal point**). **FIGURE 6.4** shows the idea for both a glass lens and an eye. The fact that parallel rays of light converge to a sharp focus explains why distant stars appear as *points* of light to our eyes or on photographs.

Light rays that are *not* parallel, such as those from a nearby object, enter a lens from different directions. These rays do not all converge at the focus, but they still follow precise rules as they bend at the lens; we will not discuss these rules in this book, but some of them are illustrated by the ray paths in **FIGURE 6.5**. The result is the bending of rays to form an **image** of the original object. The place where the image appears in focus is called the **focal plane** of the lens. In an eye with perfect vision, the focal plane is on the retina. (The retina actually is curved, rather than a flat plane, but we will ignore this detail.) Note that the image formed by a lens is upside down. In other words, our eyes actually form upside-down images, which are then flipped right-side up by our brains.

Recording Images If we want to keep an image or study it in detail, it's useful to record it with a camera (**FIGURE 6.6**). The basic operation of a camera is quite similar to that of an eye. The camera has a small opening for light to enter, much like the pupil of the eye. The camera lens bends the light, bringing it to a focus on a **detector** that makes a permanent record of the image. Today, detectors are nearly always electronic, but older cameras used photographic film. Cameras also have a *shutter* that is analogous to an eyelid: Light can reach the detector only when the shutter is open. We can use the shutter to control the **exposure time** of an image, the amount of time during which light collects on the detector. A longer exposure time means that more photons reach the detector, allowing the detector to record details that might be too faint to be seen in shorter exposures.

Modern detectors use electronic chips that are physically divided into grids of *picture elements*, or **pixels** for short. When a photon strikes a pixel, it causes a bit of electric charge to accumulate. Each subsequent photon striking the same pixel adds to this accumulated electric charge. After an exposure is complete, a computer measures the total electric charge in each pixel, thereby determining how many photons have struck each one. The overall image is stored on a memory chip as an array of numbers representing the results from each pixel. Most consumer camera chips now have 10 million or more pixels, and professional cameras can have significantly more.

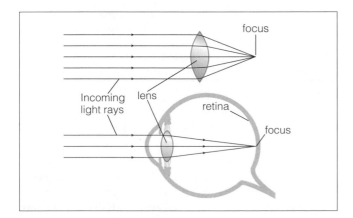

FIGURE 6.4 A glass lens bends parallel rays of light to a point called the *focus* of the lens. In an eye with perfect vision, rays of light are bent to a focus on the retina.

FIGURE 6.6 A camera works much like an eye. When the shutter is open, light passes through the lens to form an image on the detector (which may be film or an electronic device).

No detector is perfect, so a variety of tradeoffs must be made when recording images. For example, a longer exposure can reveal fainter details, but it may also cause bright regions of the image to become overexposed, meaning that so many photons collect that they can no longer be counted accurately. Nevertheless, today's detectors are able to record light over a wide range of brightness levels much more accurately than the photographic film of the past, and the technology continues to improve.

Image Processing The photographs you see in most media and in science today are not the original images recorded by cameras; instead, these images have been combined and manipulated through techniques of *image processing*. Some of the images we see today have been manipulated to change what the camera actually recorded, something that's easily done with almost any photo software. In science, image processing is often used to bring out details that might otherwise remain hidden. For example, image processing can be used to sharpen or adjust colors, to correct over- or under-exposure, or to remove artifacts or defects from an image. We will not discuss image processing much in this book, but you should be aware of its tremendous power both to increase the amount of information we can glean from images and, when misused, to distort what we see.

(MA) Telescopes Tutorial, Lessons 1–2

6.2 TELESCOPES: GIANT EYES

Telescopes are essentially giant eyes that can collect far more light than our own eyes. By combining this light-collecting capacity with cameras and other instruments that can record and analyze light in detail, modern telescopes have become extremely powerful scientific instruments.

What are the two most important properties of a telescope?

Let's begin by investigating the two most fundamental properties of any telescope: its *light-collecting area* and its *angular resolution*.

A telescope's **light-collecting area** tells us how much total light it can collect at one time. Telescopes are generally round, so we usually characterize a telescope's size by the *diameter* of its light-collecting area. For example, a "10-meter telescope" has a light-collecting area that is 10 meters in diameter. Note that, because area is proportional to the *square* of diameter, a relatively small increase in diameter can mean a big increase in light-collecting area. A 10-meter telescope has five times the diameter of a 2-meter telescope, so its light-collecting area is $5^2 = 25$ times as great. The 10-meter telescope also has

FIGURE 6.7 interactive figure Angular separation depends on distance. The headlights on the car have the same physical separation in both cases, but their angular separation is larger when the car is closer. Similarly, two stars separated by a particular distance will have a larger angular separation if they are nearby than if they are farther away.

more than 1000 times the diameter of the pupil of your eye, which means it can collect more light than $1000^2 = 1$ million human eyes.

Angular resolution is the *smallest* angle over which we can tell that two dots—or two stars—are distinct. The human eye has an angular resolution of about 1 arcminute ($\frac{1}{60}°$), meaning that two stars can appear distinct only if they have at least this much angular separation in the sky. If the stars are separated by less than 1 arcminute, our eyes will not be able to distinguish them individually and they will look like a single star. The angular separation between two points of light depends both on their actual separation and on their distance from us [**Section 2.1**]; **FIGURE 6.7** shows the idea.

Large telescopes can have amazing angular resolution. For example, the 2.4-meter Hubble Space Telescope has an angular resolution of about 0.05 arcsecond (for visible light), which would allow you to read this book from a distance of almost 1 kilometer. Larger telescopes can have even better (smaller) angular resolution, though Earth's atmosphere usually prevents ground-based telescopes from achieving their theoretical limits.

The ultimate limit to a telescope's resolving power comes from the properties of light. Because light is an electromagnetic wave [**Section 5.2**], beams of light can interfere with one

FIGURE 6.8 This computer-generated image represents interference between overlapping sets of ripples on a pond. (The colors are for visual effect only.) Where peaks or troughs meet, the effects add to make the water rise extra high or fall extra low. Where peak meets trough, the effects cancel to make the water surface flat. Light waves also exhibit interference.

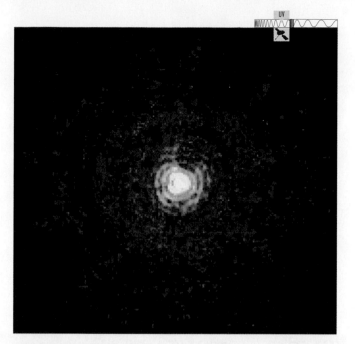

FIGURE 6.9 When examined in detail, a Hubble Space Telescope image of a star has rings (represented as green and purple in the figure) resulting from the wave properties of light. With higher angular resolution, the rings would be smaller.

another like overlapping sets of ripples on a pond (**FIGURE 6.8**). This *interference* limits a telescope's angular resolution even when all other conditions are perfect. That is why even a high-quality telescope in space cannot have perfect angular resolution (**FIGURE 6.9**).

The angular resolution that a telescope could achieve if it were limited only by the interference of light waves is called its **diffraction limit**. (*Diffraction* is a technical term for the effects of interference that limit telescope resolution.) The diffraction limit depends on both the diameter of the telescope's primary mirror and the wavelength of the light being observed (see Mathematical Insight 6.2). For any particular wavelength of light, a larger telescope has a smaller diffraction limit, meaning it can achieve a better (smaller) angular resolution. For any particular telescope, the diffraction limit is larger (poorer angular resolution) for longer-wavelength light. That is why, for example, a radio telescope must be far larger than a visible-light telescope to achieve the same angular resolution.

COMMON MISCONCEPTIONS

Magnification and Telescopes

Many people guess that magnification is the most important function of a telescope. However, even though telescopes can magnify images—much like telephoto camera lenses or binoculars—the amount of magnification a telescope can provide is *not* one of its crucial properties. No matter how much a telescope image is magnified, you cannot see details if the telescope does not collect enough light to show them or if they are smaller than the angular resolution of the telescope. Magnifying an image too much just makes it look blurry, which is why a telescope's light-collecting area and angular resolution are much more important than its magnification.

What are the two basic designs of telescopes?

Telescopes come in two basic designs: *refracting* and *reflecting*. A **refracting telescope** operates much like an eye, using transparent glass lenses to collect and focus light (**FIGURE 6.10**).

FIGURE 6.10 A refracting telescope collects light with a large transparent lens (diagram). The photo shows the 1-meter refractor at the University of Chicago's Yerkes Observatory, the world's largest refracting telescope.

The earliest telescopes, including those that Galileo built, were refracting telescopes. The world's largest refracting telescope, completed in 1897, has a lens that is 1 meter (40 inches) in diameter and a telescope tube that is 19.5 meters (64 feet) long.

A **reflecting telescope** uses a precisely curved *primary mirror* to gather light (**FIGURE 6.11**). This mirror reflects the gathered light to a *secondary mirror* that lies in front of it. The secondary mirror then reflects the light to a focus at a place where the eye or instruments can observe it—sometimes through a hole in the primary mirror and sometimes through the side of the telescope (often with the aid of additional small mirrors). The fact that the secondary mirror prevents some light from reaching the primary mirror might seem like a drawback to reflecting telescopes, but in practice it is not a problem because only a small fraction of the incoming light is blocked.

Nearly all telescopes used in current astronomical research are reflectors, mainly for two practical reasons. First, because light passes *through* the lens of a refracting telescope, lenses must be made from clear, high-quality glass with precisely shaped surfaces on both sides. In contrast, only the reflecting surface of a mirror must be precisely shaped, and the quality of the underlying glass is not a factor. Second, large glass lenses are extremely heavy and can be held in place only by their edges. Because the large lens is at the top of a refracting telescope, it is difficult to stabilize refracting telescopes and to prevent large lenses from deforming. The primary mirror of a reflecting telescope is mounted at the bottom, where its weight presents a far less serious problem. (A third problematic feature of lenses, called *chromatic aberration,* occurs because a lens brings different colors of light into focus at slightly different places. This problem can be minimized by using combinations of lenses.)

For a long time, the main factor limiting the size of reflecting telescopes was the sheer weight of the glass needed for their primary mirrors. Recent technological innovations have made it possible to build lighter-weight mirrors, such as the one in the Gemini telescope shown in Figure 6.11b, or to make many small

MATHEMATICAL INSIGHT 6.1

Angular Resolution

We often want to know whether a telescope can resolve (see as distinct) two points, such as two stars, based on their physical separation and distance. From Mathematical Insight 2.1, the angular separation of two points is given by

$$\text{angular separation} = \text{physical separation} \times \frac{360°}{2\pi \times \text{distance}}$$

This formula gives an answer in degrees. Because there are 3600 arcseconds in 1 degree (see Figure 2.8), we can rewrite the formula to give a result in arcseconds by multiplying the right side by $\frac{3600''}{1°}$ (recall that $''$ is the symbol for arcseconds). The right side will then read (physical separation)/(distance) times the numbers $360° \times \frac{3600''}{1°} \div 2\pi$, which a calculator shows to be approximately $206{,}265''$. Therefore, the formula becomes

$$\text{angular separation} = 206{,}265'' \times \frac{\text{physical separation}}{\text{distance}}$$

EXAMPLE 1: A binary star system is 20 light-years away and its two stars are separated by 200 million kilometers. Can the Hubble Space Telescope resolve the two stars? Assume an angular resolution of 0.05 arcsecond.

SOLUTION:

Step 1 Understand: The telescope can resolve the two stars if their angular separation is larger than the angular resolution of 0.05 arcsecond. We can calculate the angular separation of the two stars because we know their distance and physical separation.

Step 2 Solve: Before we can use the angular separation formula, we must have the physical separation and distance in the same units. It's easiest to convert the light-years to kilometers from the fact that 1 light-year $\approx 10^{13}$ km (see Mathematical Insight 1.1); the 20-light-year distance becomes about 20×10^{13} km $= 2 \times 10^{14}$ km. Writing the physical separation in scientific notation as 2×10^8 km, we find

$$\text{angular separation} = 206{,}265'' \times \frac{\text{physical separation}}{\text{distance}}$$

$$= 206{,}265'' \times \frac{2 \times 10^8 \text{ km}}{2 \times 10^{14} \text{ km}} = 0.2''$$

Step 3 Explain: The angular separation of the two stars is 0.2 arcsecond. Because this is larger than the telescope's angular resolution of 0.05 arcsecond, the two stars can be distinguished and studied individually.

EXAMPLE 2: If you looked at this book with a telescope that has Hubble's angular resolution of 0.05 arcsecond, how far away could you place the book and still be able to read it?

SOLUTION:

Step 1 Understand: We can read the book if we can resolve its individual letters, so answering this question hinges on determining the relevant physical and angular separations. The letters in this book are about 2 millimeters tall, so one way to think about it is to ask how closely spaced a set of dots would have to be to look like the letters in this book. If you do some test cases, you'll find that letters made 10 dots tall (and 10 dots wide) are clearly identifiable, so 2-mm tall letters would have to be composed of dots separated by 0.2 millimeter. We can use this value as the physical separation of the dots. The dots will be resolved if their angular separation is greater than or equal to the telescope's resolution of 0.05 arcsecond, so we use this value as the angular separation. We can then use the angular separation formula to calculate the distance at which the telescope could read the book.

Step 2 Solve: We solve the angular separation formula for distance:

$$\text{distance} = 206{,}265'' \times \frac{\text{physical separation}}{\text{angular separation}}$$

We substitute 0.05 arcsecond for the angular separation and 0.2 millimeter for the physical separation:

$$\text{distance} = 206{,}265'' \times \frac{0.2 \text{ mm}}{0.05''} \approx 825{,}000 \text{ mm}$$

Step 3 Explain: The distance of 825,000 millimeters, or 825 meters, is the distance at which the angular separation of the dots composing the letters would equal the angular resolution of the telescope. Therefore, the book would be readable at a distance of up to 825 meters, or a little less than 1 kilometer.

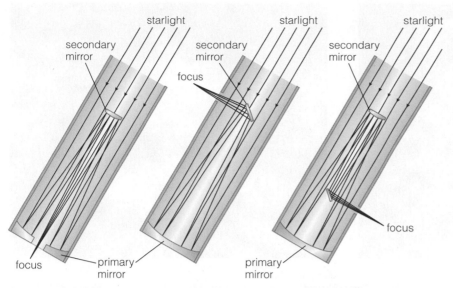

starlight · starlight · starlight

secondary mirror · focus · secondary mirror · secondary mirror

focus

focus · primary mirror · primary mirror · focus

Cassegrain Focus · **Newtonian Focus** · **Nasmyth/Coudé Focus**

a Three variations on the basic design of a reflecting telescope. In all cases, a reflecting telescope collects light with a precisely curved primary mirror that reflects light back upward to the secondary mirror. In the *Cassegrain* design, the secondary mirror reflects the light through a hole in the primary mirror, so that the light can be observed with cameras or instruments beneath the telescope. In the *Newtonian* design, the secondary mirror reflects the light out to the side of the telescope. In the *Nasmyth* and *Coudé* designs, a third mirror is used to reflect light out the side but lower down than in the Newtonian design.

b The Gemini North telescope, located on the summit of Mauna Kea, Hawaii, is a reflecting telescope with the Cassegrain design. The primary mirror, visible at the bottom of the large lattice tube, is 8 meters in diameter. The secondary mirror, located in the smaller central lattice, reflects light back down through the hole visible in the center of the primary mirror.

FIGURE 6.11 Reflecting telescopes.

MATHEMATICAL INSIGHT 6.2

The Diffraction Limit

A simple formula gives the diffraction limit of a telescope in arcseconds:

$$\text{diffraction limit} \approx 2.5 \times 10^{5\,''} \times \frac{\text{wavelength of light}}{\text{diameter of telescope}}$$

EXAMPLE 1: What is the diffraction limit of the 2.4-meter Hubble Space Telescope for visible light with a wavelength of 500 nanometers?

SOLUTION:

Step 1 Understand: We are given the wavelength of light and the telescope diameter, so we have all the information we need.

Step 2 Solve: We plug in the wavelength (500 nm = 500×10^{-9} m) and Hubble's diameter (2.4 m):

$$\text{diffraction limit} \approx 2.5 \times 10^{5\,''} \times \frac{\text{wavelength}}{\text{telescope diameter}}$$

$$= 2.5 \times 10^{5\,''} \times \frac{500 \times 10^{-9}\ \text{m}}{2.4\ \text{m}} = 0.05''$$

Step 3 Explain: The Hubble Space Telescope has a diffraction-limited angular resolution of 0.05 arcsecond for visible light with a wavelength of 500 nanometers. Therefore, it can in principle resolve objects separated by more than 0.05 arcsecond, while objects separated by less will be blurred together.

EXAMPLE 2: How large a telescope would you need to achieve a diffraction limit of 0.001 arcsecond for visible light (wavelength 500 nm)?

SOLUTION:

Step 1 Understand: We are given the diffraction limit and wavelength, so we simply need to solve the formula for telescope diameter. You should confirm that it becomes

$$\frac{\text{telescope}}{\text{diameter}} \approx 2.5 \times 10^{5\,''} \times \frac{\text{wavelength}}{\text{diffraction limit}}$$

Step 2 Solve: We substitute the given values:

$$\frac{\text{telescope}}{\text{diameter}} \approx 2.5 \times 10^{5\,''} \times \frac{500 \times 10^{-9}\ \text{m}}{0.001''} = 125\ \text{m}$$

Step 3 Explain: A telescope would need a diameter of 125 meters—longer than a football field—to achieve an angular resolution of 0.001 arcsecond for visible light. Note that this would be 50 times the diameter of the Hubble Space Telescope and give an angular resolution 50 times better.

FIGURE 6.12 (Left) The two Keck telescopes on Mauna Kea, photographed from above. Notice the primary mirrors through the openings in the domes. (Right) The primary mirror of one of the telescopes, with a man in the center for scale. If you look closely, you can see the honeycomb pattern of the 36 smaller, hexagonal mirrors that make up the primary mirror.

mirrors work together as one large one. **FIGURE 6.12** shows the primary mirror of one of the 10-meter Keck telescopes, which consists of 36 smaller mirrors that function together as one.

These new mirror-building technologies are fueling a revolution in the building of large telescopes. Before the 1990s, the 5-meter Hale telescope on Mount Palomar (outside San Diego) reigned for more than 40 years as the most powerful telescope in the world. Today, it does not even make the top-10 list for telescope size (**TABLE 6.1**). Several much larger telescopes are currently in various stages of planning or construction, including the Giant Magellan Telescope (effective size of 21 meters), the Thirty Meter Telescope (30 meters), and the European Extremely Large Telescope (39 meters).

What do astronomers do with telescopes?

Every astronomical observation is unique, and astronomers use many different kinds of instruments and detectors to extract the information contained in the light collected by a telescope. Nevertheless, most observations fall into one of three basic categories: *imaging*, which yields photographs (images) of astronomical objects; *spectroscopy*, in which astronomers obtain and study spectra; and *time monitoring*, which tracks how an object changes with time.* Let's look at each category in a little more detail.

Imaging At its most basic, an imaging instrument is simply a camera. Astronomers often place *filters* in front of a camera to allow only particular colors or wavelengths of light to pass through. In fact, most of the richly hued astronomical images that you see are made by combining images recorded through different filters (**FIGURE 6.13**).

Today, many astronomical images are made from invisible light—light that our eyes cannot see but that can be captured by specialized detectors. You can understand the idea by thinking

---

*Some astronomers include a fourth general category called *photometry*, which is the accurate measurement of light intensity from a particular object at a particular time. We do not list this as a separate category because today's detectors can generally perform photometry at the same time that they are being used for imaging, spectroscopy, or time monitoring.

TABLE 6.1 Largest Optical (Visible-Light) Telescopes

Size	Name	Location	Opened*
10.4 m	Gran Telescopio Canarias	Canary Islands	2007
10.2 m	South African Large Telescope	South Africa	2005
10 m	Keck I and Keck II	Mauna Kea, HI	1993/1996
9.2 m	Hobby-Eberly	Mt. Locke, TX	1997
2 × 8.4 m	Large Binocular Telescope	Mt. Graham, AZ	2005
4 × 8.2 m	Very Large Telescope	Cerro Paranal, Chile	1998/1999/2000/2001
8.3 m	Subaru	Mauna Kea, HI	1999
8 m	Gemini North and South	Mauna Kea, HI (North); Cerro Pachon, Chile (South)	1999/2002
6.5 m	Magellan I and II	Las Campanas, Chile	2000/2002
6.5 m	MMT	Mt. Hopkins, AZ	2000

*The year of "first light," when the telescope began operating.

FIGURE 6.13 **interactive figure** Astronomical images are usually made by combining several images taken through different filters.

The actual light collected...

...is combined to show a full-color image.

blue filter

green filter

red filter

VIS

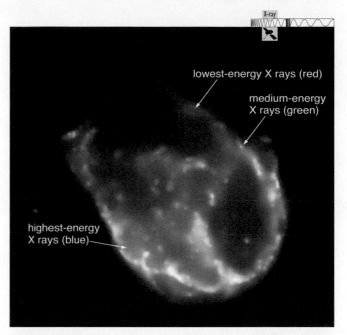

lowest-energy X rays (red)

medium-energy X rays (green)

highest-energy X rays (blue)

FIGURE 6.14 X rays are invisible, but we can color-code the information recorded by an X-ray detector to make an image of the object as it would appear in X rays. This image, from NASA's Chandra X-Ray Observatory, shows X-ray emission from the debris of a stellar explosion (a supernova remnant named N132D). Different colors represent X rays of different energy.

about X rays at a doctor's office. When the doctor "takes an X ray" of your arm, he or she uses a machine that sends X rays through your arm. The X rays that pass through are recorded with an X-ray-sensitive detector. Astronomical images work in much the same way. For example, **FIGURE 6.14** shows an X-ray image from the Chandra X-Ray Observatory (which is in space); the telescope collected X rays and the image was recorded with an X-ray-sensitive detector. In other words, what we see in Figure 6.14 is not the X rays themselves, but a picture that shows where X rays hit the detector.

Images made with invisible light cannot have any natural color, because "color" is a property only of visible light. However, we can use color-coding to help us interpret them. For example, the colors in Figure 6.14 correspond to X rays of different energy. In other cases, images may be color-coded according to the intensity of the light or to physical properties of the objects in the image.

As we discussed in Chapter 5, a spectrum can reveal a wealth of information about an object, including its chemical composition, temperature, and motion. However, just as the amount of information we can glean from an image depends on the angular resolution, the information we can glean from a

focused starlight

primary mirror

1. *Slit:* Allows only the light from the object of interest to pass through.

2. *Collimating mirror:* Makes all the reflected rays parallel.

3. *Diffraction grating:* Disperses reflected light into a spectrum.

4. *Camera mirror:* Focuses the spectrum onto a detector.

5. *Detector:* Records an image of the spectrum.

FIGURE 6.15 The basic design of a spectrograph. In this diagram, the spectrograph is attached to the bottom of a reflecting telescope, with light entering the spectrograph through a hole in the primary mirror. A narrow slit (or small hole) at the entrance to the spectrograph allows only light from the object of interest to pass through.

THINK ABOUT IT

Medical images from CT scans and MRIs are usually displayed in color, even though neither type of imaging uses visible light. What do you think the colors mean in CT scans and MRIs? How are the colors useful to doctors?

Spectroscopy Instruments called **spectrographs** use diffraction gratings (or other devices) to separate the various colors of light into spectra, which are then recorded with a detector (**FIGURE 6.15**).

spectrum depends on the **spectral resolution**: The higher the spectral resolution, the more detail we can see (**FIGURE 6.16**).

In principle, astronomers would always like the highest possible spectral resolution. However, higher spectral resolution comes at a price. A telescope collects only so much light in a given amount of time, and the spectral resolution depends on how widely the spectrograph spreads out this light. The more the light is spread out, the more total light we need in order for the spectrograph to record it successfully. Making a spectrum of an object therefore requires a longer exposure time than making an image, and high-resolution spectra require longer exposures than low-resolution spectra.

Time Monitoring Many astronomical objects vary with time. For example, some stars undergo sudden outbursts, and most stars (including our Sun) vary in brightness as starspots (or sunspots) cover more or less of their surfaces. Some objects vary periodically; for example, small, periodic changes in a star's brightness can reveal the presence of an orbiting planet [**Section 13.1**]. Time monitoring allows us to carefully study such variations.

For a slowly varying object, time monitoring may be as simple as comparing images or spectra obtained at different times. For more rapidly varying sources, time monitoring may require instruments that make rapid multiple exposures, in some cases recording the arrival time of every individual photon.

The results of time monitoring are often shown as **light curves**: graphs that show how an object's intensity varies with time. **FIGURE 6.17** shows a light curve for the star Mira. Notice that Mira's light output varies by more than a factor of 100 as it rises and falls with a period of a little less than one year.

An Astronomer's Job Although many people picture astronomers spending most of their time in late-night observing sessions, very few professional astronomers spend more than a small fraction of their time actually making observations. Some astronomers make no observations at all, instead focusing on the development of models to explain observations. Others devote their time to analyzing the wealth of data in the online archives of the world's major observatories.

For those who sometimes make observations, the day-to-day life of an astronomer goes something like this: After identifying an important unanswered question, the astronomer proposes a set of observations to an organization that manages a large telescope. The astronomer must write the proposal clearly and persuasively, explaining exactly how she or he will carry out the observations and why these observations would be a good use of telescope time. Often, several astronomers with similar interests collaborate on a proposal (and later work together to make the observations and analyze the data). A committee of other astronomers evaluates all the proposals that have been submitted, deciding which ones are worthy of being granted telescope time and which ones are not.

In most cases, the amount of telescope time requested in worthy proposals is much larger than the amount of time available. For example, if 100 worthy proposals each require 10 nights of observing time in the next year with a particular telescope, there's only enough time for about one-third of them. The selection committee therefore decides which of the worthy proposals are actually awarded telescope time, a process that can involve factors such as the importance of the question being addressed, the cost of collecting and analyzing the data, and the likelihood of success.

Because telescope time is so precious, it's crucial to use it efficiently. For working astronomers, the real work begins after an observing proposal is accepted. They must then prepare carefully to make sure the observations are successful, and later they will spend far more time analyzing and interpreting the data than they spent collecting it.

FIGURE 6.16 These two ultraviolet spectra show the same object in the same wavelength band. However, we see far more detail with higher spectral resolution, including individual spectral lines that appear merged together at lower spectral resolution. (The spectrum shows absorption lines created when interstellar gas absorbs light from a more distant star.)

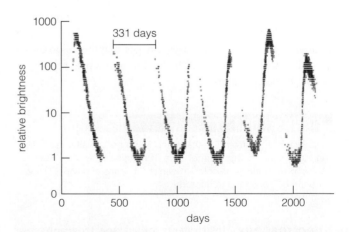

FIGURE 6.17 This graph shows a light curve for the variable star Mira (in the constellation Cetus), with data spanning several years. Centuries of observation show that Mira's brightness varies with an average period of 331 days.

FIGURE 6.18 This composite image, made from hundreds of satellite photos, shows the bright lights of cities around the world as they appear from Earth orbit at night. The lights are pretty, but they represent light pollution for astronomers.

6.3 TELESCOPES AND THE ATMOSPHERE

From the time Galileo first turned his telescope to the heavens in 1609 [**Section 3.3**] until the dawn of the space age, all astronomical telescopes were located on the ground. Even today, the vast majority of observatories are ground based, and that will probably hold true long into the future. Telescopes on the ground are much less expensive to build, operate, and maintain than telescopes in space. Nevertheless, Earth's surface is far from ideal as an observing site. In this section, we'll explore some of the problems that Earth's atmosphere poses for astronomical observations and learn why, despite the higher costs, dozens of telescopes have been lofted into Earth orbit or beyond.

How does Earth's atmosphere affect ground-based observations?

Daylight and weather are the most obvious problems with observing from the ground. Our daytime sky is bright because the atmosphere scatters sunlight, and this brightness drowns out the dim light of most astronomical objects. That is why most astronomical observations are practical only at night. Even then, we can observe only when the sky is clear rather than cloudy. (The atmosphere does not scatter most radio waves, so radio telescopes can operate day and night and under cloudy skies.)

The constraints of daylight and weather affect the timing of observations, but by themselves do not hinder observations on clear nights. However, our atmosphere creates three other problems that inevitably affect astronomical observations: the scattering of human-made light, the blurring of images by atmospheric motion, and the fact that most forms of light cannot reach the ground at all.

Light Pollution Just as our atmosphere scatters sunlight in the daytime, it also scatters the bright lights of cities at night (**FIGURE 6.18**). Because this scattered light is human-made and obscures our view of the night sky, we call it **light pollution**. Light pollution explains why you cannot see as many stars from a big city as you can from an unpopulated area, and it can seriously hinder astronomical observations.

Light pollution has become an increasing problem as cities have grown, encroaching into areas that were once remote enough to be chosen as sites for major observatories. For example, the 2.5-meter telescope at Mount Wilson, the world's largest when it was built in 1917, would be much more useful today if it weren't located so close to the lights of what was once the small town of Los Angeles. Similar but less severe light pollution affects many other telescopes, including those on Mount Palomar near San Diego and on Kitt Peak near Tucson. Fortunately, many communities are working to reduce light pollution, with benefits not only to astronomers but to everyone who enjoys our ancient connections with the night sky.

Twinkling and Atmospheric Turbulence The second major problem is the distortion of light by the atmosphere. Winds and other air currents ensure that air in our atmosphere is continually moving and mixing around, a phenomenon that we call **turbulence**. The ever-changing motions of air due to turbulence continually change the atmosphere's light-bending properties, so light rays passing through the atmosphere are continually being bent by slightly different amounts. As a result, our view of things outside Earth's atmosphere appears to jiggle around, in much the same way as your view of things outside the water when you look up from the bottom of a swimming pool. This jiggling causes the familiar twinkling of stars, which may be beautiful to the naked eye but also blurs astronomical images.

As a general rule, the blurring of images by turbulence tends to limit the angular resolution of ground-based telescopes to no better than about 0.5 arcsecond, even if a telescope's diffraction limit is much smaller than that. Today, however, a remarkable technology called **adaptive optics** can eliminate much of this blurring and allow telescopes to achieve angular resolution close to their diffraction limit. The technology works this way: Turbulence essentially causes rays of light from a star to dance around as they reach a telescope. Adaptive optics essentially makes the telescope's mirrors do an opposite dance, canceling out the atmospheric distortions (**FIGURE 6.19**). The shape of a mirror (often the secondary or even a third or fourth mirror) is changed slightly many times each second to compensate for the rapidly changing atmospheric distortions. A computer calculates the necessary changes by monitoring distortions in the image of a bright star near the object under study. In some cases, if there is no bright star near the object of interest, the observatory shines a laser into the sky to create an *artificial star* (a point of light in Earth's atmosphere) that it can monitor for distortions.

 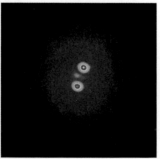

a Atmospheric distortion makes this ground-based image of a double star look like that of a single star.

b When the same telescope is used with adaptive optics, the two stars can be clearly distinguished. The angular separation between the two stars is 0.28 arcsecond.

FIGURE 6.19 The technology of adaptive optics can enable a ground-based telescope to overcome most of the blurring caused by Earth's atmosphere. (Both these images were taken in near-infrared light with the Canada-France-Hawaii telescope; colors represent infrared intensity.)

Locating Ground-Based Observatories When choosing a site for ground-based telescopes, astronomers seek to mitigate the effects of weather, light pollution, and atmospheric distortion as much as possible. This leads to the following key criteria for an observatory site: It should be *dark* in the sense of having minimal light pollution, *dry* to limit clouds and rain, *calm* to limit turbulence, and *high* so that it is above the densest parts of the atmosphere. Astronomers have found a handful of sites around the world that meet these criteria particularly well, and these sites are therefore home to many of the world's largest major telescopes. Three particularly important sites are the 4300-meter (14,000-foot) summit of Mauna Kea on the Big Island of Hawaii (**FIGURE 6.20**), a 2400-meter-high site on the island of La Palma in Spain's Canary Islands, and, for the southern hemisphere, the 2600-meter-high Paranal Observatory site in Chile.

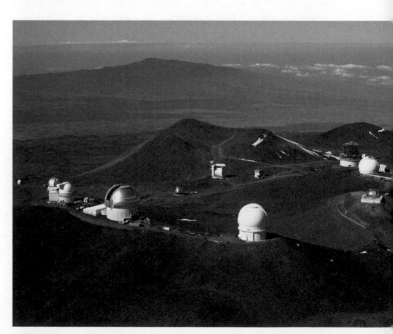

FIGURE 6.20 Observatories on the summit of Mauna Kea in Hawaii. Mauna Kea meets all the key criteria for an observing site: It is far from big-city lights, high in altitude, and in an area where the air tends to be calm and dry.

FIGURE 6.21 The Hubble Space Telescope orbits Earth. Its position above the atmosphere allows it an undistorted view of space. Hubble can observe infrared and ultraviolet light as well as visible light.

Why do we put telescopes into space?

The ultimate solution to the problems faced by ground-based observatories is to put telescopes into space, where they are above the atmosphere and unaffected by daylight, weather, light pollution, and atmospheric turbulence. That is one reason why the Hubble Space Telescope (**FIGURE 6.21**) was built and why it has been so successful despite the relatively small size of its 2.4-meter primary mirror.

COMMON MISCONCEPTIONS

Closer to the Stars?

Many people mistakenly believe that space telescopes are advantageous because their locations above Earth put them closer to the stars. You can see why this is wrong by thinking about scale. On the scale of the Voyage model solar system (see Section 1.1), the Hubble Space Telescope is so close to the surface of the millimeter-diameter Earth that you would need a microscope to resolve its altitude, while the nearest stars are thousands of kilometers away. The distances to the stars are effectively the same whether a telescope is on the ground or in space. The real advantages of space telescopes all arise from their being above Earth's atmosphere and the observational problems it presents.

However, while Hubble's visible-light images are far superior to anything that was possible from the ground at the time it was launched in 1990, in some cases adaptive optics now allows ground-based observatories to equal or better Hubble's image quality. So does that mean that the Hubble Space Telescope and other space observatories are becoming obsolete? No, and here's why: Earth's atmosphere poses one major problem that no Earth-bound technology can overcome— *our atmosphere prevents most forms of light from reaching the ground at all.*

If we studied only visible light, we'd be missing much of the story that light brings to us from the cosmos. Planets are relatively cool and emit primarily infrared light. The hot

SPECIAL TOPIC

Would You Like Your Own Telescope?

Just a couple decades ago, a decent personal telescope would have set you back a few thousand dollars and taken weeks of practice to learn to use. Today, you can get a good-quality telescope for a few hundred dollars, and built-in computer drives can make it easy to use.

Before you consider buying a telescope, you should understand what a personal telescope can and cannot do. A telescope will allow you to look for yourself at light that has traveled vast distances through space to reach your eyes. This can be a rewarding experience, but the images in your telescope will *not* look like the beautiful photographs in this book, which were obtained with much larger telescopes and sophisticated cameras. In addition, while your telescope can in principle let you see many distant objects, including star clusters, nebulae, and galaxies, it won't allow you to find anything unless you first set it up properly. Even computer-driven telescopes (sometimes called "go to" telescopes) typically take 15 minutes to a half-hour to set up for each use, and longer when you are first learning.

If your goal is just to see the Moon and a few other objects with relatively little effort, you may want to skip the telescope in favor of a good pair of binoculars, which are usually less expensive. Binoculars are generally described by two numbers, such as 7×35 or 12×50. The first number is the magnification; for example, "7×" means that objects will look seven times closer through the binoculars than to your eye. The second number is the diameter of each lens in millimeters. As with telescopes, larger lenses mean more light and better views. However, larger lenses also tend to be heavier and more difficult to hold steady, which means you may need to set them up on a tripod.

If you decide to go ahead with a telescope, the first rule to remember is that magnification is *not* the key factor, and telescopes advertised only by their magnification (such as "650 power") are rarely high quality. Instead, focus on three factors when choosing your telescope:

1. *The light-collecting area* (also called *aperture*). Most personal telescopes are reflectors, so a "6-inch" telescope has a primary mirror that is 6 inches in diameter.

2. *Optical quality.* A poorly made telescope won't do you much good. If you cannot do side-by-side comparisons, stick with a major telescope manufacturer (such as Meade, Celestron, or Orion).

3. *Portability.* A large, bulky telescope can be great if you plan to keep it on a deck, but it will be difficult to carry on camping trips. Depending on how you plan to use your telescope, you'll need to make trade-offs between size and portability.

Most important, remember that a telescope is an investment that you will keep for many years. As with any investment, learn all you can before you settle on a particular model. Read reviews of telescopes in magazines such as *Astronomy*, *Mercury*, and *Sky and Telescope*. Talk to knowledgeable salespeople at stores that specialize in telescopes. And find a nearby astronomy club that holds observing sessions at which you can try out some telescopes and learn from experienced telescope users.

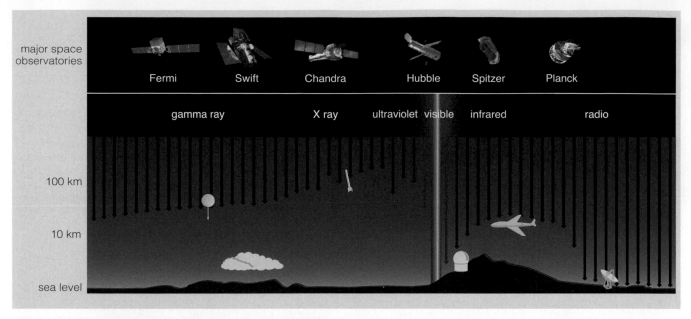

FIGURE 6.22 interactive figure This diagram shows the approximate depths to which different wavelengths of light penetrate Earth's atmosphere. Note that most of the electromagnetic spectrum—except for visible light (and the very longest wavelengths of ultraviolet light), a small portion of the infrared, and radio waves—can be observed only from very high altitudes or from space.

upper layers of stars like the Sun emit ultraviolet light and X rays. Some violent cosmic events produce bursts of gamma rays. Indeed, most astronomical objects emit light over a broad range of wavelengths. If we want to understand the universe, we must observe light all across the electromagnetic spectrum. We simply can't do this from the ground.

FIGURE 6.22 shows the approximate depths to which different forms of light penetrate Earth's atmosphere. Only radio waves, visible light (and the very longest wavelengths of ultraviolet light), and small parts of the infrared spectrum can be observed from the ground. Therefore, the most important reason for putting telescopes into space is to allow us to

observe the rest of the electromagnetic spectrum. Indeed, the Hubble Space Telescope often observes in ultraviolet or infrared wavelengths that do not reach the ground, which is why it would remain a valuable observatory even if ground-based telescopes could match all its visible-light capabilities.

The Hubble Space Telescope is the most famous observatory in space, but there are many others. Most of these observe parts of the electromagnetic spectrum that do not reach the ground. For example, the Chandra X-Ray Observatory is in space because an X-ray telescope would be completely useless on the ground. **TABLE 6.2** lists some of the most important space telescopes.

TABLE 6.2   Selected Major Space Observatories

Name	Launch Year	Lead Space Agency	Special Features
Hubble Space Telescope	1990	NASA	Optical, infrared, and ultraviolet observations
Chandra X-Ray Observatory	1999	NASA	X-ray imaging and spectroscopy
XMM–Newton	1999	ESA*	European-led mission for X-ray spectroscopy
Galaxy Evolution Explorer (GALEX)	2003	NASA	Ultraviolet observations of galaxies
Spitzer Space Telescope	2003	NASA	Infrared observations of the cosmos
Swift	2004	NASA	Study of gamma-ray bursts
Fermi Gamma-Ray Telescope	2008	NASA	Gamma-ray imaging, spectroscopy, and timing
Kepler	2009	NASA	Transit search for extrasolar Earth-like planets
Planck	2009	ESA	Study of the cosmic microwave background
Herschel	2009	ESA	Far-infrared imaging and spectroscopy
Nuclear Spectroscopic Telescope Array (NuStar)	2012	NASA	Imaging of high-energy X rays

*European Space Agency.

6.4 TELESCOPES AND TECHNOLOGY

Astronomers today are making new discoveries at an astonishing rate, driven primarily by the availability of more and larger telescopes, including space telescopes that can observe previously inaccessible portions of the electromagnetic spectrum. However, larger telescopes are not the only fuel for the current astronomical revolution.

Some new technologies make it possible to obtain better images or spectra with existing telescopes. For example, electronic detectors are constantly improving. As a result, a relatively small telescope equipped with the latest camera technology can record images as good as those that could be captured only by much larger telescopes in the past. In addition, adaptive optics allows telescopes to overcome problems of atmospheric distortion.

Other technologies make it possible to record and analyze data more efficiently. For example, obtaining spectra of distant galaxies used to be a very time-consuming and labor-intensive task. Today, astronomers can obtain hundreds of spectra simultaneously in a single telescopic observation, and then analyze this vast amount of data with the help of computers.

In this section, we'll focus our attention on two important areas of modern telescope technology that we have not yet discussed: the technology used to observe different parts of the electromagnetic spectrum and a remarkable technology that allows many small telescopes to work together to obtain higher-quality images.

How do we observe invisible light?

The basic idea behind all telescopes is the same: to collect as much light as possible with as much resolution as possible. Nearly all telescopes used by professional astronomers (except some gamma-ray telescopes) are essentially reflecting telescopes, using mirrors to bring light to a focus. Nevertheless, telescopes for most invisible wavelengths require variations on the basic design used for visible-light telescopes. Let's investigate, going in order of decreasing wavelength.

Radio Telescopes A specialized kind of radio telescope is now the most common type of telescope in the world, so common that you'll see dozens of them on a drive through almost any neighborhood. You may even own one yourself, because every satellite dish is essentially a small radio telescope designed to collect radio waves from a satellite in Earth orbit. Just by looking at a satellite dish, you can see that it operates by the same basic principles as a reflecting telescope (**FIGURE 6.23**). The metal dish is the mirror, and it is shaped to bring the radio waves to a focus in front of the dish; that's where you see the receiver, which is located where a secondary mirror would be. The receiver collects the radio waves reflected by the primary mirror and sends them to the television (or other device).

The primary differences between satellite dishes and astronomical radio telescopes are in where they look in the sky and their sizes. Communication satellites have geostationary orbits,

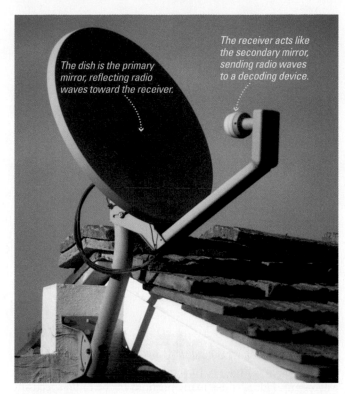

The dish is the primary mirror, reflecting radio waves toward the receiver.

The receiver acts like the secondary mirror, sending radio waves to a decoding device.

FIGURE 6.23 A satellite TV dish is essentially a small radio telescope.

which means they orbit above Earth's equator in exactly the same amount of time Earth takes to rotate (see Mathematical Insight 4.3), so that a dish aimed at a particular satellite can always point to the same spot in the local sky. In contrast, astronomical radio telescopes point toward cosmic radio sources that, like the Sun and stars, rise and set with Earth's rotation. Astronomical radio telescopes are also larger than satellite dishes, both because they need a large light-collecting area to detect the faint radio waves from cosmic sources and because they are used to make images and therefore require decent angular resolution. (Angular resolution is unimportant for satellite dishes, because they are not used to make images of the satellites in space; radio and television signals are encoded in the radio waves themselves, so the dish needs only to collect the radio waves and send them to a decoding device like a television.)

The long wavelengths of radio waves mean that very large telescopes are necessary to achieve reasonable angular resolution. For example, the world's largest single radio telescope, the Arecibo radio dish, stretches 305 meters (1000 feet) across a natural valley in Puerto Rico (**FIGURE 6.24**). Despite its large size, Arecibo's angular resolution is only about 1 arcminute at commonly observed radio wavelengths—a few hundred times worse than the visible-light resolution of the Hubble Space Telescope. Fortunately, through an amazing technique that we'll discuss shortly (interferometry), radio telescopes can work together to achieve much better angular resolution.

If you look again at Figure 6.22, you'll see that radio waves are the only form of light besides visible light that we can observe easily from the ground. Moreover, because the atmosphere does not distort radio waves the way it distorts visible light, there's no inherent advantage to observing from space.

FIGURE 6.24 The Arecibo radio telescope stretches across a natural valley in Puerto Rico. At 305 meters across, it is the world's largest single telescope.

FIGURE 6.25 This photograph shows NASA's airborne observatory, SOFIA, with its 2.5-meter infrared telescope.

However, "radio-wave pollution" is an even more serious impediment to radio astronomy than light pollution is to visible-light astronomy. Humans use many portions of the radio spectrum so heavily that radio signals from cosmic sources are almost completely drowned out. Astronomers hope someday to put radio telescopes into deep space or on the far side of the Moon, where the Moon itself would block out any radio interference from Earth. In addition, because radio telescopes can be made to work together, putting them into space in principle can allow them to be spread out over a much greater distance.

Infrared Telescopes Most of the infrared portion of the spectrum is close enough in wavelength to visible light to behave quite similarly, so infrared telescopes generally look much the same as visible-light telescopes. In fact, visible-light telescopes can in principle collect and focus much of the infrared spectrum; the practical limitation comes from Earth's atmosphere.

As you can see in Figure 6.22, most infrared wavelengths do not reach the ground. A few portions of the infrared spectrum can be observed from the tops of high mountains, such as Mauna Kea. The higher you go in the atmosphere, the more infrared light becomes accessible. NASA's airborne observatory called SOFIA (Stratospheric Observatory for Infrared Astronomy) carries a 2.5-meter infrared telescope that looks out through a large hole cut in the body of a Boeing 747 airplane (**FIGURE 6.25**). Advanced technologies make it possible for the airplane to fly smoothly despite its large hole and to keep the telescope pointed accurately at observing targets during flight.

Extreme infrared light (the longest wavelengths of infrared) poses more difficult observing challenges. Remember that all objects emit thermal radiation characteristic of their temperatures [**Section 5.4**]. At ordinary temperatures, the ground, an airplane, and even a telescope itself emit enough long-wavelength infrared light to interfere with any attempt to observe these wavelengths from the cosmos. The only

solution to this problem is to put the telescopes into space, so they get away from Earth's heat, and to cool the telescopes, so they emit less infrared light. NASA's Spitzer Space Telescope (**FIGURE 6.26**), launched in 2003, was cooled with liquid helium to just a few degrees above absolute zero.

NASA is currently working on a much more powerful infrared observatory, called the James Webb Space Telescope (scheduled for launch in 2018), which is designed as the successor mission to the Hubble Space Telescope. The James Webb Space Telescope will be kept far from Earth's heat by being put in an orbit around the Sun at a greater distance than Earth, while an attached Sun shield will prevent sunlight from heating the telescope.

Ultraviolet Telescopes Like infrared light, much of the ultraviolet spectrum is close enough in wavelength to visible light to behave similarly, so in principle it can be collected and focused by visible-light mirrors. However, Earth's atmosphere almost completely absorbs ultraviolet light, making most

FIGURE 6.26 This painting shows the Spitzer Space Telescope. The background is an artistic rendition of infrared emission from a star-forming cloud.

a Artist's illustration of the Chandra X-Ray Observatory, which orbits Earth.

X rays

The first set of nested mirrors deflects X rays to the second set . . .

X rays

10 meters

focus

. . . and the second set deflects X rays a second time to the focus point.

b This diagram shows the arrangement of Chandra's nested, cylindrical X-ray mirrors. Each mirror is 0.8 meter long and between 0.6 and 1.2 meters in diameter.

FIGURE 6.27 The Chandra X-Ray Observatory focuses X rays that enter the front of the telescope by deflecting them twice so that they end up focused at the back of the telescope.

ultraviolet observations impossible from the ground. (Very short-wavelength ultraviolet light, sometimes called *extreme ultraviolet*, behaves like X rays, which we'll discuss below.)

At present, there are two major ultraviolet observatories in space: the Galaxy Evolution Explorer (GALEX), a relatively small telescope that is conducting an ultraviolet survey of the entire sky, and the Hubble Space Telescope, which is capable of high-resolution imaging and spectroscopy in ultraviolet light. Hubble's continued operation has required periodic Space Shuttle missions to repair it, update it, and boost it back up as atmospheric drag lowers its orbit. The last servicing mission occurred in 2009, and no more are planned.

X-Ray Telescopes Besides needing to take place in space, X-ray observations pose another challenge: X rays have sufficient energy to penetrate many materials, including living tissue and ordinary mirrors. While this property makes X rays useful to medical doctors, it poses a challenge for astronomers.

Trying to focus X rays is somewhat like trying to focus a stream of bullets. If the bullets are fired directly at a metal sheet, they will puncture or damage the sheet. However, if the metal sheet is angled so that the bullets barely graze its surface, then it will slightly deflect the bullets. Specially designed mirrors can deflect X rays in much the same way. Such mirrors are called **grazing incidence mirrors** because X rays merely graze their surfaces as the rays are deflected toward the focal plane. X-ray telescopes, such as NASA's Chandra X-Ray Observatory, generally consist of several nested grazing incidence mirrors (**FIGURE 6.27**).

Chandra offers the best angular resolution of any X-ray telescope yet built, but a European X-ray telescope called XMM–Newton has a larger light-collecting area. Astronomers therefore use the two observatories in the way best suited to their science goals. For example, Chandra is better for making images of X-ray sources, while XMM–Newton's larger light-collecting area allows it to obtain more detailed X-ray spectra. The most recent X-ray telescope in space, NuStar, is optimized for imaging of higher-energy X rays than Chandra or XMM.

Gamma-Ray Telescopes Gamma rays can penetrate even grazing incidence mirrors and therefore cannot be focused in the traditional sense. Indeed, it takes a massive detector to capture gamma-ray photons at all. For example, the Large Area Telescope on the Fermi Gamma-Ray Observatory weighs 3 tons (**FIGURE 6.28**).

Gamma rays come from a number of different types of astronomical objects, but the most mysterious sources produce short bursts of gamma rays that quickly fade away [**Section 18.4**]. NASA currently has two gamma-ray observatories that can study these bursts. Swift, launched in 2004, carries X-ray and visible-light telescopes in addition to its

FIGURE 6.28 This artist's rendering shows the Fermi Gamma-Ray Telescope operating in space.

gamma-ray detectors. These telescopes help pinpoint the location of a gamma-ray burst within seconds after it is initially detected. The Fermi Gamma-Ray Observatory, launched in 2008, studies the sources of these bursts and other gamma-ray sources with much higher sensitivity and resolution.

Looking Beyond Light We have learned virtually everything we know about distant objects by observing light. However, light is not the only form of information that travels through the universe, and astronomers have begun to build and use telescopes designed to observe at least three other types of "cosmic messengers." First, there's an extremely lightweight type of subatomic particle known as the *neutrino* [**Sections S4.2, 14.2**] that is produced by nuclear reactions, including nuclear fusion in the Sun and the reactions that accompany the explosions of distant stars. Astronomers have already had some success with "neutrino telescopes"—typically located in deep mines or under water or ice—that have provided valuable insights about the Sun and stellar explosions. Second, Earth is continually bombarded by very high-energy subatomic particles from space known as *cosmic rays* [**Section 19.2**]. We still know relatively little about the origin of cosmic rays, but astronomers are now using both satellites and ground-based detectors to catch and study them. Third, Einstein's general theory of relativity predicts the existence of something called *gravitational waves* [**Section S3.4**], which are different in nature from light but travel at the speed of light. For decades, we've had indirect evidence that gravitational waves really exist, but until recently, direct detection of them was beyond our technological capabilities. Today, the first gravitational wave telescopes are up and running, and astronomers hope they will be able to detect gravitational waves from exotic objects like orbiting pairs of neutron stars and black holes.

How can multiple telescopes work together?

Individual telescopes always face limits on their capabilities. Even in space, the diffraction limit places a fundamental constraint on the angular resolution of a telescope of any particular size. In addition, while astronomers would always like larger telescopes, the current state of technology and budgetary considerations place practical limits on telescope size.

These constraints ultimately limit the amount of light that we can collect with telescopes. Even if we put a group of telescopes together, there's no getting around the fact that their total light-collecting area is simply the sum of their individual areas. However, remember that the two key properties of a telescope are light-collecting area and angular resolution. Amazingly, there is a way to make the angular resolution of a group of telescopes far better than that of any individual telescope.

In the 1950s, radio astronomers developed an ingenious technique for improving the angular resolution of radio telescopes: They learned to link two or more individual telescopes to achieve the angular resolution of a much larger telescope (**FIGURE 6.29**). This technique is called **interferometry** because

Interferometry allows these two small telescopes to work together...

...to obtain the angular resolution that would be achieved by a single, much larger telescope.

FIGURE 6.29 This diagram shows the basic idea behind interferometry: Smaller telescopes work together to obtain the angular resolution of a much larger telescope. Note that interferometry improves angular resolution but does not affect the total light-collecting area, which is simply the sum of the light-collecting areas of the individual telescopes.

it works by taking advantage of the wavelike properties of light that cause interference (see Figure 6.8). The procedure relies on precisely timing when radio waves reach each dish and using computers to analyze the resulting interference patterns.

One famous center for radio interferometry, the Very Large Array (VLA) near Socorro, New Mexico, consists of 27 individual radio dishes that can be moved along railroad tracks laid down in the shape of a Y (**FIGURE 6.30**). The light-gathering capability of the VLA's 27 dishes is simply equal to their combined area, equivalent to that of a single telescope 130 meters across. But the VLA's angular resolution is equivalent to that of a much larger telescope: When the 27 dishes are spaced as widely as possible, the VLA can achieve an angular resolution that otherwise would require a single radio telescope with a diameter of almost 40 kilometers. Today, astronomers can achieve even higher angular resolution by linking radio telescopes around the world.

Interferometry is more difficult for shorter-wavelength (higher-frequency) light, but astronomers are rapidly learning to use the technique beyond the radio portion of the spectrum. One spectacular example is the Atacama Large Millimeter/submillimeter Array (ALMA), which when complete, will combine light from 66 individual telescopes working at millimeter and submillimeter wavelengths. This portion of the spectrum has not been studied much in the past, because most of it is blocked by Earth's atmosphere; however, this light can be detected in the high (5000-meter-altitude), dry desert in which ALMA is located.

Interferometry is also now possible at shorter infrared and visible wavelengths. Indeed, new telescopes are now often built

FIGURE 6.30 The Very Large Array (VLA) in New Mexico consists of 27 telescopes that can be moved along train tracks. The telescopes work together through interferometry and can achieve an angular resolution equivalent to that of a single radio telescope almost 40 kilometers across.

in pairs (such as the Keck and Magellan telescope pairs) or with more than one telescope on a common mount (such as the Large Binocular Telescope) so that they can be used for infrared and visible-light interferometry. In addition, astronomers are testing technologies that may allow interferometry to be extended all the way to X rays. Someday, astronomers may use telescopes in space or on the Moon as giant interferometers, offering views of distant objects that may be as detailed in comparison to Hubble Space Telescope images as Hubble's images are in comparison to those of the naked eye.

The Big Picture

Putting Chapter 6 into Context

In this chapter, we've focused on the technological side of astronomy: the telescopes that we use to learn about the universe. Keep in mind the following "big picture" ideas as you continue to learn about astronomy.

- Technology drives astronomical discovery. Every time we build a bigger telescope, develop a more sensitive detector, or open up a new wavelength region to study, we learn more about the universe.

- Telescopes work much like giant eyes, enabling us to see the universe in great detail. New technologies for making larger telescopes, along with advances in adaptive optics and interferometry, are making ground-based telescopes more powerful than ever.

- For the ultimate in observing the universe, space is the place! Telescopes in space allow us to detect light from across the entire spectrum while also avoiding the distortion caused by Earth's atmosphere.

6.1 EYES AND CAMERAS: EVERYDAY LIGHT SENSORS

■ **How do eyes and cameras work?** Your eye brings rays

of light to a **focus** (or **focal point**) on your retina. Glass lenses work similarly, so distant objects form an **image** that is in focus on the **focal plane.** A camera has a detector at the focal plane, which can make a permanent record of an image.

6.2 TELESCOPES: GIANT EYES

■ **What are the two most important properties of a telescope?** A telescope's most important properties are its **light-collecting area**, which determines how much light it gathers, and its **angular resolution**, which determines how much detail we can see in its images.

■ **What are the two basic designs of telescopes?** A refract-

ing telescope forms an image by bending light through a lens. A **reflecting telescope** forms an image by focusing light with mirrors.

■ **What do astronomers do with telescopes?** The three primary uses of telescopes are **imaging** to create pictures of distant objects, **spectroscopy** to study the spectra of distant objects, and **time monitoring** to study how a distant object's brightness changes with time.

6.3 TELESCOPES AND THE ATMOSPHERE

■ **How does Earth's atmosphere affect ground-based observations?** Earth's atmosphere limits visible-light observations to nighttime and clear weather. **Light pollution** can lessen the quality of observations, and atmospheric **turbulence** makes stars twinkle, blurring their images. The technology of **adaptive optics** can overcome some of the blurring due to turbulence.

■ **Why do we put telescopes into space?** Telescopes in space are above Earth's atmosphere and the problems it causes for observations. Most important, telescopes in space can observe

all wavelengths of light, while telescopes on the ground can observe only visible light, radio waves, and small portions of the infrared.

6.4 TELESCOPES AND TECHNOLOGY

■ **How do we observe invisible light?** Telescopes for other than

visible light often use variations on the basic design of a reflecting telescope. Radio telescopes use large metal dishes as their primary mirrors. Infrared telescopes are sometimes cooled to very low temperature. X-ray telescopes use **grazing incidence** reflections rather than direct reflections.

■ **How can multiple telescopes work together?** The technique of **interferometry** links multiple telescopes in a way that allows them to obtain the angular resolution of a much larger telescope.

Use the following questions to check your understanding of some of the many types of visual information used in astronomy. Answers are provided in Appendix J. For additional practice, try the Chapter 6 Visual Quiz at MasteringAstronomy®.

The figure above, which repeats Figure 6.22, shows the approximate depths to which different wavelengths of light penetrate Earth's atmosphere. Use this figure to answer the following questions.

1. Only very small amounts of infrared and ultraviolet light can penetrate all the way to the ground. Based on the diagram, which statement is true?
   a. A small percentage of the incoming light at every infrared and ultraviolet wavelength reaches the ground, while the remaining light at the same wavelengths does not reach the ground.
   b. Most infrared and ultraviolet wavelengths do not reach the ground at all; the only wavelengths that do are the ones closest to the visible portion of the spectrum.
   c. Most infrared and ultraviolet wavelengths do not reach the ground at all; the only wavelengths that do are the ones closest to the radio and X-ray portions of the spectrum.

2. (Choose all that apply.) Observatories on mountaintops can detect
   a. visible light
   b. X rays
   c. a small portion of the infrared spectrum
   d. very long-wavelength infrared light
   e. radio waves

3. (Choose all that apply.) An observatory in space could in principle detect
   a. visible light
   b. X rays
   c. infrared light
   d. ultraviolet light
   e. radio waves
   f. gamma rays

4. What kind of light can be detected from an airplane but not from the ground?
   a. most infrared light
   b. only the shortest-wavelength infrared light
   c. radio waves
   d. X rays

5. The Planck spacecraft observes
   a. long wavelength infrared light
   b. X rays
   c. visible light
   d. ultraviolet light

EXERCISES AND PROBLEMS

MasteringAstronomy®

For instructor-assigned homework go to MasteringAstronomy®.

REVIEW QUESTIONS

Short-Answer Questions Based on the Reading

1. How does your eye focus light? How is a glass lens similar? What do we mean by the *focal plane* of a lens?

2. How does a camera record light? How are images affected by *exposure time*? What are *pixels*?

3. What are the two key properties of a telescope, and why is each important?

4. What is the *diffraction limit*, and how does it depend on a telescope's size and the wavelength of light being observed?

5. How do *reflecting telescopes* differ from *refracting telescopes*? Which type is more commonly used by professional astronomers, and why?

6. What are the three basic categories of astronomical observation, and how is each conducted?

7. What do we mean when we speak of images made from *invisible* light, such as X-ray or infrared images? What do the colors in these images mean?

8. What do we mean by *spectral resolution*? Why is higher spectral resolution more difficult to achieve?

9. List at least three ways in which Earth's atmosphere can hinder astronomical observations. What problem can *adaptive optics* help with?

10. Describe how deeply each portion of the electromagnetic spectrum penetrates Earth's atmosphere. Based on your answers, why are space telescopes so important to our understanding of the universe?

11. How do telescopes for invisible wavelengths differ from those for visible light? Answer for each major wavelength band and give examples of important observatories in those bands.

12. What is *interferometry*, and how can it improve astronomical observations?

TEST YOUR UNDERSTANDING

Does It Make Sense?

Decide whether the statement makes sense (or is clearly true) or does not make sense (or is clearly false). Explain clearly; not all of these have definitive answers, so your explanation is more important than your chosen answer.

13. The image was blurry because the detector was not placed at the focal plane.

14. I wanted to see faint details in the Andromeda Galaxy, so I photographed it with a very short exposure time.

15. I have a reflecting telescope in which the secondary mirror is bigger than the primary mirror.

16. The photograph shows what appear to be just two distinct stars, but each of those stars is actually a binary star system.

17. My 14-inch telescope has a lower diffraction limit than most large professional telescopes.

18. Now that I've bought a spectrograph, I can use my home telescope for spectroscopy as well as imaging.

19. If you lived on the Moon, you'd never see stars twinkle.

20. New technologies will soon allow astronomers to use X-ray telescopes on Earth's surface.

21. Thanks to adaptive optics, telescopes on the ground can now make ultraviolet images of the cosmos.

22. Thanks to interferometry, a properly spaced set of 10-meter radio telescopes can achieve the angular resolution of a single 100-kilometer radio telescope.

Quick Quiz

Choose the best answer to each of the following. Explain your reasoning with one or more complete sentences.

23. How much greater is the light-collecting area of a 6-meter telescope than that of a 3-meter telescope? (a) two times (b) four times (c) six times

24. Suppose that two stars are separated in the sky by 0.1 arcsecond. If you look at them with a telescope that has an angular resolution of 0.5 arcsecond, what will you see? (a) two distinct stars (b) one point of light that is the blurred image of both stars (c) nothing at all

25. The diffraction limit is a limit on a telescope's (a) size. (b) angular resolution. (c) spectral resolution.

26. The Hubble Space Telescope obtains higher-resolution images than most ground-based telescopes because it is (a) larger. (b) closer to the stars. (c) above Earth's atmosphere.

27. What does it mean if you see the color red in an X-ray image from the Chandra X-Ray Observatory? (a) The object is red in color. (b) The red parts are hotter than the blue parts. (c) It depends; the colors are chosen arbitrarily to represent something about the X rays recorded by the telescope.

28. The twinkling of stars is caused by (a) variations in stellar brightness with time. (b) light pollution. (c) motion of air in our atmosphere.

29. To achieve the same angular resolution as a visible-light telescope, a radio telescope would need to be (a) much larger. (b) slightly larger. (c) in space.

30. Where should you put a telescope designed for ultraviolet observations? (a) in Earth orbit (b) on an airplane (c) on a high mountaintop

31. Which technology can allow a single ground-based telescope to achieve images as sharp as those from the Hubble Space Telescope? (a) adaptive optics (b) grazing incidence mirrors (c) interferometry

32. Interferometry uses two or more telescopes to achieve (a) a light-collecting area equivalent to that of a much larger telescope. (b) an angular resolution equivalent to that of a much larger telescope. (c) both the light-collecting area and the angular resolution of a much larger telescope.

PROCESS OF SCIENCE

Examining How Science Works

33. *Science and Technology.* This chapter has discussed how the advance of science is intertwined with advances in technology. Choose one technology described in this chapter and summarize how its development (or improvement) has allowed us to learn more about the universe. Then project the changes you expect in this technology during the next few decades, and name at least one question about the universe that these changes should allow us to answer but that we cannot answer today.

34. *Type of Observation.* For each of the following, decide what type of observation (imaging, spectroscopy, timing) you would need to make. Explain clearly.
    a. Studying how a star's hot upper atmosphere changes with time.
    b. Learning the composition of a distant star **c.** Determining how fast a distant galaxy is moving away from Earth

GROUP WORK EXERCISE

35. *Which Telescope Would You Use?* Your job in this exercise is to choose the best telescope for observing matter around a black hole. You can assume that the matter is emitting photons at all wavelengths. Before you begin, assign the following roles to the people in your group: *Scribe* (takes notes on the group's activities), *Proposer* (proposes explanations to the group), *Skeptic* (points out weaknesses in proposed explanations), and *Moderator* (leads group discussion and makes sure everyone contributes). Then discuss the following four telescopes and rank them from best to worst for this particular observing task, explaining why you ranked each telescope where you did:
    a. an X-ray telescope, 2 meters in diameter, located at the South Pole. **b.** an infrared telescope, 2 meters in diameter, on a spacecraft in orbit around Earth and observing at a wavelength of micrometers (2×10^{-6} m). **c.** an infrared telescope, 10 meters in diameter, equipped with adaptive optics, located on Mauna Kea in Hawaii and observing at a wavelength of micrometers (10^{-5} m). **d.** a radio telescope, 300 meters in diameter, located in Puerto Rico.

INVESTIGATE FURTHER

In-Depth Questions to Increase Your Understanding

Short-Answer/Essay Questions

36. *Image Resolution.* What happens if you take a photograph and blow it up to a larger size? Does it contain more detail than it did before? Explain clearly, and relate your answer to the concepts of magnification and angular resolution in astronomical observations.

37. *Telescope Location.* In light of the problems faced by ground-based observatories, is the place where you live a good location for an astronomical observatory? Why or why not?

38. *Telescope Technology.* Suppose you were building a space-based observatory consisting of five individual telescopes. Which would be the better way to use these telescopes: as five individual telescopes with adaptive optics or as five telescopes linked together for interferometry (without adaptive optics)? Explain your reasoning clearly.

39. *Filters.* What would an American flag look like if you viewed it through a filter that transmits only red light? What would it look like through a filter that transmits only blue light?

40. *Project: Twinkling Stars.* Using a star chart, identify a few bright stars that should be visible in the early evening. On a clear night, observe each of these stars for a few minutes. Note the date and time, and for each star record the following information: approximate altitude and direction in your sky, brightness compared to other stars, color, and how much the star twinkles compared to other stars. Study your record. Can you draw any conclusions about how brightness and position in your sky affect twinkling? Explain.

41. *Project: Personal Telescope Review.* Find three telescopes that you could buy for under $1000 and evaluate each on the following criteria: light-collecting area, angular resolution, construction quality, and portability. Give each telescope a rating of 1 to 4 stars (4 is best) and state which one you would recommend for purchase.

Quantitative Problems

Be sure to show all calculations clearly and state your final answers in complete sentences.

42. *Light-Collecting Area.*
    a. How much greater is the light-collecting area of one of the 10-meter Keck telescopes than that of the 5-meter Hale telescope?
    b. Suppose astronomers built a 100-meter telescope. How much greater would its light-collecting area be than that of the 10-meter Keck telescope?

43. *Close Binary System.* Suppose that two stars in a binary star system are separated by a distance of 100 million kilometers and are located at a distance of 100 light-years from Earth. What is the angular separation of the two stars? Give your answer in both degrees and arcseconds. Can the Hubble Space Telescope resolve the two stars?

44. *Finding Planets.* Suppose you were looking at our own solar system from a distance of 10 light-years.
    a. What angular resolution would you need to see the Sun and Jupiter as distinct points of light? **b.** What angular resolution would you need to see the Sun and Earth as distinct points of light? **c.** How do the angular resolutions you found in parts a and b compare to the angular resolution of the Hubble Space Telescope? Comment on the challenge of making images of planets around other stars.

45. *Diffraction Limit of the Eye.*
    a. Calculate the diffraction limit of the human eye, assuming a wide-open pupil so that your eye acts like a lens with a diameter of 0.8 centimeter, for visible light of 500-nanometer wavelength. How does this compare to the diffraction limit of a 10-meter telescope? **b.** Now remember that humans have two eyes that are approximately 7 centimeters apart. Estimate the diffraction limit for human vision, assuming that your "optical interferometer" is just as good as one eyeball as large as the separation of two regular eyeballs.

46. *The Size of Radio Telescopes.* What is the diffraction limit of a 100-meter radio telescope observing radio waves with a wavelength of 21 centimeters? Compare this to the diffraction limit of the Hubble Space Telescope for visible light. Use your results to explain why, to be useful, radio telescopes must be much larger than optical telescopes.

47. *Your Satellite Dish.* Suppose you have a satellite dish that is 0.5 meter in diameter and you want to use it as a radio telescope. What is the diffraction limit on the angular resolution of your dish, assuming that you want to observe radio waves with a wavelength of 21 centimeters? Would it be very useful as an astronomical radio telescope?

48. *Hubble's Field of View.* Large telescopes often have small fields of view. For example, the advanced camera of the Hubble Space Telescope (HST) has a field of view that is roughly square and about 0.06 degree on a side.
    a. Calculate the angular area of the HST's field of view in square degrees. **b.** The angular area of the entire sky is about 41,250 square degrees. How many pictures would the HST have to take with its camera to obtain a complete picture of the entire sky?

49. *Hubble Sky Survey?* In Problem 48, you found out how many pictures the HST would require to photograph the entire sky. If you assume that it would take 1 hour to produce each picture, how many years would the HST need to obtain photos of the entire sky? Use your answer to explain why astronomers would like to have more than one large telescope in space.

50. *Visible-Light Interferometry.* Technological advances are now making it possible to link visible-light telescopes so that they can achieve the same angular resolution as a single telescope over 300 meters in size. What is the angular resolution (diffraction limit) of such a system of telescopes for observations at a wavelength of 500 nanometers?

Discussion Questions

51. *Science and Technology Funding.* Technological innovation clearly drives scientific discovery in astronomy, but the reverse is also true. For example, Newton made his discoveries in part because he wanted to explain the motions of the planets, but his discoveries have had far-reaching effects on our civilization. Congress often must decide between funding programs with purely scientific purposes (basic research) and programs designed to develop new technologies. If you were a member of Congress, how would you allocate spending between basic research and technology? Why?

52. *A Lunar Observatory.* Do the potential benefits of building an astronomical observatory on the Moon justify its costs at the present time? If it were up to you, would you recommend that Congress begin funding such an observatory? Defend your opinions.

Web Projects

53. *Major Ground-Based Observatories.* Take a virtual tour of one of the world's major astronomical observatories. Write a short report on why the observatory is so useful.

54. *Space Observatory.* Visit the website of a major space observatory, either existing or under development. Write a short report about the observatory, including its purpose, its orbit, and how it operates.

55. *Really Big Telescopes.* Learn about one of the projects to build a very large telescope (such as the Giant Magellan Telescope, the Thirty Meter Telescope, or the European Extremely Large Telescope). Write a short report about the telescope's current status and potential capabilities.

COSMIC CONTEXT PART II The Universality of Physics

One of Isaac Newton's great insights was that physics is universal—the same physical laws govern both the motions of heavenly objects and the things we experience in everyday life. This illustration shows some of the key physical principles used in the study of astronomy, with examples of how they apply both on Earth and in space.

EXAMPLES ON EARTH

(1) Conservation of Energy: Energy can be transferred from one object to another or transformed from one type to another, but the total amount of energy is always conserved [Section 4.3].

kinetic energy

radiative energy

potential energy

Plants transform the energy of sunlight into food containing chemical potential energy, which our bodies can convert into energy of motion.

(2) Conservation of Angular Momentum: An object's angular momentum cannot change unless it transfers angular momentum to another object. Because angular momentum depends on the product of mass, velocity, and radius, a spinning object must spin faster as it shrinks in size and an orbiting object must move faster when its orbital distance is smaller [Section 4.3].

Conservation of angular momentum explains why a skater spins faster as she pulls in her arms.

(3) Gravity: Every mass in the universe attracts every other mass through the force called gravity. The strength of gravity between two objects depends on the product of the masses divided by the square of the distance between them [Section 4.4].

The force of gravity between a ball and Earth attracts both together, explaining why the ball accelerates as it falls.

(4) Thermal Radiation: Large objects emit a thermal radiation spectrum that depends on the object's temperature. Hotter objects emit photons with a higher average energy and emit radiation of greater intensity at all wavelengths [Section 5.4].

The glow you see from a hot fireplace poker is thermal radiation in the form of visible light.

(5) Electromagnetic Spectrum: Light is a wave that affects electrically charged particles and magnets. The wavelength and frequency of light waves range over a wide spectrum, consisting of gamma rays, X rays, ultraviolet light, visible light, infrared light, and radio waves. Visible light is only a small fraction of the entire spectrum [Section 5.2].

X-ray machines

light bulb

We encounter many different kinds of electromagnetic radiation in our everyday lives.

microwave oven

gamma rays | X rays | ultraviolet | infrared | radio

visible

microwaves

A contracting gas cloud in space heats up because it transforms gravitational potential energy into thermal energy.

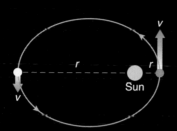

Conservation of angular momentum also explains why a planet's orbital speed increases when it is closer to the Sun.

Gravity also operates in space—its attractive force can act across great distances to pull objects closer together or to hold them in orbit.

Sunlight is also a visible form of thermal radiation. The Sun is much brighter and whiter than a fireplace poker because its surface is much hotter.

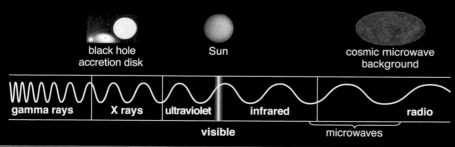

Many different forms of electromagnetic radiation are present in space. We therefore need to observe light of many different wavelengths to get a complete picture of the universe.

Earth, photographed from the outskirts of our solar system by the *Voyager* spacecraft. The "sunbeam" surrounding Earth is an artifact of light scattering in the camera.

7

OUR PLANETARY SYSTEM

LEARNING GOALS

7.1 STUDYING THE SOLAR SYSTEM

- What does the solar system look like?
- What can we learn by comparing the planets to one another?

7.2 PATTERNS IN THE SOLAR SYSTEM

- What features of our solar system provide clues to how it formed?

7.3 SPACECRAFT EXPLORATION OF THE SOLAR SYSTEM

- How do robotic spacecraft work?

We succeeded in taking that picture [left], and, if you look at it, you see a dot. Look again at that dot. That's here. That's home. That's us. On it everyone you love, everyone you know, everyone you ever heard of, every human being who ever was, lived out their lives. ... saint and sinner in the history of our species lived there—on a mote of dust suspended in a sunbeam.

—Carl Sagan

Our ancestors long ago recognized the motions of the planets through the sky, but it has been only a few hundred years since we learned that Earth is also a planet that orbits the Sun. Even then, we knew little about the other planets until the advent of large telescopes. More recently, the dawn of space exploration has brought us far greater understanding of other worlds. We've lived in this solar system all along, but only now are we getting to know it.

In this chapter, we'll explore our solar system like newcomers to the neighborhood. We'll begin by discussing what we hope to learn by studying the solar system, and in the process take a brief tour of major features of the Sun and planets. We'll also explore the major patterns we observe in the solar system—patterns that we will explain in subsequent chapters. Finally, we'll discuss the use of spacecraft to explore the solar system, examining how we are coming to learn so much more about our neighbors.

7.1 STUDYING THE SOLAR SYSTEM

Galileo's telescopic observations began a new era in astronomy in which the Sun, Moon, and planets could be studied for the first time as *worlds,* rather than as mere lights in the sky. Since that time, we have studied these worlds in different ways. Sometimes we study them individually—for example, when we map the geography of Mars or the atmospheric structure of Jupiter. Other times we compare the worlds to one another, seeking to understand their similarities and differences. This latter approach is called **comparative planetology**. Note that astronomers use the term *planetology* broadly to include moons, asteroids, and comets as well as planets.

We will use the comparative planetology approach for most of our study of the solar system in this book. Before we can compare the planets, however, we must have a general idea of the nature of our solar system and of the characteristics of individual worlds.

 Scale of the Universe Tutorial, Lesson 1

What does the solar system look like?

The first step in getting to know our solar system is to visualize what it looks like as a whole. Imagine having the perspec-tive of an alien spacecraft making its first scientific survey of our solar system. What would we see as we viewed the solar system from beyond the orbits of the planets?

Without a telescope, the answer would be "not much." Remember that the Sun and planets are all quite small compared to the distances between them [**Section 1.1**]—so small that if we viewed them from the outskirts of our solar system, the planets would be only pinpoints of light, and even the Sun would be just a small bright dot in the sky. But if we magnify the sizes of the planets by about a million times compared to their distances from the Sun and show their orbital paths, we get the central picture in **FIGURE 7.1**.

The ten pages that follow Figure 7.1 offer a brief tour through our solar system, beginning at the Sun, continuing to each of the planets, and concluding with dwarf planets such as Pluto and Eris. The tour highlights a few of the most important features of each world we visit—just enough information so that you'll be ready for the comparative study we'll undertake in later chapters. The side of each page shows the objects to scale, using the 1-to-10-billion scale introduced in Chapter 1. The map along the bottom of each page shows the locations of the Sun and each of the planets in the Voyage scale model solar system (see Figures 1.5 and 1.6), so that you can see their relative distances from the Sun. **TABLE 7.1** follows the tour and summarizes key data.

As you study Figure 7.1, the tour pages, and Table 7.1, you'll quickly see that our solar system is *not* a random collection of worlds, but a system that exhibits many clear patterns. For example, Figure 7.1 shows that all the planets orbit the Sun in the same direction and in nearly the same plane, and the tour pages show that the planets fall into two distinct groups. In science, the existence of patterns like these demands an explanation, and in Chapter 8 we will study the modern theory that explains them quite well. First, however, we need to investigate these patterns in greater detail.

 Formation of the Solar System Tutorial, Lesson 1

What can we learn by comparing the planets to one another?

The essence of comparative planetology lies in the idea that we can learn more about an individual world, including our own Earth, by studying it in the context of other objects in our solar system. It is much like learning more about a person by getting to know his or her family, friends, and culture.

(continued on page 205)

The solar system's layout and composition offer four major clues to how it formed. The main illustration below shows the orbits of planets in the solar system from a perspective beyond Neptune, with the planets themselves magnified by about a million times relative to their orbits.

1 **Large bodies in the solar system have orderly motions.** All planets have nearly circular orbits going in the same direction in nearly the same plane. Most large moons orbit their planets in this same direction, which is also the direction of the Sun's rotation.

Seen from above, planetary orbits are nearly circular.

Neptune

Saturn

Jupiter

Uranus

Mercury

Venus

Earth

Mars

White arrows indicate the rotation direction of the planets and Sun.

Red circles indicate the orbital direction of major moons around their planets.

Each planet's axis tilt is shown, with small circling arrows to indicate the direction of the planet's rotation.

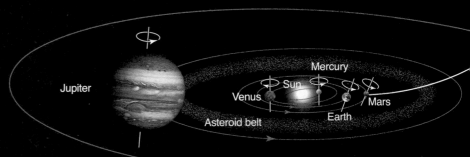

Jupiter

Orbits are shown to scale, but planet sizes are exaggerated about 1 million times relative to orbits. The Sun is not shown to scale.

Mercury

Venus

Sun

Earth

Mars

Asteroid belt

Neptune

Orange arrows indicate the direction of orbital motion.

2 **Planets fall into two major categories:** Small, rocky terrestrial planets and large, hydrogen-rich jovian planets.

terrestrial planet   jovian planet

Terrestrial Planets:
- small in mass and size
- close to the Sun
- made of metal and rock
- few moons and no rings

Jovian Planets:
- large mass and size
- far from the Sun
- made of H, He, and hydrogen compounds
- rings and many moons

3 **Swarms of asteroids and comets populate the solar system.** Vast numbers of rocky asteroids and icy comets are found throughout the solar system, but are concentrated in three distinct regions.

Asteroids are made of metal and rock, and most orbit in the **asteroid belt** *between Mars and Jupiter.*

Comets are ice-rich, and many are found in the **Kuiper belt** *beyond Neptune's orbit.*

Even more comets orbit the Sun in the distant, spherical region called the **Oort cloud**, *and only a rare few ever plunge into the inner solar system.*

Kuiper belt

4 **Several notable exceptions to these trends stand out.** Some planets have unusual axis tilts, unusually large moons, or moons with unusual orbits.

Uranus's odd tilt

Earth's relatively large moon

Saturn

Uranus

Uranus rotates nearly on its side compared to its orbit, and its rings and major moons share this "sideways" orientation.

Our own Moon is much closer in size to Earth than most other moons in comparison to their planets.

FIGURE 7.2 The Sun contains more than 99.8% of the total mass in our solar system.

Earth shown for size comparison

a A visible-light photograph of the Sun's surface. The dark splotches are sunspots—each large enough to swallow several Earths.

b This ultraviolet photograph, from the *SOHO* spacecraft, shows a huge streamer of hot gas on the Sun. The image of Earth was added for size comparison.

The Sun

- Radius: 696,000 km = 108R_{Earth}

- Mass: 333,000M_{Earth}

- Composition (by mass): 98% hydrogen and helium, 2% other elements

The Sun is by far the largest and brightest object in our solar system. It contains more than 99.8% of the solar system's total mass, making it nearly a thousand times as massive as everything else in the solar system combined.

The Sun's surface looks solid in photographs (**FIGURE 7.2**), but it is actually a roiling sea of hot (about 5800 K, or 5500°C or 10,000°F) hydrogen and helium gas. The surface is speckled with *sunspots* that appear dark in photographs only because they are slightly cooler than their surroundings. Solar storms sometimes send streamers of hot gas soaring far above the surface.

The Sun is gaseous throughout, and the temperature and pressure both increase with depth. The source of the Sun's energy lies deep in its core, where the temperatures and pressures are so high that the Sun is a nuclear fusion power plant.

Each second, fusion transforms about 600 million tons of the Sun's hydrogen into 596 million tons of helium. The "missing" 4 million tons becomes energy in accord with Einstein's famous formula, $E = mc^2$ [**Section 4.3**]. Despite converting 4 million tons of mass to energy each second, the Sun contains so much hydrogen that it has already shone steadily for almost 5 billion years and will continue to shine for another 5 billion years.

The Sun is the most influential object in our solar system. Its gravity governs the orbits of the planets. Its heat is the primary influence on the temperatures of planetary surfaces and atmospheres. It is the source of virtually all the light in our solar system—planets and moons shine by virtue of the sunlight they reflect. In addition, charged particles flowing outward from the Sun (the *solar wind*) help shape planetary magnetic fields and influence planetary atmospheres. Nevertheless, we can understand almost all the present characteristics of the planets without knowing much more about the Sun than we have just discussed. We'll therefore save more detailed study of the Sun for Chapter 14, where we will study it as our prototype for understanding other stars.

The Voyage scale model solar system represents sizes and distances in our solar system at one ten-billionth of their actual values (see Figure 1.6). The strip along the side of the page shows the sizes of the Sun and planets on this scale, and the map above shows their locations in the Voyage model on the National Mall in Washington, D.C. The Sun is about the size of a large grapefruit on this scale.

FIGURE 7.3 The left image shows that Mercury's surface is heavily cratered but also has smooth volcanic plains and long, steep cliffs. The inset shows a global composite. (Images from the *MESSENGER* spacecraft.)

Mercury

- Average distance from the Sun: 0.39 AU

- Radius: 2440 km = $0.38R_{Earth}$

- Mass: $0.055M_{Earth}$

- Average density: 5.43 g/cm^3

- Composition: rocks, metals

- Average surface temperature: 700 K (day), 100 K (night)

- Moons: 0

Mercury is the innermost planet of our solar system, and the smallest of the eight official planets. It is a desolate, cratered world with no active volcanoes, no wind, no rain, and no life. Because there is virtually no air to scatter sunlight or color the sky, you could see stars even in the daytime if you stood on Mercury with your back toward the Sun.

You might expect Mercury to be very hot because of its closeness to the Sun, but in fact it is a world of both hot and cold extremes. Tidal forces from the Sun have forced Mercury into an unusual rotation pattern [**Section 4.5**]: Its 58.6-day rotation period means it rotates exactly three times for every two of its 87.9-day orbits of the Sun. This combination of rotation and orbit gives Mercury days and nights that last about 3 Earth months each. Daytime temperatures reach 425°C, nearly as hot as hot coals. At night or in shadow, the temperature falls below −150°C, far colder than Antarctica in winter.

Mercury's surface is heavily cratered (**FIGURE 7.3**), much like the surface of our Moon. But it also shows evidence of past geological activity, such as plains created by ancient lava flows and tall, steep cliffs that run hundreds of kilometers in length. As we'll discuss in Chapter 9, these cliffs may be wrinkles from an episode of "planetary shrinking" early in Mercury's history. Mercury's high density (calculated from its mass and volume) indicates that it has a very large iron core, perhaps because it once suffered a huge impact that blasted its outer layers away.

FIGURE 7.4 The image above shows an artistic rendition of the surface of Venus as scientists think it would appear to our eyes. The surface topography is based on data from NASA's *Magellan* spacecraft. The inset (left) shows the full disk of Venus photographed by NASA's *Pioneer Venus Orbiter* with cameras sensitive to ultra-violet light. (Image above from the Voyage scale model solar system, developed by the Challenger Center for Space Science Education, the Smithsonian Institution, and NASA. Image by David P. Anderson, Southern Methodist University © 2001.)

Venus

- Average distance from the Sun: 0.72 AU

- Radius: 6051 km = $0.95R_{Earth}$

- Mass: $0.82M_{Earth}$

- Average density: 5.24 g/cm^3

- Composition: rocks, metals

- Average surface temperature: 740 K

- Moons: 0

Venus, the second planet from the Sun, is nearly identical in size to Earth. Before the era of spacecraft visits, Venus stood out largely for its strange rotation: It rotates on its axis very slowly and in the opposite direction of Earth, so days and nights are very long and the Sun rises in the west and sets in the east instead of rising in the east and setting in the west. Its surface is completely hidden from view by dense clouds, so we knew little about it until a few decades ago, when spacecraft began to map Venus with cloud-penetrating radar, discovering mountains, valleys, craters, and extensive evidence of past volcanic activity (**FIGURE 7.4**). Because we knew so little about it, some science fiction writers used its Earth-like size, thick atmosphere, and closer distance to the Sun to speculate that it might be a lush, tropical paradise—a "sister planet" to Earth.

The reality is far different. We now know that an extreme *greenhouse effect* bakes Venus's surface to an incredible 470°C (about 880°F), trapping heat so effectively that nighttime offers no relief. Day and night, Venus is hotter than a pizza oven, and the thick atmosphere bears down on the surface with a pressure equivalent to that nearly a kilometer (0.6 mile) beneath the ocean's surface on Earth. Far from being a beautiful sister planet to Earth, Venus resembles a traditional view of hell.

The fact that Venus and Earth are so similar in size and composition but so different in surface conditions suggests that Venus could teach us important lessons. In particular, Venus's greenhouse effect is caused by carbon dioxide, the same gas that is primarily responsible for global warming on Earth. Perhaps further study of Venus may help us better understand and solve some of the problems we face here at home.

a This image (left), computer generated from satellite data, shows the striking contrast between the day and night hemispheres of Earth. The day side reveals little evidence of human presence, but at night our presence is revealed by the lights of human activity. (From the Voyage scale model solar system, developed by the Challenger Center for Space Science Education, the Smithsonian Institution, and NASA. Image created by ARC Science Simulations © 2001.)

b Earth and the Moon, shown to scale. The Moon is about 1/4 as large as Earth in diameter, while its mass is about 1/80 of Earth's mass. To show the distance between Earth and Moon on the same scale, you'd need to hold these two photographs about 1 meter (3 feet) apart.

FIGURE 7.5 Earth, our home planet.

Earth

- Average distance from the Sun: 1.00 AU
- Radius: 6378 km = $1R_{Earth}$
- Mass: $1.00M_{Earth}$
- Average density: 5.52 g/cm^3
- Composition: rocks, metals
- Average surface temperature: 290 K
- Moons: 1

Beyond Venus, we next encounter our home planet, Earth, the only known oasis of life in our solar system. Earth is also the only planet in our solar system with oxygen to breathe, ozone to shield the surface from deadly solar radiation, and abundant surface water to nurture life. Temperatures are pleasant because Earth's atmosphere contains just enough carbon dioxide and water vapor to maintain a moderate greenhouse effect.

Despite Earth's small size, its beauty is striking (**FIGURE 7.5a**). Blue oceans cover nearly three-fourths of the surface, broken by the continental land masses and scattered islands. The polar caps are white with snow and ice, and white clouds are scattered above the surface. At night, the glow of artificial lights reveals the presence of an intelligent civilization.

Earth is the first planet on our tour with a moon. The Moon is surprisingly large compared with Earth (**FIGURE 7.5b**); although it is not the largest moon in the solar system, almost all other moons are much smaller relative to the planets they orbit. As we'll discuss in Chapter 8, the leading hypothesis holds that the Moon formed as a result of a giant impact early in Earth's history.

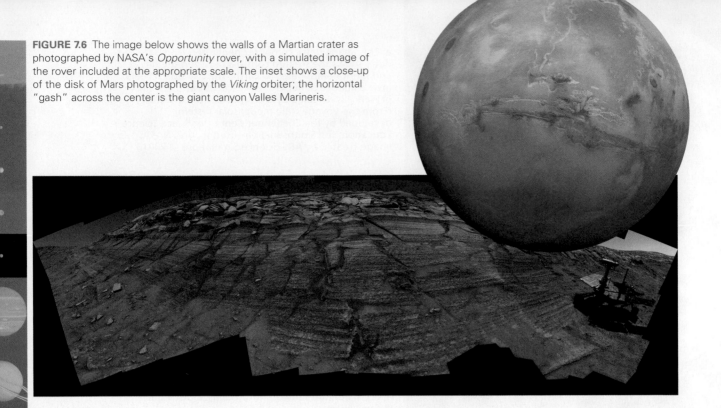

FIGURE 7.6 The image below shows the walls of a Martian crater as photographed by NASA's *Opportunity* rover, with a simulated image of the rover included at the appropriate scale. The inset shows a close-up of the disk of Mars photographed by the *Viking* orbiter; the horizontal "gash" across the center is the giant canyon Valles Marineris.

Mars

- Average distance from the Sun: 1.52 AU

- Radius: 3397 km = $0.53R_{Earth}$

- Mass: $0.11M_{Earth}$

- Average density: 3.93 g/cm^3

- Composition: rocks, metals

- Average surface temperature: 220 K

- Moons: 2 (very small)

The next planet on our tour is Mars, the last of the four inner planets of our solar system (**FIGURE 7.6**). Mars is larger than Mercury and the Moon but only about half Earth's size in diameter; its mass is about 10% that of Earth. Mars has two tiny moons, Phobos and Deimos, which probably once were asteroids that were captured into Martian orbit early in the solar system's history.

Mars is a world of wonders, with ancient volcanoes that dwarf the largest mountains on Earth, a great canyon that runs nearly one-fifth of the way around the planet, and polar caps made of frozen carbon dioxide ("dry ice") and water. Although Mars is frozen today, the presence of dried-up riverbeds, rock-strewn floodplains, and minerals that form in water offers clear evidence that Mars had at least some warm and wet periods in the past. Major flows of liquid water probably ceased at least 3 billion years ago, but some liquid water could persist underground, perhaps flowing to the surface on occasion.

Mars's surface looks almost Earth-like, but you wouldn't want to visit without a spacesuit. The air pressure is far less than that on top of Mount Everest, the temperature is usually well below freezing, the trace amounts of oxygen would not be nearly enough to breathe, and the lack of atmospheric ozone would leave you exposed to deadly ultraviolet radiation from the Sun.

More than a dozen spacecraft have flown past, orbited, or landed on Mars, and plans are in the works for more. We may even send humans to Mars within the next few decades. By overturning rocks in ancient riverbeds or chipping away at ice in the polar caps, explorers will help us learn whether Mars has ever been home to life.

FIGURE 7.7 This image shows what it would look like to be orbiting near Jupiter's moon Io as Jupiter comes into view. Notice the Great Red Spot to the left of Jupiter's center. The extraordinarily dark rings discovered during the *Voyager* missions are exaggerated to make them visible. This computer visualization was created using data from both NASA's *Voyager* and *Galileo* missions. (From the Voyage scale model solar system, developed by the Challenger Center for Space Science Education, the Smithsonian Institution, and NASA. Image created by ARC Science Simulations © 2001.)

Jupiter

- Average distance from the Sun: 5.20 AU

- Radius 71,492 km = $11.2R_{Earth}$

- Mass: $318M_{Earth}$

- Average density: 1.33 g/cm^3

- Composition: mostly hydrogen and helium

- Cloud-top temperature: 125 K

- Moons: at least 67

To reach the orbit of Jupiter from Mars, we must traverse a distance that is more than double the total distance from the Sun to Mars, passing through the *asteroid belt* along the way. Upon our arrival, we find a planet much larger than any we have seen so far (**FIGURE 7.7**).

Jupiter is so different from the planets of the inner solar system that we must adopt an entirely new mental image of the term *planet*. Its mass is more than 300 times that of Earth, and its volume is more than 1000 times that of Earth. Its most famous feature—a long-lived storm called the Great Red Spot—is itself large enough to swallow two or three Earths. Like the Sun, Jupiter is made primarily of hydrogen and helium and has no solid surface. If we plunged deep into Jupiter, the increasing gas pressure would crush us long before we ever reached its core.

Jupiter reigns over dozens of moons and a thin set of rings (too faint to be seen in most photographs). Most of the moons are very small, but four are large enough that we'd call them planets or dwarf planets if they orbited the Sun independently. These four moons—Io, Europa, Ganymede, and Callisto—are often called the *Galilean moons* (because Galileo discovered them), and they display varied and interesting geology. Io is the most volcanically active world in the solar system. Europa has an icy crust that may hide a subsurface ocean of liquid water, making it a promising place to search for life. Ganymede and Callisto may also have subsurface oceans, and their surfaces have many features that remain mysterious.

FIGURE 7.8 *Cassini*'s view of Saturn. We see the shadow of the rings on the upper right portion of Saturn's sunlit face, and the rings become lost in Saturn's shadow on the night side. The inset shows an infrared view of Saturn's largest moon, Titan, which is shrouded in a thick, cloudy atmosphere.

Saturn

- Average distance from the Sun: 9.54 AU

- Radius: 60,268 km = $9.4R_{Earth}$

- Mass: $95.2M_{Earth}$

- Average density: 0.70 g/cm^3

- Composition: mostly hydrogen and helium

- Cloud-top temperature: 95 K

- Moons: at least 62

The journey from Jupiter to Saturn is a long one: Saturn orbits nearly twice as far from the Sun as Jupiter. Saturn, the second-largest planet in our solar system, is only slightly smaller than Jupiter in diameter, but its lower density makes it considerably less massive (about one-third of Jupiter's mass). Like Jupiter, Saturn is made mostly of hydrogen and helium and has no solid surface.

Saturn is famous for its spectacular rings (**FIGURE 7.8**). Although all four of the giant outer planets have rings, only Saturn's can be seen easily. The rings look solid from a distance, but in reality they are made of countless small particles, each of which orbits Saturn like a tiny moon. These particles of rock and ice range in size from dust grains to city blocks. We are rapidly learning more about Saturn and its rings through observations made by the *Cassini* spacecraft, which has orbited Saturn since 2004.

Cassini has also taught us more about Saturn's moons, and has revealed that at least two are geologically active today: Enceladus, which has ice fountains spraying out from its southern hemisphere, and Titan, the only moon in the solar system with a thick atmosphere. Saturn and its moons are so far from the Sun that Titan's surface temperature is a frigid −180°C, making it far too cold for liquid water to exist. However, studies by *Cassini* and its *Huygens* probe, which landed on Titan in 2005, have revealed a remarkably Earth-like landscape—including an erosion-carved surface with riverbeds and lakes—except that the features are shaped by extremely cold liquid methane or ethane rather than liquid water.

Uranus

- Average distance from the Sun: 19.2 AU

- Radius: 25,559 km = $4.0R_{Earth}$

- Mass: $14.5M_{Earth}$

- Average density: 1.32 g/cm³

- Composition: hydrogen, helium, hydrogen compounds

- Cloud-top temperature: 60 K

- Moons: at least 27

It's another long journey to the next stop on our tour, as Uranus lies twice as far from the Sun as Saturn. Uranus (normally pronounced *YUR-uh-nus*) is much smaller than either Jupiter or Saturn but much larger than Earth. It is made largely of hydrogen, helium, and *hydrogen compounds* such as water (H_2O), ammonia (NH_3), and methane (CH_4). Methane gas gives Uranus its pale blue-green color (**FIGURE 7.9**). Like the other giants of the outer solar system, Uranus lacks a solid surface. More than two dozen moons orbit Uranus, along with a set of rings somewhat similar to those of Saturn but much darker and more difficult to see.

The entire Uranus system—planet, rings, and moon orbits—is tipped on its side compared to the rest of the planets. This extreme axis tilt may be the result of a cataclysmic collision that Uranus suffered as it was forming, and it gives Uranus the most extreme seasonal variations of any planet in our solar system. If you lived on a platform floating in Uranus's atmosphere near its north pole, you'd have continuous daylight for half of each orbit, or 42 years. Then, after a very gradual sunset, you'd enter into a 42-year-long night.

Only one spacecraft has visited Uranus: *Voyager 2*, which flew past all four of the giant outer planets before heading out of the solar system. Much of our current understanding of Uranus comes from that mission, though powerful new telescopes are also capable of studying it. Scientists hope it will not be too long before we can send another spacecraft to study Uranus and its rings and moons in greater detail.

FIGURE 7.10 This image shows what it would look like to be orbiting Neptune's moon Triton as Neptune itself comes into view. The dark rings are exaggerated to make them visible in this computer simulation using data from NASA's *Voyager 2* mission. (From the Voyage scale model solar system, developed by the Challenger Center for Space Science Education, the Smithsonian Institution, and NASA. Image created by ARC Science Simulations © 2001.)

Neptune

- Average distance from the Sun: 30.1 AU

- Radius: 24,764 km = $3.9 R_{Earth}$

- Mass: $17.1 M_{Earth}$

- Average density: 1.64 g/cm^3

- Composition: hydrogen, helium, hydrogen compounds

- Cloud-top temperature: 60 K

- Moons: at least 13

The journey from the orbit of Uranus to the orbit of Neptune is the longest yet in our tour, calling attention to the vast emptiness of the outer solar system. Nevertheless, Neptune looks nearly like a twin of Uranus, although it is more strikingly blue (**FIGURE 7.10**). It is slightly smaller than Uranus in size, but a higher density makes it slightly more massive even though the two planets share very similar compositions. Like Uranus, Neptune has been visited only by the *Voyager 2* spacecraft.

Neptune has rings and numerous moons. Its largest moon, Triton, is larger than Pluto and is one of the most fascinating moons in the solar system. Triton's icy surface has features that appear to be somewhat like geysers, although they spew nitrogen gas rather than water into the sky. Even more surprisingly, Triton is the only large moon in the solar system that orbits its planet "backward"—that is, in a direction opposite to the direction in which Neptune rotates. This backward orbit makes it a near certainty that Triton once orbited the Sun independently before somehow being captured into Neptune's orbit.

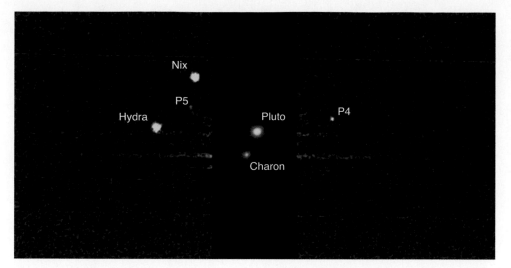

FIGURE 7.11 Pluto and its five moons, as imaged by the Hubble Space Telescope. Aside from Pluto and its moons, the other blue light in the image is scattered light within the telescope.

Dwarf planets: Pluto, Eris, and more

Pluto Data:

- Average distance from the Sun: 39.5 AU

- Radius: 1160 km = $0.18R_{Earth}$

- Mass: $0.0022M_{Earth}$

- Average density: 2.0 g/cm^3

- Composition: ices, rock

- Average surface temperature: 40 K

- Moons: 5

We conclude our tour at Pluto (**FIGURE 7.11**), which reigned for some 75 years as the "ninth planet" in our solar system. However, the 2005 discovery of the slightly larger Eris, and the fact that dozens of other recently discovered objects are not much smaller than Pluto and Eris, led scientists to reconsider the definition of "planet." The result was that we now consider Pluto and Eris to be *dwarf planets,* too small to qualify as official planets but large enough to be round in shape. Several other solar system objects also qualify as dwarf planets, including the largest asteroid of the asteroid belt, named Ceres.

Pluto and Eris belong to a collection of thousands of icy objects that orbit the Sun beyond Neptune, making up what we call the *Kuiper belt*. As you can see in Figure 7.1, the Kuiper belt is much like the asteroid belt, except it is farther from the Sun and composed of comet-like objects rather than rocky asteroids.

Pluto's characteristics help us to think about what it would be like to visit this distant realm. Pluto's average distance from the Sun lies as far beyond Neptune as Neptune lies beyond Uranus, making Pluto extremely cold and quite dark even in daytime. From Pluto, the Sun would be little more than a bright light among the stars. Pluto's largest moon, Charon, is locked together with it in synchronous rotation [**Section 4.5**], so Charon would dominate the sky on one side of Pluto but never be seen from the other side.

The great distances and small sizes of Pluto and other dwarf planets make them difficult to study, but we are rapidly learning more. Scientists are particularly excited about two upcoming events: In February 2015, the *Dawn* spacecraft will arrive at Ceres, giving us our first close-up views of this world; then, in July 2015, after a 9-year journey, the *New Horizons* spacecraft will fly past Pluto.

TABLE 7.1 The Planetary Data[a]

Photo	Planet	Relative Size	Average Distance from Sun (AU)	Average Equatorial Radius (km)	Mass (Earth = 1)	Average Density (g/cm³)	Orbital Period	Rotation Period	Axis Tilt	Average Surface (or Cloud-Top) Temperature[b]	Composition	Known Moons (2012)	Rings?
	Mercury	·	0.387	2440	0.055	5.43	87.9 days	58.6 days	0.0°	700 K (day) 100 K (night)	Rocks, metals	0	No
	Venus	•	0.723	6051	0.82	5.24	225 days	243 days	177.3°	740 K	Rocks, metals	0	No
	Earth	•	1.00	6378	1.00	5.52	1.00 year	23.93 hours	23.5°	290 K	Rocks, metals	1	No
	Mars	·	1.52	3397	0.11	3.93	1.88 years	24.6 hours	25.2°	220 K	Rocks, metals	2	No
	Jupiter	●	5.20	71,492	318	1.33	11.9 years	9.93 hours	3.1°	125 K	H, He, hydrogen compounds[c]	67	Yes
	Saturn	●	9.54	60,268	95.2	0.70	29.5 years	10.6 hours	26.7°	95 K	H, He, hydrogen compounds[c]	62	Yes
	Uranus	●	19.2	25,559	14.5	1.32	83.8 years	17.2 hours	97.9°	60 K	H, He, hydrogen compounds[c]	27	Yes
	Neptune	●	30.1	24,764	17.1	1.64	165 years	16.1 hours	29.6°	60 K	H, He, hydrogen compounds[c]	13	Yes
	Pluto	·	39.5	1160	0.0022	2.0	248 years	6.39 days	112.5°	44 K	Ices, rock	5	No
	Eris	·	67.7	1200	0.0028	2.3	557 years	1.08 days	78°	43 K	Ices, rock	1	No

[a]Including the dwarf planets Pluto and Eris; Appendix E gives a more complete list of planetary properties.
[b]Surface temperatures for all objects except Jupiter, Saturn, Uranus, and Neptune, for which cloud-top temperatures are listed.
[c]Include water (H_2O), methane (CH_4), and ammonia (NH_3).

(continued from page 191)

While we still can learn much by studying planets individually, the comparative planetology approach has demonstrated its value in at least three key ways:

- Comparative study has revealed similarities and differences among the planets that have helped guide the development of our theory of solar system formation, thereby giving us a better understanding of how we came to exist here on Earth.

- Comparative study has given us new insights into the physical processes that have shaped Earth and other worlds—insights that can help us better understand and manage our own planet.

- Comparative study has allowed us to apply lessons from our solar system to the study of the many planetary systems now known around other stars. These lessons help us understand both the general principles that govern planetary systems and the specific circumstances under which Earth-like planets—and possibly life—might exist elsewhere.

The comparative planetology approach should also benefit you as a student by helping you stay focused on *processes* rather than on a collection of facts. We now know so many individual facts about the worlds of our solar system and others that even planetary scientists have trouble keeping track of them all. By concentrating on the processes that shape planets, you'll gain a deeper understanding of how planets, including Earth, actually work.

 Orbits and Kepler's Laws Tutorial, Lessons 2–4

7.2 PATTERNS IN THE SOLAR SYSTEM

Our goal in studying the solar system as a whole is to look for clues that might help us develop a theory that could explain how it formed. In this section, we'll explore the patterns of our solar system in more depth, and organize these patterns into a set of general features that tell us about our solar system's formation.

What features of our solar system provide clues to how it formed?

We have already seen that our solar system is not a random collection of worlds, but rather a family of worlds exhibiting many traits that would be difficult to attribute to coincidence. We could make a long list of such traits, but it is easier to develop a scientific theory by focusing on the more general structure of our solar system. For our purposes, four major features stand out, each corresponding to one of the numbered steps in Figure 7.1:

1. **Patterns of motion among large bodies.** The Sun, planets, and large moons generally orbit and rotate in a very organized way.

2. **Two major types of planets.** The eight planets divide clearly into two groups: the small, rocky planets that are close together and close to the Sun, and the large, gas-rich planets that are farther apart and farther from the Sun.

3. **Asteroids and comets.** Between and beyond the planets, vast numbers of asteroids and comets orbit the Sun; some are large enough to qualify as dwarf planets. The locations, orbits, and compositions of these asteroids and comets follow distinct patterns.

4. **Exceptions to the rules.** The generally orderly solar system also has some notable exceptions. For example, among the inner planets only Earth has a large moon, and the planet Uranus is tipped on its side. A successful theory must make allowances for such exceptions even as it explains the general rules.

Because these four features are so important to our study of the solar system, let's investigate them in a little more detail.

Feature 1: Patterns of Motion Among Large Bodies If you look back at Figure 7.1, you'll notice several clear patterns of motion among the large bodies of our solar system. (In this context, a "body" is simply an individual object such as the Sun, a planet, or a moon.) For example:

- All planetary orbits are nearly circular and lie nearly in the same plane.

- All planets orbit the Sun in the same direction: counterclockwise as viewed from high above Earth's North Pole.

- Most planets rotate in the same direction in which they orbit, with fairly small axis tilts. The Sun also rotates in this direction.

- Most of the solar system's large moons exhibit similar properties in their orbits around their planets, such as orbiting in their planet's equatorial plane in the same direction as the planet rotates.

We consider these orderly patterns together as the first major feature of our solar system. As we'll see in Chapter 8, our theory of solar system formation explains these patterns as consequences of processes that occurred during the early stages of the birth of our solar system.

Feature 2: Two Types of Planets Our brief planetary tour showed that the four inner planets are quite different from the four outer planets. We say that these two groups represent two distinct planetary classes: *terrestrial* and *jovian*.

The **terrestrial planets** (*terrestrial* means "Earth-like") are the four planets of the inner solar system: Mercury, Venus, Earth, and Mars. These planets are relatively small and dense, with rocky surfaces and an abundance of metals in their cores. They have few moons, if any, and no rings. We count our Moon as a fifth terrestrial world, because its history has been shaped by the same processes that have shaped the terrestrial planets.

TABLE 7.2 Comparison of Terrestrial and Jovian Planets

Terrestrial Planets	Jovian Planets
Smaller size and mass	Larger size and mass
Higher average density	Lower average density
Made mostly of rocks and metals	Made mostly of hydrogen, helium, and hydrogen compounds
Solid surface	No solid surface
Few (if any) moons and no rings	Rings and many moons
Closer to the Sun (and closer together), with warmer surfaces	Farther from the Sun (and farther apart), with cool temperatures at cloud tops

FIGURE 7.13 Comet Hale-Bopp, photographed over Boulder, Colorado, during its appearance in 1997.

The **jovian planets** (*jovian* means "Jupiter-like") are the four large planets of the outer solar system: Jupiter, Saturn, Uranus, and Neptune. The jovian planets are much larger in size and lower in average density than the terrestrial planets, and they have rings and many moons. They lack solid surfaces and are made mostly of hydrogen, helium, and **hydrogen compounds**—compounds containing hydrogen, such as water (H_2O), ammonia (NH_3), and methane (CH_4). Because these substances are gases under earthly conditions, the jovian planets are sometimes called "gas giants." **TABLE 7.2** contrasts the general traits of the terrestrial and jovian planets.

Feature 3: Asteroids and Comets The third major feature of our solar system is the existence of vast numbers of small objects orbiting the Sun. These objects fall into two major groups: asteroids and comets.

Asteroids are rocky bodies that orbit the Sun much like planets, but they are much smaller (**FIGURE 7.12**). Most known asteroids are found within the **asteroid belt** between the orbits of Mars and Jupiter (see Figure 7.1).

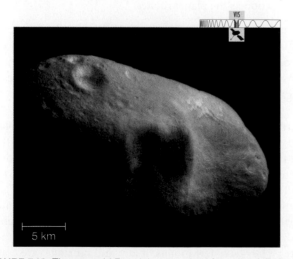

5 km

FIGURE 7.12 The asteroid Eros (photographed from the *NEAR* spacecraft). Its appearance is probably typical of most asteroids. Eros is about 40 kilometers in length. Like other small objects in the solar system, it is not spherical.

Comets are also small objects that orbit the Sun, but they are made largely of ices (such as water ice, ammonia ice, and methane ice) mixed with rock. You are probably familiar with the occasional appearance of comets in the inner solar system, where they may become visible to the naked eye with long, beautiful tails (**FIGURE 7.13**). These visitors, which may delight sky watchers for a few weeks or months, are actually quite rare among comets. The vast majority of comets never visit the inner solar system. Instead, they orbit the Sun in one of the two distinct regions shown as Feature 3 in Figure 7.1. The first is a donut-shaped region beyond the orbit of Neptune that we call the **Kuiper belt** (*Kuiper* rhymes with *piper*). The Kuiper belt contains at least 100,000 icy objects, of which Pluto and Eris are the largest known. Kuiper belt objects all orbit the Sun in the same direction as the planets. The second cometary region, called the **Oort cloud** (*Oort* rhymes with *court*), is much farther from the Sun and may contain a trillion comets; its most distant comets may sometimes reside nearly one-quarter of the distance to the nearest stars. Comets of the Oort cloud have orbits randomly inclined to the ecliptic plane, giving the Oort cloud a roughly spherical shape.

Feature 4: Exceptions to the Rules The fourth key feature of our solar system is that there are a few notable exceptions to the general rules. For example, while most of the planets rotate in the same direction as they orbit, Uranus rotates nearly on its side, and Venus rotates "backward" (clockwise as viewed from high above Earth's North Pole). Similarly, while most large moons orbit their planets in the same direction as their planets rotate, many small moons have much more unusual orbits.

One of the most interesting exceptions concerns our own Moon. While the other terrestrial planets have either no moons (Mercury and Venus) or very tiny moons (Mars), Earth has one of the largest moons in the solar system.

Summary Now that you have read through the tour of our solar system and the description of its four major features, review them again in Figure 7.1. You should now see clearly that these features hold key clues to the origin of our solar system—the main topic of the next chapter.

7.3 SPACECRAFT EXPLORATION OF THE SOLAR SYSTEM

How have we learned so much about the solar system? Much of our knowledge comes from telescopic observations, using both ground-based telescopes and telescopes in Earth orbit such as the Hubble Space Telescope. In one case—our Moon—we have learned a lot by sending astronauts to explore the terrain and bring back rocks for laboratory study. In a few other cases, we have studied samples of distant worlds that have come to us as meteorites. But most of the data fueling the recent revolution in our understanding of the solar system have come from robotic spacecraft. To date, we have sent robotic spacecraft to all the terrestrial and jovian planets, as well as to many moons, asteroids, and comets. In this section, we'll briefly investigate how we use robotic spacecraft to explore the solar system.

How do robotic spacecraft work?

The spacecraft we send to explore the planets are robots designed for scientific study. All spacecraft have computers used to control their major components, power sources such as solar cells, propulsion systems, and scientific instruments to study their targets. Robotic spacecraft operate primarily with preprogrammed instructions, but also carry radios that allow them to communicate with controllers on Earth. Most robotic spacecraft make one-way trips, never physically returning to Earth but sending their data back from space in the same way we send radio and television signals.

Broadly speaking, the robotic missions to other worlds fall into four major categories:

- **Flyby.** A spacecraft on a flyby goes past a world just once and then continues on its way.

SPECIAL TOPIC

How Did We Learn the Scale of the Solar System?

This chapter presents the layout of the solar system as we know it today, when we have precise measurements of planetary sizes and distances. But how did we learn the scale of the solar system?

By the middle of the 17th century, Kepler's laws [**Section 3.3**] had provided planetary distances in astronomical units (AU), or distances *relative* to the Earth-Sun distance, but no one yet knew the value of the AU in absolute units like miles or kilometers. A number of 17th-century astronomers proposed ideas for measuring the Earth-Sun distance, but none were practical. Then, in 1716, Edmond Halley (best known for the comet named after him) hit upon the idea that would ultimately solve the problem: He realized that during a planetary *transit*, when a planet appears to pass across the face of the Sun [**Section S1.1**], observers in different locations on Earth would see the planet trace slightly different paths across the Sun. Comparison of these paths could allow calculation of the planet's distance—which would in turn allow determination of the AU—through the simple geometry shown in **FIGURE 1**.

Only Mercury and Venus can produce transits visible from Earth. Halley realized that although Mercury transits occur more often, the measurements would be easier with Venus because its closer distance to Earth means greater separation between the paths in Figure 7.1. Unfortunately, Venus transits are rare, occurring in pairs 8 years apart about every 120 years. Halley did not live to see a Venus transit, but later astronomers followed his plan, mounting expeditions to observe transits in 1761 and 1769.

The transit observations turned out to be quite difficult in practice, partly from the inherent challenge that long expeditions posed at that time, and partly because getting the geometry right required very precise timing of the beginning of the transit. Astronomers discovered that this timing was more difficult than Halley had guessed, because of optical effects that occur during a transit. Nevertheless, astronomers studied the data from the 1761 and 1769 transits for many decades, and by the middle of the 1800s the value of the AU had been pinned down to within about 5% of its modern value of 149.6 million kilometers. The next Venus transits occurred in 1874 and 1882. Photography had been invented by then, making observations more reliable, so in principle those transits could have allowed refinement of the AU. However, by that time photography and better telescopes had also made it possible to observe parallax of

planets against stars, and by 1877 such observations had given us the value of the AU to within 0.2% of its modern value.

The most recent transits occurred in 2004 and 2012. While they were amazing spectacles to observe, they weren't important for interplanetary measurements. Nowadays, we measure the distance to Venus very precisely by bouncing radio waves off its surface with radar, a technique known as *radar ranging*. Because we know the speed of light, measuring the time it takes for the radio waves to make the round trip from Earth to Venus tells us the precise distance. We then use this distance and Venus's known distance in AU to calculate the actual value of the AU. Once we know the value of the AU, we can determine the actual distances of all the planets from the Sun, and we can determine their actual sizes from their angular sizes and distances. Indeed, we now know the layout of the solar system so well that we can launch spacecraft from Earth and send them to precise places on or around distant worlds.

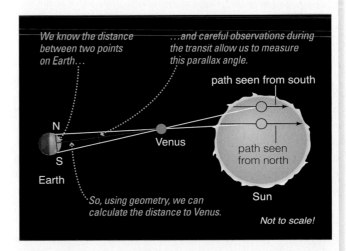

FIGURE 1 During a transit of Venus, observers at different places on Earth will see it trace slightly different paths across the Sun. The precise geometry of these events therefore allows computation of Venus's true distance, which in turn allows computation of the AU distance. (Adapted from *Sky and Telescope*.)

- **Orbiter.** An orbiter is a spacecraft that orbits the world it is visiting, allowing longer-term study.

- **Lander or probe.** These spacecraft are designed to land on a planet's surface or probe a planet's atmosphere by flying through it. Some landers carry rovers to explore wider regions.

- **Sample return mission.** A sample return mission makes a round trip to return a sample of the world it has studied to Earth.

The choice of spacecraft type depends on both scientific objectives and cost.

Flybys

Flybys tend to be cheaper than other missions because they are generally less expensive to launch into space. Launch costs depend largely on weight, and onboard fuel is a significant part of the weight of a spacecraft heading to another planet. Once a spacecraft is on its way, the lack of friction or air drag in space means that it can maintain its orbital trajectory through the solar system without using any fuel at all. Fuel is needed only when the spacecraft must change from one trajectory (orbit) to another.

Moreover, some flybys gain more "bang for the buck" by visiting multiple planets. For example, *Voyager 2* flew past Jupiter, Saturn, Uranus, and Neptune before continuing on its way out of our solar system (**FIGURE 7.14**). This trajectory allowed additional fuel savings by using the gravity of each planet along the spacecraft's path to help boost it onward to the next planet. This technique, known as a *gravitational slingshot*, can not only bend the spacecraft's path but also speed it up by essentially stealing a tiny bit of the planet's orbital energy, though the effect on the planet is unnoticeable.

THINK ABOUT IT

Study the *Voyager 2* trajectory in Figure 7.14. Given that Saturn orbits the Sun every 29 years, Uranus orbits the Sun every 84 years, and Neptune orbits the Sun every 165 years, would it be possible to send another flyby mission to all four jovian planets if we launched it now? Explain.

Although a flyby offers only a relatively short period of close-up study, it can provide valuable scientific information. Spacecraft on flybys generally carry small telescopes, cameras, and spectrographs. Because these instruments are brought relatively close (typically thousands of kilometers or less) to other worlds, they can obtain much higher-resolution images and spectra than the largest telescopes on Earth or in Earth orbit. In addition, flybys sometimes give us information that would be very difficult to obtain from Earth. For example, *Voyager 2* helped us discover Jupiter's rings and learn about the rings of Saturn, Uranus, and Neptune through views in which the rings were backlit by the Sun. Such views are possible only from beyond each planet's orbit.

Spacecraft on flybys may also carry instruments to measure local magnetic field strength or to sample interplanetary dust. The gravitational effects of the planets and their moons on the spacecraft itself provide information about object masses and

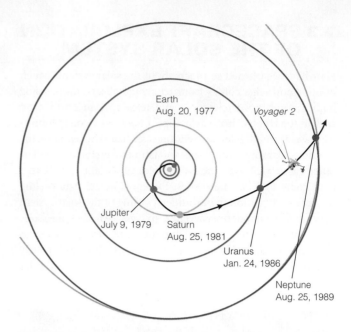

FIGURE 7.14 interactive figure The trajectory of *Voyager 2*, which made flybys of the four jovian planets in our solar system.

densities. Like the backlit views of the rings, these types of data cannot be gathered from Earth. Indeed, most of what we know about the masses and compositions of moons comes from data gathered by spacecraft that have flown past them.

Orbiters

An orbiter can study another world for a much longer period of time than a flyby. Like the spacecraft used for flybys, orbiters often carry cameras, spectrographs, and instruments for measuring the strength of magnetic fields. Some missions also carry radar, which can be used to make precise altitude measurements of surface features. Radar has proven especially valuable for the study of Venus and Titan, because it provides our only way of "seeing" through their thick, cloudy atmospheres.

An orbiter is generally more expensive than a flyby for an equivalent weight of scientific instruments, primarily because it must carry added fuel to change from an interplanetary trajectory to a path that puts it into orbit around another world. Careful planning can minimize the added expense. For example, recent Mars orbiters have saved on fuel costs by carrying only enough fuel to enter highly elliptical orbits around Mars. The spacecraft then settled into the smaller, more circular orbits needed for scientific observations by skimming the Martian atmosphere at the low point of every elliptical orbit. Atmospheric drag slowed the spacecraft with each orbit and, over several months, circularized the spacecraft orbit. (This technique is sometimes called *aerobraking*.) We have sent orbiters to the Moon, to the planets Venus, Mars, Jupiter, and Saturn, and to two asteroids.

Landers and Probes

The most "up close and personal" study of other worlds comes from spacecraft that send probes into the atmospheres or landers to the surfaces. For example, in 1995, the *Galileo* spacecraft dropped a probe into Jupiter's atmosphere [**Section 11.1**]. The probe collected temperature, pressure, composition, and radiation measurements for about an hour as it descended; it was then destroyed by the heat and pressure of Jupiter's interior.

On planets with solid surfaces, a lander can offer close-up surface views, local weather monitoring, and the ability to carry out automated experiments. Landers have successfully reached the surfaces of the Moon, the planets Venus and Mars, and Saturn's moon Titan. Several of our Mars landers have included rovers to explore wider areas of the surface, including the *Spirit* and *Opportunity* rovers that landed on Mars in 2004, and the *Curiosity* rover that landed in August 2012. Because of its weight, *Curiosity*'s landing required a particularly spectacular feat of engineering (**FIGURE 7.15**). The spacecraft carrying the lander first used a parachute to slow it down in the Martian atmosphere and then fired rockets that slowed it to a halt about 7 meters above the surface. Finally, a "sky crane" lowered the rover to the surface.

Sample Return Missions Although probes and landers can carry out experiments on surface rock or atmospheric samples, the experiments must be designed in advance and must fit inside the spacecraft. One way around these limitations is to design missions in which samples from other worlds can be scooped up and returned to Earth for more detailed study. To date, the only sample return missions have been to the Moon (brought back by the *Apollo* astronauts and by robotic spacecraft sent in the 1970s by the then–Soviet Union) and to an asteroid (Japan's *Hayabusa* mission). Many scientists are working toward a sample return mission to Mars, and they hope to launch such a mission within the next decade or so. A slight variation on the theme of a sample return mission is the *Stardust* mission, which collected comet dust on a flyby and returned to Earth in 2006.

Combination Spacecraft Many missions combine more than one type of spacecraft. For example, the *Galileo* mission to Jupiter included an orbiter that studied Jupiter and its moons as well as the probe that entered Jupiter's atmosphere. The *Cassini* spacecraft included flybys of Venus, Earth, and Jupiter during its 7-year trip to Saturn. The spacecraft itself is an orbiter that is studying Saturn and its moons, but it also carried the *Huygens* probe, which descended through the atmosphere and landed on the surface of Saturn's moon Titan.

Exploration—Past, Present, and Future Over the past several decades, studies using both telescopes on Earth and robotic spacecraft have allowed us to learn the general characteristics of all the major planets and moons in our solar system as well as the general characteristics of asteroids and comets. Telescopes will continue to play an important role in future observations, but for detailed study we will probably continue to depend on spacecraft.

TABLE 7.3 lists some significant robotic missions of the past and present. The next few years promise many new discoveries as missions arrive at their destinations. Over the longer term, all the world's major space agencies have hopes of launching numerous and diverse missions to answer many specific questions about the nature of our solar system and its numerous worlds.

SEE IT FOR YOURSELF

It can be easy with a book full of planetary images to forget that these are real objects, many of which you can see in the night sky. Search the Web for "planets tonight" and then go out and see if you can find any of the planets in tonight's sky. Which planets can you see? Why can't you see the others?

1 Friction slows spacecraft as it enters Mars atmosphere.

2 Parachute slows spacecraft to about 350 km/hr.

3 Rockets slow spacecraft to halt; "sky crane" tether lowers rover to surface.

4 Tether released, the rocket heads off to crash a safe distance away.

As it flew overhead, the *Mars Reconnaissance Orbiter* took this photo of the spacecraft with its parachute deployed.

FIGURE 7.15 An artist's conception of the landing sequence that brought the *Curiosity* rover to Mars, along with a photo of its descent taken from orbit.

TABLE 7.3 Selected Robotic Missions to Other Worlds

Destination	Mission	Arrival Year	Agency*
Mercury	*MESSENGER* orbiter studies surface, atmosphere, and interior	2011	NASA
Venus	*Magellan* orbiter mapped surface with radar	1990	NASA
	Venus Express focuses on atmosphere studies	2006	ESA
Moon	The United States, China, Japan, India, and Russia all have current or planned robotic missions to explore the Moon	—	—
Mars	*Spirit* and *Opportunity* rovers learn about water on ancient Mars	2004	NASA
	Mars Express orbiter studies Mars's climate, geology, and polar caps	2004	ESA
	Mars Reconnaissance Orbiter takes very high-resolution photos	2006	NASA
	Phoenix lander studied soil near the north polar cap	2008	NASA
	Curiosity rover explores Gale Crater to understand prospects for life	2012	NASA
	MAVEN orbiter to study how Mars has lost atmospheric gas over time	2014	NASA
Asteroids	*Hayabusa* orbited and landed on asteroid Itokawa; returned sample to Earth in 2010	2005	JAXA
	Dawn visited asteroid Vesta and will visit the dwarf planet Ceres	2011/2015	NASA
Jovian planets	*Voyagers 1* and *2* visited all the jovian planets and left the solar system	1979	NASA
	Galileo's orbiter studied Jupiter and its moons; probe entered Jupiter's atmosphere	1995	NASA
	Cassini orbits Saturn; its *Huygens* probe (built by ESA) landed on Titan	2004	NASA
	Juno orbiter to study Jupiter's deep interior	2016	NASA
Pluto and comets	*New Horizons,* the first mission to Pluto, passed Jupiter in 2007	2015	NASA
	Stardust flew through the tail of Comet Wild 2; returned comet dust in 2006	2004	NASA
	Deep Impact observed its "lander" impacting Comet Tempel 1 at 10 km/s	2005	NASA
	Rosetta to orbit Comet Churyumov-Gerasimenko and release a lander	2014	ESA

*ESA = European Space Agency; JAXA = Japan Aerospace Exploration Agency.

The Big Picture

Putting Chapter 7 into Perspective

This chapter introduced the major features of our solar system and discussed some important patterns and trends that provide clues to its formation. As you continue your study of the solar system, keep in mind the following "big picture" ideas:

■ Our solar system is not a random collection of objects moving in random directions. Rather, it is highly organized, with clear patterns of motion and with most objects falling into just a few basic categories.

■ Each planet has its own unique and interesting features. Becoming familiar with the planets is an important first step in understanding the root causes of their similarities and differences.

■ Much of what we now know about the solar system comes from spacecraft exploration. Choosing the type of mission to send to a planet involves many considerations, from the scientific to the purely political. Many missions are currently under way, offering us hope of learning much more in the near future.

SUMMARY OF KEY CONCEPTS

7.1 STUDYING THE SOLAR SYSTEM

■ **What does the solar system look like?** The planets are

tiny compared to the distances between them. Our solar system consists of the Sun, the planets and their moons, and vast numbers of asteroids and comets. Each world has its own unique character, but there are many clear patterns among the worlds.

■ **What can we learn by comparing the planets to one another?** Comparative studies reveal the similarities and

differences that give clues to solar system formation and highlight the underlying processes that give each planet its unique appearance.

7.2 PATTERNS IN THE SOLAR SYSTEM

■ **What features of our solar system provide clues to how it formed?** Four major features provide clues: (1) The Sun, planets, and large moons generally rotate and orbit in a

very organized way. (2) The eight official planets divide clearly into two groups: terrestrial and jovian. (3) The solar system contains huge numbers of asteroids and comets. (4) There are some notable exceptions to these general patterns.

7.3 SPACECRAFT EXPLORATION OF THE SOLAR SYSTEM

■ **How do robotic spacecraft work?** Spacecraft can be categorized as flyby, orbiter, lander or probe, or sample return mission. In all cases, robotic spacecraft carry their own propul-

sion, power, and communication systems, and can operate under preprogrammed control or with updated instructions from ground controllers.

VISUAL SKILLS CHECK

Use the following questions to check your understanding of some of the many types of visual information used in astronomy. Answers are provided in Appendix J. For additional practice, try the Chapter 7 Visual Quiz at MasteringAstronomy®.

The plots above show the masses of the eight major planets on the vertical axis and their radii on the horizontal axis. The plot on the left shows the information on a linear scale, meaning that each tick mark indicates an increase by the same amount. The plot on the right shows the same information plotted on an exponential scale, meaning that each tick mark represents another factor-of-ten increase. Before proceeding, convince yourself that the points on each plot are the same.

1. Based on the information given in Table 7.1, which dots on each plot correspond to which planets? Which correspond to the terrestrial planets, and which to the jovian planets?
2. Notice how the eight planets group roughly into pairs on the graphs. Which planets are in each pair?
3. Which statement most accurately describes the relationship between the largest and smallest planets?
   a. The largest planet is 6000 times as wide (in diameter) and 30 times as massive as the smallest.
   b. The largest planet is 6000 times as wide (in diameter) and 6000 times as massive as the smallest.
   c. The largest planet is 30 times as wide (in diameter) and 30 times as massive as the smallest.
   d. The largest planet is 30 times as wide (in diameter) and 6000 times as massive as the smallest.

4. Answer each of the following questions to compare the two plots.
   a. Which plot, if either, best shows mass and radius information for all the planets?
   b. Which plot, if either, best emphasizes the differences between Jupiter and Saturn?
   c. Which plot, if either, could most easily be extended to show a planet with twice Jupiter's mass or radius?

For instructor-assigned homework go to MasteringAstronomy®.

REVIEW QUESTIONS

Short-Answer Questions Based on the Reading

1. What do we mean by *comparative planetology*? Does it apply only to planets?

2. What would the solar system look like to your naked eye if you could view it from beyond the orbit of Neptune?

3. Briefly describe the overall layout of the solar system as it is shown in Figure 7.1. What are the four major features of our solar system that provide clues to how it formed?

4. For each of the objects in the solar system tour (pages 194–203), describe at least two features that you find interesting.

5. Briefly describe the patterns of motion that we observe among the planets and moons of our solar system.

6. What are the basic differences between the *terrestrial* and *jovian* planets? Which planets fall into each group?

7. What do we mean by *hydrogen compounds*? In what kinds of planets or small bodies are they major ingredients?

8. What kind of object is Pluto? Explain.

9. What are *asteroids*? What are *comets*? Describe the basic differences between the two, and where we find them in our solar system.

10. What is the *Kuiper belt*? What is the *Oort cloud*? How do the orbits of comets differ in the two regions?

11. Describe at least two "exceptions to the rules" that we find in our solar system.

12. Describe and distinguish between space missions that are *flybys*, *orbiters*, *landers* or *probes*, and *sample return missions*. What are the advantages and disadvantages of each type?

TEST YOUR UNDERSTANDING

Does It Make Sense?

Decide whether the statement makes sense (or is clearly true) or does not make sense (or is clearly false). Explain clearly; not all of these have definitive answers, so your explanation is more important than your chosen answer.

13. Pluto orbits the Sun in a direction opposite that of all the other planets.

14. If Pluto were as large as the planet Mercury, we would classify it as a terrestrial planet.

15. Comets in the Kuiper belt and Oort cloud have long, beautiful tails that we can see when we look through telescopes.

16. Our Moon is about the same size as moons of the other terrestrial planets.

17. The mass of the Sun compared to the mass of all the planets combined is like the mass of an elephant compared to the mass of a cat.

18. On average, Venus is the hottest planet in the solar system—even hotter than Mercury.

19. The weather conditions on Mars today are much different than they were in the distant past.

20. Moons cannot have atmospheres, active volcanoes, or liquid water.

21. Saturn is the only planet in the solar system with rings.

22. We could probably learn more about Mars by sending a new spacecraft on a flyby than by any other method of studying the planet.

Quick Quiz

Choose the best answer to each of the following. Explain your reasoning with one or more complete sentences.

23. The largest terrestrial planet and jovian planet are, respectively, (a) Venus and Jupiter. (b) Earth and Jupiter. (c) Earth and Saturn.

24. Which terrestrial planets have had volcanic activity at some point in their histories? (a) only Earth (b) Earth and Mars (c) all of them

25. Large moons orbit their planets in the same direction as the planet rotates (a) rarely. (b) half of the time. (c) most of the time.

26. Which of the following three kinds of objects resides closest to the Sun on average? (a) comets (b) asteroids (c) jovian planets

27. What's unusual about our Moon? (a) It's the only moon that orbits a terrestrial planet. (b) It's by far the largest moon in the solar system. (c) It's surprisingly large relative to the planet it orbits.

28. Planetary orbits are (a) very eccentric (stretched-out) ellipses and in the same plane. (b) fairly circular and in the same plane. (c) fairly circular but oriented in every direction.

29. Which have more moons on average? (a) jovian planets (b) terrestrial planets (c) Terrestrial and jovian planets both have about the same number of moons.

30. The most abundant ingredient of the Sun and Jupiter is (a) ionized metal. (b) hydrogen. (c) ammonia.

31. Are there any exceptions to the rule that planets rotate with small axis tilts and in the same direction as they orbit the Sun? (a) No (b) Venus is the only exception. (c) Venus and Uranus are exceptions.

32. The *Cassini* spacecraft (a) flew past Pluto. (b) landed on Mars. (c) is orbiting Saturn.

PROCESS OF SCIENCE

Examining How Science Works

33. *Why Wait?* To explore a planet, we often first send a flyby, then an orbiter, then a probe or a lander. There's no doubt that probes and landers give the most close-up detail, so why don't we send this type of mission first? For the planet of your choice, based just on the information in this chapter, give an example of why such a strategy might cause a mission to provide incomplete information about the planet or to fail outright.

Group Work Exercise

34. *Comparative Planetology.* This chapter advocates learning about how planets work by comparing the planets in general, as opposed to studying the individual planets in great depth. Compare this approach with any previous study you might have made of the planets—for example, in grade school or an earth sciences class. Before you begin, assign the following roles to the people in your group: *Scribe* (takes notes on the group's activities), *Proposer* (proposes explanations to the group), *Skeptic* (points out weaknesses in proposed explanations), and *Moderator* (leads group discussion and makes sure everyone contributes). Make a list of advantages and any disadvantages of the comparative approach for planets. Then describe in a few sentences how a comparative approach might be used in a completely different field, such as another branch of science or social science.

INVESTIGATE FURTHER

In-Depth Questions to Increase Your Understanding

Short-Answer/Essay Questions

35. *Planetary Tour.* Based on the brief planetary tour in this chapter, which planet besides Earth do you think is the most interesting, and why? Defend your opinion clearly in two or three paragraphs.

36. *Patterns of Motion.* In one or two paragraphs, explain why the existence of orderly patterns of motion in our solar system suggests that the Sun and the planets all formed at one time from one cloud of gas, rather than as individual objects at different times.

37. *Solar System Trends.* Study the planetary data in Table 7.1 to do each of the following.
   a. Notice the relationship between distance from the Sun and surface temperature. Describe the trend, explain why it exists, and explain any notable exceptions to the trend. **b.** The text says that planets can be classified as either terrestrial or jovian, with Pluto fitting neither category. Describe in general how the columns for density, composition, and distance from the Sun support this classification. **c.** Describe the trend you see in orbital periods, and explain it in terms of Kepler's third law.

38. *Comparing Planetary Conditions.* Use the planetary data in Table 7.1 and Appendix E to answer each of the following.
   a. Which column of data would you use to find out which planet has the shortest days? Do you see any notable differences in the length of a day for the different types of planets? Explain. **b.** Which planets should not have seasons? Why? **c.** Which column tells you how much a planet's orbit deviates from a perfect circle? Based on that column, are there any planets for which you would expect the surface temperature to vary significantly over its orbit? Explain.

Quantitative Problems

Be sure to show all calculations clearly and state your final answers in complete sentences.

39. *Size Comparisons.* How many Earths could fit inside Jupiter (assuming you could fill up all the volume)? How many Jupiters could fit inside the Sun? The equation for the volume of a sphere is $V = \frac{4}{3}\pi r^3$.

40. *Asteroid Orbit.* Ceres, the largest asteroid, has an orbital semimajor axis of 2.77 AU. Use Kepler's third law to find its orbital period. Compare your answer with the value in Table 7.1, and name the planets that orbit just inside and outside Ceres's orbit.

41. *Density Classification.* Imagine that a new planet is discovered in our solar system with a mass of 5.97×10^{25} kilograms and a radius of 12,800 kilometers. Based just on its density, would we consider it the largest terrestrial planet or the smallest jovian planet? Explain. (*Hint:* Be careful to convert your density to units of grams per cubic centimeter in order to compare it with the terrestrial and jovian planet data in this chapter.)

42. *Escape Velocity.* By studying the data in Appendix E, briefly describe how escape velocity is related to mass and radius. Is the trend what you expect based on what you learned about escape velocity in Chapter 4?

43. *Comparative Weight.* Suppose you weigh 100 pounds. How much would you weigh on each of the other planets in our solar system? Assume you can stand either on the surface or in an airplane in the planet's atmosphere. (*Hint:* Recall from Chapter 4 that weight is mass times the acceleration of gravity. The surface gravity column in Appendix E tells you how the acceleration of gravity on other planets compares to Earth's.)

44. *Mission to Pluto.* The *New Horizons* spacecraft will take about 9 years to travel from Earth to Pluto. About how fast is it traveling on average? Assume that its trajectory is close to a straight line. Give your answer in AU per year and kilometers per hour.

45. *Planetary Parallax.* Suppose observers at Earth's North Pole and South Pole use a transit of the Sun by Venus to discover that the angular size of Earth as viewed from Venus would be 62.8 arcseconds. Earth's radius is 6378 kilometers. Estimate the distance between Venus and Earth in kilometers and AU. Compare your answer with information from the chapter.

Discussion Questions

46. *Where Would You Go?* Suppose you could visit any one of the planets or moons in our solar system for 1 week. Which object would you choose to visit, and why?

47. *Planetary Priorities.* Suppose you were in charge of developing and prioritizing future planetary missions for NASA. What would you choose as your first priority for a new mission, and why?

Web Projects

48. *Current Mission.* Visit the website for one of the current missions listed in Table 7.3. Write a one- to two-page summary of the mission's basic design, goals, and current status.

49. *Mars Missions.* Go to the home page for NASA's Mars Exploration Program. Write a one- to two-page summary of the plans for future exploration of Mars.

8 FORMATION OF THE SOLAR SYSTEM

The evolution of the world may be compared to a display of fireworks that has just ended: some few red wisps, ashes and smoke. Standing on a cooled cinder, we see the slow fading of the suns, and we try to recall the vanished brilliance of the origin of the worlds.

—G. Lemaître (1894–1966),
astronomer and Catholic priest

How did Earth come to be? How old is it? Is it unique? Our ancestors could do little more than guess at the answers to these questions, but today we are able to address them scientifically. As we'll discuss in this chapter, careful study of the major features of our solar system has enabled scientists to put together a detailed theory of how Earth and our solar system were born.

Our theory of solar system formation is important not only because it helps us understand our cosmic origins, but also because it holds the key to understanding the nature of planets. If the planets in our solar system all formed together, then their differences must be attributable to physical processes that occurred during the birth and subsequent evolution of the solar system. Our study of the solar system's birth will therefore form the basis for our comparative study of the planets in subsequent chapters. It will also help us extend these ideas to the myriad of other planetary systems now known to exist, a topic we will study in Chapter 13.

8.1 THE SEARCH FOR ORIGINS

The development of any scientific theory is an interplay between observations and attempts to explain those observations [**Section 3.4**]. Hypotheses that seem to make sense at one time might later be dismissed because they fail to explain new data. For example, ancient Greek ideas about Earth's origins probably seemed quite reasonable when people assumed that Earth was the center of the universe, but they no longer made sense after Kepler and Galileo proved that Earth is a planet orbiting the Sun.

By the end of the 17th century, the Copernican revolution [**Section 3.3**] and Newton's discovery of the universal law of gravitation [**Section 4.4**] had given us a basic understanding of the layout and motion of the planets and moons in our solar system. It was only natural that scientists would begin to speculate about how this system came to be.

How did we arrive at a theory of solar system formation?

Recall that a hypothesis can rise to the status of a scientific theory only if it offers a detailed physical model that explains a broad range of observed facts. For our solar system, the most important facts to explain are the four major features discussed in Chapter 7. If a hypothesis fails to explain even one of the four features, then it cannot be correct. If it successfully explains all four, then we might reasonably assume it is on the right track. We therefore arrive at the following four criteria for the success of a solar system formation theory:

1. It must explain the patterns of motion discussed in Chapter 7.

2. It must explain why planets fall into two major categories: small, rocky *terrestrial* planets near the Sun and large, hydrogen-rich *jovian* planets farther out.

3. It must explain the existence of huge numbers of asteroids and comets and why these objects reside primarily in the regions we call the *asteroid belt*, the *Kuiper belt*, and the *Oort cloud*.

4. It must explain the general patterns while at the same time making allowances for exceptions to the general rules, such as the odd axis tilt of Uranus and the existence of Earth's large Moon.

From Hypothesis to Theory Although it took time to learn all four of the major features, scientists began speculating about how our solar system formed as soon as the basic patterns of motion were known. We generally credit two 18th-century scientists with proposing the hypothesis that ultimately blossomed into our modern scientific theory of the origin of the solar system. Around 1755, German philosopher Immanuel Kant proposed that our solar system formed from the gravitational collapse of an interstellar cloud of gas. About 40 years later, French mathematician Pierre-Simon Laplace put forth the same idea independently. Because an interstellar cloud is usually called a *nebula* (Latin for "cloud"), their idea became known as the *nebular hypothesis.*

The nebular hypothesis remained popular throughout the 19th century. By the early 20th century, however, scientists had found a few aspects of our solar system that the nebular hypothesis did not seem to explain well—at least in its original form as described by Kant and Laplace. While some scientists sought to modify the nebular hypothesis, others looked for different ways to explain how the solar system might have formed.

During much of the first half of the 20th century, the nebular hypothesis faced stiff competition from a hypothesis proposing that the planets represent debris from a near-collision between the Sun and another star. According to this *close encounter hypothesis,* the planets formed from blobs of gas that had been gravitationally pulled out of the Sun during the near-collision.

Today, the close encounter hypothesis has been discarded. It began to lose favor when calculations showed that it could not account for either the observed orbital motions of the planets or the neat division of the planets into two major categories (terrestrial and jovian). Moreover, the close encounter hypothesis required a highly improbable event: a near-collision between our Sun and another star. Given the vast separation between star systems in our region of the galaxy, the chance of such an encounter is so small that it would be difficult to imagine it happening even once in order to form our solar system. It certainly could not account for the many other planetary systems that we have discovered in recent years.

While the close encounter hypothesis was losing favor, new discoveries about the physics of planet formation led to modifications of the nebular hypothesis. Using more sophisticated models of the processes that occur in a collapsing cloud of gas, scientists found that the nebular hypothesis offered natural explanations for all four general features of our solar system. By the latter decades of the 20th century, so much evidence had accumulated in favor of the nebular hypothesis that it achieved the status of a scientific *theory* [**Section 3.4**]—the **nebular theory** of our solar system's birth.

Putting the Theory to the Test Recall that in science, a theory is never really complete and we must put it to continual tests and modify it as necessary. In the case of a theory that claims to explain the origin of *our* solar system, one critical set of tests involves its ability to predict and explain the characteristics of other solar systems. The nebular theory has clearly passed the most important of these tests: Because it claims that planets are a natural outgrowth of the star formation process, it predicts that other planetary systems ought to be common, a prediction that has now been borne out by observations. Other observations have presented greater challenges; many of the recently discovered planetary systems are organized in ways somewhat different from our own. Nevertheless, as we'll discuss in Chapter 13, scientists have not found any major flaws in the nebular theory, and with relatively minor modifications it seems capable of explaining the diversity of planetary systems that we observe. As a result, the nebular theory today stands on stronger ground than ever. We'll therefore devote the rest of this chapter to understanding the basic theory and how it explains the major features of our solar system.

Where did the solar system come from?

The nebular theory begins with the idea that our solar system was born from the gravitational collapse of an interstellar cloud of gas, called the **solar nebula**, that collapsed under its own gravity. As we'll discuss in more detail in the next section, this cloud gave birth to the Sun at its center and the planets in a spinning disk that formed around the young Sun.

Where did the gas that made up the solar nebula come from? According to modern science, it was the product of billions of years of galactic recycling that occurred before the Sun and planets were born. Recall that the universe as a whole is thought to have been born in the Big Bang [**Section 1.2**], which essentially produced only two chemical elements: hydrogen and helium. Heavier elements were produced later by massive stars, and released into space when the stars died. The heavy elements then mixed with other interstellar gas that formed new generations of stars (**FIGURE 8.1**).

Although this process of creating heavy elements in stars and recycling them within the galaxy has probably gone on

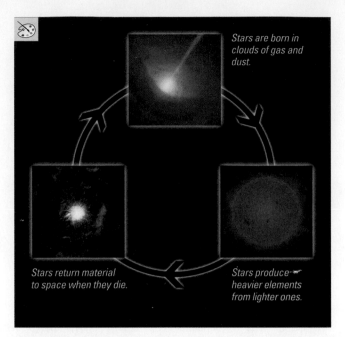

Stars are born in clouds of gas and dust.

Stars return material to space when they die.

Stars produce heavier elements from lighter ones.

FIGURE 8.1 This figure, which is a portion of Figure 1.10, summarizes the galactic recycling process.

for most of the 14-billion-year history of our universe, only a small fraction of the original hydrogen and helium has been converted into heavy elements. By studying the composition of the Sun, other stars of the same age, and interstellar gas clouds, we have learned that the gas that made up the solar nebula contained (by mass) about 98% hydrogen and helium and 2% all other elements combined. The Sun and planets were born from this gas, and Earth and the other terrestrial worlds were made primarily from the heavier elements mixed within it. As we discussed in Chapter 1, we are "star stuff" because we and our planet are made of elements forged in stars that lived and died long ago.

THINK ABOUT IT

Could a solar system like ours have formed with the first generation of stars after the Big Bang? Explain.

Strong observational evidence supports this scenario. Spectroscopy shows that old stars have a smaller proportion of heavy elements than younger ones, just as we would expect if they were born at a time before many heavy elements had been manufactured. Moreover, visible and infrared telescopes allow us to study stars that are in the process of formation today. **FIGURE 8.2** shows the Orion Nebula, in which many stars are in various stages of formation. Just as our scenario predicts, the forming stars are embedded within gas clouds like our solar nebula, and the characteristics of these clouds match what we expect if they are collapsing due to gravity.

FIGURE 8.2 **interactive figure** The Orion Nebula, an interstellar cloud in which new star systems are forming. Over the next few million years, thousands of stars will be born in this gas cloud. Some of these stars may end up with their own planetary systems. This image combines an infrared view from the Spitzer Space Telescope and visible data from the Hubble Space Telescope.

MA Formation of the Solar System Tutorial, Lessons 1–2

8.2 EXPLAINING THE MAJOR FEATURES OF THE SOLAR SYSTEM

We are now ready to look at the nebular theory in somewhat more detail. In the process, we'll see how it successfully accounts for all four major features of our solar system.

What caused the orderly patterns of motion?

The solar nebula probably began as a large, roughly spherical cloud of very cold and very low-density gas. Initially, this gas was probably so spread out—perhaps over a region a few light-years in diameter—that gravity alone may not have been strong enough to pull it together and start its collapse. Instead, the collapse may have been triggered by a cataclysmic event, such as the impact of a shock wave from the explosion of a nearby star (a supernova).

Once the collapse started, gravity enabled it to continue. Remember that the strength of gravity follows an inverse square law with distance [**Section 4.4**]. The mass of the cloud remained the same as it shrank, so the strength of gravity increased as the diameter of the cloud decreased.

Because gravity pulls inward in all directions, you might at first guess that the solar nebula would have remained spherical as it shrank. Indeed, the idea that gravity pulls in all directions explains why the Sun and the planets are spherical. However, we must also consider other physical laws that apply to a collapsing gas cloud in order to understand how orderly motions arose in the solar nebula.

Heating, Spinning, and Flattening As the solar nebula shrank in size, three important processes altered its density, temperature, and shape, changing it from a large, diffuse (spread-out) cloud to a much smaller spinning disk (**FIGURE 8.3**):

- **Heating.** The temperature of the solar nebula increased as it collapsed. Such heating represents energy conservation in action [**Section 4.3**]. As the cloud shrank, its gravitational potential energy was converted to the kinetic energy of individual gas particles falling inward. These particles crashed into one another, converting the kinetic energy of their inward fall to the random motions of thermal energy (see Figure 4.14b). The Sun formed in the center, where temperatures and densities were highest.

- **Spinning.** Like an ice skater pulling in her arms as she spins, the solar nebula rotated faster and faster as it shrank in radius. This increase in rotation rate represents conservation of angular momentum in action [**Section 4.3**]. The rotation of the cloud may have been imperceptibly slow before its collapse began, but the cloud's shrinkage made fast rotation inevitable. The rapid rotation helped ensure that not all the material in the solar nebula collapsed into the center: The greater the angular momentum of a rotating cloud, the more spread out it will be.

- **Flattening.** The solar nebula flattened into a disk. This flattening is a natural consequence of collisions between particles in a spinning cloud. A cloud may start with any size or shape, and different clumps of gas within the cloud may be moving in random directions at random speeds. These clumps collide and merge as the cloud collapses, and each new clump has the average velocity of the clumps that formed it. The random motions of the original cloud therefore become more orderly as the cloud collapses, changing the cloud's original lumpy shape into a rotating, flattened disk. Similarly, collisions between clumps of material in highly elliptical orbits reduce their eccentricities, making the orbits more circular.

The formation of the spinning disk explains the orderly motions of our solar system today. The planets all orbit the Sun in nearly the same plane because they formed in the flat disk. The direction in which the disk was spinning became the direction of the Sun's rotation and the orbits of the planets. Computer models show that planets would have tended to rotate in this same direction as they formed—which is why most planets rotate the same way—though the small sizes of planets compared to the entire disk allowed some exceptions to arise. The fact that collisions in the disk tended to make orbits more circular explains why the planets in our solar system have nearly circular orbits.

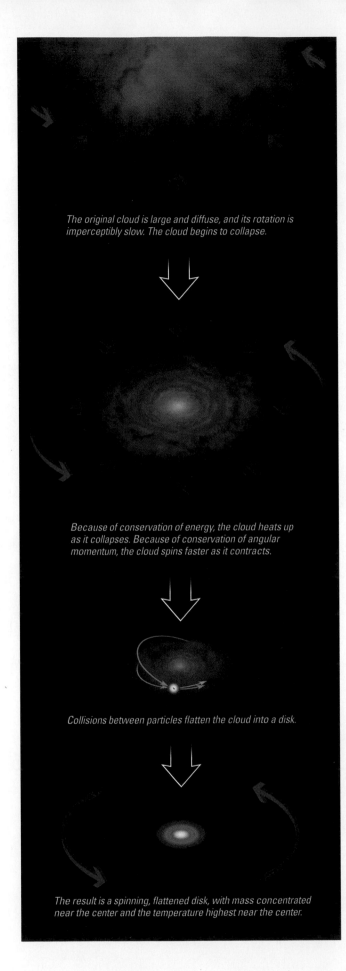

The original cloud is large and diffuse, and its rotation is imperceptibly slow. The cloud begins to collapse.

Because of conservation of energy, the cloud heats up as it collapses. Because of conservation of angular momentum, the cloud spins faster as it contracts.

Collisions between particles flatten the cloud into a disk.

The result is a spinning, flattened disk, with mass concentrated near the center and the temperature highest near the center.

FIGURE 8.3 interactive figure This sequence of illustrations shows how the gravitational collapse of a large cloud of gas causes it to become a spinning disk of matter. The hot, dense central bulge becomes a star, while planets can form in the surrounding disk.

SEE IT FOR YOURSELF

You can demonstrate the development of orderly motion by sprinkling pepper into a bowl of water and stirring it quickly in random directions. The water molecules constantly collide with one another, so the motion of the pepper grains will tend to settle down into a slow rotation representing the average of the original, random velocities. Try the experiment several times, stirring the water differently each time. Do the random motions ever cancel out exactly, resulting in no rotation at all? Describe what occurs, and explain how it is similar to what took place in the solar nebula.

Testing the Model Because the same processes should affect other collapsing gas clouds, we can test our model by searching for disks around other forming stars. Observational evidence does indeed support our model of spinning, heating, and flattening.

The heating that occurs in a collapsing cloud of gas means the gas should emit thermal radiation [**Section 5.4**], primarily in the infrared. We've detected infrared radiation from many nebulae where star systems appear to be forming. More direct evidence comes from flattened, spinning disks around other stars (**FIGURE 8.4**), some of which appear to be ejecting jets of material perpendicular to their disks [**Section 16.2**]. These jets are thought to result from the flow of material from the disk onto the forming star, and they may influence the solar system formation processes.

Other support for the model comes from computer simulations of the formation process. A simulation begins with a set of data representing the conditions we observe in interstellar clouds. Then, with the aid of a computer, we apply the laws of physics to predict the changes that should occur over time. Computer simulations successfully reproduce most of the general characteristics of motion in our solar system, suggesting that the nebular theory is on the right track.

Additional evidence that our ideas about the formation of flattened disks are correct comes from many other structures in the universe. We expect flattening to occur anywhere orbiting particles can collide, which explains why we find so many cases of flat disks, including the disks of spiral galaxies like the Milky Way, the disks of planetary rings, and the *accretion disks* that surround neutron stars and black holes in close binary star systems [**Section 18.3**].

Why are there two major types of planets?

The planets began to form after the solar nebula had collapsed into a flattened disk of perhaps 200 AU in diameter (about twice the present-day diameter of Pluto's orbit). The churning and mixing of gas in the solar nebula should have ensured that the nebula had the same composition throughout: 98% hydrogen and helium plus 2% heavier elements.

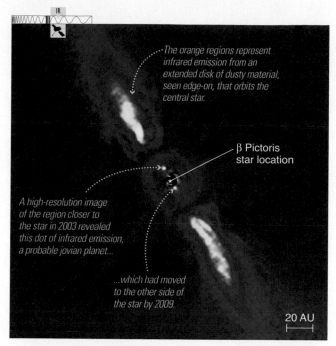

The orange regions represent infrared emission from an extended disk of dusty material, seen edge-on, that orbits the central star.

β Pictoris star location

A high-resolution image of the region closer to the star in 2003 revealed this dot of infrared emission, a probable jovian planet...

...which had moved to the other side of the star by 2009.

20 AU

a This infrared image composite from the European Southern Observatory Very Large Telescope shows a large debris disk orbiting the star Beta Pictoris and a probable jovian planet that has formed in the disk. Images were taken with the star itself blocked; the star's position has been added digitally.

nearby companion stars

Wisps and clumps may show the influence of the neighboring stars or unseen planets.

orbiting disk of material

central star blocked out

200 AU

b This Hubble Space Telescope image shows a disk around the star HD141569A. The colors are not real; a black-and-white image has been tinted red to bring out faint detail.

FIGURE 8.4 These images show flattened, spinning disks of material around other stars.

How, then, did the terrestrial planets end up so different in composition from the jovian planets? The key clue comes from their locations: Terrestrial planets formed in the warm, inner regions of the swirling disk, while jovian planets formed in the colder, outer regions.

Condensation: Sowing the Seeds of Planets

In the center of the collapsing solar nebula, gravity drew together enough material to form the Sun. In the surrounding disk, however, the gaseous material was too spread out for gravity alone to clump it together. Instead, material had to begin clumping in some other way and then grow in size until gravity could start pulling it together into planets. In essence, planet formation required the presence of "seeds"—solid bits of matter from which gravity could ultimately build planets.

The basic process of seed formation was probably much like the formation of snowflakes in clouds on Earth: When the temperature is low enough, some atoms or molecules in a gas may bond and solidify. The general process in which solid (or liquid) particles form in a gas is called **condensation**—we say that the particles *condense* out of the gas. (Pressures in the solar nebula were generally too low to allow the condensation of liquid droplets.) These particles start out microscopic in size, but they can grow larger with time.

Different materials condense at different temperatures. As summarized in **TABLE 8.1**, the ingredients of the solar nebula fell into four major categories:

- **Hydrogen and helium gas (98% of the solar nebula).** These gases never condense in interstellar space.

TABLE 8.1 Materials in the Solar Nebula

A summary of the four types of materials present in the solar nebula. The squares represent the relative proportions of each type (by mass).

	Examples	Typical Condensation Temperature	Relative Abundance (by mass)
Hydrogen and Helium Gas	hydrogen, helium	do not condense in nebula	98%
Hydrogen Compounds	water (H_2O), methane (CH_4), ammonia (NH_3)	<150 K	1.4%
Rock	various minerals	500–1300 K	0.4%
Metal	iron, nickel, aluminum	1000–1600 K	0.2%

- **Hydrogen compounds (1.4% of the solar nebula).** Materials such as water (H_2O), methane (CH_4), and ammonia (NH_3) can solidify into **ices** at low temperatures (below about 150 K under the low pressure of the solar nebula).

- **Rock (0.4% of the solar nebula).** Rocky material is gaseous at very high temperatures, but condenses into solid bits of mineral at temperatures between about 500 K and 1300 K, depending on the type of rock. (A *mineral* is a type of rock with a particular chemical composition and structure.)

- **Metal (0.2% of the solar nebula).** Metals such as iron, nickel, and aluminum are also gaseous at very high temperatures, but condense into solid form at higher temperatures than rock—typically in the range of 1000 K to 1600 K.

Because hydrogen and helium gas made up 98% of the solar nebula's mass and did not condense, the vast majority of the nebula remained gaseous at all times. However, other materials could condense wherever the temperature allowed (**FIGURE 8.5**). Close to the forming Sun, where the temperature was above 1600 K, it was too hot for any material to condense. Near what is now Mercury's orbit, the temperature was low enough for metals and some types of rock to condense into tiny, solid particles, but other types of rock and all the hydrogen compounds remained gaseous. More types of rock could condense, along with the metals, at the distances from the Sun where Venus, Earth, and Mars would form. In the region where the asteroid belt would eventually be located, temperatures were low enough to allow dark, carbon-rich minerals to condense, along with minerals containing small amounts of water. Hydrogen compounds could condense into ices only beyond the **frost line**—the distance at which it was cold enough for ices to condense—which lay between the present-day orbits of Mars and Jupiter.

> ### THINK ABOUT IT
>
> Consider a region of the solar nebula in which the temperature was about 1300 K. Based on the data in Table 8.1, what fraction of the material in this region was gaseous? What were the solid particles in this region made of? Answer the same questions for a region with a temperature of 100 K. Would the 100 K region be closer to or farther from the Sun? Explain.

The frost line marked the key transition between the warm inner regions of the solar system where terrestrial planets formed and the cool outer regions where jovian planets formed. Inside the frost line, only metal and rock could condense into solid "seeds," which is why the terrestrial planets ended up being made of metal and rock. Beyond the frost line, where it was cold enough for hydrogen compounds to condense into ices, the solid seeds were built of ice along with metal and rock. Moreover, because hydrogen compounds were nearly three times as abundant in the nebula as metal

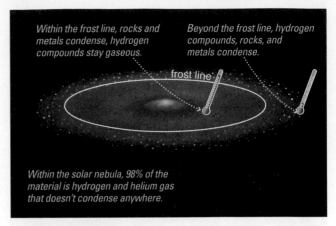

FIGURE 8.5 interactive figure Temperature differences in the solar nebula led to different kinds of condensed materials at different distances from the Sun, sowing the seeds for two kinds of planets.

and rock combined (see Table 8.1), the total amount of solid material was far greater beyond the frost line than within it. The stage was set for the birth of two types of planets: planets born from seeds of metal and rock in the inner solar system and planets born from seeds of ice (as well as metal and rock) in the outer solar system.

Building the Terrestrial Planets From this point, the story of the inner solar system seems fairly clear. The solid seeds of metal and rock in the inner solar system ultimately grew into the terrestrial planets we see today, but these planets ended up relatively small in size because rock and metal made up such a small amount of the material in the solar nebula.

The process by which small "seeds" grew into planets is called **accretion** (**FIGURE 8.6**). Accretion began with the microscopic solid particles that condensed from the gas of the solar nebula. These particles orbited the forming Sun with the same orderly, circular paths as the gas from which they condensed. Individual particles therefore moved at nearly the same speed as neighboring particles, so "collisions" were more like gentle touches. Although the particles were far too small to attract each other gravitationally at this point, they were able to stick together through electrostatic forces—the same "static electricity" that makes hair stick to a comb. Small particles thereby began to combine into larger ones. As the particles grew in mass, gravity began to aid the process of

COMMON MISCONCEPTIONS

Solar Gravity and the Density of Planets

Some people guess that it was the Sun's gravity that pulled the dense rocky and metallic materials to the inner part of the solar nebula, or that gases escaped from the inner nebula because gravity couldn't hold them. But this is not the case—all the ingredients were orbiting the Sun together under the influence of the Sun's gravity. The orbit of a particle or a planet does *not* depend on its size or density, so the Sun's gravity cannot be the cause of the different kinds of planets. Rather, the different temperatures in the solar nebula are the cause.

Early in the acccretion process, there are many relatively large planetesimals on crisscrossing orbits.

As time passes, a few planetesimals grow larger by accreting smaller ones, while others shatter in collisions.

Ultimately, only the largest planetesimals avoid shattering and grow into full-fledged planets.

Not to scale!

FIGURE 8.6 These diagrams show how planetesimals gradually accrete into terrestrial planets.

their sticking together, accelerating their growth into boulders large enough to count as **planetesimals**, which means "pieces of planets."

The planetesimals grew rapidly at first. As they grew larger, they had both more surface area to make contact with other planetesimals and more gravity to attract them. Some planetesimals probably grew to hundreds of kilometers in size in only a few million years—a long time in human terms, but only about *one-thousandth* of the present age of the solar system. However, once the planetesimals reached these relatively large sizes, further growth became more difficult.

Gravitational encounters [**Section 4.5**] between planetesimals tended to alter their orbits, particularly those of the smaller planetesimals. With different orbits crossing each other, collisions between planetesimals tended to occur at higher speeds and hence became more destructive. Such collisions tended to shatter planetesimals rather than help them grow. Only the largest planetesimals avoided being shattered and could grow into terrestrial planets.

Theoretical evidence in support of this model comes from computer simulations of the accretion process. Observational evidence comes from **meteorites**, rocks that have fallen to Earth from space. Meteorites that appear to be surviving fragments from the period of condensation contain metallic grains embedded in rocky minerals (**FIGURE 8.7**), just as we would expect if metal and rock condensed in the inner solar system. Meteorites thought to come from the outskirts of the asteroid belt contain abundant carbon-rich materials, and some contain water—again as we would expect for planetesimals that formed in that region.

Making the Jovian Planets Accretion should have occurred similarly in the outer solar system, but condensation of ices meant both that there was more solid material and that this material contained ice in addition to metal and rock. The solid objects that reside in the outer solar system today, such as comets and the moons of the jovian planets,

still show this ice-rich composition. However, the growth of icy planetesimals cannot be the whole story of jovian planet formation, because the jovian planets contain large amounts of hydrogen and helium gas.

The leading model for jovian planet formation holds that these planets formed as gravity drew gas around ice-rich planetesimals much more massive than Earth. Because of their large masses, these planetesimals had gravity strong enough to capture some of the hydrogen and helium gas that made up the vast majority of the surrounding solar nebula. This added gas made their gravity even stronger, allowing them to capture even more gas. Ultimately, the jovian planets accreted so much gas that they bore little resemblance to the icy seeds from which they grew.

FIGURE 8.7 Shiny flakes of metal are clearly visible in this slice through a meteorite (a few centimeters across), mixed in among the rocky material. Such metallic flakes are just what we would expect to find if condensation really occurred in the solar nebula as described by the nebular theory.

As planetesimals accrete to form terrestrial planets in the inner solar system...

×5

...farther out, huge icy planetesimals gather hydrogen and helium to form jovian planets...

Sun

spinning disk around forming jovian planet

...while the disk forms large moons by condensation and accretion, and captures small moons.

2 AU

FIGURE 8.8 The young jovian planets were surrounded by disks of gas, much like the disk of the entire solar nebula but smaller in size. According to the leading model, the planets grew as large, ice-rich planetesimals captured hydrogen and helium gas from the solar nebula. This painting shows the gas and planetesimals surrounding one jovian planet in the larger solar nebula.

This model also explains most of the large moons of the jovian planets. The same processes of heating, spinning, and flattening that made the disk of the solar nebula should also have affected the gas drawn by gravity to the young jovian planets. Each jovian planet came to be surrounded by its own disk of gas, spinning in the same direction as the planet rotated (**FIGURE 8.8**). Moons that accreted from ice-rich planetesimals within these disks ended up with nearly circular orbits going in the same direction as their planet's rotation and lying close to their planet's equatorial plane.

Clearing the Nebula The vast majority of the hydrogen and helium gas in the solar nebula never became part of any planet. So what happened to it? Apparently, it was cleared away by a combination of intense radiation from the young Sun and the **solar wind**—a stream of charged particles (such as protons and electrons) continually blown outward in all directions from the Sun. Although the solar wind is fairly weak today, observations show that stars tend to have much stronger winds when they are young. The young Sun therefore also should have had a strong solar wind—strong enough to have swept huge quantities of gas out of the solar system.

The clearing of the gas sealed the compositional fate of the planets. If the gas had remained longer, it might have continued to cool until hydrogen compounds could have condensed into ices even in the inner solar system. In that case, the terrestrial planets might have accreted abundant ice, and perhaps hydrogen and helium gas as well, changing their basic nature. At the other extreme, if the gas had been blown out earlier, the raw materials of the planets might have been swept away before the planets could fully form. Although these extreme scenarios did not occur in our solar system, they may sometimes occur around other stars. Planet formation may also sometimes be interrupted when radiation from hot neighboring stars drives away material in a solar nebula.

The clearing of the nebula also helps explain what was once considered a surprising aspect of the Sun's rotation. According to the law of conservation of angular momentum, the spinning disk of the solar nebula should have spun fastest near its center, where most of the mass became concentrated. We therefore expect the young Sun to have rotated very fast. But the Sun rotates quite slowly today, with each full rotation taking about a month. If the young Sun really did rotate fast, as the nebular theory seems to demand, how did its rotation slow down?

Angular momentum cannot simply disappear, but it can be transferred from one object to another—and then the other object can be pushed away. A spinning skater can slow her spin by grabbing her partner and then pushing him away. The Sun probably lost angular momentum in a similar way.

The young Sun's rapid rotation would have generated a magnetic field far stronger than that of the Sun today, which in turn would have led to the strong solar wind and to strong surface activity (such as large sunspots and frequent solar flares [**Section 14.3**]) that would have caused the young Sun to emit intense ultraviolet and X-ray radiation. This high-energy radiation ionized gas in the solar nebula, creating many charged particles, while the magnetic field swept through the nebula with the Sun's rapid rotation. Because charged particles and magnetic fields tend to move together, the magnetic field dragged the charged particles along faster than the rest of the nebula, effectively slowing the Sun's rotation and transferring some of the Sun's angular momentum to the nebula. When the nebula was cleared into interstellar space, the gas carried the angular momentum away, leaving the Sun with the greatly diminished angular momentum and slow rotation that we see today (**FIGURE 8.9**).

Although we cannot prove that the young Sun really did lose angular momentum in this way, support for the idea comes

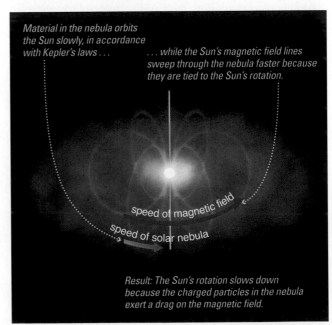

Material in the nebula orbits the Sun slowly, in accordance with Kepler's laws . . .

. . . while the Sun's magnetic field lines sweep through the nebula faster because they are tied to the Sun's rotation.

speed of magnetic field

speed of solar nebula

Result: The Sun's rotation slows down because the charged particles in the nebula exert a drag on the magnetic field.

FIGURE 8.9 The young Sun should have rotated rapidly, but today the Sun rotates quite slowly. As shown in this diagram, the Sun probably lost angular momentum because of drag between slow-moving charged particles in the solar nebula and the Sun's rotating magnetic field (represented by purple loops).

from observations of other stars. When we look at young stars in interstellar clouds, we find that nearly all of them rotate rapidly and have strong magnetic fields and strong winds [**Section 16.2**]. Older stars almost invariably rotate slowly, like our Sun. This suggests that nearly all stars have their original rapid rotations slowed by transferring angular momentum to charged particles in their disks—particles that are later swept away—just as we suspect happened with the Sun.

Where did asteroids and comets come from?

The process of planet formation also explains the origin of the many asteroids and comets that populate our solar system (including those large enough to qualify as dwarf planets): They are "leftovers" from the era of planet formation. Asteroids are the rocky leftover planetesimals of the inner solar system, while comets are the icy leftover planetesimals of the outer solar system. We'll see in Chapter 12 why most asteroids ended up grouped in the asteroid belt while most comets ended up split between two regions (the Kuiper belt and the Oort cloud).

Evidence that asteroids and comets are leftover planetesimals comes from analysis of meteorites, spacecraft visits to comets and asteroids, and theoretical models of solar system formation. In fact, the nebular theory allowed scientists to make predictions about the locations of comets that weren't verified until decades later, when we discovered large comets orbiting in the vicinity of Neptune and Pluto.

The asteroids and comets that exist today probably represent only a small fraction of the leftover planetesimals that roamed the young solar system. The rest are now gone. Some of these "lost" planetesimals may have been flung into deep space by gravitational encounters, but many others must have collided with the planets. When impacts occur on solid worlds, they leave behind **impact craters** as scars. Impacts have thereby transformed planetary landscapes and, in the case of Earth, have altered the course of evolution. For example, an impact is thought to have been responsible for the death of the dinosaurs [**Section 12.4**].

Although impacts occasionally still occur, the vast majority of these collisions occurred in the first few hundred million years of our solar system's history, during the period we call the **heavy bombardment**. Every world in our solar system must have been pelted by impacts during the heavy bombardment (**FIGURE 8.10**), and most of the craters we see on the Moon and other worlds date from this period. These impacts did more than just batter the planets. They also brought materials from other regions of the solar system—a fact that is critical to our existence on Earth today.

The metal and rock planetesimals that built the terrestrial planets probably contained no water or other hydrogen compounds at all, because it was too hot for these compounds to condense in our region of the solar nebula. How, then, did Earth come to have the water that makes up our oceans and the gases that first formed our atmosphere? The likely answer is that water, along with other hydrogen compounds, was brought to Earth and other terrestrial planets by the impacts of water-bearing planetesimals that formed farther from

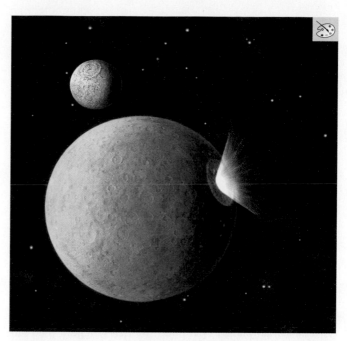

FIGURE 8.10 Around 4 billion years ago, Earth, its Moon, and the other planets were heavily bombarded by leftover planetesimals. This painting shows the young Earth and Moon, with an impact in progress on Earth.

the Sun. We don't yet know whether these planetesimals came primarily from the outer asteroid belt, where rocky planetesimals contained small amounts of water and other hydrogen compounds, or whether they were comets containing huge amounts of ice. Either way, the water we drink and the air we breathe probably originated beyond the orbit of Mars.

How do we explain "exceptions to the rules"?

We have now explained all the major features of our solar system except the exceptions to the rules, including our surprisingly large Moon. Today, we think that most of these exceptions arose from collisions or close gravitational encounters.

Captured Moons We have explained the orbits of most large jovian planet moons by their formation in a disk that swirled around the forming planet. But how do we explain moons with less orderly orbits, such as those that go in the "wrong" direction (opposite their planet's rotation) or that have large inclinations to their planet's equator? These moons are probably leftover planetesimals that originally orbited the Sun but were then captured into planetary orbit.

It's not easy for a planet to capture a moon. An object cannot switch from an unbound orbit (for example, an asteroid whizzing by Jupiter) to a bound orbit (for example, a moon orbiting Jupiter) unless it somehow loses orbital energy [**Section 4.5**]. For the jovian planets, captures probably occurred when passing planetesimals lost energy to drag in the extended and relatively dense gas that surrounded these planets as they formed. The planetesimals would have been slowed by friction with the gas, just as artificial satellites are slowed by drag in encounters with Earth's atmosphere. If friction reduced its orbital energy enough, a planetesimal

a Phobos   b Deimos

FIGURE 8.11 The two moons of Mars are probably captured asteroids. Phobos is only about 13 kilometers across, and Deimos is only about 8 kilometers across—making each of these two moons small enough to fit within the boundaries of a typical large city. (Images from the *Mars Reconnaissance Orbiter*.)

could have become an orbiting moon. Because of the random nature of the capture process, captured moons would not necessarily orbit in the same direction as their planet or in its equatorial plane. Computer models suggest that this capture process would have worked only on objects up to a few kilometers in size. Most of the small moons of the jovian planets are a few kilometers across, supporting the idea that they were captured in this way. Mars may have similarly captured its two small moons, Phobos and Deimos, at a time

when the planet had a much more extended atmosphere than it does today (**FIGURE 8.11**).

The Giant Impact Formation of Our Moon Capture processes cannot explain our own Moon, because it is much too large to have been captured by a small planet like Earth. We can also rule out the possibility that our Moon formed simultaneously with Earth, because if both had formed together, they would have accreted from planetesimals of the same type and would therefore have approximately the same composition and density. But this is not the case: The Moon's density is considerably lower than Earth's, indicating that it has a very different average composition. So how did we get our Moon? Today, the leading hypothesis suggests that it formed as the result of a **giant impact** between Earth and a huge planetesimal.

According to models, a few leftover planetesimals may have been as large as Mars. If one of these Mars-size objects struck a young planet, the blow might have tilted the planet's axis, changed the planet's rotation rate, or completely shattered the planet. The giant impact hypothesis holds that a Mars-size object hit Earth at a speed and angle that blasted Earth's outer layers into space. According to computer simulations, this material could have collected into orbit around our planet, and accretion within this ring of debris could have formed the Moon (**FIGURE 8.12**).

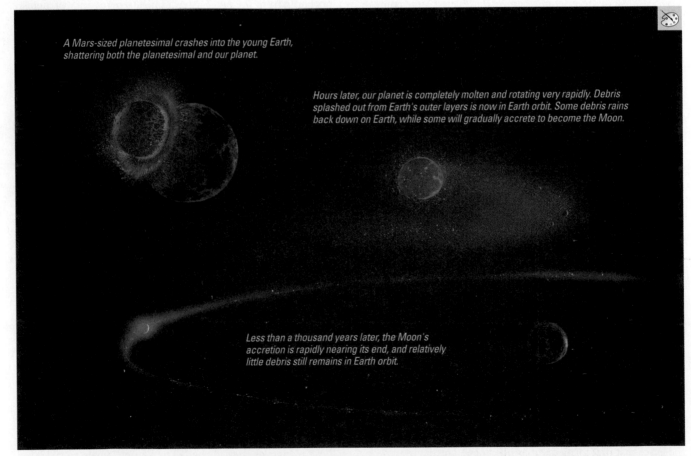

A Mars-sized planetesimal crashes into the young Earth, shattering both the planetesimal and our planet.

Hours later, our planet is completely molten and rotating very rapidly. Debris splashed out from Earth's outer layers is now in Earth orbit. Some debris rains back down on Earth, while some will gradually accrete to become the Moon.

Less than a thousand years later, the Moon's accretion is rapidly nearing its end, and relatively little debris still remains in Earth orbit.

FIGURE 8.12 Artist's conception of the giant impact hypothesis for the formation of our Moon. The fact that ejected material came mostly from Earth's outer rocky layers explains why the Moon contains very little metal. The impact must have occurred more than 4.4 billion years ago, since that is the age of the oldest Moon rocks. As shown, the Moon formed quite close to a rapidly rotating Earth, but over billions of years, tidal forces have slowed Earth's rotation and moved the Moon's orbit outward (see Figure 4.26).

Strong support for the giant impact hypothesis comes from two features of the Moon's composition. First, the Moon's overall composition is quite similar to that of Earth's outer layers—just as we would expect if it were made from material blasted away from those layers. Second, the Moon has a much smaller proportion of easily vaporized ingredients (such as water) than Earth. This fact supports the hypothesis because the heat of the impact would have vaporized these ingredients. As gases, they would not have participated in the subsequent accretion of the Moon.

Other Exceptions Giant impacts may also explain other exceptions to the general trends. For example, Pluto's moon Charon shows signs of having formed in a giant impact similar to the one thought to have formed our Moon. Mercury has a metal core extending to 85% of its radius—the largest of any planet—which may be the result of a giant impact that blasted away its outer, rocky layers. Giant impacts could have been responsible for tilting the axes of many planets (including Earth) and perhaps for tipping Uranus on its side. Venus's slow and backward rotation could also be the result of a giant impact, though some scientists suspect it is a consequence of processes attributable to Venus's thick atmosphere.

Although we cannot definitively explain these exceptions to the general rules, the overall lesson is clear: The chaotic processes that accompanied planet formation, including the many collisions that surely occurred, are *expected* to have led to at least a few exceptions. We therefore conclude that the nebular theory can account for all four of the major features of our solar system. **FIGURE 8.13** summarizes what we have discussed. Computer models of planetary formation suggest that the entire process took no more than about 50 million years, which is only about 1% of the current age of the solar system.

Planetary Destiny We've now seen that the nebular theory accounts for all the major features of our solar system. Of course, that does not mean that it explains *everything* about the formation of our solar system. As we'll discuss in Chapter 13, discoveries of other planetary systems have already forced us to revise parts of the theory to allow for a wider range of possibilities than we observe in our own solar system. Nevertheless, the basic theory is widely accepted today because it is so strongly supported by the available evidence.

Assuming the nebular theory is correct, was it inevitable that our solar nebula would form the solar system we see today? Probably not. The first stages of planet formation were orderly and inevitable according to the nebular theory. The creation of a spinning disk, condensation within that disk, and the first stages of accretion were relatively gradual processes that probably would happen again if we turned back the clock. However, the final stages of accretion, and giant impacts in particular, are inherently random and probably would not happen again in the same way. A larger or smaller planet might form at Earth's location and might suffer from a larger giant impact or from no giant impact at all. We don't yet know whether these differences would fundamentally alter the solar system or simply change a few "minor" details—such as the possibility of life on Earth. What we do know is that, among the billions of star systems in our galaxy, it's unlikely that any other is exactly like ours.

8.3 THE AGE OF THE SOLAR SYSTEM

The nebular theory seems to explain *how* our solar system was born. But *when* was it born, and how do we know? The answer is that the planets began to form through accretion just over $4\frac{1}{2}$ billion years ago, a fact we learned by determining the age of the oldest rocks in the solar system.

How do we measure the age of a rock?

The first step in understanding how we learned the age of our solar system is to understand exactly what we mean by the age of a rock. A rock is a collection of a great many atoms held together in solid form. The atoms must be older than Earth, having been forged in the Big Bang or in stars that lived long ago. We cannot determine the ages of the individual atoms, because old atoms are indistinguishable from young ones. However, some atoms undergo changes with time that allow us to determine how long they have been held in place within the rock's solid structure. In other words, the age of a rock is the time since its atoms became locked together in their present arrangement, which in most cases means the time *since the rock last solidified*.

Radiometric Dating The method by which we measure a rock's age is called **radiometric dating**, and it relies on careful measurement of the rock's proportions of various atoms and isotopes. Recall that each chemical element is uniquely characterized by the number of protons in its nucleus, and that different *isotopes* of the same element differ in their number of neutrons [**Section 5.3**]. The key to radiometric dating lies in the fact that some isotopes are radioactive, which is just a fancy way of saying that their nuclei tend to undergo some type of spontaneous change (also called *decay*) with time, such as breaking into two pieces or having a neutron turn into a proton. Decay can change one element into an entirely different one, with different chemical properties.

As a specific example, consider the radioactive isotope potassium-40 (nuclei consisting of 19 protons and 21 neutrons), which decays when one of its protons turns into a neutron, thereby transforming it into argon-40 (18 protons and 22 neutrons). We say that potassium-40 is the *parent isotope*, because it is the original isotope before the decay, and argon-40 is the *daughter isotope* left behind by the decay process.

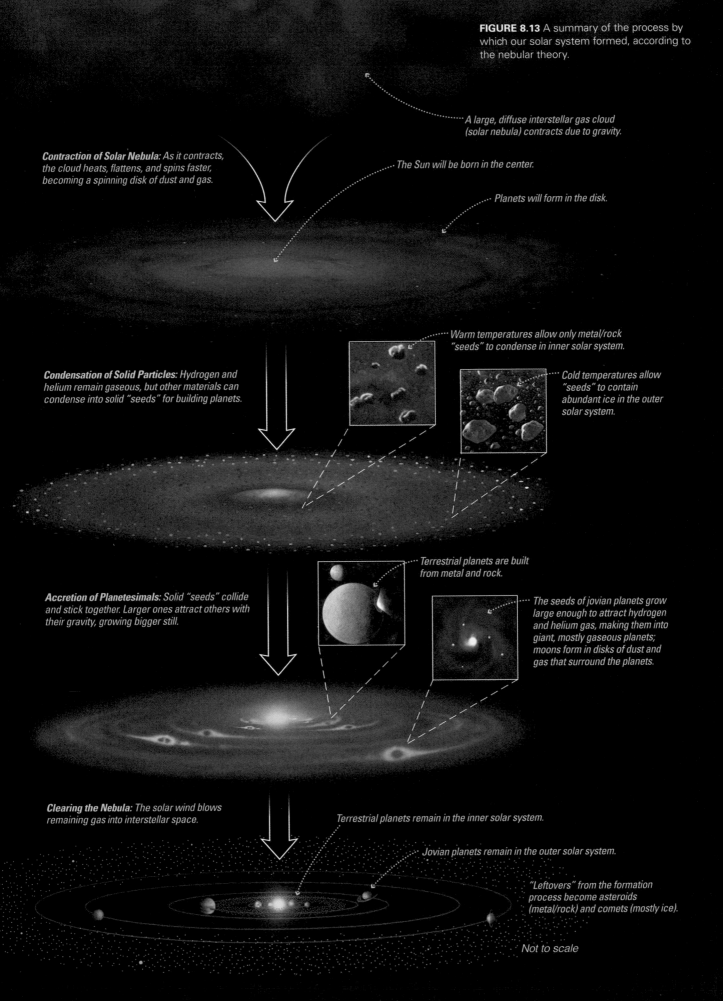

FIGURE 8.13 A summary of the process by which our solar system formed, according to the nebular theory.

A large, diffuse interstellar gas cloud (solar nebula) contracts due to gravity.

Contraction of Solar Nebula: As it contracts, the cloud heats, flattens, and spins faster, becoming a spinning disk of dust and gas.

The Sun will be born in the center.

Planets will form in the disk.

Condensation of Solid Particles: Hydrogen and helium remain gaseous, but other materials can condense into solid "seeds" for building planets.

Warm temperatures allow only metal/rock "seeds" to condense in inner solar system.

Cold temperatures allow "seeds" to contain abundant ice in the outer solar system.

Accretion of Planetesimals: Solid "seeds" collide and stick together. Larger ones attract others with their gravity, growing bigger still.

Terrestrial planets are built from metal and rock.

The seeds of jovian planets grow large enough to attract hydrogen and helium gas, making them into giant, mostly gaseous planets; moons form in disks of dust and gas that surround the planets.

Clearing the Nebula: The solar wind blows remaining gas into interstellar space.

Terrestrial planets remain in the inner solar system.

Jovian planets remain in the outer solar system.

"Leftovers" from the formation process become asteroids (metal/rock) and comets (mostly ice).

Not to scale

(Potassium-40 also decays by other paths, but we will focus only on decay into argon-40 to keep the discussion simple.)

While the decay of any single nucleus is an instantaneous event, laboratory studies show that a modest amount (millions of atoms or more, which is still only a tiny fraction of a gram) of any radioactive parent isotope will gradually transform itself into the daughter isotope at a very steady rate. You don't have to watch for all that long—rarely longer than a few years at most—before you can measure the rate. You can then use this rate to calculate what we call the **half-life,** which is the time it would take for half of the parent nuclei to decay. Every radioactive isotope has its own unique half-life, which may be anywhere from a fraction of a second to many billions of years. Laboratory measurements show that the half-life for the transformation of potassium-40 into argon-40 is 1.25 billion years.

To understand the idea of half-life better, imagine a small piece of rock that contained 1 microgram of potassium-40 and no argon-40 when it solidified long ago. The half-life of 1.25 billion years means that half the original potassium-40 had decayed into argon-40 by the time the rock was 1.25 billion years old, so at that time the rock contained $\frac{1}{2}$ microgram of potassium-40 and $\frac{1}{2}$ microgram of argon-40. Half of this remaining potassium-40 had then decayed by the end of the next 1.25 billion years, so after 2.5 billion years the rock contained $\frac{1}{4}$ microgram of potassium-40 and $\frac{3}{4}$ microgram of argon-40. After three half-lives, or 3.75 billion years, only $\frac{1}{8}$ microgram of potassium-40 remained, while $\frac{7}{8}$ microgram had become argon-40. **FIGURE 8.14** summarizes the gradual

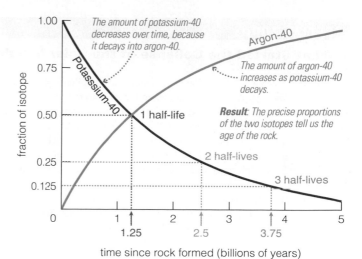

FIGURE 8.14 Potassium-40 is radioactive, decaying into argon-40 with a half-life of 1.25 billion years. The red line shows the decreasing amount of potassium-40, and the blue line shows the increasing amount of argon-40. The remaining amount of potassium-40 drops by half with each successive half-life.

decrease in the amount of potassium-40 and the corresponding rise in the amount of argon-40.

You can now see the essence of radiometric dating. Suppose you find a rock that contains equal numbers of atoms of potassium-40 and argon-40. If you assume that all the argon came from potassium decay (and if the rock shows no evidence of subsequent heating that could have allowed any argon to escape), then it must have taken precisely one half-life for the rock to

MATHEMATICAL INSIGHT 8.1

Radiometric Dating

The amount of a radioactive substance decreases by half with each half-life, so we can express the decay process with a simple formula relating the current and original amounts of a radioactive substance in a rock:

$$\frac{\text{current amount}}{\text{original amount}} = \left(\frac{1}{2}\right)^{t/t_{\text{half}}}$$

where t is the time since the rock formed and t_{half} is the half-life of the radioactive material. We can solve this equation for the age t by taking the base-10 logarithm of both sides and rearranging the terms. The resulting general equation for the age is

$$t = t_{\text{half}} \times \frac{\log_{10}\left(\dfrac{\text{current amount}}{\text{original amount}}\right)}{\log_{10}\left(\dfrac{1}{2}\right)}$$

EXAMPLE: You heat and chemically analyze a small sample of a meteorite. Potassium-40 and argon-40 are present in a ratio of approximately 0.85 unit of potassium-40 atoms to 9.15 units of gaseous argon-40 atoms. (The units are unimportant, because only the relative amounts of the parent and daughter materials matter.) How old is the meteorite?

SOLUTION:

Step 1 Understand: The formula above allows us to find the meteorite's age from the current and original amounts of a parent isotope. We are given that the current amount of the parent isotope potassium-40 is 0.85 unit. We are not told the original amount, but we can figure it out from the fact that the sample now has 9.15 units of the daughter isotope argon-40. As discussed in the text, we can assume that *all* the trapped argon-40 in the meteorite is a decay product of potassium-40, which means that the 9.15 units of argon-40 must originally have been potassium-40. Therefore, the original amount of potassium-40 was $0.85 + 9.15 = 10$ units.

Step 2 Solve: We use the formula above with current amount = 0.85 unit and original amount = 10 units:

$$t = 1.25 \text{ billion yr} \times \frac{\log_{10}\left(\dfrac{0.85}{10}\right)}{\log_{10}\left(\dfrac{1}{2}\right)}$$

$$= 1.25 \text{ billion yr} \times \left(\frac{-1.07}{-0.301}\right)$$

$$= 4.45 \text{ billion yr}$$

Step 3 Explain: We have used radiometric dating to determine that the meteorite's age is 4.45 billion years, which means it solidified 4.45 billion years ago. (*Note:* This example captures the essence of radiometric dating, but real cases generally require more detailed analysis.)

What Started the Collapse of the Solar Nebula?

The solar nebula probably could not have started collapsing on its own, because gravity was too weak when the gas was still spread out over a large region of space. We therefore suspect that some cataclysmic event triggered the collapse. But what type of event? Although we may never know for sure, radioactive elements and their decay products give us some clues. Radioactive elements are made in the violent stellar explosions called *supernovae*, so their presence in meteorites and on Earth underscores the fact that our solar system is made from the remnants of past generations of stars.

Some meteorites contain the rare isotope xenon-129. Xenon is gaseous even at extremely low temperatures, so it could not have condensed and become trapped in the meteorites as they condensed in the solar nebula. Instead, it must be a product of radioactive decay, and laboratory studies show that its parent isotope is iodine-129. This is a crucial clue, because iodine-129 has a relatively short half-life of 17 million years. We conclude that the xenon-129 we observe in meteorites today is the product of radioactive decay of iodine-129 produced in a supernova that occurred no more than a few tens of millions of years before the collapse of the solar nebula began.

The relatively short time between a supernova and the birth of our solar system suggests that a shock wave from the exploding star may have been the trigger for the nebula's collapse. Once this shock wave got the collapse started, gravity took over and made the rest of the collapse inevitable.

---

end up with equal amounts of the two isotopes. You can therefore conclude that the rock is 1.25 billion years old. The only question is whether you are right in assuming that the rock lacked argon-40 when it formed. In this case, knowing a bit of "rock chemistry" helps. Potassium-40 is a natural ingredient of many minerals in rocks, but argon-40 is a gas that does not combine with other elements and did not condense in the solar nebula. Therefore, if you find argon-40 gas trapped inside minerals, it must have come from radioactive decay of potassium-40.

Validity of the Method Radiometric dating is possible with many other radioactive isotopes as well. In many cases, a rock contains more than one radioactive isotope, and agreement between the ages calculated from the different isotopes gives us confidence that we have dated the rock correctly. We can also check results from radiometric dating against those from other methods of measuring or estimating ages. For example, some archaeological artifacts have original dates printed on them, and these dates agree with ages found by radiometric dating. The same agreement is found for wood artifacts that can be dated both by counting tree rings and by the radiometric method.

We can validate the $4\frac{1}{2}$-billion-year radiometric age for the solar system as a whole by comparing it to an age based on detailed study of the Sun. Theoretical models of the Sun, along with observations of other stars, show that stars slowly expand and brighten as they age. The model ages are not nearly as precise as radiometric ages, but they confirm that the Sun is between about 4 and 5 billion years old. Overall, the technique of radiometric dating has been checked in so many ways and relies on such basic scientific principles that there is no serious scientific debate about its validity.

How do we know the age of the solar system?

Radiometric dating tells us how long it has been since a rock solidified, which is not the same as the age of a planet as a whole. For example, we find rocks of many different ages on Earth. Some rocks are quite young because they formed recently from molten lava; others are much older. The oldest Earth rocks are about 4 billion years old, and some small mineral grains date to almost 4.4 billion years ago, but even these are not as old as Earth itself, because Earth's entire surface has been reshaped through time.

Moon rocks brought back by the *Apollo* astronauts have also been dated, and many are older than the oldest Earth rocks, demonstrating that parts of the Moon's surface have not changed since very early in its history. The oldest Moon rocks contain minerals with a small amount of uranium-238, which decays (in several steps) into lead-206 with a half-life of about 4.5 billion years. Lead and uranium have very different chemical behaviors, and some minerals start with virtually no lead. Laboratory analysis of such minerals in lunar rocks shows that they now contain almost equal proportions of atoms of uranium-238 and lead-206, which means they are about one half-life old. More precise work shows them to be about 4.4 billion years old.

Although the oldest Moon rocks are older than the oldest Earth rocks, the rocks must still be younger than the Moon and Earth as a whole. In fact, their age tells us that the giant impact thought to have created the Moon must have occurred more than 4.4 billion years ago. But how do we determine when the planets first began to form?

To go all the way back to the origin of the solar system, we must find rocks that have not melted or vaporized since they first condensed in the solar nebula. Meteorites that have fallen to Earth are our source of such rocks. Many meteorites appear to have remained unchanged since they condensed and accreted in the early solar system. Careful analysis of radioactive isotopes in these meteorites shows that the oldest ones formed about 4.55 billion years ago, so this time must mark the beginning of accretion in the solar nebula. Because the planets apparently accreted within about 50 million (0.05 billion) years after that, Earth and the other planets had formed by about 4.5 billion years ago. In other words, the age of our solar system is only about a third of the 14-billion-year age of our universe.

Putting Chapter 8 into Perspective

This chapter described the current scientific theory of our solar system's formation, and how this theory explains the major features we observe. As you continue your study of the solar system, keep in mind the following "big picture" ideas:

- The nebular theory of solar system formation gained wide acceptance because of its success in explaining the major characteristics of our solar system.

- Most of the general features of the solar system were determined by processes that occurred very early in the solar system's history.

- Chance events may have played a large role in determining how individual planets turned out. No one knows how different our solar system might be if the process started over.

- We have learned the age of our solar system—about $4\frac{1}{2}$ billion years—from radiometric dating of the oldest meteorites. This age agrees with ages estimated through a variety of other techniques, making it clear that we are recent arrivals on a very old planet.

SUMMARY OF KEY CONCEPTS

8.1 THE SEARCH FOR ORIGINS

- **How did we arrive at a theory of solar system formation?** A successful theory must explain four major features of our solar system: patterns of motion, the existence of two types of planets (terrestrial and jovian), the presence of asteroids and comets, and exceptions to the rules. Developed over a period of more than two centuries, the **nebular theory** explains all four features and also can account for other planetary systems.

- **Where did the solar system come from?** The nebular theory holds that the solar system formed from the gravitational collapse of an interstellar cloud known as the **solar nebula**. This cloud was the product of recycling of gas through many generations of stars within our galaxy. This material consisted of 98% hydrogen and helium and 2% all other elements combined.

8.2 EXPLAINING THE MAJOR FEATURES OF THE SOLAR SYSTEM

- **What caused the orderly patterns of motion?** As the solar nebula collapsed under gravity, natural processes caused it to heat up, spin faster, and flatten out as it shrank. The orderly motions we observe today all came from the orderly motion of this spinning disk.

- **Why are there two major types of planets?** The inner regions of the solar nebula were relatively hot, so only metal and rock could condense into tiny, solid grains; these grains accreted into larger **planetesimals** that ultimately merged to make the terrestrial planets. Beyond the frost line, cooler temperatures also allowed more abundant **hydrogen compounds** to condense into ice, building ice-rich planetesimals; some of these grew large enough for their gravity to draw in hydrogen and helium gas, forming the jovian planets.

- **Where did asteroids and comets come from?** Asteroids are the rocky leftover planetesimals of the inner solar system, and comets are the ice-rich leftover planetesimals of the outer solar system. These objects still occasionally collide with planets or moons, but the vast majority of impacts occurred during the **heavy bombardment** in the solar system's first few hundred million years.

- **How do we explain "exceptions to the rules"?** Most of the exceptions probably arose from collisions or close encounters with leftover planetesimals. Our Moon is most likely the result of a **giant impact** between a Mars-size planetesimal and the young Earth.

8.3 THE AGE OF THE SOLAR SYSTEM

- **How do we measure the age of a rock?** Radiometric **dating** is based on carefully measuring the proportions of radioactive isotopes and their decay products within rocks. The ratio of the isotopes changes with time in a steady and predictable way that we characterize by an isotope's **half-life**, the time it takes for half the atoms in a collection to decay.

- **How do we know the age of the solar system?** Radiometric dating of the oldest meteorites tells us that accretion began in the solar nebula about 4.55 billion years ago, with the planets forming by about 4.5 billion years ago.

Use the following questions to check your understanding of some of the many types of visual information used in astronomy. Answers are provided in Appendix J. For additional practice, try the Chapter 8 Visual Quiz at MasteringAstronomy®.

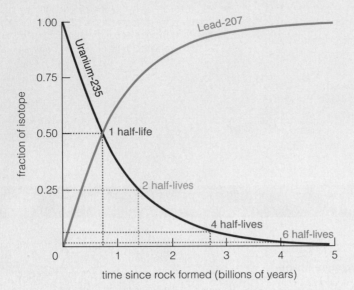

The graph above, similar to Figure 8.14, shows the radioactive decay of uranium-235 to lead-207. (Uranium-235 also decays to lead-206, but we'll ignore that reaction for the sake of clarity in this example.)

1. Compare the graph above to Figure 8.14, which shows the decay of potassium-40. Which element is more radioactive (undergoes radioactive decay more quickly)?
   a. uranium-235
   b. potassium-40
   c. Both are equally radioactive.
2. What fraction of the original uranium-235 should be left after 3.5 billion years?
   a. $\frac{1}{2}$
   b. $\frac{1}{4}$
   c. $\frac{1}{8}$
   d. $\frac{1}{32}$
   e. $\frac{1}{64}$

3. You find a mysterious rock on the ground and determine that 60% of its uranium-235 has been converted into lead-207. What is the most likely origin of the rock, based on its radiometric age?
   a. It's older than our solar system, so it must have come from another solar system.
   b. It's a meteorite dating back to the formation of the solar system.
   c. It's a volcanic rock nearly a billion years old.
   d. It was just formed this year during the eruption of a nearby volcano.

MasteringAstronomy®

For instructor-assigned homework go to MasteringAstronomy®.

REVIEW QUESTIONS

Short-Answer Questions Based on the Reading

1. Briefly describe the four major features of our solar system that a formation theory must explain.
2. What is the *nebular theory,* and why is it widely accepted by scientists today?
3. What do we mean by the *solar nebula*? What was it made of, and where did it come from?
4. Describe the three key processes that led the solar nebula to take the form of a spinning disk. What observational evidence supports this scenario?
5. List the approximate condensation temperature and abundance for each of the four categories of materials in the solar nebula. Which ingredients are present in terrestrial planets? In jovian planets? In comets and asteroids?
6. What was the *frost line*? Which ingredients condensed inside and outside the frost line? What role did it play in the formation of two distinct types of planets?
7. Briefly describe the process by which terrestrial planets are thought to have formed.
8. How was the formation of jovian planets similar to that of the terrestrial planets? How was it different? Why did the jovian planets end up with so many moons?

9. What is the *solar wind*, and what roles did it play in the early solar system?

10. How did planet formation lead to the existence of asteroids and comets?

11. What was the *heavy bombardment*, and when did it occur?

12. What is the leading hypothesis for the Moon's formation, and what evidence supports this hypothesis?

13. Describe the technique of *radiometric dating*. What is a *half-life*?

14. How old is the solar system, and how do we know?

TEST YOUR UNDERSTANDING

Surprising Discoveries?

Suppose we found a solar system with the property described. (These are not real discoveries.) In light of what you've learned about the formation of our own solar system, decide whether the discovery should be considered reasonable or surprising. Explain your reasoning.

15. A solar system has five terrestrial planets in its inner solar system and three jovian planets in its outer solar system.

16. A solar system has four large jovian planets in its inner solar system and seven small terrestrial planets in its outer solar system.

17. A solar system has ten planets that all orbit the star in approximately the same plane. However, five planets orbit in one direction (e.g., counterclockwise), while the other five orbit in the opposite direction (e.g., clockwise).

18. A solar system has 12 planets that all orbit the star in the same direction and in nearly the same plane. The 15 largest moons in this solar system orbit their planets in nearly the same direction and plane as well. However, several smaller moons have highly inclined orbits around their planets.

19. A solar system has six terrestrial planets and four jovian planets. Each of the six terrestrial planets has at least five moons, while the jovian planets have no moons at all.

20. A solar system has four Earth-size terrestrial planets. Each of the four planets has a single moon that is nearly identical in size to Earth's Moon.

21. A solar system has many rocky asteroids and many icy comets. However, most of the comets orbit in the inner solar system, while the asteroids orbit in far-flung regions much like the Kuiper belt and Oort cloud of our solar system.

22. A solar system has several planets similar in composition to the jovian planets of our solar system but similar in mass to the terrestrial planets of our solar system.

23. A solar system has several terrestrial planets and several larger planets made mostly of ice. (*Hint:* What would happen if the solar wind started earlier or later than in our solar system?)

24. Radiometric dating of the oldest meteorites from another solar system shows that they are a billion years younger than rocks from the terrestrial planets of the same system.

Quick Quiz

Choose the best answer to each of the following. Explain your reasoning with one or more complete sentences.

25. How many of the planets orbit the Sun in the same direction as Earth does? (a) a few (b) most (c) all

26. What fraction of the large moons of the planets orbit in the same direction as their planets rotate? (a) some (b) most (c) all

27. The solar nebula was 98% (a) rock and metal. (b) hydrogen compounds. (c) hydrogen and helium.

28. Which of the following did *not* occur during the collapse of the solar nebula? (a) spinning faster (b) heating up (c) concentrating denser materials nearer the Sun

29. What is Jupiter's main ingredient? (a) rock and metal (b) hydrogen compounds (c) hydrogen and helium

30. Which lists the major steps of solar system formation in the correct order? (a) collapse, accretion, condensation (b) collapse, condensation, accretion (c) accretion, condensation, collapse

31. Which of the following is *not* true about the young Sun? (a) It rotated much faster than it does today. (b) It was much brighter and hotter than it is today. (c) It had a much stronger solar wind than it does today.

32. Leftover ice-rich planetesimals are called (a) comets. (b) asteroids. (c) meteorites.

33. What's the leading hypothesis for the origin of the Moon? (a) It formed along with Earth. (b) It formed from material ejected from Earth in a giant impact. (c) It split out of a rapidly rotating Earth.

34. About how old is the solar system? (a) 4.5 million years (b) 4.5 billion years (c) 4.5 trillion years

PROCESS OF SCIENCE

Examining How Science Works

35. *Explaining the Past.* Is it really possible for science to inform us about things that may have happened billions of years ago? To address this question, test the nebular theory against each of the three hallmarks of science discussed in Chapter 3. Be as detailed as possible in explaining whether the theory does or does not satisfy these hallmarks. Use your findings to decide whether the theory can really tell us about how our solar system formed. Defend your opinion.

36. *Unanswered Questions.* As discussed in this chapter, the nebular theory explains many but not all questions about the origin of our solar system. Choose one important but unanswered question about the origin of our solar system and write two or three paragraphs in which you discuss how we might answer this question in the future. Be as specific as possible, focusing on the type of evidence necessary to answer the question and how the evidence could be gathered. What are the benefits of finding answers to this question?

GROUP WORK EXERCISE

37. *A Cold Solar Nebula.* In this chapter, you've learned that the nature of the planets in our solar system was determined by the location of the frost line when the era of planet formation ended. In our solar system, the frost line was located between Mars and Jupiter, but study of other solar systems suggests that our solar system could have turned out differently. Here we examine what would have happened if the entire solar nebula had cooled to 50 K before the solar wind cleared it away. Before you begin, assign the following roles to the people in your group: *Scribe* (takes notes on the group's activities), *Proposer* (proposes explanations to the group), *Skeptic* (points out weaknesses in proposed explanations), and *Moderator* (leads group discussion and makes sure everyone contributes).
a. Make a list of ingredients that will condense at 50 K, and another list of the main ingredients of a typical jovian planet and terrestrial planet. **b.** Use your lists to suggest ways in which jovian and/or terrestrial planets would be different from the ones in our solar system under this alternative formation scenario. Be sure to consider differences in size and composition, and assess whether any differences would be minor or major. **c.** Discuss the likelihood that your predicted changes would match the actual characteristics of this alternative solar system. What factors limit the degree to which you can make successful predictions? **d.** Come up with additional "what if" scenarios, discussing various ways in which planets might have turned out differently.

In-Depth Questions to Increase Your Understanding

Short-Answer/Essay Questions

38. *An Early Solar Wind.* Suppose the solar wind had cleared away the solar nebula before the seeds of the jovian planets could gravitationally draw in hydrogen and helium gas. How would the planets of the outer solar system be different? Would they still have many moons? Explain your answer in a few sentences.

39. *Angular Momentum.* Suppose our solar nebula had begun with much more angular momentum than it did. Do you think planets could still have formed? Why or why not? What if the solar nebula had started with zero angular momentum? Explain your answers in one or two paragraphs.

40. *Two Kinds of Planets.* The jovian planets differ from the terrestrial planets in a variety of ways. Using sentences that members of your family would understand, explain why the jovian planets differ from the terrestrial planets in each of the following: composition, size, density, distance from the Sun, and number of satellites.

41. *Two Kinds of Planetesimals.* Why are there two kinds of "leftovers" from solar system formation? List as many differences between comets and asteroids as you can.

42. *History of the Elements.* Our bodies (and most living things) are made mostly of water: H_2O. Summarize the "history" of a typical hydrogen atom from its creation to the formation of Earth. Do the same for a typical oxygen atom. (*Hint:* Which elements were created in the Big Bang, and where were the others created?)

43. *Understanding Radiometric Dating.* Imagine you had the good fortune to find a rocky meteorite in your backyard. Qualitatively, how would you expect its ratio of potassium-40 to argon-40 to be different from that of other rocks in your yard? Explain why, in a few sentences.

44. *Rocks from Other Solar Systems.* Many "leftovers" from planetary formation were likely ejected from our solar system, and the same has presumably happened in other star systems. Given that fact, should we expect to find meteorites that come from other star systems? How rare or common would you expect them to be? (Be sure to consider the distances between stars.) Suppose that we *did* find a meteorite identified as a leftover from another stellar system. What could we learn from it?

Quantitative Problems

Be sure to show all calculations clearly and state your final answers in complete sentences.

45. *Radiometric Dating.* You are dating rocks by their proportions of parent isotope potassium-40 (half-life 1.25 billion years) and daughter isotope argon-40. Find the age for each of the following. **a.** A rock that contains equal amounts of potassium-40 and argon-40 **b.** A rock that contains three times as much argon-40 as potassium-40

46. *Lunar Rocks.* You are dating Moon rocks based on their proportions of uranium-238 (half-life of about 4.5 billion years) and its ultimate decay product, lead. Find the age for each of the following.
a. A rock for which you determine that 55% of the original uranium-238 remains, while the other 45% has decayed into lead
b. A rock for which you determine that 63% of the original uranium-238 remains, while the other 37% has decayed into lead

47. *Carbon-14 Dating.* The half-life of carbon-14 is about 5700 years.
a. You find a piece of cloth painted with organic dye. By analyzing the dye, you find that only 77% of the carbon-14 originally in the dye remains. When was the cloth painted? **b.** A well-preserved piece of wood found at an archaeological site has 6.2% of the carbon-14 it must have had when it was living. Estimate when the wood was cut. **c.** Is carbon-14 useful for establishing Earth's age? Why or why not?

48. *Collapsing Cloud.* The time it takes for a cloud 100,000 AU in diameter to collapse in "free fall" to form a new star is *half* the time it would take an object to orbit the star on an elliptical orbit with a semimajor axis of 50,000 AU (half the 100,000 AU diameter). Use Kepler's third law to find the collapse time, assuming the star has the same mass as the Sun.

49. *Icy Earth.* How massive would Earth have been if it had accreted hydrogen compounds in addition to rock and metal? Assume the same proportions of the ingredients as listed in Table 8.1. What would Earth's mass be if it had been possible to capture hydrogen and helium gas in the proportion listed in the table?

50. *What Are the Odds?* The fact that all the planets orbit the Sun in the same direction is cited as support for the nebular hypothesis. Imagine that there's a different hypothesis in which planets can be created orbiting the Sun in either direction. Under this hypothesis, what is the probability that eight planets would end up traveling in the same direction? (*Hint:* It's the same probability as that of flipping a coin eight times and getting all heads.)

51. *Spinning Up the Solar Nebula.* The orbital speed of the material in the solar nebula at Pluto's average distance from the Sun was about 5 km/s. What was the orbital speed of this material when it was 40,000 AU from the Sun (before it fell inward with the collapse of the nebula)? Use the law of conservation of angular momentum (see Section 4.3).

Discussion Questions

52. *Theory and Observation.* Discuss the interplay between theory and observation that has led to our modern theory of solar system formation. What role does technology play in allowing us to test this theory?

53. *Lucky to Be Here?* Considering the overall process of solar system formation, do you think formation of a planet like Earth was likely? Could random events in the early history of the solar system have prevented our being here today? What implications do your answers have for the possibility of Earth-like planets around other stars? Defend your opinions.

Web Projects

54. *Spitzer Space Telescope.* NASA's Spitzer Space Telescope operates at the infrared wavelengths especially useful for studies of star and planet formation. Visit the Spitzer website to see what such studies have told us about how planets form. Summarize your findings in a one- to two-page report.

55. *The Stardust Puzzle.* Search the Internet for recent progress in explaining the *Stardust* mission's discovery of minerals from a comet nucleus that must have formed very close to the Sun. Write a one- to two-page report, describing how the result has affected our ideas about the processes in the early solar nebula.

56. *Dating the Past.* The method of radiometric dating that tells us the age of our solar system is also used to determine when many other past events occurred. For example, it is used to determine ages of fossils that tell us when humans first evolved and ages of relics that teach us about the rise of civilization. Research one key aspect of human history in which radiometric dating helps us piece the story together. Write two or three paragraphs on how radiometric dating is used in this case (such as what materials are dated and what radioactive elements are used) and what these studies have concluded. Does your understanding of the method lead you to accept the results? Why or why not?

9 PLANETARY GEOLOGY
EARTH AND THE OTHER TERRESTRIAL WORLDS

Nothing is rich but the inexhaustible wealth of nature. She shows us only surfaces, but she is a million fathoms deep.

—Ralph Waldo Emerson

It's easy to take for granted the qualities that make Earth so suitable for human life: a temperature neither boiling nor freezing, abundant water, a protective atmosphere, and a relatively stable environment. But we need look only as far as our neighboring terrestrial worlds to see how fortunate we are. The Moon is airless and barren, and Mercury is much the same. Venus is a searing hothouse, while Mars has an atmosphere so thin and cold that liquid water cannot last on its surface today.

How did the terrestrial worlds come to be so different, and why did Earth alone develop conditions that permit abundant life? We'll explore these questions through careful, comparative study of the planets, focusing on the geology of the terrestrial worlds in this chapter and on their atmospheres in Chapter 10. As we will see, the histories of the terrestrial worlds have been determined largely by properties endowed at their births.

(MA) Formation of the Solar System Tutorial, Lesson 1

9.1 CONNECTING PLANETARY INTERIORS AND SURFACES

Earth's surface seems solid and steady, but every so often it offers us a reminder that nothing about it is permanent. If you live in Alaska or California, you've probably felt the ground shift beneath you in an earthquake. In Washington State, you may have witnessed the rumblings of Mount St. Helens. In Hawaii, a visit to the active Kilauea volcano will remind you that you are standing on mountains of volcanic rock protruding from the ocean floor.

Volcanoes and earthquakes are not the only processes acting to reshape Earth's surface. They are not even the most dramatic: Far greater change can occur on the rare occasions when an asteroid or a comet slams into Earth. More gradual processes can also have spectacular effects. The Colorado River causes only small changes in the landscape from year to year, but its unrelenting flow over the past few million years carved the Grand Canyon. The Rocky Mountains were once twice as tall as they are today, but they have been cut down in size through tens of millions of years of erosion by wind, rain, and ice. Entire continents even move slowly about, completely rearranging the map of Earth every few hundred million years.

Earth is not alone in having undergone tremendous change since its birth. The surfaces of all five terrestrial worlds—Mercury, Venus, Earth, the Moon, and Mars—must have looked quite similar when they were young. All five were made of rocky material that condensed in the solar nebula, and all five were subjected early on to the impacts of the heavy bombardment [**Section 8.2**]. The great differences in their present-day appearances must therefore be the result of changes that have occurred through time. Ultimately, these changes must be traceable to fundamental properties of the planets.

FIGURE 9.1 shows global views of the terrestrial surfaces to scale, as well as sample surface views from orbit. Profound differences among these worlds are immediately obvious. Mercury and the Moon show the scars of their battering during the heavy bombardment: They are densely covered by craters, except in areas that appear to be volcanic plains. Bizarre bulges and odd volcanoes dot the surface of Venus. Mars, despite its middling size, has the solar system's largest volcanoes and a huge canyon cutting across its surface, along with numerous features that appear to have been shaped by liquid water. Earth has surface features similar to all those on the other terrestrial worlds, and more—including a

Venus   Earth

Mars

Mercury   Earth's Moon

50 km   100 km   100 km   100 km   50 km

Heavily cratered Mercury has long steep cliffs (arrow).

Cloud-penetrating radar revealed this twin-peaked volcano on Venus.

A portion of Earth's surface as it appears without clouds.

The Moon's surface is heavily cratered in most places.

Mars has features that look like dry riverbeds; note the impact craters.

FIGURE 9.1 Global views to scale, along with sample close-ups viewed from orbit, of the five terrestrial worlds. All the images were taken with visible light except those for Venus, which are based on radar data.

rocky crust
(lower density)

mantle
(medium density)

metal core
(highest density)

rigid lithosphere
(crust and
part of mantle)

lithosphere

lithosphere

lithosphere

lithosphere

Key: ▮ crust ▮ mantle ⊥ lithosphere ▮ core

FIGURE 9.2 Interior structures of the terrestrial worlds, with sizes shown to scale from largest to smallest. Color coding shows the core-mantle-crust layering by density; a dashed circle represents the inner boundary of the lithosphere, defined by the strength of the rock rather than by density. The thicknesses of the crust and lithosphere on Venus and Earth are exaggerated to make them visible in this figure.

unique layer of living organisms that covers almost the entire surface of the planet.

Our goal in this chapter is to understand how these differences among the terrestrial planets came to be. This type of study is called **planetary geology**; geology is literally the study of Earth (*geo* means "Earth"), so planetary geology is the extension of this science to other worlds. Note that we use the term *planetary* geology for the study of any solid world, even if it is a moon rather than a planet. Because most geological features are shaped by processes that take place deep beneath the surface, we'll begin our study of planetary geology by examining what planets are like on the inside.

What are terrestrial planets like on the inside?

We cannot see inside Earth or any other planet, but a variety of clues tell us about their internal structures. For Earth, our most detailed information comes from **seismic waves**, vibrations that travel both through the interior and along the surface after an earthquake (*seismic* comes from the Greek word for "shake"). In much the same way that shaking a box containing a gift offers clues about what's inside, seismic vibrations offer clues about what's inside Earth. We also have seismic data for the Moon, thanks to monitoring stations left on the Moon by the *Apollo* astronauts.

For other worlds, we can learn about their interiors in a variety of ways. For example, comparing a world's overall density (found by dividing its total mass by its total volume) to the density of its surface rock tells us how much more dense it must be inside; precise measurements of a world's gravity from spacecraft can tell us about how its mass is distributed inside it; study of magnetic fields tells us about the interior layers in which these fields are generated; and observations of surface rocks that have emerged from the interior (such as volcanic rock) can tell us about interior composition.

Layering by Density All the terrestrial worlds have layered interiors. You are probably familiar with the idea of dividing the interior into three layers by density:

- **Core.** The highest-density material, consisting primarily of metals such as nickel and iron, resides in the central core.

- **Mantle.** Rocky material of moderate density—mostly minerals that contain silicon, oxygen, and other elements—forms the thick mantle that surrounds the core.

- **Crust.** The lowest-density rock, such as granite and basalt (a common form of volcanic rock), forms the thin crust, essentially representing the world's outer skin.

FIGURE 9.2 shows these layers for the five terrestrial worlds. Although not shown in the figure, Earth's metallic core actually consists of two distinct regions: a solid *inner core* and a molten (liquid) *outer core*.* Venus may have a similar core structure, but without seismic data we cannot be sure.

We can understand *why* the interiors are layered by thinking about what happens in a mixture of oil and water: Gravity pulls the denser water to the bottom, driving the less dense oil to the top, in a process called **differentiation** (because it results in layers made of *different* materials). The layered interiors of the terrestrial worlds tell us that they underwent differentiation at some time in the past, which means all these worlds must once have been hot enough inside for their interior rock and metal to melt. Dense metals like iron sank toward the center, driving less dense rocky material toward the surface.

Comparing the terrestrial worlds' interiors provides important clues about their early histories. Models indicate that the relative proportions of metal and rock should have been similar throughout the inner solar system at the time the terrestrial

*Because temperature increases with depth in a planet, it may seem surprising that Earth's inner core is solid while its outer core is molten. The inner core is kept solid by the higher pressure at its greater depth, even though the temperature is also higher.

planets formed, which means we should expect smaller worlds to have correspondingly smaller metal cores. We do indeed see this general pattern in Figure 9.2, but the pattern is not perfect: Mercury's core seems surprisingly big, while the Moon's core seems surprisingly small. These surprises are a major reason scientists suspect that giant impacts affected both worlds [**Section 8.2**]. In Mercury's case, a giant impact that blasted away its outer rocky layers while leaving its core intact could explain why the core is so large compared to the rest of the planet. We can explain the Moon's small core by assuming the Moon formed from debris blasted out of Earth's rocky outer layers (see Figure 8.12): The debris would have contained relatively little high-density metal and therefore would have accreted into an object with a very small metal core.

Layering by Rock Strength In geology, it's often more useful to categorize interior layers by rock strength than by density. The idea that rock can vary in strength may seem surprising, since we often think of rock as the very definition of strength. However, like all matter built of atoms, rock is mostly empty space; its apparent solidity arises from electrical bonds between its atoms and molecules [**Section 5.3**]. Although these bonds are strong, they can still break and re-form when subjected to heat or sustained stress, which means that even solid rock can slowly deform and flow over millions and billions of years. The long-term behavior of rock is much like that of Silly Putty, which breaks like a brittle solid when you pull it sharply but deforms and stretches when you pull it slowly (**FIGURE 9.3**). Also like Silly Putty, rock becomes softer and easier to deform when it is warmer.

In terms of rock strength, a planet's outer layer consists of relatively cool and rigid rock, called the **lithosphere** (*lithos* is Greek for "stone"), that essentially "floats" on the warmer, softer rock beneath. As shown by the dashed circles in Figure 9.2, the lithosphere encompasses the crust and part of the mantle of each world.

FIGURE 9.3 Silly Putty stretches when pulled slowly but breaks cleanly when pulled rapidly. Rock behaves just the same way, but on a longer time scale.

Notice that lithospheric thickness is closely related to a world's size: Smaller worlds tend to have thicker lithospheres. The two largest terrestrial planets, Earth and Venus, have thin lithospheres that extend only a short way into their upper mantles. The smaller worlds—Mars, Mercury, and the Moon—have thick lithospheres that extend nearly to their cores. The thickness of the lithosphere is very important to geology. A thin lithosphere is brittle and can crack easily. A thick lithosphere is much stronger and inhibits the passage of molten rock from below, making volcanic eruptions and the formation of mountain ranges less likely.

Why Big Worlds Are Round The fact that rock can deform and flow also explains why large worlds are spherical while small moons and asteroids are "potato-shaped." The weak gravity of a small object is unable to overcome the rigidity of its rocky material, so the object retains the shape it had when it was born. For a larger world, gravity can overcome the strength of solid rock, slowly deforming and molding it into a spherical shape. Gravity will make any rocky object bigger than about 500 kilometers in diameter into a sphere within about 1 billion years. Larger worlds become spherical more quickly, especially if they are molten (or gaseous) at some point in their history.

What causes geological activity?

The most interesting aspects of planetary geology are those that cause the surfaces of the terrestrial worlds to change with time. We use the term **geological activity** to describe ongoing changes. For example, we say that Earth is geologically active, because volcanic eruptions, earthquakes, erosion, and other geological processes continually reshape its surface. In contrast, the Moon and Mercury have virtually no geological activity, which is why their surfaces today look essentially the same as they did billions of years ago.

Interior heat is the primary driver of geological activity. For example, volcanoes can erupt only if the interior is hot enough to melt at least some rock into molten lava. But what makes some planetary interiors hotter than others? To find the answer, we must investigate how interiors heat up and cool off. As we'll see, we can ultimately trace a planet's internal heat, and hence its geological activity, back to its size.

How Interiors Get Hot A hot interior contains a lot of thermal energy, and the law of conservation of energy tells us that this energy must have come from somewhere [**Section 4.3**]. Although you might first guess that the Sun would be the heat source, this is not the case: Sunlight is the primary heat source for the *surfaces* of the terrestrial planets, but virtually none of this solar energy penetrates more than a few meters into the ground. Internal heat is a product of the planets themselves, not of the Sun. Three sources of energy explain nearly all the interior heat of the terrestrial worlds (**FIGURE 9.4**):

■ **Heat of accretion.** Accretion deposits energy brought in from afar by colliding planetesimals. As a planetesimal approaches a forming planet, its gravitational potential energy is converted to kinetic energy, causing it to accelerate. Upon impact, much of the kinetic energy is converted to heat, adding to the thermal energy of the planet.

■ **Heat from differentiation.** When a world undergoes differentiation, the sinking of dense material and rising of less-dense material means that mass moves inward, losing gravitational potential energy. This energy is converted to thermal energy by the friction generated as materials separate by density. The same thing happens when you drop a brick into a pool: As the brick sinks to the bottom, friction with the surrounding water heats the pool—though the amount of heat from a single brick is too small to be noticed.

■ **Heat from radioactive decay.** The rock and metal that built the terrestrial worlds contained radioactive isotopes of elements such as uranium, potassium, and thorium. When radioactive nuclei decay, subatomic particles fly off at high speeds, colliding with neighboring atoms and heating them. In essence, this transfers some of the mass-energy ($E = mc^2$) of the radioactive nuclei to the thermal energy of the planetary interior.

SPECIAL TOPIC

How Do We Know What's Inside Earth?

Our deepest drills have barely pricked Earth's surface, penetrating much less than 1% of the way into the interior. How, then, can we claim to know what our planet is like on the inside?

For Earth, much of our information about the interior comes from *seismic waves,* vibrations created by earthquakes. Seismic waves come in two basic types, analogous to the two ways you can generate waves in a Slinky (**FIGURE 1**). Pushing and pulling on one end of a Slinky (while someone holds the other end still) generates a wave in which the Slinky is bunched up in some places and stretched out in others. Waves like this in rock are called P waves. The *P* stands for *primary,* because these waves travel fastest and are the first to arrive after an earthquake, but it is easier to think of *P* as meaning *pressure* or *pushing.* P waves can travel through almost any material—whether solid, liquid, or gas—because molecules can always push on their neighbors no matter how weakly they are bound together. (Sound travels as a pressure wave quite similar to a P wave.)

Shaking a Slinky slightly up and down or side to side generates a different type of wave; in rock, such waves are called S waves. The *S* stands for *secondary* but is easier to remember as meaning *shear* or *side to side.* S waves travel only through solids, because the bonds between neighboring molecules in a liquid or gas are too weak to transmit up-and-down or sideways forces.

The speeds and directions of seismic waves depend on the composition, density, pressure, temperature, and phase (solid or liquid) of the material they pass through. For example, P waves reach the side of the world opposite an earthquake, but S waves do not. This tells us that a liquid layer has stopped the S waves, which is how we know that Earth has a liquid outer core (**FIGURE 2**). More careful analysis of seismic waves has allowed geologists to develop a detailed picture of Earth's interior structure.

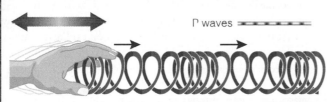

P waves result from compression and stretching in the direction of travel.

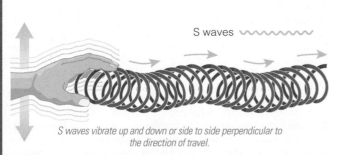

S waves vibrate up and down or side to side perpendicular to the direction of travel.

FIGURE 1 Slinky examples demonstrating P and S waves.

The liquid outer core bends P waves . . .

. . . but stops S waves.

FIGURE 2 Because S waves do not reach the side of Earth opposite an earthquake, we infer that part of Earth's core is liquid.

Accretion:
Gravitational potential energy is converted into kinetic energy . . .

. . . which upon impact is converted to thermal energy.

Differentiation:
Light materials rise to the surface . . .

. . . while dense materials fall to the core, converting gravitational potential energy into thermal energy.

Radioactive Decay:
Mass-energy contained in nuclei is converted into thermal energy.

FIGURE 9.4 The three main heat sources for terrestrial planet interiors are accretion, differentiation, and radioactive decay. Only the last is still a major heat source today.

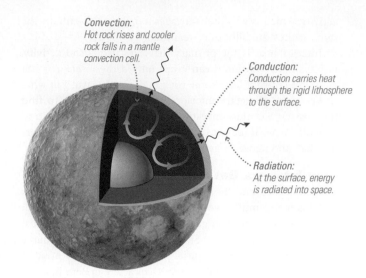

Convection:
Hot rock rises and cooler rock falls in a mantle convection cell.

Conduction:
Conduction carries heat through the rigid lithosphere to the surface.

Radiation:
At the surface, energy is radiated into space.

FIGURE 9.5 The three main cooling processes in planets. Convection can occur only in a planet that is still hot inside.

Note that accretion and differentiation deposited heat into planetary interiors only when the planets were very young. In contrast, radioactive decay provides an ongoing source of heat. Over billions of years, the total amount of heat deposited by radioactive decay has been comparable to or greater than the amount that was deposited initially by accretion and differentiation. However, the rate of radioactive decay has declined with time, so it was an even more significant heat source when the planets were young than it is today.

The combination of the three heat sources explains how the terrestrial interiors ended up with their core-mantle-crust structures. The many violent impacts that occurred during the latter stages of accretion deposited so much energy that the outer layers of the young planets began to melt. This started the process of differentiation, which then released its own additional heat. This heat, along with the substantial heat from early radioactive decay, made the interiors hot enough to melt and differentiate throughout.

How Interiors Cool Off Cooling a planetary interior requires transporting heat outward. Just as there are three basic heating processes for planetary interiors, there are also three basic cooling processes (**FIGURE 9.5**):

- **Convection.** Convection is the process by which hot material expands and rises while cooler material contracts and falls. It therefore transfers heat upward and can occur whenever there is strong heating from below. You can see convection in a pot of soup on a hot burner, and you may be familiar with it in weather: Warm air near the ground tends to rise while cool air above tends to fall.

- **Conduction.** Conduction is the transfer of heat from hot material to cooler material through contact; it is operating when a hot potato transfers its heat to your cooler hand when you pick it up. Conduction occurs through the microscopic collisions of individual atoms or molecules. Molecules of materials in close contact are constantly

colliding with one another, so the faster-moving molecules in hot material tend to transfer some of their energy to the slower-moving molecules of cooler material.

- **Radiation.** Planets ultimately lose heat to space through radiation. Remember that objects emit *thermal radiation* characteristic of their temperatures [**Section 5.4**]; this radiation (light) carries energy away and therefore cools an object. Because of their relatively low temperatures, planets radiate primarily in the infrared.

For Earth, convection is the most important heat transfer process in the interior. Hot rock from deep in the mantle gradually rises, slowly cooling as it makes its way upward. By the time it reaches the top of the mantle, the rock has transferred its excess heat to its surroundings, so it is now cool and it begins to fall. This ongoing process creates individual **convection cells** within the mantle, shown as small circles in Figure 9.5; arrows indicate the direction of flow. Keep in mind that mantle convection primarily involves solid rock, not molten rock. Because solid rock flows quite slowly, mantle convection takes a long time. At the typical rate of mantle convection on Earth—about 1 centimeter per year—it would take about 100 million years for a piece of rock to be carried from the base of the mantle to the top.

COMMON MISCONCEPTIONS

Pressure and Temperature

Some people think that Earth's interior should be hot just because the internal pressure is so high. After all, if we compress a gas from low pressure to high pressure, it heats up. But the same is not necessarily true of rock. High pressure compresses rock only slightly, so the compression causes little increase in temperature. Although high pressures and temperatures sometimes go together in planets, they don't have to. In fact, after all the radioactive elements decay (billions of years from now), Earth's deep interior will become quite cool even though the pressure will be the same as it is today. The temperatures inside Earth and the other planets can remain high only if there is a source of heat for the interior, such as accretion, differentiation, or radioactive decay.

Mantle convection stops at the base of the lithosphere, where the rock is too strong to flow as readily as it does lower down. From the base of the lithosphere to the surface, heat continues upward primarily through conduction. (Some heat also reaches the surface through volcanic eruptions that directly carry hot rock upward.) When the heat finally reaches Earth's surface, it is radiated away into space.

Planetary Size Controls Geological Activity Size is the single most important factor in planetary cooling: Just as a hot potato remains hot inside much longer than a hot pea, a large planet can stay hot inside much longer than a small one. You can see why size is the critical factor by picturing a large planet as a smaller planet wrapped in extra layers of rock. The extra rock acts as insulation, so it takes much longer for interior heat to reach the surface.

SEE IT FOR YOURSELF

The fact that large objects stay warm longer than small objects is easy to see with food and drink. The next time you are eating something large and hot, cut off a small piece; notice how much more quickly the small piece cools than the rest of it. A similar approach can be used to explore the time it takes a cold object to warm up: Find two ice cubes of the same size, crack one into small pieces with a spoon, and then compare the rates of melting. Explain your observations in terms that your friends would understand.

Size is therefore the primary factor in determining geological activity. The relatively small sizes of the Moon and Mercury probably allowed their interiors to cool within a billion years or so after they formed. As they cooled, their lithospheres thickened and mantle convection was confined to deeper and deeper layers. Ultimately, the mantle convection probably stopped altogether. With insufficient internal heat to drive any further movement of interior rock, the Moon and Mercury are now geologically "dead," meaning they have little or no heat-driven geological activity today.

In contrast, the much larger size of Earth has allowed our planet to stay quite hot inside. Mantle convection keeps interior rock in motion and the heat keeps the lithosphere thin, which is why geological activity can continually reshape the surface. Venus probably remains nearly as active as Earth, thanks to its very similar size. Mars, with a size in between those of the other terrestrial worlds, probably represents an intermediate case: It has cooled significantly during its history, but it probably still retains enough internal heat for at least some geological activity.

THINK ABOUT IT

In general, what kind of planet would you expect to have the *thickest* lithosphere: the largest planet, the smallest planet, the planet closest to the Sun, or the planet farthest from the Sun? Why? (Assume the planet has the same composition in all four cases.)

MATHEMATICAL INSIGHT 9.1

The Surface Area–to–Volume Ratio

The total amount of heat contained in a planet depends on its volume, but this heat can escape into space only from its *surface*. As heat escapes, more heat flows upward from the interior to replace it until the interior is no hotter than the surface. The time it takes a planet to lose its internal heat is related to the ratio of the *surface area* through which it loses heat to the *volume* that contains heat, or the *surface area–to–volume ratio*:

$$\text{surface area–to–volume ratio} = \frac{\text{surface area}}{\text{volume}}$$

A spherical planet (radius r) has surface area $4\pi r^2$ and volume $\frac{4}{3}\pi r^3$, so the ratio becomes

$$\underset{\text{(for a sphere)}}{\text{surface area–to–volume ratio}} = \frac{4\pi r^2}{\frac{4}{3}\pi r^3} = \frac{3}{r}$$

Because r appears in the denominator, we conclude that *larger objects have smaller surface area–to–volume ratios*. Note that this idea holds for objects of any shape, which is why the larger of two objects that start at the same temperature retains heat longer. It also explains why crushed ice cools a drink more quickly than an equal amount of ice cubes: The crushed ice pieces have a greater combined surface area than the ice cubes, which means faster cooling because more ice surface is in direct contact with the surrounding liquid.

EXAMPLE: Compare and interpret the surface area–to–volume ratios of the Moon and Earth.

SOLUTION:

Step 1 Understand: We compare the two surface area–to–volume ratios by dividing the Moon's ratio (which is larger, because the Moon is smaller) by Earth's. The result tells us the relative rates at which the two worlds would lose heat if they started with the same internal temperature.

Step 2 Solve: We've already found that the surface area–to–volume ratio for a spherical world is $3/r$. Therefore, we find

$$\frac{\text{surface area–to–volume ratio (Moon)}}{\text{surface area–to–volume ratio (Earth)}} = \frac{3/r_{\text{Moon}}}{3/r_{\text{Earth}}} = \frac{r_{\text{Earth}}}{r_{\text{Moon}}}$$

From Appendix E, the radii are $r_{\text{Moon}} = 1738$ km and $r_{\text{Earth}} = 6378$ km. Substituting these values into our equation, we find

$$\frac{\text{surface area–to–volume ratio (Moon)}}{\text{surface area–to–volume ratio (Earth)}} = \frac{6378 \text{ km}}{1738 \text{ km}} = 3.7$$

Step 3 Explain: The Moon's surface area–to–volume ratio is nearly four times as large as Earth's, which means the Moon would cool four times as fast if both worlds started with the same temperature. However, Earth's larger size also gave it much more internal heat to begin with, which amplifies the difference in heat loss found from the surface area–to–volume ratio alone. Overall, we expect the Moon to have cooled many times faster than Earth, which explains why the Moon's interior is so much cooler than Earth's interior today.

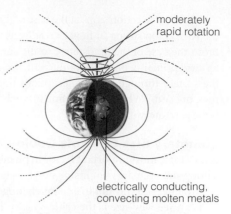

moderately
rapid rotation

electrically conducting,
convecting molten metals

a This photo shows how a bar magnet influences iron filings (small black specks) around it. The *magnetic field lines* (red) represent this influence graphically.

b A similar magnetic field is created by an electromagnet, which is essentially a wire wrapped around a metal bar and attached to a battery. The field is created by the battery-forced motion of charged particles (electrons) along the wire.

c Earth's magnetic field also arises from the motion of charged particles. The charged particles move within Earth's liquid outer core, which is made of electrically conducting, convecting molten metals.

FIGURE 9.6 Sources of magnetic fields.

Why do some planetary interiors create magnetic fields?

Interior heat plays another important role: It can help create a global **magnetic field.** Earth's magnetic field determines the direction in which a compass needle points, but it also plays many other important roles. As we'll discuss in Chapter 10, the magnetic field helps create a *magnetosphere* (see Figure 10.11) that surrounds our planet and diverts the paths of high-energy charged particles coming from the Sun. The magnetic field therefore protects Earth's atmosphere from being stripped away into space by these particles; many scientists suspect that this protection has been crucial to the long-term habitability of Earth, and hence to our own existence.

You are probably familiar with the pattern of the magnetic field created by an iron bar (**FIGURE 9.6a**). Earth's magnetic field is generated by a process more like that of an *electromagnet,* in which the magnetic field arises as a battery forces charged particles (electrons) to move along a coiled wire (**FIGURE 9.6b**). Earth does not contain a battery, but charged particles move with the molten metal in its liquid outer core (**FIGURE 9.6c**). Internal heat causes the liquid metal to rise and fall (convection), while Earth's rotation twists and distorts the convection pattern. The result is that electrons in the molten metal move within Earth's outer core in much the same way as they move in an electromagnet, generating Earth's magnetic field.

We can generalize these ideas to other worlds. There are three basic requirements for a global magnetic field:

1. An interior region of electrically conducting fluid (liquid or gas), such as molten metal

2. Convection in that layer of fluid

3. At least moderately rapid rotation

Earth is the only terrestrial world that meets all three requirements, which is why it is the only terrestrial world with a strong magnetic field. The Moon has no magnetic field, presumably because its core has long since cooled and ceased

convecting. Mars's core probably still retains some heat, but not enough to drive core convection, which is why it also lacks a magnetic field today. Venus probably has a molten core layer much like that of Earth, but either its convection or its 243-day rotation period is too slow to generate a magnetic field. Mercury remains an enigma: It possesses a measurable magnetic field despite its small size and slow, 59-day rotation. The reason for this may be Mercury's huge metal core, which may still be partially molten and convecting.

The same three requirements for a magnetic field also apply to jovian planets and stars. For example, Jupiter's strong magnetic field comes from its rapid rotation and its layer of convecting metallic hydrogen that conducts electricity [**Section 11.1**]. The Sun's magnetic field is generated by the combination of convection of ionized gas (plasma) in its interior and rotation.

THINK ABOUT IT

Recall that the Sun had a strong solar wind when it was young (which cleared out the remaining gas of the solar nebula [**Section 8.2**]) because its magnetic field was much stronger than it is today. What single factor explains why the young Sun's magnetic field was so strong?

MA **Shaping Planetary Surfaces Tutorial, Lessons 1–3**

9.2 SHAPING PLANETARY SURFACES

Now that we have discussed how Earth and other terrestrial worlds work on the inside, we are ready to turn to their surfaces. The surfaces of the planets are remarkably flat. For example, the tallest mountains on Earth rise only about 10 kilometers above sea level, less than 1/600 of Earth's radius—which means they could be represented by grains of sand on a typical globe. Nevertheless, surface features tell us a great deal about the histories of the planets.

FIGURE 9.7 **interactive figure** Artist's conception of the impact process.

What processes shape planetary surfaces?

Although we find a huge variety of geological surface features on Earth and the other terrestrial worlds, nearly all of them can be explained by just four major geological processes:

- **Impact cratering**: the creation of bowl-shaped *impact craters* by asteroids or comets striking a planet's surface
- **Volcanism**: the eruption of molten rock, or *lava*, from a planet's interior onto its surface
- **Tectonics**: the disruption of a planet's surface by internal stresses
- **Erosion**: the wearing down or building up of geological features by wind, water, ice, and other phenomena of planetary weather

Let's examine each of these processes in a little more detail.

Impact Cratering The scarred faces of the Moon and Mercury (see Figure 9.1) attest to the battering that the terrestrial worlds have taken from leftover planetesimals, such as comets and asteroids. They also immediately reveal an important feature of impact cratering: Small craters far outnumber large ones, confirming that many more small asteroids and comets orbit the Sun than large ones. While the Moon and Mercury bear the most obvious scars, all the terrestrial worlds have suffered similar impacts.

An impact crater forms when an asteroid or comet slams into a solid surface (**FIGURE 9.7**). Impacting objects typically hit the surface at a speed between about 40,000 and 250,000 km/hr (10–70 km/s). At such tremendous speed, the impact releases enough energy to vaporize solid rock and blast out a *crater* (the Greek word for "cup"). Debris from the blast shoots high above the surface and then rains down over a large area. If the impact is large enough, some of the ejected material can escape into space.

Craters are usually circular, because an impact blasts out material in all directions, regardless of the incoming object's direction. Laboratory experiments show that craters are typically about 10 times as wide as the objects that create them and about 10–20% as deep as they are wide. For example, an asteroid 1 kilometer in diameter will blast out a crater about 10 kilometers wide and 1–2 kilometers deep. A large crater may have a central peak, which forms when the center rebounds after impact in much the same way that water rebounds after you drop a pebble into it. **FIGURE 9.8** shows two typical impact craters, one on Earth and one on the Moon.

a Meteor Crater in Arizona is more than a kilometer across and almost 200 meters deep. It was created around 50,000 years ago by the impact of a metallic asteroid about 50 meters across.

FIGURE 9.8 Impact craters.

b This photo shows a crater, named Tycho, on the Moon. Note the classic shape and central peak.

A simple bowl-shaped crater, showing a sharp rim . . .

. . . and a ring of ejected debris.

1 km

a A crater with a typical bowl shape.

Unusual ridges suggest the impact debris was muddy.

10 km

b This crater was probably made by an impact into icy ground.

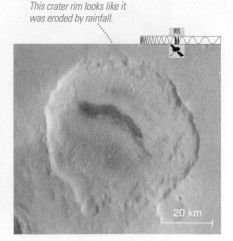

This crater rim looks like it was eroded by rainfall.

20 km

c This crater shows evidence of erosion.

FIGURE 9.9 Crater shapes on Mars tell us about Martian geology. (Photos from *Mars Global Surveyor* (a) and *Viking* spacecraft (b and c).)

Details of crater shapes provide clues about geological conditions. For example, **FIGURE 9.9** contrasts three craters on Mars. The crater in Figure 9.9a has a simple bowl shape, as we expect from the basic cratering process. The crater in Figure 9.9b has an extra large bump in its center and appears to be surrounded by mud flows, suggesting that underground water (or ice) melted or vaporized on impact; the muddy debris then flowed across the surface and hardened into the pattern we see today. The crater in Figure 9.9c shows obvious signs of erosion: It lacks a sharp rim and its floor no longer has a well-defined bowl shape. This suggests that ancient rainfall eroded the crater and that the crater bottom was once a lake. Studies of crater shapes on other worlds provide similar clues to their surface conditions and history.

Volcanism We use the term *volcanism* to refer to any eruption of molten lava, whether the lava comes from a tall volcano or simply rises to the surface through a crack in a planet's lithosphere. Volcanism occurs when underground molten rock, technically called *magma*, finds a path to the surface (**FIGURE 9.10**).

Molten rock tends to rise for three main reasons. First, molten rock is generally less dense than solid rock, and lower-density materials tend to rise when surrounded by higher-density materials. Second, because most of Earth's interior is *not* molten, the solid rock surrounding a chamber of molten rock (a *magma chamber*) can squeeze the molten rock, driving it upward under pressure. Third, molten rock often contains trapped gases that expand as it rises, which can make it rise much faster and lead to dramatic eruptions. The same molten rock that is called magma when it is underground is called *lava* once it erupts onto the surface.

The result of an eruption depends on how easily the lava flows across the surface. Lava that is "runny" can flow far before cooling and solidifying, while "thick" lava tends to collect in one place.* Broadly speaking, lava can shape three different types of volcanic features:

- The runniest lavas flow far and flatten out before solidifying, creating vast **volcanic plains** (**FIGURE 9.11a**).

- Somewhat thicker lavas tend to solidify before they completely spread out, creating **shield volcanoes** (so named because of their shape). Shield volcanoes can be very tall

molten rock in upper mantle

FIGURE 9.10 Volcanism. This photo shows the eruption of an active volcano on the flanks of Kilauea on the Big Island of Hawaii. The inset shows the underlying process: Molten rock collects in a magma chamber and can erupt upward.

---

*The "thickness" of a liquid is described by the technical term *viscosity*, which refers to how much internal friction it has. For example, liquid water has a low viscosity that allows it to flow quickly and easily; honey and molasses flow more slowly because they have higher viscosities.

Lava plains (maria) on the Moon

10 km

a Very runny lava makes flat lava plains like these on the Moon. The long, winding channel near the upper left was made by a river of molten lava.

Olympus Mons (Mars)

50 km

b Slightly thicker lava makes shallow-sloped shield volcanoes, such as Olympus Mons on Mars.

Mount Hood (Earth)

c The thickest lavas make steep-sloped stratovolcanoes like Oregon's Mount Hood.

FIGURE 9.11 The shapes of volcanic structures depend on whether the lava is runny or thick.

but are not very steep. Examples include the mountains of the Hawaiian Islands on Earth and Olympus Mons on Mars (**FIGURE 9.11b**).

- The thickest lavas cannot flow very far before solidifying and therefore build up tall, steep **stratovolcanoes** (**FIGURE 9.11c**). Examples include Mount Fuji (Japan), Mount Kilimanjaro (Tanzania), and Mount Hood (Oregon).

Lava plains and shield volcanoes are made of **basalt,** a mixture of many different minerals that erupts from volcanoes as a high-density but fairly runny lava. All the terrestrial worlds and some jovian moons show evidence of volcanic plains and shield volcanoes, telling us that basalt is common throughout the solar system. Stratovolcanoes are made by lower-density volcanic rock that erupts as a thicker lava; this type of lava is common on Earth but rare in the rest of the solar system.

Volcanic mountains are the most obvious result of volcanism, but volcanism has had a much more profound effect on our planet: It explains the existence of our atmosphere and oceans. Recall that Earth accreted from rocky and metallic planetesimals, while water and other ices were brought in by planetesimals from more distant reaches of the solar system that crashed into the growing planets [**Section 8.2**]. Water and gases became trapped in the interiors of the planets in much the same way that the gas in a carbonated beverage is trapped in a pressurized bottle. Volcanic eruptions later released some of this gas into the atmosphere in a process known as **outgassing**. Outgassing can range from dramatic, as during a volcanic eruption (**FIGURE 9.12a**), to more gradual, as when gas escapes from volcanic vents (**FIGURE 9.12b**). The same type of outgassing also occurred on other terrestrial worlds. That is, virtually all the gas that made the atmospheres of Venus, Earth, and Mars—and the water vapor that rained down to form Earth's oceans—originally was released from the planetary interiors by outgassing.

Tectonics The term *tectonics* comes from the Greek word *tekton*, for "builder"; notice the same root in the word *architect*, which means "master builder." In geology, tectonics refers to the "building" of surface features by stretching, compression, or other forces acting on the lithosphere.

Tectonic features arise in a variety of ways. For example, the weight of a volcano can bend or crack the lithosphere beneath it, while a rising plume of hot material can push up on the lithosphere to create a bulge. However, most tectonic activity is a direct or indirect result of mantle convection (**FIGURE 9.13**). The crust can be compressed in places where adjacent convection cells push rock together; this type of compression helped create the Appalachian Mountains of the eastern United States. Cracks and valleys form in places where adjacent convection cells pull the crust apart; examples of such cracks and valleys include the Guinevere Plains on Venus, the Ceraunius Valleys on Mars, and New Mexico's Rio Grande Valley. (The river named the Rio Grande came *after* the valley formed from tectonic processes.)

On Earth, the ongoing stress of mantle convection ultimately fractured Earth's lithosphere into more than a dozen pieces, or **plates**. These plates move over, under, and around each other in a process we call **plate tectonics**. The movements of the plates explain nearly all Earth's major geological features, including the arrangement of the continents, the nature of the seafloor, and the origin of earthquakes. Because plate tectonics appears to be unique to Earth, we'll save further discussion of it for Section 9.6. (Because the term *plate tectonics* refers to a process, it is generally considered to be singular rather than plural, despite the *-s* on *tectonics*.)

a The eruption of Mount St. Helens, May 18, 1980.

b More gradual outgassing from a volcanic vent in Volcanoes National Park, Hawaii.

FIGURE 9.12 Examples of outgassing, which released the gases that ultimately made the atmospheres of Venus, Earth, and Mars.

Erosion The last of the four major geological processes is *erosion*, which refers to the breakdown or transport of surface rock through the action of ice, liquid, or gas (**FIGURE 9.14**). The shaping of valleys by glaciers (ice), the carving of canyons by rivers (liquid), and the shifting of sand dunes by wind (gas) are all examples of erosion. Note that although we often associate erosion with breakdown, it also builds things such as sand dunes, river deltas, and lake-bed deposits. Indeed, the most common type of rock on Earth's surface is **sedimentary rock**, which has been built over long periods of time as erosion piled sediments into layers on the floors of oceans and seas. You can see layered sedimentary rock in almost any canyon wall, including the walls of the Grand Canyon, which were built up from sediments long before the Colorado River carved the canyon.

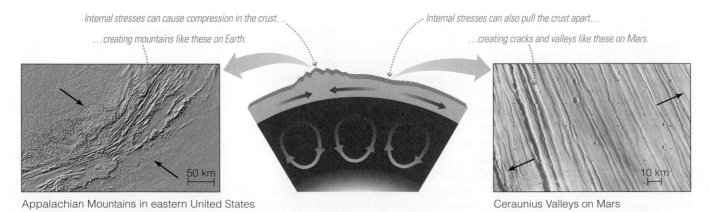

Internal stresses can cause compression in the crust…

…creating mountains like these on Earth.

Internal stresses can also pull the crust apart…

…creating cracks and valleys like these on Mars.

50 km

10 km

Appalachian Mountains in eastern United States

Ceraunius Valleys on Mars

FIGURE 9.13 interactive figure Tectonic forces produce a wide variety of features, including mountain ranges and fractured plains. The Mars image is a visible-light photo from orbit. The Earth image was computer generated from topographical data.

a The Colorado River has been carving the Grand Canyon for millions of years.

b Glaciers created Yosemite Valley during ice ages.

c Wind erosion wears away rocks and builds up sand dunes.

d This river delta is built from sediments worn away by wind and rain and then carried downstream.

FIGURE 9.14 A few examples of erosion on Earth.

How do impact craters reveal a surface's geological age?

Notice that impact cratering is the only one of the four processes with an external cause: the random impacts of objects from space. This fact leads to one of the most useful insights in planetary geology: *We can estimate the geological age of any surface region from its number of impact craters, with more craters indicating an older surface.* By "geological age" we mean the age of the surface as it now appears; a geologically young surface is dominated by features that have formed relatively recently in the history of the solar system, while a geologically old surface still looks about the same today as it did billions of years ago.

You can understand this idea by thinking about why the Moon has so many more impact craters than Earth. Recall that all the terrestrial worlds, regardless of their size or distance from the Sun, were battered by impacts during the heavy bombardment that occurred early in our solar system's history [**Section 8.2**]. Most impact craters were made during that time, and relatively few impacts have occurred since. In places where we see numerous craters, such as on much of the Moon's surface, we must be looking at a surface that has stayed virtually unchanged for billions of years. In contrast, when we see very few craters, as we do on Earth, we must be looking at a younger surface, one on which the scars of ancient impacts have been erased over time by other geological processes, such as volcanic eruptions or erosion.

Careful studies of the Moon have allowed planetary scientists to be more precise about surface ages. The degree of crowding among craters varies greatly from place to place on the Moon (**FIGURE 9.15**). In the **lunar highlands**, craters

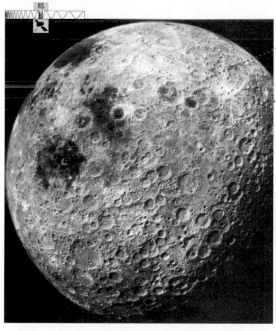

a This *Apollo* photograph of the Moon shows that some areas are much more heavily cratered than others. (This view of the Moon is *not* the one we see from Earth.)

Lunar maria are huge impact basins that were flooded by lava. Only a few small craters appear on the maria.

Lunar highlands are ancient and heavily cratered.

b This map shows the entire surface of the Moon in the same way a flat map of Earth represents the entire globe. Radiometric dating shows the heavily cratered lunar highlands to be a half-billion or more years older than the darkly colored lunar maria, telling us that the impact rate dropped dramatically after the end of the heavy bombardment.

FIGURE 9.15 These photos of the Moon show that crater crowding is closely related to surface age.

are so crowded that we see craters on top of other craters. In the **lunar maria**, we see only a few craters on top of generally smooth volcanic plains. Radiometric dating of rocks brought back by the *Apollo* astronauts indicates that those from the lunar highlands are about 4.4 billion years old, telling us that the heavy cratering occurred early in the solar system's history. Rocks from the maria date to 3.0–3.9 billion years ago, telling us that the lava flows that made these volcanic plains had occurred by that time. Because the maria contain only about 3% as many craters as the highlands, we conclude that the heavy bombardment must have ended by about 4 billion years ago, and relatively few impacts have occurred since that time.

In fact, radiometric dating of Moon rocks has allowed scientists to reconstruct the rate of impacts during much of the Moon's history. Because impacts are essentially random events, the same changes in impact rate over time must apply to all the terrestrial worlds. The degree of crater crowding therefore allows us to estimate the geological age of a planetary surface to within a few hundred million years. Although this is nowhere near as accurate as radiometric dating, it has the advantage of requiring nothing more than orbital photos.

THINK ABOUT IT

Mercury is heavily cratered, much like the lunar highlands. Venus has relatively few craters. Which planet has the older surface? Is there any difference between the ages of the two planets as a whole? Explain how the surfaces of Mercury and Venus can have different ages while the planets themselves are the same age.

Why do the terrestrial planets have different geological histories?

The same four geological processes operate on all solid worlds, yet the terrestrial worlds have had very different geological histories. For example, the surfaces of the Moon and Mercury are dominated by impact craters, but show virtually no features of erosion. Volcanic and tectonic features are found on all the worlds, but we find tall volcanoes only on Earth, Venus, and Mars. And Earth is the only terrestrial world on which features of erosion dominate the landscape. Why do the processes operate so differently on different worlds? Ultimately, we can trace the answer back to three fundamental planetary properties: size, distance from the Sun, and rotation rate (**FIGURE 9.16**).

Planetary Properties Controlling Volcanism and Tectonics
Volcanism and tectonics both require internal heat, which means they depend on planetary size. As shown in Figure 9.16, larger planets have more internal heat and hence more volcanic and tectonic activity.

All the terrestrial worlds probably had some degree of volcanism and tectonics when their interiors were young and hot. As an interior cools, volcanic and tectonic activity subsides. The Moon and Mercury lack ongoing volcanism and tectonics because their small sizes allowed their interiors to cool long ago. Earth has active volcanism and tectonics because it is large enough to still have a hot interior. Venus, nearly the same size as Earth, must still be hot inside and probably also

has active volcanism and tectonics, though we lack conclusive evidence of volcanic eruptions in the past few tens of millions of years. Mars, with its smaller size, has much less volcanism and tectonic activity today than it did in the distant past.

Planetary Properties Controlling Erosion
Erosion arises from weather phenomena such as wind and rain. Figure 9.16 shows that erosion therefore has links with all three fundamental planetary properties. Planetary size is important because erosion requires an atmosphere, and a terrestrial world can have an atmosphere only if it is large enough to have had significant volcanic outgassing and if its gravity is strong enough to have prevented the gas from escaping to space. Distance from the Sun is important because of its role in temperature: If all else is equal (such as planetary size), the higher temperatures on a world closer to the Sun will make it easier for atmospheric gases to escape into space, while the colder temperatures on a world farther from the Sun may cause atmospheric gases to freeze out. Distance is also important because water erosion is much more effective with liquid water than with water vapor or ice and therefore is strongest when moderate temperature allows water vapor to condense into liquid form. Rotation rate is important because it is the primary driver of winds and other weather [**Section 10.2**]: Faster rotation means stronger winds and storms.

The Moon and Mercury lack significant atmospheres and erosion because they lack outgassing today, and any atmospheric gases they had in the distant past have been lost to space. Mars has limited erosion at the present time; it has only a thin atmosphere because much of the water vapor and carbon dioxide outgassed in its past either escaped into space or lies frozen in its polar caps or beneath the surface. Venus and Earth probably had similar amounts of outgassing, but cooler temperatures on Earth led to condensation and the formation of oceans, allowing wind and weather to drive strong erosion. Most of Venus's gas remained in its atmosphere, making its atmosphere much thicker than Earth's, but Venus has little erosion because its slow rotation rate means that it has very little surface wind and its high temperature means that rain never falls to the surface.

Planetary Properties Controlling Impact Cratering
Impacts are random events and therefore the *creation* of craters is not "controlled" by fundamental planetary properties. However, because impact craters can be *destroyed* over time, the number of remaining impact craters on a world's surface *is* controlled by fundamental properties. The primary factor is size: Larger worlds have more volcanism and tectonics (and in some cases erosion), processes that tend to cover up or destroy ancient impact craters over time. Mercury and the Moon are heavily cratered today because their small sizes have supported relatively little geological activity, so we can still see most of the impact craters that formed during the heavy bombardment. Mars has had more geological activity and therefore has somewhat fewer impact craters remaining today. The active geologies of Venus and Earth have erased nearly all ancient craters, leaving only those that have formed relatively recently (within the last billion years or so).

The Role of Planetary Size

Small Terrestrial Planets

Interior cools rapidly...

...so that tectonic and volcanic activity cease after a billion years or so. Many ancient craters therefore remain.

Lack of volcanism means little outgassing, and low gravity allows gas to escape more easily; no atmosphere means no erosion.

Large Terrestrial Planets

Warm interior causes mantle convection...

...leading to ongoing tectonic and volcanic activity; most ancient craters have been erased.

Outgassing produces an atmosphere and strong gravity holds it, so that erosion is possible.

Core may be molten, producing a magnetic field if rotation is fast enough.

The Role of Distance from the Sun

Planets Close to the Sun

Sun

Surface is too hot for rain, snow, or ice, so little erosion occurs.

High atmospheric temperature allows gas to escape more easily.

Planets at Intermediate Distances from the Sun

Moderate surface temperatures can allow for oceans, rain, snow, and ice, leading to substantial erosion.

Gravity can more easily hold atmospheric gases.

Planets Far from the Sun

Low surface temperatures can allow for ice and snow, but not rain or oceans, limiting erosion.

Atmosphere may exist, but gases can more easily condense to make surface ice.

The Role of Planetary Rotation

Slow Rotation

Less wind and weather means less erosion, even with a substantial atmosphere.

Slow rotation means weak magnetic field, even with a molten core.

Rapid Rotation

More wind and weather means more erosion.

Rapid rotation is necessary for a global magnetic field.

FIGURE 9.16 A planet's fundamental properties of size, distance from the Sun, and rotation rate are responsible for its geological history. This illustration shows the role of each key property separately, but a planet's overall geological evolution depends on the combination of all these effects.

9.3 GEOLOGY OF THE MOON AND MERCURY

In the rest of this chapter we will investigate the geological histories of the terrestrial worlds, with the goal of explaining why each ended up as it did. Because smaller worlds have simpler geological histories, we'll start with the smallest worlds, the Moon and Mercury (FIGURE 9.17).

The similar appearances of the Moon and Mercury should not be surprising: Their small sizes mean that most of their internal heat was lost long ago, leaving them without ongoing volcanism or tectonics. They therefore still show the craters made by most of the objects that have hit them during the past $4\frac{1}{2}$ billion years. Small size also explains their lack of significant atmospheres and erosion: Their gravity is too weak to hold gas for long periods of time, and without ongoing volcanism they lack the outgassing needed to replenish gas lost in the past. Nevertheless, both worlds have a few volcanic or tectonic features, indicating that they had geological activity in the past. This also makes sense: Long ago, before they had a chance to cool, even these relatively small worlds were hot enough inside to support some volcanism and tectonics.

What geological processes shaped our Moon?

The familiar face of the full moon reveals some of the surface distinctions we've already discussed (FIGURE 9.18). The bright, heavily cratered regions are lunar highlands. The smooth dark regions are the lunar maria, which got their name because they look much like oceans when seen from afar; *maria* (singular, *mare*) is Latin for "seas." But the entire surface of the Moon should have been packed with craters during the heavy bombardment. So what happened to the craters that must once have been located in the regions where we now see the maria?

Moon   Mercury

FIGURE 9.17 Views of the Moon and Mercury, shown to scale. The Mercury photo is from *MESSENGER*.

FIGURE 9.18 The familiar face of the Moon shows numerous dark smooth maria.

Volcanism and Tectonics in the Lunar Maria

FIGURE 9.19 shows how the maria probably formed. During the heavy bombardment, craters covered the Moon's entire surface. The largest impacts were violent enough to fracture the Moon's lithosphere beneath the huge craters they created. However, the Moon's interior had already cooled since its formation, so there was no molten rock to flood these craters immediately. Instead, the lava floods came hundreds of millions of years later, thanks to heat released by the decay of radioactive elements in the Moon's interior. This heat gradually built up during the Moon's early history, until mantle material melted between about 3 and 4 billion years ago. Molten rock then welled up through the cracks in the lithosphere, flooding the largest impact craters with lava. The maria are generally circular because they are essentially flooded craters (and craters are almost always round). Their dark color comes from the dense, iron-rich rock (basalt) that rose up from the lunar mantle as molten lava. The flat surfaces of the maria tell us that the lunar lava spread easily and far, which means that it must have been among the runniest lava in the solar system. It was so runny that in places it flowed like rivers of molten rock, carving out long, winding channels (see Figure 9.11a). The relatively few craters within the maria today were made by impacts that occurred after the maria formed, when the heat from radioactive decay was no longer sufficient to produce lava flows.

The Moon's relatively few tectonic features are also found within the maria. These features, which look much like surface wrinkles, probably were created by small-scale tectonic stresses that arose when the lava that flooded the maria cooled and contracted (FIGURE 9.20).

Surprisingly, the Moon's far side looks quite different from the side facing Earth. The far side landscape consists almost entirely of heavily cratered highlands, with very few maria (see Figure 9.15); indeed, the entire far side essentially has a higher altitude than the near side. No one knows what caused this difference in altitude between the two sides, but it apparently allowed lava to well up and create maria much more easily on the near side.

a This illustration shows the Mare Humorum region as it probably looked about 4 billion years ago, when it would have been completely covered in craters.

b Around that time, a huge impact excavated the crater that would later become Mare Humorum. The impact fractured the Moon's lithosphere and erased the many craters that had existed earlier.

c A few hundred million years later, heat from radioactive decay built up enough to melt the Moon's upper mantle. Molten lava welled up through the lithospheric cracks, flooding the impact crater.

d This photo shows Mare Humorum as it appears today, and the inset shows its location on the Moon.

FIGURE 9.19 The lunar maria formed between 3 and 4 billion years ago, when molten lava flooded large craters that had formed hundreds of millions of years earlier. This sequence of diagrams represents the formation of Mare Humorum.

The Moon Today The Moon's era of geological activity is long gone. Today, the Moon is a desolate and nearly unchanging place. Impacts may occur from time to time, but we are unlikely ever to witness a major one. The only ongoing geological change on the Moon is a very slow "sandblasting" of the surface by *micrometeorites*, sand-size particles from space. These tiny particles burn up as meteors in the atmospheres of Earth, Venus, and Mars but rain directly onto the surface of the airless Moon. The micrometeorites gradually pulverize the surface rock, which explains why the lunar surface is covered by a thin layer of powdery "soil." You can see this powdery surface in photos from the *Apollo* missions to the Moon (**FIGURE 9.21**). Pulverization by micrometeorites is a very slow process, so the rover tracks and astronaut footprints could remain for millions of years.

Aside from the micrometeorite sandblasting and rare larger impacts, the Moon has been "geologically dead" since the maria formed more than 3 billion years ago. Nevertheless, the Moon remains a prime target of exploration, because study of lunar geology can help us understand both the Moon's history and the history of our solar system. Several robotic missions are currently studying the Moon, and more are planned. In the future, humans may return to the Moon, and lunar resources may someday prove valuable.

THINK ABOUT IT

It has now been more than 40 years since the last time humans walked on the Moon. Do you favor a return of humans to the Moon? Why or why not?

FIGURE 9.20 The wrinkles within the maria were probably made when the lava that flooded the maria cooled and contracted. (This photo shows Mare Imbrium.)

a Astronaut Gene Cernan takes the Lunar Roving Vehicle for a spin during the final *Apollo* mission to the Moon (*Apollo 17*, December 1972).

b The *Apollo* astronauts left footprints, like this one, in the Moon's powdery "soil." Micrometeorites will eventually erase the footprints, but not for millions of years.

FIGURE 9.21 The Moon today is geologically dead, but it can still tell us a lot about the history of our solar system.

50 km

100 km

2 km

a A close-up view of Mercury's surface, showing impact craters and smooth regions where lava apparently covered up craters.

b The Rembrandt Basin, a large impact crater on Mercury.

c The light-colored "hollows" on this crater floor (shown with colors enhanced) are thought to have formed as easy-to-vaporize materials escaped over millions of years, causing the rock to crumble and make hollowed-out pits.

FIGURE 9.22 *MESSENGER* images of geological features of Mercury.

What geological processes shaped Mercury?

Mercury looks so much like the Moon that it's often difficult to tell which world you are looking at in surface photos. Nevertheless, the two worlds have a few important differences.

Impact Craters and Volcanism on Mercury Impact craters are visible almost everywhere on Mercury, indicating an ancient surface. However, Mercury's craters are less crowded together than the craters in the most ancient regions of the Moon, suggesting that molten lava later covered up some of the craters that formed on Mercury during the heavy bombardment (**FIGURE 9.22**). As on the Moon, these lava flows probably occurred when heat from radioactive decay accumulated enough to melt part of the mantle. Although we have not found evidence of lava flows as large as those that created the lunar maria, the lesser crater crowding and the many smaller lava plains suggest that Mercury had at least as much volcanism as the Moon.

The largest surface features on Mercury are huge impact craters called *basins*. The impacts were so large and so violent that the craters filled with molten rock heated by the impact. The *Caloris Basin* spans more than half of Mercury's radius. Many of these basins have few craters within them, indicating that they must have formed at a time when the heavy bombardment was already subsiding.

Tectonic Evidence of Planetary Shrinking The most surprising features of Mercury are its many tremendous cliffs—evidence of a type of tectonics quite different from anything we have found on any other terrestrial world (**FIGURE 9.23**). Mercury's cliffs have vertical faces up to 3 or more kilometers high and typically run for hundreds of kilometers across the surface. They probably formed when tectonic forces compressed the crust, causing the surface to crumple. Because crumpling would have shrunk the portions of the surface it affected, Mercury as a whole could not have stayed the same size unless other parts of the surface expanded. However, we find no evidence of similar global-scale "stretch marks" on Mercury. Could the whole planet have shrunk?

Apparently so. Recall that in addition to being larger than the Moon, Mercury also has a surprisingly large iron core; Mercury therefore gained and retained more internal heat from accretion and differentiation than the Moon. This heat caused Mercury's core to swell in size. Later, as the core cooled,

Mercury's core and mantle shrank . . .

Some portions of the crust were forced to slide under others.

Today we see long, steep cliffs created by this crustal movement.

Not to scale!

. . . causing Mercury's crust to contract.

100 km

a This diagram shows how Mercury's cliffs probably formed as the core shrank and the surface crumpled.

b This cliff extends about 100 kilometers in length, and its vertical face is as much as 2 kilometers tall. (Photo from *Mariner 10*.)

FIGURE 9.23 Long cliffs on Mercury offer evidence that the entire planet shrank early in its history, perhaps by as much as 20 kilometers in radius.

it contracted by perhaps as much as 20 kilometers in radius. The mantle and lithosphere must have contracted along with the core, generating the tectonic stresses that created the great cliffs. The contraction probably also closed off any remaining volcanic vents, ending Mercury's period of volcanism.

Mercury Today Because Mercury is larger than the Moon, it probably retained its internal heat a bit longer. Nevertheless, it is still small enough that its interior should have cooled relatively quickly. Crater counts on Mercury suggest that most of its volcanic and tectonic activity ceased within 1 to 2 billion years after its formation, though some features may be more recent. For most practical purposes, Mercury is geologically "dead" today, a fact traceable to its small size.

Although there is no active volcanism on Mercury today, its surface still displays one odd form of geological activity: Some crater floors appear to be releasing easily vaporized materials from the rock, causing the rock to crumble and make pits nicknamed "hollows" (see Figure 9.22c). The release of the vaporized gases leaves behind a light-colored coating, whose composition remains unknown. Scientists hope that ongoing *MESSENGER* studies will tell us more about its nature.

Mercury has always seemed an unusual case in our solar system, but the discovery of numerous planets in other solar systems that are rocky and very close to their stars suggests that Mercury may be representative of a class of planets more common elsewhere.

9.4 GEOLOGY OF MARS

As we turn our attention to Mars, it's worth noting that Mars has long held a special place in the human imagination. Its bright reddish color made it a part of many ancient myths and legends. Then, starting more than 200 years ago, Mars became a prime example of scientific enthusiasm gone astray.

Have you ever wondered why people often speak of Martians but rarely of, say, Venusians or Jupiterians? The story begins with observations of Mars made in the late 18th century by brother and sister astronomers William and Caroline Herschel, who are also famous for discovering the planet Uranus. They discovered several features of Mars that resemble features of Earth, including a similar axis tilt, a day just slightly longer than 24 hours, the presence of polar caps, and seasonal variations in appearance over the course of the Martian year (about 1.9 Earth years). By 1784, taking these resemblances to an unwarranted extreme, William Herschel had begun speaking with confidence about "inhabitants" of Mars.

The hypothetical Martians got a bigger break about a century later. In 1879, Italian astronomer Giovanni Schiaparelli reported seeing a network of linear features on Mars through his telescope. He named these features *canali,* by which he meant the Italian word for "channels," but it was frequently translated as "canals." Excited by what sounded like evidence of intelligent life, wealthy American astronomer Percival Lowell commissioned the building of an observatory for the study of Mars.

The Lowell Observatory opened in 1894 in Flagstaff, Arizona. Barely a year later, Lowell published detailed maps of the Martian canals and the first of three books in which

FIGURE 9.24 The image at left is a telescopic photo of Mars. The image at right is a drawing of Mars made by Percival Lowell. As you can see, Lowell had a vivid imagination. Nevertheless, if you blur your eyes while looking at the photo, you might see how some of the features resemble what Lowell thought he saw.

he argued that the canals were the work of an advanced civilization. He suggested that Mars was falling victim to unfavorable climate changes and that the canals had been built to carry water from the poles to thirsty cities elsewhere. Lowell's work drove rampant speculation about Martians and inspired science fiction fantasies such as H. G. Wells's *The War of the Worlds* (published in 1898). Other scientists were skeptical of Lowell's claims, but the public imagination kept them alive until 1965, when NASA's *Mariner 4* spacecraft flew past Mars and sent back photos of a barren, cratered surface. Lowell's canals were nowhere to be seen, confirming that they had been the work of a vivid imagination upon blurry telescopic images (**FIGURE 9.24**).

The reality may have dampened the enthusiasm of science fiction writers, but scientists find Mars as intriguing as ever. Indeed, although we now know there have never been Martian cities, it's possible that Mars has harbored more primitive life. The search for past or present life on Mars is one of the major goals of Mars exploration, and a first step in this search is understanding Martian geology.

What geological processes have shaped Mars?

Mars is much larger than the Moon or Mercury (see Figure 9.1), so we expect it to have had a more interesting and varied geological history. However, it is much smaller than Venus or Earth, so we expect it to be less geologically active than our own world. Observations confirm this basic picture, though its geology has also been influenced by the fact that it is about 50% farther than Earth from the Sun.

Interestingly, although Mars has only about one-fourth the surface area of Earth, both planets have nearly the same amount of *land* area, because Earth's surface is about three-fourths covered by water. For context, **FIGURE 9.25** shows the full surface of Mars (and the locations of particular features we will discuss later), which reveals extensive evidence of all four geological processes.

Impact Cratering on Mars Aside from the polar caps, the most striking feature of Figure 9.25 is the dramatic difference in terrain around different parts of Mars. Much of

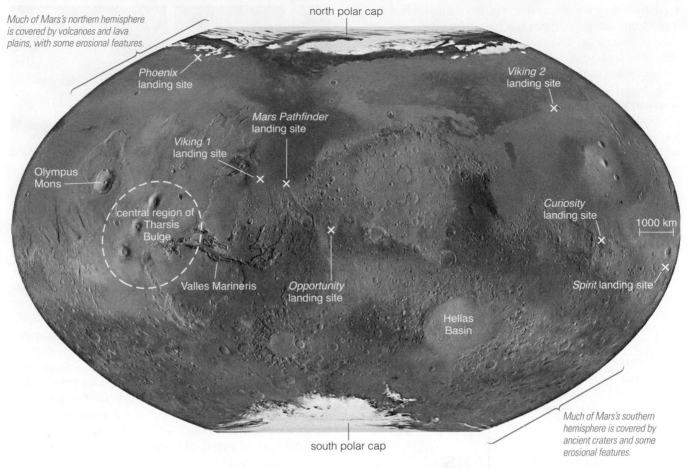

Much of Mars's northern hemisphere is covered by volcanoes and lava plains, with some erosional features.

north polar cap

Phoenix landing site

Viking 2 landing site

Mars Pathfinder landing site

Viking 1 landing site

Olympus Mons

Curiosity landing site

central region of Tharsis Bulge

1000 km

Valles Marineris

Opportunity landing site

Spirit landing site

Hellas Basin

south polar cap

Much of Mars's southern hemisphere is covered by ancient craters and some erosional features.

FIGURE 9.25 interactive figure This image showing the full surface of Mars is a composite made by combining more than 1000 images with more than 200 million altitude measurements from the *Mars Global Surveyor* mission. Several key geological features are labeled, and landing sites of Mars missions are marked.

the southern hemisphere has relatively high elevation and is scarred by numerous large impact craters, including the very large crater known as the Hellas Basin. In contrast, the northern plains tend to be below the average Martian surface level and show few impact craters. The differences in cratering tell us that the southern highlands are an older surface than the northern plains, which must have had their early craters erased by other geological processes.

THINK ABOUT IT

Which fundamental planetary property (size, distance from the Sun, or rotation rate) explains why Earth does not have any terrain that is as heavily cratered as the southern highlands of Mars?

Volcanism and Olympus Mons Further study suggests that volcanism was the most important process in erasing craters on the northern plains, although tectonics and erosion also played a part. However, no one knows why volcanism affected the northern plains so much more than the southern highlands or why the two regions differ so much in elevation.

More dramatic evidence of volcanism on Mars comes from several towering shield volcanoes. One of these, Olympus Mons, is the tallest known volcano in the solar system (**FIGURE 9.26**; see also Figure 9.11b). Its base is some 600 kilometers across, large enough to cover an area the size of Arizona. Its peak stands about 26 kilometers above the average Martian surface level, or about three times as high as Mount Everest stands above sea level on Earth. Much of Olympus Mons is rimmed by a cliff that in places is 6 kilometers high.

Olympus Mons and several other large volcanoes are concentrated on or near the continent-size *Tharsis Bulge* (see Figure 9.25). Tharsis is some 4000 kilometers across, and most of it rises several kilometers above the average Martian surface level. It was probably created by a long-lived plume of rising mantle material that bulged the surface upward and provided the molten rock for the eruptions that built the giant volcanoes.

Is there any ongoing volcanism on Mars? Until recently, we didn't think so. We expect Mars to be much less volcanically active than Earth, because its smaller size has allowed its interior to cool much more, and Martian volcanoes show enough impact craters on their slopes to suggest that they have been inactive for at least tens of millions of years. However, geologically speaking, tens of millions of years is

FIGURE 9.26 Olympus Mons, photographed from orbit; note the tall cliff around its rim and the central volcanic crater from which lava erupted. The inset shows a 3-D perspective on this immense volcano.

FIGURE 9.27 Valles Marineris is a huge system of valleys on Mars created in part by tectonic stresses. The inset shows a perspective view looking north across the center of the canyon, obtained by *Mars Express.*

not that long. In addition, radiometric dating of meteorites that appear to have come from Mars (so-called *Martian meteorites* [**Section 24.2**]) shows some of them to be made of volcanic rock that solidified from molten lava as little as 180 million years ago—quite recently in the $4\frac{1}{2}$-billion-year history of the solar system. Given this evidence of geologically recent volcanic eruptions, it is likely that Martian volcanoes will erupt again someday. In addition, scientists suspect the presence of methane in the Martian atmosphere, which might also point to ongoing volcanic activity [**Section 24.2**]. Nevertheless, the Martian interior is presumably cooling and its lithosphere thickening. Within a few billion years, Mars will become as geologically dead as the Moon and Mercury.

Tectonics and Valles Marineris Mars also has tectonic features, though none on a global scale like the plate tectonics of Earth. The most prominent tectonic feature is the long, deep system of valleys called *Valles Marineris* (**FIGURE 9.27**). Named for the *Mariner 9* spacecraft that first imaged it, Valles Marineris extends almost a fifth of the way along the planet's equator. It is as long as the United States is wide and almost four times as deep as Earth's Grand Canyon. No one knows exactly how Valles Marineris formed. Parts of the canyon are completely enclosed by high cliffs on all sides, so neither flowing lava nor water could have been responsible. However, extensive cracks on its western end run up against the Tharsis Bulge (see Figure 9.25), suggesting a connection between the two features. Perhaps it formed through tectonic stresses accompanying the uplift that created Tharsis, cracking the surface and leaving the tall cliff walls of the valleys.

Erosion on Mars Impacts, volcanism, and tectonics explain most of the major geological features of Mars, but closer examination shows extensive evidence of erosion by liquid water. For example, **FIGURE 9.28** shows what appears to be a dried-up riverbed (see also Figure 9.1) that was almost certainly carved by running water, although we cannot yet say whether the water came from runoff after rainfall, from erosion by water-rich debris flows, or from an underground source. Regardless of the specific mechanism, water is the only substance that could have been liquid under past Martian conditions and that is sufficiently abundant to have created such extensive erosion features. We must therefore look more carefully at the evidence suggesting that Mars once had abundant surface water.

What geological evidence tells us that water once flowed on Mars?

No liquid water exists anywhere on the surface of Mars today. We know this in part because orbiting spacecraft have photographed most of the surface in high resolution and have seen

FIGURE 9.28 This photo, taken by *Mars Reconnaissance Orbiter,* shows what appears to be a dried-up meandering riverbed, now filled with dunes of windblown dust. Notice the numerous small impact craters scattered about the image.

no lakes, rivers, or even puddles. More important, current surface conditions do not allow liquid water to remain stable on Mars. In most places and at most times, Mars is so cold that any liquid water would immediately freeze into ice. Even when the temperature rises above freezing, as it often does at midday near the equator, the air pressure is so low that liquid water would quickly evaporate [**Section 5.3**]. If you put on a spacesuit and took a cup of water outside your pressurized spaceship, the water would rapidly either freeze or boil away (or some combination of both).

When we combine the clear evidence of water erosion with the fact that liquid water is unstable on Mars today, we conclude that Mars must once have had very different surface conditions—conditions such as warmer temperatures and greater air pressure that would have allowed water to flow and rain to fall. Geological evidence indicates that this warmer and wetter period must have ended long ago. For example, notice the impact craters that lie on top of the channels in Figure 9.28. Counts of these craters suggest that no water has flowed through these channels for at least about 3 billion years.

Evidence for Ancient Water Flows Strong evidence that Mars had rain and surface water in the distant past comes from both orbital and surface studies. **FIGURE 9.29a** shows a broad region of the ancient, heavily cratered southern highlands. Notice the indistinct rims of many large craters and the relative lack of small craters. Both facts argue for ancient rainfall, which would have eroded crater rims and erased small craters altogether. **FIGURE 9.29b** shows a three-dimensional perspective of the surface that suggests water once flowed between two ancient crater lakes. **FIGURE 9.29c** shows what

looks like a river delta where water flowed into an ancient crater. Further evidence that the crater was once a lake comes from images and spectra indicating the presence of clay minerals on the crater floor, presumably deposited by sediments flowing down the river. Some studies even suggest that the northern plains may once have held a vast ocean, though the evidence is less definitive than the evidence for smaller lakes. The evidence for the ocean comes from features that look like an ancient shoreline, along with radar data suggesting that the rock along the proposed shoreline is sedimentary rather than volcanic, just as we would expect if there had once been an ocean; this rock may even contain water ice in its pores and cracks.

Surface studies further strengthen the case for past water. In 2004, the robotic rovers *Spirit* and *Opportunity* landed on opposite sides of Mars. *Spirit* landed in Gusev Crater, the site of a possible ancient lake (see Figure 9.29b). *Opportunity* landed in the Meridiani Plains, where orbital spacecraft had detected spectroscopic hints of minerals that form in water. The rovers carried cameras, instruments to identify rock composition, and a grinder to expose fresh rock for analysis. The rovers long outlasted their design lifetime of 3 months, with *Spirit* lasting more than 6 years and *Opportunity* still going as this book was being written, more than 8 years after arrival.

The rovers provided clinching evidence through identification of several minerals known to form only in the presence of liquid water. Immediately after landing, *Opportunity* found tiny spheres—nicknamed "blueberries" — that contain the iron-rich mineral hematite, and nearby rocks containing the sulfur-rich mineral jarosite. Terrestrial examples of these minerals form in standing water or water seeping through rock (**FIGURE 9.30**). Detailed chemical analysis suggests these

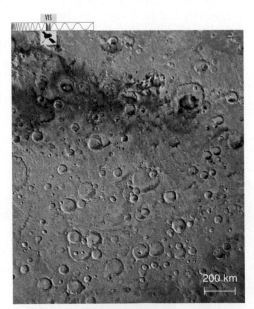

a This photo shows a broad region of the southern highlands on Mars. The eroded rims of large craters and the relative lack of small craters suggest erosion by rainfall.

b This computer-generated perspective view shows how a Martian valley forms a natural passage between two possible ancient lakes (shaded blue). Vertical relief is exaggerated 14 times to reveal the topography.

landing site of *Spirit* rover

100 km

c Combined visible/infrared image of an ancient river delta that formed where water flowing down a valley emptied into a lake filling a large crater (portions of the crater wall are identified). Clay minerals are identified in green.

portion of crater wall

river delta deposits

1 km

FIGURE 9.29 More evidence of past water on Mars.

Mars (Endurance Crater)   **Earth (Utah)**

FIGURE 9.30 *"Blueberries"* on two planets. In both cases the foreground shows hematite "blueberries," which formed within sedimentary rock layers like those in the background, then eroded out and rolled downhill; the varying tilts of the rock layers hint at changing wind or waves during formation. The background rocks are about twice as far away from the camera in the Earth photo as in the Mars photo (taken by the *Opportunity* rover).

FIGURE 9.31 This enhanced-color image shows a layer of the mineral gypsum filling a crack in Martian rock. As on Earth, the gypsum formed from mineral-rich water flowing through a crack in the rock.

where it landed in 2012 (**FIGURE 9.32**). The landing site was chosen in large part because orbital images of the crater's central mountain—the 5-kilometer-tall Mount Sharp—indicate that it is built from sedimentary rock layers dating to many different times over the past several billion years. *Curiosity* carries the most powerful suite of scientific instruments ever landed on another world, including cameras, drills, microscopes, rock and soil analyzers, a laser to vaporize rock, and a spectrograph to analyze the vaporized material.

minerals formed in a salty environment such as a sea or ocean. In 2012, after 20 miles of roving, *Opportunity* came across mineral outcrops made of gypsum, another sulfur-rich mineral that only forms in water (**FIGURE 9.31**); on Earth it's the main ingredient in drywall building material. *Opportunity* team leaders called the discovery a "slam-dunk story that water flowed through underground fractures in the rock."

Scientists expect to learn much more about the history of Martian water as the *Curiosity* rover explores Gale Crater,

FIGURE 9.32 The view of Mount Sharp in Gale Crater, taken from the Bradbury landing site by NASA's *Curiosity* rover (see Figure 9.25). In the main image, parts of the rover and its shadow are visible in the foreground. The zoom-in on Mount Sharp, more than 20 km away, shows the tilted sedimentary rock layers that scientists hope will reveal whether Mars was once habitable.

Evidence for More Recent Water Flows Taken together, the orbital and surface studies provide convincing evidence for abundant liquid water in Mars's distant past. But could water have flowed more recently? As we'll discuss in Chapter 10, abundant water ice still remains on Mars at the polar caps and underground over much of the planet. Although this ice probably represents only a fraction of the water that once existed on Mars, it is still enough that if it all melted, it could make an ocean about 10 meters deep over the entire planet. The atmospheric pressure on Mars is too low today for significant melting to occur on the surface, but it is possible that some liquid water exists underground, near sources of volcanic heat. If so, this liquid water could potentially provide a home for microscopic life. Moreover, if there is enough volcanic heat, ice might occasionally melt and flow along the surface for the short time until it freezes or evaporates. Although we have found no geological evidence to suggest that any large-scale water flows have occurred on Mars in the past billion years, orbital photographs offer tantalizing hints of smaller-scale water flows in much more recent times.

The strongest evidence comes from photos of gullies on crater and channel walls (FIGURE 9.33). These gullies look strikingly similar to the gullies we see on almost any eroded slope on Earth, and spacecraft images show that new gullies are still forming. One hypothesis suggests that the gullies form when snow accumulates on the crater walls in winter and melts away from the base of the snowpack in spring. If

this hypothesis is correct, the water at the base could melt (rather than sublimating directly to water vapor, as ice normally does on Mars) because of the angle of sunlight and the pressure of the overlying snow; the melting may have occurred during a time within the past million years or so when Mars's axis tilt was slightly different than it is today. Alternatively, the gullies may be formed by landslides, which have been seen to occur elsewhere on Mars with the change of seasons [**Section 10.4**].

If any liquid water does sometimes flow on Mars, the amount must be a tiny fraction of the water that flowed when riverbeds and lakes were formed long ago. Mars clearly was much warmer and wetter in the past than it is today. Ironically, Percival Lowell's supposition that Mars was drying up has turned out to be basically correct, although in a very different way than he imagined. We'll discuss the reasons for Mars's dramatic climate change in Chapter 10.

9.5 GEOLOGY OF VENUS

The surface of Venus is searing hot with brutal pressure [**Section 7.1**], making it seem quite unlike the "sister planet" to Earth it is sometimes called. However, beneath the surface, Venus and Earth must be quite similar. The two planets are nearly the same size—Venus is only about 5% smaller than Earth in radius—and their similar densities (see Table 7.1) suggest similar overall compositions. We therefore expect the interiors of Venus and Earth to have the same structure and to retain about the same level of internal heat today. Nevertheless, we see ample evidence of "skin-deep" differences in surface geology.

What geological processes have shaped Venus?

Venus's thick cloud cover prevents us from seeing through to its surface with visible light, but we can study its geological features with radar because radio waves can pass through clouds. **Radar mapping** bounces radio waves off the surface and uses the reflections to create three-dimensional images of the surface. From 1990 to 1993, the *Magellan* spacecraft used radar to map Venus's surface with resolution good enough to permit identification of features as small as 100 meters across (**FIGURE 9.34**). Scientists have named most of the major geological features on Venus for goddesses and famous women. Three elevated "continents" (the three "Terra" in Figure 9.34) are the largest surface features.

Impact Cratering on Venus Figure 9.34 shows that Venus has many geological similarities to Earth. Like Earth, Venus has a relatively small number of impact craters, indicating that its ancient craters were erased by other geological processes. Moreover, while Venus has a few large craters, it lacks the small craters that are most common on other worlds—probably because the small objects that could make such craters burn up completely as they enter Venus's thick atmosphere. But even this thick atmosphere has little effect on objects large enough to make the craters we can see in global views like Figure 9.34.

FIGURE 9.33 This *Mars Reconnaissance Orbiter* image supports the hypothesis that running water has etched gullies into crater walls. The image below shows the crater, and the close-up shows details of a gully network that has carried sediments downward. The lack of small impact craters on the sediment deposits indicates that the gullies formed within the last million years or so.

100 m

These two volcanic peaks are probably much like the shield volcanoes that make up the Hawaiian Islands on Earth.

100 km

The round blobs are steep stratovolcanoes, apparently built from a "thick" lava.

10 km

Ishtar Terra

Aphrodite Terra

Tectonic forces have fractured and twisted the crust.

10 km

Lada Terra

This central image shows the full surface of Venus, 98% of which was mapped by Magellan. Notice the three large, elevated "continents" called Ishtar Terra, Lada Terra, and Aphrodite Terra.

Impact craters, like this one, are relatively rare on Venus and are distributed uniformly over the surface.

10 km

This round corona was probably made by a mantle plume. It is dotted with small volcanoes (the round dots) and surrounded by tectonic stress marks.

50 km

FIGURE 9.34 These images show the surface of Venus as revealed by radar observations from the *Magellan* spacecraft. Bright regions in the radar images represent rough areas or higher altitudes.

Volcanic and Tectonic Features Volcanism has clearly been important on Venus. Like Earth, Venus shows evidence of having had volcanic flows with a variety of lava types, since we see both lava plains and volcanic mountains. Some mountains are shield volcanoes, indicating eruptions in which the lava was about as runny as that which formed the Hawaiian Islands on Earth. A few volcanoes have steeper sides, indicating eruptions of a thicker lava.

Tectonics has also been very important on Venus, as the entire surface appears to have been extensively contorted and fractured by tectonic forces. Some of these features, including the large circular *coronae* (Latin for "crowns"), provide strong evidence for mantle convection beneath the lithosphere. The coronae were probably pushed upward by hot, rising plumes of rock in the mantle. The plumes probably also forced lava to the surface, which would explain why numerous volcanoes are found near coronae.

Venus almost certainly remains geologically active today, since it should still retain nearly as much internal heat as Earth. Although we have not witnessed any volcanic eruptions, at least two lines of evidence suggest recent volcanic activity. First, Venus's clouds contain sulfuric acid. Sulfuric acid is made from sulfur dioxide (SO_2) and water, and the sulfur dioxide must have entered the atmosphere through volcanic outgassing. Because sulfur dioxide is gradually removed from the atmosphere by chemical reactions with surface rocks, the existence of sulfuric acid clouds means that volcanic outgassing must have occurred within about the past 100 million years. The second line of evidence narrows the time scale further. The European Space Agency's *Venus Express* spacecraft (orbiting Venus since 2006) detected an infrared spectral feature on three volcanoes that suggests their rocks are "fresh," erupted within about the past 250,000 years (**FIGURE 9.35**).

IR DATA

100 km

FIGURE 9.35 This composite image shows a volcano called Idunn Mons on Venus. Surface topography details are from NASA's *Magellan* radar mapper (enlarged about 30 times to make the volcano easier to see), and colors represent infrared data from the *Venus Express* spacecraft. Red colors indicate relatively new rock that has not been chemically altered by Venus's harsh atmosphere, suggesting that lava flows occurred within the past 250,000 years.

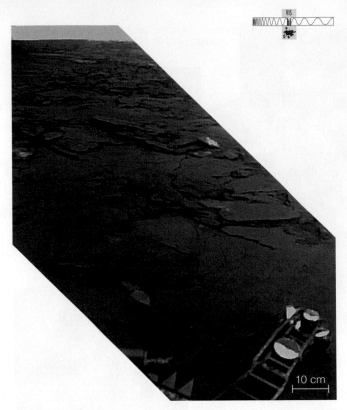

10 cm

FIGURE 9.36 This photo from one of the former Soviet Union's *Venera* landers shows Venus's surface; part of the lander is in the foreground, with sky in the background. Many volcanic rocks are visible, hardly affected by erosion despite their presumed age of about 750 million years (the age of the entire surface).

Weak Erosion We might naively expect Venus's thick atmosphere to produce strong erosion, but the view both from orbit and on the surface suggests otherwise. The former Soviet Union sent several landers to Venus in the 1970s and early 1980s. Before the intense surface heat destroyed them, the probes returned images of a bleak, volcanic landscape with little evidence of erosion (**FIGURE 9.36**). We can trace the lack of erosion on Venus to two simple facts. First, Venus is far too hot for any type of rain or snow on its surface. Second, as we discussed earlier, Venus's slow rotation means it has very little surface wind.

THINK ABOUT IT

Suppose Venus rotated as fast as Earth. How would this change its relative levels of volcanism, tectonics, and erosion? Explain.

Does Venus have plate tectonics?

We can easily explain the lack of erosion on Venus, but another "missing feature" of its geology is more surprising: Venus shows no evidence of Earth-like plate tectonics. As we'll discuss in the next section, plate tectonics shapes nearly all of Earth's major geological features, including mid-ocean ridges, deep ocean trenches, and long mountain ranges like the Rockies and Himalayas. Venus lacks any similar features.

Instead, Venus shows evidence of a very different type of global geological change. On Earth, plate tectonics resculpts the surface gradually, so different regions have different ages.

In contrast, Venus's relatively few impact craters are distributed fairly uniformly over the entire planet, suggesting that the surface is about the same age everywhere; crater counts suggest a surface age of about 750 million years. We therefore conclude that the entire surface of Venus was somehow "repaved" at that time, erasing all craters that formed earlier.

We do not know how much of the repaving was due to tectonic processes and how much was due to volcanism, but both probably were important. It is even possible that plate tectonics played a role before and during the repaving, only to stop after the repaving episode was over. Either way, the absence of present-day plate tectonics on Venus poses a major mystery.

Earth's lithosphere was broken into plates by forces due to the underlying mantle convection. The lack of plate tectonics on Venus therefore suggests either that it has weaker mantle convection or that its lithosphere somehow resists fracturing. The first possibility seems unlikely: Venus's similar size to that of Earth means it should have a similar level of mantle convection. Most scientists therefore suspect that Venus's lithosphere resists fracturing into plates because it is thicker and stronger than Earth's lithosphere, though we have no direct evidence to support this hypothesis.

Even if a thicker and stronger lithosphere explains the lack of plate tectonics on Venus, we are still left with the question of why the lithospheres of Venus and Earth differ. One possible answer is Venus's high surface temperature. Venus is so hot that any water in its crust and mantle has probably been baked out over time. Water tends to soften and lubricate rock, so its loss would have tended to thicken and strengthen Venus's lithosphere. If this idea is correct, then Venus might have had plate tectonics if it had not become so hot in the first place.

The mystery of plate tectonics is just one of many reasons scientists would like to explore Venus in greater detail. Unfortunately, the harsh surface conditions make it difficult for robotic landers to survive, and sending astronauts to Venus seems almost completely out of the question. We are therefore limited to studies from orbit.

9.6 THE UNIQUE GEOLOGY OF EARTH

Most aspects of Earth's geology should make sense in light of what we have found for other terrestrial worlds. Earth's size—largest of the five terrestrial worlds—explains why our planet retains internal heat and remains volcanically and tectonically active. Earth's rampant erosion by water and wind is explained by the combination of our planet's size, distance from the Sun, and rotation rate: Earth is large enough for volcanism and outgassing to have produced an atmosphere while its distance from the Sun allowed water vapor to condense and fall to the surface as rain, and Earth's moderately rapid rotation drives wind and other weather.

However, Earth's geology stands apart from that of the other terrestrial worlds in one important way: Earth is the only planet shaped primarily by ongoing plate tectonics. The importance of this fact goes far beyond basic geology. As we'll discuss in

FIGURE 9.37 This relief map shows known plate boundaries (solid yellow lines), with arrows to represent directions of plate motion. Color represents elevation, progressing from blue (lowest) to red (highest). Labels identify some of the geological features discussed in the text.

Chapter 10, plate tectonics plays a crucial role in the stability of Earth's climate. Without this climate stability, it is unlikely that we could be here today.

How is Earth's surface shaped by plate tectonics?

The term *plate tectonics* refers to the scientific theory that explains much of Earth's surface geology as a result of the slow motion of *plates* that essentially "float" over the mantle, gradually moving over, under, and around each other as convection moves Earth's interior rock. Earth's lithosphere is broken into more than a dozen plates (**FIGURE 9.37**). The plate motions are barely noticeable on human time scales: On average, plates move only a few centimeters per year—about the rate at which your fingernails grow. Nevertheless, over millions of years these motions can completely reshape the surface of our planet.

The Discovery of Continental Motion We usually trace the origin of the theory of plate tectonics to a slightly different hypothesis proposed in 1912 by German meteorologist and geologist Alfred Wegener: *continental drift*, the idea that continents gradually drift across the surface of Earth. Wegener got his idea in part from the puzzle-like fit of continents such as South America and Africa (**FIGURE 9.38**). He also noted that similar types of distinctive rocks and rare fossils were found in eastern South America and western Africa, suggesting that these two regions once had been close together.

Despite this evidence, no one at the time knew of a mechanism that could allow the continents to move about. Wegener suggested that Earth's gravity and tidal forces from the Sun and Moon were responsible, but other scientists quickly showed that these forces were too weak to move entire continents. As a result, Wegener's idea of continental drift was widely rejected by geologists for decades after he proposed it.

In the mid-1950s, scientists began to observe geological features that suggested a mechanism for continental motion. In particular, they discovered *mid-ocean ridges* (such as the Mid-Atlantic Ridge shown in Figure 9.37) along which mantle material erupts onto the ocean floor, pushing apart the existing seafloor on either side. This **seafloor spreading** helped explain how the continents could move apart with time. In addition, as more fossil evidence was gathered, it became clear that the continents really were arranged differently in the past, and Wegener's idea of a "continental fit" for Africa and South America ultimately gained acceptance.

Today, geologists can directly measure the slow plate motions by comparing readings taken with the global positioning system (GPS) [**Section S1.3**] on either side of plate boundaries. We also now understand that continental motion is coupled to the underlying mantle convection, driven by the heat released from Earth's interior. Because this idea is quite different from Wegener's original notion of continents plowing through the solid rock beneath them, geologists no longer use the term *continental drift* and instead consider continental motion within the context of plate tectonics.

Seafloor Crust and Continental Crust Another key piece of evidence for plate tectonics came with the discovery that Earth's surface has two distinct types of crust, one type found on seafloors and the other on continents (**FIGURE 9.39**). **Seafloor crust** is thinner, denser, and younger than **continental crust**. No other planet shows evidence of such distinct differences in crust from place to place.

Seafloor crust is typically 5–10 kilometers thick and is made primarily of the relatively high-density mineral mixture called basalt. Recall that basalt lava is very runny, so it spreads outward when it erupts from volcanoes along mid-ocean

FIGURE 9.38 The puzzle-like fit of South America and Africa.

The relatively dense, young seafloor crust is 5–10 km thick.

The less dense, older continental crust is 20–70 km thick.

upper mantle

FIGURE 9.39 Earth has two distinct kinds of crust.

ridges. Radiometric dating shows that seafloor basalt is quite young—usually less than 200 million years old—indicating that it erupted to the surface relatively recently in geological history. Further evidence of the young age of seafloors comes from studies of impact craters. Large impacts should occur more or less uniformly over Earth's surface, and the oceans are not deep enough to prevent a large asteroid or comet from making a seafloor crater. However, we find far fewer large craters on the seafloor than on the continents, which means that seafloor crust must have been created more recently.

Continental crust is much thicker—typically between 20 and 70 kilometers thick—but it sticks up only slightly higher than seafloor crust because its sheer weight presses it down farther into the mantle below. It is made mostly of granite and other types of rock that are less dense than basalt. Continental crust spans a wide range of ages and in some places contains rocks that are up to 4 billion years old.

The Conveyor Belt of Plate Tectonics The theory of plate tectonics explains continental motion, seafloor spreading, and the existence of two types of crust as direct results of the way plates move about on Earth. Over millions of years, the movements involved in plate tectonics act like a giant conveyor belt for Earth's lithosphere (**FIGURE 9.40**).

Mid-ocean ridges occur at places where mantle material rises upward, creating new seafloor crust and pushing plates apart. The newly formed basaltic crust cools and contracts as it spreads sideways from the central ridge, giving seafloor spreading regions their characteristic ridged shape. Along the mid-ocean ridges worldwide, new crust covers an area of about 2 square kilometers every year, enough to replace the entire seafloor within about 200 million years—and thereby explaining the less-than-200-million-year age of seafloor crust.

Over tens of millions of years, any piece of seafloor crust gradually makes its way across the ocean bottom, then finally gets recycled into the mantle in the process we call **subduction**. Subduction occurs where a seafloor plate meets a continental plate, which is generally somewhat offshore at the edge of a sloping continental shelf. As the dense seafloor crust of one plate pushes under the less dense continental crust of another plate, it can pull the entire surface downward to form a deep *ocean trench*. At some trenches, the ocean depth is more than 8 kilometers (5 miles).

Beneath a subduction zone, the descending seafloor crust heats up and may begin to melt as it moves deeper into the mantle. If enough melting occurs, the molten rock may erupt upward. This melting tends to occur under the edges of the continents, which is why so many active volcanoes are found along those edges. Moreover, the lowest-density material tends to melt more easily, which is why the continental crust emerging from these landlocked volcanoes is lower in density than seafloor crust. This fact also explains why steep-sided stratovolcanoes, made from low-density but thick lava (see Figure 9.11c), are common on Earth but rare on other worlds: Without plate tectonics to recycle crust, other worlds generally have only volcanic plains or shield volcanoes made from basalt.

The conveyor-like process of plate tectonics is undoubtedly driven by the heat flow from mantle convection, although the precise relationship between the convection cells and the plates remains an active topic of research. For example, it is not clear whether mantle convection simply pushes plates apart at

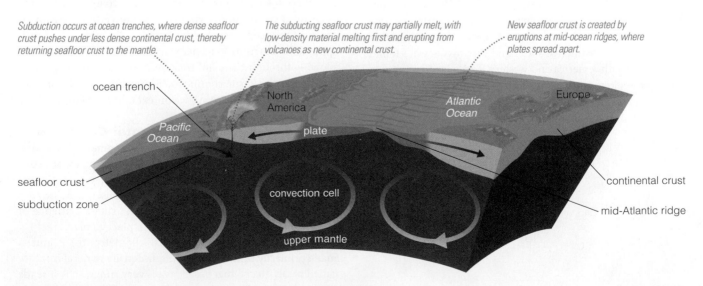

Subduction occurs at ocean trenches, where dense seafloor crust pushes under less dense continental crust, thereby returning seafloor crust to the mantle.

The subducting seafloor crust may partially melt, with low-density material melting first and erupting from volcanoes as new continental crust.

New seafloor crust is created by eruptions at mid-ocean ridges, where plates spread apart.

ocean trench

North America

Atlantic Ocean

Europe

Pacific Ocean

plate

seafloor crust

subduction zone

convection cell

continental crust

mid-Atlantic ridge

upper mantle

FIGURE 9.40 Plate tectonics acts like a giant conveyor belt for Earth's lithosphere.

the sites of seafloor spreading or whether the plates are denser than the underlying material and pull themselves down at subduction zones.

Building Continents Unlike seafloor crust, continental crust does not get recycled back into the mantle. As a result, present-day continents have been built up over billions of years. However, their histories are not just a simple "piling up" of continental crust. Instead, the continents are continually reshaped by volcanism and stresses associated with plate tectonics, as well as by erosion. **FIGURE 9.41** shows some of the complex geological history of North America, which we can use as an example of how the continents have been built.

The west coast regions of Alaska, British Columbia, Washington, Oregon, and most of California began as numerous volcanic islands in the Pacific. Long ago, plate motion carried these islands from their ocean birthplaces toward the North American mainland, where subduction continues to take the seafloor crust downward into the mantle (see Figure 9.40). Because these islands were made of lower-density continental crust, they remained on the surface and attached themselves to the edge of North America, even as the seafloor crust beneath them slid back into the mantle. Similar processes affect other islands today. Alaska's Aleutian Islands, a string of volcanoes located over a region where one seafloor plate is subducting under another (see the plate boundaries in Figure 9.37), are gradually growing and will someday merge together. Japan and the Philippines represent a later stage in this process; each once contained many small islands that merged into the fewer islands we see today. As these islands continue to grow and merge, they may eventually create a new continent or merge with an existing one.

Other portions of North America have been shaped by erosion. The Great Plains and the Midwest once were ancient seas. Erosion gradually filled these seas with sediment that hardened into sedimentary rock. The Deep South formed from the buildup of sediments that were carried to these regions after eroding from other parts of the continent. Northeastern Canada features some of the oldest continental crust on Earth, worn down over time by erosion.

Tectonic processes are responsible for long mountain ranges. The Sierra Nevada range in California, which lies over the region where the Pacific plate subducts under North America, formed when partially molten granite rose up from the mantle and pushed the surface rock higher. The original surface rock was sedimentary, but erosion gradually wore it away and left the granite exposed, which is why the Sierra Nevada mountains are made largely of granite.

Mountain ranges can also form through collisions of continent-bearing plates. The Himalayas, the tallest mountains on Earth, offer the most famous example of this process (**FIGURE 9.42**). They are still growing taller as the plate carrying the Indian subcontinent rams into the plate carrying Eurasia. Because both colliding plates are made of low-density continental crust, neither plate can subduct under the other. Instead, they push into each other and the resulting compression pushes the land upward to make mountains. The Appalachian range in the eastern United States was built through multiple collisions of continental plates: Over a period of a few hundred million years, North America apparently collided twice with South America and then with western Africa. The Appalachians probably were once as tall as the Himalayas are now, but tens of millions of years of erosion transformed them into the fairly modest mountain range we see today. Similar processes contributed to the formation of the Rocky Mountains in the United States and Canada.

Rifts, Faults, and Earthquakes So far, we have discussed features formed in places where plates are colliding, either with a seafloor plate subducting under a continental

This is ancient continental crust, heavily eroded.

Volcanic and tectonic mountains formed over subduction zones.

Former Pacific Islands merged into the continent during subduction.

Sediments were deposited in ancient seas; they later turned to rock.

Mountains formed by repeated collisions with other continents.

Deep sedimentary layer formed by erosion of the continent.

FIGURE 9.41 The major geological features of North America record the complex history of plate tectonics. Only the basic processes behind the largest features are shown.

FIGURE 9.42 This satellite photo shows the Himalayas, which are still slowly growing as the plate carrying India pushes into the Eurasian plate. Arrows indicate directions of plate motion.

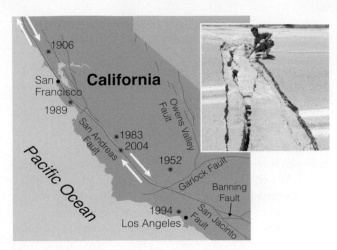

FIGURE 9.44 California's San Andreas Fault marks a boundary where plates are sliding sideways, as shown by the white arrows; asterisks indicate recent earthquakes. The inset photo shows a place along the San Andreas Fault where the painted lines in a road allow us to see how far the two sides of the fault moved in an earthquake.

plate or with two continental plates pushing into each other. Different types of features occur in places where plates pull apart or slide sideways relative to each other.

In places where continental plates are pulling apart, the crust thins and can create a large *rift valley*. The East African rift zone is an example (see Figure 9.37). This rift is slowly growing and will eventually tear the African continent apart. At that point, rock rising upward with mantle convection will begin to erupt from the valley floor, creating a new zone of seafloor spreading. A similar process tore the Arabian Peninsula from Africa, creating the Red Sea (FIGURE 9.43).

Places where plates slip sideways relative to each other are marked by what we call **faults**—fractures in the lithosphere.

The San Andreas Fault in California marks a line where the Pacific plate is moving northward relative to the continental plate of North America (FIGURE 9.44). In about 20 million years, this motion will bring Los Angeles and San Francisco together. The two plates do not slip smoothly against each other; instead, their rough surfaces catch. Stress builds up until it is so great that it forces a rapid and violent shift, causing an earthquake. In contrast to the usual motion of plates, which proceeds at a few centimeters per year, an earthquake can move plates by several *meters* in a few seconds. The movement can raise mountains, level cities, set off destructive tsunamis, and make the whole planet vibrate with seismic waves.

THINK ABOUT IT

By studying the plate boundaries in Figure 9.37, explain why California, Oregon, and Washington have more earthquakes and volcanoes than the rest of the continental United States. Find the locations of recent earthquakes and volcanic eruptions worldwide. Do the locations fit the pattern you expect?

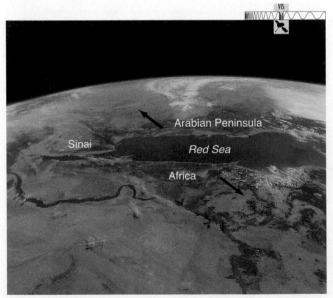

FIGURE 9.43 When continental plates pull apart, the crust thins and deep rift valleys form. This process tore the Arabian Peninsula from Africa, forming the Red Sea. Arrows indicate directions of plate motion.

Hot Spots Not all volcanoes occur near plate boundaries. Sometimes, a plume of hot mantle material rises in what we call a **hot spot**. The Hawaiian Islands are the result of a hot spot that has been erupting basaltic lava for tens of millions of years. Plate tectonics has gradually carried the Pacific plate over the hot spot, thereby forming a chain of volcanic islands (FIGURE 9.45). Today, most of the lava erupts on the Big Island of Hawaii, giving this island a young, rocky surface. About a million years ago, the Pacific plate lay farther to the southeast (relative to its current location), and the hot spot built the island of Maui. Before that, the hot spot created other islands, including Oahu (3 million years ago), Kauai (5 million years ago), and Midway (27 million years ago). The older islands are more heavily eroded. Midway has eroded so much that it barely rises above sea level. If plate tectonics were not moving

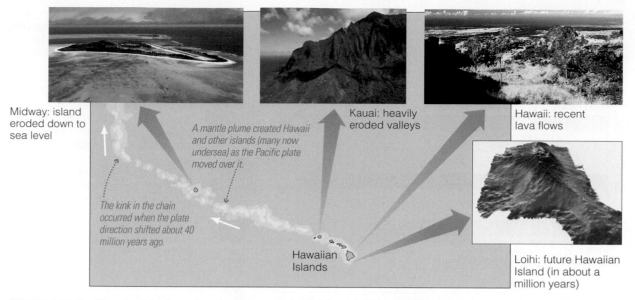

Midway: island eroded down to sea level

A mantle plume created Hawaii and other islands (many now undersea) as the Pacific plate moved over it.

Kauai: heavily eroded valleys

Hawaii: recent lava flows

The kink in the chain occurred when the plate direction shifted about 40 million years ago.

Hawaiian Islands

Loihi: future Hawaiian Island (in about a million years)

FIGURE 9.45 The Hawaiian Islands are just the most recent of a very long string of volcanic islands made by a mantle hot spot. The image of Loihi (lower right) was obtained by sonar, as it is still entirely under water.

the plate relative to the hot spot, a single huge volcano would have formed—perhaps looking somewhat like Olympus Mons on Mars (see Figure 9.26).

The movement of the plate over the hot spot continues today, building underwater volcanoes that eventually will rise above sea level to become new Hawaiian Islands. The growth of a future island, named Loihi, is already well under way—prime beach real estate should be available there in a million years or so. Hot spots can also occur beneath continental crust. For example, a hot spot is responsible for the geysers and hot springs of Yellowstone National Park.

THINK ABOUT IT

Find your hometown or current home on the map in Figure 9.37 or Figure 9.41 (or both). Based on what you have learned about Earth's geology, describe the changes your home has undergone over millions (or billions) of years.

Plate Tectonics Through Time We can use the current motions of the plates to project the arrangement of continents millions of years into the past or the future. For example, at a speed of 2 centimeters per year, a plate will travel 2000 kilometers in 100 million years. **FIGURE 9.46** shows two past arrangements of the continents, along with one future arrangement. Note that about 200 million years ago the present-day continents were together in a single "supercontinent," sometimes called *Pangaea* (which means "all lands").

Studies of magnetized rocks (which record the orientation of ancient magnetic fields) and comparisons of fossils found in different places around the world have allowed geologists to map the movement of the continents even further into the past. It seems that, over the past billion years or more, the continents have slammed together, pulled apart, spun around, and changed places on the globe. Central Africa once lay at

Earth's South Pole, and Antarctica once was near the equator. The continents continue to move, and their current arrangement is no more permanent than any past arrangement.

Was Earth's geology destined from birth?

Now that we have completed our geological tour of the terrestrial worlds, let's consider whether fundamental planetary properties shape all geological destinies. If so, then we should be able to predict the geological features of terrestrial worlds we find around other stars. If not, then we still have more to learn before we'll know whether other worlds could have Earth-like geology.

FIGURE 9.47 summarizes the key trends we've seen among the terrestrial worlds. All the worlds were heavily cratered during the heavy bombardment, but the extent of volcanism and tectonics has depended on planetary size. The interiors of the smallest worlds cooled quickly, so they have not had volcanism or tectonics for billions of years. The interiors of the larger worlds have stayed hot longer, allowing volcanism and tectonics to continue for much longer time periods.

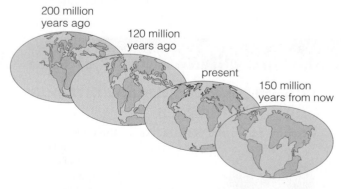

200 million years ago

120 million years ago

present

150 million years from now

FIGURE 9.46 Past, present, and future arrangements of Earth's continents. The present continents were all combined into a single "supercontinent" about 200 million years ago.

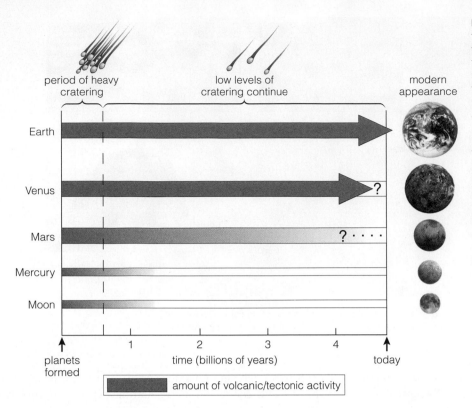

FIGURE 9.47 This diagram summarizes the geological histories of the terrestrial worlds. The brackets along the top indicate that impact cratering has affected all worlds similarly. The arrows represent volcanic and tectonic activity. A thicker and darker arrow means more volcanic/tectonic activity, and the arrow length tells us how long this activity persisted. Notice this trend's relationship to planetary size: Earth remains active today; Venus has also been active, though we are uncertain whether it remains so; Mars has had an intermediate level of activity and might still have low-level volcanism; Mercury and the Moon have had very little volcanic/tectonic activity. Erosion is not shown, because it has played a significant role only on Earth (ongoing) and Mars (in the past and at low levels today).

The major remaining question concerns whether fundamental properties can also explain Earth's unique plate tectonics. The lack of plate tectonics on Venus tells us that we cannot attribute plate tectonics solely to size, since Venus is so similar in size to Earth. However, if we are correct in guessing that Venus lacks plate tectonics because its high surface temperature led to loss of water and lithospheric thickening, then a lower surface temperature might have led to plate tectonics on Venus as well as Earth. As we'll discuss in the next chapter, we can ultimately trace Venus's high surface temperature to its nearness to the Sun. While plenty of uncertainty still remains, it seems likely that the broad geological histories of Earth and the other terrestrial worlds were indeed destined from birth by the fundamental properties of size, distance from the Sun, and rate of rotation.

The Big Picture

Putting Chapter 9 into Perspective

In this chapter, we have explored the geology of the terrestrial worlds. As you continue your study of the solar system, keep the following "big picture" ideas in mind:

- The terrestrial worlds all looked much the same when they were born, so their present-day differences are a result of geological processes that occurred in the ensuing $4\frac{1}{2}$ billion years.

- The extent to which different geological processes operate on the different worlds depends largely on their fundamental properties, especially their size.

- A planet's geology is largely destined from its birth—which means we should be able to predict the geology of as-yet-undiscovered planets once we know their fundamental properties.

- Earth has been affected by the same geological processes affecting the other terrestrial worlds. However, erosion is far more important on Earth than on any other terrestrial world, and Earth's unique plate tectonics may be very important to our existence.

SUMMARY OF KEY CONCEPTS

9.1 CONNECTING PLANETARY INTERIORS AND SURFACES

- **What are terrestrial planets like on the inside?**

In order of decreasing density and depth, the interior structure consists of **core, mantle,** and **crust.** The crust and part of the mantle together make up the rigid **lithosphere.** In general, a thinner lithosphere allows more geological activity.

- **What causes geological activity?** Interior heat drives

geological activity by causing mantle convection, keeping the lithosphere thin, and keeping the interior partially molten. All the terrestrial interiors were once hot, but larger planets cool slowly, retaining more interior heat and staying geologically active longer.

- **Why do some planetary interiors create magnetic fields?** A planetary **magnetic field** requires three things: an interior layer of electrically conducting fluid, convection of that fluid, and rapid rotation. Among the terrestrial planets, only Earth has all three characteristics.

9.2 SHAPING PLANETARY SURFACES

- **What processes shape planetary surfaces?** The four major geological processes are **impact cratering, volcanism, tectonics,** and **erosion.**

- **How do impact craters reveal a surface's geological age?** More craters indicate an older surface. All the terrestrial worlds were battered by impacts when they were young, so those that still have many impact craters must look much the same as they did long ago. Those with fewer impact craters must have had their ancient craters erased by other geological processes.

- **Why do the terrestrial planets have different geological histories?** Fundamental planetary properties, especially size, determine a planet's geological history. Larger worlds have more volcanism and tectonics, and these processes erase more of the world's ancient impact craters. Erosion depends on a planet's size, distance from the Sun, and rotation rate.

9.3 GEOLOGY OF THE MOON AND MERCURY

- **What geological processes shaped our Moon?** The lunar surface is a combination of extremely ancient, heavily cratered terrain and somewhat younger lava plains called the lunar **maria.** Some small tectonic features are also present. The Moon lacks erosion because it has so little atmosphere.

- **What geological processes shaped Mercury?** Mercury's surface resembles that of the Moon in being shaped by impact cratering and volcanism. It also has tremendous tectonic cliffs that probably formed when the whole planet cooled and contracted in size.

9.4 GEOLOGY OF MARS

- **What geological processes have shaped Mars?** Mars shows evidence of all four geological processes. It has the tallest volcano and the biggest canyon in the solar system, evidence of a period of great volcanic and tectonic activity. It also has abundant craters and evidence of erosion by wind and flowing water.

- **What geological evidence tells us that water once flowed on Mars?** Images of dry river channels and eroded craters, along with chemical analysis of Martian rocks, show that water once flowed on Mars. Any periods of rainfall seem to have ended at least 3 billion years ago. Mars still has water ice underground and in its polar caps and could possibly have pockets of underground liquid water.

9.5 GEOLOGY OF VENUS

- **What geological processes have shaped Venus?** Venus's surface shows evidence of major volcanic and tectonic activity in the past billion years, as expected for a planet nearly as large as Earth. This activity explains the relative lack of craters. Despite Venus's thick atmosphere, erosion is only a minor factor because of the high temperature (no rainfall) and slow rotation (little wind). Venus almost certainly remains geologically active today.

- **Does Venus have plate tectonics?** Venus appears to have undergone planetwide resurfacing, but we do not see evidence of ongoing plate tectonics. The lack of plate tectonics probably means that Venus has a stiffer and stronger lithosphere than Earth, perhaps because the high surface temperature has baked out any water that might have softened the lithospheric rock.

9.6 THE UNIQUE GEOLOGY OF EARTH

- **How is Earth's surface shaped by plate tectonics?** On Earth, the lithosphere is broken into **plates** that move around in the special type of tectonics that we call **plate tectonics.** Plate tectonics works much like a giant conveyor belt. New seafloor **crust** emerges from **mid-ocean ridges** and is recycled into the mantle at **subduction zones,** near which lower-density **continental crust** can erupt to build up the continents. Over time, the shifting of plates rearranges the continents on Earth's surface.

- **Was Earth's geology destined from birth?** It seems likely that the geological histories of Earth and the other terrestrial worlds were indeed destined from birth by the fundamental properties of size, distance from the Sun, and rate of rotation.

Use the following questions to check your understanding of some of the many types of visual information used in astronomy. Answers are provided in Appendix J. For additional practice, try the Chapter 9 Visual Quiz at MasteringAstronomy®.

This image from a MESSENGER flyby shows evidence of impact cratering, volcanism, and tectonic activity on Mercury. Answer the following questions based on the image. Remember that craters are bowl-shaped and rough-floored when they form, and wipe out any preexisting features in the area. Lava on Mercury appears to be fairly runny and makes flat smooth plains as it spreads out.

1. Label 1a lies on the rim of a large crater, and label 1b lies on the rim of a smaller one. Which crater must have formed first?
   a. crater 1a
   b. crater 1b
   c. Cannot be determined
2. The region around 2b has far fewer craters than the region around 2c. The crater floor at 2a is also flat and smooth, without many smaller craters on it. Why are regions 2a and 2b so smooth?
   a. Few small craters ever formed in these regions.
   b. Erosion erased craters that once existed in these regions.
   c. Lava flows covered craters that once existed in these regions.
3. A tectonic ridge appears to connect points 3a and 3b, crossing several craters. From its appearance, we can conclude that it must have formed
   a. before the area was cratered
   b. after the area was cratered
   c. at the same time the area was cratered
4. Using your answers to questions 1–3, list the following features in order from oldest to youngest:
   a. the tectonic ridge from 3a to 3b
   b. crater 1a
   c. the smooth floor of crater 1b

MasteringAstronomy®

For instructor-assigned homework go to MasteringAstronomy®.

REVIEW QUESTIONS

Short-Answer Questions Based on the Reading

1. Describe the core-mantle-crust structures of the terrestrial worlds. What is *differentiation*? What do we mean by the *lithosphere*? How does lithospheric thickness vary among the five terrestrial worlds?
2. Summarize the processes by which planetary interiors get hot and cool off. Why do large planets retain internal heat longer than smaller planets?
3. Why does Earth have a global magnetic field? Why don't the other terrestrial worlds have similarly strong magnetic fields?
4. Define each of the four major geological processes, and give examples of features shaped by each process.
5. What is *outgassing*, and why is it so important to our existence?
6. Why is the Moon so much more heavily cratered than Earth? Explain how crater counts tell us the age of a surface.
7. Summarize the ways in which a terrestrial world's size, distance from the Sun, and rotation rate each affect its relative level of impact cratering, volcanism, tectonics, and erosion.
8. Briefly summarize the geological history of the Moon. How did the lunar maria form?
9. Briefly summarize the geological history of Mercury. How are Mercury's great cliffs thought to have formed?
10. Choose five features on the global map of Mars (Figure 9.25) and explain the nature and likely origin of each.
11. Why isn't liquid water stable on Mars today, and why do we nonetheless think it flowed on Mars in the distant past?
12. Choose at least three major geological features of Venus and explain how we think each one formed.
13. What evidence tells us that Venus was "repaved" about 750 million years ago? What might account for the lack of plate tectonics on Venus?
14. Describe the conveyor-like action of plate tectonics on Earth, and how it explains the differences between seafloor and continental crust.
15. Briefly explain how each of the following geological features of Earth is formed: seafloors, continents, islands, mountain ranges, rift valleys, and faults.
16. To what extent do we think the geologies of the terrestrial worlds were destined from their birth? Explain.

TEST YOUR UNDERSTANDING

Surprising Discoveries?

Suppose we were to make the following discoveries. (These are not real discoveries.) In light of your understanding of planetary geology, decide whether the discovery should be considered reasonable or surprising. (In some cases, both views can be defended.) Explain your answer, if possible tracing your logic back to the terrestrial worlds' fundamental properties of size, distance from the Sun, and rotation rate.

17. New photographs reveal sand dunes on Mercury.
18. Seismographs placed on the surface of Mercury record frequent and violent earthquakes.

19. A new orbiter observes a volcanic eruption on Venus.

20. A Venus radar mapper discovers extensive regions of layered sedimentary rocks.

21. Radiometric dating of rocks brought back from one lunar crater shows that the crater was formed only a few tens of millions of years ago.

22. New high-resolution orbital photographs of Mars show many crater bottoms filled with pools of liquid water.

23. Drilling into the surface, a robotic spacecraft discovers liquid water beneath the slopes of a Martian volcano.

24. Clear-cutting in the Amazon rain forest on Earth exposes terrain that is as heavily cratered as the lunar highlands.

25. Seismic studies on Earth reveal a "lost continent" that held great human cities just a few thousand years ago but is now buried deep under water off the western coast of Europe.

26. We find a planet in another solar system that has Earth-like plate tectonics; the planet is the size of the Moon and orbits 1 AU from its star.

27. We find a planet in another solar system that is as large as Earth but as heavily cratered as the Moon.

28. We find a planet in another solar system that has Earth-like seafloor crust and continental crust but apparently lacks plate tectonics or any other kind of crustal motion.

Quick Quiz

Choose the best answer to each of the following. Explain your reasoning with one or more complete sentences.

29. What is the longest-lasting internal heat source responsible for geological activity? (a) accretion (b) radioactive decay (c) sunlight

30. In general, what kind of planet would you expect to have the thickest lithosphere? (a) a large planet (b) a small planet (c) a planet far from the Sun

31. Which of a planet's fundamental properties has the greatest effect on its level of volcanic and tectonic activity? (a) size (b) distance from the Sun (c) rotation rate

32. What is the name of the rigid outer layer of a planet? (a) crust (b) mantle (c) lithosphere

33. Which describes our understanding of flowing water on Mars? (a) It was never important. (b) It was important once, but is no longer. (c) It is a major process on the Martian surface today.

34. What do we conclude if a planet has few impact craters of any size? (a) The planet was never bombarded by asteroids or comets. (b) Its atmosphere stopped impactors of all sizes. (c) Other geological processes have wiped out craters.

35. How many of the five terrestrial worlds have surfaces being continually reshaped by plate tectonics? (a) one (b) two (c) three or more

36. On how many of the five terrestrial worlds has erosion been an important process? (Be sure that you explain *why* erosion is important on this many worlds and not more.) (a) one (b) two (c) three or more

37. How many of the terrestrial worlds have lava plains or shield volcanoes? (a) one (b) three (c) five

38. How many of the five terrestrial worlds are considered "geologically dead"? (a) none (b) two (c) four

PROCESS OF SCIENCE
Examining How Science Works

39. *Mars Attracts.* William Herschel, Giovanni Schiaperelli, and Percival Lowell were all respected astronomers who made important and widely accepted discoveries, yet all three jumped to incorrect conclusions about Mars. Scientifically, where do you think each went wrong, and how seriously? Try to frame your answer in terms of the hallmarks of science discussed in Chapter 3. Do you think their mistakes had any long-term impact on the study of Mars?

40. *What Is Predictable?* Briefly explain why much of a planet's geological history is destined from its birth, and discuss the level of detail that is predictable. For example, was Mars's general level of volcanism predictable? Could we have predicted a mountain as tall as Olympus Mons or a canyon as long as Valles Marineris? Explain.

41. *Unanswered Questions.* Our current understanding of the terrestrial worlds allows us to paint a broad-brush overview of their geological histories, but we still have much more to learn. Choose one important but unanswered question related to the geology of the terrestrial worlds, and describe why the question is important and how we might answer it in the future. Be as specific as possible, focusing on the type of evidence necessary to answer the question and how the evidence could be gathered.

GROUP WORK EXERCISE

42. *Real and Imaginary Planets.* Before you begin, assign the following roles to the people in your group: *Scribe* (takes notes on the group's activities), *Proposer* (proposes explanations to the group), *Skeptic* (points out weaknesses in proposed explanations), and *Moderator* (leads group discussion and makes sure everyone contributes).
    a. Figure 9.16 shows the general principles for planetary behavior for "generic" planets. For each of the seven example worlds shown, decide which planet or planets in our solar system match the listed characteristics (large or small; close, intermediate, or far; etc.). **b.** For at least one of the blue annotations on each of the seven worlds in this figure, find an image in the chapter that exemplifies the principle. For example, which image supports the principle that slow rotation leads to less erosion? **c.** Next, select one of the planet examples you chose in part a, and imagine a significant change in the property listed. How would the planet's geological history have been different? **d.** Now propose two imaginary planets, whose size, distance from the Sun, and rotation rates are different from each other and different from those of any terrestrial world. Give them names, and predict their geological histories, emphasizing comparisons with the other planets.

INVESTIGATE FURTHER
In-Depth Questions to Increase Your Understanding
Short-Answer/Essay Questions

43. *Dating Planetary Surfaces.* We have discussed two basic techniques for determining the age of a planetary surface: studying the abundance of impact craters and radiometric dating of surface rocks. Which technique seems more reliable? Which technique is more practical? Explain.

44. *Comparative Erosion.* Of Mercury, Venus, the Moon, and Mars, which world has the greatest erosion? Why? Write a paragraph explaining each world's level of erosion, tracing it back to fundamental properties.

45. *Miniature Mars.* Suppose Mars had turned out to be significantly smaller than its current size—say, the size of our Moon. How would this have affected the number of geological features due to each of the four major geological processes? Do you think Mars would be a better or worse candidate for harboring extraterrestrial life? Summarize your answers in two or three paragraphs.

46. *Change in Fundamental Properties.* Choose one property of Earth—either size or distance from the Sun—and suppose that it had been different (for example, smaller size or greater distance). Describe

how this change might have affected Earth's subsequent geological history and the possibility of our existence today on Earth.

47. *Predictive Geology.* Suppose another star system has a rocky terrestrial planet twice as large as Earth but at the same distance from its star (which is just like our Sun) and with a similar rotation rate. In one or two paragraphs, describe the type of geology you would expect it to have.

48. *Mystery Planet.* It's the year 2098, and you are designing a robotic mission to a newly discovered planet around a nearby star that is nearly identical to our Sun. The planet is as large in radius as Venus, rotates with the same daily period as Mars, and lies 1.2 AU from its star. Your spacecraft will orbit but not land on the planet. **a.** Some of your colleagues believe that the planet has no metallic core. How could you support or refute their hypothesis? **b.** Other colleagues suspect that the planet has no atmosphere, but the instruments designed to study the planet's atmosphere fail because of a software error. However, the spacecraft can still photograph geological features. How could you use the spacecraft's photos of geological features to determine whether a significant atmosphere is (or was) present on this planet?

49. *Experiment: Planetary Cooling in a Freezer.* Fill two small plastic containers of similar shape but different sizes with cold water and put both into the freezer at the same time. Every hour or so, record the time and estimate the thickness of the "lithosphere" (the frozen layer) in the two tubs. How long does it take the water in each tub to freeze completely? Describe the relevance of your experiment to planetary geology. Extra credit: Plot your results on a graph with time on the *x*-axis and lithospheric thickness on the *y*-axis. What is the ratio of the two freezing times?

50. *Amateur Astronomy: Observing the Moon.* Any amateur telescope has resolution adequate for identifying geological features on the Moon. The light highlands and dark maria should be evident, and shadowing should be visible near the line between night and day. Try to observe the Moon near the first- or third-quarter phase. Sketch or photograph the Moon at low magnification, and then zoom in on a region of interest. Again sketch or photograph your field of view, labeling its features and identifying the geological process that created them. Look for craters, volcanic plains, and tectonic features. Estimate the size of each feature by comparing it to the size of the entire Moon (radius = 1738 km).

Quantitative Problems

Be sure to show all calculations clearly and state your final answers in complete sentences.

51. *Surface Area-to-Volume Ratio.* Compare the surface area-to-volume ratios of the following: **a.** the Moon and Mars; **b.** Earth and Venus. In each case, use your answer to discuss differences in internal heat on the two worlds.

52. *Doubling Your Size.* Just as the surface area-to-volume ratio depends on size, so can other properties. To see how, suppose that your size suddenly doubled—that is, your height, width, and depth all doubled. (For example, if you were 5 feet tall before, you now are 10 feet tall.) **a.** By what factor has your waist size increased? **b.** How much more material will be required for your clothes? (*Hint:* Clothes cover the surface *area* of your body.) **c.** By what factor has your weight increased? (*Hint:* Weight depends on the *volume* of your body.) **d.** The pressure on your weight-bearing joints depends on how much *weight* is supported by the *surface area* of each joint. How has this pressure changed?

53. *Lunar Footprints.* Assume that the Moon is hit by about 25 million micrometeorite impacts each day (this number comes from observa-

tions of meteors in Earth's atmosphere) and that these impacts strike randomly around the Moon's surface. Also assume that it takes about 20 such impacts to destroy a footprint. About how long would it take for one of the footprints left by the *Apollo* astronauts to be erased? (*Hints:* Use the Moon's surface area to determine the impact rate per square centimeter, and estimate the size of a footprint.)

54. *Geological Proportions.* Express the approximate height and width of Olympus Mons (26 km tall and 600 km wide) as percentages of Mars's radius. Repeat for Valles Marineris (7 km deep and 4000 km long). Then compare the height of Mt. Everest (9 km tall) and the height and width of the Grand Canyon (1.8 km deep and 450 km long) to Earth's radius, and comment on the relative sizes of geological features on the two planets.

55. *Internal versus External Heating.* In daylight, Earth's surface absorbs about 400 watts per square meter. Earth's internal radioactivity produces a total of 30 trillion watts that leak out through our planet's entire surface. Calculate the amount of heat from radioactive decay that flows outward through each square meter of Earth's surface (your answer should have units of watts per square meter). Compare quantitatively to solar heating, and comment on why internal heating drives geological activity.

56. *Plate Tectonics.* Typical motion of one plate relative to another is 1 centimeter per year. At this rate, how long would it take for two continents 3000 kilometers apart to collide? What are the global consequences of motion like this?

57. *More Plate Tectonics.* Consider a seafloor spreading zone creating 1 centimeter of new crust over its entire 2000-kilometer length every year. How many square kilometers of surface will be created in 100 million years? What fraction of Earth's surface does this constitute?

Discussion Questions

58. *Worth the Effort?* Politicians often argue over whether planetary missions are worth the expense involved. If you were in Congress, would you support more or fewer missions? Why?

59. *Evidence of Our Civilization.* Discuss how the geological processes will affect the evidence of our current civilization in the distant future. For example, what evidence of our current civilization will survive in 100,000 years? in 100 million years? Do you think that future archaeologists or alien visitors will be able to know that we existed here on Earth?

Web Projects

60. *"Coolest" Surface Photo.* Visit the Astronomy Picture of the Day website, and search for past images of the terrestrial worlds. Look at many of them, and choose the one you think is the "coolest." Write a short description of what it shows, and explain what you like about it.

61. *Water on Mars.* Go to the home page for NASA's Mars Exploration Program, and look for the latest evidence concerning recent water flows on Mars. Write a few paragraphs describing the new evidence and what it tells us. How will future missions help resolve the questions?

62. *Planetary Geology.* Choose one of the terrestrial worlds, and do a Web search to learn about at least two geological features on it that were *not* highlighted in this chapter. Are the features consistent with what we would expect based on the ideas of "geological destiny" discussed in this chapter? Explain.

63. *Volcanoes and Earthquakes.* Learn about one major earthquake or volcanic eruption that occurred during the past decade. Report on the geological conditions that led to the event, as well as on its geological and human consequences.

10 PLANETARY ATMOSPHERES
EARTH AND THE OTHER TERRESTRIAL WORLDS

LEARNING GOALS

10.1 ATMOSPHERIC BASICS
- What is an atmosphere?
- How does the greenhouse effect warm a planet?
- Why do atmospheric properties vary with altitude?

10.2 WEATHER AND CLIMATE
- What creates wind and weather?
- What factors can cause long-term climate change?
- How does a planet gain or lose atmospheric gases?

10.3 ATMOSPHERES OF THE MOON AND MERCURY
- Do the Moon and Mercury have any atmosphere?

10.4 THE ATMOSPHERIC HISTORY OF MARS
- What is Mars like today?
- Why did Mars change?

10.5 THE ATMOSPHERIC HISTORY OF VENUS
- What is Venus like today?
- How did Venus get so hot?

10.6 EARTH'S UNIQUE ATMOSPHERE
- How did Earth's atmosphere end up so different?
- Why does Earth's climate stay relatively stable?
- How is human activity changing our planet?

For the first time in my life, I saw the horizon as a curved line. It was accentuated by a thin seam of dark blue light—our atmosphere. Obviously this was not the ocean of air I had been told it was so many times in my life. I was terrified by its fragile appearance.

—Ulf Merbold, astronaut (Germany)

Life as we know it would be impossible on Earth without our atmosphere. This thin layer of gas supplies the oxygen we breathe, shields us from harmful ultraviolet and X-ray radiation from the Sun, protects us from continual bombardment by micrometeorites, generates rain-giving clouds, and traps just enough heat to keep Earth habitable.

How did Earth end up with atmospheric conditions that are so favorable to life? In this chapter, we'll explore the answers—and learn why the other terrestrial worlds ended up so different, despite having formed under similar conditions. We'll also discuss why Earth's climate remains relatively stable and how human activity may be threatening that stability, with potential consequences that we are only beginning to understand.

10.1 ATMOSPHERIC BASICS

The atmospheres of the terrestrial worlds are even more varied than their geologies. **FIGURE 10.1** shows global and surface views of each world. **TABLE 10.1** lists general characteristics of their atmospheres. You'll notice vast differences between the worlds.

The Moon and Mercury have so little atmosphere that it's reasonable to call them "airless"; they have no wind or weather of any kind. At the other extreme, Venus is enshrouded by a thick atmosphere composed almost entirely of carbon dioxide, giving it surface conditions so hot and harsh that not even robotic space probes can survive there for long. Mars also has a carbon dioxide atmosphere, but its air

is so thin that your body tissues would bulge painfully if you stood on the surface without a pressurized spacesuit. Only Earth has the "just right" conditions that allow liquid water on the surface, making it hospitable to life.

Despite these great differences among the terrestrial atmospheres, the same basic processes are at work in all cases. In fact, the same processes are at work on any world with an atmosphere, and in later chapters we'll see how they apply to the atmospheres of the jovian planets, of moons with atmospheres, and even of planets around other stars. Let's begin our comparative study of planetary atmospheres by discussing the basic nature of an atmosphere.

What is an atmosphere?

An *atmosphere* is a layer of gas that surrounds a world. In most cases, it is a surprisingly thin layer. On Earth, for example, about two-thirds of the air in the atmosphere lies within 10 kilometers of the surface (**FIGURE 10.2**). You could represent this air on a standard globe (to scale) with a layer only as thick as a dollar bill.

Atmospheric air is a mixture of gases that may consist either of individual atoms or of molecules. Temperatures in the terrestrial atmospheres are generally low enough (even on Venus) for atoms to combine into molecules. For example, the air we breathe consists of *molecular* nitrogen (N_2) and oxygen (O_2), as opposed to individual atoms (N or O). Other common molecules in terrestrial atmospheres include water (H_2O) and carbon dioxide (CO_2).

Atmospheric Pressure Collisions of individual atoms or molecules in an atmosphere create *pressure* [**Section 5.3**] that pushes in all directions. On Earth, for example, the nitrogen and oxygen molecules in the air fly around at average speeds of about 500 meters per second—fast enough to cross your bedroom a hundred times in 1 second. Given that a single breath of air contains more than a billion trillion molecules, you can imagine how frequently molecules collide. On average,

FIGURE 10.1 Views of the terrestrial worlds and their atmospheres from orbit and from the surface. The surface views for Mercury and Venus are artists' conceptions; the others are photos. The global views are visible-light photos taken from spacecraft. (Venus appears in gibbous phase as it was seen by the *Galileo* spacecraft during its Venus flyby en route to Jupiter.)

TABLE 10.1   Atmospheres of the Terrestrial Worlds

World	Composition of Atmosphere	Surface Pressure*	Average Surface Temperature	Winds, Weather Patterns	Clouds, Hazes
Mercury	helium, sodium, oxygen	10^{-14} bar	day: 425°C (797°F) night: −175°C (−283°F)	none: too little atmosphere	none
Venus	96% carbon dioxide (CO_2) 3.5% nitrogen (N_2)	90 bars	470°C (878°F)	slow winds, no violent storms, acid rain	sulfuric acid clouds
Earth	77% nitrogen (N_2) 21% oxygen (O_2) 1% argon H_2O (0.4%, but variable) 0.04% carbon dioxide (CO_2)	1 bar	15°C (59°F)	winds, hurricanes, rain, snow	H_2O clouds, pollution
Moon	helium, sodium, argon	10^{-14} bar	day: 125°C (257°F) night: −175°C (−283°F)	none: too little atmosphere	none
Mars	95% carbon dioxide (CO_2) 2.7% nitrogen (N_2) 1.6% argon	0.007 bar	−50°C (−58°F)	winds, dust storms	H_2O and CO_2 clouds, dust

*1 bar = the atmospheric pressure at sea level on Earth.

each molecule in the air around you will suffer a million collisions in the time it takes to read this paragraph. These collisions create pressure that pushes in all directions, and this pressure holds up the atmosphere so that it does not collapse under its own weight.

A balloon offers a good example of how pressure works in a gas. The air molecules inside a balloon exert pressure, pushing outward as they constantly collide with the balloon's inside surface. At the same time, outside air molecules collide with the balloon's outer surface, exerting pressure that by itself would make the balloon collapse. A balloon

FIGURE 10.2 Earth's atmosphere, visible in this photograph from the Space Shuttle, is a very thin layer over Earth's surface. Most of the air is in the lowest 10 kilometers of the atmosphere, visible along the edge of the planet.

stays inflated when the inward and outward pressures are balanced (**FIGURE 10.3a**). (We are ignoring the tension in the rubber of the balloon walls.) Imagine that you blow more air into the balloon (**FIGURE 10.3b**). The extra molecules inside mean more collisions with the balloon wall, momentarily making the pressure inside greater than the pressure outside. The balloon therefore expands until the inward and outward pressures are again in balance.

If you heat the balloon (**FIGURE 10.3c**), the gas molecules begin moving faster and collide harder and more frequently with the inside surface, which also momentarily increases the inside pressure until the balloon expands. As it expands, the pressure inside it decreases and the balloon comes back into pressure balance. Conversely, cooling a balloon makes it contract, because the outside pressure momentarily exceeds the inside pressure.

SEE IT FOR YOURSELF

Find an empty plastic bottle with a screwtop that makes a good seal. Warm the air inside by filling the bottle partway with hot water and then shaking and emptying the bottle. Seal the bottle and place it in the refrigerator or freezer for about 15 minutes. What happens, and why? Explain in terms of the individual air molecules.

We can understand **atmospheric pressure** by applying similar principles. Gas in an atmosphere is held down by gravity. The atmosphere above any given altitude therefore has some weight that presses downward, tending to compress the atmosphere beneath it. At the same time, the fast-moving molecules exert pressure in all directions, including upward, which tends to make the atmosphere expand. Planetary atmospheres exist in a perpetual balance between the downward weight of their gases and the upward push of their gas pressure.

The higher you go in an atmosphere, the less the weight of the gas above you, and less weight means less pressure. That is why the pressure decreases as you climb a mountain or ascend

a A balloon stays inflated when the inside and outside pressures are balanced.

b Adding air molecules temporarily increases the pressure inside the balloon, so the balloon expands until the pressure balance is restored.

c Heating the balloon increases the speeds of air molecules inside it, thereby increasing the inside pressure. Again, the balloon expands until the pressure balance is restored.

FIGURE 10.3 Gas pressure in a balloon depends on both density and temperature.

in an airplane. You can visualize this concept by imagining the atmosphere as a very big stack of pillows (**FIGURE 10.4**). The pillows at the bottom are highly compressed because of the weight of all the pillows above. Going upward, the pillows are less and less compressed because less weight lies on top of them.

In planetary science, we usually measure atmospheric pressure in a unit called the **bar** (as in *barometer*). One bar is roughly equal to Earth's atmospheric pressure at sea level. It is also equivalent to 1.03 kilograms per square centimeter or

14.7 pounds per square inch. In other words, if you gathered up all the air directly above any 1 square inch of Earth's surface, you would find that it weighed about 14.7 pounds.

You might wonder why you don't feel Earth's atmospheric weight bearing down on you. After all, if we placed 14.7 pounds of weight on every square inch of your shoulders, you'd certainly feel the downward pressure. You don't notice the atmospheric pressure for two reasons: First, the pressure pushes in all directions, so it pushes upward and inward on you as well as downward. Second, the fluids in your body push outward with an equivalent pressure, so there is no net pressure trying to compress or expand your body. You can tell that the pressure is there, however, because you'll quickly notice any pressure changes. For example, even slight changes in pressure as you go up or down in altitude can cause your ears to "pop." More extreme changes, such as those that affect deep-sea divers when they rise too rapidly, can be deadly. You'd face even greater pressure differences if you visited other planets; as shown in Table 10.1, atmospheric pressure varies by a factor of 10^{16} among the terrestrial worlds.

Where Does an Atmosphere End? If you study Figure 10.4, you'll see that there is no clear boundary between the atmosphere and space above, because pressure and density decrease gradually with increasing altitude. At some point, the density becomes so low that we can't really think of the gas as "air" anymore. Collisions between atoms or molecules are rare at these altitudes, and the gas is so thin that it would

The weight of the material up here . . .

. . . creates higher pressure and density down here.

FIGURE 10.4 Atmospheric density and pressure decrease with altitude for the same reason that a giant stack of pillows would be more compressed at the bottom than at the top. The pressure in each layer is enough to hold up the weight of layers above, counteracting the force of gravity.

FIGURE 10.5 Low-density gas extends into what we often think of as "space," as you can see from the visible glow of gas around a Space Shuttle's tail in this photograph. Notice the aurora in the background, created by interactions between charged particles and atmospheric gas at altitudes above 100 kilometers, well below the Shuttle. Most of Earth's atmospheric gas lies below 10 kilometers.

look and feel as if you had already entered space. On Earth, this occurs at an altitude of about 100 kilometers; above that, the sky is black even in the daytime, much like the sky on the Moon. This altitude is often described as "the edge of space."

However, some gas is still present above 100 kilometers. Earth's tenuous upper atmosphere extends for several hundred kilometers above Earth's surface. The Space Station and many satellites orbit Earth within these outer reaches of the atmosphere (FIGURE 10.5). The low-density gas may be barely noticeable under most conditions, but it still exerts drag on orbiting spacecraft. That is why satellites in low-Earth orbit slowly spiral downward, eventually burning up as they reenter the denser layers of the atmosphere [Section 4.5].

How Atmospheres Affect Planets Earth's atmosphere is obviously important to our existence, but the full range of atmospheric effects is greater than most people realize. To provide context for our study of planetary atmospheres, here is a brief list of key effects of atmospheres that we will study in this chapter:

- As we've seen, atmospheres create pressure that determines whether liquid water can exist on the surface.

- Atmospheres absorb and scatter light. Scattering can make daytime skies bright and absorption can prevent dangerous radiation from reaching the ground.

- Atmospheres can create wind and weather and play a major role in long-term climate change.

- Interactions between atmospheric gases and the solar wind can create a protective *magnetosphere* around planets with strong magnetic fields.

- Atmospheres can make planetary surfaces warmer than they would be otherwise via the **greenhouse effect**.

The greenhouse effect is arguably the most important effect that an atmosphere can have on its planet, and we'll therefore focus our attention on it first.

How does the greenhouse effect warm a planet?

You've probably heard of the greenhouse effect, because it is an important part of the environmental problem known as *global warming* [**Section 10.6**]. But you may be surprised to learn that the greenhouse effect is also critical to the existence of life on Earth. Without the greenhouse effect, Earth's surface would be too cold for liquid water to flow and for life to flourish. Let's explore how the greenhouse effect can warm planetary surfaces.

How the Greenhouse Effect Works FIGURE 10.6 shows the basic idea behind the greenhouse effect. Some of the visible light that reaches the ground is reflected and some is absorbed. The absorbed energy must be returned to space, but planetary surfaces are too cool to emit visible light. (Recall that the type of thermal radiation an object emits depends on its temperature [**Section 5.4**].) Instead, planetary surface temperatures are in the range in which they emit mostly infrared light. The greenhouse effect works by temporarily "trapping" some of this infrared light, slowing its return to space.

The greenhouse effect occurs only when an atmosphere contains gases that can absorb the infrared light. Gases that are particularly good at absorbing infrared light are called **greenhouse gases**, and they include water vapor (H_2O), carbon dioxide (CO_2), and methane (CH_4). These gases absorb

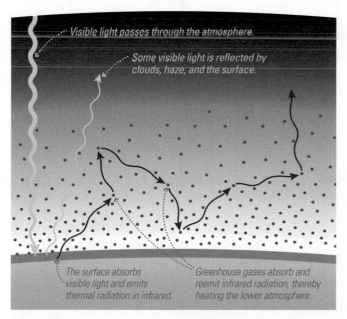

Visible light passes through the atmosphere.

Some visible light is reflected by clouds, haze, and the surface.

The surface absorbs visible light and emits thermal radiation in infrared.

Greenhouse gases absorb and reemit infrared radiation, thereby heating the lower atmosphere.

FIGURE 10.6 The greenhouse effect. The lower atmosphere becomes warmer than it would be if it had no greenhouse gases such as water vapor, carbon dioxide, and methane.

infrared light effectively because their molecular structures begin rotating or vibrating when they absorb an infrared photon (see Figure 5.18).

A greenhouse gas molecule that absorbs an infrared photon does not retain this energy for long; instead, it quickly reemits it as another infrared photon, which may head off in any random direction. This photon can then be absorbed by another greenhouse gas molecule, which does the same thing. The net result is that greenhouse gases tend to slow the escape of infrared radiation from the lower atmosphere, while their molecular motions heat the surrounding air. In this way, the greenhouse effect makes the surface and the lower atmosphere warmer than they would be from sunlight alone. The more greenhouse gases present, the greater the degree of surface warming.

Note that the greenhouse effect does *not* alter a planet's overall energy balance. The total amount of energy that a planet receives from the Sun must be precisely balanced with the amount of energy it returns to space through reflection and radiation. Otherwise, the planet would either rapidly heat up (if it received more energy than it returned) or rapidly cool down (if it returned more energy than it received). All the energy "trapped" by the greenhouse effect still escapes to space, just not as directly as it would otherwise. A blanket offers a good analogy: You stay warmer under a blanket not because the blanket itself provides any heat, but because it slows the escape of your body heat into the cold outside air. In summary, the greenhouse effect makes the total amount of energy in the lower atmosphere greater than it would be otherwise—making it warmer—while leaving the total amount of energy that escapes to space unchanged.

Incidentally, the name *greenhouse effect* is a bit of a misnomer. The name comes from botanical greenhouses, but greenhouses actually trap heat through a different mechanism than planetary atmospheres: Rather than absorbing infrared radiation, greenhouses stay warm by preventing warm air from rising.

Greenhouse Warming of the Terrestrial Worlds

We can better appreciate the importance of the greenhouse effect by comparing each planet's average surface temperature—or **global average temperature**—with and without it. Recall that a terrestrial planet's interior heat has very little effect on its surface temperature, so sunlight is the only significant energy source for the surface [**Section 9.1**]. Therefore, without the greenhouse effect, a planet's global average surface temperature would depend on only two things:

- *The planet's distance from the Sun,* which determines the amount of energy received from sunlight. The closer a planet is to the Sun, the greater the intensity of the incoming sunlight.

- *The planet's overall reflectivity,* which determines the relative proportions of incoming sunlight that the planet reflects and absorbs. The higher the reflectivity, the less light absorbed and the cooler the planet.

MATHEMATICAL INSIGHT 10.1

"No Greenhouse" Temperatures

The "no greenhouse" temperature of a planet depends only on its distance from the Sun, its overall reflectivity, and the Sun's brightness. It can be calculated with the following formula:

$$T_{\text{"no greenhouse"}} = 280 \text{ K} \times \sqrt[4]{\frac{(1 - \text{reflectivity})}{d^2}}$$

where d is the distance from the Sun in AU; reflectivity should be stated as a fraction (such as 0.2 rather than 20%). The symbol $\sqrt[4]{}$ means the fourth root, or $\frac{1}{4}$ power.

We will not derive the formula here, but you can see how it makes sense. The term (1 − reflectivity) is the proportion of sunlight that the planet absorbs, which is the light that heats its surface. This term is divided by d^2 because the amount of energy from sunlight (per unit area) declines with the square of distance from the Sun. Thus, the term (1 − reflectivity)/d^2 represents the total amount of energy that the planet absorbs from sunlight (per unit area) each second. This energy warms the surface, which returns the energy to space as thermal radiation. Because the energy emitted by thermal radiation depends on temperature raised to the fourth power (see Mathematical Insight 5.2), calculating the planet's temperature requires taking the fourth root of the absorbed energy. The 280 K in the formula is the temperature of a perfectly black planet at 1 AU from the Sun.

EXAMPLE: Calculate the "no greenhouse" temperature of Mercury.

SOLUTION:

Step 1 Understand: We can use the "no greenhouse" formula with Mercury's distance and reflectivity from Table 10.2; its distance is 0.387 AU and we write the reflectivity of 12% as 0.12.

Step 2 Solve: We substitute the values to find

$$T_{\text{"no greenhouse"}} = 280 \text{ K} \times \sqrt[4]{\frac{(1 - \text{reflectivity})}{d^2}}$$

$$= 280 \text{ K} \times \sqrt[4]{\frac{(1 - 0.12)}{0.387^2}} = 436 \text{ K}$$

Step 3 Explain: Mercury's "no greenhouse" temperature is 436 K. We subtract 273 to convert it to Celsius, finding that 436 K is equivalent to 436 − 273 = 163°C. Note that this agrees with the value in Table 10.2. It also makes sense: Because Mercury has no greenhouse effect, its "no greenhouse" temperature should be roughly midway between its day and night temperatures. The precise halfway point of the day and night temperatures shown in Table 10.2 is 125°C; although this is not an exact match to 163°C, it is close enough for us to be confident that we have the right general idea.

TABLE 10.2 The Greenhouse Effect on the Terrestrial Worlds

World	Average Distance from Sun (AU)	Reflectivity	"No Greenhouse" Average Surface Temperature*	Actual Average Surface Temperature	Greenhouse Warming (actual temperature minus "no greenhouse" temperature)
Mercury	0.387	12%	163°C	day: 425°C night: −175°C	—
Venus	0.723	75%	−40°C	470°C	510°C
Earth	1.00	29%	−16°C	15°C	31°C
Moon	1.00	12%	−2°C	day: 125°C night: −175°C	—
Mars	1.524	16%	−56°C	−50°C	6°C

*The "no greenhouse" temperature is calculated by assuming no change to the atmosphere other than lack of greenhouse warming. For example, Venus has a lower "no greenhouse" temperature than Earth even though it is closer to the Sun, because the high reflectivity of its bright clouds means that it absorbs less sunlight than Earth.

Note that a planet's reflectivity (sometimes called its *albedo*) depends on its composition and color; darker colors reflect less light. For example, clouds, snow, and ice reflect more than 70% of the light that hits them, absorbing only about 30%, while rocks typically reflect only about 20% of the light that hits them and absorb the other 80%.

Both distance from the Sun and reflectivity have been measured for all the terrestrial worlds. With a little mathematics, these measurements can be used to calculate the "no greenhouse" temperature that each world would have without greenhouse gases (see Mathematical Insight 10.1). **TABLE 10.2** shows the results. The "no greenhouse" temperatures for Mercury and the Moon lie between their actual day and night temperatures, since they have little atmosphere and hence no greenhouse effect. Mars has a weak greenhouse effect that makes its global average temperature only 6°C higher than its "no greenhouse" temperature. Venus is the extreme case, with a greenhouse effect that bakes its surface to a temperature more than 500°C hotter than it would be otherwise.

We can also see why the greenhouse effect is so important to life on Earth. Without the greenhouse effect, our planet's global average temperature would be a chilly −16°C (+3°F), well below the freezing point of water. With it, the global average temperature is about 15°C (59°F), or about 31°C warmer than the "no greenhouse" temperature. This greenhouse warming is even more remarkable when you realize that it is caused by gases, such as water vapor and carbon dioxide, that are only trace constituents of Earth's atmosphere. Most of the atmosphere consists of nitrogen (N_2) and oxygen (O_2) molecules, which have no effect on infrared light and do not contribute to the greenhouse effect. (Molecules with only two

atoms, especially those with two of the same kind of atom, such as N_2 and O_2, are poor infrared absorbers because they have very few ways to vibrate and rotate.)

Why do atmospheric properties vary with altitude? The greenhouse effect can warm a planet's surface and lower atmosphere, but other processes affect the temperature at higher altitudes. The way in which temperature varies with altitude determines what is often called the **atmospheric structure**. Earth's atmospheric structure has four basic layers (**FIGURE 10.7**), each of which affects the planet in a distinct way:

- The **troposphere** is the lowest layer, in which temperature drops with altitude (something you've probably noticed if you've ever climbed a mountain).

- The **stratosphere** begins where the temperature stops dropping and instead begins to rise with altitude. High in the stratosphere, the temperature falls again.*

- The **thermosphere** begins where the temperature again starts to rise at high altitude.

- The **exosphere** is the uppermost region, in which the atmosphere gradually fades away into space.

Interactions Between Light and Atmospheric Gases The key to understanding atmospheric structure lies in interactions between atmospheric gases and energy from the Sun. Although most light coming from the Sun is visible light, the Sun also emits significant amounts of ultraviolet light

---

*Technically, the region where the temperature falls again is called the *mesosphere*, but we will not make that distinction in this book.

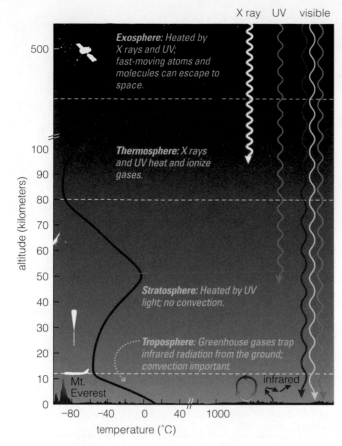

FIGURE 10.7 Earth's average atmospheric structure. The curve shows average temperature for each altitude. The squiggly arrows to the right represent light of different wavelengths and show where it is typically absorbed.

FIGURE 10.8 The primary effects when light of different energies strikes common atmospheric gases.

and X rays. In addition, the planetary surface emits infrared light. Atmospheric gases interact with each of these forms of light in different ways (**FIGURE 10.8**):

■ X rays have enough energy to *ionize* (knock electrons from) almost any atom or molecule. They can therefore be absorbed by virtually all atmospheric gases.

■ Ultraviolet photons generally do not have enough energy to cause ionization, but they can sometimes break molecules apart. For example, ultraviolet photons can split water (H_2O) molecules, and are even more likely to be absorbed by weakly bonded molecules, such as ozone (O_3), which split apart in the process.

■ Visible-light photons generally pass through atmospheric gases without being absorbed, but some are *scattered* so that their direction changes.

■ As we've already discussed, infrared photons can be absorbed by greenhouse gases, which are molecules that easily begin rotating and vibrating.

With these ideas in mind, we are ready to examine the reasons for Earth's atmospheric structure, working our way from the ground up.

Visible Light: Warming the Surface and Coloring the Sky Most visible sunlight reaches the ground and warms the surface, but the small amount that is scattered has two important effects.

First, scattering makes the daytime sky bright, which is why we can't see stars in the daytime. Without scattering, sunlight would travel only in perfectly straight lines, which means we'd see the Sun against an otherwise black sky, just as it appears on the Moon. Scattering also prevents shadows on Earth from being pitch black. On the Moon, shadows receive little scattered sunlight and are extremely cold and dark.

Second, scattering explains why our sky is blue. Visible light consists of all the colors of the rainbow, but not all the colors are scattered equally. Gas molecules scatter blue light (higher energy) much more effectively than red light (lower energy). The difference in scattering is so great that, for practical purposes, we can imagine that only the blue light gets scattered. When the Sun is overhead, this scattered blue light reaches our eyes from all directions and the sky appears blue (**FIGURE 10.9**).

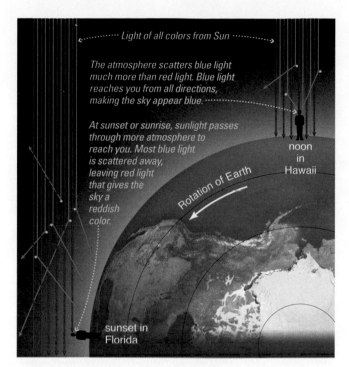

Light of all colors from Sun

The atmosphere scatters blue light much more than red light. Blue light reaches you from all directions, making the sky appear blue.

At sunset or sunrise, sunlight passes through more atmosphere to reach you. Most blue light is scattered away, leaving red light that gives the sky a reddish color.

noon in Hawaii

Rotation of Earth

sunset in Florida

FIGURE 10.9 This diagram summarizes why the daytime sky is blue and sunsets (and sunrises) are red.

At sunset or sunrise, the sunlight must pass through a greater amount of atmosphere on its way to us. Most of the blue light is scattered away, leaving only red light to color the sky.

Infrared Light and the Troposphere The ground returns the energy it absorbs by radiating in the infrared. Greenhouse gases absorb this infrared light and warm the troposphere. Because the infrared light comes from the surface, more is absorbed closer to the ground than at higher altitudes, which is why the temperature drops with altitude in the troposphere. (The relatively small amount of infrared light coming from the Sun does not have a significant effect on the atmosphere.)

The drop in temperature with altitude, combined with the relatively high density of air in the troposphere, explains why the troposphere is the only layer of the atmosphere with storms. The primary cause of storms is the churning of air by convection [**Section 9.1**], in which warm air rises and cool air falls. Recall that convection occurs only when there is strong heating from below; in the troposphere, the heating from the ground can drive convection. In fact, the troposphere gets its name from convection; *tropos* is Greek for "turning."

Ultraviolet Light and the Stratosphere Above the troposphere, the air density is too low for greenhouse gases to have much effect, so infrared light from below can travel unhindered through higher layers of the atmosphere and into space. Heating from below therefore has little effect on the stratosphere. Instead, the primary source of heating in the stratosphere is the absorption of solar ultraviolet light by ozone.

Most of this ultraviolet absorption and heating occurs at moderately high altitudes in the stratosphere, which is why

temperature tends to *increase* with altitude as we go upward from the base of the stratosphere. This temperature structure prevents convection in the lower stratosphere, because heat cannot rise if the air above is hotter. The lack of convection makes the air relatively stagnant and *stratified* (layered), with layers of warm air overlying cooler air; this stratification explains the name *stratosphere*. The lack of convection also means that the stratosphere has essentially no weather and no rain. Pollutants that reach the stratosphere, including the ozone-destroying chemicals known as chlorofluorocarbons (CFCs), remain there for decades.

Note that a planet can have a stratosphere *only* if its atmosphere contains molecules that are particularly good at absorbing ultraviolet photons. Ozone (O_3) plays this role on Earth, but the lack of oxygen in the atmospheres of the other terrestrial worlds means that they also lack ozone. As a result, Earth is the only terrestrial world with a stratosphere, at least in our solar system. (The jovian planets have stratospheres due to other ultraviolet-absorbing molecules [**Section 11.1**].)

X Rays and the Thermosphere Because nearly all gases are good X-ray absorbers, X rays from the Sun are absorbed by the first gases they encounter as they enter the atmosphere. The density of gas in the exosphere is too low for it to absorb significant amounts of these X rays, so most X rays are absorbed in the thermosphere. The absorbed energy makes temperatures quite high in the thermosphere (*thermos* is Greek for "hot"), but you wouldn't feel much heat because the density and pressure are so low [**Section 4.3**]. Virtually no X rays penetrate beneath the thermosphere, which is why X-ray telescopes are useful only on very high-flying balloons, rockets, and spacecraft.

Solar X rays also ionize a small but important fraction of the thermosphere's gas. The portion of the thermosphere that contains most of the ionized gas is called the *ionosphere*. The ionosphere is very important to radio communication, because it reflects most radio broadcasts back to Earth's surface. Without this reflection, radio communication would work only between locations in sight of each other.

The Exosphere The exosphere is the extremely low-density gas that forms the gradual and fuzzy boundary between the atmosphere and space (*exo* means "outermost" or "outside"). The gas density in the exosphere is so low that collisions between

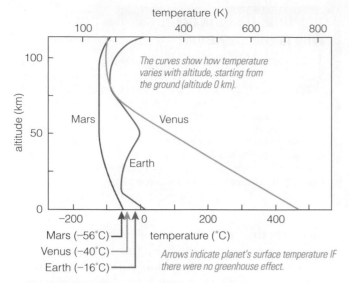

The curves show how temperature varies with altitude, starting from the ground (altitude 0 km).

Mars

Venus

Earth

temperature (K)

altitude (km)

temperature (°C)

Mars (−56°C)
Venus (−40°C)
Earth (−16°C)

Arrows indicate planet's surface temperature IF there were no greenhouse effect.

FIGURE 10.10 Venus, Earth, and Mars all have tropospheres and thermospheres (and exospheres, not shown), but only Earth has middle-atmosphere heating to make a stratosphere. The graph also shows that Venus and Earth are considerably warmer than they would be without the greenhouse effect.

atoms or molecules are very rare, although the high temperature means that gas particles move quite rapidly. Lightweight atoms and molecules sometimes reach escape velocity [**Section 4.5**] and fly off into space.

Comparative Structures of Terrestrial Atmospheres

We can now understand the structures of all the terrestrial atmospheres. The Moon and Mercury have so little gas that they essentially contain only an exosphere and have no structure to speak of. **FIGURE 10.10** contrasts the atmospheric

structures of Venus, Earth, and Mars. Notice that all three planets have a troposphere warmed by the greenhouse effect, and all three have a thermosphere heated by solar X rays. However, only Earth has the extra "bump" of a stratosphere, because it is the only planet with a layer of ultraviolet-absorbing gas (ozone). Without ozone, the middle altitudes of Earth's atmosphere would be almost as cold as those on Mars.

Magnetospheres and the Solar Wind There is one other important type of energy coming from the Sun: the low-density flow of subatomic charged particles called the *solar wind* [**Section 8.2**]. The solar wind does not significantly affect atmospheric structure, but it has other important effects.

On the Moon and Mercury, solar wind particles hit the surface, where they can blast atoms free. On Venus and Mars, solar wind particles can strip away atmospheric gas. In contrast, Earth's strong magnetic field creates a **magnetosphere** that acts like a protective bubble surrounding our planet, deflecting most solar wind particles around our planet (**FIGURE 10.11a**).

The magnetosphere still allows a few solar wind particles to get through, especially near the magnetic poles. Once inside the magnetosphere, these particles move along magnetic field lines, collecting in **charged particle belts** (or *Van Allen belts*, after their discoverer) that encircle our planet. The high energies of the particles in these belts can be hazardous to spacecraft and astronauts passing through them.

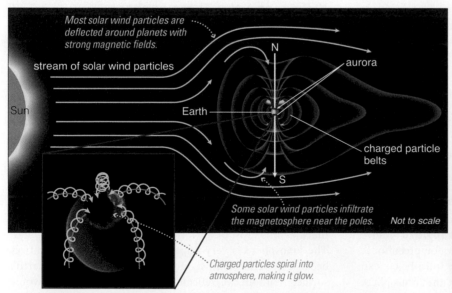

Most solar wind particles are deflected around planets with strong magnetic fields.

stream of solar wind particles

Sun

N

Earth

aurora

charged particle belts

S

Some solar wind particles infiltrate the magnetosphere near the poles. *Not to scale*

Charged particles spiral into atmosphere, making it glow.

a This diagram shows how Earth's magnetosphere deflects solar wind particles. Some particles accumulate in charged particle belts encircling our planet. The inset is a photo of a ring of auroras around the North Pole; the bright crescent at its left is part of the day side of Earth.

b This photograph shows the aurora near Yellowknife, Northwest Territories, Canada. In a video, you would see these lights dancing about in the sky.

FIGURE 10.11 Earth's magnetosphere acts like a protective bubble, shielding our planet from the charged particles of the solar wind.

Charged particles trapped in the magnetosphere also create the beautiful light of the **aurora** (**FIGURE 10.11b**). Variations in the solar wind can buffet the magnetosphere and give energy to particles trapped there. If a trapped particle gains enough energy, it can follow the magnetic field all the way down to Earth's atmosphere, where it collides with atmospheric atoms and molecules. These collisions cause the atoms and molecules to radiate and produce the moving lights of the aurora. Because the charged particles follow the magnetic field, auroras are most common near the magnetic poles and are best viewed at high latitudes. In the Northern Hemisphere, the aurora is often called the *aurora borealis,* or northern lights. In the Southern Hemisphere, it is called the *aurora australis,* or southern lights. The aurora can also be seen from space, where the lights look much like surf in the upper atmosphere (see Figure 10.5).

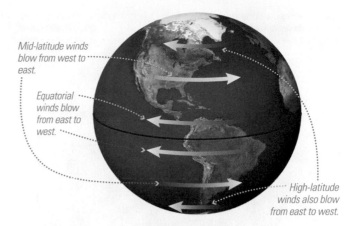

FIGURE 10.12 Schematic of Earth's global wind patterns. Notice that mid-latitude winds blow from west to east, explaining why storms generally move across the United States from the West Coast toward the East Coast.

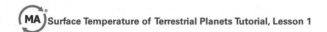

Surface Temperature of Terrestrial Planets Tutorial, Lesson 1

10.2 WEATHER AND CLIMATE

So far, we've talked mostly about *average* conditions in planetary atmospheres, such as the global average temperature and the average atmospheric structure. However, experience tells us that surface and atmospheric conditions constantly change through weather and climate.

Weather and climate are closely related, but they are not quite the same thing. **Weather** refers to the ever-varying combination of winds, clouds, temperature, and pressure that makes some days hotter or cooler, clearer or cloudier, or calmer or stormier than others. **Climate** refers to the average of weather over many years. For example, we say that Antarctic deserts have a cold, dry climate, even though it might sometimes rain or snow. Geological records show that climates can change, but these changes usually occur gradually over decades, centuries, or millennia.

Weather and climate can be hard to distinguish on a human time scale. For example, a drought lasting a few years may be the result of either random weather fluctuations or the beginning of a gradual change to a drier climate. The difficulty in distinguishing between random weather and real climate trends is an important part of the debate about human influences on the climate. We'll return to this topic in Section 10.6; for now, let's focus on understanding weather and climate generally.

What creates wind and weather?

Wind, rain, and other weather phenomena are all driven by energy in the atmosphere, which means only planets with atmospheres can have weather. Even then, weather varies dramatically among different worlds. Earth has the most diverse weather of the terrestrial planets, so we'll use it as our example of how weather works.

The complexity of weather makes it difficult to predict, and at best the local weather can be predicted only a week or so in advance (see Special Topic, page 280). Nevertheless, when we look at Earth as a whole, we can identify certain general characteristics of weather.

Global Wind Patterns The wind's direction and strength can change rapidly in any particular place, but we find distinctive patterns on a more global scale. For example, winds generally cause storms moving in from the Pacific to hit the West Coast of the United States first and then make their way eastward across the Rocky Mountains and the Great Plains, heading to the East Coast.

FIGURE 10.12 shows Earth's major **global wind patterns** (or *global circulation*). Notice that the wind direction varies with latitude: Equatorial winds blow from east to west, mid-latitude winds blow from west to east, and high-latitude winds blow like equatorial winds from east to west. Two factors explain this pattern: atmospheric heating and planetary rotation. Let's examine each in turn.

THINK ABOUT IT

Hurricanes usually form over warm, equatorial waters. Based on the wind patterns shown in Figure 10.12, in which direction will a hurricane move after forming off the northeastern coast of South America? Explain why these wind patterns make the Caribbean and the southeastern United States especially vulnerable to hurricanes.

Atmospheric Heating and Circulation Cells

Atmospheric heating affects global wind patterns because equatorial regions receive more heat from the Sun than polar regions. Warm equatorial air therefore rises upward and flows toward the poles, where cool air descends and flows toward the equator. If Earth's rotation did not influence this process, the result would be two huge **circulation cells** (or *Hadley cells,* after the man who first suggested their existence), one in each hemisphere (**FIGURE 10.13**).

The circulation cells transport heat both from lower to higher altitudes and from the equator to the poles. They therefore make Earth's polar regions much warmer than they would be in the absence of circulation. The same idea applies to different extents on Venus and Mars. On Venus, the dense atmosphere allows the circulation cells to transport so much thermal energy that temperatures are nearly the same at the

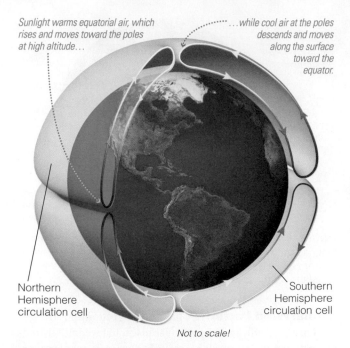

Sunlight warms equatorial air, which rises and moves toward the poles at high altitude...

...while cool air at the poles descends and moves along the surface toward the equator.

Northern Hemisphere circulation cell

Southern Hemisphere circulation cell

Not to scale!

FIGURE 10.13 interactive figure Atmospheric heating creates a circulation cell in each hemisphere that carries warm equatorial air toward the poles while cool polar air moves toward the equator. This diagram shows how the circulation cells would work if our planet's rotation didn't affect them. The vertical extent of the circulation cells is greatly exaggerated; in reality, the tops of the cells are only a few kilometers above Earth's surface.

equator and the poles. On Mars, the circulation cells transport very little heat because the atmosphere is so thin, so the poles remain much colder than the equator.

Rotation and the Coriolis Effect Planetary rotation affects global wind patterns through the **Coriolis effect** (named for the French physicist who first explained it), which you can understand by thinking about a spinning merry-go-round (**FIGURE 10.14**). The outer parts of the merry-go-round move at a faster speed than the inner parts, because they have a greater distance to travel around the axis with each rotation. If you sit near the edge and roll a ball toward the center, the ball begins with your relatively high speed around the axis. As it rolls inward, the ball's high speed makes it move ahead of the slower-moving inner regions. On a merry-go-round rotating counterclockwise, the ball therefore deviates *to the right* instead of heading straight inward. The Coriolis effect also makes the ball deviate to the right if you roll it outward from a position near the center. In that case, the ball starts with your slower speed around the center and lags behind as it rolls outward to faster-moving regions, again deviating to the right. (If the merry-go-round rotates clockwise rather than counterclockwise, the deviations go to the left instead of the right.)

The Coriolis effect alters the path of air on the rotating Earth in much the same way (**FIGURE 10.15**). Equatorial regions circle around Earth's rotation axis faster than polar regions (see Figure 1.12). Air moving away from the equator therefore

Weather and Chaos

Scientists today have a very good understanding of the physical laws and mathematical equations that govern the behavior of the atmosphere and oceans. Why, then, do we have so much trouble predicting the weather? To understand the answer, we must look at the nature of scientific prediction.

Prediction always requires knowing two things: (1) the current state of a system, sometimes called its *initial conditions*, and (2) how the system is changing. This is easy for something simple, like a car; if you know where a car is and how fast and in what direction it is traveling, you can predict where it will be a few minutes from now. Weather prediction is more difficult because weather is created by the motions of countless individual atoms and molecules. Scientists therefore attempt to predict the weather by creating a "model world." For example, suppose you overlay a globe of Earth with graph paper and specify the current temperature, pressure, cloud cover, and wind within each square; these are your initial conditions, which you can input into a computer along with a set of equations (physical laws) describing the processes that can change weather from one moment to the next. You can now use your computer model to predict the weather for the next month in New York City. The model might tell you that tomorrow will be warm and sunny, with cooling during the next week and a major storm passing through a month from now.

Now, suppose you run the model again, but you make one minor change in the initial conditions—say, a small change in the wind speed somewhere over Brazil. This slightly different initial condition will not change the weather prediction for tomorrow in New York City, and

may only slightly affect the prediction for next week's weather. For next month's weather, however, the two predictions may not agree at all!

The disagreement between the two predictions arises because the laws governing weather can cause tiny changes in initial conditions to be greatly magnified over time. This extreme sensitivity to initial conditions is sometimes called the *butterfly effect:* If initial conditions change by as much as the flap of a butterfly's wings, the resulting long-term prediction may be very different. That is why it is possible to predict the weather with reasonable accuracy a few days in advance, but not a few months in advance.

The butterfly effect is a hallmark of *chaotic systems*. Simple systems are described by linear equations in which, for example, increasing a cause produces a proportional increase in an effect. In contrast, chaotic systems are described by nonlinear equations, which allow for subtler and more intricate interactions. For example, the economy is nonlinear because a rise in interest rates does not automatically produce a corresponding change in consumer spending. Weather is nonlinear because a change in the wind speed in one location does not automatically produce a corresponding change in another location.

Despite their name, chaotic systems are not necessarily random. In fact, many chaotic systems have a kind of underlying order that explains the general features of their behavior even though details at any particular moment remain unpredictable. In a sense, many chaotic systems—like the weather—are "predictably unpredictable."

FIGURE 10.14 **interactive figure** The Coriolis effect on a merry-go-round rotating counterclockwise. Notice that the ball's path deviates to the *right* regardless of whether the ball is rolled inward or outward. If the merry-go-round were rotating clockwise, in both cases the ball would veer to the left.

has "extra" speed that causes it to move ahead of Earth's rotation to the east, while air moving toward the equator lags behind Earth's rotation to the west. In either case, moving air turns to the *right* in the Northern Hemisphere and to the *left* in the Southern Hemisphere, which explains why storms circulate in opposite directions in the two hemispheres. Uneven heating and cooling of Earth's surface creates regions of slightly higher pressure ("H" on weather maps) or lower pressure ("L" on weather maps) than average. Storms generally occur around low-pressure regions, which draw air inward from surrounding regions; as shown in Figure 10.15a, the Coriolis effect makes the inward-flowing air rotate counterclockwise in the Northern Hemisphere and clockwise in the Southern Hemisphere. This rotation around a low-pressure

zone can be quite stable, which is why storms can persist for days or weeks while being carried across the globe.

THINK ABOUT IT

The Coriolis effect also is important for long-range missiles. Suppose you are in the Northern Hemisphere on Earth and are trying to send a missile to hit a target that lies 5000 kilometers due north of you. Should you aim directly at the target, somewhat to the left, or somewhat to the right? Explain.

The Coriolis effect plays an even more important role in shaping Earth's global wind patterns: It splits each of the two huge circulation cells shown in Figure 10.13 into three

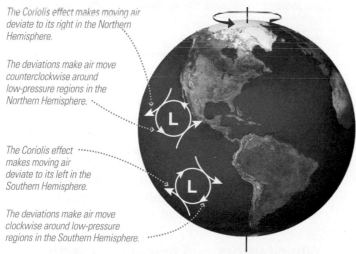

a Low-pressure regions ("L") draw in air from surrounding areas, and the Coriolis effect causes this air to circulate counterclockwise in the Northern Hemisphere and clockwise in the Southern Hemisphere.

b This photograph shows the opposite directions of storm circulation in the two hemispheres.

FIGURE 10.15 **interactive figure** The Coriolis effect works on the rotating Earth much as it does on a merry-go-round, because regions near the equator move at faster speed around the axis than regions near the poles.

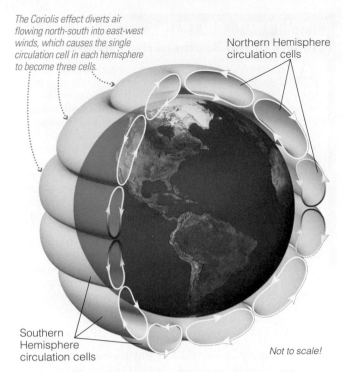

Northern Hemisphere circulation cells

Southern Hemisphere circulation cells

Not to scale!

FIGURE 10.16 interactive figure On Earth, the Coriolis effect causes each of the two large circulation cells that would be present without rotation (see Figure 10.13) to split into three smaller cells.

smaller circulation cells (**FIGURE 10.16**). You can understand why by considering air flowing southward along the surface from the North Pole. Without rotation, this air would travel 10,000 kilometers due south to the equator. But as Earth rotates, the Coriolis effect diverts this air to the right (westward) well before it reaches the equator, forcing the single large circulation cell to split. The three resulting cells circulate the air somewhat like three interlocking gears. These motions explain the global wind directions: Notice that surface air moves toward the equator in the cells near the equator and near the poles, so the Coriolis effect diverts this air into westward winds (see Figure 10.12). In contrast, surface air moves toward the poles in the mid-latitude cells, so the Coriolis effect diverts it into winds that blow eastward. In essence, the Coriolis effect on a rotating planet tends to divert air moving north or south into east-west winds.

The Coriolis effect operates to some extent on all planets. Its strength depends on a planet's size and rotation rate: Larger size and faster rotation both contribute to a stronger Coriolis effect. Among the terrestrial planets, Earth is the only one with a Coriolis effect strong enough to split the two large circulation cells. Venus has a very weak Coriolis effect because of its slow rotation, while Mars has a weak Coriolis effect because of its small size. The Coriolis effect is much stronger on the large and fast-rotating jovian planets, and it can therefore split their circulation cells into numerous smaller cells [**Section 11.1**].

Clouds and Precipitation In addition to winds, the other key components of weather are rain, snow, and hail, which together are called **precipitation** in weather reports.

FIGURE 10.17 The cycle of water on Earth's surface and in the atmosphere.

Precipitation requires clouds. We may think of clouds as imperfections on a sunny day, but they have profound effects on Earth and other planets, despite being made from minor ingredients of the atmosphere (see Table 10.1). Besides being the source of precipitation, clouds can alter a planet's energy balance. Clouds reflect sunlight back to space, thereby reducing the amount of sunlight that warms a planet's surface, but they also tend to be made from greenhouse gases that contribute to planetary warming.

On Earth, clouds are made from tiny droplets of liquid water or flakes of ice, which you can feel if you walk through a cloud on a mountaintop. Clouds are produced by condensation of water vapor (**FIGURE 10.17**). The water vapor enters the atmosphere through evaporation of surface water (or sublimation of ice and snow). Convection then carries the water vapor to high, cold regions of the troposphere, where it can condense to form clouds. Clouds can also form as winds blow over mountains, because the mountains force the air high enough to allow condensation. The condensed droplets or ice flakes start out very small, but gradually grow larger. If they get large enough that the upward convection currents can no longer hold them aloft, then they begin to fall toward the surface as rain, snow, or hail.

Stronger convection means more clouds and precipitation. That is why thunderstorms are common on summer afternoons, when the sunlight-warmed surface drives strong convection. The linkage between clouds and convection also explains why Earth has lush jungles near the equator and deserts at latitudes of 20°–30° north or south. Equatorial regions experience high rainfall because they receive more sunlight, which causes more convection. This high rainfall depletes the air of moisture as Earth's circulation cells carry it away from the equator (see Figure 10.16), leaving little moisture to fall as rain at the latitudes of the deserts.

What factors can cause long-term climate change?

We now turn our attention to climate, which varies much more slowly than weather. Scientists have identified four major factors that can lead to long-term climate change on the terrestrial worlds:

Solar brightening: As the Sun brightens with time, the increasing sunlight tends to warm the planets.

Changes in reflectivity: Higher reflectivity tends to cool a planet, while lower reflectivity leads to warming.

Changes in axis tilt: Greater tilt makes more extreme seasons, while smaller tilt keeps polar regions colder.

Changes in greenhouse gas abundance: An increase in greenhouse gases slows escape of infrared radiation, warming the planet, while a decrease leads to cooling.

FIGURE 10.18 Four major factors affecting long-term climate change.

- **Solar brightening.** The Sun has grown gradually brighter with time, increasing the amount of solar energy reaching the planets.

- **Changes in axis tilt.** The tilt of a planet's axis may change over long periods of time.

- **Changes in reflectivity.** An increase in a planet's reflectivity means a decrease in the amount of sunlight it absorbs, and vice versa.

- **Changes in greenhouse gas abundance.** More greenhouse gases tend to make a planet warmer, and less make it cooler.

FIGURE 10.18 summarizes these four factors, which we now investigate in a little more detail. Keep in mind that more than one factor may be acting at any given time.

Solar Brightening Both theoretical models of the Sun and observations of other Sun-like stars tell us that the Sun has gradually brightened with age [**Section 14.2**]: The Sun today is probably about 30% brighter than it was when our solar system was young. This brightening tends to warm climates with time, though we'd expect the effects to be noticeable only over periods of tens of millions of years or longer. Moreover, the brightening is so gradual that other climate change factors can easily overwhelm it. For example, while solar brightening alone would mean that all planets should be warmer today than they were in the past, both Mars and Earth are cooler than they were at times in their early histories.

Changes in Axis Tilt Small gravitational tugs from moons, other planets, and the Sun can change a planet's axis tilt over thousands or millions of years. For example, while Earth's current axis tilt is about $23\frac{1}{2}°$, the tilt has varied

over tens of thousands of years between about 22° and 25° (**FIGURE 10.19**). These small changes affect the climate by making seasons more or less extreme. Greater tilt means more extreme seasons, with warmer summers and colder winters. The extra summer warmth tends to prevent ice from building up, which reduces the planet's reflectivity and thereby makes the whole planet warmer. Conversely, a smaller tilt means less extreme seasons, which can allow ice to build up and make a planet cooler.

Earth's past periods of smaller axis tilt correlate well with the times of past ice ages, especially when considered along with other small changes in Earth's rotation and orbit. They

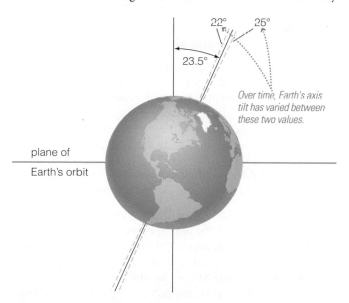

FIGURE 10.19 Earth's axis tilt is currently about $23\frac{1}{2}°$, but it has varied between about 22° and 25° over tens of thousands of years.

are therefore thought to be a primary factor in climate changes on Earth. (The cyclical changes in Earth's axis tilt and orbit are often called *Milankovitch cycles*, after the Serbian scientist who suggested their role in climate change.) As we'll discuss shortly, Mars probably experiences much more extreme changes in axis tilt and climate than Earth, and similar climate cycles may occur on some of the icy moons of the outer solar system.

Changes in Reflectivity Changes in reflectivity affect climate because they change the proportions of sunlight absorbed and reflected. If a planet reflects more sunlight, it must absorb less, which can lead to planetwide cooling. Microscopic dust particles (called *aerosols*) released by volcanic eruptions can reflect sunlight and cool a planet. Small but measurable planetwide cooling has been detected on Earth following a major volcanic eruption. The cooling can continue for years if the dust particles reach the stratosphere.

Human activity is currently altering Earth's reflectivity, although we are not sure in which direction or by precisely how much. Smog particles can act like volcanic dust, reflecting sunlight before it reaches the ground. Deforestation also increases reflectivity because it removes sunlight-absorbing plants. On the other hand, roads and cities tend to decrease reflectivity, which is why they tend to be hotter than surrounding areas of vegetation.

Changes in Greenhouse Gas Abundance Perhaps the most important factor in long-term climate change is a change in the abundance of greenhouse gases, which strengthens or weakens the greenhouse effect. If the abundance of greenhouse gases increases, the planet generally will warm. If the planet warms enough, increased evaporation and sublimation may add substantial amounts of gas to the planet's atmosphere, leading to an increase in atmospheric pressure. Conversely, if the abundance of greenhouse gases decreases, the planet generally will cool, and atmospheric pressure may decrease as gases freeze.

How does a planet gain or lose atmospheric gases?

Of the factors that affect planetary climate, changes in greenhouse gas concentrations appear to have had the greatest effect on the long-term climates of Venus, Earth, and Mars. Such changes generally occur as part of more general changes in the abundances of atmospheric gases. We must therefore investigate how atmospheres gain and lose gas.

Sources of Atmospheric Gas Terrestrial atmospheres can gain gas in three basic ways (**FIGURE 10.20**):

■ **Outgassing.** Volcanic *outgassing* has been the primary source of gases for the atmospheres of Venus, Earth, and Mars. Recall that the terrestrial worlds were built primarily of metal and rock, but impacts of ice-rich planetesimals (from beyond the frost line [**Section 8.2**]) brought in water and gas that became trapped in their interiors during accretion. Studies of volcanic eruptions show that the most common gases released by outgassing are water (H_2O),

carbon dioxide (CO_2), nitrogen (N_2), and sulfur-bearing gases (H_2S and SO_2).*

■ **Evaporation/sublimation.** After outgassing creates an atmosphere, some atmospheric gases may condense to become surface liquids or ices. The subsequent evaporation or sublimation of these surface liquids and ices therefore represents a secondary source of atmospheric gas. For example, if a planet warms, the rates of evaporation and sublimation will increase, adding gas to the atmosphere.

■ **Surface ejection.** The tiny impacts of micrometeorites, solar wind particles, and high-energy solar photons can knock individual atoms or molecules free from the surface. This *surface ejection* process explains the small amounts of gas that surround the Moon and Mercury. It is not a source process for planets that already have substantial atmospheres, because the atmospheres prevent small particles and high-energy solar photons from reaching the surface.

Losses of Atmospheric Gas Planets can also lose atmospheric gas, and there are four major loss processes (**FIGURE 10.21**). Note that the first two of these processes simply recycle gas from the atmosphere to the planet's surface or interior, while the latter two lead to permanent loss of gas.

■ **Condensation.** The condensation of gases that then fall as rain, hail, or snow is essentially the reverse of the release of gas by evaporation or sublimation. On Mars, for example, it is cold enough for carbon dioxide to condense into dry ice (frozen carbon dioxide), especially at the poles.

■ **Chemical reactions.** Some chemical reactions incorporate gas into surface metal or rock. Rusting is a familiar example: Iron rusts when it reacts with oxygen, thereby removing oxygen from the atmosphere and incorporating it into the metal.

■ **Solar wind stripping.** For any world without a protective magnetosphere, particles from the solar wind can gradually strip away gas particles into space.

■ **Thermal escape.** If an atom or a molecule of gas in a planet's exosphere achieves escape velocity [**Section 4.5**], it will fly off into space. The relative importance of thermal escape on any world depends on its size, distance from the Sun, and atmospheric composition. In general, more thermal escape will occur if a planet is small (so that it has a low escape velocity) or close to the Sun (which makes it hotter, so that atoms and molecules of atmospheric gas are moving faster). Lightweight gases, such as hydrogen and helium, escape more easily than heavier gases, such as carbon dioxide, nitrogen, and oxygen.

SEE IT FOR YOURSELF

You can see some atmospheric gain and loss processes in your freezer. Look for loose ice cubes that are gradually shrinking away, or look for the buildup of frost. What process is occurring in each case, and where do these processes occur in the solar system?

---

*Among these materials, only water existed in modest quantities in the solar nebula. The others were created by chemical reactions that occurred *inside* the planets after gas became trapped.

Outgassing is the primary source of gas for terrestrial atmospheres.

Evaporation/sublimation adds gas released by surface liquids or ices.

Impacts of particles and photons on worlds without significant atmospheres can eject small amounts of gas from surface.

FIGURE 10.20 Three processes that can provide gas to terrestrial atmospheres.

How Atmospheres Lose Gas

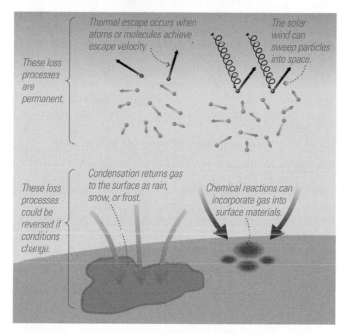

These loss processes are permanent.

Thermal escape occurs when atoms or molecules achieve escape velocity.

The solar wind can sweep particles into space.

These loss processes could be reversed if conditions change.

Condensation returns gas to the surface as rain, snow, or frost.

Chemical reactions can incorporate gas into surface materials.

FIGURE 10.21 Four processes that can remove gas from terrestrial atmospheres.

All the terrestrial planets are small enough and warm enough for hydrogen and helium to escape. That is why they were unable to hold on to any hydrogen or helium gas that they may have captured from the solar nebula when they were very young. Venus, Earth, and Mars have been able to retain heavier gases released later by outgassing, which is why they have substantial atmospheres today. However, the compositions and densities of their atmospheres have changed with time because of other source and loss processes.

10.3 ATMOSPHERES OF THE MOON AND MERCURY

We have now covered the basic ideas needed to understand the atmospheric histories of the terrestrial worlds. In the rest of this chapter, we'll use these to learn how and why each world ended up with its current atmosphere. As we did in Chapter 9, we'll begin with the smallest worlds, the Moon and Mercury.

Do the Moon and Mercury have any atmosphere?

We usually don't think of the Moon and Mercury as having atmospheres, but they are not totally devoid of gas. However, their gas densities are far too low for sunlight to be scattered or absorbed. The lack of scattering means that, even in broad daylight, you would see a pitch-black sky surrounding the bright Sun. The lack of absorption means that their atmospheres do not have a troposphere, stratosphere, or thermosphere. In essence, the Moon and Mercury have only extremely low-density exospheres, without any other atmospheric layers.

The total amount of gas in the exospheres of the Moon and Mercury is very small. If you could condense the entire atmosphere of either the Moon or Mercury into solid form, you would have so little material that you could almost store it in a dorm room. The low density of the gas means that collisions between atoms or molecules are rare. The gas particles therefore can rise as high as their speeds allow—sometimes even escaping to space. As a result, the exospheres of the Moon and Mercury extend thousands of kilometers above their surfaces (**FIGURE 10.22**).

Source and Loss Processes on the Moon and Mercury The Moon and Mercury may once have had some gas released by volcanic outgassing, but they no longer have any volcanic activity, and any gas released in the distant past is long gone. Some of the gas released long ago was probably lost through stripping by the solar wind, but it would have been a The Moon's exosphere, which extends high above the surface.

b Mercury's exosphere, much of which is escaping in this image.

lost to thermal escape anyway. Mercury cannot hold much of an atmosphere because its small size and high daytime temperature mean that nearly all gas particles eventually achieve escape velocity. The Moon is cooler than Mercury,

FIGURE 10.22 These "images" of the atmospheres of the Moon and Mercury—which are essentially exospheres only—are based on data collected with instruments sensitive to emission lines from sodium atoms; the colors represent gas density from highest (red) to lowest (blue). The inset photos of the Moon and Mercury are on the same scale as the images.

MATHEMATICAL INSIGHT 10.2

Thermal Escape from an Atmosphere

Thermal escape depends on the speeds of gas particles (atoms or molecules), because a particle can escape to space only if it reaches or exceeds a world's escape velocity. The fact that gas particles continually collide with one another means that individual particles move at a wide range of speeds. However, these speeds are not totally random; instead, their distribution has a characteristic shape that depends on the temperature and the particle masses. **FIGURE 1** shows this characteristic shape for sodium atoms in the Moon's daytime exosphere ($T = 400$ K). The peak in the figure represents the most common speed, or *peak thermal velocity*, which is given by the following formula:

$$v_{\text{thermal}} = \sqrt{\frac{2kT}{m}}$$

where m is the mass of a single gas particle, T is the temperature on the Kelvin scale, and $k = 1.38 \times 10^{-23}$ joule/K is *Boltzmann's constant*. Note that the peak thermal velocity increases with temperature (because higher temperature means higher average kinetic energy for gas particles), and decreases with particle mass (because at any particular temperature, particles of lighter gases, with smaller m, move at faster speeds than particles of heavier gases).

If a gas's peak thermal velocity is greater than the escape velocity, most of the gas particles will quickly escape to space. However, some gas particles can escape even if the peak thermal velocity is much lower. For example, Figure 1 shows that a small fraction of the sodium atoms have speeds exceeding the Moon's escape velocity

(about 2.4 km/s), even though the peak thermal velocity (about 0.5 km/s) is well below it. These atoms can escape if they are moving in the right direction (upward) and don't collide with other particles on their way out. (Once these atoms escape, ongoing collisions redistribute the speeds of the remaining atoms, so that the shape of the distribution remains the same and the same small fraction always exceeds escape velocity.) The time it takes for a gas to completely escape depends on how its peak thermal velocity compares to the escape velocity. As a rule of thumb, a peak thermal velocity above about 20% of the escape velocity will allow the gas to be completely lost to space within a few billion years—which is less than the age of our solar system.

EXAMPLE: Why does the Moon's exosphere contain sodium atoms but virtually no hydrogen gas? Useful data: The Moon's daytime temperature is 400 K and its escape velocity is about 2.4 km/s; the mass of a hydrogen atom is 1.67×10^{-27} kg; the mass of a sodium atom is 3.84×10^{-26} kg (about 23 times the mass of a hydrogen atom).

SOLUTION:

Step 1 Understand: The rate of thermal escape depends on how each gas's peak thermal velocity compares to the escape velocity at the Moon's daytime temperature of 400 K. From Figure 1, we know the peak thermal velocity for sodium atoms (about 0.5 km/s). We need to find the peak thermal velocity for hydrogen atoms at the same temperature.

Step 2 Solve: We use the given formula and data to calculate the peak thermal velocity for hydrogen atoms at a temperature of 400 K:

$$v_{\text{thermal (H atoms)}} = \sqrt{\frac{2kT}{m_{\text{H atom}}}}$$

$$= \sqrt{\frac{2 \times (1.38 \times 10^{-23} \frac{\text{joule}}{\text{K}}) \times (400 \text{ K})}{1.67 \times 10^{-27} \text{ kg}}}$$

$$\approx 2600 \text{ m/s} = 2.6 \text{ km/s}$$

Step 3 Explain: The peak thermal velocity of the hydrogen atoms is 2.6 km/s, which is slightly *greater* than the Moon's escape velocity of 2.4 km/s. This tells us that hydrogen atoms quickly escape, which is why the Moon cannot retain hydrogen in its atmosphere. In contrast, the peak thermal velocity for sodium atoms (about 0.5 km/s) is only about 20% of the Moon's escape velocity, so their escape rate is slow, and they are continually replenished as micrometeorites, solar wind particles, and high-energy photons eject new sodium atoms from the Moon's surface. That is why the Moon has a thin sodium exosphere.

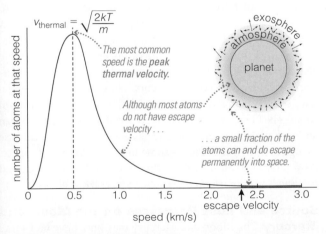

FIGURE 1 interactive figure In a gas at a given temperature, different atoms are always moving at different speeds. This plot shows the range of speeds of sodium atoms in the lunar atmosphere.

but its smaller size gives it a lower escape velocity; as a result, gases escape about as easily on both worlds.

The only ongoing source of gas on the Moon and Mercury is the surface ejection that occurs when micrometeorites, solar wind particles, or high-energy solar photons knock free surface atoms and molecules. This gas never accumulates because it is lost as quickly as it is gained. Some of the relatively few particles released from the surface are blasted upward fast enough to achieve escape velocity and therefore escape directly to space. The rest bounce around like tiny rubber balls, arcing hundreds of kilometers into the sky before crashing back down to the surface. Each gas particle typically bounces a few dozen times before being absorbed back into the surface.

Ice in Polar Craters There cannot be liquid water on the Moon or Mercury, because the lack of significant atmospheric pressure means that water cannot remain stable in liquid form. Daytime temperatures on both worlds are generally high enough that any water ice would long ago have sublimated into gas and escaped to space. However, scientists long wondered whether water ice might reside in permanently shadowed craters near the poles, deposited by comet impacts over millions of years.

For the Moon, the presence of polar ice was confirmed in 2009, when scientists sent the rocket from the *LCROSS* spacecraft crashing into a crater near the south pole. The debris that splashed upward revealed the presence of water vaporized from ice in the lunar soil. A radar sensor aboard India's *Chandrayaan-1* spacecraft detected ice deposits in similar craters near the Moon's north pole (**FIGURE 10.23**). More

FIGURE 10.23 This radar map shows a region near the Moon's north pole, imaged by a NASA instrument on India's *Chandrayaan-1* spacecraft. The green circles represent craters in which water ice was detected. The ice lies at the bottoms of craters that are in perpetual shadow. Shadows are not visible since radar was used to take the image.

surprisingly, other missions detected small amounts of water mixed into the upper layer of lunar soil over much of the lunar surface; the origin of this water is unknown. The evidence for ice in polar craters on Mercury is more limited, but scientists hope to learn more from ongoing observations by the *MESSENGER* spacecraft, which has been orbiting Mercury since 2011.

10.4 THE ATMOSPHERIC HISTORY OF MARS

One of our key goals in this chapter is to understand how and why the atmospheres of Venus, Earth, and Mars came to differ so profoundly, despite the fact that all three worlds must have had similar early atmospheres supplied by outgassing. We are now ready to focus on this question as we continue our atmospheric tour with Mars. Recall that Mars is only about 40% larger in radius than Mercury, but its surface reveals a much more fascinating and complex atmospheric history.

What is Mars like today?

The present-day surface of Mars looks much like deserts or volcanic plains on Earth (for example, see Figure 7.6). However, its thin atmosphere makes Mars quite different. The low atmospheric pressure—less than 1% of that on Earth's surface—explains why liquid water is unstable on the Martian surface [**Section 9.4**] and why visiting astronauts could not survive without a pressurized spacesuit. The atmosphere is made mostly of carbon dioxide, but the total amount of gas is so small that it creates only a weak greenhouse effect. The temperature is usually well below freezing, with a global average of about −50°C (−58°F). The lack of oxygen means that Mars lacks an ozone layer, so much of the Sun's damaging ultraviolet radiation passes unhindered to the surface.

While axis tilt is the only important influence on Earth's seasons, Mars's seasons are also affected by its orbit (**FIGURE 10.24**). Mars's more elliptical orbit puts it significantly closer to the Sun during southern hemisphere summer (and farther from the Sun during southern hemisphere winter), giving its southern hemisphere more extreme seasons—that is, shorter, warmer summers and longer, colder winters—than its northern hemisphere.

Seasons on Mars

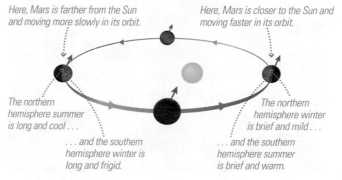

Here, Mars is farther from the Sun and moving more slowly in its orbit.

Here, Mars is closer to the Sun and moving faster in its orbit.

The northern hemisphere summer is long and cool . . .

. . . and the southern hemisphere winter is long and frigid.

The northern hemisphere winter is brief and mild . . .

. . . and the southern hemisphere summer is brief and warm.

FIGURE 10.24 The ellipticity of Mars's orbit makes seasons more extreme in the southern hemisphere than in the northern hemisphere.

FIGURE 10.25 This image from the *Mars Global Surveyor* shows the residual south polar cap during summer. A layer of frozen carbon dioxide around 8 meters thick overlies a much thicker cap of water ice. In winter, the whole area shown in the image is covered in CO_2 frost.

Seasonal changes lead to several major features of Martian weather. Temperatures at the winter pole drop so low (about −130°C) that carbon dioxide condenses into "dry ice" at the winter polar cap. Meanwhile, frozen carbon dioxide at the summer pole sublimates into carbon dioxide gas, and by the peak of summer only a residual cap of water ice remains (FIGURE 10.25). The atmospheric pressure therefore increases at the summer pole and decreases at the winter pole. Overall, as much as one-third of the total carbon dioxide of the Martian atmosphere moves seasonally between the north and south polar caps.

The Martian Winds The strong winds associated with the cycling of carbon dioxide gas can initiate huge dust storms, particularly when the more extreme summer approaches in the southern hemisphere (FIGURE 10.26). At times, the Martian surface becomes almost completely obscured by airborne dust. As the dust settles out, it can change the surface appearance over vast areas (for example, by covering dark regions with brighter dust); such changes fooled astronomers of the past into thinking they were seeing seasonal changes in vegetation.

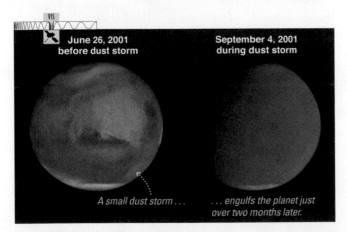

June 26, 2001
before dust storm

September 4, 2001
during dust storm

A small dust storm . . .

. . . engulfs the planet just over two months later.

FIGURE 10.26 These two Hubble Space Telescope photos contrast the appearance of the same face of Mars in the absence (left) and presence (right) of a global dust storm.

FIGURE 10.27 **interactive figure** This photograph shows a dust devil on Mars, photographed by the *Spirit* rover.

Martian winds can also spawn *dust devils,* swirling winds that you may have seen over desert sands or dry dirt on Earth. Dust devils look much like miniature tornadoes, but they rise up from the ground rather than coming down from the sky. The air in dust devils is heated from below by the sunlight-warmed ground; it swirls because of the way it interacts with prevailing winds. Dust devils on Mars are especially common during summer in either hemisphere. While many are quite small (FIGURE 10.27), some can be far larger than their counterparts on Earth.

Martian winds and dust storms leave Mars with perpetually dusty air, which helps explain the colors of the Martian sky. The air on Mars is so thin that, without suspended dust, the sky would be essentially black even in daytime. However, light scattered by the suspended dust tends to give the sky a yellow-brown color. Different hues can occur as the amount of suspended dust varies, and in the mornings and evenings.

Water Ice on Mars Although there is no liquid water on Mars today, there is a fair amount of water ice. The polar caps are made mostly of water ice, overlaid with a thin layer (at most a few meters thick) of CO_2 ice (see Figure 10.25). Radar instruments on *Mars Express* and the *Mars Reconnaissance Orbiter* have found substantial quantities of water frozen in vast layers of dusty ice surrounding both poles. Radar has even detected some icy glaciers well away from the poles; a protective layer of rocks and dust prevents the ice from sublimating away. Scientists sent the *Phoenix* lander to explore the north polar region in 2008 (FIGURE 10.28), and were surprised to discover that the spacecraft's landing rockets had exposed a patch of water ice, despite being hundreds of kilometers away from the polar cap itself. If all the water ice now known on Mars melted, it could make an ocean averaging about 10 meters deep over the planet. The fact that plenty of frozen water still remains on Mars makes it conceivable that some water is present in liquid form in underground locations where it is kept warm by volcanic heat.

Mars Climate and Axis Tilt The weather on Mars does not change much from one year to the next. However,

a The *Phoenix* lander used a robotic arm to scoop up Martian soil for analysis by on-board instruments. Results will help scientists determine whether and when Mars's polar regions may have been habitable.

b The robotic arm camera of the *Phoenix* lander found a bright patch of water ice underneath it. The spacecraft's landing rockets (visible at top) blasted away an overlying layer of dust.

FIGURE 10.28 The *Phoenix* lander.

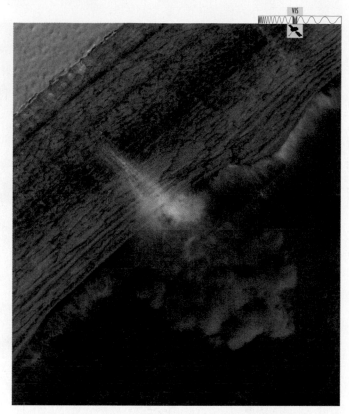

FIGURE 10.29 The *Mars Reconnaissance Orbiter* captured this image of a landslide in layered terrain in the north polar region. Despite the dark appearance, water makes up the bulk of the material. Layers of dusty ice more than 700 meters thick built up over many cycles of climate change. During the northern spring of 2010, warming conditions apparently weakened the cliff walls and triggered landslides.

changes in its axis tilt probably cause Mars to undergo longer-term cycles of climate change. Theoretical calculations suggest that Mars's axis tilt varies far more than Earth's—from as little as 0° to 60° or more on time scales of hundreds of thousands to millions of years. This extreme variation arises for two reasons. First, Jupiter's gravity has a greater effect on the axis of Mars than on that of Earth, because Mars's orbit is closer to Jupiter's orbit. Second, Earth's axis is stabilized by the gravity of our relatively large Moon. Mars's two tiny moons, Phobos and Deimos (see Figure 8.11), are far too small to offer any stabilizing influence on its axis.

As we discussed earlier, changes in axis tilt affect both the severity of the seasons and the global average temperature. When Mars's axis tilt is small, the poles may stay in a perpetual deep freeze for tens of thousands of years. With more carbon dioxide frozen at the poles, the atmosphere becomes thinner, lowering the pressure and weakening the greenhouse effect. When the axis is highly tilted, the summer pole becomes much warmer, allowing substantial amounts of water ice to sublimate, along with carbon dioxide, into the atmosphere. The pressure therefore increases, and Mars becomes warmer as the greenhouse effect strengthens—although probably not by enough to allow liquid water to become stable at the surface. Changes in axis tilt may therefore destroy existing glaciers and create new ones at different latitudes. The Martian polar regions show layering of dust and

ice that probably reflects changes in climate due to the changing axis tilt (**FIGURE 10.29**). Small changes in the tilt over the last few million years may also be responsible for the formation of the gullies seen on many crater walls (see Figure 9.33).

Why did Mars change?

Mars's varying axis tilt may have had important effects on its climate, but it does not explain the most dramatic climate change in Martian history: the change from a world of flowing water to the cold and dry world we see today.

Past Climate on Mars The geological evidence discussed in Chapter 9 leaves little doubt that Mars had wetter and warmer periods, probably with rainfall, before about 3 billion years ago. The full extent of these periods is a topic of considerable scientific debate. Some scientists think that Mars may have been continuously warm and wet for much of its first billion years of existence. Others think that Mars may have had only intermittent periods of rainfall, perhaps triggered by the heat of large impacts, and that ancient lakes, ponds, or oceans may have been completely ice-covered. Either way, rain could have fallen only if temperatures were warm enough to keep water from freezing and the atmospheric pressure were high enough for liquid water to be stable. We conclude that Mars once must have had a much thicker atmosphere, with a much stronger greenhouse effect.

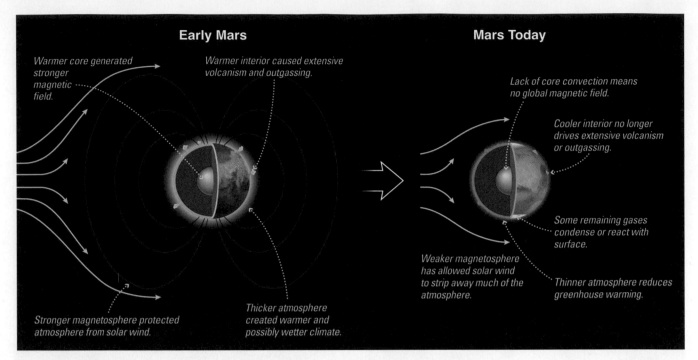

Early Mars

Warmer core generated stronger magnetic field.

Warmer interior caused extensive volcanism and outgassing.

Stronger magnetosphere protected atmosphere from solar wind.

Thicker atmosphere created warmer and possibly wetter climate.

Mars Today

Lack of core convection means no global magnetic field.

Cooler interior no longer drives extensive volcanism or outgassing.

Some remaining gases condense or react with surface.

Weaker magnetosphere has allowed solar wind to strip away much of the atmosphere.

Thinner atmosphere reduces greenhouse warming.

FIGURE 10.30 Some 3 billion years ago, Mars underwent dramatic climate change, ensuring that rain could never fall again.

The idea that Mars once had a much thicker atmosphere and stronger greenhouse effect makes sense. Calculations suggest that Martian volcanoes should have outgassed enough carbon dioxide to make the atmosphere about 400 times as dense as it is today, and enough water to fill oceans tens or even hundreds of meters deep. Moreover, computer simulations show that if Mars had that much carbon dioxide today, its greenhouse effect would make it warm enough to allow for liquid water, and the pressure would be great enough for the water to remain stable. However, because the Sun was dimmer in the distant past (see Figure 10.18), even more greenhouse warming would have been needed to allow for liquid water when Mars was young. Current models are unable to account for the necessary additional warming with carbon dioxide gas alone, but additional greenhouse warming may have been provided by carbon dioxide ice clouds or atmospheric methane.

Loss of Atmospheric Gas The bigger question is not whether Mars once had a denser atmosphere, but what happened to it. Mars must somehow have lost most of its carbon dioxide gas. This loss would have weakened the greenhouse effect until the planet essentially froze over. Some of the carbon dioxide condensed and became part of the polar caps. Some may be chemically bound into **carbonate rocks** (rocks rich in carbon and oxygen), according to recent finds by Mars orbiters and rovers. But the bulk of the gas was probably lost to space.

The precise way in which Mars lost its carbon dioxide gas is not clear, but the leading hypothesis suggests a close link to a change in Mars's magnetic field (**FIGURE 10.30**). Early in its history, Mars probably had molten, convecting metals in its core, much like Earth today. The combination of this convecting metal with Mars's rotation should have

produced a magnetic field and a protective magnetosphere. However, the magnetic field would have weakened as the small planet cooled and core convection ceased, leaving atmospheric gases vulnerable to being stripped into space by solar wind particles. This hypothesis will be put to the test by the orbiting *MAVEN* mission, scheduled for launch in late 2013, which will measure the escape of gases from Mars's atmosphere today and thereby help us understand past gas loss as well.

With regard to water, although some lies frozen in the polar caps and underground, much of the water once present on Mars is also probably gone for good. Like the carbon dioxide, some water vapor may have been stripped away by the solar wind. However, Mars also lost water in another way. Because Mars lacks an ultraviolet-absorbing stratosphere, atmospheric water molecules would have been easily broken apart by ultraviolet photons. The hydrogen atoms that broke away from the water molecules would have been lost rapidly to space through thermal escape. With these hydrogen atoms gone, the water molecules could not be made whole again. Initially, oxygen from the water molecules would have remained in the atmosphere, but over time this oxygen was lost, too. Some was probably stripped away by the solar wind, and the rest was drawn out of the atmosphere through chemical reactions with surface rock. This process literally rusted the Martian rocks, giving the "red planet" its distinctive tint.

THINK ABOUT IT

Some people have proposed "terraforming" Mars—that is, making it more Earth-like—by finding a way to release all the carbon dioxide frozen in its polar caps into its atmosphere. If Mars had oceans in the distant past, could this release of gas allow it to have oceans again? Why or why not?

Size as the Critical Factor Summarizing these ideas, we conclude that Mars's fate was shaped primarily by its relatively small size. It was big enough for volcanism and outgassing to release plenty of water and atmospheric gas early in its history, but too small to maintain the internal heat needed to keep this water and gas. As its interior cooled, its volcanoes quieted and released far less gas, while its relatively weak gravity and the loss of its magnetic field allowed existing gas to be stripped away to space. If Mars had been as large as Earth, so that it could still have outgassing and a global magnetic field, it might still have a moderate climate today. Mars's distance from the Sun helped seal its fate: Even with its small size, Mars might still have some flowing water if it were significantly closer to the Sun, where the extra warmth could melt the water that remains frozen underground and at the polar caps.

The history of the Martian atmosphere holds important lessons for us on Earth. Mars apparently was once a world with more moderate temperatures and streams, rain, glaciers, lakes, and possibly oceans. It had all the necessities for life as we know it. But this once-hospitable planet turned into a frozen and barren desert at least 3 billion years ago, and it is unlikely that Mars will ever again be warm enough for its frozen water to flow. Any life that may have existed on Mars is either extinct or hidden away in a few choice locations, such as in underground water near not-quite-dormant volcanoes. As we consider the possibility of future climate change on Earth, Mars presents us with an ominous example of how drastically things can change.

10.5 THE ATMOSPHERIC HISTORY OF VENUS

Venus (**FIGURE 10.31**) presents a stark contrast to Mars, but it is easy to understand why: Its larger size allowed it to retain more interior heat, leading to greater volcanism. The associated outgassing released the vast quantities of carbon dioxide that create Venus's strong greenhouse effect.

Venus becomes more mysterious when we compare it to Earth. Because Venus and Earth are so similar in size, we might naively expect both planets to have had similar atmospheric histories. Clearly, this is not the case. In this section, we'll explore how Venus's atmosphere ended up so different from Earth's, and in the process we'll learn important lessons about the habitability of our own planet.

What is Venus like today?

If you stood on the surface of Venus, you'd feel a searing heat hotter than that of a self-cleaning oven and a tremendous pressure 90 times greater than that on Earth. A deep-sea diver would have to go nearly 1 kilometer (0.6 mile) beneath the ocean surface on Earth to feel comparable pressure. Venus's atmosphere consists almost entirely of carbon dioxide (CO_2). It has virtually no molecular oxygen (O_2), so you could not breathe the air even if you cooled it to a comfortable temperature.

Moving through the thick air near Venus's surface would feel like a cross between swimming and flying: Its density is

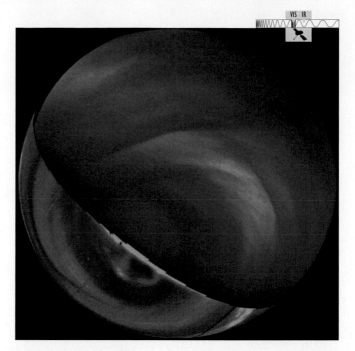

FIGURE 10.31 This composite image from the *Venus Express* spacecraft combines a visible-wavelength image of the day side (left, shaded red) and an infrared image of the night side (right, shaded blue). Venus's south pole lies at the center. At most wavelengths, clouds completely prevent any view of the surface.

about 10% that of water. Looking upward, you'd see a perpetually overcast sky, with only weak sunlight filtering through the thick clouds above. Because the thick atmosphere scatters nearly all the blue light away, the dimly lit sky appears reddish-orange in color.

The weather forecast for the surface of Venus today and every day is dull, dull, dull. Venus's slow rotation (243 Earth days) means a very weak Coriolis effect. As a result, Venus has little wind on its surface and never has hurricane-like storms. The top wind speeds measured by the Soviet Union's *Venera* landers were only about 6 kilometers per hour. No rain falls, because droplets that form and fall from the cool upper atmosphere evaporate long before they reach the ground. The weak Coriolis effect also means that Venus's atmosphere has just two large circulation cells, much like what Earth would have if rotation didn't split its cells (see Figure 10.13). The thick atmosphere makes the circulation so efficient at transporting heat from the equator to the poles that the surface temperature is virtually the same everywhere: The poles are no cooler than the equator, and night is just as searingly hot as day. Moreover, Venus has no seasons because it has virtually no axis tilt,* so temperatures are the same year-round.

The weather is much more interesting at high altitudes. Strong convection drives hot air upward; high in the troposphere, where the temperature is 400°C cooler than on the surface (see Figure 10.10), sulfuric acid (H_2SO_4) condenses into droplets that create Venus's bright, reflective clouds. The droplets sometimes fall through the upper troposphere as

---

*In tables (such as Table 7.1), Venus's axis tilt is usually written as 177.3°. This may sound large, but notice that it is nearly 180°—the same tilt as 0° but "upside down." It is written this way because Venus rotates backward compared to its orbit, and backward rotation is equivalent to forward rotation that is upside down.

sulfuric acid rain, but they evaporate at least 30 kilometers above the surface. Moreover, high-altitude winds circle the planet in just 4 days—much faster than the planet rotates. No one knows why these fast winds blow, but they are responsible for the dynamic cloud patterns visible in Figure 10.31.

How did Venus get so hot?

It's tempting to attribute Venus's high surface temperature solely to the fact that Venus is closer than Earth to the Sun, but Venus would actually be quite cold without its strong greenhouse effect, because its bright clouds reflect much more sunlight than Earth (see Table 10.2). The real question is why Venus has such a strong greenhouse effect.

The simple answer is that Venus has a huge amount of carbon dioxide in its atmosphere—nearly 200,000 times as much as in Earth's atmosphere. However, a deeper question still remains. Given their similar sizes and compositions, we expect Venus and Earth to have had similar levels of volcanic outgassing, and the released gas ought to have had about the same composition on both worlds. Why, then, is Venus's atmosphere so different from Earth's?

The Fate of Outgassed Water and Carbon Dioxide

We expect that huge amounts of water vapor and carbon dioxide should have been outgassed into the atmospheres of both Venus and Earth. Venus's atmosphere does indeed have an enormous amount of carbon dioxide, but it has virtually no water. Earth's atmosphere has very little of either gas. We conclude that Venus must have somehow lost its outgassed water, while Earth lost both water vapor and carbon dioxide. But how were these gases lost?

We can easily account for the missing gases on Earth. The huge amounts of water vapor released into our atmosphere condensed into rain, forming our oceans. In other words, the water is still here, but mostly in liquid rather than gaseous form. The huge amount of carbon dioxide released into our atmosphere is also still here, but in solid form: Carbon dioxide dissolves in water, where it can undergo chemical reactions to make carbonate rocks such as limestone. Earth has about 170,000 times as much carbon dioxide locked up in rocks as in its atmosphere—which means that Earth does indeed have almost as much total carbon dioxide as Venus. Of course, the fact that Earth's carbon dioxide is mostly in rocks rather than in the atmosphere makes all the difference in the world: If this carbon dioxide were in our atmosphere, our planet would be nearly as hot as Venus and certainly uninhabitable.

We are left with the question of what happened to Venus's water. Venus today is incredibly dry. It is far too hot to have any liquid water or ice on its surface; it is even too hot for water to be chemically bound in surface rock, and any water deeper in its crust or mantle was probably baked out long ago. Measurements also show very little water in the atmosphere. Overall, the total amount of water on Venus is about 10,000 times smaller than the total amount on Earth, a fact that explains why Venus retains so much carbon dioxide in its atmosphere: Without oceans, carbon dioxide cannot dissolve or become locked away in carbonate rocks. If it is true that a huge amount of water was outgassed on Venus, it has somehow disappeared.

The leading hypothesis for the disappearance of Venus's water invokes one of the same processes thought to have removed water from Mars. Ultraviolet light from the Sun broke apart water molecules in Venus's atmosphere. The hydrogen atoms then escaped to space (through thermal escape), ensuring that the water molecules could never re-form. The oxygen from the water molecules was lost to a combination of chemical reactions with surface rocks and stripping by the solar wind; Venus's lack of a magnetic field leaves its atmosphere vulnerable to the solar wind.

Acting over billions of years, the breakdown of water molecules and the escape of hydrogen can easily explain the loss of an ocean's worth of water from Venus, and careful study of Venus's atmospheric composition offers evidence that such a loss really occurred. Recall that most hydrogen nuclei contain just a single proton, but a tiny fraction of all hydrogen atoms (about 1 in 6400 measured in Earth's oceans) contain a neutron in addition to the proton, making the isotope of hydrogen that we call *deuterium.* Water molecules that contain one or two atoms of deuterium instead of hydrogen (called *heavy water*) behave chemically just like ordinary water and can be broken apart by ultraviolet light just as easily. However, a deuterium atom is twice as heavy as an ordinary hydrogen atom and therefore does not escape to space as easily when the water molecule is broken apart. If Venus lost a huge amount of hydrogen from water molecules to space, the rare deuterium atoms would have been more likely to remain behind than the ordinary hydrogen atoms. Measurements show that this is the case: The fraction of deuterium among hydrogen atoms is a hundred times higher on Venus than on Earth, suggesting that a substantial amount of water was lost by having its molecules broken apart and its hydrogen lost to space. We cannot determine exactly how much water Venus has lost, but it seems plausible that Venus really did outgas as much water as Earth and then lost virtually all of it.

Of course, Venus could have lost all this water only if it had been in the atmosphere as water vapor, where ultraviolet light could break the molecules apart, rather than in liquid oceans like the water on Earth. Our quest to understand Venus's high temperature therefore leads to one more question: Why didn't Venus, like Earth, end up with oceans to trap its carbon dioxide in carbonate rocks and prevent its water from being lost to space?

The Runaway Greenhouse Effect To understand why Venus does not have oceans, we need to consider the role of **feedback processes**—processes in which a change in one property amplifies (positive feedback) or counteracts (negative feedback) the behavior of the rest of the system. You are probably familiar with feedback processes in daily life. For example, if someone brings a microphone too close to a loudspeaker, it picks up and amplifies small sounds from the speaker. These amplified sounds are again picked up by the microphone and further amplified, causing a loud screech.

If Earth moved to Venus's orbit

More intense sunlight...

...would raise surface temperature by about 30°C.

Higher temperature increases evaporation, and warmer air holds more water vapor.

Runaway greenhouse effect

Additional water vapor further strengthens the greenhouse effect.

Result: Oceans evaporate and carbonate rocks decompose, releasing CO_2...

...making Earth hotter than Venus.

FIGURE 10.32 This diagram shows how, if Earth were placed at Venus's distance from the Sun, the runaway greenhouse effect would cause the oceans to evaporate completely.

This sound feedback is an example of *positive feedback,* because it automatically amplifies itself. The screech usually leads to a form of *negative feedback:* The embarrassed person holding the microphone moves away from the loudspeaker, thereby stopping the positive sound feedback.

THINK ABOUT IT

Think of at least one other everyday example each for positive feedback and negative feedback.

With the idea of feedback in mind, let's consider what would happen if we could magically move Earth to the orbit of Venus (**FIGURE 10.32**). The greater intensity of sunlight would almost immediately raise Earth's global average temperature by about 30°C, from its current 15°C to about 45°C (113°F). Although this is still well below the boiling point of water, the higher temperature would lead to increased evaporation of water from the oceans. The higher temperature would also allow the atmosphere to hold more water vapor before the vapor condensed to make rain. The combination of more evaporation and greater atmospheric capacity for water vapor would substantially increase the total amount of water vapor in Earth's atmosphere. Now, remember that water vapor, like carbon dioxide, is a greenhouse gas. The added water vapor would therefore strengthen the greenhouse effect, driving temperatures a little higher. The higher temperatures, in turn, would lead to even more ocean evaporation and more water vapor in the atmosphere, strengthening the greenhouse effect even further. In other words, we'd have a positive feedback process in which each little bit of additional water vapor in the atmosphere would lead to higher temperature and even more water vapor. The process would rapidly spin out of control, resulting in a **runaway greenhouse effect**.

The runaway greenhouse effect would cause Earth to heat up until the oceans were completely evaporated and the carbonate rocks had released all their carbon dioxide back into the atmosphere. By the time the runaway process was complete, temperatures on our "moved Earth" would be even higher than they are on Venus today, thanks to the combined greenhouse effects of carbon dioxide and water vapor in the atmosphere. The water vapor would then gradually disappear, as ultraviolet light broke

water molecules apart and the hydrogen escaped to space. In short, moving Earth to Venus's orbit would essentially turn our planet into another Venus.

We have arrived at a simple explanation of why Venus is so much hotter than Earth. Even though Venus is only about 30% closer to the Sun than Earth is, this difference was critical. On Earth, it was cool enough for water to rain down to make oceans. The oceans dissolved carbon dioxide and chemical reactions locked it away in carbonate rocks, leaving our atmosphere with only enough greenhouse gases to make our planet pleasantly warm. On Venus, the greater intensity of sunlight made it just warm enough that oceans either never formed or soon evaporated, leaving Venus with a thick atmosphere full of greenhouse gases.

The next time you see Venus shining brightly as the morning or evening "star," consider the radically different path it has taken from Earth—and thank your lucky star. If Earth had formed a bit closer to the Sun, or if the Sun had been slightly hotter, our planet might have suffered the same greenhouse-baked fate.

THINK ABOUT IT

We've seen that moving Earth to Venus's orbit would cause our planet to become Venus-like. If we could somehow move Venus to Earth's orbit, would it become Earth-like? Why or why not?

A Pleasant Early Venus? Venus's closeness to the Sun may have sealed its ultimate fate, but it's possible that Venus might have been more moderate in its early history. Recall that the Sun has gradually brightened with age; some 4 billion years ago, the intensity of sunlight shining on the young Venus was not much greater than it is on Earth today. Rain might have fallen, and oceans could have formed. It's even conceivable that life could have arisen on the young Venus.

As the Sun gradually brightened, however, any liquid water or life on Venus was doomed. The runaway greenhouse effect raised the temperature so high that all the water evaporated. In the upper atmosphere, ultraviolet light broke apart the water molecules, and the hydrogen escaped to space. If Venus had oceans in its youth, the water is now gone forever.

We'll probably never know for sure whether Venus ever had oceans. The global "repaving" of Venus's surface by tectonics and volcanism [**Section 9.5**] would long ago have covered up any shorelines or other geological evidence of past oceans, and the high surface temperatures would have baked out any gases that might once have been incorporated into surface rock. If the climate once was pleasant, it's unlikely that any evidence survives to tell the tale.

10.6 EARTH'S UNIQUE ATMOSPHERE

Earth's atmosphere, composed mostly of nitrogen and oxygen, makes our lives possible. It provides our planet with just enough warmth and pressure to enable water to cycle among all three phases (solid ice, liquid water, and gaseous water vapor), it protects us from harmful solar radiation, and it produces the weather patterns that variously bring us days of sunshine, clouds, and rain or snow. In this section, we'll discuss how and why our atmosphere is so hospitable—and how we humans may be altering the very balances upon which we depend for survival.

How did Earth's atmosphere end up so different?

As we've seen, most of the major physical differences in the atmospheres of the terrestrial planets are easy to understand. For example, the Moon and Mercury lack substantial atmospheres because of their small sizes, while a very strong greenhouse effect causes the high surface temperature on Venus. In contrast, differences in atmospheric composition seem more surprising, because outgassing should have released the same gases on Venus, Earth, and Mars. How, then, did Earth's atmosphere end up so different? We can break down this general question into four separate questions:

1. Why did Earth retain most of its outgassed water—enough to form vast oceans—while Venus and Mars lost theirs?

2. Why does Earth have so little carbon dioxide (CO_2) in its atmosphere compared to Venus, when Earth should have outgassed about as much of it as Venus?

3. Why is Earth's atmosphere composed primarily of nitrogen (N_2) and oxygen (O_2), when these gases are only trace constituents in the atmospheres of Venus and Mars?

4. Why does Earth have an ultraviolet-absorbing stratosphere, while Venus and Mars do not?

Water and Carbon Dioxide We have already answered the first two questions in our discussions of the atmospheres of Mars and Venus. On Mars, some of the outgassed water was lost after solar ultraviolet light broke water vapor molecules apart, and the rest froze and may remain in the polar caps or underground. On Venus, it was too hot for water vapor to condense, so virtually all the water molecules were ultimately broken apart, allowing the hydrogen atoms to escape to space. Earth retained its outgassed water because temperatures were low enough for water vapor to condense into rain and form oceans. Evidence from tiny mineral grains suggests that Earth may have had oceans as early as 4.3–4.4 billion years ago. The oceans, in turn, explain the low level of carbon dioxide in our atmosphere. Most of the carbon dioxide outgassed by volcanism on Earth dissolved in the oceans, where chemical reactions turned it into carbonate rocks. Even today, about 60 times as much carbon dioxide is dissolved in the oceans as is present in the atmosphere, and carbonate rocks contain some 170,000 times as much CO_2 as the atmosphere.

Nitrogen, Oxygen, and Ozone Turning our attention to the third question, it's relatively easy to explain the substantial nitrogen content (77%) of our atmosphere. Nitrogen is the third most common gas released by outgassing, after water vapor and carbon dioxide. Because most of Earth's water ended up in the oceans and most of the carbon dioxide ended up in rocks, our atmosphere was left with nitrogen as its dominant ingredient.

The oxygen content (21%) is a little more mysterious. Molecular oxygen (O_2) is not a product of outgassing or any other geological process. Moreover, oxygen is a highly reactive chemical that is easily removed from the atmosphere. Fire, rust, and the discoloration of freshly cut fruits and vegetables are everyday examples of chemical reactions that remove oxygen from the atmosphere (called *oxidation reactions*). Similar reactions between oxygen and surface materials (especially iron-bearing minerals) give rise to the reddish appearance of much of Earth's rock and clay, including the beautiful reds of Arizona's Grand Canyon. Without continual replenishment, these types of chemical reactions would remove all the oxygen in Earth's atmosphere in just a few million years. We must therefore explain not only how oxygen got into Earth's atmosphere in the first place, but how it is replenished as chemical reactions remove it.

The answer to the oxygen mystery is *life* (**FIGURE 10.33**). Plants and many microorganisms release oxygen through photosynthesis. Photosynthesis takes in CO_2 and, through a complex chain of chemical reactions, releases O_2. Because chemical reactions can remove oxygen, it took a long time for oxygen to accumulate in Earth's atmosphere. According

Photosynthetic life in oceans creates O_2.

O_2 + UV creates O_3 (ozone).

O_3 layer makes surface safe for living things.

Photosynthetic life on land adds O_2.

FIGURE 10.33 The origin of oxygen and ozone in Earth's atmosphere.

to present evidence, it took at least a billion years of photosynthesis before the buildup of atmospheric oxygen began, and Earth's atmosphere probably has had enough oxygen for us to breathe only for the past few hundred million years [**Section 24.1**]. If you had a time machine and spun the dial to arrive randomly at any point in Earth's past, there's only about a 1 in 10 chance that you would arrive at a time when you'd be able to breathe the air. Today, plants and single-celled photosynthetic organisms return oxygen to the atmosphere in approximate balance with the rate at which animals and chemical reactions consume oxygen, keeping the oxygen levels relatively steady.

THINK ABOUT IT

Suppose that, somehow, all photosynthetic life on Earth died out. What would happen to the oxygen in our atmosphere? Could animals, including us, still survive?

Life and oxygen also explain the presence of Earth's ultraviolet-absorbing stratosphere. In the upper atmosphere, chemical reactions involving solar ultraviolet light transform some of the O_2 into molecules of O_3, or ozone. The O_3 molecule is more weakly bound than the O_2 molecule, which allows it to absorb solar ultraviolet energy even better. The absorption of solar energy by ozone heats the upper atmosphere, creating the stratosphere. This ozone layer prevents harmful ultraviolet radiation from reaching the surface. Mars and Venus lack photosynthetic life and therefore have too little O_2, and consequently too little ozone, to form a stratosphere.

Maintaining Balance We have answered all four basic questions about Earth's atmosphere. However, a deeper look at what we have learned still leaves us with a major mystery. Our oceans exist because Earth has a greenhouse effect that is "just right" to keep them from either freezing or boiling away, and the presence of liquid water allowed life to arise and produce oxygen. But *why* does the amount of carbon dioxide in our atmosphere stay "just right"? For example, why didn't the chemical reactions either remove *all* the carbon dioxide or leave so much of it that the oceans would boil away? The long-term existence of Earth's oceans tells us that our planet has enjoyed remarkable climate stability—presenting a stark contrast to the dramatic climate changes that apparently occurred on Mars and Venus.

Why does Earth's climate stay relatively stable?

Earth's long-term climate stability has clearly been important to the ongoing evolution of life—and hence to our own relatively recent arrival as a species (see Figure 1.11). Had our planet undergone a runaway greenhouse effect like Venus, life would certainly have been extinguished. If Earth had suffered loss of atmosphere and a global freezing like Mars, any surviving life would have been driven to hide in underground pockets of liquid water.

Earth's climate is not perfectly stable—our planet has endured numerous ice ages and warm periods in the past. Nevertheless, even in the coldest ice ages and warmest warm periods, Earth's temperature has remained in a range in

which some liquid water could still exist and harbor life. This long-term climate stability is even more remarkable when you remember that the Sun has brightened substantially over the past 4 billion years, yet Earth's temperature has managed to stay in nearly the same range throughout this time. Apparently, the strength of the greenhouse effect self-adjusts to keep the climate stable. How does it do this?

The Carbon Dioxide Cycle The mechanism by which Earth self-regulates its temperature is called the **carbon dioxide cycle**, or the **CO_2 cycle** for short. Let's follow the cycle as illustrated in **FIGURE 10.34**, starting at the top center:

- Atmospheric carbon dioxide dissolves in rainwater, creating a mild acid.
- The mildly acidic rainfall erodes rocks on Earth's continents, and rivers carry the broken-down minerals to the oceans.
- In the oceans, calcium from the broken-down minerals combines with dissolved carbon dioxide and falls to the ocean floor, making carbonate rocks such as limestone.*
- Over millions of years, the conveyor belt of plate tectonics (see Figure 9.39) carries the carbonate rocks to subduction zones, where they are carried downward.
- As they are pushed deeper into the mantle, some of the subducted carbonate rocks melt and release their carbon dioxide, which then outgasses back into the atmosphere through volcanoes.

---

*During the past half billion years or so, the carbonate minerals have been made by shell-forming sea animals, falling to the bottom in the seashells left after the animals die. Without the presence of animals, chemical reactions would do the same thing—and apparently did for most of Earth's history.

Atmospheric CO_2 dissolves in rainwater.

Rainfall erodes rock on land; rivers carry broken down minerals to the sea.

Volcanoes outgas CO_2.

Broken-down minerals react with dissolved CO_2 to form carbonate rocks.

Carbonate rocks subduct and melt, releasing CO_2.

FIGURE 10.34 This diagram shows how the CO_2 cycle continually moves carbon dioxide from the atmosphere to the ocean to rock and back to the atmosphere. Note that plate tectonics (subduction in particular) plays a crucial role in the cycle.

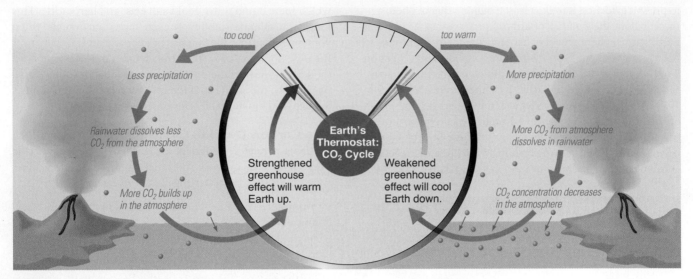

FIGURE 10.35 The carbon dioxide cycle acts as a thermostat for Earth through negative feedback processes. Cool temperatures cause atmospheric CO_2 to increase, and warm temperatures cause atmospheric CO_2 to decline.

The CO_2 cycle acts as a long-term thermostat for Earth, because it has a built-in form of negative feedback that returns Earth's temperature toward "normal" whenever it warms up or cools down (**FIGURE 10.35**). The negative feedback occurs because the overall rate at which carbon dioxide is pulled from the atmosphere is very sensitive to temperature: the higher the temperature, the higher the rate at which carbon dioxide is removed.

Consider first what happens if Earth warms up a bit. The warmer temperature means more evaporation and rainfall, pulling more CO_2 out of the atmosphere. The reduced atmospheric CO_2 concentration leads to a weakened greenhouse effect, which counteracts the initial warming and cools the planet back down. Similarly, if Earth cools a bit, precipitation decreases and less CO_2 is dissolved in rainwater, allowing the CO_2 released by volcanism to build back up in the atmosphere. The increased CO_2 concentration strengthens the greenhouse effect and warms the planet. Overall, the natural thermostat of the carbon dioxide cycle has allowed the greenhouse effect to strengthen or weaken just enough to keep Earth's climate fairly stable, regardless of what other changes have occurred on our planet.

COMMON MISCONCEPTIONS

Ozone—Good or Bad?

Ozone often generates confusion, because in human terms it is sometimes good and sometimes bad. Near the ground, ozone is one of the main ingredients in urban air pollution, produced as a by-product of automobiles and industry. This is bad for us, because ozone is poisonous to most living creatures. In contrast, the ozone that exists naturally high above the ground, in the stratosphere, serves as a protective shield from the Sun's ultraviolet radiation. This ozone is therefore crucial to our survival, which is why the release of chemicals (such as chlorofluorocarbons, or CFCs) that destroy stratospheric ozone has been a significant environmental problem.

Ice Ages and Other Long-Term Climate Change

While Earth's climate has remained stable enough for the oceans to stay at least partly liquid throughout history, significant variations have still occurred. Such variations are possible because the CO_2 cycle does not act instantly. When something begins to change the climate, it takes time for the feedback mechanisms of the CO_2 cycle to come into play because of their dependence on the gradual actions of plate tectonics and mineral formation in the oceans. Calculations show that it takes hundreds of thousands of years for atmospheric CO_2 to stabilize through the CO_2 cycle. That is, if the amount of CO_2 in the atmosphere were to rise as a result of, say, increased volcanism, it would take some 400,000 years for the CO_2 cycle to restore temperatures to their current values.[*]

Ice ages occur when the global average temperature drops by a few degrees. The slightly lower temperatures lead to increased snowfall, which may cover continents with ice down to fairly low latitudes. For example, the northern United States was entirely covered with glaciers during the peak of the most recent ice age, which ended only about 10,000 years ago. The causes of ice ages are complex and not fully understood. Over periods of tens or hundreds of millions of years, the Sun's gradual brightening and the changing arrangement of the continents around the globe have at least in part influenced the climate. During the past few million years—a period too short for solar changes or continental motion to have a significant effect—the ice ages appear to have been strongly influenced by small changes in Earth's axis tilt and other characteristics of Earth's rotation and orbit (the Milankovitch cycles noted earlier).

---

[*]This time scale applies to ocean/atmosphere equilibrium only. The time scale for crust recycling is much longer, while shorter-term climate variations in atmospheric CO_2 concentration can occur through factors besides the inorganic CO_2 cycle, such as cycling of carbon dioxide by life.

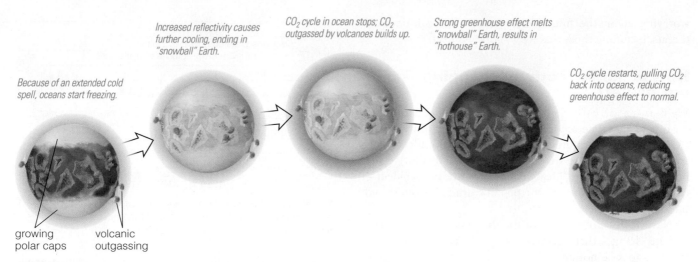

Because of an extended cold spell, oceans start freezing.

Increased reflectivity causes further cooling, ending in "snowball" Earth.

CO_2 cycle in ocean stops; CO_2 outgassed by volcanoes builds up.

Strong greenhouse effect melts "snowball" Earth, results in "hothouse" Earth.

CO_2 cycle restarts, pulling CO_2 back into oceans, reducing greenhouse effect to normal.

growing polar caps

volcanic outgassing

FIGURE 10.36 The CO_2 cycle rescues Earth from a "snowball" phase.

Geological evidence also points to several particularly long and deep ice ages between about 750 and 580 million years ago. During these periods, glaciers appear to have advanced all the way to the equator. Because ice can reflect up to about 90% of the sunlight hitting it, this increase in global ice would have set up a positive feedback process that would have cooled Earth even further. Geologists suspect that in this way our planet may have entered the periods we now call **snowball Earth** (**FIGURE 10.36**). We do not know why these episodes occurred or precisely how extreme the cold became. Some models suggest the positive feedback may have driven the global average temperature as low as –50°C (–58°F), causing the oceans to freeze to a depth of 1 kilometer or more. Other models suggest the oceans never froze completely, making Earth more of a "slushball" than a snowball. Either way, it seems that Earth became far colder during these periods than in more recent ice ages.

How did Earth recover from a "snowball" phase? The drop in surface temperature would not have affected Earth's interior heat, so volcanic outgassing would have continued to add CO_2 to the atmosphere. Oceans covered by ice would have been unable to absorb this CO_2 gas, which therefore would have accumulated in the atmosphere and strengthened the greenhouse effect. Eventually (perhaps after as long as 10 million years), the strengthening greenhouse effect would have warmed Earth enough to start melting the ice. The feedback processes that started the snowball Earth episode then moved in reverse. As the ice melted, more sunlight would have been absorbed (because liquid water absorbs more and reflects less sunlight than ice), warming the planet further. In fact, because the CO_2 concentration was so high, the warming would have continued well past current temperatures— perhaps taking the global average temperature to higher than 50°C (122°F). In just a few centuries, Earth would have emerged from a "snowball" phase into a "hothouse" phase. Geological evidence supports the occurrence of dramatic increases in temperature at the end of each snowball Earth episode. Earth then recovered over hundreds of thousands of years as the CO_2 cycle removed carbon dioxide from the atmosphere.

THINK ABOUT IT

Suppose Earth did not have plate tectonics. Could the planet ever recover from a "snowball" phase? Explain.

The snowball Earth episodes must have had severe consequences for any life on Earth at the time. Indeed, the end of the snowball Earth episodes roughly coincides with a dramatic increase in the diversity of life on Earth (the *Cambrian explosion* [**Section 24.1**]). Some scientists suspect that the environmental pressures caused by the snowball Earth periods may have led to a burst of evolution. If so, we might not be here today if not for Earth's dramatic climate changes.

Earth's Long-Term Future Climate Despite going through periods of ice ages, snowball Earth episodes, and hothouse phases, Earth's climate has remained suitable for life for some 4 billion years. However, the continuing brightening of the Sun will eventually overheat our planet.

According to some climate models, the warming Sun could cause Earth to begin losing its water as soon as a billion or so years from now. If so, then life on Earth has already completed about 75% of its history on this planet. However, there are enough uncertainties in the models to make it possible that they are wrong, and that the CO_2 cycle will keep the climate steady much longer. Either way, by about 3–4 billion years from now, the Sun will have grown so warm that sunlight on Earth will be as intense as it is on Venus today. The effect will be the same as if we moved Earth to Venus's orbit (see Figure 10.32)—a runaway greenhouse effect. Our planet will become a Venus-like hothouse, with temperatures far too high for liquid water to exist and all the CO_2 baked out of the rocks into the atmosphere.

In summary, Earth's habitability will cease between 1 and 4 billion years from now. Although this may seem depressing, remember that a billion years is a very long time— equivalent to some 10 million human lifetimes and far longer than humans have existed so far. If you want to lose sleep worrying about the future, there are far more immediate threats, including those that we will discuss next.

How is human activity changing our planet?

We humans are well adapted to the present-day conditions on our planet. The amount of oxygen in our atmosphere, the average temperature of our planet, and the ultraviolet-absorbing ozone layer are just what we need to survive. We have seen that these "ideal" conditions are no accident—they are consequences of our planet's unique geology and biology.

Nevertheless, the stories of the dramatic and permanent climate changes that occurred on Venus and Mars should teach us to take nothing for granted. Our planet may regulate its own climate quite effectively over long time scales, but fossil and geological evidence tells us that substantial and rapid changes in global climate can occur on shorter ones. In some cases, Earth's climate appears to have warmed several degrees Celsius in just decades. Evidence also shows that these past climate changes have had dramatic effects on local climates by raising or lowering sea level as much as tens of meters, altering ocean currents that keep coastlines warm, and transforming rainforests into deserts.

These past climate changes have been due to "natural" causes, such as small changes in Earth's tilt, major volcanic eruptions, the release of trapped carbon dioxide from oceans, and a variety of other geological processes. Today, however, Earth is apparently undergoing climate change for a new reason: Human activity is rapidly increasing the atmospheric concentration of carbon dioxide and other greenhouse gases. (The carbon dioxide cycle operates much too slowly to absorb these emissions on human time scales.) Effects of this increase in greenhouse gas concentration are already apparent: Global average temperatures have risen by about 0.8°C (1.4°F) in the past century (**FIGURE 10.37**). This **global warming** is one of the most important issues of our time.

Global Warming Global warming has been a hot political issue, both because some people debate its cause and because efforts to slow or stop the warming would require finding

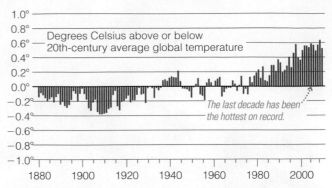

FIGURE 10.37 Average global temperatures from 1860 through 2011. Notice the clear global warming trend of the past few decades. (Data from the National Climate Data Center.)

new energy sources and making other changes that would dramatically affect the world's economy. However, a major research effort has gradually added to our understanding of the potential threat, particularly in the past two decades. The case linking global warming with human activity rests on three basic facts:

1. The greenhouse effect is a simple and well-understood scientific model. We can be confident in our understanding of it because it so successfully explains the observed surface temperatures of other planets. Given this basic model, there is no doubt that a rising concentration of greenhouse gases would make our planet warm up more than it would otherwise; the only debate is about how soon and how much.

2. The burning of fossil fuels and other human activities are clearly increasing the amounts of greenhouse gases in the atmosphere. Observations show that currently the atmospheric concentration of carbon dioxide is significantly higher (about 30%) than it has been at any time during the past million years, and it is rising rapidly (**FIGURE 10.38**). We can be confident that the rise is a result of human activity, because the atmosphere is becoming enriched in molecules of CO_2 carrying the distinct ratio of isotopes present in fossil fuels.

3. Climate models that ignore human activity fail to match the observed rise in global temperatures. In contrast, climate models that include the enhanced greenhouse effect from human production of greenhouse gases match the observed temperature trend quite well (**FIGURE 10.39**). Comparisons between observations and models therefore clearly indicate that global warming results from human activity.

These facts, summarized in **FIGURE 10.40**, offer convincing evidence that we humans are now tinkering with the climate in a way that may cause major changes not just in the distant future, but in our own lifetimes. The same models that convince scientists of the reality of human-induced global warming tell us that if current trends in the greenhouse gas concentration continue—that is, if we do nothing to slow our emissions of carbon dioxide and other greenhouse gases—the warming trend will continue to accelerate. By the end of this century, the models predict that the global average temperature would be

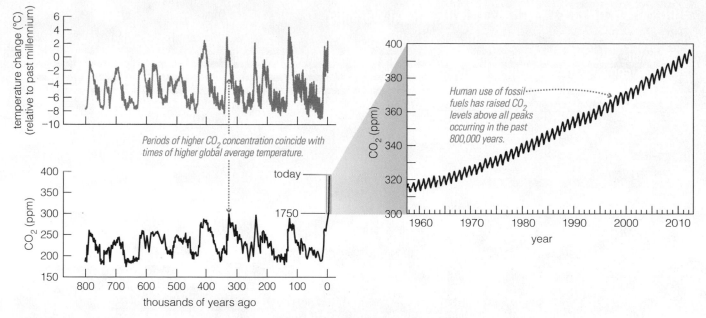

FIGURE 10.38 This diagram shows the atmospheric concentration of carbon dioxide and global average temperature over the past 800,000 years. Data for the past few decades come from direct measurements; earlier data come from studies of air bubbles trapped in Antarctic ice (ice core samples). The CO_2 concentration is measured in parts per million (ppm), which is the number of CO_2 molecules among every 1 million air molecules.

3°C–5°C (6°F–10°F) higher than it is now, giving our children and grandchildren the warmest climate that any generation of *Homo sapiens* has ever experienced.

THINK ABOUT IT

Use Figure 10.38 to estimate the current rate at which the carbon dioxide concentration is rising. If the rise continues at the same rate, how long will it be until we reach a doubling of the pre-industrial-age concentration of 280 ppm? What if the rate of increase also rises, as it has over the past few years? Discuss the implications of your answer.

While the models tell us what general trends to expect, we do not expect them to precisely match Earth's behavior because of the complexity of the climate system. In particular, numerous feedback mechanisms can counter or enhance greenhouse warming on time scales of years, decades, and even centuries, and we don't yet understand all these mechanisms. For example, increased evaporation of ocean water can enhance the greenhouse effect because water vapor is also a greenhouse gas, but it can counter global warming if it causes the formation of more clouds (which can block sunlight from reaching the ground). Melting of polar ice, on the other hand, makes global warming even worse, because it changes the reflectivity of part of Earth's surface from high to low, leading to more absorption of solar energy.

Consequences of Global Warming A temperature increase of a few degrees might not sound so bad, but small changes in *average* temperature can lead to much more dramatic changes in climate patterns. These changes will cause some regions to warm much more than the average, while other

regions may actually cool. Some regions might experience more rainfall or might become deserts.

Polar regions will warm the most, causing ice to melt. This is clearly threatening to the species of these regions (polar bears, which depend on an abundance of ice floes, are already endangered), but it also warms the oceans everywhere and changes their salt content as melting ice pours more fresh water into the sea. The fact that the waters of the Gulf of

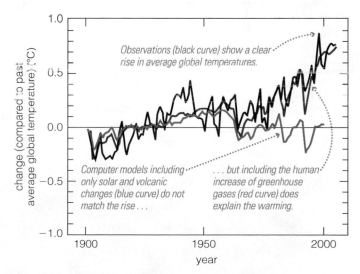

FIGURE 10.39 This graph compares observed temperature changes (black curve) with the predictions of climate models that include only natural factors such as changes in the brightness of the Sun and effects of volcanoes (blue curve) and models that also include the human contribution to increasing greenhouse gas concentration (red curve). Only the red curve matches the observations well. (The red and blue model curves are averages of many scientists' independent models of global warming, which generally agree with each other within 0.1°C–0.2°C.)

Scientific studies of global warming apply the same basic approach used in all areas of science: We create models of nature, compare the predictions of those models with observations, and use our comparisons to improve the models. We have found that climate models agree more closely with observations if they include human production of greenhouse gases like carbon dioxide, making scientists confident that human activity is indeed causing global warming.

① The greenhouse effect makes a planetary surface warmer than it would be otherwise because greenhouse gases such as carbon dioxide, methane, and water vapor slow the escape of infrared light radiated by the planet. Scientists have great confidence in models of the greenhouse effect because they successfully predict the surface temperatures of Venus, Earth, and Mars.

② Human activity is adding carbon dioxide and other greenhouse gases to the atmosphere. While the carbon dioxide concentration also varies naturally, its concentration is now much higher than it has been at any time in the previous million years, and it is continuing to rise rapidly.

Visible light passes through the atmosphere.

Some visible light is reflected by clouds, haze, and the surface.

The surface absorbs visible light and emits thermal radiation in infrared.

Greenhouse gases absorb and reemit infrared radiation, thereby heating the lower atmosphere.

The graph shows that today's CO_2 levels are higher than at any point in the past 800,000 years.

Global Average Surface Temperature

Planet	Temperature Without Greenhouse Effect	Temperature With Greenhouse Effect
Venus	−40°C	470°C
Earth	−16°C	15°C
Mars	−56°C	−50°C

This table shows planetary temperatures as they would be without the greenhouse effect and as they actually are with it. The greenhouse effect makes Earth warm enough for liquid water and Venus hotter than a pizza oven.

(3) Observations show that Earth's average surface temperature has risen during the last several decades. Computer models of Earth's climate show that an increased greenhouse effect triggered by CO_2 from human activities can explain the observed temperature increase.

Observations (white curve) show a clear rise in average global temperatures.

Computer models including only solar and volcanic changes (blue curve) do not match the rise

. . . but including the human increase of greenhouse gases (red curve) does explain the warming.

(y-axis) change (compared to past average global temperature) (°C)

1.0 / 0.5 / 0.0 / -0.5 / -1.0

(x-axis) 1900 / 1950 / 2000

year

(4) Models can also be used to predict the consequences of a continued rise in greenhouse gas concentrations. These models show that, without significant reductions in greenhouse gas emissions, we should expect further increases in global average temperature, rising sea levels, and more intense and destructive weather patterns.

Tallahassee

Orlando

Miami

Key West

HALLMARK OF SCIENCE **Science progresses through creation and testing of models of nature that explain the observations as simply as possible.** Observations showing a rise in Earth's temperature demand a scientific explanation. Models that include an increased greenhouse effect due to human activity explain those observations better than models without human activity.

This diagram shows the change in Florida's coastline that would occur if sea levels rose by 1 meter. Some models predict that this rise could occur within a century. The light blue regions show portions of the existing coastline that would be flooded.

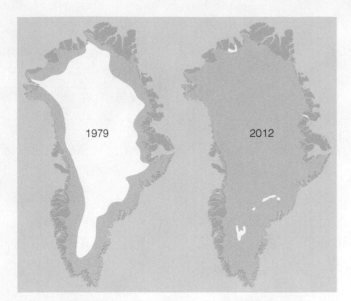

FIGURE 10.41 These maps contrast the extent of the year-round Greenland ice sheet, shown in white, in 1979 and 2012. The pink area indicates the region in which at least some melting occurred during the warm season. The melt region spanned at least 97% of the ice sheet during the extreme warming of 2012.

Mexico are at their warmest level in at least a century may be contributing to the greater strength of hurricanes that have recently blown through the Caribbean, though it is difficult to cite causes for specific storms. More generally, the greater overall warmth of the atmosphere will increase evaporation from the oceans, leading to more numerous and more intense storms; ironically, this fact means that global warming can lead to more severe winter blizzards. In some regions, spring is arriving earlier, which can lead to drier summers and greater wildfire risk. Some researchers also worry that the influx of large quantities of fresh water into the oceans may alter major ocean currents, such as the Gulf Stream—a "river" within the ocean that regulates the climate of western Europe and parts of the United States.

Melting polar ice may significantly increase sea level in the future, but global warming has already caused a rise in sea level for a different reason. Water expands very slightly as it warms—so slightly that we don't notice the change in a glass of water, but enough that sea level has risen some 20 centimeters in the past hundred years. This effect alone is likely to cause sea level to rise another 30 centimeters during this century, with potentially devastating effect on coastal communities and low-lying countries such as Bangladesh. The oceans are slow to

warm and expand; depending on how long it takes to stabilize greenhouse gases, the expansion could raise sea levels another 3 meters in subsequent centuries. The added effect of melting ice could increase sea level much more. While the melting of ice in the Arctic Ocean does not affect sea level—it is already floating—melting of landlocked ice does. Such melting appears to be occurring already. For example, the famous "snows" (glaciers) of Mount Kilimanjaro are rapidly retreating and may be gone within the next decade or so. More ominously, recent data suggest that the Greenland ice sheet is melting much more rapidly than models have predicted (**FIGURE 10.41**). If this trend continues, sea level could rise as much as several *meters*—enough to flood most of Florida—by the end of this century (see item 4 in Figure 10.40). Looking further ahead, complete melting of the polar ice caps would increase sea level by some 70 meters (more than 200 feet). Although such melting would probably take centuries or millennia, it suggests the disconcerting possibility that future generations will have to send deep-sea divers to explore the underwater ruins of many of our major cities.

Fortunately, most scientists believe that we still have time to avert the most serious consequences of global warming, provided that we dramatically and rapidly curtail our greenhouse gas emissions. The most obvious way to reduce these emissions is to improve energy efficiency. Doubling the average gas mileage of cars—which we could do easily with current technology—would immediately cut automobile-related carbon dioxide emissions in half. Other tactics could include replacing fossil fuels with alternative energy sources—such as biofuels, solar, wind, or nuclear energy—or finding ways to bury the carbon dioxide by-products of the fossil fuels that we still use. The key idea to keep in mind is that global warming is a global problem, and significant international cooperation will be required if we hope to solve it. But there is precedent for success: In the 1980s and 1990s, as we learned that human-produced chemicals (known as CFCs) were causing great damage to the ozone layer, the nations of the world agreed on a series of treaties that ultimately phased those chemicals out of production. As a result, the ozone layer is beginning to recover from its earlier damage, and we learned that people can be moved to act in the face of a threat to the environment on which we depend for survival.

THINK ABOUT IT

If you were a political leader, how would *you* deal with the threat of global warming?

The Big Picture

Putting Chapter 10 into Perspective

This chapter and the previous chapter have given us a complete "big picture" view of how the terrestrial worlds started out so similar yet ended up so different. As you continue your studies, keep in mind the following important ideas:

- Atmospheres affect planets in many ways. They absorb and scatter light, distribute heat, and create the weather that can lead to erosion. Perhaps most important, the greenhouse effect

allows atmospheres to make a planet warmer than it would be otherwise.

- Atmospheric properties differ widely among the terrestrial worlds, but we can trace these differences to root causes. For example, only the larger worlds have significant atmospheres. The smaller worlds lack the internal heat needed for volcanism and the outgassing that releases atmospheric gases, and they also lack the gravity necessary to retain these gases.

- The histories of Venus and Mars suggest that major climate change is the rule, not the exception. Mars was once warm and wet, but its small size and lack of a magnetic field caused it to lose gas and freeze over 3 billion or more years ago. Venus may once have had oceans, but its proximity to the Sun doomed it to a runaway greenhouse effect.

- We are here to talk about these things today only because Earth has managed to be the exception, a planet whose climate has remained relatively stable. We humans are ideally adapted to Earth today, but we have no guarantee that Earth will remain as hospitable in the future, especially as we tinker with the balance of greenhouse gases that has kept our climate stable.

SUMMARY OF KEY CONCEPTS

10.1 ATMOSPHERIC BASICS

- **What is an atmosphere?** An atmosphere is a layer of gas that surrounds a world. It can create pressure, absorb and scatter sunlight, create wind and weather, interact with the solar wind to create a magnetosphere, and cause a **greenhouse effect** that can make a planet's surface warmer than it would be otherwise.

- **How does the greenhouse effect warm a planet?** **Greenhouse gases** such as carbon dioxide, methane, and water vapor absorb infrared light emitted from a planet's surface. The absorbed photons are quickly reemitted, but in random directions. The result acts much like a blanket, warming the planet's surface.

- **Why do atmospheric properties vary with altitude?** **Atmospheric structure** is determined by the way atmospheric gases interact with sunlight. On Earth, the basic structure consists of the **troposphere**, where most greenhouse warming occurs; the **stratosphere**, where **ozone** absorbs ultraviolet light from the Sun; the **thermosphere**, where solar X rays are absorbed; and the **exosphere**, the extremely low-density outer layer of the atmosphere.

10.2 WEATHER AND CLIMATE

- **What creates wind and weather?** **Global wind patterns** are shaped by atmospheric heating and the **Coriolis effect** caused by a planet's rotation. Convection in the troposphere can lead to the formation of clouds and rain, hail, or snow.

- **What factors can cause long-term climate change?** The four factors that cause climate change are solar brightening (noticeable only over many millions of years), changes in axis tilt, changes in reflectivity, and changes in greenhouse gas abundance.

- **How does a planet gain or lose atmospheric gases?** Three sources of atmospheric gas are outgassing, evaporation or sublimation, and—on worlds with little atmosphere—surface ejection by tiny impacts of particles and photons. Four loss processes

are condensation, chemical reactions with surface materials, the stripping of gas by the solar wind, and **thermal escape**.

10.3 ATMOSPHERES OF THE MOON AND MERCURY

- **Do the Moon and Mercury have any atmosphere?** The Moon and Mercury have only very thin exospheres consisting of gas particles released through surface ejection by micrometeorites, solar wind particles, and high-energy solar photons.

10.4 THE ATMOSPHERIC HISTORY OF MARS

- **What is Mars like today?** Mars is cold and dry, with an atmospheric pressure so low that liquid water is unstable; however, a substantial amount of water is frozen in and near the polar caps. Martian weather is driven largely by seasonal changes that cause carbon dioxide alternately to condense and sublimate at the poles, creating pole-to-pole winds and sometimes leading to huge dust storms.

- **Why did Mars change?** Mars's atmosphere must once have been much thicker with a stronger greenhouse effect, so change must have occurred due to loss of atmospheric gas. Much of the gas probably was stripped away by the solar wind, which was able to reach the atmosphere as Mars cooled and lost its magnetic field. Water was probably lost as ultraviolet light broke apart water molecules in the atmosphere, and the lightweight hydrogen then escaped to space.

10.5 THE ATMOSPHERIC HISTORY OF VENUS

- **What is Venus like today?** Venus has a thick carbon dioxide atmosphere that creates its strong greenhouse effect, explaining why it is so hot. It rotates slowly and therefore has a weak Coriolis effect and weak winds, and is far too hot for rain to fall. Its atmospheric circulation keeps temperatures about the same day and night, and its lack of axis tilt means no seasonal changes.

- **How did Venus get so hot?** Venus's distance from the Sun ultimately led to a **runaway greenhouse effect**: Venus became too hot to develop liquid oceans like those on Earth. Without oceans to dissolve outgassed carbon dioxide and lock it away in carbonate rocks, all of Venus's carbon dioxide remained in its atmosphere, creating its intense greenhouse effect.

10.6 EARTH'S UNIQUE ATMOSPHERE

■ **How did Earth's atmosphere end up so different?**
Temperatures on Earth were just right for outgassed water vapor to condense and form oceans. The oceans dissolve carbon dioxide and ultimately lock it away in carbonate rocks, keeping the greenhouse effect moderate. Nitrogen from outgassing remained in the atmosphere. Oxygen and ozone were produced by photosynthesis, which was possible because the moderate conditions allowed the origin and evolution of abundant life.

■ **Why does Earth's climate stay relatively stable?**

Earth's long-term climate is remarkably stable because of feedback processes that tend to counter any warming or cooling that occurs. The most important feedback process is the **carbon dioxide cycle**, which naturally regulates the strength of the greenhouse effect.

■ **How is human activity changing our planet?** Human activity is releasing carbon dioxide and other greenhouse gases into the atmosphere, and scientific evidence confirms that this

is causing **global warming**. This warming may have many consequences, including a rise in sea level, an increase in severity of storms, and dramatic changes in local climates.

VISUAL SKILLS CHECK

Use the following questions to check your understanding of some of the many types of visual information used in astronomy. Answers are provided in Appendix J. For additional practice, try the Chapter 10 Visual Quiz at MasteringAstronomy®.

The graph above, a portion of Figure 10.38, shows the atmospheric concentration of carbon dioxide for the past 800,000 years. Use the information in the graph to answer the following questions.

1. Based on the graph, when was the maximum abundance of CO_2 over the past 800,000 years?
   a. about 300,000 years ago
   b. about 125,000 years ago
   c. 1750
   d. now

2. Today's CO_2 abundance is approximately ____ times the average value over the last 800,000 years.
   a. 17   b. 1.7   c. 1.17   d. 0.17

3. How would you describe the rapid variations in CO_2 on the inset graph on the right?
   a. The variations occur randomly every few years.
   b. The variations occur randomly every few months.

   c. The variations occur regularly every few years.
   d. The variations occur regularly every year.
   e. The variations occur regularly every few months.

4. Which physical mechanism is consistent with your answer to question 3?
   a. Major volcanic eruptions every decade or so place additional CO_2 in the atmosphere.
   b. Every few years, another huge coal-powered power plant opens up and produces more CO_2.
   c. Seasonal changes in plant growth produce a regular yearly variation in CO_2 levels, with the Northern Hemisphere (with much more land area) causing almost all the variation.

The graph below, the same as Figure 10.37, shows average global temperatures from 1860 through 2011 relative to the average temperature from the last century. Compare the graph below with the graph of CO_2 levels from questions 1–4.

5. What is the relationship between the graph of global average temperatures and the graph of CO_2 levels? Focus on the period since about 1960.
   a. The warmest years correlate exactly with the years of greatest CO_2 abundance, indicating that CO_2 is the single most important factor controlling global temperature changes.
   b. The warmest years correlate roughly with the years of greatest CO_2 abundance, indicating that CO_2 is important but other factors also contribute to global temperature changes.
   c. The warmest years are not correlated with years of high CO_2 abundance, indicating that factors other than CO_2 have the most important effect on global temperature changes.

EXERCISES AND PROBLEMS

For instructor-assigned homework go to MasteringAstronomy®.

REVIEW QUESTIONS

Short-Answer Questions Based on the Reading

1. Briefly describe the basic atmospheric characteristics of each of the five terrestrial worlds.
2. Use the balloon analogy to explain the origin of gas pressure. What is *atmospheric pressure*, and why does it decrease with altitude? What is 1 *bar* of pressure?
3. Is there any atmosphere at the orbital altitude of the Space Station? Explain.
4. What is the *greenhouse effect*? Describe how it warms a planet.
5. What factors determine a world's "no greenhouse" surface temperature? Explain the difference between "no greenhouse" and actual temperatures for each of the terrestrial worlds.
6. Describe Earth's basic *atmospheric structure*, from the ground up. How do interactions of sunlight and gases explain the existence of each of the atmospheric layers?
7. Why is the sky blue? Why are sunrises and sunsets red?
8. Why does convection occur in the *troposphere*, leading to active weather, but not in the *stratosphere*?
9. What is ozone? How does the absence of ozone on Venus and Mars explain why these planets lack a stratosphere?
10. What is a *magnetosphere*? Describe its role in protecting any atmosphere from the solar wind and in creating *auroras*.
11. What is the difference between *weather* and *climate*?
12. Describe Earth's *global wind patterns* and the role of *circulation cells*. How does rotation affect these cells?
13. What are clouds made of? How does rain or snow form?
14. Describe each of the four factors that can lead to long-term climate change.
15. Describe each process by which atmospheres gain or lose gas. What factors control *thermal escape*?
16. Why do the Moon and Mercury have so little atmospheric gas?
17. How and why do seasons on Mars differ from seasons on Earth?
18. Describe the leading hypothesis for how Mars lost atmospheric gas. What role does Mars's size play in this process?
19. What do we mean by a *runaway greenhouse effect*? Explain why this process occurred on Venus but not on Earth.
20. Describe four ways in which Earth's atmosphere is unique among the terrestrial worlds, and how each is important to our existence.
21. What is the *carbon dioxide cycle*, and why is it so crucial to life on Earth?
22. Briefly summarize the evidence linking human activity to *global warming*. What are its potential consequences?

TEST YOUR UNDERSTANDING

Does It Make Sense?

Decide whether the statement makes sense (or is clearly true) or does not make sense (or is clearly false). Explain clearly; not all these have definitive answers, so your explanation is more important than your chosen answer.

23. If Earth's atmosphere did not contain molecular nitrogen, X rays from the Sun would reach the surface.
24. If the molecular oxygen content of Earth's atmosphere increases, it will cause our planet to warm up.
25. Earth's oceans must have formed at a time when no greenhouse effect operated on Earth.
26. In the distant past, when Mars had a thicker atmosphere, it also had a stratosphere.
27. If Earth rotated faster, hurricanes would be more common and more severe.
28. Mars would still have seasons even if its orbit around the Sun were perfectly circular rather than elliptical.
29. Mars once may have been warmer than it is today, but it could never have been warmer than Earth.
30. If the solar wind were much stronger, Mercury might develop a carbon dioxide atmosphere.
31. If Earth had as much carbon dioxide in its atmosphere as Venus, our planet would not have oceans.
32. A planet in another solar system has no life but has an Earth-like atmosphere with plentiful oxygen.

Choose the best answer to each of the following. Explain your reasoning with one or more complete sentences.

33. Which terrestrial world has the most atmosphere? (a) Venus (b) Earth (c) Mars

34. The greenhouse effect occurs in the (a) troposphere. (b) stratosphere. (c) lithosphere.

35. What kind of light warms the stratosphere? (a) infrared (b) visible (c) ultraviolet

36. Which of the following is a strong greenhouse gas? (a) nitrogen (b) water vapor (c) oxygen

37. In which direction do hurricanes in the Southern Hemisphere rotate? (a) clockwise (b) counterclockwise (c) either direction

38. What is the leading hypothesis for Venus's lack of water? (a) Venus formed closer to the Sun and accreted very little water. (b) Its water is locked away in the crust. (c) Its water molecules were broken apart, and hydrogen was lost to space.

39. What kind of gas is most affected by thermal escape? (a) greenhouse gases (b) light gases (c) all gases equally

40. About what fraction of Earth's atmosphere is CO_2? (a) 90% (b) 1% (c) less than 0.1%

41. What causes the release of oxygen into Earth's atmosphere? (a) outgassing (b) evaporation/sublimation (c) photosynthesis

42. Where is most of the CO_2 that has outgassed from Earth's volcanoes? (a) in the atmosphere (b) in space (c) in rocks

PROCESS OF SCIENCE

Examining How Science Works

43. *Science with Consequences.* A small but vocal group of people still dispute that humans are causing global warming. Do some research to find the basis of their claims. Then defend or refute their findings based on your own studies and your understanding of the hallmarks of science discussed in Chapter 3.

44. *Unanswered Questions.* Choose one important but unanswered question about Mars's past, and write two or three paragraphs in which you discuss how we might answer this question in the future. Be as specific as possible, focusing on the type of evidence necessary to answer the question and how the evidence could be gathered. What are the benefits of finding answers to this question?

GROUP WORK EXERCISE

45. *Are We Causing Global Warming?* This exercise is intended to help you understand the scientific evidence behind the issue of global warming. Before you begin, assign the following roles to the people in your group: *Scribe* (takes notes on the group's activities), *Advocate* (argues in favor of the claim that human activity is causing global warming), *Skeptic* (points out weaknesses in the arguments made by the *Advocate*), and *Moderator* (leads group discussion and makes sure everyone contributes).
a. Work together to make a list of scientific observations that have been proposed as evidence that humans are causing global warming. Your list should include, but is not limited to, the evidence in Figures 10.37–10.40. **b.** *Advocate* presents the case that humans are causing global warming, drawing on the evidence from part a. **c.** *Skeptic* attempts to refute *Advocate*'s case using scientific arguments. **d.** After hearing these arguments, *Moderator* and *Scribe* decide whose arguments were more persuasive and explain their reasoning. **e.** Each person in the group writes up a summary of the discussion.

INVESTIGATE FURTHER

In-Depth Questions to Increase Your Understanding

Short-Answer/Essay Questions

46. *Clouds of Venus.* Table 10.2 shows that Venus's surface temperature in the absence of the greenhouse effect is lower than Earth's, even though Venus is closer to the Sun.
a. Explain this unexpected result in one or two sentences. **b.** Suppose Venus had neither clouds nor greenhouse gases. What do you think would happen to the surface temperature of Venus? Why? **c.** How are clouds and volcanoes linked on Venus? What change in volcanism might result in the disappearance of clouds? Explain.

47. *Atmospheric Structure.* Study Earth's average atmospheric structure (Figure 10.7). Sketch a similar curve for each of the following cases, and explain how and why the structure would be different in each case.
a. Suppose Earth had no greenhouse gases. **b.** Suppose the Sun emitted no ultraviolet light. **c.** Suppose the Sun had a higher output of X rays.

48. *Magic Mercury.* Suppose we could magically give Mercury the same atmosphere as Earth. Assuming this magical intervention happened only once, would Mercury be able to keep its new atmosphere? Explain.

49. *A Swiftly Rotating Venus.* Suppose Venus rotated as rapidly as Earth. Briefly explain how and why you would expect it to be different in terms of each of the following: geological processes, atmospheric circulation, magnetic field, and climate history.

50. *Coastal Winds.* During the daytime, heat from the Sun tends to make the air temperature warmer over land near the coast than over the water offshore. At night, land cools off faster than the sea, so temperatures tend to be cooler over land. Use these facts to predict the directions in which winds generally blow during the day and at night in coastal regions. Explain your reasoning in a few sentences. (Diagrams might help.)

51. *Sources and Losses.* Choose one process by which atmospheres can gain gas and one by which they can lose gas. For each process, write a few sentences that describe it and how it depends on each of the following fundamental planetary properties: size, distance from the Sun, and rotation rate.

52. *Two Paths Diverged.* Briefly explain how the different atmospheric properties of Earth and Venus can be explained by the fundamental properties of size and distance from the Sun.

53. *Change in Fundamental Properties.* Choose one property of Earth—either size or distance from the Sun—and suppose that it had been different (for example, smaller size or greater distance). Describe how this change would have affected Earth's subsequent atmospheric history and the possibility of life on Earth.

54. *Feedback Processes in the Atmosphere.* As the Sun gradually brightens in the future, how can the CO_2 cycle respond to reduce the warming effect? Which parts of the cycle will be affected? Is this an example of positive or negative feedback?

55. *Earth to Mars.* Section 10.5 discusses what might happen to Earth if it were suddenly moved to the orbit of Venus. What do you think would happen to Earth if it were suddenly moved to the orbital distance of Mars? Write a few sentences explaining your answer.

56. *Global Warming.* What, if anything, should we be doing that we are not doing already to alleviate the threat of global warming? Write a one-page editorial summarizing and defending your opinion.

Quantitative Problems

57. *The Mass of an Atmosphere.* What is the total mass of Earth's atmosphere? You may use the fact that 1 bar is the pressure exerted

by 10,000 kilograms pushing down on a square meter in Earth's gravity. Remember that every square meter of Earth experiences this pressure from the atmosphere above it. Alternatively, you may start with the English unit value for pressure of 14.7 pounds per square inch and convert to kilograms for your final answer. Remember that the surface area of a sphere of radius r is $4\pi r^2$.

58. *The Role of Reflectivity.* By assuming 0% and 100% reflectivity (respectively), find the maximum and minimum possible "no greenhouse" temperatures for a planet at 1 AU. What reflectivity would be necessary to keep the average temperature exactly at the freezing point? Compare to Earth's actual reflectivity in Table 10.2.

59. *The Cooling Clouds of Venus.* Table 10.2 shows that Venus's temperature in the absence of the greenhouse effect is lower than Earth's, even though Venus is closer to the Sun. What would Venus's "no greenhouse" temperature be if its clouds were more transparent, giving a reflectivity the same as Earth's? What would the actual surface temperature be in this case if the greenhouse effect increased the surface temperature by the same number of degrees that it does today?

60. *Mars's Elliptical Orbit.* Mars's distance from the Sun varies from 1.38 AU to 1.66 AU. How much does this affect its "no greenhouse" surface temperature at different times of year? Comment on how this affects Mars's seasons.

61. *Escape from Venus.*
    a. Calculate the escape velocity from Venus's exosphere, which begins about 200 kilometers above the surface. (*Hint:* See Mathematical Insight 4.4.) **b.** Calculate and compare the thermal speeds of hydrogen and deuterium atoms at the exospheric temperature of 350 K. The mass of a hydrogen atom is 1.67×10^{-27} kilogram, and the mass of a deuterium atom is about twice the mass of a hydrogen atom. **c.** In a few sentences, comment on the relevance of these calculations to the question of whether Venus has lost large quantities of water.

Discussion Questions

62. *Lucky Earth.* The climate histories of Venus and Mars make it clear that getting a pleasant climate like that of Earth isn't easy. How does this affect your opinion about whether Earth-like planets might exist around other stars? Explain.

63. *Terraforming Mars.* Some people have suggested that we might be able to engineer Mars in a way that would cause its climate to warm and its atmosphere to thicken. This type of planet engineering is called *terraforming,* because its objective is to make a planet more Earth-like and easier for humans to live on. Discuss possible ways to terraform Mars. Do any of these ideas seem practical? Do they seem like good ideas? Defend your opinions.

64. *Terraforming Venus.* Can you think of ways in which it would be possible to terraform Venus? Discuss the possibilities as well as their practicality.

Web Projects

65. *Human Threats to Earth.* Write a three- to five-page research report about current understanding and controversy regarding global warming. Be sure to address both the latest knowledge about the issue and proposals for alleviating any dangers associated with it. End your report by making your own recommendations about what, if anything, needs to be done to prevent damage to Earth.

66. *Spacecraft Study of Atmospheres.* Learn about a current or planned mission to study the atmosphere of one of the terrestrial worlds (including Earth). Write a one- to two-page essay describing the mission and what we hope to learn from it.

67. *Martian Weather.* Find the latest weather report for Mars from spacecraft and other satellites. What season is it in the northern hemisphere? When was the most recent dust storm? What surface temperature was most recently reported from Mars's surface, and at what location? Summarize your findings by writing a 1-minute script for a television news update on Martian weather.

The photo of Saturn was taken by the *Cassini* spacecraft while it was in Saturn's shadow. The small blue dot of light just inside Saturn's rings at the left (about the 10 o'clock position) is Earth, far in the distance.

11 JOVIAN PLANET SYSTEMS

LEARNING GOALS

11.1 A DIFFERENT KIND OF PLANET

- Are jovian planets all alike?
- What are jovian planets like on the inside?
- What is the weather like on jovian planets?
- Do jovian planets have magnetospheres like Earth's?

11.2 A WEALTH OF WORLDS: SATELLITES OF ICE AND ROCK

- What kinds of moons orbit the jovian planets?
- Why are Jupiter's Galilean moons so geologically active?

- What is remarkable about Titan and other major moons of the outer solar system?
- Why are small icy moons more geologically active than small rocky planets?

11.3 JOVIAN PLANET RINGS

- What are Saturn's rings like?
- How do other jovian ring systems compare to Saturn's?
- Why do the jovian planets have rings?

Do there exist many worlds, or is there but a single world? This is one of the most noble and exalted questions in the study of Nature.

—St. Albertus Magnus (1206–1280)

In Roman mythology, the namesakes of the jovian planets are rulers among gods: Jupiter is the king of the gods, Saturn is Jupiter's father, Uranus is the lord of the sky, and Neptune rules the sea. However, our ancestors could not have foreseen the true majesty of the four jovian planets. The smallest, Neptune, is large enough to contain the volume of more than 50 Earths. The largest, Jupiter, has a volume some 1400 times that of Earth. These worlds are totally unlike the terrestrial planets. They are essentially giant balls of gas, with no solid surface.

Why should we care about a set of worlds so different from our own? Apart from satisfying natural curiosity, studies of the jovian planets and their moons help us understand the birth and evolution of our solar system—which in turn helps us understand our own planet Earth. In addition, the jovian planets provide stepping stones to understanding other planetary systems, which may have a wider range of planetary types than we find in our own solar system. In this chapter, we'll explore the jovian planet systems, focusing first on the planets themselves, then on their many moons, and finally on their beautifully complex rings.

 Formation of the Solar System Tutorial, Lesson 1

11.1 A DIFFERENT KIND OF PLANET

The great differences between the terrestrial and the jovian planets are a relatively recent discovery in human history. Jupiter and Saturn are easily visible in the night sky, but the naked eye cannot discern any details about their nature. Uranus and Neptune were discovered in 1781 and 1846, respectively (see Special Topic, page 312), so they were unknown even during the Copernican revolution.

Astronomers first recognized the immense sizes of Jupiter and Saturn about 250 years ago. Recall that we need to know both angular size and distance to calculate an object's true size (see Mathematical Insight 2.1). Although Copernicus had figured out the *relative* distances to the known planets in astronomical units (AU), scientists did not establish the absolute scale of the solar system until the 1760s, when they were able to measure the true length of an astronomical unit with data from a transit of Venus (see Special Topic, page 207). Only then could scientists calculate the true sizes of Jupiter and Saturn from their distances and angular sizes (measured through telescopes). Knowing the scale of the solar system also told scientists the distances of orbiting moons, and because the orbital periods of the moons were easy to measure, the orbital distances allowed scientists to use Newton's version of Kepler's third law [**Section 4.4**] to calculate the masses of the jovian planets. Together, the measurements of size and mass revealed the low densities of the jovian planets, proving that these worlds are very different in nature from Earth.

The real revolution in understanding jovian planet systems began with the first spacecraft sent to visit them: *Pioneer 10* and *Pioneer 11*, which flew past Jupiter and Saturn in the early 1970s. The *Voyager* missions followed soon thereafter; *Voyager 1* flew past Jupiter and Saturn, and *Voyager 2* flew past all four jovian planets. More recently, we've learned much more about Jupiter from the *Galileo* spacecraft, which orbited Jupiter from 1995 to 2003, and *New Horizons*, which flew past Jupiter in 2007 on its way to Pluto; the *Juno* mission is scheduled to begin orbiting Jupiter in August 2016. Much of our current understanding of Saturn comes from *Cassini*, which has been orbiting Saturn since 2004.

Are jovian planets all alike?

FIGURE 11.1 shows a montage of the jovian planets compiled by the *Voyager* spacecraft, along with basic data; Earth is included for scale. The immense sizes of the jovian worlds are apparent. But while all four are enormous, there are important differences between them. In particular, they differ substantially in mass, density, and overall composition.

Jovian Planet Composition Jupiter and Saturn are made almost entirely of hydrogen and helium, with just a few percent of their masses in the form of hydrogen compounds and even smaller amounts of rock and metal. In fact, their overall compositions are much more similar to the composition of the Sun than to the compositions of the terrestrial planets. Some people even call Jupiter a "failed star" because it has a starlike composition but lacks the nuclear fusion needed to make it shine. This is a consequence of its size: Although Jupiter is large for a planet, it is much less massive than any star. As a result, its gravity is too weak to compress its interior to the extreme temperatures and densities needed for nuclear fusion. (Jupiter would have needed to grow to about 80 times its current mass to have become a star.) Of course, where some people see a failed star, others see an extremely successful planet.

Uranus and Neptune are much smaller than Jupiter and Saturn, and also contain smaller proportions of hydrogen and helium gas. Instead, they are made primarily of hydrogen compounds such as water (H_2O), methane (CH_4), and ammonia (NH_3), along with smaller amounts of metal and rock.

The differences in composition among the jovian planets can probably be traced to their origins. Recall that the jovian planets formed in the outer solar system, where it was cold enough for hydrogen compounds to condense into ices [**Section 8.2**]. Because hydrogen compounds were so much more abundant than metal and rock, some of the ice-rich planetesimals of the outer solar system grew to great size. Once these planetesimals became sufficiently massive, their gravity allowed them to draw in the hydrogen and helium gas that surrounded them. All four jovian planets are thought to have grown from ice-rich planetesimals of about the same mass—roughly 10 times the mass of Earth—but they captured different amounts of hydrogen and helium gas from the surrounding solar nebula.

Earth

Jupiter
Distance from Sun = 5.20 AU
Mass = $318 M_{Earth}$
Density = 1.33 g/cm^3
Composition: mostly H, He

Saturn
Distance from Sun = 9.54 AU
Mass = $95 M_{Earth}$
Density = 0.71 g/cm^3
Composition: mostly H, He

Uranus
Distance from Sun = 19.2 AU
Mass = $14 M_{Earth}$
Density = 1.24 g/cm^3
Composition: H compounds,
  rock, H and He

Neptune
Distance from Sun = 30.1 AU
Mass = $17 M_{Earth}$
Density = 1.67 g/cm^3
Composition: H compounds,
  rock, H and He

FIGURE 11.1 Jupiter, Saturn, Uranus, and Neptune, shown to scale with Earth for comparison.

Jupiter and Saturn captured so much hydrogen and helium gas that these gases now make up the vast majority of their masses. The ice-rich planetesimals from which they grew now represent only about 3% of Jupiter's mass and about 10% of Saturn's mass.

Uranus and Neptune pulled in much less hydrogen and helium gas. Notice in Figure 11.1 that Uranus's mass is about 14 times Earth's mass. Since Uranus is thought to have grown around an ice-rich planetesimal that had about 10 times Earth's mass, the captured hydrogen and helium gas must make up only about a third of Uranus's total mass. The bulk of its mass consists of material from the original ice-rich planetesimal: hydrogen compounds mixed with smaller amounts of rock and metal. The same is true for Neptune, though its higher density suggests that it may have formed around a slightly more massive ice-rich planetesimal.

Why did the different planets capture different amounts of gas? The answer probably lies in their distances from the Sun as they formed. The solid particles that condensed farther from the Sun should have been more widely spread out than those that condensed nearer to the Sun, which means it would have taken longer for them to accrete into large, icy planetesimals. As the nearest jovian planet to the Sun, Jupiter would have been the first to get a planetesimal large enough for its gravity to start drawing in gas, followed by Saturn, Uranus, and Neptune. Because all the planets stopped accreting gas at the same time—when the solar wind blew all the remaining gas into interstellar space—the more distant planets had less time to capture gas and ended up smaller in size.

Density Differences Figure 11.1 shows that Saturn is considerably less dense than Uranus or Neptune. This should make sense when you compare compositions, because the hydrogen compounds, rock, and metal that make up Uranus and Neptune are normally much more dense than hydrogen or helium gas. By the same logic, we'd expect Jupiter to be even less dense than Saturn—but it's not. To understand Jupiter's surprisingly high density, we need to think about how massive planets are affected by their own gravity.

Building a planet of hydrogen and helium is a bit like making one out of fluffy pillows. Imagine assembling a planet pillow by pillow. As each new pillow is added, those on the bottom are compressed more by those above. As the lower layers are forced closer together, their mutual gravitational attraction increases, compressing them even further. At first the stack grows substantially with each pillow, but eventually the growth slows until adding pillows barely increases the height of the stack (**FIGURE 11.2a**).

This analogy explains why Jupiter is only slightly larger than Saturn in radius even though it is more than three times as massive. The extra mass of Jupiter compresses its interior to a much higher density. More precise calculations show that Jupiter's radius is almost the maximum possible radius for a jovian planet. If much more gas were added to Jupiter, its weight would actually compress the interior enough to make the planet *smaller* rather than larger (**FIGURE 11.2b**). Some extrasolar planets that are larger in mass than Jupiter are therefore smaller in size. In fact, the smallest stars are significantly smaller in radius than Jupiter, even though they are at least 80 times as massive.

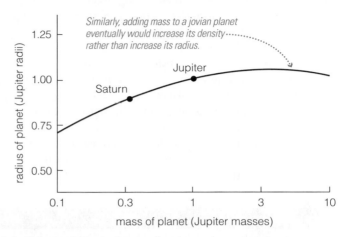

a Adding pillows to a stack may increase its height at first but eventually just compresses the stack, making its density greater. Similarly, adding mass to a jovian planet eventually will increase its density rather than increasing its radius.

b This graph shows how radius depends on mass for a hydrogen/helium planet. Notice that Jupiter is only slightly larger in radius than Saturn, despite being three times as massive. Gravitational compression of a planet much more massive than Jupiter would actually make it smaller in size.

FIGURE 11.2 The relationship between mass and radius for a planet made of hydrogen and helium.

Jovian Planet Rotation and Shape The jovian planets rotate much more rapidly than any of the terrestrial worlds. However, precisely defining their rotation rates can be difficult because they lack solid surfaces. We can observe the rotation of clouds, but these observations can be deceptive, because the speeds at which we observe clouds to move are affected by winds as well as by planetary rotation. Nevertheless, observations of clouds at different latitudes suggest that the jovian planets do not rotate like solid balls. Instead, their rotation rates vary with latitude: Equatorial regions complete each rotation in less time than polar regions. (The Sun rotates similarly [**Section 14.1**].)

We can measure the rotation rates of the jovian *interiors* by tracking emissions from charged particles trapped in their magnetospheres. This technique tells us the rotation period of the magnetosphere, which should be the same as the rotation period deep in the interior, where the magnetic field is generated. These measurements show that the jovian "day" ranges from about 10 hours on Jupiter and Saturn to 16–17 hours on Uranus and Neptune. Note that, because the jovian planets are much larger in radius than Earth, their surface rotation speeds are much greater compared to Earth than the periods alone would suggest.

The rapid rotation rates of the jovian planets affect their shapes (**FIGURE 11.3**). Gravity alone would make the jovian planets into perfect spheres. Rotation makes them less like perfect spheres, because it makes material bulge outward. Material near the equator, where speeds around the rotation axis are highest, is flung outward in the same way you feel yourself flung outward when you ride on a merry-go-round. The size of the equatorial bulge depends on the balance between the strength of gravity (which pulls material inward) and the rate of rotation (which pushes material outward).

The balance tips most strongly toward flattening on Saturn: With its rapid 10-hour rotation period and relatively weak surface gravity, Saturn is about 10% larger in diameter at its equator than from pole to pole. In addition to altering a planet's shape, the equatorial bulge exerts an extra gravitational pull that helps keep moons and rings aligned with the equator.

What are jovian planets like on the inside?

The jovian planets are often called "gas giants," making it sound as if they were entirely gaseous like air on Earth. Based on their compositions, Jupiter and Saturn may seem to deserve this name; after all, they became giants primarily by capturing

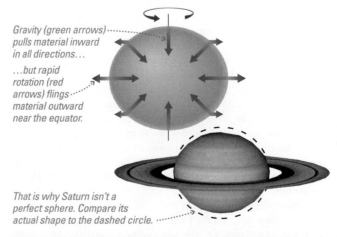

FIGURE 11.3 interactive figure Because of their rapid rotation, the jovian planets are not quite spherical. Saturn shows the biggest difference between its actual shape and a perfect sphere.

so much hydrogen and helium gas. The name may seem less fitting for Uranus and Neptune, since they are made mostly of materials besides pure hydrogen and helium. However, closer inspection shows that the name is a little misleading even for Jupiter and Saturn, because their strong gravity compresses most of the "gas" into forms of matter quite unlike anything we are familiar with in everyday life on Earth.

You may wonder how we can claim to know what the jovian planets are like on the inside. The answer is a combination of theoretical modeling and laboratory experiments. Just as with the terrestrial planets, detailed observations of the strength of a planet's gravity and of its magnetic field can help us put together a model of the planet's interior structure. We then use the model to predict observable phenomena, such as a planet's average density, and modify it until the predictions match the data. Today, advanced computer models

successfully explain the observed sizes, densities, atmospheric compositions, and precise shapes of the jovian planets. We are therefore confident that the interior structure found by the models is fairly close to reality. Laboratory studies also provide important data, such as showing us how hydrogen and helium behave under the tremendous temperatures and pressures that exist deep beneath the jovian cloudtops.

Inside Jupiter Let's begin discussion of the jovian interiors by using Jupiter as a prototype. Jupiter's lack of a solid surface makes it tempting to think of the planet as "all atmosphere," but you could not fly through Jupiter's interior in the way airplanes fly through air. A spacecraft plunging into Jupiter would find increasingly higher temperatures and pressures as it descended. The *Galileo* spacecraft dropped a scientific probe into Jupiter (in 1995) that collected measurements for about

---

SPECIAL TOPIC

How Were Uranus, Neptune, and Pluto Discovered?

The planets Mercury, Venus, Mars, Jupiter, and Saturn are all easily visible to the naked eye and hence were well known to ancient people. In contrast, Uranus, Neptune, and the solar system's many small bodies were discovered relatively recently.

Uranus is actually visible to the naked eye, but it is so faint and moves so slowly in its 84-year orbit of the Sun that ancient people did not recognize it as a planet. It was even recorded as a star on some detailed early sky charts. Uranus was discovered as a planet in 1781 by brother-and-sister astronomers William and Caroline Herschel. William is the more famous of the pair by far, partly because Caroline's gender made many scientists of the time unwilling to take her seriously, but also because he was nearly 12 years older than his sister and had been the one who drew her into astronomical research. William Herschel originally suggested naming the new planet *Georgium Sidus,* Latin for "George's Star," in honor of his patron, King George III. Fortunately, the idea of "Planet George" never caught on. Instead, many 18th- and 19th-century astronomers referred to the new planet as Herschel. The modern name, Uranus (after the mythological father of Saturn), was first suggested by one of Herschel's contemporaries, astronomer Johann Bode. This name was generally accepted by the middle of the 19th century.

Neptune's discovery more than 60 years later represented an important triumph for Newton's universal law of gravitation and the young science of astrophysics. By that time, careful observations of Uranus had shown its orbit to be slightly inconsistent with that predicted by Newton's law—at least if it were being influenced only by the Sun and the other known planets. In the early 1840s in England, a student named John Adams suggested that the inconsistency could be explained if there were an unseen "eighth planet" orbiting the Sun beyond Uranus. According to the official story, he used Newton's theory to predict the location of the planet but was unable to convince British astronomers to carry out a telescopic search. However, documents from the time suggest his prediction may not have been as precise as the official history claims. Meanwhile, in the summer of 1846, French astronomer Urbain Leverrier independently made very precise calculations. He sent a letter to Johann Galle of the Berlin Observatory, suggesting a search for the eighth planet. On the night of September 23, 1846, Galle pointed his telescope to the position suggested by Leverrier. There,

within 1° of its predicted position, he saw the planet Neptune. Hence, Neptune's discovery truly was made by mathematics and physics and was only confirmed with a telescope.

As a side note to this story, Leverrier had such faith in Newton's universal law of gravitation that he also suggested a second unseen planet, this one orbiting closer to the Sun than Mercury. He got this idea because other astronomers had identified slight discrepancies between Mercury's actual orbit and the orbit predicted by Newton's theory. He assumed that no one had yet seen the planet, which he called Vulcan, because it was so close to the Sun. Leverrier died in 1877, still believing that Vulcan would someday be discovered. But Vulcan does not exist. Mercury's actual orbit does not match the orbit predicted by Newton's law because Newton's theory is not the whole story of gravity. About 40 years after Leverrier's death, Einstein showed that Newton's theory is only an approximation of a broader theory of gravity, known today as Einstein's *general theory of relativity.* Einstein's theory predicts an orbit for Mercury that matches its actual orbit. This match was one of the first key pieces of evidence in favor of Einstein's theory [**Section S3.4**].

Pluto was discovered in 1930 by American astronomer Clyde Tombaugh, culminating a search that began when astronomers analyzed the orbit of Neptune. The story of Pluto's discovery at first seemed similar to Neptune's. Just as discrepancies between the predicted orbit and the actual orbit of Uranus led to the prediction that Neptune must exist, lingering discrepancies between the updated prediction for Uranus's orbit and its actual orbit suggested the existence of an even more distant "ninth planet." Tombaugh found Pluto just 6° from the position in the sky where this ninth planet had been predicted to lie, so it seemed that the search had been successful. However, while initial estimates suggested that Pluto was much larger than Earth, we now know that Pluto has a radius of only 1160 kilometers and a mass of just 0.002 Earth mass—making it far too small to affect the orbit of Neptune. In retrospect, the supposed orbital irregularities of Neptune appear to have been errors in measurement; no "ninth planet" is needed to explain the orbits of Uranus or Neptune. As we now know from the fact that many similar objects orbit in the same region of the solar system [**Section 12.3**], Pluto's discovery was just a coincidence, and its small size is why we now call it a *dwarf planet.*

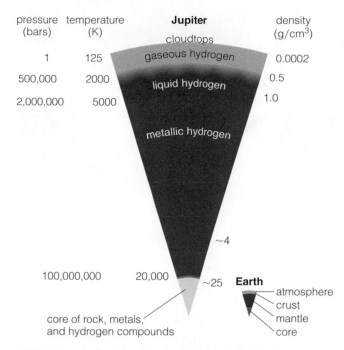

pressure (bars)	temperature (K)	**Jupiter** cloudtops	density (g/cm³)
1	125	gaseous hydrogen	0.0002
500,000	2000	liquid hydrogen	0.5
2,000,000	5000		1.0
		metallic hydrogen	
		~4	
100,000,000	20,000	~25 **Earth**	atmosphere crust mantle core

core of rock, metals, and hydrogen compounds

FIGURE 11.4 Jupiter's interior structure, labeled with the pressure, temperature, and density at various depths. Earth's interior structure is shown to scale for comparison. (The thickness of Earth's crust and atmosphere are exaggerated.) Note that Jupiter's core is only slightly larger than Earth but is about 10 times as massive.

an hour before the ever-increasing pressures and temperatures destroyed it. *Galileo* provided valuable data about Jupiter's atmosphere, but didn't last long enough to sample the interior: It survived only to a depth of about 200 kilometers, or about 0.3% of Jupiter's radius.

Computer models tell us that Jupiter has fairly distinct interior layers (**FIGURE 11.4**). The layers do not differ much in composition—all except the core are mostly hydrogen and helium. Instead they differ in the phase (such as liquid or gas) of their hydrogen. To get a better sense of this layering, imagine plunging head-on into Jupiter in a futuristic spacesuit that allows you to survive the extreme interior conditions.

Near the cloudtops, you'll find the temperature to be a brisk 125 K (–148°C), the density to be a low 0.0002 g/cm³, and the atmospheric pressure to be about 1 bar (the same as the pressure at sea level on Earth [**Section 10.1**]). As you plunge downward, conditions quickly become more extreme.

By a depth of 7000 kilometers, about 10% of the way to the center, you'll find that the temperature has increased to a scorching 2000 K and the pressure has reached 500,000 bars. The density is about 0.5 g/cm³, or about half that of water. Under these conditions, hydrogen acts more like a liquid than a gas, which is why the layer that begins at this depth is labeled *liquid hydrogen* in Figure 11.4. Of course, like the rest of Jupiter, the layer also contains helium and hydrogen compounds.

At a depth of 14,000 kilometers, about 20% of the way to the center, the density has become about as high as that of water (1.0 g/cm³). The temperature is near 5000 K, almost as hot as the surface (but not the interior) of the Sun. The pressure has reached 2 million bars. This extreme pressure forces hydrogen into a compact, metallic form. Just as is the case with everyday

metals, electrons are free to move around in *metallic hydrogen,* so it conducts electricity quite well. This layer extends through most of the rest of Jupiter's interior and, as we'll see shortly, it is where Jupiter's magnetic field is generated.

You'll finally reach Jupiter's core at a depth of 60,000 kilometers, about 10,000 kilometers from the center. The core temperature is some 20,000 K and the pressure is about 100 million bars. The density of the core is about 25 g/cm³, much denser than any material you'll find on Earth's surface. The core is a mix of hydrogen compounds, rock, and metals, but these materials bear little resemblance to familiar solids or liquids because of the combination of high temperature and extreme pressure. Moreover, while the rock and metal of the terrestrial planets have separated into layers by composition (that is, core, mantle, and crust), Jupiter's core materials are probably all mixed together. The core contains about 10 times as much mass as the entire Earth, but it is only about the same size as Earth because it is compressed to such high density.

Comparing Jovian Interiors Because all four jovian planets have cores of about the same mass, their interiors differ mainly in the hydrogen/helium layers that surround their cores. **FIGURE 11.5** contrasts the four jovian interiors. Remember that while the outer layers are named for the phase of their hydrogen, they also contain helium and hydrogen compounds.

Saturn is the most similar to Jupiter, just as we should expect given its similar size and composition. Its four interior layers differ from those of Jupiter only because of its lower mass and weaker gravity. The lower mass makes the weight of the overlying layers less on Saturn than on Jupiter, so you must look deeper into Saturn to find each level where pressure changes hydrogen from one phase to another. That is, Saturn has thicker layers of gaseous and liquid hydrogen and a thinner and more deeply buried layer of metallic hydrogen.

Pressures within Uranus and Neptune are not high enough to form liquid or metallic hydrogen at all. Each of these two planets has only a thick layer of gaseous hydrogen surrounding its core of hydrogen compounds, rock, and metal. This core material may be liquid, making for very odd "oceans" buried deep inside Uranus and Neptune. The cores of Uranus and Neptune are larger in radius than the cores of Jupiter and Saturn, even though they have about the same mass, because they are less compressed by their lighter-weight overlying layers. The less extreme interior conditions also allowed Uranus's and Neptune's cores to differentiate, so hydrogen compounds reside in a layer around a center of rock and metal.

Internal Heat Recall that internal heat drives surface geology on the terrestrial worlds. The jovian planets do not have surface geology, because they have no solid surfaces, but internal heat still plays an important role.

Jupiter has a tremendous amount of internal heat, and like any hot object, it gradually loses this heat to space by emitting thermal radiation [**Section 5.4**]. In fact, Jupiter emits almost twice as much energy as it receives from the Sun. This heat contributes significant energy to Jupiter's upper atmosphere. (For comparison, Earth's internal heat contributes only 0.005% as much energy to the surface as sunlight does.)

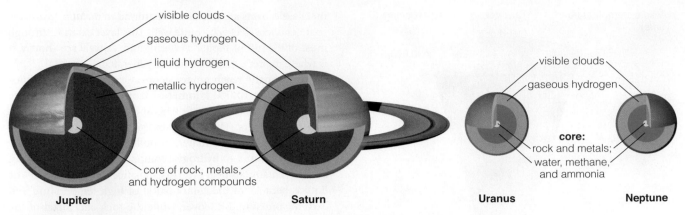

visible clouds
gaseous hydrogen
liquid hydrogen
metallic hydrogen

core of rock, metals,
and hydrogen compounds

Jupiter

visible clouds

Saturn

visible clouds
gaseous hydrogen

core:
rock and metals;
water, methane,
and ammonia

Uranus

Neptune

FIGURE 11.5 These diagrams compare the interior structures of the jovian planets, shown approximately to scale. All four planets have cores of rock, metals, and hydrogen compounds, with masses about 10 times the mass of Earth's core. They differ primarily in the thicknesses of the hydrogen/helium layers that surround their cores. Notice that the cores of Uranus and Neptune are differentiated into a layer of hydrogen compounds around a center of rock and metals.

What keeps Jupiter so hot inside? Jupiter's large size means it loses internal heat very slowly (see Mathematical Insight 9.1), and calculations show that the remaining heat of accretion and differentiation is not enough to explain Jupiter's present-day heat loss. Radioactive decay also adds heat, but still not enough to account for all the internal heat. The most likely explanation for Jupiter's excess heat is that the planet is still slowly contracting. Contraction converts gravitational potential energy to thermal energy, so continued contraction would be an ongoing source of internal heat. Although we have not measured any such contraction, theoretical models suggest that it is probably occurring. Moreover, calculations tell us that contraction could easily explain Jupiter's internal heat, even if the contraction is so gradual that we have little hope of ever measuring it directly.

Regardless of the specific mechanism, Jupiter has undoubtedly lost substantial heat during the 4.5 billion years since its formation. Jupiter's interior must have been much warmer in the distant past, and this heat would have "puffed up" its atmosphere. With a larger size, Jupiter would have reflected more sunlight, so in the distant past Jupiter must have been even more prominent in Earth's sky than it is today.

Saturn also emits nearly twice as much energy as it receives from the Sun, suggesting that it, too, must have some ongoing source of heat. However, Saturn's mass is too small for it to be generating heat by contracting like Jupiter. Instead, Saturn's pressure and its lower interior temperatures may allow helium to condense into liquid form at relatively high levels within the interior. The helium droplets slowly rain down to the deeper interior. This gradual helium rain represents a sort of ongoing *differentiation* [**Section 9.1**], because it means that higher-density material (the liquid helium) is still sinking inside the planet. Spacecraft measurements have confirmed that Saturn's atmosphere has less helium than Jupiter's, just as we would expect if helium has been raining down into Saturn's interior for billions of years.

Neither Uranus nor Neptune has internal conditions that should allow helium rain to form, and most of their original heat from accretion should have escaped long ago. This explains why Uranus emits virtually no excess internal energy. Neptune, however, is more mysterious. Like Jupiter and Saturn, Neptune emits nearly twice as much energy as it receives from the Sun. The only reasonable explanation for this internal heat is that Neptune is somehow still contracting, somewhat like Jupiter, thereby converting gravitational potential energy into thermal energy. However, we do not yet know why a planet of Neptune's size would still be contracting more than 4 billion years after its formation.

What is the weather like on jovian planets?

Jovian atmospheres, by which we mean the portions of the planets that are visible to telescopes and spacecraft, have dynamic winds and weather, with colorful clouds and enormous storms. These atmospheres are made mostly of hydrogen and helium gas, mixed with small amounts of various hydrogen compounds. Because oxygen, carbon, and nitrogen are among the most common elements in the universe besides hydrogen and helium, the most common hydrogen compounds are methane (CH_4), ammonia (NH_3), and water (H_2O). Spectroscopy also reveals the presence of small amounts of more complex hydrogen compounds, such as acetylene (C_2H_2), ethane (C_2H_6), and propane (C_3H_8). Although all these hydrogen compounds together make up only a minuscule fraction of the jovian planets' atmospheres, they are responsible for virtually all aspects of their appearances. Some of these compounds condense to form the clouds that are so prominent in telescope and spacecraft images. Others are responsible for the great variety of colors we see among the jovian planets. Without these compounds in their atmospheres, the jovian planets would be uniform, colorless balls of gas.

THINK ABOUT IT

Several gases in Jupiter's atmosphere—including methane, propane, and acetylene—are highly flammable here on Earth. Jupiter has plenty of lightning to provide sparks, so why don't these gases ignite in Jupiter's atmosphere? (*Hint:* What's missing from Jupiter's atmosphere that's necessary for ordinary fire?)

We can understand the jovian atmospheres using the same basic principles that we used with the terrestrial atmospheres. As we did for the jovian planet interiors, let's examine different aspects of the jovian atmospheres by starting with Jupiter as the prototype for each feature.

Atmospheric Structure and Clouds Telescopic observations of Jupiter, along with data returned from the *Galileo* probe during its 1995 plunge into Jupiter, tell us that the temperature structure of Jupiter's atmosphere is very similar to that of Earth's atmosphere (see Figure 10.7). High above the cloudtops, Jupiter's *thermosphere* consists of very low-density gas heated to about 1000 K by solar X rays and by energetic particles from Jupiter's magnetosphere. Below the thermosphere but still above the clouds, we find Jupiter's *stratosphere*. Recall that a planet can have a stratosphere only if it has a gas that can absorb ultraviolet light from the Sun. Ozone plays this role on Earth. Jupiter lacks molecular oxygen and ozone, but has a few minor atmospheric ingredients that absorb solar ultraviolet photons. This absorption gives the stratosphere a peak temperature of around 200 K (–73°C). Chemical reactions driven by the solar ultraviolet photons create a smog-like haze that masks the color and sharpness of the clouds below. Below the stratosphere lies Jupiter's *troposphere,* where the temperature rises with depth because greenhouse gases trap both solar heat and Jupiter's own internal heat. **FIGURE 11.6** shows how temperature varies with altitude in Jupiter's atmosphere.

In the troposphere, the warmer temperatures below drive strong convection. This convection is responsible for the thick clouds that enshroud Jupiter. Recall that clouds form when a gas condenses to make tiny liquid droplets or solid flakes. Water vapor is the only gas that can condense in Earth's atmosphere, which is why clouds on Earth are made of water droplets or ice flakes that can produce rain or snow. In contrast, Jupiter's atmosphere has several gases that can condense to form clouds. Each of these gases condenses at a different temperature, leading to distinctive cloud layers at different altitudes.

Jupiter has three primary cloud layers, which are also shown in Figure 11.6. We can understand them by imagining that we could watch gases rising through the troposphere with convection. Deep in the troposphere, the gases include three ingredients that will condense when temperatures are low enough: water (H_2O), ammonium hydrosulfide (NH_4SH), and ammonia (NH_3). The rising gas first encounters temperatures cool enough for water vapor to condense into liquid water but not cool enough for the other gases to condense. Therefore, the lowest layer of clouds contains water droplets. As the remaining gas continues its rise, it next reaches an altitude at which ammonium hydrosulfide condenses to make the second cloud layer. After rising another 50 kilometers, the gas reaches an altitude at which ammonia condenses to make the upper cloud layer.

Data from telescopic and spacecraft observations tell us that the atmospheric structures of the other three jovian planets are quite similar to that of Jupiter (**FIGURE 11.7**). The primary difference among them is that the atmospheres get progressively cooler with increasing distance from the Sun. These temperature differences lead the planets to have their cloud layers at different altitudes.

FIGURE 11.6 This graph shows the temperature structure of Jupiter's atmosphere. Jupiter has at least three distinct cloud layers because different atmospheric gases condense at different temperatures and hence at different altitudes.

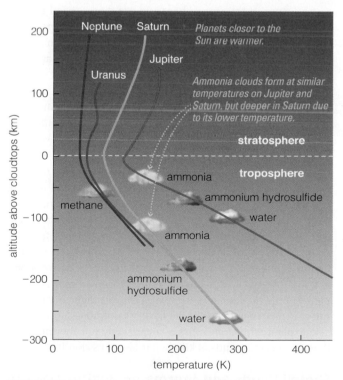

FIGURE 11.7 The figure contrasts the atmospheric structures and cloud layers of the four jovian planets. Note that the altitude scale is different on this figure than on Figure 11.6, allowing us to see the troposphere and stratospheres only.

Saturn has the same set of three cloud layers as Jupiter, but the lower overall temperatures cause these layers to lie deeper in Saturn's atmosphere. For example, to find the relatively warm temperatures at which water vapor can condense to form water clouds, we must look about 200 kilometers deeper into Saturn than into Jupiter. Saturn's cloud layers are also separated by greater vertical distances than Jupiter's, because Saturn's weaker gravity causes less atmospheric compression.

Uranus and Neptune are so cold that any cloud layers similar to those of Jupiter or Saturn would be buried too deep in their atmospheres for us to see. Therefore, we do not know whether such clouds exist on these two planets. However, high in the atmospheres of Uranus and Neptune, we see clouds made from flakes of methane snow. Methane can condense in the very cold upper tropospheres of Uranus and Neptune but not in the warmer tropospheres of Jupiter and Saturn.

Clouds and Colors

The spectacular colors of the jovian planets are probably the first thing that jumps out at you when you look at the photos in Figure 11.1. Many mysteries remain about precisely why the jovian planets are so colorful, but clouds are responsible for many of the color features. Different types of clouds reflect light of different colors. For example, Earth's clouds look white from space because they are made of water that reflects the white light of the Sun.

Water clouds on Jupiter and Saturn probably also reflect white light, as do the upper-layer clouds made of ammonia. In contrast, the mid-layer clouds of ammonium hydrosulfide reflect brown and red light, although no one knows exactly why. Most likely, the colors in these clouds come from as-yet-unidentified compounds that are produced by chemical reactions deeper in the atmosphere and are then carried upward by convection. The fact that Saturn's clouds lie deeper in its atmosphere than Jupiter's probably explains Saturn's more subdued colors: Less light penetrates to the depths at which Saturn's clouds are found, and the light they reflect is more obscured by the atmosphere above them.

The blue colors of Uranus and Neptune come from methane gas, which is at least 20 times as abundant (by percentage) on these planets as on Jupiter or Saturn. Methane gas in the upper atmospheres of Uranus and Neptune absorbs red light, allowing only blue light to penetrate to the level at which the methane clouds exist. The methane clouds reflect this blue light upward, giving the planets their blue colors (**FIGURE 11.8**). Uranus has a lighter blue color than Neptune, probably because it has more smog-like haze to scatter sunlight before it reaches the level of the methane clouds. The extra haze is probably a result of Uranus's "sideways" axis tilt, which leads to extreme seasons. With one hemisphere remaining sunlit for decades, gases have plenty of time to interact with solar ultraviolet light and to make the chemical ingredients of the haze. Continuous sunlight may also explain why Uranus has a surprisingly hot thermosphere that extends thousands of kilometers above its cloudtops.

Global Winds and Storms on Jupiter

Jupiter has dynamic weather, with strong winds and powerful storms. To understand Jupiter's weather, we must think about how its rapid rotation affects its global winds.

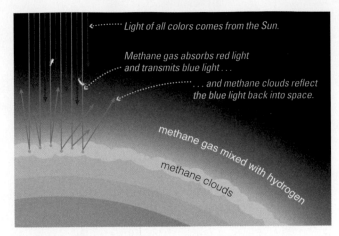

FIGURE 11.8 Neptune and Uranus look blue because methane gas absorbs red light but transmits blue light. Clouds of methane snowflakes reflect the transmitted blue light back to space.

Jupiter has planetwide circulation cells similar to those on Earth (see Figure 10.16). As on Earth, solar heat causes equatorial air to expand and spill toward the poles, while cooler polar air flows toward the equator. Also as on Earth, the Coriolis effect splits the large equator-to-pole circulation cells into smaller cells. However, Jupiter's greater size and faster rotation make the Coriolis effect much stronger on Jupiter than on Earth. Instead of being split into just three smaller cells encircling each hemisphere, Jupiter's circulation cells split into many alternating bands of rising and falling air. These bands are visible as the stripes of alternating color in photographs of Jupiter.

The alternating colors come from differences in the clouds that form in rising and falling air. As shown in **FIGURE 11.9**, the entire planet is blanketed with the reddish-brown ammonium hydrosulfide clouds that make up Jupiter's middle cloud layer. In contrast, the upper layer of white ammonia clouds forms only in places where rising air carries ammonia gas to altitudes high enough and cold enough for the ammonia to condense. In other words, the white bands (sometimes called *zones*) represent regions of rising air with cold ammonia clouds, while the darker bands (sometimes called *belts*) represent regions of falling air in which we can see down to a lower, warmer cloud level. This distinction between Jupiter's bright and dark bands is analogous to the difference on Earth between cloudy, rainy equatorial regions (regions of generally rising air) and the clear desert skies found at latitudes roughly 20°–30° north and south of the equator (regions of descending air). The infrared photo of Jupiter in Figure 11.9 confirms this idea: Warmer air emits more infrared light, so the warmer reddish bands appear brighter in the infrared photograph than the cooler white bands.

You might wonder why ammonia clouds form only in the rising air and not in the falling air. The answer goes back to the clouds themselves. Jupiter's lower atmosphere has ammonia throughout, so there's always ammonia in the air rising upward. However, once the ammonia condenses to form clouds of ammonia ice, these ice flakes fall back down as ammonia "snow," so the air that continues to rise above the clouds has very little ammonia left in it. Therefore, when

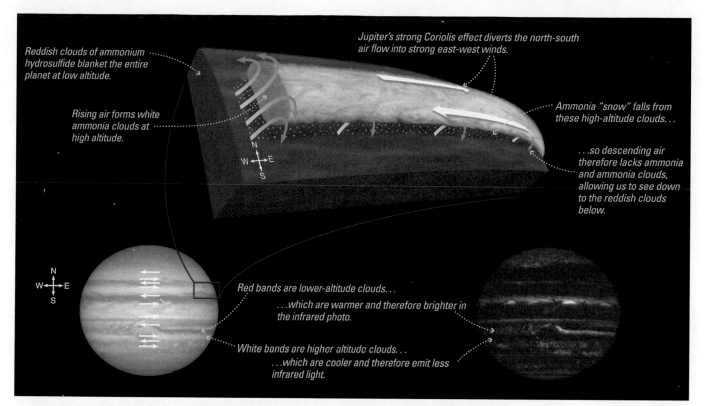

FIGURE 11.9 interactive figure Jupiter's bands of color represent alternating regions of rising and falling air: We see white ammonia clouds in regions of rising air, and see down to the underlying layer of reddish ammonium hydrosulfide clouds in regions of falling air. The white arrows on the visible-light photo indicate wind directions. The infrared photo at the right, taken nearly simultaneously with the visible-light photo, confirms that the red bands represent warmer, deeper clouds and the white bands represent cooler, higher-altitude clouds.

this ammonia-depleted air spills to the north and south and returns downward in the bands of falling air, there is not enough ammonia to form clouds. It's analogous to the way that air descending over Earth's desert latitudes is usually depleted of water by precipitation over the equator, leaving it dry.

The alternating bands of rising and falling air shape Jupiter's global wind patterns. As on Earth, the rising and falling air drives slow winds that are directed north or south, but Jupiter's strong Coriolis effect diverts these winds into fast east or west winds. The east-west winds have peak speeds above 400 km/hr, making hurricane winds on Earth seem mild by comparison. The winds are generally strongest at the equator and at the boundaries between bands of rising and falling air.

Jupiter's global wind patterns are sometimes interrupted by powerful storms, much as storms can interrupt Earth's global wind patterns, but Jupiter's storms dwarf those that we see on Earth. Jupiter's most famous feature—the **Great Red Spot**—is a giant storm more than twice as wide as all of planet Earth. It is somewhat like a hurricane, except that its winds circulate around a high-pressure region rather than a low-pressure region (**FIGURE 11.10**). It is also extremely long-lived: Astronomers have seen it throughout the three centuries during which telescopes have been powerful enough to detect it. Other, smaller storms are constantly brewing in Jupiter's atmosphere. Brown ovals are low-pressure storms with their cloudtops deeper in Jupiter's atmosphere, and white ovals are high-pressure storms topped with ammonia clouds.

No one knows what drives Jupiter's storms, why Jupiter has only one Great Red Spot, or why the Great Red Spot has persisted so much longer than storms on Earth. Storms on Earth lose their strength when they pass over land, so perhaps Jupiter's biggest storms last for centuries simply because no solid surface is present to sap their energy. In fact, two long-lived storms have recently (one in 2006 and one in 2008)

FIGURE 11.10 interactive photo This photograph shows Jupiter's Great Red Spot, a huge high-pressure storm that is large enough to swallow two Earths. The overlaid diagram shows a weather map of the region.

a This Hubble Space Telescope image shows Jupiter's southern hemisphere with the Great Red Spot, "Baby Red" (to its left), and "Red Jr." (below). Baby Red was torn apart by the Great Red Spot a few days later.

b Saturn's atmosphere, photographed by *Voyager 1*. Its banded appearance is very similar to that of Jupiter, but it has even faster winds.

c This infrared image of Uranus from the Keck Telescope shows several storms (the bright blotches) and Uranus's thin rings (red).

d Neptune's atmosphere, viewed from *Voyager 2*, shows bands and occasional strong storms. The large storm (white arrow) was called the Great Dark Spot.

FIGURE 11.11 Selected views of weather patterns on the four jovian planets.

been observed to undergo mysterious change, turning red. The more recent one was torn apart by the Great Red Spot as it passed nearby (**FIGURE 11.11a**). Scientists hope to learn more about Jupiter's weather by studying storms like these.

The bright colors of the Great Red Spot also pose a mystery: We might expect its high-altitude clouds to be white like the high-altitude ammonia clouds elsewhere, but instead, of course, they are red. The colors may be the result of chemicals formed by interactions between the storm's high-altitude gas and solar ultraviolet light, but no one really knows for sure.

Jupiter does not have seasons, because it has no appreciable axis tilt. In fact, Jupiter's polar temperatures are quite similar to its equatorial temperatures, presumably because heat from Jupiter's interior keeps the planet uniformly warm.

Global Winds and Storms on Saturn, Uranus, and Neptune The other jovian planets also have dramatic weather patterns (**FIGURE 11.11b–d**). As on Jupiter, Saturn's rapid rotation creates alternating bands of rising and falling air, along with rapid east-west winds. In fact, Saturn's winds are even faster than Jupiter's—a surprising finding that scientists have yet to explain. We might expect seasons on Saturn, because it has an axis tilt similar to that of Earth. Some seasonal weather changes have been observed, but Saturn's internal heat keeps temperatures about the same year-round and planetwide.

Neptune's atmosphere is also banded, and we have seen a high-pressure storm, called the *Great Dark Spot,* similar to Jupiter's Great Red Spot. However, the Great Dark Spot did not last as long, disappearing from view just 6 years after its discovery. Like Saturn, Neptune has an axis tilt similar to Earth's, but it has relatively little seasonal change because of its internal heat.

The greatest surprise in jovian weather comes from Uranus. When *Voyager 2* flew past Uranus in 1986, photographs revealed virtually no clouds and no banded structures

like those found on the other jovian planets. Scientists attributed the lack of weather to the lower internal heat of Uranus. However, more recent observations from the Hubble Space Telescope and ground-based adaptive optics telescopes show storms raging in Uranus's atmosphere. The storms may be brewing because of the changing seasons: Thanks to Uranus's extreme axis tilt and 84-year orbit of the Sun [**Section 7.1**], its northern hemisphere began to see sunlight only in 2007 after decades of night.

Do jovian planets have magnetospheres like Earth's?

Recall that Earth has a global magnetic field generated by the movements of charged particles in our planet's metallic outer core (see Figure 9.6c). The jovian planets also have global magnetic fields generated by motions of charged particles deep in their interiors. Just as with Earth, these magnetic fields create bubble-like *magnetospheres* that surround the planets and shield them from the solar wind.

Jupiter's Magnetosphere Of the four jovian planets, Jupiter has by far the strongest magnetic field; it is some 20,000 times as strong as Earth's magnetic field. As discussed in Chapter 9, a planet can have a global magnetic field if it has (1) an interior region of electrically conducting fluid, (2) convection in that layer of fluid, and (3) at least moderately rapid rotation. In Jupiter's case, the electrically conducting fluid region is its thick layer of metallic hydrogen (see Figure 11.4). The great extent of this region, combined with Jupiter's rapid rotation, explains Jupiter's strong magnetic field.

Jupiter's strong magnetic field gives it an enormous magnetosphere that begins to deflect the solar wind some 3 million kilometers (about 40 Jupiter radii) in front of Jupiter (**FIGURE 11.12**). If we could see Jupiter's magnetosphere, it would be larger than the full moon in our sky.

solar
wind

magnetic
axis

axis of
rotation

Gases escaping from Io's volcanoes create a ring of charged particles encircling Jupiter.

200,000 km

5,000,000 km

Earth's magnetosphere
magnified 10X

These ultraviolet images show auroral zones in Jupiter's polar regions, much like those on Earth.

FIGURE 11.12 Jupiter's strong magnetic field gives it an enormous magnetosphere. Gases escaping from Io feed the donut-shaped torus, and particles entering Jupiter's atmosphere near its magnetic poles contribute to auroras on Jupiter. The image at the right is a composite of ultraviolet images of the polar regions overlaid on a visible image of the whole planet, all taken by the Hubble Space Telescope.

Jupiter's magnetosphere traps far more charged particles than Earth's, largely because it has a source of particles that Earth's lacks. Nearly all the charged particles in Earth's magnetosphere come from the solar wind, but in Jupiter's case its volcanically active moon, Io, contributes many additional particles. These particles help create auroras on Jupiter. They also create belts of intense radiation around Jupiter, which can cause damage to orbiting spacecraft.

Jupiter's magnetosphere, in turn, has important effects on Io and the other moons of Jupiter. The charged particles bombard the surfaces of Jupiter's icy moons, with each particle blasting away a few atoms or molecules. This process alters the surface materials and can even generate thin atmospheres, much like the thin atmospheres of Mercury and the Moon [**Section 10.3**]. On Io, bombardment by charged particles leads to the continuous escape of gases that were released by volcanic outgassing. The escaping gases are ionized and feed a donut-shaped charged particle belt, called the *Io torus,* that approximately traces Io's orbit (the red "donut" in Figure 11.12). Scientists hope to learn much more about Jupiter's interior, atmosphere, and magnetosphere through NASA's *Juno* mission, which will orbit Jupiter for about a year after its arrival in 2016.

Comparing Jovian Magnetospheres The other jovian planets also have magnetic fields and magnetospheres, but theirs are much weaker than Jupiter's (although still much stronger than Earth's). The strength of each planet's magnetic field depends primarily on the size of the electrically conducting layer buried in its interior. Saturn's magnetic field is weaker than Jupiter's because it has a thinner layer of electrically conducting metallic hydrogen. Uranus and Neptune, smaller still, have no metallic hydrogen at all. Their relatively weak magnetic fields must be generated in their core "oceans" of hydrogen compounds, rock, and metals.

The size of a planet's magnetosphere depends not only on the magnetic field strength, but also on the pressure of the solar wind against it. The pressure of the solar wind is weaker at greater distances from the Sun, so the magnetospheric "bubbles" surrounding more distant planets are larger than they would be if these planets were closer to the Sun. That is why Uranus and Neptune have moderate-size magnetospheres despite their weak magnetic fields (**FIGURE 11.13**). No other magnetosphere is as full of charged particles as Jupiter's, primarily because no other jovian planet has a satellite like Io. Because trapped particles generate auroras, Jupiter has the brightest auroras, while those of the more distant jovian planets are progressively weaker.

We still have much to learn about the jovian magnetic fields and magnetospheres. For example, we generally expect magnetic fields to be closely aligned with planetary rotation, because the magnetic fields are generated within the rotating interiors of planets. However, while this is the case for Jupiter and Saturn (whose magnetic fields are inclined to their rotation axes by 10° and 0°, respectively), it is not the case for Uranus and Neptune. *Voyager* observations showed that the magnetic field of Uranus is tipped by a whopping 60° relative to its rotation axis, and the magnetic field's center is also significantly offset from the planet's center. Neptune's magnetic field is inclined by 46° to its rotation axis. No one has yet explained these surprising observations.

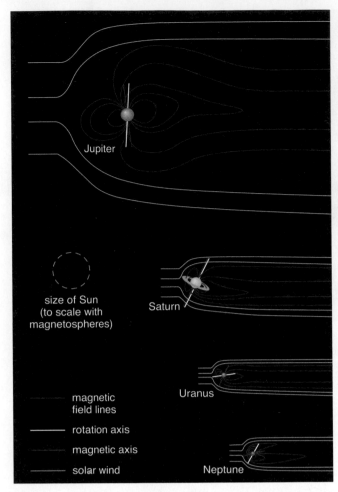

FIGURE 11.13 Comparison of jovian planet magnetospheres. Note the significant tilts of the magnetic fields of Uranus and Neptune compared to their rotation axes. (Planet sizes exaggerated compared to scale of magnetospheres.)

11.2 A WEALTH OF WORLDS: SATELLITES OF ICE AND ROCK

The jovian planets are majestic and fascinating, but they are only the beginning of our exploration of jovian planet *systems*. Each of the four jovian systems includes numerous moons and a set of rings. The total mass of all the moons and rings put together is minuscule compared to that of any one of the jovian planets, but the remarkable diversity of these satellites makes up for their lack of size. In this section, we'll explore a few of the most interesting aspects of the jovian moons.

What kinds of moons orbit the jovian planets?

We now know of more than 170 moons orbiting the jovian planets. Jupiter and Saturn have the most, each with more than 60 moons known to date. It's helpful to organize these moons into three groups by size: small moons less than about 300 kilometers in diameter, medium-size moons ranging from about 300 to 1500 kilometers in diameter, and large moons more than 1500 kilometers in diameter. These categories

are useful because size relates to geological activity. In general, larger moons are more likely to show evidence of past or present geological activity.

FIGURE 11.14 shows a montage of all the medium-size and large moons. These moons resemble the terrestrial planets in many ways. Each is spherical with a solid surface and its own unique geology. Some possess atmospheres, hot interiors, and even magnetic fields. The two largest—Jupiter's moon Ganymede and Saturn's moon Titan—are larger than the planet Mercury, while four others (Jupiter's moons Io, Europa, and Callisto and Neptune's moon Triton) are larger than the largest known dwarf planets, Pluto and Eris. However, they differ from terrestrial worlds in their compositions: Because they formed in the cold outer solar system, most of these worlds contain substantial amounts of ice in addition to metal and rock.

Most of the medium-size and large moons probably formed by accretion within the disks of gas surrounding individual jovian planets [Section 8.2]. That explains why their orbits are almost circular and lie close to the equatorial plane of their parent planet, and also why these moons orbit in the same direction in which their planet rotates. Nearly all

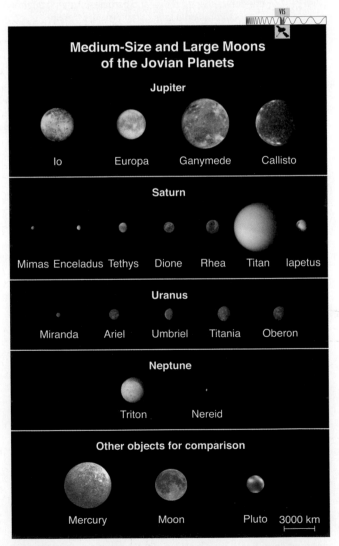

FIGURE 11.14 The medium-size and large moons of the jovian planets, with sizes shown to scale. The moons are shown in order of distance from their planet, from left (nearest) to right (farthest).

VIS

Calypso
(22 km)

Pandora
(84 km)

Telesto
(24 km)

Hyperion
(180 km)

Phoebe
(220 km)

Janus
(180 km)

FIGURE 11.15 These photos from the *Cassini* spacecraft show six of Saturn's smaller moons. All are much smaller than the smallest moons shown in Figure 11.14. Their irregular shapes are due to their small size, which makes their gravities too weak to force them into spheres. The sizes in parentheses represent approximate lengths along their longest axes.

these moons also share an uncanny trait: They always keep the same face turned toward their planet, just as our Moon always shows the same face to Earth. This synchronous rotation arose from the strong tidal forces [**Section 4.5**] exerted by the jovian planets, which caused each moon to end up with equal periods of rotation and orbit regardless of how fast the moon rotated when it formed.

The medium-size and large moons tend to have more interesting geology, but small moons are more numerous. Many small moons are probably captured asteroids or comets, and thus do not follow any particular orbital patterns. Dozens of the smallest moons have been discovered only within the past few years, and many more may yet be discovered. (See Table E.3 in Appendix E for a current list.)

The small moons have irregular shapes, much like potatoes (**FIGURE 11.15**), because their gravities are too weak to force their rigid material into spheres. We have not studied these moons in depth, but we expect their small sizes to allow for little if any geological activity. For the most part, the small moons are just chunks of ice and rock held captive by the gravity of a massive jovian planet.

In case you are wondering, scientists originally named Jupiter's moons for the mythical lovers of the Roman god Jupiter. However, Jupiter has so many moons that more recent discoveries are named for more distant mythological relations. Saturn's moon Titan is named for the Greek gods called the Titans, who ruled before Zeus (Jupiter); other moons of Saturn are named for individual Titans. Moons of Uranus all take their names from characters in the works of William Shakespeare and Alexander Pope, while moons of Neptune are all characters related to the sea in Greek and Roman mythology.

Why are Jupiter's Galilean moons so geologically active?

We are now ready to embark on a brief tour of the most interesting moons of the jovian planets. Our first stop is Jupiter. Jupiter's four largest moons, known as the *Galilean moons* because Galileo discovered them [**Section 3.3**], are all large enough that they would count as planets or dwarf planets if they orbited the Sun (**FIGURE 11.16**).

Io: The Most Volcanically Active World in the Solar System
For anyone who thinks of moons as barren, geologically dead places like our own Moon, Io shatters the stereotype. When the *Voyager* spacecraft first photographed Io up close about three decades ago, we discovered a world with a surface so young that not a single impact crater has survived. Moreover, *Voyager* cameras recorded volcanic eruptions in progress as the spacecraft passed by. We now know that Io is by far the most volcanically active world in our solar system. Large volcanoes pockmark its entire surface

VIS

1000 km

Io

Europa

Ganymede

Callisto

FIGURE 11.16 This set of photos, taken by the *Galileo* spacecraft, shows global views of the four Galilean moons. Sizes are shown to scale; Io is about the size of Earth's Moon. Colors have been enhanced to show detail.

x12

50 km

This close-up shows the glow of intensely hot lava from a volcanic eruption.

VIS

500 km

x50

jets of SO₂ gas

SO_2 frost

basalt lava flow

10 km

An 80-km-high gas plume was created when hot lava flowed over sulfur dioxide frost, causing it to sublimate explosively into gas.

FIGURE 11.17 interactive photo Io is the most volcanically active body in the solar system. Most of the black, brown, and red spots on the surface are recently active volcanic features. White and yellow areas are sulfur dioxide and sulfur deposits, respectively, from volcanic gases. (Photographs from the *Galileo* spacecraft; some colors slightly enhanced or altered.)

(FIGURE 11.17), and eruptions are so frequent that they constantly repave the surface. Io probably also has tectonic activity, because tectonics and volcanism generally go hand in hand. However, debris from volcanic eruptions has probably buried most tectonic features.

Many of Io's frequent eruptions are surprisingly similar to those of volcanoes on Earth. Some of the taller volcanoes appear to be shield volcanoes built from flows of basalt lava, and the eruptions are accompanied by outgassing. The primary gases released from Io's volcanoes are sulfur dioxide (SO_2), sulfur, and a hint of sodium. Some of this gas escapes into space, where it supplies ionized gas (plasma) to the Io torus and Jupiter's magnetosphere, and some of it gives Io a very thin atmosphere. Much of the gas condenses and falls back to the surface, where the sulfur gives Io its distinctive red and orange colors and the sulfur dioxide makes a white frost (FIGURE 11.18). As hot lava flows across the surface, it can re-vaporize the sulfur dioxide surface ice in much the same way that lava flowing into the ocean vaporizes water on Earth. Io's low gravity and thin atmosphere allow tall plumes of this vaporized sulfur dioxide to rise upward to altitudes of

hundreds of kilometers. Over a period of a few months, fallout from these tall volcanic plumes can blanket an area the size of Arizona.

Io's active volcanoes tell us that it must be quite hot inside. However, Io is only about the size of our geologically dead Moon, so it should have long ago lost any heat from its birth and it is too small for radioactivity to provide much ongoing heat. How, then, can Io be so hot inside? The only possible answer is that some other ongoing process must be heating Io's interior. Scientists have identified this process and call it **tidal heating**, because it arises from effects of tidal forces exerted by Jupiter.

Just as Earth exerts a tidal force that causes the Moon to keep the same face toward us at all times [**Section 4.5**], a tidal force from Jupiter makes Io keep the same face toward Jupiter as it orbits. But Jupiter's mass makes this tidal force far larger than the tidal force that Earth exerts on the Moon. Moreover, Io's orbit is slightly elliptical, so its orbital speed and distance from Jupiter vary. This variation means that the strength and direction of the tidal force change slightly as Io moves through each orbit, which in turn changes the size and orientation of Io's tidal bulges (**FIGURE 11.19a**). The result is that Io is continuously being flexed in different directions, which generates friction inside it. The flexing heats the interior in the same way that flexing warms Silly Putty. Tidal heating generates tremendous heat on Io—more than 200 times as much heat (per gram of mass) as the radioactive heat driving much of Earth's geology. This heat explains Io's volcanic activity. The energy for tidal heating ultimately comes from Jupiter's rotation, which is gradually slowing, although at a rate too small to be observed.

VIS IR

Three large plumes lit by the Sun blanket the surface in sulfur-rich snow . . .

. . . while many more eruptions are visible on Io's night side glowing in the infrared.

FIGURE 11.18 Two views of Io's volcanoes taken by *New Horizons* on its way to Pluto. Io rotated a little between these images, so the eruption at the top is the same one in both cases.

SEE IT FOR YOURSELF

Pry apart the overlapping ends of a paper clip so that you can hold one end in each hand. Flex the ends apart and together until the paper clip breaks. Lightly touch the broken end to your finger or lips—can you feel the warmth produced by flexing? How is this heating similar to the tidal heating of Io?

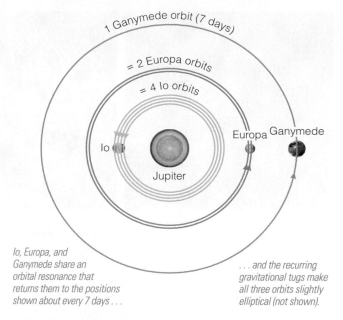

Io's elliptical orbit means continual changes in the strength and direction of the tidal force from Jupiter . . .

. . . and the changing tides flex Io's interior and cause tidal heating.

close to Jupiter: large tidal bulges

far from Jupiter: small tidal bulges

Jupiter

a Tidal heating arises because Io's elliptical orbit (exaggerated in this diagram) causes varying tides.

FIGURE 11.19 interactive figure These diagrams explain the cause of tidal heating on Io. Tidal heating has a weaker effect on Europa and Ganymede, because they are farther from Jupiter and tidal forces weaken with distance.

1 Ganymede orbit (7 days)

= 2 Europa orbits

= 4 Io orbits

Io   Europa Ganymede

Jupiter

Io, Europa, and Ganymede share an orbital resonance that returns them to the positions shown about every 7 days . . .

. . . and the recurring gravitational tugs make all three orbits slightly elliptical (not shown).

b Io's orbit is elliptical because of the orbital resonance Io shares with Europa and Ganymede.

However, we are still left with a deeper question: Why is Io's orbit slightly elliptical, when almost all other large satellites have nearly circular orbits? The answer lies in an interesting dance executed by Io and its neighboring moons (**FIGURE 11.19b**). During the time Ganymede takes to complete one orbit of Jupiter (about 7 days), Europa completes exactly two orbits and Io completes exactly four orbits. The three moons therefore line up periodically, and the gravitational tugs they exert on one another add up over time.* Because the tugs are always in the same direction with each alignment, they tend to stretch out the orbits, making them slightly elliptical. The effect is much like that of pushing a child on a swing. If timed properly, a series of small pushes can add up to a *resonance* that causes the child to swing quite high. For the three moons, the resonance that makes their orbits elliptical comes from the small gravitational tugs that repeat at each alignment.

The phenomenon of orbital periods falling into a simple mathematical relationship is called an **orbital resonance**. Resonances are quite common in planetary systems. For example, they affect not only the Galilean moons but also planetary rings, the asteroid belt, and the Kuiper belt in our solar system, and they affect the orbits of some of the planets known around other stars.

Europa: The Water World? Europa offers a striking contrast to Io. Instead of having active volcanoes dotting its surface, Europa is covered by water ice (see Figure 11.16). Nevertheless, its fractured, frozen surface must hide an interior made hot by the same type of tidal heating that powers Io's volcanoes, though tidal heating is weaker on Europa because it lies farther from Jupiter.

Key evidence about how the hot interior affects the icy surface comes from the fact that Europa has only a handful of impact craters. This fact tells us that some type of ongoing geological activity must have erased the evidence of nearly all past impacts. But what is doing the erasing?

Scientists suspect that the answer is either liquid water rising up from an ocean that lies beneath the icy crust or interior water ice that is just warm enough to undergo convection, so that some of it rises up and flows across the surface. Close-up photos of the surface, combined with the fact that Europa has enough internal heat to melt subsurface ice into liquid water, support the ocean hypothesis. For example, **FIGURE 11.20** shows some of the many double-ridged cracks visible on Europa's surface. As the model at the right shows, these cracks are best explained by assuming that the icy crust can slide on a softer or liquid layer below. The double ridges may form as tidal stresses force parts of the icy crust to scrape past each other, warming and possibly melting the ice along the fault.

Other *Galileo* data provide additional support for the ocean hypothesis. Theoretical models based on *Galileo* measurements of the strength of gravity over different parts of the surface indicate that Europa has a metallic core and a rocky mantle surrounded by enough water to make a layer of ice about 100 kilometers thick. The models suggest that the upper 5 to 25 kilometers of the water should be solidly frozen as an icy crust, but that tidal heating should provide enough warmth to turn the underlying ice into a layer of either liquid water or relatively warm, convecting ice. Evidence favoring the idea that the underlying layer consists of liquid water comes from magnetic field data collected by the *Galileo* spacecraft. Europa is one of only a few moons in the solar system to have a magnetic field, and its magnetic field changes as Jupiter rotates. The simplest way to explain this change is if Europa's magnetic field is created (or *induced*)

---

*You may wonder why periodic alignments occur. Like synchronous rotation, they are not a coincidence but rather a consequence of feedback from the tides that the moons raise on Jupiter. The periodic tugs the moons exert on one another actually work to sustain the recurring alignments.

Tidal stresses cause parts of Europa's icy crust to slowly slide past each other.

Frictional heating expands ice here, forming the ridge...

...and may melt ice here, collapsing the ridge center.

Europa's surface appears heavily cracked even from a distance.

Close-up photos show double-ridged cracks, best explained by an icy crust moving upon a soft or liquid layer below.

FIGURE 11.20 This sequence shows how some of the many double-ridged cracks found on Europa's surface may have formed. (Photos are from the *Galileo* spacecraft; colors are enhanced in the global view.)

in response to the rotation of Jupiter's strong magnetic field. This type of response is possible only if Europa has a *liquid* layer of electrically conducting material. A salty ocean would fit the bill, but convecting ice would not. The data also suggest that Europa's liquid ocean must be global in extent, and that it is about as salty as Earth's oceans. Further support for this idea comes from another of *Galileo's* instruments, which found evidence for salty compounds on Europa's surface—possible seepage from a briny deep.

Taken together, the evidence from surface photos, gravitational measurements, and magnetic fields makes a strong case for a deep ocean of liquid water on Europa (**FIGURE 11.21**). Because Europa must have a hot interior from tidal heating, it seems reasonable to imagine that volcanic vents dot the seafloor and sometimes erupt, creating rising plumes of warm water. These plumes may lead to the formation of subsurface lakes within the icy crust, which could cause cracking of the icy

surface above. Surface features that look like jumbled icebergs, such as the region shown on the right of Figure 11.21, may be explained by such cracking.

If these ideas are correct, then Europa may have a hidden ocean containing more than twice as much liquid water as all of Earth's oceans combined, along with a seafloor dotted with the same types of volcanic vents around which life thrives in Earth's oceans. That is why scientists are so interested in the possibility that Europa might also be home to life—a possibility we'll explore further in Chapter 24.

Ganymede: King of the Moons Ganymede is the largest moon in the solar system, and its surface tells of a complex geological history. Like Europa, Ganymede has a surface of water ice. However, while Europa's surface appears relatively young everywhere, Ganymede's surface appears to have a dual personality (**FIGURE 11.22**). Some regions are dark

metallic core

rocky interior

H_2O layer

Europa may have a 100-km-thick ocean under an icy crust.

Rising plumes of warm water may sometimes create lakes within the ice, causing the crust above to crack . . .

. . . explaining surface terrain that looks like a jumble of icebergs suspended in a place where liquid or slushy water froze.

FIGURE 11.21 This sequence shows a model of Europa's interior structure based on *Galileo* data, along with a *Galileo* surface photo that may tell us about processes occurring in the ice and ocean below. (The layer thicknesses are not to scale in the central panel.)

FIGURE 11.22 Ganymede, the largest moon in the solar system, has both old and young regions on its surface of water ice. The dark regions are heavily cratered and must be billions of years old, while the light regions are younger landscapes where eruptions of water have presumably erased ancient craters; the long grooves in the light regions were probably formed by water erupting along surface cracks. Notice that the boundary between the two types of terrain can be quite sharp.

and densely cratered, suggesting that they look much the same today as they did billions of years ago. Other regions are light-colored with very few craters. In some cases, fairly sharp boundaries separate the two types of terrain.

The young terrain argues for occasional upwelling of liquid water or icy slush to the surface. This material would cover craters before refreezing, explaining why there are so few craters in this terrain. The long grooves are probably made by the eruption of water or slush along a crack in the surface. As the water in the crack freezes, it expands and pushes outward, creating the groove.

If liquid water occasionally wells up to the surface, could it mean that Ganymede has a subsurface ocean like that thought to exist on Europa? The case for an ocean on Ganymede is less strong than the case for an ocean on Europa, but magnetic field measurements indicate that such an ocean is possible. Like Europa, Ganymede has a magnetic field that varies with Jupiter's rotation, suggesting the presence of a salty ocean beneath the surface.

One difficulty with the idea of an ocean on Ganymede is figuring out the source of the heat needed to melt subsurface ice. Because Ganymede is farther from Jupiter than Europa or Io, its tidal heating is weaker and could not by itself supply enough heat to melt ice today. However, Ganymede's larger size means it should retain more heat from radioactive decay. Perhaps tidal heating and radioactive decay together provide enough heat to make a liquid layer beneath the icy surface. If so, the lesser heating on Ganymede means that its ocean probably lies at least 150 kilometers beneath the surface—much deeper than Europa's potential ocean.

Callisto: Last of the Galilean Moons The outermost Galilean moon, Callisto, looks most like what scientists originally expected for an outer solar system satellite: a heavily cratered iceball (FIGURE 11.23). The bright, circular patches on its surface are impact craters. They are probably bright because large impacts blast out "clean" ice from deep underground, and this ice reflects more light than the dirtier ice that has been on the surface longer.

Craters make sense on an old surface like Callisto's, but other features are more difficult to interpret. For example, close-up photos show a dark, powdery substance concentrated in low-lying areas, leaving ridges and crests bright white (small photo in Figure 11.23). The dark powder may be debris left behind when ice sublimates into gas from Callisto's surface, much as dark material is left behind on a comet's nucleus when ice sublimates away [**Section 12.2**].

Despite its relatively large size (the third-largest moon in the solar system), Callisto lacks volcanic and tectonic features. This tells us that it lacks any significant source of internal heat. In fact, gravity measurements by the *Galileo* spacecraft showed that Callisto never underwent differentiation: Dense rock and lighter ice are mixed throughout most of its interior, which means that its interior never warmed significantly. The lack of heat is not surprising: Callisto has no tidal heating because it does not participate in the orbital resonances that affect the other three Galilean moons, and we do not expect its icy interior to contain enough radioactive material to supply much heat through radioactive decay.

Nevertheless, it is possible that Callisto, too, may hide a subsurface ocean. Like Europa and Ganymede, Callisto has a magnetic field that suggests the presence of a salty interior ocean. Perhaps Callisto's surface provides just enough

Callisto is heavily cratered, indicating an old surface that nonetheless may hide a deeply buried ocean.

Close-up photo shows a dark powder overlaying the low areas of the surface.

FIGURE 11.23 Callisto, the outermost of the four Galilean moons, has a heavily cratered, icy surface.

insulation for heat from radioactive decay to melt water beneath the thick, icy crust. If so, we face the intriguing possibility that there could be oceans on three worlds orbiting Jupiter—with far more total ocean than we find here on Earth.

What is remarkable about Titan and other major moons of the outer solar system?

Leaving Jupiter behind, we will continue our tour of moons by discussing the major moons of Saturn, Uranus, and Neptune. We'll go in planetary order, first focusing special attention on Saturn's enigmatic Titan, the second-largest moon in the solar system after Ganymede.

Saturn's Moon Titan: Piercing the Veil
Titan is unique among the moons of our solar system in having a thick atmosphere—so thick that it hides the surface from view, except at a few specific wavelengths of light (**FIGURE 11.24**). Titan's color comes from chemicals in its atmosphere, which act much like those that make smog over cities on Earth. The atmosphere is about 90% molecular nitrogen (N_2), not that different from the 77% nitrogen content of Earth's atmosphere. However, on Earth the rest of the atmosphere is mostly oxygen, while the rest of Titan's atmosphere consists of argon, methane (CH_4), ethane (C_2H_6), and other hydrogen compounds.

Titan's atmospheric composition can be understood in terms of our general understanding of atmospheric production and loss processes [**Section 10.2**]. Titan's icy composition supplies methane and ammonia gas through evaporation, sublimation, and possibly volcanic eruptions. Solar ultraviolet light breaks down some of those molecules, releasing hydrogen atoms and leaving highly reactive compounds containing carbon and nitrogen. The hydrogen atoms can leave Titan forever by thermal escape, while the remaining molecular fragments can react to make the other ingredients of Titan's atmosphere. For example, the abundant molecular nitrogen is made after ultraviolet light breaks down ammonia (NH_3) molecules, and ethane is made from methane.

The methane and ethane in Titan's atmosphere are both greenhouse gases and therefore give Titan an appreciable greenhouse effect [**Section 10.1**] that makes it warmer than it would be otherwise. Still, because of its great distance from the Sun, its surface temperature is a frigid 93 K ($-180°C$). The surface pressure on Titan is about 1.5 times the sea level pressure on Earth, which would be fairly comfortable if not for the lack of oxygen and the cold temperatures.

A moon with a thick atmosphere would be intriguing enough, but we have at least two other reasons for special interest in Titan. First, its complex atmospheric chemistry probably produces numerous organic chemicals—the chemicals that are the basis of life. Second, although it is far too cold for liquid water to exist on Titan's surface, conditions are right for methane or ethane rain, perhaps making rivers that feed into lakes or oceans.

Titan is so intriguing that NASA and the European Space Agency (ESA) combined forces to explore it, with NASA's *Cassini* "mother ship" releasing the ESA-built probe called *Huygens* (pronounced "Hoy-guns"), which parachuted to Titan's surface in 2005 (**FIGURE 11.25**). During its descent, the probe photographed river valleys merging together, flowing down to what looks like a shoreline. On landing, instruments on the probe discovered that the surface has a hard crust but is a bit squishy below, perhaps like sand with liquid mixed in, and photos showed "ice boulders" rounded by erosion. All these results support the idea of a wet climate—but wet with liquid methane rather than liquid water.

Cassini observations of Titan have also taught us a lot. The brighter regions in the photographs are icy hills, perhaps made by ice volcanoes. The dark valleys were probably created when methane rain carried down "smog particles" that concentrated on river bottoms. The vast plains into which the valleys appear to empty are low-lying regions, but they do not appear to be liquid. Instead, they are probably covered by smog particles carried down by the rivers and then sculpted into vast dune fields by Titan's global winds. All in all, conditions in Titan's equatorial regions appear to be analogous to those of the desert southwest of the United States, where infrequent rainfall carves valleys and creates vast dry lakes called *playas* where the water evaporates or soaks into the ground.

The polar regions of Titan, revealed by *Cassini* radar, contain numerous lakes of liquid methane or ethane (**FIGURE 11.26**). Images also reveal polar storm clouds and riverbeds leading into the lakes, suggesting that Titan has a methane/ethane cycle resembling the water cycle on Earth. Interestingly, while almost all of the lakes lie at high northern latitudes, most of the storm clouds have been seen in the south. The reason is probably traceable to Titan's seasons. Titan shares Saturn's 27° axis tilt, which gives it seasons, and Saturn's 29-year orbit of the Sun makes the four seasons last about 7 years each. So far, *Cassini*'s observations have been mostly during Titan's southern hemisphere summer, and the summer warmth explains the rising air responsible for the methane storm clouds seen there. On Earth, the warmth of a summer afternoon or of

VIS

900 km

FIGURE 11.24 Titan, as photographed by the *Cassini* spacecraft, is enshrouded by a thick atmosphere with clouds and haze. *Cassini* was outfitted with filters designed to peer through the atmosphere at specific near-infrared wavelengths of light. The inset shows sunlight reflecting off Kraken Mare, Titan's largest lake.

FIGURE 11.25 interactive photo This sequence zooms in on the *Huygens* landing site on Titan. Left: A global view taken by the *Cassini* orbiter. Center: An aerial view from the descending probe. Right: A surface view taken by the probe after landing; the "rocks," which are 10 to 20 centimeters across, are presumably made of ices. Keep in mind that you are looking at the surface of a world more than a billion kilometers away.

Earth's equatorial region creates storm clouds that produce intense rainfall, but Titan's southern hemisphere storms apparently don't produce enough rainfall to create many lakes in the south. The situation may differ for the northern hemisphere, because Saturn's elliptical orbit brings it (and Titan) closer to the Sun during northern summer. The more intense northern summer may create stronger circulation and storms with enough rainfall to fill the northern hemisphere lakes. This hypothesis is being tested now: Spring has begun in Titan's northern hemisphere, and *Cassini* is monitoring the storms and lake depths.

FIGURE 11.26 Radar image of Ligeia Mare near Titan's north pole showing lakes of liquid methane and ethane at a temperature of −180°C. Most solid surfaces reflect radar well, and these regions are artificially shaded tan to suggest land. The liquid surfaces reflect radar poorly, and these regions are shaded blue and black to suggest lakes.

Perhaps the most astonishing result from the *Cassini/Huygens* mission is how familiar the landscape looks in this alien environment with these unfamiliar materials. Instead of liquid water, Titan has liquid methane and ethane. Instead of rock, Titan has ice. Instead of molten lava, Titan has a slush of water ice mixed with ammonia. Instead of surface dirt, Titan's surface has smog-like particles that rain out of the sky and accumulate on the ground. Evidently, the similarities between the physical processes that occur on Titan and Earth are far more important in shaping the landscapes than the fact that the two worlds have very different compositions and temperatures.

THINK ABOUT IT

What other geological features might you expect on Titan, given its similarities to Earth? How might those features be different, given its differences?

The abundance of organic molecules on Titan, along with the similarity of its surface geology to that of Earth, makes it natural to wonder whether Titan could be home to life. Although we cannot rule it out, most scientists think it unlikely that Titan's surface could support life because of the combination of the very low temperatures and the fact that chemical properties of liquid methane and ethane make them seem less suited to supporting biology than liquid water [**Section 24.2**]. Another possibility is subsurface life. *Cassini* observations of small changes in Titan's surface gravity and shape as it orbits Saturn indicate that it probably has a subsurface ocean of liquid water, much like the subsurface ocean on Europa. However, theoretical models of Titan's interior suggest that the bottom of its ocean would be in contact with a layer of high-density ice, rather than with a rocky layer like that thought to exist at the bottom of Europa's ocean. Scientists are unsure whether this difference would have any effect on the potential habitability of Titan's subsurface ocean.

| Mimas (398 km) | Enceladus (498 km) | Tethys (1060 km) | Dione (1118 km) | Rhea (1528 km) | Iapetus (1436 km) |

FIGURE 11.27 Portraits taken by the *Cassini* spacecraft of Saturn's medium-size moons (not to scale). All but Mimas show evidence of past volcanism and/or tectonics.

Saturn's Medium-Size Moons The *Cassini* mission is also teaching us more about Saturn's other moons, especially its six medium-size moons (**FIGURE 11.27**). The photographs suggest that these moons have had a complex history.

Only Mimas, the smallest of these six moons, shows little evidence of past volcanism or tectonics. It is essentially a heavily cratered iceball. The huge crater visible at Mimas's upper right in Figure 11.27 is sometimes called "Darth Crater" because of Mimas's resemblance to the Death Star in the *Star Wars* movies. (The crater's official name is *Herschel*.) The impact that created this crater probably came close to breaking Mimas apart.

Most of Saturn's other medium-size moons also have heavily cratered surfaces, confirming that they lack global geological activity today. However, we find abundant evidence of volcanism and/or tectonics that must have occurred more recently. Smooth regions appear to be places where icy lava once flowed, and close-up views of the bright streaks (such as the long streaks visible on Dione) show them to be vast sets of tectonic cliffs running parallel to one another.

Iapetus is particularly bizarre (**FIGURE 11.28**). It has an astonishing ridge more than 10 kilometers high that spans nearly half its circumference, curiously aligned along the equator. No one knows its origin, but it is likely the result of tectonic activity. Moreover, half of Iapetus appears coated in very dark dust (some is visible at the right of the Iapetus image in Figure 11.27). The dust apparently comes from Phoebe, a dark and dusty moon which orbits Saturn in a backwards orbit at a great distance (see Figure 11.15). Small impacts easily eject dust because of Phoebe's low gravity, and the dust forms a ring as it spirals inward toward Iapetus.

Enceladus provided an even bigger surprise: This moon is barely 500 kilometers across—small enough to fit inside the borders of Colorado—and yet it shows clear evidence of *ongoing* geological activity (**FIGURE 11.29**). Scientists knew

FIGURE 11.29 *Cassini* photo of Saturn's moon Enceladus. The blue "tiger stripes" near the bottom of the main photo are regions of fresh ice that must have recently emerged from below. The colors are exaggerated; the image is a composite made at near-ultraviolet, visible, and near-infrared wavelengths. The inset shows Enceladus backlit by the Sun, with fountains of ice particles (and water vapor) clearly visible as they spray out of the south polar region.

FIGURE 11.28 Saturn's moon Iapetus has a 10-kilometer equatorial ridge (white arrow) that spans nearly half its circumference. The inset shows a portion of the ridge in perspective.

that Enceladus undergoes some tidal heating through an orbital resonance, but were surprised to learn that the heating is enough to make this moon active today. Its surface has very few impact craters—and some regions have none at all—telling us that recent geological activity has erased older craters. Moreover, the strange grooves near its south pole are measurably warmer than the surrounding terrain, and photographs show this region venting huge clouds of water vapor and ice crystals. These fountains must have some subsurface source, which could potentially mean the existence of an ocean beneath the icy crust. In that case, there would be at least a slim possibility that Enceladus could harbor life [**Section 24.2**].

The Medium-Size Moons of Uranus

Uranus does not have any moons that fall into our large category, but it has five medium-size moons: Miranda, Ariel, Umbriel, Titania, and Oberon (see Figure 11.14). Like other jovian moons, these moons are made largely of ice. Because of Uranus's great distance from the Sun, this ice includes a great deal of ammonia and methane as well as water.

Our only close look at these moons came during the *Voyager* flyby in 1986, which left us with many unanswered questions. For example, Ariel and Umbriel are virtual twins in size, yet Ariel shows evidence of volcanism and tectonics, while the heavily cratered surface of Umbriel suggests a lack of geological activity. Titania and Oberon also are twins in size, but Titania appears to have had much more geological activity than Oberon. No one knows why these two pairs of similar-size moons should vary so greatly in geological activity.

Miranda, the smallest of Uranus's medium-size moons, is the most surprising (**FIGURE 11.30**). We might expect Miranda to be a cratered iceball like Saturn's similar-size moon, Mimas. Instead, *Voyager* images of Miranda show tremendous tectonic features and relatively few craters. Why should Miranda be so much more geologically active than Mimas? Our best guess is that Miranda had an episode of tidal heating

billions of years ago, perhaps during a temporary orbital resonance with another moon of Uranus. Mimas apparently never had such an episode of tidal heating and thus lacks similar features.

Neptune's Triton: A Captured Moon

The last stop on our tour is Neptune, so distant that only two of its moons (Triton and Nereid) were known to exist before the *Voyager 2* spacecraft visited it. Nereid is Neptune's only medium-size moon, and Triton is its only large moon. Triton is one of the coldest worlds in the solar system, even colder than Pluto because it reflects more of the weak sunlight that reaches the outskirts of the solar system.

Triton may appear to be a typical moon, but it is not. It orbits Neptune "backward" (opposite to Neptune's rotation) and at a high inclination to Neptune's equator. These are telltale signs of a moon that was captured rather than having formed in the disk of gas around its planet. No one knows how a moon as large as Triton could have been captured, but models suggest one possible mechanism: Triton may have once been a member of a binary Kuiper belt object that passed so close to Neptune that Triton lost energy and was captured while its companion gained energy and was flung away at high speed.

Triton's geology is just as surprising as its origin. Triton is smaller than our own Moon, yet its surface shows evidence of relatively recent geological activity (**FIGURE 11.31**). Some regions show evidence of past volcanism, perhaps an icy equivalent to the volcanism that shaped the lunar maria. Other regions show wrinkly ridges (nicknamed "cantaloupe terrain") that appear tectonic in nature; apparently, blobs of ice of different density have risen and fallen, contorting the crust. Triton even has a very thin atmosphere that has left wind streaks on its surface.

What could have generated the heat needed to drive Triton's geological activity? The best guess is tidal heating. Triton probably had a very elliptical orbit and a more rapid rotation when Neptune first captured it. Tidal forces would have circularized its orbit, brought it into synchronous rotation, and perhaps heated its interior enough to cause its geological activity. This idea would also explain why the geological activity has subsided with time.

FIGURE 11.30 The surface of Miranda shows astonishing tectonic activity despite its small size. The cliff walls seen in the inset are higher than those of the Grand Canyon on Earth.

cantaloupe terrain

frost deposits?

wind streaks

200 km

Triton's southern hemisphere as seen by *Voyager 2*.

50 km

This close-up shows lava-filled impact basins similar to the lunar maria, but the lava was water or slush rather than molten rock.

FIGURE 11.31 Neptune's moon Triton shows evidence of a surprising level of past geological activity.

Terrestrial Planet Geology

- Internal heat, primarily from radioactive decay, can cause volcanic and tectonic activity.
- Only large planets retain enough internal heat to stay geologically active today.
- Example: Mars (photo above) probably retains some internal heat. If it had been smaller, like Mercury, it would be geologically "dead" today. If it had been larger, like Earth, it would probably have much more active and ongoing tectonics and volcanism.

Jovian Moon Geology

- Tidal heating can cause tremendous geological activity on moons with elliptical orbits around massive planets.
- Even without tidal heating, icy materials can melt and deform at lower temperatures than rock, increasing the likelihood of geological activity.
- Together, these effects explain why icy moons are much more likely to have ongoing geological activity than rocky terrestrial worlds of the same size.
- Example: Ganymede (photo above) shows evidence of recent geological activity, even though it is similar in size to the geologically dead terrestrial planet Mercury.

FIGURE 11.32 Jovian moons can be much more geologically active than terrestrial worlds of similar size because of their icy compositions and tidal heating, which is not important on the terrestrial worlds.

Why are small icy moons more geologically active than small rocky planets?

Based on what we learned when studying the geology of the terrestrial worlds, the active geology of the jovian moons seems out of character with their sizes. Numerous jovian moons remained geologically active far longer than Mercury or our Moon, yet they are no bigger and in many cases are much smaller in size. However, there are two crucial differences between the jovian moons and the terrestrial worlds: icy compositions and tidal heating.

Because they formed far from the Sun, most of the jovian moons contain ices that can melt or deform at far lower temperatures than rock. As a result, they can experience geological activity even when their interiors have cooled to temperatures far below those of rocky worlds. Indeed, except on Io, most of the volcanism that has occurred in the outer solar system probably did not produce any hot lava at all. Instead, it produced icy lava that was essentially liquid water, perhaps mixed with methane and ammonia.

The major lesson, then, is that "ice geology" is possible at far lower temperatures than "rock geology." This fact, combined in many cases with tidal heating, explains how the jovian moons have had such interesting geological histories despite their small sizes. In essence, the same physical properties that

allowed hydrogen compounds to condense as ices in the outer solar system also allowed the worlds made from these ices to stay geologically active for long periods of time. **FIGURE 11.32** summarizes the differences between the geology of jovian moons and that of the terrestrial worlds.

11.3 JOVIAN PLANET RINGS

The jovian planet systems have three major components: the planets themselves, the moons, and the rings that encircle the planets. We have already studied the planets and their moons, so we now turn our attention to their amazing rings. For a long time, Saturn's rings were thought to be unique in the solar system. We now know that all four jovian planets have rings. We'll begin by exploring the rings of Saturn, since they are by far the most spectacular.

What are Saturn's rings like?

Saturn's rings have dazzled and puzzled astronomers since Galileo first saw them through his small telescope and suggested that they resembled "ears" on Saturn. You can see the rings through a backyard telescope, but learning their nature requires higher resolution (**FIGURE 11.33**). Earth-based views make the rings appear to be continuous, concentric sheets of material separated by a large gap (called the

a This Earth-based telescopic view of Saturn makes the rings look like large, concentric sheets. The dark gap within the rings is called the *Cassini division*.

b This image of Saturn's rings from the *Cassini* spacecraft reveals many individual rings separated by narrow gaps.

c Artist's conception of particles in a ring system. Particles clump together because of gravity, but small random velocities cause collisions that break them up.

FIGURE 11.33 interactive figure Zooming in on Saturn's rings.

Cassini division). Spacecraft images reveal these "sheets" to be made of many individual rings, each separated from the next by a narrow gap. But even these appearances are somewhat deceiving. If we could wander into Saturn's rings, we'd find that they are made of countless icy particles ranging in size from dust grains to large boulders, sometimes clumped together by their mutual gravity. All are far too small to be photographed even from spacecraft passing nearby.

Ring Particle Characteristics Spectroscopy reveals that Saturn's ring particles are made of relatively reflective water ice. The rings look bright where they contain enough particles to intercept sunlight and scatter it back toward us. We see gaps in places where there are few particles to reflect sunlight.

Each individual ring particle orbits Saturn independently in accord with Kepler's laws, so the rings are much like myriad tiny moons. The individual ring particles are so close together that they collide frequently. In the densest parts of the rings, each particle collides with another every few hours. However, the collisions are fairly gentle: Despite the high orbital speeds of the ring particles, nearby ring particles are moving at nearly the same speed and touch only gently when they collide.

THINK ABOUT IT

Which ring particles travel faster: those closer to Saturn or those farther away? Explain why. (*Hint*: Review Kepler's third law.)

The frequent collisions explain why Saturn's rings are perhaps the thinnest known astronomical structure. They span more than 270,000 kilometers in diameter but are only a few tens of *meters* thick. The rings are so thin that they disappear from view when we see Saturn edge-on, as we do around the equinoxes of its 29.5-year orbit of the Sun.

To understand how collisions keep the rings thin, imagine what would happen to a ring particle on an orbit slightly inclined to the central ring plane. The particle would collide with other particles every time its orbit intersected the ring plane, and its orbital tilt would be reduced with every collision. Before long, these collisions would force the particle to conform to the orbital pattern of the other particles, and any particle that moved away from the narrow ring plane would

soon be brought back within it. A similar idea explains why ring particles have almost perfectly circular orbits: Any particle with an elliptical orbit would quickly suffer enough collisions to force it into an orbit matching that of its neighbors.

Rings and Gaps Close-up photographs show an astonishing number of rings, gaps, ripples, and other features—as many as 100,000 altogether. Scientists are still struggling to explain all the features, but some general ideas are now clear.

Rings and gaps are caused by particles bunching up at some orbital distances and being forced out at others. This bunching happens when gravity nudges the orbits of ring particles in some particular way. One source of nudging comes from small moons located within the gaps in the rings themselves, sometimes called *gap moons*. The gravity of a gap moon can effectively keep the gap clear of smaller ring particles while creating ripples in the ring edges (**FIGURE 11.34a**). The ripples appear to move in opposite directions on the two sides of the gap, because ring particles on the inner side orbit Saturn slightly faster than the gap moon, while those on the outer side orbit slightly slower than the gap moon. In some cases, two nearby gap moons can force particles between them into a very narrow ring (**FIGURE 11.34b**). The gap moons are often called *shepherd moons* in those cases, because they act like a shepherd forcing particles into line.

Ring particles also may be nudged by the gravity from larger, more distant moons. For example, a ring particle orbiting about 120,000 kilometers from Saturn's center will circle the planet in exactly half the time it takes the moon Mimas to orbit. Every time Mimas returns to a certain location, the ring particle will also be at its original location and therefore will experience the same gravitational nudge from Mimas. The periodic nudges reinforce one another and clear a gap in the rings—in this case, the large gap visible from Earth (the Cassini division). This type of reinforcement due to repeated gravitational tugs is another example of an *orbital resonance*, much like the orbital resonances that make Io's orbit elliptical (see Figure 11.19). Some resonances create vast numbers of ripples and waves that travel great distances through the rings (**FIGURE 11.35**). Other orbital resonances, caused by moons both within the rings and farther out from Saturn, probably explain most of the intricate structures visible in ring photos.

A gap moon (white dot) creates ripples as its gravity nudges particles that orbit faster than the moon (inside the gap) or slower (outside).

100 km

x13

The 8-km-wide moon Daphnis clears a gap in the rings.

2000 km

a Some small moons create gaps within the rings.

These two moons are shepherding a narrow ringlet around Saturn.

200 km

x35

The 100-km-wide moon Prometheus might be supplying or attracting ring material.

10,000 km

b Some moons force particles between them into a very narrow ring, as is the case with the two shepherd moons shown here. The inset shows a close-up of one of them.

FIGURE 11.34 Small moons within the rings have important effects on ring structure (*Cassini* photos).

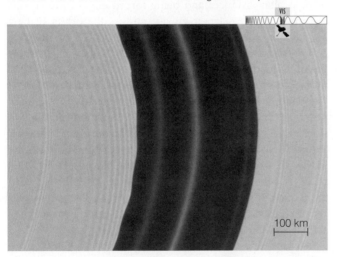

100 km

FIGURE 11.35 This *Cassini* image of a gap in Saturn's rings shows how gravitational tugs from the moon Pan create waves that shape the rings over vast distances.

How do other jovian ring systems compare to Saturn's?

The rings of Jupiter, Uranus, and Neptune are so much fainter than Saturn's that it took almost four centuries longer to discover them. Ring particles in these three systems are far less numerous, generally smaller, and much darker. Despite these differences, a family portrait of the jovian ring systems shows many similarities (**FIGURE 11.36**). All rings lie in their planet's equatorial plane. Particle orbits are nearly circular, with small orbital tilts relative to the equator. Individual rings and gaps are probably shaped by gap moons, shepherd moons, and orbital resonances.

Uranus's rings were discovered in 1977 during observations of a *stellar occultation*—a star passing behind Uranus as

seen from Earth. During the occultation, the star "blinked" on and off nine times before it disappeared behind Uranus and nine more times as it emerged. Scientists concluded that these nine "blinks" were caused by nine thin rings encircling Uranus. Similar observations of stars passing behind Neptune yielded more confounding results: Rings appeared to be present at some times but not at others. Could Neptune's rings be incomplete or transient?

The *Voyager* spacecraft provided some answers. *Voyager* cameras first discovered thin rings around Jupiter in 1979. After next providing incredible images of Saturn's rings, *Voyager 2* photographed the rings of Uranus as it flew past in 1986. In 1989, *Voyager 2* passed by Neptune and found that it does, in fact, have partial rings—at least when seen from Earth. The space between the ring segments is filled with dust not detectable from Earth. In addition to the dust within the rings of Neptune, *Voyager 2* detected vast dust sheets between the widely separated rings of both Uranus and Neptune.

The differences among the ring systems present us with unsolved mysteries. The larger size, higher reflectivity, and much greater number of particles in Saturn's rings compel us to wonder whether different processes might be at work there. We do not know why Uranus's thin rings have a slight tilt and slightly eccentric orbits, nor do we know why Neptune's rings contain dusty regions that make them appear as partial rings when viewed from Earth. Scientists hope that we may someday learn the answers with new missions to visit Uranus and Neptune, but none are yet under development. Meanwhile, advances in ground-based observing, including the new technology of adaptive optics [**Section 6.3**], are providing views that in some cases rival those obtained from the *Voyager* spacecraft.

FIGURE 11.36 Four ring systems (not to scale). The rings differ in appearance and in the composition and sizes of the ring particles. (Jupiter: Keck Telescope, infrared; Saturn: *Cassini*, visible; Uranus: Hubble Space Telescope, infrared; Neptune: *Voyager*, visible.)

Why do the jovian planets have rings?

Where do rings come from? Did they form together with the planets out of the solar nebula, or are they a more recent phenomenon? An important clue is that the rings lie close to their planet in a region where tidal forces are very strong.

Within two to three planetary radii of any planet, the tidal forces tugging apart an object become comparable to the gravitational forces holding it together. (This region is often called the *Roche tidal zone.*) Only relatively small objects held together by nongravitational forces—such as the electromagnetic forces that hold solid rock, spacecraft, and human beings together—can avoid being ripped apart by the strong tidal forces in this region. Because a large moon would be ripped apart by tidal forces in the region of the rings, scientists once suspected that Saturn's rings formed after a large moon came too close to the planet. However, moons don't simply "wander" away from their orbits, and a large moon could have been deflected toward Saturn only by a close gravitational encounter with an unusually large asteroid or comet that happened to pass nearby. Such encounters should be extremely rare. Back in the days when Saturn was the only planet known to have rings, it might have seemed possible that such a rare encounter was responsible for the rings. The discovery of rings around the other jovian planets all but ruled out this possibility, because it seems inconceivable that such an unlikely event could have occurred in all four cases.

If the rings did not originate with the destruction of a large moon, where did all the ring particles come from? Another idea that once seemed reasonable was that the ring particles might be leftover chunks of rock and ice that condensed in the disks of gas that orbited each jovian planet when it was young. This would explain why all four jovian planets have rings, because tidal forces near each planet would have prevented these chunks from accreting into a full-fledged moon. However, we now know that the ring particles cannot be leftovers from the births of the planets, because they could not have survived for billions of years. Ring particles are continually being ground down in size, primarily by the impacts of the countless sand-size particles that orbit the Sun—the same types of particles that become meteors in Earth's atmosphere and cause micrometeorite impacts on the Moon [**Section 9.3**]. Millions of years of such tiny impacts would have ground the existing ring particles to dust long ago. The dust particles could not have survived either, because over time they are slightly slowed in their orbits by the pressure of sunlight, and this slowing eventually causes them to spiral into their planet. Ring particles may also be lost in other ways. For example, the thermosphere of Uranus extends up into the ring region, exerting atmospheric drag that also causes ring particles to spiral slowly into the planet. Between the effects of micrometeorite impacts and other processes, *none* of the abundant small particles that now occupy the jovian rings can have been there since the solar system formed more than 4 billion years ago.

We are left with only one reasonable possibility: New particles must be continually supplied to the rings to replace those that are destroyed. These new particles must come from a source that lies in each planet's equatorial plane. The most likely source is numerous small "moonlets"—moons the size

jovian planet

Tidal forces near the planet prevent small moonlets from accreting into larger moons.

Moonlets are occasionally disrupted by impacts.

Ongoing small impacts blast off dust and debris to form the rings.

FIGURE 11.37 This illustration summarizes the origin of rings around the jovian planets.

of gap moons (see Figure 11.34)—that formed in the disks of material orbiting the young jovian planets. As with the ring particles themselves, tiny impacts are gradually grinding away these small moons, but they are large enough to still exist despite $4\frac{1}{2}$ billion years of such sandblasting.

The small moons contribute ring particles in two ways. First, each tiny impact releases particles from a small moon's surface, and these released particles become new, dust-size ring particles. Ongoing impacts ensure that some ring particles are present at all times. Second, occasional larger impacts can shatter a small moon completely, creating a supply of boulder-size ring particles. The frequent tiny impacts then slowly grind these boulders into smaller ring particles. Some of these particles are "recycled" by forming into small clumps, only to come

apart again later on; others are ground down to dust and slowly spiral onto their planet. In summary, all ring particles ultimately come from the gradual dismantling of small moons that formed during the birth of the solar system (**FIGURE 11.37**).

The collisions that shatter small moons and generate large ring particles must occur only occasionally and at essentially random times, which means that the numbers and sizes of particles in any particular ring system may vary dramatically over millions and billions of years. Rings could be broad and bright when they are full of particles and almost invisible when particles are few. The brilliant spectacle of Saturn's rings could be a special treat of our epoch, one that might not have been seen a billion years ago and that might not last long on the time scale of our solar system.

The Big Picture

Putting Chapter 11 into Context

In this chapter, we saw that the jovian planets really are a different kind of planet and, indeed, a different kind of planet system. As you continue your study of the solar system, keep in mind the following "big picture" ideas:

- The jovian planets dwarf the terrestrial planets. Even some of their moons are as large as terrestrial worlds.

- The jovian planets may lack solid surfaces on which geology can occur, but they are interesting and dynamic worlds with rapid winds, huge storms, strong magnetic fields, and interiors in which common materials behave in unfamiliar ways.

- Despite their relatively small sizes and frigid temperatures, many jovian moons are geologically active by virtue of their icy compositions—a result of their formation in the outer regions of the solar nebula—and tidal heating.

- Ring systems probably owe their existence to small moons formed in the disks of gas that produced the jovian planets billions of years ago. The rings we see today are composed of particles liberated from those moons quite recently.

- Understanding the jovian planet systems forced us to modify many of our earlier ideas about the solar system by adding the concepts of ice geology, tidal heating, and orbital resonances. Each new set of circumstances that we discover offers new opportunities to learn how our universe works.

11.1 A DIFFERENT KIND OF PLANET

- **Are jovian planets all alike?** Jupiter and Saturn are made almost entirely of hydrogen and helium, while Uranus and Neptune are made mostly of hydrogen compounds mixed with metals and rock. These differences arose because all four planets started from ice-rich planetesimals of about the same size, but captured different amounts of hydrogen and helium gas from the solar nebula.

- **What are jovian planets like on the inside?** The jovian planets have layered interiors with very high internal tempera- tures and pressures. All have a core about 10 times as massive as Earth, consisting of hydrogen compounds, metals, and rock. They differ mainly in their surrounding layers of hydrogen and helium, which can take on unusual forms under the extreme internal conditions of the planets.

- **What is the weather like on jovian planets?** The jovian planets all have multiple cloud layers that give them distinctive colors, fast winds, and large storms. Some storms, such as Jupiter's **Great Red Spot,** can apparently rage for centuries or longer.

- **Do jovian planets have magnetospheres like Earth's?** Jupiter has a magnetic field 20,000 times stronger than Earth's, which leads to an enormous magnetosphere. Many of the particles in Jupiter's magnetosphere come from volcanic eruptions on Io. Other jovian planets also have magnetic fields and magnetospheres, but they are weaker and smaller than Jupiter's.

11.2 A WEALTH OF WORLDS: SATELLITES OF ICE AND ROCK

- **What kinds of moons orbit the jovian planets?** We can categorize the more than 170 known moons as small, medium-size, or large. Most of the medium-size and large moons probably formed with their planet in the disks of gas that surrounded the jovian planets when they were young. Smaller moons are often captured asteroids or comets.

- **Why are Jupiter's Galilean moons so geologically active?** Io is the most volcanically active object in the solar system, thanks to an interior kept hot by **tidal heating**—which occurs because Io's close orbit is made elliptical by **orbital resonances** with other moons. Europa (and possibly Ganymede) may have a deep, liquid water ocean under its icy crust, also thanks to tidal heating. Callisto is the least geologically active, since it has no orbital resonance or tidal heating.

- **What is remarkable about Titan and other major moons of the outer solar system?** Many medium-size and large moons show a surprisingly high level of past or present volcanism or tectonics. Titan has a thick atmosphere and ongoing erosion, and Enceladus is also geologically active today. Triton, which apparently was captured by Neptune, also shows signs of recent geological activity.

- **Why are small icy moons more geologically active than small rocky planets?** Ices deform and melt at much lower temperatures than rock, allowing icy volcanism and tectonics at surprisingly low temperatures. In addition, some jovian moons have a heat source—tidal heating—that is not important for the terrestrial worlds.

11.3 JOVIAN PLANET RINGS

- **What are Saturn's rings like?** Saturn's rings are made up of countless individual particles, each orbiting Saturn independently like a tiny moon. The rings lie in Saturn's equatorial plane, and they are extremely thin. Moons within and beyond the rings create many ringlets and gaps in part through orbital resonances.

- **How do other jovian ring systems compare to Saturn's?** The other jovian planets have ring systems that are much fainter in photographs. Their ring particles are generally smaller, darker, and less numerous than Saturn's ring particles.

- **Why do the jovian planets have rings?** Ring particles probably come from the dismantling of small moons formed in the disks of gas that surrounded the jovian planets billions of years ago. Small ring particles come from countless tiny impacts on the surfaces of these moons, while larger ones come from impacts that shatter the moons.

Use the following questions to check your understanding of some of the many types of visual information used in astronomy. Answers are provided in Appendix J. For additional practice, try the Chapter 11 Visual Quiz at MasteringAstronomy®.

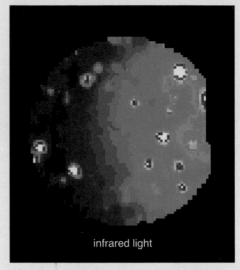

Left: Approximate colors of Io in visible light; black spots are volcanoes that are active or have recently gone inactive. Right: Infrared thermal emission from Io; bright spots are active volcanoes. (Both images are from Galileo *data, but taken at different times.)*

1. What do the colors in the right image represent?
   a. the actual colors of Io's surface
   b. the colors we would see if we had infrared vision
   c. the intensity of the infrared light
   d. regions of different chemical composition on the surface
2. Which color in the right image represents the highest temperatures?
   a. blue   b. green   c. orange   d. red   e. white
3. The right image was obtained when only part of Io was in sunlight. Based on the colors which part of the surface was in sunlight?
   a. the left side
   b. the right side
   c. only the peaks of the volcanoes

4. By comparing the two images, what can you conclude about Io's volcanoes?
   a. Every black spot in the visible image has a bright spot in the infrared image, so all of Io's volcanoes were active when the photos were taken.
   b. There are more black spots in the visible image than bright spots in the infrared image, so many of Io's volcanoes were inactive when the photos were taken.
   c. There are more bright spots in the infrared image than black spots in the visible image, so new eruptions must have started after the visible photo was taken.

For instructor-assigned homework go to MasteringAstronomy®.

MasteringAstronomy®

REVIEW QUESTIONS

Short-Answer Questions Based on the Reading

1. Briefly describe how differences in composition among the jovian planets can be traced to their formation.
2. Why is Jupiter so much more dense than Saturn? Could a planet be smaller in size than Jupiter but greater in mass?
3. Briefly describe the interior structure of Jupiter and why it is layered in this way. How do the interiors of the other jovian planets compare to that of Jupiter?
4. Why does Jupiter have so much internal heat? What generates internal heat on other jovian planets?
5. Briefly describe Jupiter's atmospheric structure and cloud layers. How do the structures and clouds differ on the other jovian planets?

6. How do clouds contribute to Jupiter's colors? Why are Saturn's colors more subdued? Why are Uranus and Neptune blue?
7. Briefly describe Jupiter's weather patterns. What is the *Great Red Spot*?
8. Why does Jupiter have such a strong magnetic field? Describe a few features of Jupiter's magnetosphere and compare it to the magnetospheres of the other jovian planets.
9. Briefly describe how we categorize jovian moons by size. What is the origin of most of the medium-size and large moons? What is the origin of many of the small moons?
10. Describe key features of Jupiter's four Galilean moons, and the role of *tidal heating* and *orbital resonances* in shaping these features.
11. Describe the atmosphere and surface features of Titan. How is Titan's landscape similar to Earth's? How is it different?

12. Summarize the evidence for and some of the mysteries of past or present geological activity on the medium-size moons of Saturn and Uranus.

13. Why do we think Triton is a captured moon?

14. Briefly explain why icy moons can have active geology at much smaller sizes than rocky worlds.

15. What are planetary rings made of, and how do they differ among the four jovian planets? Briefly describe the effects of gap moons and orbital resonances on ring systems.

16. Explain why we think that ring particles must be replenished over time. Will the jovian planet rings always look the same?

TEST YOUR UNDERSTANDING

Surprising Discoveries?

Suppose someone claimed to make the discoveries described below. (These are not real discoveries.) Decide whether each discovery should be considered reasonable or surprising. More than one right answer may be possible, so explain your answer clearly.

17. Saturn's core is pockmarked with impact craters and dotted with volcanoes erupting basaltic lava.

18. Neptune's deep blue color is not due to methane, as previously thought, but instead is due to its surface being covered with an ocean of liquid water.

19. A jovian planet in another star system has a moon as big as Mars.

20. An extrasolar planet is made primarily of hydrogen and helium. It has approximately the same mass as Jupiter but is the same size as Neptune.

21. A previously unknown moon orbits Jupiter outside the orbits of other known moons. It is the smallest of Jupiter's moons but has several large, active volcanoes.

22. A previously unknown moon orbits Neptune in the planet's equatorial plane and in the same direction that Neptune rotates, but it is made almost entirely of metals such as iron and nickel.

23. An icy, medium-size moon orbits a jovian planet in a star system that is only a few hundred million years old. The moon shows evidence of active tectonics.

24. A jovian planet is discovered in a star system that is much older than our solar system. The planet has no moons but has a system of rings as spectacular as the rings of Saturn.

25. Future observations discover rainfall of liquid water on Titan.

26. During a future mission to Uranus, scientists discover that it is orbited by another 20 previously unknown moons.

Quick Quiz

Choose the best answer to each of the following. Explain your reasoning with one or more complete sentences.

27. Which lists the jovian planets in order of increasing distance from the Sun? (a) Jupiter, Saturn, Uranus, Pluto (b) Saturn, Jupiter, Uranus, Neptune (c) Jupiter, Saturn, Uranus, Neptune

28. Why does Neptune appear blue and Jupiter red? (a) Neptune is hotter, which gives bluer thermal emission. (b) Methane in Neptune's atmosphere absorbs red light. (c) Neptune's air molecules scatter blue light, much as Earth's atmosphere does.

29. Why is Jupiter denser than Saturn? (a) It has a larger proportion of rock and metal. (b) It has a larger proportion of hydrogen. (c) Its higher mass and gravity compress its interior.

30. Some jovian planets give off more energy than they receive because of (a) fusion in their cores. (b) tidal heating. (c) ongoing contraction or differentiation.

31. The main ingredients of most satellites of the jovian planets are (a) rock and metal. (b) hydrogen compound ices. (c) hydrogen and helium.

32. Why is Io more volcanically active than our moon? (a) Io is much larger. (b) Io has a higher concentration of radioactive elements. (c) Io has a different internal heat source.

33. What is unusual about Triton? (a) It orbits its planet backward. (b) It does not keep the same face toward its planet. (c) It is the only moon with its own rings.

34. Which moon shows evidence of rainfall and erosion by some liquid substance? (a) Europa (b) Titan (c) Ganymede

35. Saturn's many moons affect its rings through (a) tidal forces. (b) orbital resonances. (c) magnetic field interactions.

36. Saturn's rings (a) have looked basically the same since they formed along with Saturn. (b) were created long ago when tidal forces tore apart a large moon. (c) are continually supplied with new particles by impacts with small moons.

PROCESS OF SCIENCE

Examining How Science Works

37. *Europan Ocean.* Scientists strongly suspect that Europa has a subsurface ocean, even though we cannot see through the surface ice. Briefly explain why scientists think the ocean exists. Is this "belief" in a Europan ocean scientific? Explain.

38. *Breaking the Rules.* As discussed in Chapter 9, the geological "rules" for the terrestrial worlds tell us that a world as small as Io should not have any geological activity. However, the *Voyager* images of Io's volcanoes proved that the old "rules" had been wrong. Based on your understanding of the nature of science [**Section 3.4**], should this be seen as a failure in the process of the science? Defend your opinion.

39. *Unanswered Questions.* Choose one unanswered question about one of the jovian planets or its moons. Write a few paragraphs discussing the question and the specific types of evidence we would need to answer it.

GROUP WORK EXERCISE

40. *Comparing Jovian Moons.* Comparing the masses, radii, and densities of jovian moons reveals clues about their composition and their history. In this exercise, your team will draw on the data in Appendix E to develop a hypothesis about the moons of Jupiter. Before you begin, assign the following roles to the people in your group: *Scribe* (collects data and takes notes on the group's activities), *Proposer* (proposes hypotheses and explanations of the data), *Skeptic* (points out weaknesses in the hypotheses and explanations), *Moderator* (leads group discussion and makes sure everyone contributes). Each person should write down the answers for each part of the exercise.
a. *Scribe* collects data on Jupiter's four largest moons from Table E.2 in Appendix E and determines which of Jupiter's moons has the greatest density. **b.** *Moderator* uses Table E.2 to determine what other solar system moon most resembles the moon from part a in mass, radius, and density. **c.** *Proposer* proposes a hypothesis about the composition of the moon from part a, based on its resemblance to the moon from part b. **d.** *Skeptic* questions the hypothesis from part c, stating concerns about its viability. **e.** *Scribe* and *Moderator* use Table E.2 to determine whether there is a trend in density with orbital distance among the major moons of Jupiter and briefly describe any trends they discover. **f.** *Proposer* offers a hypothesis that accounts for any trend found in part e. **g.** *Skeptic* raises questions about the revised hypothesis, stating any reasons to doubt it. **h.** Together, the team develops and describes an experiment that could test the hypotheses in parts c and f.

INVESTIGATE FURTHER

In-Depth Questions to Increase Your Understanding

Short-Answer/Essay Questions

41. *The Importance of Rotation.* Suppose the material that formed Jupiter came together without any rotation- so that no "jovian nebula" formed and the planet today wasn't spinning. How else would the jovian system be different? Think of as many effects as you can, and explain each in a sentence.

42. *The Great Red Spot.* Based on the infrared and visible images in Figure 11.9, is Jupiter's Great Red Spot warmer or cooler than nearby clouds? Is it higher or lower in altitude than the nearby clouds? Explain.

43. *Comparing Jovian Planets.* You can do comparative planetology armed only with telescopes and an understanding of gravity.
    a. The small moon Amalthea orbits Jupiter at about the distance in kilometers at which Mimas orbits Saturn, yet Mimas takes almost twice as long to orbit. From this observation, what can you conclude about how Jupiter and Saturn differ? Explain.
    b. Jupiter and Saturn are not very different in radius. When you combine this information with your answer to part a, what can you conclude? Explain.

44. *Minor Ingredients Matter.* Suppose the jovian planets' atmospheres were composed only of hydrogen and helium, with no hydrogen compounds at all. How would the atmospheres be different in terms of clouds, color, and weather? Explain.

45. *Galilean Moon Formation.* Look up the densities of Jupiter's four Galilean moons in Appendix E, and notice that they follow a trend with distance from Jupiter. Based on what you've learned about condensation in the solar nebula, can you suggest a reason for this trend among the Galilean moons? Next, compare the densities of the moons with the planetary densities in Table E.1. Based on the comparison, do you think it was as hot toward the center of the nebula surrounding Jupiter as it was at the center of the solar nebula? Explain.

46. *Super Jupiter.* Suppose that the solar wind had not cleared the solar nebula until much later in our solar system's history, so that Jupiter accumulated 40 times as much mass as it actually has today.
    a. Make a sketch showing how the interior layers of this "Super Jupiter" would compare to the interior layers of the real Jupiter shown in Figure 11.4. **b.** How would you expect the cloud layers and colors to differ on Super Jupiter? Explain. **c.** How would you expect the wind speeds to be different on Super Jupiter? Explain. **d.** How would you expect the magnetic field to be different on Super Jupiter? Explain. **e.** Do you think Super Jupiter would have more or less excess heat than the real Jupiter? Explain.

47. *Observing Project: Jupiter's Moons.* Using binoculars or a small telescope, view the moons of Jupiter. Make a sketch of what you see, or take a photograph. Repeat your observations several times (nightly, if possible) over a period of a couple of weeks. Can you determine which moon is which? Can you measure the moons' orbital periods? Can you determine their approximate distances from Jupiter? Explain.

48. *Observing Project: Saturn's Rings.* Using binoculars or a small telescope, view the rings of Saturn. Make a sketch of what you see, or take a photograph. What season is it in Saturn's northern hemisphere? How far do the rings extend above Saturn's atmosphere? Can you identify any gaps in the rings? Describe any other features you notice.

Quantitative Problems

Be sure to show all calculations clearly and state your final answers in complete sentences.

49. *Disappearing Moon.* Io loses about a ton (1000 kilograms) of sulfur dioxide per second to Jupiter's magnetosphere.
    a. At this rate, what fraction of its mass would Io lose in 4.5 billion years? **b.** Suppose sulfur dioxide currently makes up 1% of Io's mass. When will Io run out of this gas at the current loss rate?

50. *Ring Particle Collisions.* Each ring particle in the densest part of Saturn's rings collides with another about every 5 hours. If a ring particle survived for the age of the solar system, how many collisions would it undergo?

51. *Prometheus and Pandora.* These two moons orbit Saturn at 139,350 and 141,700 kilometers, respectively.
    a. Using Newton's version of Kepler's third law, find their two orbital periods. Find the percent difference in their distances from Saturn and in their orbital periods. **b.** Consider the two in a race around Saturn: In one Prometheus orbit, how far behind is Pandora (in units of time)? In how many Prometheus orbits will Pandora have fallen behind by one of its own orbital periods? Convert this number of periods back into units of time. This is how often the satellites pass by each other.

52. *Orbital Resonances.* Using the data in Appendix E, identify the orbital resonance relationship between Titan and Hyperion. (*Hint:* If the orbital period of one were 1.5 times the orbital period of the other, we would say that they were in a 3:2 resonance.) Which medium-size moon is in a 2:1 resonance with Enceladus?

53. *Titanic Titan.* What is the ratio of Titan's mass to that of the other satellites of Saturn whose masses are listed in Appendix E? Calculate the strength of gravity on Titan compared to that on Mimas. Comment on how this affects the possibility of an atmosphere on each.

54. *Titan's Evolving Atmosphere.* Titan's exosphere lies nearly 1400 kilometers above its surface. What is the escape velocity from this altitude? What is the thermal speed of a hydrogen atom at the exospheric temperature of about 200 K? Use these answers (and the method of Mathematical Insight 10.2) to comment on whether thermal escape of hydrogen is likely to be important for Titan.

55. *Saturn's Thin Rings.* Saturn's ring system is more than 270,000 kilometers wide and only a few tens of meters thick; let's assume 50 meters thick for this problem. Assuming the rings could be shrunk down so that their diameter was the width of a dollar bill (6.6 cm), how thick would the rings be? Compare your answer to the actual thickness of a dollar bill (0.01 cm).

Discussion Questions

56. *Jovian Planet Mission.* We can study terrestrial planets up close by landing on them, but jovian planets have no surfaces to land on. Suppose that you were in charge of planning a long-term mission to "float" in the atmosphere of a jovian planet. Describe the technology you would use and how you would ensure survival for any people assigned to this mission.

57. *Pick a Moon.* Suppose you could choose any one moon to visit in the solar system. Which one would you pick, and why? What dangers would you face in your visit to this moon? What kinds of scientific instruments would you want to bring along for studies?

Web Projects

58. *News from Cassini.* Find the latest news about the *Cassini* mission to Saturn. What is the current mission status? Write a short report about the mission's status and recent findings.

59. *Ocean of Europa.* Investigate plans for future study of Europa's ocean, either from Earth or with spacecraft. Write a short summary of the plans, how they might help us learn whether Europa really has an ocean, and what the ocean might contain if it exists.

Comet McNaught and the Milky Way over Patagonia, Argentina (2007). The fuzzy patches above the comet tail are the Magellanic Clouds, satellite galaxies of the Milky Way.

12 ASTEROIDS, COMETS, AND DWARF PLANETS
THEIR NATURE, ORBITS, AND IMPACTS

LEARNING GOALS

12.1 ASTEROIDS AND METEORITES
- What are asteroids like?
- Why is there an asteroid belt?
- How are meteorites related to asteroids?

12.2 COMETS
- What are comets like?
- Where do comets come from?

12.3 PLUTO: LONE DOG NO MORE
- How big can a comet be?
- What are the large objects of the Kuiper belt like?

12.4 COSMIC COLLISIONS: SMALL BODIES VERSUS THE PLANETS
- Have we ever witnessed a major impact?
- Did an impact kill the dinosaurs?
- Is the impact threat a real danger or media hype?
- How do the jovian planets affect impact rates and life on Earth?

> *As we look out into the Universe and identify the many accidents of physics and astronomy that have worked to our benefit, it almost seems as if the Universe must in some sense have known that we were coming.*
>
> —*Freeman Dyson*

Asteroids and comets might at first seem insignificant compared to the planets and moons we've discussed so far, but there is strength in numbers, and small bodies probably number in the trillions. The appearance of a comet has more than once altered the course of human history when our ancestors acted on superstitions related to the sighting. More profoundly, asteroids or comets falling to Earth have scarred our planet with impact craters and have altered the course of biological evolution. Asteroids and comets are also important scientifically: As remnants from the birth of our solar system, they teach us about how our solar system formed.

In this chapter, we will explore the small bodies of our solar system. We'll consider asteroids and the pieces of them that fall to Earth as meteorites, comets, and dwarf planets like Pluto and Eris. We'll also explore the dramatic effects of the occasional collisions between small bodies and large planets.

12.1 ASTEROIDS AND METEORITES

The objects we've studied so far—terrestrial planets, jovian planets, and large moons—have changed dramatically since their formation. We have used these objects to infer much about the history of our solar system, but they no longer hold direct clues to what the solar system was like when it was born. In contrast, many small bodies remain much as they were when they first formed, some 4.5 billion years ago. Comets, asteroids, and meteorites therefore have the story of our solar system's birth encoded in their compositions, locations, and numbers.

Using these small bodies to understand planetary formation is a bit like picking through the trash in a carpenter's shop to see how furniture is made. By studying the scraps, sawdust, paint chips, and glue, we can develop some understanding of how the carpenter builds furniture. Asteroids, comets, and meteorites are the "scraps" left over from the formation of our solar system, so their study can give us an understanding of how the planets and larger moons came to exist.

We'll begin our study of small bodies by focusing on asteroids and meteorites, which are closely related. As we discussed in Chapter 8, asteroids are planetesimals (or fragments of planetesimals) left over from the birth of our solar system. Asteroids and meteorites are therefore closely related: Most meteorites are pieces of asteroids that orbited the Sun for billions of years before falling to Earth.

What are asteroids like?

The word *asteroid* means "starlike," but asteroids are starlike only in their appearance through a telescope. Asteroids are virtually undetectable by the naked eye and remained unnoticed for almost two centuries after the invention of the telescope. The first asteroids were discovered about 200 years ago, and it took 50 years to discover the first 10. Today, advanced telescopes can discover far more than that in a single night, and around 600,000 asteroids have been cataloged. Asteroids can be recognized in telescopic images because they move relative to the stars (**FIGURE 12.1**).

Newly discovered asteroids first get a provisional name based on the discovery year and month and order of discovery; for example, the first asteroid discovered in January 2015 would be called Asteroid 2015 AA. (Once letters run out, numbers are added after the letters.) An asteroid's orbit can be calculated from the law of gravity [**Section 4.4**], even after observation of only a fraction of a complete orbit, but less than half of the asteroids identified in images have been tracked long enough for their orbits to be calculated. In those cases, the discoverer may choose a name for the asteroid, subject to approval by the International Astronomical Union. The earliest discovered asteroids bear the names of mythological figures. More recent discoveries often carry names of scientists, cartoon heroes, pets, rock stars, or textbook authors.

Asteroid Sizes, Numbers, and Shapes Asteroids come in a wide range of sizes. The largest, Ceres, is just under 1000 kilometers in diameter, which is a little over a quarter of the Moon's diameter. About a dozen others are large enough that we would call them medium-size moons if they orbited a planet. Smaller asteroids are far more numerous. There are probably more than a million asteroids with diameters greater than 1 kilometer, and many more even smaller in size. Despite their large numbers, asteroids don't add up to much in total mass. If we could put all the asteroids together and allow gravity to compress them into a sphere, they'd make an object less than 2000 kilometers in diameter—just over half the diameter of our Moon.

FIGURE 12.1 interactive photo Because asteroids orbit the Sun, they move through our sky relative to the stars. In this long-exposure photograph, stars show up as distinct dots, while the motion of an asteroid relative to the stars makes it show up as a short streak.

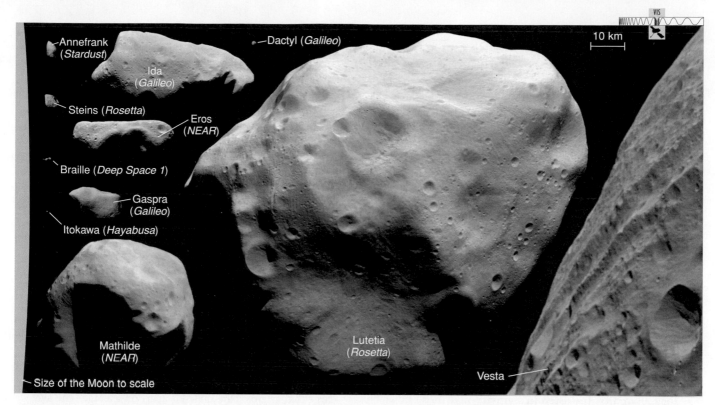

FIGURE 12.2 Images of all the asteroids visited by spacecraft as of 2013 shown to scale. Some objects were the destinations of dedicated missions; others were imaged on the way to different targets. The size of Earth's Moon is included for comparison.

Asteroid shapes depend on the strength of the asteroid's gravity. If an object is large enough, over time gravity can mold it into a spherical shape, even if the object is solid throughout. Some large bodies may also have had enough heat to have been molten inside at one time, allowing gravity to work faster. Among asteroids, Ceres is large enough for gravity to have compressed it into a roughly spherical shape—allowing it to qualify as a *dwarf planet* (see Special Topic, page 8). The next two largest asteroids are somewhat "squashed" (Pallas, with equatorial diameter 580 kilometers and polar diameter 500 kilometers; and Vesta, with equatorial diameter 570 kilometers and polar diameter 460 kilometers). For smaller asteroids, gravity is too weak to reshape the rocky material. Much like potatoes, these asteroids have a variety of shapes. Some objects that appear to be single asteroids are probably two or more distinct objects held in contact by a weak gravitational attraction, while other small asteroids are little more than weakly bound piles of rubble.

Measuring Sizes and Shapes Given that most asteroids appear even to powerful telescopes as little more than moving dots of light similar to that in Figure 12.1, you may wonder how we can make general statements about their sizes and shapes. In about a dozen cases, we've sent spacecraft to visit asteroids, giving us clear pictures of what they look like (**FIGURE 12.2**). Note the numerous craters, confirming that asteroids, like planets and moons, have been battered by impacts.

For the majority of asteroids, which we can observe only through telescopes, we can learn about sizes through careful measurements of brightness. Asteroids shine with reflected sunlight, so an asteroid's brightness in our sky depends on its size, distance, and reflectivity. For example, if two asteroids at the same distance have the same reflectivity, the one that appears brighter must be larger in size. We can determine an asteroid's distance from its position in its orbit. Determining reflectivity is more difficult, but we can do it by comparing the asteroid's brightness in visible light to its brightness in infrared light. The visible light is reflected sunlight, while the infrared light is thermal radiation emitted by the asteroid itself [**Section 5.4**]; the latter depends on the asteroid's temperature, which in turn depends on how much sunlight it absorbs. Comparing the visible and infrared brightness therefore tells us the proportions of incoming sunlight that an asteroid reflects and absorbs. Once we know these proportions, the asteroid's brightness and distance tell us its size. This technique has been used to make size estimates for more than a thousand asteroids.

THINK ABOUT IT

Suppose you discover two asteroids of similar size that are equally bright in visible light, but infrared observations tell you that Asteroid 1 is more reflective than Asteroid 2. Which one is farther away? Explain.

Determining asteroid shapes without close-up photographs is difficult but not impossible. In some cases, we can determine shape by monitoring brightness variations as an asteroid rotates: A nonspherical asteroid with a uniformly bright surface will reflect more light when it presents its larger side toward the Sun and our telescopes. In other cases, astronomers have determined shapes by bouncing radar signals off asteroids that have passed close to Earth.

Asteroid Compositions Analysis of light can also tell us about asteroid compositions. Recall that spectra of distant objects contain spectral lines that are essentially "fingerprints" left by the objects' chemical constituents [**Section 5.4**]. Thousands of asteroids have been analyzed through spectroscopy. The results are consistent with what we expect from our theory of solar system formation: Asteroids are made mostly of metal and rock, because they condensed within the frost line in the solar nebula. Those near the outskirts of the asteroid belt contain larger proportions of dark, carbon-rich material, because this material was able to condense at the relatively cool temperatures found in this region of the solar nebula but not in the regions closer to the Sun; some even contain small amounts of water, telling us that they formed close to the frost line. A few asteroids appear to be made mostly of metals, such as iron, suggesting that they may be fragments of the metal cores of shattered worlds.

Asteroid Masses and Densities The only direct way to measure a distant object's mass is to observe its gravitational effect on another object. We've been able to observe the gravitational effects of asteroids in only a few cases: those in which a spacecraft has passed by close enough to have its trajectory altered by the asteroid's gravity, and those in which asteroids have even smaller asteroids as tiny, orbiting "moons." For asteroids with moons, such as the asteroid Ida with its tiny moon Dactyl (see Figure 12.2) and the asteroid Eugenia in **FIGURE 12.3**, we can determine the mass of the central asteroid by applying Newton's version of Kepler's third law to the orbital characteristics of the moon [**Section 4.4**]. Astronomers have found a few dozen asteroid masses by this method.

Once we have measured an asteroid's mass and size, we can calculate its density. The density can offer valuable insights into the asteroid's origin and makeup. For example, the asteroid Eros (see Figure 12.2) has a density of 2.4 g/cm³, which is close to the value expected for solid rock; notice the subtle pattern of surface ridges and troughs that hints at a set of parallel faults and fractures in its interior. In contrast, the asteroid Mathilde has a density of only 1.5 g/cm³, which is so low that it must be a loosely bound "rubble pile," held together by its weak gravity, rather than a solid chunk of rock. Support for this idea comes from the fact that Mathilde has a huge, central crater (see Figure 12.2): The impact that formed this crater would have shattered a solid object, but a "rubble pile" could absorb the shock and remain intact.

Asteroid Orbits The vast majority of the asteroids that have been cataloged are located in the asteroid belt between the orbits of Mars and Jupiter (**FIGURE 12.4**). All asteroids orbit the Sun in the same direction as the planets, but their orbits tend to be more elliptical and more highly inclined to the ecliptic plane (up to 20°–30°) than those of planets.

Science fiction movies often show the asteroid belt as a crowded and hazardous place, but the asteroid belt is so large that asteroids are quite far apart on average. The average distance between asteroids in the asteroid belt is millions of kilometers—the equivalent of grains of sand separated by kilometers.

Not all asteroids are located in the asteroid belt. Two sets of asteroids, called *Trojan asteroids*, share Jupiter's 12-year orbit around the Sun. One clump of Trojan asteroids always stays 60° ahead of Jupiter in its orbit and the other clump always

FIGURE 12.3 This set of images, made by a telescope using adaptive optics [**Section 6.3**], shows a small moon observed in five positions in its orbit around the asteroid Eugenia. The moon completes one orbit around the asteroid every 4.7 days. The asteroid itself was blocked out during the observations and is shown as the gray oval in the image.

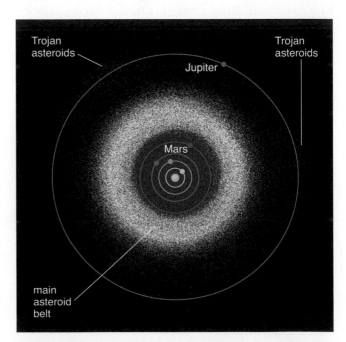

FIGURE 12.4 This figure shows the positions of more than 150,000 asteroids on a single night. To scale, the asteroids themselves would be much smaller than shown. The asteroids that share Jupiter's orbit, found 60° ahead of and behind Jupiter, are called *Trojan asteroids.*

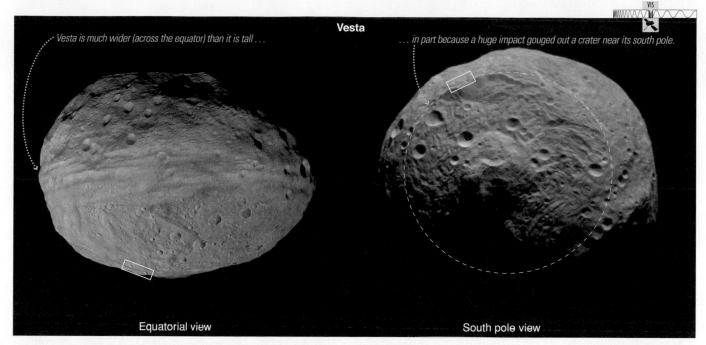

Vesta

Vesta is much wider (across the equator) than it is tall ...

... in part because a huge impact gouged out a crater near its south pole.

Equatorial view

South pole view

FIGURE 12.5 Global images of Vesta taken by the *Dawn* spacecraft. The white boxes show the same pair of craters in the two views, and the dashed circle is the outline of the huge polar crater.

stays 60° behind. The number of Trojan asteroids could be as large as the number in the asteroid belt, but the greater distance to the Trojan asteroids (and their darker surfaces) makes them more difficult to study or even count from Earth.

A relatively small number of asteroids have orbits that pass through the inner solar system—including the *near-Earth asteroids* that pass near Earth's orbit. These asteroids are probably "impacts waiting to happen." Many asteroids with similar orbits must have hit our planet in the past. In Section 12.4, we will discuss both past impacts and the future dangers these asteroids may pose.

The *Dawn* Mission Our most detailed knowledge of asteroids comes from those few that have been visited by spacecraft. The most in-depth studies are under way with the *Dawn* spacecraft, which spent a year orbiting Vesta, the second-most-massive asteroid (Pallas is slightly larger in size but smaller in mass), in 2011/12. *Dawn* is scheduled to enter orbit of the largest asteroid, Ceres, in 2015.

COMMON MISCONCEPTIONS

Dodge Those Asteroids!

Science fiction movies often show brave spacecraft pilots navigating through crowded fields of asteroids, dodging this way and that as they heroically pass through with only a few bumps and bruises. It's great drama, but not very realistic. The asteroid belt looks crowded in Figure 12.4, but in reality it is an enormous region of space. Despite their large numbers, asteroids are thousands to millions of kilometers apart on average—so far apart that it would take incredibly bad luck to crash into one by accident. Indeed, spacecraft must be carefully guided to fly close enough to an asteroid to take a decent photograph. Future space travelers will have plenty of dangers to worry about, but dodging asteroids is not likely to be one of them.

It's long been suspected that Vesta is massive enough to have undergone differentiation, giving it a metallic core, a low-density rocky crust, and a mantle in between. *Dawn* images (**FIGURE 12.5**) revealed that Vesta is relatively far from being a perfect sphere, as it is nearly 25% wider (equatorial) than it is tall (polar). This irregular shape is probably in large part a result of the impact that gouged out the huge impact crater at Vesta's south pole. Nevertheless, the nonspherical shape makes it debatable whether Vesta should qualify as a dwarf planet, and as of 2012 no official decision had been made.

Vesta's south polar crater is remarkable in many ways. Like many large craters, it has a central mountain formed from the rebound after the impact. But at 23 kilometers tall, this mountain may be the largest in the solar system. The impact that formed this crater excavated so deeply into Vesta's interior that it should have blasted out substantial amounts of rock from the crust and mantle. Measurements from the *Dawn* spacecraft and from the Hubble Space Telescope confirm that the composition of rock near the crater's center is unlike that of surface rocks anywhere else in the solar system, but matches the composition of rock expected deep inside terrestrial worlds. The specific spectral signature of this rock matches that of many small asteroids and of many meteorites that have been found on Earth (including the meteorite shown on the right in Figure 12.9b), suggesting that these asteroids and meteorites are pieces of Vesta that were blasted away by the impact that formed the south polar crater.

Vesta still presents numerous mysteries. For example, the spectral signature of basalt, a volcanic rock, is present on Vesta and in the asteroids and meteorites that appear to have come from it, but *Dawn* scientists have not yet identified any ancient volcanoes or lava flows. Perhaps they've been hidden by billions of years of cratering. In addition, Vesta's equator is ringed with ridges that currently defy explanation.

Why is there an asteroid belt?

The asteroid belt between Mars and Jupiter gets its name from the fact that it is where we find the majority of asteroids. But why are asteroids concentrated in this region, and why didn't a full-fledged planet form here? The answers lie in *orbital resonances* much like those that explain the elliptical orbits of Jupiter's moons Io, Europa, and Ganymede [**Section 11.2**] and the gaps and other structures in Saturn's rings [**Section 11.3**].

Recall that an orbital resonance occurs whenever two objects periodically line up with each other. Because gravity tugs at the objects in the same direction at each alignment, the effects can build up over time. Objects will periodically line up—and hence have an orbital resonance—whenever one object's orbital period is a simple ratio of another object's period, such as $\frac{1}{2}$, $\frac{1}{4}$, or $\frac{2}{5}$.

In the asteroid belt, orbital resonances occur between asteroids and Jupiter, the most massive planet by far. These resonances tend to clear gaps in the asteroid belt, much as resonances clear gaps in Saturn's rings. For example, any asteroid in an orbit in which it takes 6 years to circle the Sun—half of Jupiter's 12-year orbital period—would receive the same gravitational nudge from Jupiter every 12 years and therefore would soon be pushed out of this orbit. The same is true for asteroids with orbital periods of 4 years ($\frac{1}{3}$ of Jupiter's period) and 3 years ($\frac{1}{4}$ of Jupiter's period). Music offers an elegant analogy: When a vocalist sings into an open piano, the strings of notes "in resonance" with the voice—not just the note being sung but also notes with a half or a quarter of the note's frequency—are "nudged" and begin to vibrate.

We can see the results of these orbital resonances on a graph of the numbers of asteroids with various orbital periods (**FIGURE 12.6**). For example, notice the lack of asteroids with periods exactly $\frac{1}{2}$, $\frac{1}{3}$, or $\frac{1}{4}$ of Jupiter's—periods for which the gravitational tugs from Jupiter have cleared gaps in the asteroid belt. (The gaps are often called *Kirkwood gaps,* after their discoverer.)

Orbital resonances with Jupiter likely also explain why no planet ever formed between Mars and Jupiter. When the solar system was forming, this region of the solar nebula probably contained enough rocky material to form another planet as large as Earth or Mars. However, resonances with the young planet Jupiter disrupted the orbits of this region's planetesimals, preventing them from accreting into a full-fledged planet. Over the ensuing 4.5 billion years, ongoing orbital disruptions gradually kicked pieces of this "unformed planet" out of the asteroid belt altogether. Once booted from the asteroid belt, these objects either crashed into a planet or moon or were flung out of the solar system. The asteroid belt thereby lost most of its original mass, which explains why the total mass of all its asteroids is now less than that of any terrestrial planet.

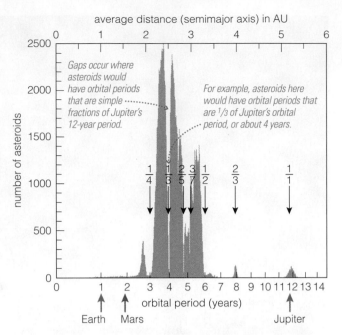

FIGURE 12.6 This graph shows the numbers of asteroids with various orbital periods, which correspond to different average distances from the Sun (labeled along the top). Notice the gaps created by orbital resonances with Jupiter.

The asteroid belt is still undergoing slow change. Jupiter's gravity continues to nudge asteroid orbits, sending asteroids on collision courses with each other and occasionally the planets. A major collision occurs somewhere in the asteroid belt every 100,000 years or so. Over long periods of time, larger asteroids therefore tend to be broken into smaller ones, along with numerous dust-size particles. The asteroid belt has been grinding itself down for more than 4 billion years and will continue to do so for as long as the solar system exists.

Most orbital resonances result in gaps, but some actually gather asteroids. Such is the case for the Trojan asteroids. Their orbits remain stable in the zones 60° ahead of and behind Jupiter, because any asteroid that wanders away from one of these zones is nudged back into the zone by Jupiter's gravity. That explains why we see a grouping of asteroids, rather than a gap, at the $\frac{1}{1}$ resonance (where orbital periods are the same as Jupiter's) in Figure 12.6.

How are meteorites related to asteroids?

Because asteroids are leftovers from the birth of our solar system, studying samples of them should teach us a lot about how Earth and the other planets formed. You might think it would be difficult to obtain such samples, but in fact, we already have tens of thousands of them—the rocks called *meteorites* that fall from the sky.

The Difference Between Meteors and Meteorites

In everyday language, people often use the terms *meteors* and *meteorites* interchangeably. Technically, however, a **meteor** (which means "a thing in the air") is only a flash of light caused by a particle entering our atmosphere at high speed, not the particle itself. Meteors are sometimes called *shooting stars* or *falling stars*, because some people once thought they really were stars falling from the sky.

The vast majority of the particles that make meteors are no larger than peas and burn up completely before reaching the ground. Only in rare cases is a meteor caused by a chunk of rock large enough to survive the plunge through our atmosphere and leave a *meteorite* (which means "associated with meteors") on the ground. Those cases make unusually bright meteors, called *fireballs*. Observers find a few meteorites each year by following the trajectories of fireballs. Meteorites found in this way can be especially valuable to science, because we can trace their trajectories back to learn exactly where they came from and because they can be collected before they have much time to be contaminated by terrestrial material.

Meteorite Falls People didn't always accept that rocks fall from the sky. Stories of such events arose occasionally in human history, and sometimes even affected it. For example, stories of "fallen stars" influenced the philosophy of the ancient Greek scientist Anaxagoras (see Figure 3.12), who concluded that planets and stars were flaming rocks in the sky. Anaxagoras's assumption that meteorites fell from the heavens made him the first person in history known to believe that the heavens and Earth are made of the same materials, even though his guess about the nature of planets and stars was not quite correct. Many later scientists regarded stories of fallen stars more skeptically. Upon hearing of a meteorite fall in Connecticut, Thomas Jefferson (who was a student of science as well as politics) reportedly said, "It is easier to believe that Yankee professors would lie than that stones would fall from heaven."

Today we know that rocks really do fall from the heavens (**FIGURE 12.7**). Meteorites are often blasted apart in their fiery

descent through our atmosphere, scattering fragments over an area several kilometers across. A direct hit on the head by a meteorite would be fatal, but there are no reliable accounts of human deaths from meteorites. Of course, most meteorites fall into the ocean, which covers three-fourths of Earth's surface.

Unless you actually see a meteorite fall, it can be difficult to distinguish a meteorite from an Earth rock. Fortunately, a few clues can help. Meteorites are usually covered with a dark, pitted crust resulting from their fiery passage through the atmosphere (**FIGURE 12.8**). Some have an unusually high metal content, enough to attract a magnet hanging on a string. The ultimate judge of extraterrestrial origin is laboratory analysis: Meteorites often contain elements such as iridium that are very rare in Earth rocks, and even common elements in meteorites tend to have different ratios among their isotopes [**Section 5.3**] than are found in rocks from Earth. If you suspect that you have found a meteorite, many museums will analyze a small chip free of charge.

Scientists have found and cataloged more than 30,000 meteorites, some found by accident and others by organized meteorite searches. Antarctica is one of the best places to hunt for meteorites, not because more fall there but because the icy surface makes them easier to find. Few terrestrial rocks end up on the ice, so a rock found on the ice has a good chance of being a meteorite. Moreover, the slow movement of Antarctic glaciers tends to carry meteorites along, concentrating them in small areas where the ice floes run into mountains. The majority of meteorites now in museums and laboratories were found in Antarctica.

Types of Meteorites The origin of meteorites was long a mystery, but in recent decades we've been able to determine where in our solar system they come from. The most direct evidence comes from the relatively few meteorites whose trajectories have been observed or filmed as they fell to the ground. In every case so far, these meteorites clearly originated in the asteroid belt.

FIGURE 12.7 On March 26, 2003, a meteorite crashed through the roof of this home in Chicago. The main photo shows the hole in the ceiling; the inset shows the meteorite at rest on the floor. No one was hurt.

FIGURE 12.8 This large meteorite, called the Ahnighito Meteorite, is located at the American Museum of Natural History in New York. Its dark, pitted surface is a result of its fiery passage through Earth's atmosphere.

Stony primitive meteorite: Made of rocky material embedded with shiny metal flakes (arrow).

Carbon-rich primitive meteorite: Also rocky but with dark carbon compounds and small whitish spheres (arrow).

a Primitive meteorites.

Metal-rich processed meteorite: Made of iron and other metals that came from a shattered asteroid's core.

Rocky processed meteorite: Resembles volcanic rocks found on Earth. This meteorite probably came from Vesta's south pole.

b Processed meteorites.

FIGURE 12.9 There are two basic types of meteorites: primitive and processed. Each also has two subtypes. They are shown slightly smaller than actual size. (The meteorites have flat faces because they have been sliced with rock saws.)

Detailed analysis of thousands of meteorites shows that they come in two basic types, primitive and processed,* each of which can be further classified into two subtypes:

■ **Primitive meteorites** (**FIGURE 12.9a**) are "primitive" in the sense of being the first (or primary) type of meteorite to have formed. Radiometric dating [**Section 8.3**] shows them to be nearly 4.6 billion years old, making them remnants from the birth of our solar system, essentially unchanged since they first accreted in the solar nebula. Primitive meteorites come in two subtypes:

 ■ *Stony primitive meteorites* are composed of rocky minerals with a small but noticeable fraction of pure metallic flakes mixed in.

 ■ *Carbon-rich primitive meteorites* are generally similar to stony primitive meteorites, but they also contain substantial amounts of carbon compounds and, sometimes, a small amount of water.

■ **Processed meteorites** (**FIGURE 12.9b**) apparently once were part of a larger object that "processed" the original material of the solar nebula into another form. Radiometric dating shows that processed meteorites are generally younger than the primitive meteorites by a few hundred million years. The processed meteorites can also be divided into two subtypes:

 ■ *Metal-rich processed meteorites* are made mostly of high-density iron and nickel mixed with smaller amounts of other metals. That is, they resemble the terrestrial planet cores in composition.

 ■ *Rocky processed meteorites* have lower densities and are made of rock with compositions resembling that of terrestrial mantles and crusts. A few have compositions remarkably close to that of the basalts [**Section 9.2**] that erupt from terrestrial volcanoes.

The Origin of Primitive Meteorites The vast majority of all meteorites are primitive, and their structures confirm what their radiometric ages tell us: They are pieces of rock

that accreted in the solar nebula and orbited the Sun for billions of years before finally falling to Earth. The individual flakes may represent the tiny particles that first condensed from the gas of the solar nebula. The small, roundish features visible in Figure 12.9a may be solidified droplets splashed out in the accretion process.

Why are some primitive meteorites stony in composition while others are carbon-rich? Both theory and observation indicate that the answer depends on where they formed in the solar nebula. The nebular theory tells us that all meteorites accreted inside the frost line and are therefore made of metal and rock. However, models tell us that, beyond about 3 AU from the Sun, temperatures in the solar nebula were low enough for condensation of carbon compounds. This suggests that carbon-rich primitive meteorites come from the outer regions of the asteroid belt, while stony primitive meteorites come from closer in. Laboratory studies of primitive meteorites confirm these ideas. The composition of carbon-rich meteorites matches the composition of dark, carbon-rich asteroids in the outer part of the asteroid belt. The composition of stony meteorites matches the composition of asteroids in the inner part of the asteroid belt.

Curiously, most of the meteorites collected on Earth are of the stony variety even though most asteroids are of the carbon-rich variety. Evidently, orbital resonances are more effective at pitching material our way from the inner part of the asteroid belt than from more distant regions.

Scientists are taking the study of the asteroid-meteorite connection to the next level with sample return missions. In 2010, the Japanese *Hayabusa* mission, overcoming a host of technical challenges, returned a small dust sample from the asteroid Itokawa (**FIGURE 12.10**), confirming that this asteroid represents material in the stony primitive category. In 2016, NASA hopes to launch the *Osiris-REX* mission, which will collect a sample from the near-Earth asteroid 1999 RQ36. This asteroid is thought to be a carbon-rich primitive object, perhaps representative of the kind that brought the ingredients for oceans, atmosphere, and life to Earth.

The Origin of Processed Meteorites The processed meteorites tell a more complex story. Their compositions look similar to those of the cores, mantles, or crusts of

---

*Primitive meteorites are also called *chondrites* because of the roundish features visible in them, known as *chondrules*. Processed meteorites lack these features and hence are called *achondrites* (*a-* means "not").

FIGURE 12.10 The top panel shows the central portion of the 500-meter-long asteroid Itokawa. The dark spot near the center is the shadow of the Japanese spacecraft *Hayabusa*, dominated by its two large solar panels. The spacecraft descended to the surface, where it picked up a small sample, at the location of the zoom-in. *Hayabusa*, appropriately, means "hawk."

the terrestrial worlds. We conclude that these meteorites are fragments of larger asteroids that underwent *differentiation* [**Section 9.1**], in which their interiors melted so that metals sank to the center and rocks rose to the surface. This idea explains why we find two types of processed meteorites.

Rocky processed meteorites are often so close in composition to volcanic rocks on Earth that they must have been made by lava flows. We conclude that these meteorites are rocks from the surfaces of large asteroids that once were volcanically active, probably chipped off by collisions with smaller asteroids. Perhaps as many as a dozen large asteroids were geologically active shortly after the formation of the solar system. The fact that the processed meteorites date to only a few tens to hundreds of millions of years after the birth of the solar system tells us that this active period was short-lived; the interiors of the geologically active asteroids must have cooled quickly. As noted earlier, spectroscopy indicates that numerous rocky processed meteorites were chipped from the surface of the asteroid Vesta by the impact that created Vesta's south polar crater.

Metal-rich processed meteorites have core-like compositions, suggesting that they are fragments of larger asteroids

that shattered in collisions. These shattered worlds not only lost a chance to grow into planets but also sent many fragments onto collision courses with other asteroids and planets, including Earth. Moreover, these processed meteorites present us with an opportunity to study a "dissected planet," and they represent a form of direct proof that large worlds really do undergo differentiation, confirming what we infer from seismic studies of Earth.

Meteorites from the Moon and Mars In a few cases, the compositions of processed meteorites do not match those of asteroids but instead appear to match either the Moon or Mars. Careful analysis of these meteorites' compositions makes us very confident that we really do have a few meteorites that were once part of the Moon or Mars. Moderately large impacts can blast surface material from terrestrial worlds into interplanetary space. Once they are blasted into space, the rocks orbit the Sun until they come crashing down on another world. Calculations show that it is not surprising that we should have found a few meteorites chipped off the Moon and Mars in this way.

Study of these *lunar meteorites* and *Martian meteorites* is providing new insights into conditions on the Moon and Mars. One Martian meteorite sparked debate over whether it contains evidence of past life on Mars [**Section 24.2**].

12.2 COMETS

Asteroids are one of the two major categories of small bodies in the solar system. The other is comets, to which we now turn our attention. Asteroids and comets have much in common. Both are leftover planetesimals from the birth of our solar system. Both come in a wide range of sizes. The primary difference between them is in composition. Asteroids are rocky because they formed in the inner solar system where only metal and rock condensed. Comets are ice-rich because they formed beyond the frost line, where abundant hydrogen compounds condensed into ice [**Section 8.2**]. Note that we refer to any icy leftover planetesimal orbiting the Sun as a comet, regardless of its size, whether it has a tail, or where it resides or comes from.

What are comets like?

For most of human history, comets were familiar only from their occasional presence in the night sky. Every few years, a comet becomes visible to the naked eye, appearing as a fuzzy ball with a long tail. Indeed, the word *comet* comes from the Greek word for "hair," because comets were recognized by the long, hair-like tails they can display in our sky (**FIGURE 12.11**).

In comet photographs, the tails make it look as if the comets were racing across the sky, but they are not. If you watch a comet for minutes or hours, you'll see it staying nearly stationary relative to the stars around it in the sky. Over many days it will rise and set just like the stars, while gradually moving relative to the constellations. You may be able to see it night after night for several weeks or more, until it finally fades from view.

a Comet Hyakutake.

b Comet Hale-Bopp, photographed over Phoenix.

FIGURE 12.11 interactive photo Brilliant comets can appear at almost any time, as demonstrated by the back-to-back appearances of Comet Hyakutake in 1996 and Comet Hale-Bopp in 1997.

Today, we know that the vast majority of comets do not have tails and never venture anywhere close to Earth. Instead, they remain in the outer reaches of our solar system, orbiting the Sun far beyond the orbit of Neptune. The comets that appear in the night sky are the rare ones that have had their orbits changed by the gravitational influences of planets, other comets, or stars passing by in the distance. Their new orbits may carry them much closer to the Sun, eventually bringing them into the inner solar system, where they may grow the tails that allow us to see them in the sky. Most of these comets will not return to the inner solar system for thousands of years, if ever. A few happen to pass near enough to a planet to have their orbits changed further, and some end up on elliptical orbits that periodically bring them close to the Sun.

Comets in Human History The occasional presence of a bright comet in the night sky was hard to miss before the advent of electric lights. In some cultures, these rare intrusions into the unchanging heavens foretold bad or good luck. The ancient Chinese, for example, believed that the appearance of a comet led to a time of major and tumultuous change.

Few ancient cultures made any attempt to explain comets in astronomical terms. In fact, comets were generally thought to be within Earth's atmosphere until 1577, when Tycho Brahe [**Section 3.3**] used observations made from different locations on Earth to prove that a comet lay far beyond the Moon. A century later, Newton correctly deduced that comets orbit the Sun. Then, in a book published in 1705, English scientist Edmond Halley (1656–1742) used Newton's law of gravitation to calculate the orbit of a comet that had been seen in 1682. He showed that it was the same comet that had been observed on numerous prior occasions, which meant that it orbited the Sun every 76 years. Halley predicted that the comet would return in 1758. Although he had died by that

time, the comet returned as he predicted and was then named in his honor. Halley's Comet last passed through the inner solar system in 1986, and it will return in 2061.

Today, both professional and amateur astronomers routinely scan the skies with telescopes and binoculars in hopes of discovering a new comet. Several comets are discovered every year. The first observers (up to three) to report a comet discovery to the International Astronomical Union have the comet named after them. In that sense, comets now bring a form of good luck to their discoverers. More generally, comets are good luck for astronomers seeking to understand the solar system. These leftover planetesimals teach us about the history of the outer solar system just as asteroids teach us about the history of the inner solar system.

The surprise champion of comet discovery is the *SOHO* spacecraft, an orbiting solar observatory. *SOHO* has detected more than 2100 "Sun-grazing" comets, most on their last pass by the Sun (**FIGURE 12.12**).

This comet's tail is at least two solar diameters long—more than 3 million km.

FIGURE 12.12 Comet SOHO-6's final blaze of glory. The *SOHO* spacecraft observed this "Sun-grazing" comet a few hours before it passed just 50,000 kilometers above the Sun's surface. The comet did not survive its passage, because of the intense solar heating and tidal forces. In this image, the large, orange disk blocks out the Sun; the Sun's size is indicated by the white circle.

b The first image shows Comet Wild 2 photographed from Earth, and the inset shows its nucleus photographed by the *Stardust* spacecraft. The irregular surface probably shows effects from a combination of impacts and uneven vaporization rates in different regions.

a These photos were taken by the *Deep Impact* spacecraft as its 370-kg impactor crashed into Comet Tempel 1. The entire sequence, which starts just before impact at the upper left, unfolded over just 67 seconds.

FIGURE 12.13 interactive photo Comet nuclei.

Comet Composition We have learned about the composition of comets by studying their spectra. The results are just what we expect for objects that formed in the cold outer solar system: Comets are basically chunks of ice mixed with rocky dust and some more complex chemicals, and hence they are often described as "dirty snowballs."

Spectra of comets confirm their distant origin because they show the presence of compounds that could have condensed only in the cold outer regions of the solar nebula. For example, comet spectra invariably show emission features from hydrogen compounds, including water. They also show emissions from carbon dioxide and carbon monoxide, gases that condensed only in the very coldest and most distant regions of the solar nebula. Comet spectra also show evidence of many more complex molecules, including some organic molecules, leading some scientists to speculate that much of the organic material that made life possible on Earth was brought by comets.

Telescopic spectroscopy tells us about the material comets shed into space, but is this material representative of their internal compositions? NASA's *Deep Impact* mission was designed to answer this and other questions about comets. The main spacecraft carried a 370-kilogram impactor that it released as it approached Comet Tempel 1. Instruments on the spacecraft then observed the results as the impactor slammed into the comet at a speed of 37,000 kilometers per hour on July 4, 2005 (**FIGURE 12.13a**). The blast created a plume of hot gas composed of material vaporized from deep within the comet. Spectroscopy showed this material to contain many complex organic molecules, as expected. The plume also contained a huge amount of dust, telling us that the surface must have been covered by dust to a depth of tens of meters. However, because images show that Comet Tempel 1 looks quite different from other comets, we cannot conclude that all comets have similar compositions or similar amounts of surface dust.

We've learned even more about comet composition by studying comet material in the laboratory. In 2004, NASA's *Stardust* spacecraft used a material called "aerogel" to capture dust particles from Comet Wild 2 (pronounced "Vilt 2") (**FIGURE 12.13b**). The aerogel was then sealed in a re-entry capsule, which returned to Earth in 2006. Scientists are looking forward to data from the European Space Agency's *Rosetta* mission, on track to enter orbit of Comet Churyumov-Gerasimenko in May 2014 and to drop a lander to the surface in November 2014.

The Flashy Lives of Comets Comets grow tails only as they enter the inner solar system, where they are heated by the warmth of the Sun. To see what happens, let's look at the comet path shown in **FIGURE 12.14a** and the comet anatomy in **FIGURE 12.14b**.

Far from the Sun, the comet is completely frozen—in essence, a "dirty snowball." It begins to shed gas and dust as it approaches the Sun, but it still has its central chunk of ice, which we call the **nucleus**. Our best views of comet nuclei have come during spacecraft flybys: The first was a flyby of Halley's Comet by the European Space Agency's *Giotto* spacecraft in 1986, followed more recently by the *Deep Impact* and *Stardust* missions. Despite their icy compositions, comet nuclei are darker than charcoal, reflecting less than 5% of the light that falls on them. Apparently, it doesn't take much rocky or carbon-rich material to darken a comet. Density measurements show that comet nuclei are not packed very solidly. For example, the density of Halley's nucleus is considerably less than the density of water (1 g/cm^3), suggesting that the nucleus is part ice and part empty space.

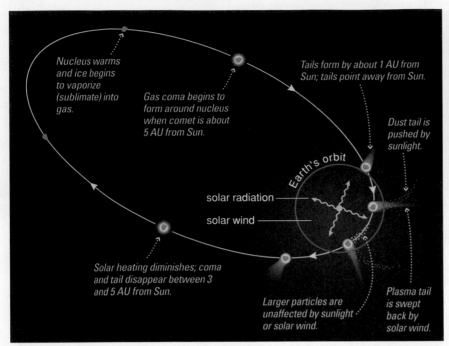

Nucleus warms and ice begins to vaporize (sublimate) into gas.

Gas coma begins to form around nucleus when comet is about 5 AU from Sun.

Tails form by about 1 AU from Sun; tails point away from Sun.

Dust tail is pushed by sunlight.

Earth's orbit

solar radiation

solar wind

Solar heating diminishes; coma and tail disappear between 3 and 5 AU from Sun.

Larger particles are unaffected by sunlight or solar wind.

Plasma tail is swept back by solar wind.

a This diagram (not to scale) shows the changes that occur when a comet's orbit takes it on a passage into the inner solar system.

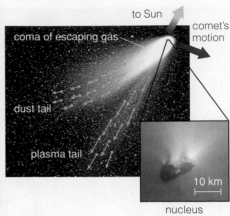

to Sun

comet's motion

coma of escaping gas

dust tail

plasma tail

nucleus

10 km

b Anatomy of a comet. The larger image is a ground-based photo of Comet Hale-Bopp. The inset shows the nucleus of Halley's Comet photographed by the *Giotto* spacecraft.

FIGURE 12.14 A comet grows a coma and tail around its nucleus only if it happens to come close to the Sun. Most comets never do this, instead remaining perpetually frozen in the far outer solar system.

The scale bars in Figures 12.13 and 12.14b show that comet nuclei are typically less than about 20 kilometers across. How can such small objects put on such spectacular shows? As a comet accelerates toward the Sun, its surface temperature increases, and ices begin to vaporize into gas that easily escapes the comet's weak gravity. Some of the escaping gas drags away dust particles from the nucleus, and the gas and dust create a huge, dusty atmosphere called a **coma**. The coma is far larger than the nucleus it surrounds. Spacecraft observations show that the release of gas from the comet's nucleus can be quite violent, with jets of gas and dust shooting out at speeds of hundreds of meters per second from the interior of the nucleus into space.

The coma grows as the comet continues into the inner solar system, and tails form as extensions of the coma. Comet tails can be hundreds of millions of kilometers in length. Because they look much like exhaust trails behind rockets, many people mistakenly guess that tails extend behind comets as they travel through their orbits. In fact, comet tails generally point away from the Sun, regardless of the direction in which the comet is traveling.

Comets have two visible tails, one made of ionized gas, or *plasma,* and the other made of dust. The **plasma tail** consists of gas escaping from the coma. Ultraviolet light from the Sun ionizes the gas, and the solar wind then carries this gas straight outward from the Sun at speeds of hundreds of kilometers per second. That is why the plasma tail extends almost directly away from the Sun at all times. The **dust tail** is made of dust-size particles escaping from the coma. These particles are not affected by the solar wind and instead are pushed away from the Sun by the much weaker pressure of sunlight itself (*radiation pressure*). The dust tail therefore points generally away from the Sun, but has a slight curve back in the direction the comet came from.

After the comet loops around the Sun and begins to head back outward, vaporization declines, the coma dissipates, and the tails disappear. Nothing happens until the comet again comes sunward—in a century, a millennium, a million years, or perhaps never.

Comets that repeatedly visit the inner solar system, like Halley's Comet, cannot last long on the time scale of our solar system. A comet probably loses about 0.1% of its ice on every pass around the Sun, so it could not make more than a few hundred passages before losing most of its original ice. Changes in composition may end the comet's "life" even faster. In a close pass by the Sun, a comet may shed a layer of material a meter thick. Dust that is too heavy to escape accumulates on the surface. This thick, dusty deposit helps make comets dark and may eventually block the escape of interior gas, preventing the comet from growing a coma or tails on future passes by the Sun. Several possible fates await a comet after its ices can no longer vaporize into gas and escape. In some cases, the remaining dust may disguise the dead comet as an asteroid. In other cases, the comet may come "unglued" and break apart, or even disintegrate along its orbit (**FIGURE 12.15**).

Comets striking Earth do not strew meteorites across the surface as asteroids do. The icy materials of a small comet vaporize on entry, and most of the rocky material takes the form of dust or pebbles that also vaporize as they are heated by friction in Earth's atmosphere.

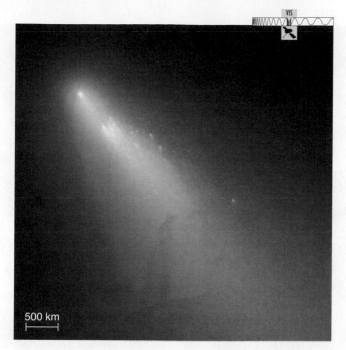

FIGURE 12.15 This photo shows a comet fragment (one of dozens that made up Comet Schwassmann-Wachmann 3) disintegrating as it passed Earth in 2006.

TABLE 12.1 Major Annual Meteor Showers

Shower Name	Approximate Date	Associated Comet
Quadrantids	January 3	?
Lyrids	April 22	Thatcher
Eta Aquarids	May 5	Halley
Delta Aquarids	July 28	?
Perseids	August 12	Swift-Tuttle
Orionids	October 22	Halley
Taurids	November 3	Encke
Leonids	November 17	Tempel-Tuttle
Geminids	December 14	Phaeton
Ursids	December 23	Tuttle

Comet Tails and Meteor Showers In addition to the plasma and dust in their two visible tails, the gas escaping from comets also carries away sand- to pebble-size pieces of rocky material. These particles are too big to be affected by either the solar wind or sunlight, so they drift away slowly and spread along the orbital path. These particles essentially form a third, invisible tail that follows the comet around its orbit. They are also the particles responsible for most meteors and meteor showers.

We see a meteor when one of these small particles (or a similar-size particle from an asteroid) burns up in our atmosphere. The sand- to pebble-size particles are much too small to be seen themselves, but they enter the atmosphere at such high speeds—up to about 250,000 kilometers per hour (70 km/s)—that they make the surrounding air glow with heat. It is this glow that we see as the brief but brilliant flash of a meteor. The small particles are vaporized by the heat and never reach the ground. An estimated 25 million particles of comet dust enter the atmosphere worldwide every day, burning up as meteors and adding hundreds of tons of comet dust to Earth daily.

Comet dust is sprinkled throughout the inner solar system, but the "third tails" of ejected particles make the dust much more concentrated along the orbits of comets. This dust rains down on our planet whenever we cross a comet's orbit, producing a **meteor shower**. A meteor shower recurs at about the same time each year, because the orbiting Earth passes through a particular comet's orbit at the same time each year. For example, the meteor shower known as the *Perseids* occurs about August 12 each year, when Earth passes through the orbit of Comet Swift-Tuttle. **TABLE 12.1** lists major annual meteor showers and their parent comet, if known.

You can typically see a few meteors each hour on any clear night, but you may see far more during one of the annual meteor showers. During a "good" shower at a dark site, you may see dozens of meteors per hour. Remember that you are watching the effects of comet dust entering our planet's atmosphere from space. The meteors generally appear to radiate from a particular direction in the sky, for essentially the same reason that snow or heavy rain seems to come from a particular direction in front of a moving car (**FIGURE 12.16**). Because more meteors hit Earth from the front than from behind (just as more snow hits the front windshield of a moving car), meteor showers are best observed in the predawn sky, when part of the sky faces in the direction of Earth's motion.

SEE IT FOR YOURSELF

Try to observe the next meteor shower (see Table 12.1); be prepared with a star chart, a marker pen, and a dim (preferably red) flashlight. Each time you see a meteor, record its path on your star chart. Record at least a dozen meteors, and try to determine the *radiant* of the shower—that is, the constellation from which the meteors appear to radiate. Does the meteor shower live up to its name?

Where do comets come from?

We've already stated that comets come from the outer solar system, but we can be much more specific. By analyzing the orbits of comets that pass close to the Sun, scientists learned that there are two major "reservoirs" of comets in the outer solar system.

Most comets that visit the inner solar system follow orbits that seem almost random. They do not orbit the Sun in the same direction as the planets, and their elliptical orbits can be tilted at any angle relative to the ecliptic plane. Moreover, their orbits show that they come from far beyond the orbits of the planets—sometimes nearly a quarter of the distance to the nearest star. These comets must come plunging sunward from a vast, spherical region of space that scientists call the **Oort cloud** (after astronomer Jan Oort; rhymes with "court"). Be sure to note that the Oort cloud is not a cloud of gas, but rather a collection of many individual comets. Based on the number of Oort cloud comets that enter the inner solar

Snowflakes and meteors appear to radiate from a single direction based on our motion relative to them.

a Meteors appear to radiate from a particular point in the sky for the same reason that we see snow or heavy rain come from a single point in front of a moving car.

b This digital composite photo, taken in Australia during the 2001 Leonid meteor shower, shows meteors as streaks of light. The large rock is Uluru, also known as Ayers Rock.

FIGURE 12.16 The geometry of meteor showers.

system each year, we estimate that the Oort cloud must contain about a trillion (10^{12}) comets orbiting the Sun at distances up to about 50,000 AU.

A smaller number of the comets that visit the inner solar system have a pattern to their orbits. They travel around the Sun in the same direction and in nearly the same plane as the planets, and their elliptical orbits carry them no more than about twice as far from the Sun as Neptune. These comets must come from a ring of comets that orbit the Sun beyond the orbit of Neptune. This ring is usually called the **Kuiper belt** (after astronomer Gerald Kuiper; rhymes with "piper"). **FIGURE 12.17** contrasts the general features of the Kuiper belt and the Oort cloud.

How did comets end up in these far-flung regions of the solar system? The only answer that makes scientific sense comes from thinking about what happened to the leftover icy planetesimals that roamed the region in which the jovian planets formed.

The leftover planetesimals that cruised the spaces between Jupiter, Saturn, Uranus, and Neptune were doomed to suffer either a collision or a close gravitational encounter with one of the young jovian planets. Recall that when a small object passes near a large planet, the planet is hardly affected but the small object may be flung off at high speed [**Section 4.5**]. The planetesimals that escaped being swallowed up by the jovian planets tended to be flung off in all directions. Some may have been cast away at such high speeds that they completely escaped the solar system and now drift through interstellar space. The rest ended up on orbits with very large average distances from the Sun, becoming the comets of the Oort cloud. The random directions in which they were flung

explain why the Oort cloud is roughly spherical in shape. Oort cloud comets are so far from the Sun that they can be nudged by the gravity of nearby stars (and even by the mass

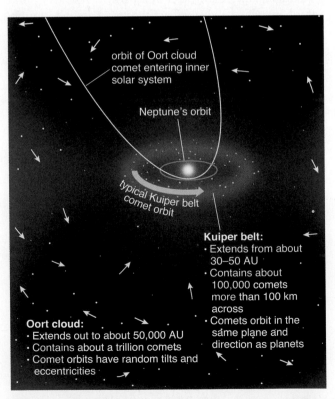

orbit of Oort cloud comet entering inner solar system

Neptune's orbit

typical Kuiper belt comet orbit

Oort cloud:
· Extends out to about 50,000 AU
· Contains about a trillion comets
· Comet orbits have random tilts and eccentricities

Kuiper belt:
· Extends from about 30–50 AU
· Contains about 100,000 comets more than 100 km across
· Comets orbit in the same plane and direction as planets

FIGURE 12.17 The comets we occasionally see in the inner solar system come from two major reservoirs in the outer solar system: the Kuiper belt and the Oort cloud.

of the galaxy as a whole), preventing some of them from ever returning near the planets and sending others plummeting toward the Sun.

Beyond the orbit of Neptune, the icy planetesimals were much less likely to be cast off by gravitational encounters. Instead, they remained in orbits going in the same directions as planetary orbits and concentrated relatively near the ecliptic plane. These are the comets of the donut-shaped Kuiper belt. Kuiper belt comets can be nudged by the gravity of the jovian planets through orbital resonances, sending some on close passes through the inner solar system. The Kuiper belt extends to only about twice Neptune's orbital distance, because at greater distances the density of the solar nebula was apparently too low to allow the accretion of planetesimals.

To summarize, the comets of the Kuiper belt seem to have originated farther from the Sun than the comets of the Oort cloud, even though the Oort cloud comets are now much more distant. The Oort cloud consists of leftover planetesimals that were flung outward after forming between the jovian planets, while the Kuiper belt consists of leftover planetesimals that formed and still remain in the outskirts of the planetary realm, roughly between the orbit of Neptune and twice Neptune's distance from the Sun.

(MA) Formation of the Solar System Tutorial, Lesson 3

12.3 PLUTO: LONE DOG NO MORE

Pluto was considered the "ninth planet" for some 75 years after its discovery in 1930, but it had long seemed a misfit among the planets. Its 248-year orbit is more elliptical and more inclined to the ecliptic plane than that of the first eight planets (FIGURE 12.18). In fact, Pluto sometimes comes closer to the Sun than Neptune, although there is no danger of collision: Neptune orbits the Sun precisely three times for every two Pluto orbits, and this stable orbital resonance means that Neptune is always a safe distance away whenever Pluto

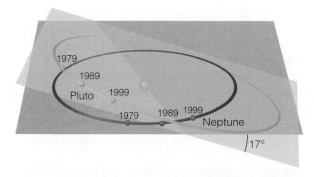

FIGURE 12.18 interactive figure Pluto's orbit is significantly elliptical and tilted relative to the ecliptic. Pluto comes closer to the Sun than Neptune for 20 years in each 248-year orbit, as was the case between 1979 and 1999. There's no danger of a collision, however, thanks to an orbital resonance in which Neptune completes three orbits for every two of Pluto's.

approaches its orbit. Neptune and Pluto will probably continue their dance of avoidance until the end of the solar system.

As we learned more about Pluto, it proved to be a misfit among the planets in size and composition as well: It is far smaller than even the terrestrial planets, and its ice-rich composition fits neither the terrestrial nor the jovian category. These seeming oddities made Pluto seem like a lone dog of the outer solar system. But it is not: We now know that Pluto is part of a huge pack of similar objects—essentially large comets—that orbit in its region of the solar system. In fact, Pluto is not even the largest member of this pack.

How big can a comet be?

We've known since the 1950s that many of the comets that visit the inner solar system come from the reservoir that we call the Kuiper belt. We've also known that Pluto orbits the Sun near the middle of this reservoir. We now realize that Pluto's location is no mere coincidence, but rather evidence of an astonishing fact: Some of the objects that populate the Kuiper belt are far larger than any of the comets that we ever see in the inner solar system. This fact forces us to reconsider just exactly what we mean by a "comet."

All the comets we've observed in the inner solar system are fairly small, with nuclei no larger than about 20 kilometers across. This shouldn't be surprising based on our understanding of comet origins, because only relatively small objects are likely to have their orbits changed enough to send them from the Kuiper belt or Oort cloud into the inner solar system. But given that the comets we see are rare and relatively small visitors from distant reaches of the solar system, could these outer realms hold much larger objects as well?

Large Comets of the Kuiper Belt According to our understanding of solar system formation, the icy planetesimals of the Kuiper belt should have been able to continue their accretion as long as other icy particles were nearby. In principle, one of these planetesimals could have grown large enough to become the seed of a fifth jovian planet, beyond Neptune, but that did not occur—probably because the density of material was too low at this great distance from the Sun. Nevertheless, it's reasonable to think that many of them grew to hundreds or even thousands of kilometers in diameter.

You might think it would be easy to observe icy objects of such sizes, but it requires powerful telescope technology. Trying to see a 1000-kilometer iceball at Pluto's distance from Earth is equivalent to looking for a snowball the size of your fist that is 600 kilometers away (and in dim light, too). Aside from Pluto, the first large objects in the Kuiper belt were discovered in the early 1990s. The discoveries have come at a rapid pace ever since.

As astronomers surveyed more and more of the sky in search of large, icy objects, the record size seemed to rise with each passing year. In 2002, the record was set by an object named Quaoar (pronounced "kwa-o-whar"), which is more than half the diameter of Pluto. Two objects announced in 2004—Orcus and Sedna—are between about two-thirds and

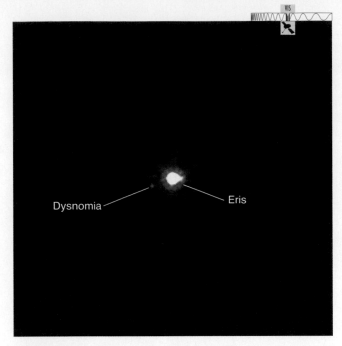

FIGURE 12.19 Eris and its moon, photographed by the Hubble Space Telescope.

FIGURE 12.20 Artist paintings of the largest known objects of the Kuiper belt (as of 2012), along with their known moons, shown to scale with Earth and the asteroids Ceres and Vesta for comparison. Pluto, Eris, Makemake, and Haumea officially qualify as dwarf planets, and others may yet join the list. The paintings are guesses about appearance, since the Kuiper belt objects have not been photographed with high resolution.

three-quarters of Pluto's size. It seemed only a matter of time until scientists would find an object larger than Pluto, and the time came in 2005, when Caltech astronomer Michael Brown (who had also discovered Quaoar, Orcus, and Sedna) announced the discovery of the object now named Eris (**FIGURE 12.19**). Eris is named for a Greek goddess who caused strife and arguments among humans (a commentary on the arguments its discovery caused about the definition of "planet"). Eris even has a moon, named Dysnomia, a mythological goddess of lawlessness and daughter of Eris.

Eris currently lies about twice as far as Pluto from the Sun. Its 557-year orbit is inclined to the ecliptic plane more than twice as much as Pluto's, and it is so eccentric that Eris must sometimes come closer to the Sun than Pluto. The latest observations place Pluto and Eris in a virtual tie for largest Kuiper belt object, although Eris is about 27% larger in mass.

Classification of Kuiper Belt Comets What should we call Pluto, Eris, and other big iceballs? The terminology has been a topic of much debate, but the science behind it seems clear. By 2012, more than 1100 icy objects had been directly observed in the Kuiper belt, allowing scientists to infer that the region contains at least 70,000 objects more than 100 kilometers across, and many more that are smaller. The new Pan-STARRS telescopes should discover tens of thousands of them. The smaller ones are easy to refer to as comets, but the larger ones pose a classification challenge, as you can see by comparing object sizes (**FIGURE 12.20**). Note that the Kuiper belt has numerous objects significantly larger than the largest asteroids, but still far smaller than Earth and the other terrestrial planets.

Under the current classification scheme (see Special Topic, page 8), four objects of the Kuiper belt (Pluto, Eris,

Makemake, and Haumea*) qualify as *dwarf planets* because they are big enough to be round but have not cleared their orbital neighborhoods. Dozens of other objects of the Kuiper belt may yet qualify as dwarf planets, though further observations will be needed to establish whether these objects are really round.

Regardless of what we call them, all the objects of the Kuiper belt—from the smallest boulders to the largest dwarf planets—probably share the same basic composition of ice and rock. In other words, they are all essentially comets of different sizes. That is why we often refer to all of them as *comets* of the Kuiper belt. However, some astronomers object to calling objects "comets" if they never venture into the inner solar system and show tails; therefore, you may also hear these objects referred to as *Kuiper belt objects* (*KBOs*) or *trans-Neptunian objects* (*TNOs*).

What are the large objects of the Kuiper belt like?

The small sizes and great distances of objects in the Kuiper belt make them very difficult to study. Even our best photographs from Earth show little more than fuzzy blobs or dots of light. As a result, we generally know little about these objects besides their orbits and the fact that their reflectivities and spectra suggest that they have ice-rich compositions. This should soon change, as scientists hope to learn much more when the *New Horizons* spacecraft flies past Pluto in the summer of 2015. In the meantime, let's look at our pre-flyby understanding of Pluto and its Kuiper belt companions.

Pluto We know more about Pluto than any other object in the Kuiper belt, largely because it was discovered back in 1930, decades before any of the others. In addition, the fact that

---

*Haumea is actually oblong, but it counts as a dwarf planet because it would be round if not for its high rotation rate.

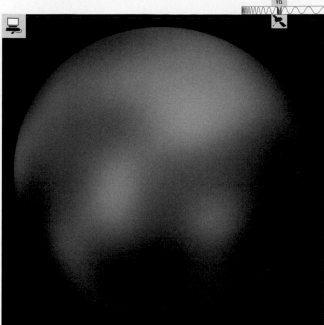

a This Hubble Space Telescope photo shows Pluto and its five known moons, along with orbital paths for the moons. Horizontal stripes are scattered light from Charon and Pluto in the long exposure.

b The surface of Pluto in approximately true color, based on a computer-processed series of Hubble Space Telescope images, each just a few pixels across. Comparison with images taken 8 years earlier reveals changes in surface markings, which may be due to the gradual change in seasons on Pluto.

FIGURE 12.21 Pluto's great distance and small size make it difficult to study.

Pluto has moons (**FIGURE 12.21a**) meant scientists could learn much more. The 1978 discovery of Pluto's largest moon, Charon, was especially important. Recall that we can determine an object's mass using Newton's version of Kepler's third law only if we can observe its gravitational effect on some other object [**Section 4.4**]. As a result, Charon's discovery allowed the first precise determination of Pluto's mass, which was a key step in learning how greatly Pluto differed from either the terrestrial or the jovian planets.

Charon soon provided another opportunity for scientists to learn about Pluto through some good luck in timing: Just a few years after Charon's discovery, Pluto and Charon entered a period of time (lasting from 1985 to 1990) during which they eclipsed each other every few days as seen from Earth—something that happens only about every 120 years. Detailed analysis of brightness variations during these eclipses allowed the calculation of accurate sizes, masses, and densities for both Pluto and Charon. These measurements confirmed that Pluto and Charon are both made of ice mixed with rock, just like comets. The eclipse data even allowed astronomers to construct rough maps of Pluto's surface markings. More recent images show that Pluto's surface is being altered, probably because of seasonal changes, which on Pluto are affected both by axis tilt and by changing distance from the Sun (**FIGURE 12.21b**).

Comparing Pluto and Charon leads to some interesting surprises. Charon is closer in distance and in relative size to its parent planet than our Moon is to Earth; it has more than half the diameter and about $\frac{1}{8}$ the mass of Pluto, and orbits only 20,000 kilometers away. (For comparison, our Moon

has a mass $\frac{1}{80}$ of Earth's and orbits 400,000 kilometers away.) Charon is also slightly lower in density than Pluto. Together, these facts have led astronomers to guess that Charon was created by a *giant impact* similar to the one thought to have formed our Moon [**Section 8.2**]. A large comet crashing into Pluto may have blasted away its low-density outer layers, which then formed a ring around Pluto and eventually re-accreted to make Charon and the smaller moons. The impact may also explain why Pluto rotates almost on its side.

Pluto is very cold, with an average temperature of only 40 K, as we would expect at its great distance from the Sun. Nevertheless, Pluto currently has a thin atmosphere of nitrogen and other gases formed by vaporization of surface ices. The atmosphere should gradually refreeze onto the surface as Pluto's 248-year orbit carries it farther from the Sun, although recent results have puzzled astronomers by showing the opposite effect. In fact, it now appears that Pluto has started to eject carbon monoxide gas, perhaps released as Pluto's frigid pole turns slowly toward the Sun.

Despite the cold, the view from Pluto would be stunning. Charon would dominate the sky, appearing almost 10 times as large in angular size as our Moon appears from Earth. The mutual tidal forces acting between Pluto and Charon long ago made them rotate synchronously with each other [**Section 4.5**], so Charon is visible from only one side of Pluto and always shows the same face to Pluto. This synchronous rotation also means that Pluto's "day" is the same length as Charon's "month" (orbital period) of 6.4 Earth days. Viewed from Pluto's surface, Charon neither rises nor sets but instead hangs motionless as it cycles through its phases every 6.4 days. The Sun would appear more than a thousand times fainter than it appears here on Earth and would be no larger in angular size than Jupiter appears in our skies.

Other Kuiper Belt Objects We don't know nearly as much about any other large comets of the Kuiper belt, though scientists hope to send the *New Horizons* spacecraft to at least one or two others after its encounter with Pluto. Nevertheless, because all Kuiper belt comets presumably formed in the same region of the solar system, we expect them to be similar in nature and composition to Pluto. One key piece of evidence of similarities comes from careful study of orbits. Like Pluto, many Kuiper belt comets have stable orbital resonances with Neptune. In fact, hundreds of Kuiper belt comets have the *same* orbital period and average distance from the Sun as Pluto itself (and are nicknamed "Plutinos").

Other evidence comes from the fact that several other Kuiper belt comets (including Eris) are also known to have moons, so in those cases we can calculate masses and densities. These results, along with spectral information, confirm the idea that these objects have comet-like compositions of ice and rock. One remarkable object, Haumea (see Figure 12.20), seems to have suffered a giant impact like the one thought to have created Charon. The impact created two moons and seems to have set Haumea spinning so fast (once every 4 hours) that its shape has become very elongated. Four other known Kuiper belt comets may also be fragments from the same impact, since their orbits resemble that of Haumea and they have similar spectral properties.

Perhaps most intriguingly, if you think back to what we learned in our tour of jovian moons in Chapter 11, you'll realize that Pluto probably will *not* be the first Kuiper belt comet to be photographed close up. Instead, that honor probably belongs to Neptune's moon Triton. Recall that Triton's "backward" orbit indicates that it must be a captured object [Section 11.2]. Its location suggests that it was most likely captured from the Kuiper belt, since many Kuiper belt comets have orbits that cross Neptune's orbit. In that case, *Voyager* images of Triton (see Figure 11.31) represent images of a *former* member of the Kuiper belt.

The fact that Triton appears to be quite similar in nature to Pluto lends further support to the idea that it once orbited the Sun within the Kuiper belt. Moreover, Triton is about 15% larger than Pluto (in diameter), which suggests that other objects bigger than Pluto were able to form in the Kuiper belt. Some may still await discovery, and others may have been destroyed in giant impacts. Those that did not gain the stability provided by being in an orbital resonance may have fallen inward in the solar system, eventually being accreted by one of the jovian planets, or may have been thrown outward and ejected from the solar system. All in all, it seems quite likely that the Kuiper belt, like the asteroid belt, once held many more large objects than it does today.

12.4 COSMIC COLLISIONS: SMALL BODIES VERSUS THE PLANETS

The hordes of small bodies orbiting the Sun are slowly shrinking in number through collisions with the planets and ejection from the solar system. Many more must have roamed the solar system in the days of the heavy bombardment, when most impact craters formed [Section 8.2]. Plenty of small bodies remain, however, and cosmic collisions still occur on occasion. These collisions have had important effects on Earth and other planets.

Have we ever witnessed a major impact?

We have never witnessed a major impact on a terrestrial world, but have seen one on Jupiter: the 1994 impact of a comet named *Shoemaker-Levy 9,* or *SL9* for short. Rather than having a single nucleus, Comet SL9 consisted of a string of nuclei (**FIGURE 12.22a**). Apparently, tidal forces from Jupiter had ripped apart a single comet nucleus during a previous close pass by the planet. Chains of craters observed on some of Jupiter's moons provide evidence that similar breakups of comets have occurred near Jupiter in the past.

Comet SL9 was discovered more than a year before it collided with Jupiter, and orbital calculations told astronomers precisely when the collision would occur. When the impacts began, they were observed with nearly every major telescope in existence, as well as by spacecraft that were in position to get a view. Each of the individual nuclei crashed into Jupiter with an energy equivalent to that of a million hydrogen bombs (**FIGURE 12.22b** and **c**). Comet nuclei barely a kilometer across left scars large enough to swallow Earth. The scars lasted for months before Jupiter's strong winds finally swept them from view.

The SL9 impacts allowed scientists to study both the impact process and material splashed out from deep inside Jupiter. Scientists felt quite fortunate, because estimates suggested that impacts as large as those of SL9 occur only about once in a thousand years. Astronomers were therefore surprised when another impact occurred on Jupiter in 2009. This impact was first noticed by an amateur astronomer from Australia, who reported a new "dark spot" on Jupiter that was soon recognized as the aftermath of an unseen impact (**FIGURE 12.22d**). Two amateur astronomers independently recorded images of an even smaller impact event in 2010. Astronomers are still not sure if these recent events signify that Jupiter impacts are more common than previously thought, or if improved monitoring of Jupiter can now capture smaller impact events, which occur more frequently than impacts of SL9's magnitude. Either way, we now know

a Jupiter's tidal forces ripped apart the single comet nucleus of SL9 into a chain of smaller nuclei.

b This painting shows how the SL9 impacts might have looked from the surface of Io. The impacts occurred on Jupiter's night side.

c This infrared photo shows the brilliant glow of a rising fireball from the impact of one SL9 nucleus in 1994. Jupiter is the round disk, with the impact occurring near the lower left.

d The black spot in this Hubble Space Telescope photo is a scar from the impact of an unknown object that struck Jupiter in July 2009.

FIGURE 12.22 The impacts of Comet Shoemaker-Levy 9 on Jupiter allowed astronomers their first direct view of a cosmic collision.

that violent impacts have happened on other planets in our lifetime. Could they also happen on Earth?

Did an impact kill the dinosaurs?

There's no doubt that major impacts have occurred on Earth in the past. Meteor Crater in Arizona (see Figure 9.8a) formed about 50,000 years ago when a metallic asteroid roughly 50 meters across crashed to Earth with the explosive power of a 20-megaton hydrogen bomb. Although the crater is only a bit more than 1 kilometer across, the blast and ejecta probably battered an area covering hundreds of square kilometers. Meteor Crater is relatively small and recent. Despite the fact

that erosion and other geological processes have erased most of Earth's impact craters, geologists have identified more than 150 impact craters on our planet. So before we consider whether an impact might occur in our lifetimes, it's worth examining the potential consequences if it did. Clearly, an impact could cause widespread physical damage. But a growing body of evidence, accumulated over the past three decades, suggests that an impact can do much more—in some cases, large impacts may have altered the entire course of evolution.

In 1978, while analyzing geological samples collected in Italy, a scientific team led by Luis and Walter Alvarez (father and son) made a startling discovery. They found that a thin layer of dark sediments deposited about 65 million years ago—about the time the dinosaurs went extinct—was unusually rich in the element iridium. Iridium is a metal that is rare on Earth's surface (because it sank to Earth's core when our planet underwent differentiation) but common in meteorites. Subsequent studies found the same iridium-rich layer in 65-million-year-old sediments around the world (**FIGURE 12.23**).* The Alvarez team suggested a stunning hypothesis: The extinction of the dinosaurs was caused by the impact of an asteroid or comet.

In fact, the death of the dinosaurs was only a small part of the biological devastation that seems to have occurred 65 million years ago. The fossil record suggests that up to 99% of all living organisms died around that time and that up to 75% of all existing *species* were driven to extinction. This makes the event a clear example of a **mass extinction**—the rapid extinction of a large fraction of all living species. Could it really have been caused by an impact?

Evidence for the Impact There's still some scientific debate about whether the impact was the sole cause of the mass extinction or just one of many causes, but there's

---

*The layer marks what geologists call the *K-T boundary*, because it separates sediments deposited in the Cretaceous and Tertiary periods (the K comes from the German word for Cretaceous, *Kreide*). The mass extinction that occurred 65 million years ago is therefore called the *K-T event*.

FIGURE 12.23 Around the world, sedimentary rock layers dating to 65 million years ago share the evidence of the impact of a comet or asteroid. Fossils of dinosaurs and many other species appear only in rocks below the iridium-rich layer.

little doubt that a major impact coincided with the death of the dinosaurs. Key evidence comes from further analysis of the sediment layer. Besides being unusually rich in iridium, this layer contains four other unusual features: (1) high abundances of several other metals, including osmium, gold, and platinum; (2) grains of "shocked" quartz, quartz crystals with a distinctive structure that indicates they experienced the high-pressure conditions of an impact; (3) spherical rock droplets of a type known to form when drops of molten rock cool and solidify in the air; and (4) soot (at some sites) that appears to have been produced by widespread forest fires.

All these features point to an impact. The metal abundances look much like what we commonly find in meteorites rather than what we find elsewhere on Earth's surface. Shocked quartz is also found at other impact sites, such as Meteor Crater in Arizona. The rock droplets presumably were made from molten rock splashed into the air by the force and heat of the impact. Some debris would have been blasted so high that it rose above the atmosphere, spreading worldwide before falling back to Earth. On their downward plunge, friction would have heated the debris particles until they became a hot, glowing rain of rock. The soot probably came from vast forest fires ignited by radiation from this impact debris.

In addition to the evidence within the sediments, scientists have identified a large impact crater that appears to match the age of the sediment layer. The crater, called the *Chicxulub crater* after a nearby fishing village, is about 200 kilometers across and is located on the coast of Mexico's Yucatán Peninsula, about half on land and half underwater (**FIGURE 12.24**). Its size indicates that it was created by the impact of an asteroid or a comet measuring about 10 kilometers across, which is large enough to account for the iridium and other metals found in the sediment layer.

The Mass Extinction If the impact was indeed the cause of the mass extinction, here's how it probably happened: On that fateful day some 65 million years ago, the asteroid or comet slammed into Mexico with the force of a hundred million hydrogen bombs (**FIGURE 12.25**). It apparently hit at a slight angle, sending a shower of red-hot debris across the continent of North America. A huge tsunami sloshed more than 1000 kilometers inland. Much of North American life may have been wiped out almost immediately. Not long after, the hot debris raining around the rest of the world ignited fires that killed many other living organisms.

The longer-term effects were even more severe. Dust and smoke remained in the atmosphere for weeks or months, blocking sunlight and causing temperatures to fall as if Earth were experiencing a global and extremely harsh winter. The reduced sunlight would have stopped photosynthesis for up to a year, killing large numbers of species throughout the food chain. This period of cold may have been followed by a period of unusual warmth. Some evidence suggests that the impact site was rich in carbonate rocks, so the impact may have released large amounts of carbon dioxide into the atmosphere. The added carbon dioxide would have strengthened the greenhouse effect, and the months of global winter may have been followed by decades or longer of global summer.

The impact probably also caused chemical reactions in the atmosphere that produced large quantities of harmful compounds, such as nitrous oxides. These compounds dissolved in the oceans, where they would have killed vast numbers of marine organisms. Acid rain may have been another by-product, killing vegetation and acidifying lakes around the world.

FIGURE 12.24 This computer-generated image, based on measurements of small local variations in the strength of gravity, shows an impact crater (dashed circle) in the northwest corner of Mexico's Yucatán Peninsula; the inset shows the location on a map. The crater measures about 200 kilometers across, and about half of it lies underwater off the coast.

FIGURE 12.25 This painting shows an asteroid or comet moments before its impact on Earth, some 65 million years ago. The impact probably caused the extinction of the dinosaurs, and if it hadn't occurred the dinosaurs might still rule Earth today.

Perhaps the most astonishing fact is not that 75% of all species died but that 25% survived. Among the survivors were a few small mammals. These mammals may have survived in part because they lived in underground burrows and managed to store enough food to outlast the global winter that immediately followed the impact.

The evolutionary impact of the extinctions was profound. For 180 million years, dinosaurs had diversified into a great many species large and small, while mammals (which had arisen at almost the same time as the dinosaurs) had generally remained small and rodent-like. With the dinosaurs gone, mammals became the new kings of the planet. Over the next 65 million years, the mammals rapidly evolved into an assortment of much larger mammals—ultimately including us.

Controversies and Other Mass Extinctions

There seems little doubt that a major impact coincided with the mass extinction of 65 million years ago, but does it tell the entire story? The jury is still out. The fossil record can be difficult to read when we are trying to understand events that happened over just a few years rather than a few million years. Some scientists suspect that the dinosaurs were already in decline and the impact was only the last straw. Others suggest that major volcanic eruptions also may have played a role.

Was the dinosaur extinction a unique event? Measuring precise extinction rates becomes more difficult as we look to older fossils, but there appear to have been at least four other mass extinctions during the past 500 million years. None of the other mass extinctions is as closely tied to an impact as the dinosaur extinction, but impacts almost certainly played a role. Sediments from the times of other mass extinctions reveal evidence similar to that found in the iridium-rich layer tied to the death of the dinosaurs. In some cases, impact craters have been found that date to about the right times. Much more research is needed, but impacts appear to have played a major role in shaping the history of life on Earth.

Is the impact threat a real danger or media hype?

On June 14, 2002, the 100-meter-wide asteroid 2002 MN passed within 120,000 kilometers of Earth (less than a third of the Earth-Moon distance). There was no advance warning; in fact, the asteroid wasn't discovered until after it had passed by. Since then, more than 30 smaller asteroids have passed even closer. Most recently, the 1-meter-wide asteroid 2011 CQ1 passed less than 6000 kilometers above the Pacific Ocean in 2011. Although asteroids of these sizes would not cause global devastation, one of the larger ones (tens of meters or more across) could kill millions of people in the unlikely event that it struck a large city. How concerned should we be about the possibility of objects like this striking Earth—or, worse yet, even larger objects such as those that may have caused past mass extinctions?

One way to gauge the threat is to look at evidence from past impacts. Geological data show that impacts large enough to cause mass extinctions happen many tens of millions of years apart on average, which means that while another such

FIGURE 12.26 This photo shows forests burned and flattened by the 1908 impact over Tunguska, Siberia.

impact will probably occur in the future, it's highly unlikely to be within the next few thousand years. We're far more likely to do ourselves in than to be done in by a large asteroid or comet.

Smaller impacts can be expected more frequently. While such impacts would not wipe out our civilization, they could kill thousands or millions of people. We know of one close call in modern times. In 1908, a tremendous explosion occurred over Tunguska, Siberia (**FIGURE 12.26**). Entire forests were flattened and set on fire, and air blasts knocked over people and furniture up to 200 kilometers away. Seismic disturbances were recorded at distances of up to 1000 kilometers, and atmospheric pressure fluctuations were detected almost 4000 kilometers away. The explosion, now estimated to have released energy equivalent to that of several atomic bombs, is thought to have been caused by a small asteroid no more than about 40 meters across. Atmospheric friction caused it to explode completely before it hit the ground, so it left no impact crater. If the asteroid had exploded over a major city instead of Siberia, it would have been the worst natural disaster in human history.

A less devastating impact occurred in 2007, when eyewitnesses saw a bright fireball streak across the mid-day sky near Carancas, Peru. A stony meteorite about 1–2 meters across slammed into the ground and excavated a crater 15 meters across (**FIGURE 12.27**). The impact spewed debris more than 300 meters and shattered windows as far as a kilometer away.

Another way to gauge the threat is to look at objects in space that might strike Earth. As of 2012, astronomers had identified nearly 9000 asteroids with orbits that pass near Earth's orbit, including more than 1300 that could potentially hit Earth someday. Orbital calculations show that none of these known objects will collide with Earth in the foreseeable future. However, while astronomers estimate that more than 90% of the "potentially hazardous asteroids" larger than 1 kilometer have been found, the estimates also suggest

FIGURE 12.27 The impact crater made by a stony meteorite 1–2 meters across near Carancas, Peru, in 2007. The crater filled with groundwater soon after the impact.

that more than 15,000 asteroids with diameters between 100 meters and 1 kilometer also pass near Earth's orbit but remain undiscovered. The threat from comets is lower, but we are unlikely to see a comet plunging in from the outer solar system until it is well on its way, giving us no more than a few years to prepare for the impact.

We can use the data from both geology and astronomy to estimate how often, on average, we expect objects of different sizes to hit Earth. The results are shown in FIGURE 12.28. Objects a few meters across probably enter Earth's atmosphere every week or so, each liberating energy equivalent to that of an atomic bomb as it burns up. Larger impacts are obviously more devastating but thankfully more rare. Hundred-meter impacts that forge craters similar to Meteor Crater in Arizona probably strike only every few thousand years or so. Objects of the size that caused the Tunguska event probably strike our planet every few hundred years

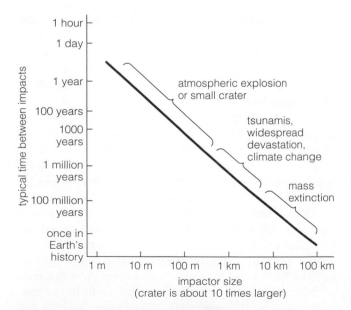

FIGURE 12.28 This graph shows that larger objects (asteroids or comets) hit Earth less frequently than smaller ones. The labels describe the effects of impacts of different sizes.

or so, making them a real threat. Fortunately, these impacts are unlikely to strike densely populated areas, because most of Earth's surface is ocean, and even on land humans are concentrated in relatively small urban areas.

If we were to find an asteroid or a comet on a collision course with Earth, could we do anything about it? Many people have proposed schemes to save Earth by using nuclear weapons or other means to demolish or divert an incoming asteroid, but no one knows whether current technology is really up to the task. We can only hope that the threat doesn't become a reality before we're ready.

While near-Earth asteroids may pose real dangers, they may also offer opportunities. These asteroids bring valuable resources tantalizingly close to Earth. Iron-rich asteroids are particularly enticing, because they probably contain many precious metals that have mostly sunk to the core on Earth. In the future, it may prove technically feasible and financially profitable to mine metals from asteroids and bring these resources to Earth. It may also be possible to gather fuel and water from asteroids for use in missions to the outer solar system.

How do the jovian planets affect impact rates and life on Earth?

Ancient people imagined that the mere movement of planets relative to the visible stars in our sky could somehow have an astrological influence on our lives. Although scientists no longer give credence to this ancient superstition, we now know that planets can have a real effect on life on Earth. By catapulting asteroids and comets in Earth's direction, other planets have caused cosmic collisions that helped shape Earth's destiny.

Jupiter and the other jovian planets have had the greatest effects because of their influence on the small bodies of the solar system. As we saw earlier in this chapter, Jupiter disturbed the orbits of rocky planetesimals outside Mars's orbit, preventing a planet from forming and creating the asteroid belt. The jovian planets ejected icy planetesimals to create the distant Oort cloud, and orbital resonances with Neptune shape the orbits of many comets in the Kuiper belt. Ultimately, every asteroid or comet that has impacted Earth since the end of the heavy bombardment was in some sense sent our way by the influence of Jupiter or one of the other jovian planets. FIGURE 12.29 summarizes the ways the jovian planets have controlled the motions of asteroids and comets.

We thereby find a deep connection between the jovian planets and the survival of life on Earth. If Jupiter did not exist, the threat from asteroids might be much smaller, since the objects that make up the asteroid belt might instead have become part of a planet. On the other hand, the threat from comets might be much greater: Jupiter probably ejected more comets to the Oort cloud than any other jovian planet, and

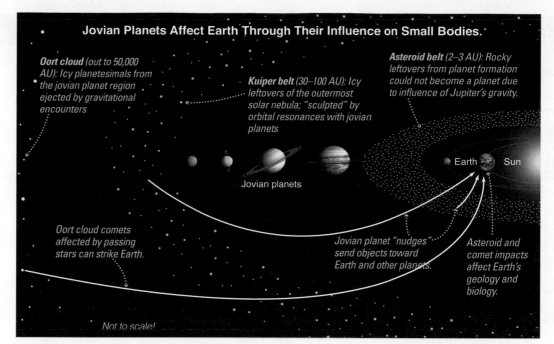

FIGURE 12.29 The connections between the jovian planets, small bodies, and Earth. The gravity of the jovian planets helped shape both the asteroid belt and the Kuiper belt, and the Oort cloud consists of comets ejected from the jovian planet region by gravitational encounters with these large planets. Ongoing gravitational influences sometimes send asteroids or comets toward Earth.

Within the figure:

Jovian Planets Affect Earth Through Their Influence on Small Bodies.

Oort cloud (out to 50,000 AU): Icy planetesimals from the jovian planet region ejected by gravitational encounters

Kuiper belt (30–100 AU): Icy leftovers of the outermost solar nebula; "sculpted" by orbital resonances with jovian planets

Asteroid belt (2–3 AU): Rocky leftovers from planet formation could not become a planet due to influence of Jupiter's gravity.

Jovian planets

Earth   Sun

Oort cloud comets affected by passing stars can strike Earth.

Jovian planet "nudges" send objects toward Earth and other planets.

Asteroid and comet impacts affect Earth's geology and biology.

Not to scale!

without Jupiter, those comets might have remained dangerously close to Earth. Of course, even if Jupiter has protected us from impacts, it's not clear whether that has been good or bad for life overall. The dinosaurs appear to have suffered from an impact, but the same impact may have paved the way for our existence. So while some scientists argue that more impacts would have damaged life on Earth, others argue that more impacts might have sped up evolution.

The role of Jupiter has led some scientists to wonder whether we could exist if our solar system had been laid out differently. Could it be that civilizations can arise only in solar systems that happen to have a Jupiter-like planet in a Jupiter-like orbit? No one yet knows the answer to this question. What we do know is that Jupiter has had profound effects on life on Earth and will continue to have effects in the future.

The Big Picture

Putting Chapter 12 into Context

In this chapter we concluded the study of our own solar system by focusing on its smallest objects, finding that these objects can have big consequences. Keep in mind the following "big picture" ideas:

- Asteroids, comets, and meteorites may be small compared to planets, but they provide evidence that has helped us understand how the solar system formed.

- The small bodies are subject to the gravitational whims of the largest. The jovian planets shaped the asteroid belt, the Kuiper belt, and the Oort cloud, and they continue to nudge objects onto collision courses with the planets.

- Collisions not only bring meteorites and leave impact craters but also can profoundly affect life on Earth. An impact probably was responsible for wiping out the dinosaurs, and future impacts pose a threat that we cannot ignore.

12.1 ASTEROIDS AND METEORITES

- **What are asteroids like?** Asteroids are rocky leftovers from the era of planetary formation. Most are small, and despite their enormous numbers their total mass is less than that of any terrestrial planet.

- **Why is there an asteroid belt?** Orbital resonances with Jupiter disrupted the orbits of planetesimals in the asteroid belt, preventing them from accreting into a planet. Many were ejected but some remained and make up the asteroid belt today. Most asteroids in other regions of the inner solar system accreted into one of the planets.

- **How are meteorites related to asteroids?** Most meteorites are pieces of asteroids. **Primitive meteorites** are essentially unchanged since the birth of the solar system. **Processed meteorites** are fragments of larger asteroids that underwent differentiation.

12.2 COMETS

- **What are comets like?** Comets are icy leftovers from the era of planet formation, and most orbit far from the Sun. If a comet approaches the Sun, its **nucleus** heats up and its ice vaporizes into gas. The escaping gases carry along some dust, forming a **coma** and two tails: a **plasma tail** of ionized gas and a **dust tail**. Larger particles can also escape, becoming the particles that cause **meteor showers** on Earth.

- **Where do comets come from?** Comets come from two reservoirs: the **Kuiper belt** and the **Oort cloud**. The Kuiper belt comets still reside in the region beyond Neptune in which they formed. The Oort cloud comets formed between the jovian planets, and were flung out to great distances from the Sun by gravitational encounters with the planets.

12.3 PLUTO: LONE DOG NO MORE

- **How big can a comet be?** In the Kuiper belt, icy planetesimals were able to grow to hundreds or thousands of kilometers in size. Eris is the largest known of these objects, and Pluto is the second largest.

- **What are the large objects of the Kuiper belt like?**

Like smaller comets, these objects are ice-rich in composition. They orbit the Sun roughly between the orbit of Neptune and twice that distance from the Sun. Their orbits tend to be more elliptical and more inclined to the ecliptic plane than those of the terrestrial and jovian planets. Many share orbital resonances with Neptune. A few, including Pluto and Eris, have moons.

12.4 COSMIC COLLISIONS: SMALL BODIES VERSUS THE PLANETS

- **Have we ever witnessed a major impact?** In 1994, we observed the impacts of Comet Shoemaker-Levy 9 on Jupiter. The comet had fragmented into a string of individual nuclei, so there was a string of impacts that left Jupiter's atmosphere scarred for months.

- **Did an impact kill the dinosaurs?** It may not have been

the sole cause, but a major impact clearly coincided with the **mass extinction** in which the dinosaurs died out, about 65 million years ago. Sediments from the time contain iridium and other clear evidence of an impact, and an impact crater of the right age lies near the coast of Mexico.

- **Is the impact threat a real danger or media hype?** Impacts certainly pose a threat, though the probability of a major impact in our lifetimes is fairly low. Impacts like the Tunguska event may occur every couple of hundred years, and would be catastrophic if they struck populated areas.

- **How do the jovian planets affect impact rates and life on Earth?** Impacts are always linked in at least some way

to the gravitational influences of Jupiter and the other jovian planets. These influences have shaped the asteroid belt, the Kuiper belt, and the Oort cloud, and continue to determine when an object is flung our way.

Use the following questions to check your understanding of some of the many types of visual information used in astronomy. Answers are provided in Appendix J. For additional practice, try the Chapter 12 Visual Quiz at MasteringAstronomy®.

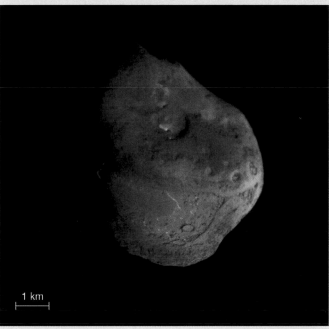

The graph above (Figure 12.28) shows how often impacts occur for objects of different sizes. The photo above shows Comet Tempel 1 moments before the Deep Impact spacecraft crashed into it.

1. Estimate Comet Tempel 1's diameter, using the scale bar in the photo above.
2. According to the graph above, how frequently do objects the size of Comet Tempel 1 strike Earth?
   a. once in Earth's history
   b. about once every hundred million years
   c. about once every million years
   d. about once every thousand years
3. Consider an object twice the size of Comet Tempel 1. What kind of damage would this object cause if it hit our planet?
   a. mass extinction
   b. widespread devastation and climate change
   c. atmospheric explosion or a small crater
4. Meteor Crater in Arizona is about 1.2 kilometers across. According to the graph, about how big was the object that made this crater? (*Note*: Be sure to read the axis labels carefully.)
   a. 1 meter
   b. 10 meters
   c. 100 meters
   d. 1 kilometer
5. How often do objects big enough to create craters like Meteor Crater impact Earth?
   a. once in Earth's history
   b. about once every ten thousand years
   c. about once every few million years
   d. about once every day, but most burn up in the atmosphere or land in the ocean

EXERCISES AND PROBLEMS

MasteringAstronomy®

For instructor-assigned homework go to MasteringAstronomy®.

REVIEW QUESTIONS

Short-Answer Questions Based on the Reading

1. Briefly describe the general characteristics of asteroids—including sizes, masses, densities, and compositions—and how we measure them.
2. How does the largest asteroid compare in size to the planets? How does the total mass of all asteroids compare to the mass of a terrestrial world?
3. Where is the *asteroid belt* located, and why? Briefly explain how orbital resonances with Jupiter have affected the asteroid belt.
4. What is the difference between a *meteor* and a *meteorite*? Distinguish between *primitive meteorites* and *processed meteorites* in terms of both composition and origin.
5. What does a comet look like when it is far from the Sun? Why do only a few comets ever enter the inner solar system?

6. What produces the *coma* and *tails* of a comet? What is the *nucleus*? Why do tails point away from the Sun?

7. How are *meteor showers* linked to comets, and why do they recur at about the same time each year?

8. Describe the *Kuiper belt* and *Oort cloud* in terms of their locations and the orbits of comets within them. How did comets come to exist in these two regions?

9. Briefly describe Pluto and Charon. Why won't Pluto collide with Neptune? How do we think Charon formed?

10. How are other Kuiper belt comets similar to Pluto?

11. Briefly describe the impact of Comet Shoemaker-Levy 9 on Jupiter.

12. Briefly describe the evidence suggesting that an impact caused the *mass extinction* that killed off the dinosaurs. How might the impact have led to the mass extinction?

13. How often should we expect impacts of various sizes on Earth? How serious a threat do we face from these impacts?

14. Summarize the role of the jovian planets in shaping the orbits of small bodies in the solar system. Why do some people think that Jupiter's existence is critical to our own?

TEST YOUR UNDERSTANDING

Surprising Discoveries?

Suppose we found a solar system with the property described. (These are not real discoveries.) In light of what you've learned about the formation of our own solar system, decide whether the discovery should be considered reasonable or surprising. Explain.

15. A small asteroid that orbits within the asteroid belt has an active volcano.

16. Scientists discover a meteorite that, based on radiometric dating, is 7.9 billion years old.

17. An object that resembles a comet in size and composition is discovered orbiting in the inner solar system.

18. Studies of a large object in the Kuiper belt reveal that it is made almost entirely of rocky (as opposed to icy) material.

19. Astronomers discover a previously unknown comet that will be brightly visible in our night sky about 2 years from now.

20. A mission to Eris finds that it has lakes of liquid water on its surface.

21. Geologists discover a crater from a 5-kilometer object that impacted Earth more than 100 million years ago.

22. Archaeologists learn that the fall of ancient Rome was caused in large part by an asteroid impact in Asia.

23. In another solar system, astronomers discover an object the size of Earth orbiting its star at the distance of the Kuiper belt.

24. Astronomers discover an asteroid with an orbit suggesting that it will collide with Earth in the year 2064.

Quick Quiz

Choose the best answer to each of the following. Explain your reasoning with one or more complete sentences.

25. The asteroid belt lies between the orbits of (a) Earth and Mars. (b) Mars and Jupiter. (c) Jupiter and Saturn.

26. Jupiter nudges the asteroids through the influence of (a) tidal forces. (b) orbital resonances. (c) magnetic fields.

27. Can an asteroid be pure metal? (a) No; all asteroids contain rock. (b) Yes; it must have formed where only metal could condense in the solar nebula. (c) Yes; it must have been the core of a shattered asteroid.

28. Did a large terrestrial planet ever form in the region of the asteroid belt? (a) No, because there was never enough mass there. (b) No, because Jupiter prevented one from accreting. (c) Yes, but it was shattered by a giant impact.

29. What does Pluto most resemble? (a) a terrestrial planet (b) a jovian planet (c) a comet

30. How big an object causes a typical shooting star? (a) a grain of sand or a small pebble (b) a boulder (c) an object the size of a car

31. Which have the most elliptical and tilted orbits? (a) asteroids (b) Kuiper belt comets (c) Oort cloud comets

32. Which are thought to have formed farthest from the Sun? (a) asteroids (b) Kuiper belt comets (c) Oort cloud comets

33. About how often does a 1-kilometer object strike Earth? (a) every year (b) every million years (c) every billion years

34. What would happen if a 1-kilometer object struck Earth? (a) It would break up in the atmosphere without causing widespread damage. (b) It would cause widespread devastation and climate change. (c) It would cause a mass extinction.

PROCESS OF SCIENCE

Examining How Science Works

35. *The Pluto Debate.* Research the decision to demote Pluto to dwarf planet. In your opinion, is this a good example of the scientific process? Does it exhibit the hallmarks of science described in Chapter 3? Compare your conclusions to opinions you find about the debate, and describe how you think astronomers should handle this or similar debates in the future.

36. *Life or Death Astronomy.* In most cases, the study of the solar system has little direct effect on our lives. But the discovery of an asteroid or comet on a collision course with Earth is another matter. How should the standards for verifiable observations described in Chapter 3 apply in this case? Is the potential danger so great that any astronomer with any evidence of an impending impact should spread the word as soon as possible? Or is the potential for panic so great that even higher standards of verification ought to be applied? What kind of review process, if any, would you set in place? Who should be informed of an impact threat, and when?

GROUP WORK EXERCISE

37. *Assessing Impact Danger.* Your task in this exercise is to assess the risks we face on Earth from meteorite and comet impacts. Before you begin, assign the following roles to the people in your group: *Scribe* (collects data and takes notes on the group's activities), *Proposer* (proposes hypotheses and explanations of the data), *Skeptic* (points out weaknesses in the hypotheses and explanations), *Moderator* (leads group discussion and makes sure everyone contributes). Each person should write down the answers from each part of the exercise.
a. Estimate the odds that human civilization will be destroyed by an impact during your lifetime. The *Moderator* should lead a discussion about how to go about making such an estimate. As a group, determine the kinds of information you would need, develop a method for making the estimate, and write down your method. **b.** *Scribe* analyzes Figure 12.28, determines whether it contains any of the necessary information, and shares his or her findings with the group. **c.** *Proposer* uses the group's method to estimate the probability that civilization will be destroyed by an impact during your lifetime, which you can assume to be 100 years for the purpose of this exercise. *Skeptic* should check *Proposer's* work. **d.** Estimate the probability that an impact will cause widespread devastation somewhere on Earth during your lifetime. *Moderator* should lead the discussion about how to make this estimate. *Scribe* should gather the information from Figure 12.28. *Proposer* should do the calculation, and *Skeptic* should check *Proposer's* work. **e.** Finding near-Earth asteroids early greatly increases our chances of deflecting them. Given the probabilities

from parts c and d and considering the damage these events would cause, decide as a group how much money per year should be spent on finding near-Earth asteroids and explain your reasoning.

INVESTIGATE FURTHER

In-Depth Questions to Increase Your Understanding

Short-Answer/Essay Questions

38. *The Role of Jupiter.* Suppose that Jupiter had never existed. Describe at least three ways in which our solar system would be different, and clearly explain why.

39. *Life Story of an Iron Atom.* Imagine that you are an iron atom in a processed meteorite made mostly of iron. Tell the story of how you got to Earth, beginning from the time you were part of the gas in the solar nebula 4.6 billion years ago. Include as much detail as possible. Your story should be scientifically accurate but also creative and interesting.

40. *Asteroid Discovery.* You have discovered two new asteroids and named them Albert and Isaac. Both lie at the same distance from Earth and have the same brightness when you look at them through your telescope, but Albert is twice as bright as Isaac at infrared wavelengths. What can you deduce about the relative reflectivities and sizes of the two asteroids? Which would make a better target for a mission to mine metal? Which would make a better target for a mission to obtain a sample of a carbon-rich planetesimal? Explain.

41. *Asteroids vs. Comets.* Contrast the compositions and locations of comets and asteroids, and explain in your own words why they have turned out differently.

42. *Comet Tails.* Describe in your own words why comets have tails. Why do most comets have two distinct visible tails, and why do the tails go in different directions? Why is the third, invisible tail of small pebbles of interest to us on Earth?

43. *Oort Cloud vs. Kuiper Belt.* Explain in your own words how and why there are two different reservoirs of comets. Be sure to discuss where the two groups of comets formed and what kinds of orbits they travel on.

44. *Project: Dirty Snowballs.* If there is snow where you live or study, make a dirty snowball. (The ice chunks that form behind tires work well.) How much dirt does it take to darken snow? Find out by allowing your dirty snowball to melt in a container and measuring the approximate proportions of water and dirt afterward.

Quantitative Problems

Be sure to show all calculations clearly and state your final answers in complete sentences.

45. *Adding Up Asteroids.* It's estimated that there are a million asteroids 1 kilometer across or larger. If a million asteroids 1 kilometer across were all combined into one object, how big would it be? How many 1-kilometer asteroids would it take to make an object as large as the Earth? (*Hint:* You can assume they're spherical. The expression for the volume of a sphere is $\frac{4}{3}\pi r^3$, where r is the radius.)

46. *Impact Energies.* A relatively small impact crater 20 kilometers in diameter could be made by a comet 2 kilometers in diameter traveling at 30 kilometers per second (30,000 m/s).
    a. Assume that the comet has a total mass of 4.2×10^{12} kilograms. What is its total kinetic energy? (*Hint:* The kinetic energy is equal to $\frac{1}{2}mv^2$, where m is the comet's mass and v is its speed. If you use mass in kilograms and velocity in m/s, the answer for kinetic energy will have units of joules.) **b.** Convert your answer from part a to an equivalent in megatons of TNT, the unit used for nuclear bombs. Comment on the degree of devastation the

impact of such a comet could cause if it struck a populated region on Earth. (*Hint:* One megaton of TNT releases 4.2×10^{15} joules of energy.)

47. *The "Near Miss" of Toutatis.* The 5-kilometer asteroid Toutatis passed a mere 1.5 million kilometers from Earth in 2004. Suppose Toutatis were destined to pass *somewhere* within 1.5 million kilometers of Earth. Calculate the probability that this "somewhere" would have meant that it slammed into Earth. Based on your result, do you think it is fair to call the 2004 passage a "near miss"? Explain. (*Hint:* You can calculate the probability by considering an imaginary dartboard of radius 1.5 million kilometers in which the bull's-eye has Earth's radius, 6378 kilometers.)

48. *Room to Roam.* It's estimated that there are a trillion comets in the Oort cloud, which extends out to about 50,000 AU. What is the total volume of the Oort cloud, in cubic AU? How much space does each comet have in cubic AU, on average? Take the cube root of the average volume per comet to find the comets' typical spacing in AU. (*Hints:* For the purpose of this calculation, you can assume the Oort cloud fills the whole sphere out to 50,000 AU. The volume of a sphere is given by $\frac{4}{3}\pi r^3$, where r is the radius.)

49. *Comet Temperatures.* Find the "no greenhouse" temperatures for a comet at distances from the Sun of 50,000 AU (in the Oort cloud), 3 AU, and 1 AU (see Mathematical Insight 10.1). Assume that the comet reflects 3% of the incoming sunlight. At which location will the temperature be high enough for water ice to vaporize (about 150 K)? How do your results explain comet anatomy? Explain.

50. *Comet Dust Accumulation.* A few hundred tons of comet dust are added to Earth daily from the millions of meteors that enter our atmosphere. Estimate the time it would take for Earth to get 0.1% heavier at this rate. Is this mass accumulation significant for Earth as a planet? Explain.

Discussion Questions

51. *Rise of the Mammals.* Suppose the impact 65 million years ago had not occurred. How do you think our planet would be different? For example, do you think that mammals still would eventually have come to dominate Earth? Would we be here? Defend your opinions.

52. *How Should Kids Count Planets?* The new definitions that have officially demoted Pluto from planet to dwarf planet have many educational implications. For example, many children learn songs in school that refer to "nine planets" and Pluto. How would you recommend that school teachers deal with the new definitions? Be sure to consider the fact that while these definitions have "official" status today (from the International Astronomical Union), it is possible that they may change again in the future.

Web Projects

53. *Asteroid and Comet Missions.* Learn about a past or present space mission to study asteroids or comets, such as *NEAR*, *Dawn*, or *Rosetta*. What did it (or will it) accomplish? Write a one- to two-page summary of your findings.

54. *The New Horizons Mission to Pluto.* Find out the current status of the *New Horizons* mission. What are its science goals? What has it done so far? Summarize your findings in a few paragraphs.

55. *Impact Hazards.* Many groups are searching for near-Earth asteroids that might impact our planet. They use something called the *Torino Scale* to evaluate the possible danger posed by an asteroid based on how well we know its orbit. What is this scale? What object has reached the highest level on this scale? What were the estimated chances of impact, and when?

b

c

e

d

20 AU
0.5"

This infrared image from the Keck Telescope shows direct detection of a four-planet system (planets marked b, c, d, e) orbiting the star HR 8799. Light from the star itself (center) was mostly blocked out during the exposure.

13 OTHER PLANETARY SYSTEMS
THE NEW SCIENCE OF DISTANT WORLDS

LEARNING GOALS

13.1 DETECTING PLANETS AROUND OTHER STARS
- Why is it so challenging to learn about extrasolar planets?
- How can a star's motion reveal the presence of planets?
- How can changes in a star's brightness reveal the presence of planets?

13.2 THE NATURE OF PLANETS AROUND OTHER STARS
- What properties of extrasolar planets can we measure?
- How do extrasolar planets compare with planets in our solar system?

13.3 THE FORMATION OF OTHER SOLAR SYSTEMS
- Can we explain the surprising orbits of many extrasolar planets?
- Do we need to modify our theory of solar system formation?

13.4 THE FUTURE OF EXTRASOLAR PLANETARY SCIENCE
- How will future observations improve our understanding?

How vast those Orbs must be, and how inconsiderable this Earth, the Theatre upon which all our mighty Designs, all our Navigations, and all our Wars are transacted, is when compared to them. A very fit consideration, and matter of Reflection, for those Kings and Princes who sacrifice the Lives of so many People, only to flatter their Ambition in being Masters of some pitiful corner of this small Spot.

—Christiaan Huygens, c. 1690

For more than 450 years after Copernicus recognized Earth as one member of our Sun's planetary system, the study of planetary systems remained limited to our own. Then, about two decades ago, a new scientific revolution began with the first discoveries of planets around other stars. This revolution is now in full bloom, with new discoveries coming at a rapid pace.

The rapidly advancing science of extrasolar planets has dramatic implications for our understanding of our place in the universe. The fact that planets are common in the universe makes it seem more likely that we might someday find life elsewhere, perhaps even intelligent life. Moreover, having many more worlds to compare to our own vastly enhances our ability to learn how planets work, which may help us better understand our home planet, Earth. It also allows us to test the nebular theory of solar system formation in new settings. In this chapter, we'll focus our attention on the exciting new science of other planetary systems.

 Detecting Extrasolar Planets Tutorial, Lessons 1–3

13.1 DETECTING PLANETS AROUND OTHER STARS

The very idea of planets around other stars, or **extrasolar planets** for short, would have shattered the world views of many people throughout history. After all, cultures of the western world long regarded Earth as the center of the universe, and nearly all ancient cultures imagined the heavens to be a realm distinct from Earth.

The Copernican revolution, which taught us that Earth is a planet orbiting the Sun, opened up the possibility that planets might also orbit other stars. Still, until the 1990s, no extrasolar planets were known. Given that the number of known or suspected extrasolar planets is now in the thousands and growing rapidly, you may wonder what took so long.

Why is it so challenging to learn about extrasolar planets?

We've known for centuries that other stars are distant suns (see Special Topic, page 368), making it natural to suspect that they would have their own planetary systems. The nebular theory of solar system formation, well established decades ago, made extrasolar planets seem even more likely.

As we discussed in Chapter 8, the nebular theory explains our planetary system as a natural consequence of processes that accompanied the birth of our Sun. The theory therefore seems to predict that planets should be common around other stars. However, it took many decades for technology to reach the point at which this prediction could be put to the test.

The first technological challenge is one we already discussed in the context of our scale model solar system in Chapter 1. Recall that on a 1-to-10-billion scale, the Sun is the size of a grapefruit, Earth is a pinhead orbiting 15 meters away, and Jupiter is a marble orbiting 80 meters away. On the same scale, the distance to the nearest stars (besides the Sun) is equivalent to the distance across the United States. Therefore, looking for planets orbiting even the nearest stars would be like looking from San Francisco for pinheads and marbles orbiting a grapefruit in Washington, D.C. Most stars are much farther away, making the challenge far greater.

A second challenge comes from the fact that a Sun-like star is a *billion times* brighter than the light reflected from any of its planets. Because even the best telescopes blur the light from stars at least a little, the glare of scattered starlight generally overwhelms any small spots of planetary light, even if those spots are bright enough to be detectable in principle. The problem is somewhat lessened if we observe in infrared light, because planets emit their own infrared light and Sun-like stars are dimmer in the infrared, but it is still significant.

As recently as the early 1990s, these challenges made even some astronomers think that we were still decades away from finding extrasolar planets. But human ingenuity proved greater than the pessimists had guessed. Thanks to technological advances, clever planet-hunting strategies, and some unexpected differences between our solar system and others, we are discovering vast numbers of planets orbiting other stars. Although it is too soon to know for sure, it seems likely that our Milky Way Galaxy is home to billions of planetary systems.

Nevertheless, the combination of the distance scale and the vast differences in brightness between stars and planets still adds up to a tremendous technological challenge as we seek to obtain images or spectra of extrasolar planets. This brings us to a key point about our current ability to study planets around other stars. In general, there are two ways to learn about distant objects: *directly*, which means by obtaining images or spectra of the objects themselves, and *indirectly*, which means by inferring existence or properties without actually seeing the objects under study.

In principle, direct study is always preferable and allows us to learn far more. In the case of extrasolar planets, direct study could mean imaging of visible light reflected by the planets or of infrared light that the planets emit, or it could mean obtaining visible or infrared spectra of the planets. We've achieved some success in direct detection, but we will need significant advances in technology before we are able to obtain images or spectra of many extrasolar planets with even moderate resolution. The best we can do today is exemplified by images such as the one on the chapter opening page (page 366), in which we see dots of infrared light for planets that are much larger and farther from their star than the planets of our own solar system. We'll discuss present and future direct detection efforts in more detail in Section 13.4.

In the meantime, the difficulty of direct study means that nearly everything we know today about extrasolar planets comes from indirect study. There are two major approaches to indirectly finding and studying extrasolar planets:

1. Observing the motion of a star to detect the subtle gravitational effects of orbiting planets.

2. Observing changes to a star's brightness that occur when one of its planets passes in front of the star as viewed from Earth.

The earliest discoveries of extrasolar planets came primarily through the first approach, but the current champion of extrasolar planet discoveries—NASA's *Kepler* mission—uses the second. We can learn even more in cases in which we can combine both indirect approaches, because each can provide different information about a distant planet.

Because these two approaches to indirect study are so important to current understanding of extrasolar planets, we will devote the rest of this section to studying them in somewhat more detail.

How can a star's motion reveal the presence of planets?

The first approach to indirect study relies on observing stars in order to detect motion that we can attribute to gravitational tugs from orbiting planets. This type of detection is indirect because we discover the planets by observing their stars without actually seeing the planets themselves.

Although we usually think of a star as remaining still while planets orbit around it, that is only approximately correct. In reality, all the objects in a star system, including the star itself, orbit the system's "balance point," or *center of mass* [**Section 4.4**]. To understand how this fact allows us to discover extrasolar planets, imagine the viewpoint of extraterrestrial astronomers observing our solar system from afar.

Let's start by considering only the influence of Jupiter, the most massive planet in our solar system. The center of mass between the Sun and Jupiter lies just outside the Sun's visible surface (**FIGURE 13.1**), so what we usually think of as Jupiter's 12-year orbit around the Sun is really a 12-year orbit around the center of mass. Because the Sun and Jupiter are always on opposite sides of the center of mass (otherwise it wouldn't be a "center"), the Sun must orbit this point with the same 12-year period. The Sun's orbit traces out only a small circle (or ellipse) with each 12-year period, because the Sun's average orbital distance is barely larger than its own radius; that is why

How Did We Learn That Other Stars Are Suns?

Today we know that stars are other suns—meaning objects that produce enough energy through nuclear fusion to supply light and heat to orbiting planets—but this fact is not obvious from looking at the night sky. After all, the feeble light of stars hardly seems comparable to the majestic light of the Sun. Most ancient observers guessed that stars were much more mundane; typical guesses suggested that they were holes in the celestial sphere or flaming rocks in the sky.

The only way to realize that stars are suns is to know that they are incredibly far away; then, a simple calculation will show that they are actually as bright as or brighter than the Sun [**Section 15.1**]. The first person to make reasonably accurate estimates of the distances to stars was Christiaan Huygens. By *assuming* that other stars were indeed suns, as some earlier astronomers had guessed, Huygens successfully estimated stellar distances. The late Carl Sagan eloquently described the technique:

Huygens drilled small holes in a brass plate, held the plate up to the Sun and asked himself which hole seemed as bright as he remembered the bright star Sirius to have been the night before. The hole was effectively $\frac{1}{28,000}$ the apparent size of the Sun. So Sirius, he reasoned, must be 28,000 times farther from us than the Sun, or about half a light-year away. It is hard to remember just how bright a star is many hours after you look at it, but Huygens remembered very well. If he had known that Sirius was intrinsically brighter than the Sun, he would have come up with [a better estimate of] the right answer: Sirius is [8.6] light-years away.[*]

Christiaan Huygens (1629–1695)

Huygens could not actually prove that stars are suns, since his method was based on the assumption that they are. However, his results explained a fact known since ancient times: Stellar parallax is undetectable to the naked eye [**Section 2.4**]. Recall that the lack of detectable parallax led many Greeks to conclude that Earth must be stationary at the center of the universe, but there was also an alternative explanation: Stars are incredibly far away. Even with his original estimate that Sirius was only half a light-year away, Huygens knew that its parallax would be far too small to observe by naked eye or with the telescopes available at the time. Huygens thereby "closed the loop" on the ancient mystery of the nature of stars, showing that their appearance and lack of parallax made perfect sense if they were very distant suns. This new knowledge apparently made a great impression on Huygens, as you can see from the quotation at the top of page 367.

[*]From *Cosmos*, by Carl Sagan (Random House, 1980). Sagan demonstrates the technique in the *Cosmos* video series, Episode 7.

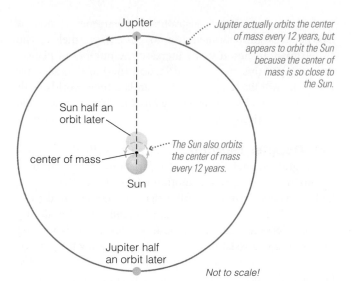

FIGURE 13.1 This diagram shows how both the Sun and Jupiter orbit around their mutual center of mass, which lies very close to the Sun. The diagram is not to scale; the sizes of the Sun and its orbit are exaggerated about 100 times compared to the size shown for Jupiter's orbit, and Jupiter's size is exaggerated even more.

In the diagram: *Jupiter actually orbits the center of mass every 12 years, but appears to orbit the Sun because the center of mass is so close to the Sun.* *The Sun also orbits the center of mass every 12 years.* Labels: Jupiter; Sun half an orbit later; center of mass; Sun; Jupiter half an orbit later; Not to scale!

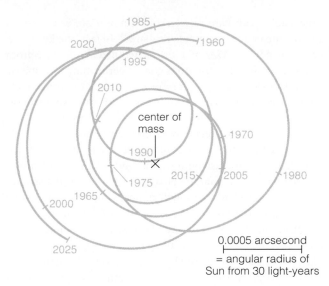

FIGURE 13.2 This diagram shows the orbital path of the Sun around the center of mass of our solar system as it would appear from a distance of 30 light-years away, for the period 1960–2025. Notice that the entire range of motion during this period is only about 0.0015 arcsecond, which is almost 100 times smaller than the angular resolution of the Hubble Space Telescope. Nevertheless, if alien astronomers could measure this motion, they could learn of the existence of planets in our solar system.

In the diagram: center of mass; 0.0005 arcsecond = angular radius of Sun from 30 light-years.

we generally don't notice the Sun's motion. Nevertheless, with sufficiently precise measurements, extraterrestrial astronomers could detect this orbital movement of the Sun and thereby deduce the existence of Jupiter, even without having observed Jupiter itself. They could even determine Jupiter's mass from the orbital characteristics of the Sun as it goes around the center of mass. A more massive planet at the same distance would pull the center of mass farther from the Sun's center, giving the Sun a larger orbit and a faster orbital speed around the center of mass.

SEE IT FOR YOURSELF

To see how a small planet can make a big star wobble, find a pencil and tape a heavy object (such as a set of keys) and a lighter object (perhaps a few coins) to opposite ends. Tie a string (or piece of floss) at the balance point—the center of mass—so that the pencil is horizontal. Then tap the lighter object into "orbit" around the heavier object. What does the heavier object do, and why? How does your setup correspond to a planet orbiting a star? You can experiment further with objects of different weights or shorter pencils; try to explain the differences you see.

Now let's add in the effects of Saturn. Saturn takes 29.5 years to orbit the Sun, so by itself it would cause the Sun to orbit their mutual center of mass every 29.5 years. However, because Saturn's influence is secondary to that of Jupiter, this 29.5-year period appears as a small effect added on top of the Sun's 12-year orbit around its center of mass with Jupiter. In other words, every 12 years the Sun would return to *nearly* the same orbital position around its center of mass with Jupiter, but the precise point of return would move around with Saturn's 29.5-year period. By measuring this motion carefully over many years, an extraterrestrial astronomer could deduce the existence and masses of both Jupiter and Saturn.

The other planets also exert gravitational tugs on the Sun, which further affect the Sun's orbital motion around the solar system's center of mass (**FIGURE 13.2**). These extra effects are smaller still, and so are even more difficult to measure in practice, but extremely precise observations could in principle allow an extraterrestrial astronomer to discover all the planets in our solar system. If we turn this idea around, it means we can search for planets in other star systems by carefully watching for the tiny orbital motion of a star around the center of mass of its star system.

Astronomers use two distinct observation methods to search for this type of orbital motion by stars. The first, called the **astrometric method** (*astrometry* means "measurement of the stars"), uses very precise measurements of stellar positions in the sky to look for the slight motion caused by orbiting planets. The second, called the **Doppler method**, takes advantage of the Doppler effect [**Section 5.4**], which allows us to measure changes in a star's velocity toward or away from us that are caused by orbiting planets.

The Astrometric Method The astrometric method has been used for decades to identify binary star systems, because two orbiting stars move periodically around their center of mass. The technique works especially well for binary systems in which the two stars are not too close together, because the stellar motions tend to be larger in those cases. In the case of planet searches, however, the expected stellar motion is much more difficult to detect.

For example, from a distance of 10 light-years, a Jupiter-size planet in a Jupiter-like orbit (5 AU from a Sun-like star) would cause its star to move slowly over a side-to-side angular distance of only about 0.003 arcsecond—approximately the width of a hair seen from a distance of 5 kilometers.

Remarkably, with careful telescope calibration, astronomers can now measure movements this small. In fact, the European Space Agency's *GAIA* mission, scheduled to operate from 2013 to 2018, has the ambitious goal of performing astrometric observations of a billion stars in our galaxy to an accuracy that in some cases will be better than 10 microarcseconds. (A microarcsecond is one-millionth of an arcsecond.)

Even with such precise measurements, two unavoidable constraints limit the number of planetary systems that can be studied with the astrometric method. The first constraint comes from the fact that the farther away a star is, the smaller its side-to-side movement will appear. For example, while Jupiter causes the Sun to move by about 0.003 arcsecond as seen from 10 light-years away, the observed motion is only half as large when seen from 20 light-years away and one-tenth as large when seen from 100 light-years away. The astrometric method therefore works best for detecting the gravitational effects of massive planets around relatively nearby stars.

The second constraint arises from the fact that, for a planet of any particular mass, a greater orbital distance means a larger astrometric effect on its star, but it also means that the effect takes longer to measure. For example, moving Jupiter farther from the Sun would increase the angular extent of the Sun's side-to-side motion as seen from a distance (because moving Jupiter outward would also move the center of mass outward from the Sun), but Kepler's third law tells us that this move would also increase the period of Jupiter's orbit. For example, if Jupiter moved to Neptune's distance from the Sun, at which the orbital period is 165 years, it might take a century or more of patient observation to be confident that stellar motion was occurring in a 165-year cycle.

As a result of these constraints, the astrometric method has been of only limited use to date. Nevertheless, the *GAIA* mission should vastly increase the number of planets detected with this method (see Section 13.4). If all is successful, *GAIA* will likely discover thousands of new worlds, with more and more discoveries as the mission goes on and is able to observe motions caused by planets in more distant orbits.

The Doppler Method The Doppler method also looks for the gravitational influence of a planet on a star, but instead of looking for side-to-side motion of the star in the sky, we look for back-and-forth motion that can be measured from Doppler shifts in a star's spectrum (**FIGURE 13.3**). Recall that the Doppler effect causes a blueshift when a star is moving toward us and a redshift when it is moving away from us, so alternating blueshifts and redshifts (relative to a star's average Doppler shift) indicate orbital motion around a center of mass.

The Doppler method has the distinction of having been used for the first discovery of an extrasolar planet around a Sun-like star*: a planet orbiting a star called 51 Pegasi, found in 1995 by Swiss astronomers Michel Mayor and Didier Queloz, and soon confirmed by a team led by American astronomers Geoffrey W. Marcy and R. Paul Butler. The discovery came

---

*The first actual detection of extrasolar planet-size objects occurred in 1992, when precise timing measurements revealed the existence of three objects with Earth-like masses orbiting a type of "dead" star known as a *pulsar* [**Section 18.2**]. Because pulsars are created when stars die in supernova explosions, these "planets" must be either the charred remains of preexisting planets or, more likely, objects that somehow formed from supernova debris. Either way, they are not planets in the same sense as those that form during star birth; in this chapter, we will focus only on planets orbiting ordinary stars like our Sun.

SPECIAL TOPIC

The Names of Extrasolar Planets

The planets in our solar system have familiar names rooted in mythology. Unfortunately, there's not yet a well-accepted scheme for naming extrasolar planets. Astronomers still generally refer to extrasolar planets by the star they orbit, such as "the planet orbiting the star named…." Worse still, the stars themselves often have confusing or even multiple names, reflecting naming schemes used in star catalogs made by different people at different times in history.

A few hundred of the brightest stars in the sky carry names from ancient times. Many of these names are Arabic—such as Betelgeuse, Algol, and Aldebaran—because of the work of the Arab scholars of the Middle Ages [**Section 3.2**]. In the early 1600s, German astronomer Johann Bayer developed a system that gave names to many more stars: Each star gets a name based on its constellation and a Greek letter indicating its ranking in brightness within that constellation. For example, the brightest star in the constellation Andromeda is called Alpha Andromedae, the second brightest is Beta Andromedae, and so on. Bayer's system worked for only the 24 brightest stars in each constellation, because there are only 24 letters in the Greek alphabet. About a century later, English astronomer John Flamsteed published a more extensive star catalog in which he used numbers once the Greek letters were exhausted. For example, 51 Pegasi gets its name from Flamsteed's catalog. (Flamsteed's numbers are based on position within a constellation rather than brightness.)

As more powerful telescopes made it possible to discover more and fainter stars, astronomers developed many new star catalogs. The names we use today usually come from one of these catalogs. For example, the star HD 209458 appears as star number 209458 in a catalog compiled by Henry Draper (HD). You may see star names consisting of numbers preceded by other catalog names, including Gliese, Ross, and Wolf; these catalogs are also named for the astronomers who compiled them. Because the same star is often listed in several catalogs, a single star can have several different names.

Objects orbiting other stars usually carry the star name plus a letter denoting their order of discovery around that star. If the second object is another star, a capital B is added to the star name; a lowercase b is added if it's a planet. For example, HD 209458b is the first planet discovered orbiting star number 209458 in the Henry Draper catalog; Upsilon Andromedae d is the third planet discovered orbiting the twentieth brightest star (because upsilon is the twentieth letter in the Greek alphabet) in the constellation Andromeda. Some recently discovered planets have been named for the observing program that discovered them. For example, Kepler 11g is the sixth planet in the eleventh planetary system announced by the *Kepler* mission. Many astronomers hope soon to devise a better naming system for these wonderful new worlds.

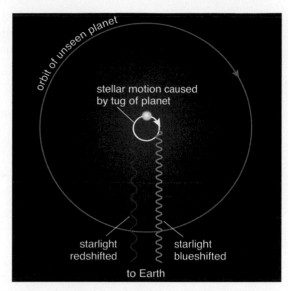

FIGURE 13.3 interactive figure The Doppler technique for discovering extrasolar planets: The star's Doppler shift alternates toward the blue and toward the red, allowing us to detect its slight motion—caused by an orbiting planet—around the center of mass.

from careful measurements showing that the spectrum of 51 Pegasi has alternating blueshifts and redshifts with a period of about 4 days (**FIGURE 13.4**).

The 4-day period of the star's motion must be the orbital period of its planet. We therefore know that the planet lies so close to the star that its "year" lasts only 4 Earth days and its surface temperature is probably over 1000 K. The Doppler data also allow us to determine the planet's approximate mass, because a more massive planet has a greater gravitational effect on the star (for a given orbital distance) and therefore causes the star to move at higher speed around the

system's center of mass. In the case of 51 Pegasi, the data show that the planet has about half the mass of Jupiter. The combination of the planet's high mass with its closeness to its star makes it an example of what we call a **hot Jupiter**, because it has a Jupiter-like mass but a much higher surface temperature.

Current techniques can measure a star's velocity to within about 1 meter per second—walking speed—which corresponds to a Doppler wavelength shift of just 0.0000003%. We can therefore find planets that exert a considerably smaller gravitational tug on their stars than the planet orbiting 51 Pegasi does on its star. More than 700 planets have been detected by the Doppler method as of 2012, many by the same teams of astronomers that made the discovery around 51 Pegasi. In fact, the Doppler data are often good enough to allow us to measure the gravitational tugs of two or more planets on a single star; as of 2012, nearly 100 multiple-planet systems had been identified by the Doppler method.

Despite the success of the Doppler method, it has some limitations. The fact that the Doppler method, like the astrometric method, searches for gravitational tugs from orbiting planets means that it is much better for finding massive planets like Jupiter than small planets like Earth. However, the Doppler method is best suited to identifying massive planets that orbit relatively *close* to their star (in contrast to the astrometric method, which works better for planets farther from their star), because being closer means a stronger gravitational tug and hence a greater velocity for the star as it orbits the center of mass. The fact that the Doppler method, like all indirect methods, requires observing long enough to see a star's orbit around the center of mass—which is equivalent to the orbital period of the planet causing the motion—means that the Doppler method is biased toward finding close-in planets with short orbital periods. That is why many of the earliest

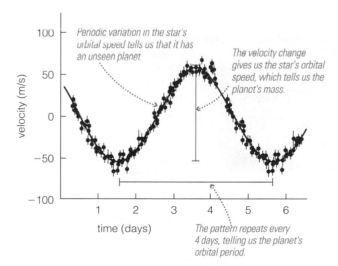

a A periodic Doppler shift in the spectrum of the star 51 Pegasi shows the presence of a large planet with an orbital period of about 4 days. Dots are actual data points; bars through dots represent measurement uncertainty.

FIGURE 13.4 The discovery of a planet orbiting 51 Pegasi.

b Artist's conception of the planet orbiting 51 Pegasi, which probably has a mass similar to that of Jupiter but orbits its star at only about one-eighth of Mercury's orbital distance from the Sun. It probably has a surface temperature above 1000 K, making it an example of what we call a hot Jupiter.

discoveries of extrasolar planets were of hot Jupiters like 51 Pegasi b. Finally, the Doppler method requires a fairly large telescope in order to obtain spectra with high enough resolution to look for very small Doppler shifts, which limits the number of stars that can be studied with this method.

How can changes in a star's brightness reveal the presence of planets?

The second general approach to detecting planets indirectly relies on searching for slight changes in a star's brightness. This approach has its greatest use in what we call the **transit method**, which is the method used by the *Kepler* mission to search for planets around other stars.

Transits If we were to examine a large sample of stars with planets, a small number of them—typically one in several hundred—would by chance be aligned in such a way that one or more of its planets would pass directly between us and the star once each orbit. In such cases, we say that a planet's orbital tilt, or *inclination,* is aligned with the direction to Earth. The result is a **transit**, in which the planet appears to move across the face of the star. We occasionally witness this effect in our own solar system when Mercury or Venus crosses in front of the Sun (see Figure S1.5). Other star systems are so far away that we cannot actually see a planetary dot set against the face of the star as we can see Mercury or Venus against the face of the Sun. Nevertheless, a transiting planet will block a little of its star's light, allowing us to detect its passage as the star appears temporarily dimmer (**FIGURE 13.5**). The larger the planet, the more dimming it will cause. Many transiting planets also offer a measurable **eclipse** as the planet goes behind the star, during which we see a dip in the system's brightness. The reason is that the system's total brightness comes from the star and planet together (perhaps with additional contributions from other planets in the system), so this brightness goes down when the planet is hidden behind the star during the eclipse. Eclipse observations are more easily accomplished in the infrared, because planets contribute a greater proportion of a system's infrared brightness than visible-light brightness.

Note that, because most stars exhibit intrinsic variations in brightness, we can be confident that we have detected a planet only if we observe the dips in brightness to repeat with a regular period, indicating that the same planet is passing in front of the star repeatedly with each orbit. (Intrinsic stellar brightness variations tend to be more random.) In that case, the period of the repeated dimmings tells us the orbital period of the transiting planet. Scientists generally require at least three repeated transits before concluding that a planet is probably responsible. Using the transit method to discover a planet therefore requires monitoring a star's brightness for at least three full orbits of the planet. For example, we would need at least about three years of monitoring with the transit method to detect a planet in an Earth-like orbit.

Putting these ideas together, we find that the first requirement for success with the transit method is that a system have planet orbits aligned "just right" so that they pass in front of their star as seen from Earth. When that is the case, the transit method will always be able to find close-in planets with short orbital periods before it can find more distant planets, simply because less time is needed to observe repeated transits of the closer-in planets. Adding the fact that larger planets create bigger dips in brightness means that transits are easiest to measure for hot Jupiters, because they are large and close in.

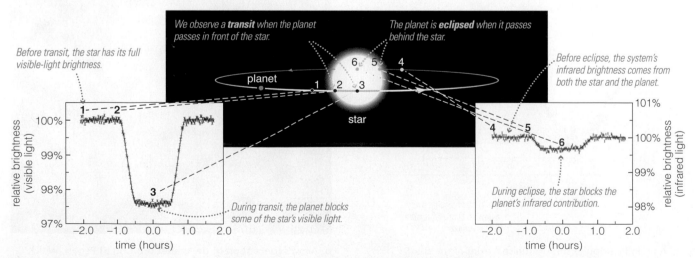

FIGURE 13.5 interactive figure This diagram represents transits and eclipses of the planet orbiting the star HD 189733. The graph shows that each transit lasts about 2 hours, during which the star's visible-light brightness dips by about 2.5%. The eclipses are observable in the infrared, because the star blocks the infrared contribution of the planet. The transits and eclipses each occur once during every 2.2-day orbit of the planet.

The first success of the transit method came in 1999, when a relatively small ground-based telescope detected transit dips from a hot Jupiter known as HD 209458b. This planet had already been detected by the Doppler method, so the transits in this case served as confirmation rather than discovery. Nevertheless, this success spurred scientists to begin searching more deliberately for transits, and many dozens of planets have now been discovered with ground-based searches. The transit method has met even greater success with space-based telescopes, beginning with the European Space Agency's *COROT* mission, launched in 2006. NASA's Spitzer Space Telescope, which observes infra-red light, has also successfully observed transits, even though it was not designed for that purpose. But the study of extraso-lar planets through transits truly came of age with the launch of the *Kepler* spacecraft in 2009.

The *Kepler* Mission The *Kepler* mission was con-ceived with the ambitious goal of discovering Earth-size planets in Earth-like orbits. This poses a much greater tech-nical challenge than detecting hot Jupiters. There are three major challenges, each of which was addressed in the mission's design.

The first challenge comes from the statistics of orbital alignments. Remember that only a small fraction of planets will by chance have their orbits oriented in such a way as to pass in front of their star from our vantage point on Earth. The chance declines with a planet's distance from its star. Calculations show that only about 1 in 200 planetary systems should by chance have an orientation that would allow us to see a transit by an Earth-size planet in an Earth-like orbit. To compensate for these low odds, *Kepler* must monitor a large number of stars in order to have reasonable expectation of a small number of successes. *Kepler* monitors about 150,000 stars by keeping its telescope continuously pointed at a rela-tively wide field of view in the constellation Cygnus.

The second challenge comes from transit timing. For an Earth-like planet in an Earth-like orbit, a transit will last no more than a few hours and occur only once a year. As a result, we may miss transits unless we watch for them both day and night over a period of years. *Kepler* was therefore placed into an orbit around the Sun rather than Earth, so that Earth does not get in the way of its observations at any time. To make sure no transit is missed, *Kepler* measures the brightnesses of all its stars about every 30 minutes. To allow time for observ-ing multiple transits, the mission was designed to last for at least four years.

The final challenge is in the need for extraordinarily precise measurement, because the dips in brightness caused by Earth-size planets are very small. For example, viewed from afar, a transit of Earth across the Sun would dim the Sun's light by only about 0.008%—not quite one part in 10,000. *Kepler*'s cameras are sensitive enough to detect transits of Earth-size planets around Sun-like stars, and transits of planets as small as Mercury around smaller, dimmer stars. As a result, *Kepler* is currently capable of finding much smaller planets than can be found with any other method, especially if the planets are at larger distances from their star.

The early success of *Kepler* has been nothing short of phenomenal. Because it was designed to detect planets as small as Earth on year-long orbits, it's had no trouble finding Jupiter- and Neptune-size planets on shorter orbits. As of mid-2012, *Kepler* mission scientists had sifted through data from only the first 16 months of observations, and had already identified planet "candidates" around nearly 1800 stars. More than 300 of these stars have two or more planet candidates, making a total of more than 2300 planet candidates. The detections are called "candidates" until they're confirmed by follow-up observa-tions (such as detection by the Doppler method), because even when three dips in a star's brightness that look like transits are observed, there's a small possibility that the dips are being caused by something besides an orbiting planet. Statistically, though, it's expected that about 90% of the candidates will eventually be confirmed. Given this early success, it's likely that the *Kepler* mission will have found hundreds or even thousands of additional planets by the time you are reading this book.

THINK ABOUT IT

Check the current status of the *Kepler* mission online. How many planets has it found? What is the most Earth-like planet discovered by the mission so far?

Advantages and Limitations of the Transit Method The most obvious limitation of the transit method is that it can work only for the small fraction of planetary sys-tems in which planet orbits are oriented edge-on to Earth. This method therefore will never be able to detect more than a very small fraction of the planets that actually exist. Nevertheless, as the *Kepler* mission demonstrates, this limitation can be coun-terbalanced by observing large numbers of stars, so that even a small fraction can represent large numbers of planets. The transit method shares the bias of the other methods toward finding planets with shorter orbital periods, simply because it takes longer to observe planets with longer periods.

The transit method also has several advantages over the astrometric and Doppler methods. Of greatest significance, the transit method can reveal planets far smaller than can currently be found with the astrometric or Doppler method. There are two reasons for this advantage. First, although small planets cause only a very small dimming of their star, the *Kepler* mission proves that such dimmings are measurable for planets as small as Earth. Second, in a multiplanet system, the astrometric and Doppler methods can detect the exist-ence of small planets only as small effects added to the much larger gravitational effects of larger planets in the system. In contrast, a small planet leaves a distinct signature with the transit method, because planets with different periods will rarely transit in front of a star at the same time.

Perhaps most remarkably, in some cases transits cause enough stellar dimming to be measurable even with small, personal telescopes. A telescope as small as 4 inches in diam-eter has been used to discover an extrasolar planet, and it's relatively easy to confirm for yourself some of the transits that have already been detected. What was once considered impossible for scientists can now be assigned as homework for students (see Problem 48 at the end of the chapter).

TABLE 13.1 Major Advantages and Limitations of Extrasolar Planet Detection Methods

	Major Advantages	Major Limitations	Major Planetary Properties Measured	Confirmed or Candidate Planets (as of mid-2012)
Direct Detection	■ Allows direct study of images or spectra	■ Currently possible only in rare cases	■ Spectra ■ Surface details (with future technology)	A few
Astrometric Method	■ Best for detecting massive planets that orbit far from their stars	■ Currently possible only in rare cases ■ Works best for nearby stars; for distant planets, requires space observatory	■ Planet mass ■ Orbital period and distance (semimajor axis) ■ Orbital eccentricity ■ Orbital inclination	A few
Doppler Method	■ Best for detecting massive planets with close-in orbits ■ Detects planets in all orbit orientations except face-on	■ Less sensitive to planets farther from their stars ■ Requires large telescopes ■ Provides only minimum mass when orbital inclination to Earth unknown	■ Minimum planet mass ■ Orbital period and distance (semimajor axis) ■ Orbital eccentricity	About 1000
Transit Method	■ Allows many stars to be observed at once ■ Can detect very small planets ■ Feasible with small telescopes ■ Can provide some atmospheric information in cases of measurable eclipses	■ Possible only for planets with edge-on orbits as viewed from Earth ■ For small planets, requires sensitivity possible only from space observatory	■ Planet size ■ Orbital period and distance (semimajor axis) ■ Orbital inclination (edge-on)	About 2000

Summary TABLE 13.1 summarizes the advantages and limitations of the major methods we have discussed for studying extrasolar planets, and Cosmic Context FIGURE 13.6 (pages 376–377) summarizes how these methods work. Note that while these will almost certainly be the most important methods over the long term, several other strategies have met with some success. More than a dozen planets have been detected by what is sometimes called *microlensing*, which is an example of gravitational lensing [**Sections S3.4, 23.2**] in which the light of a star is temporarily magnified as another star passes in front of it and bends its light. Careful study of the microlensing event can reveal whether the foreground star has planets. The major drawback to this method is that the special alignment necessary for microlensing is a one-time event, which generally means that there is no opportunity for confirmation or follow-up observations. A different strategy takes advantage of the fact that many stars are surrounded by disks of dust. A planet within such a disk can exert small gravitational tugs on dust particles to produce gaps, waves, or ripples that may be detectable. Some astronomers are searching for thermal emission attributable to the heat of large impacts on planets in young planetary systems, and others are searching for the special kinds of emission expected from the magnetospheres of jovian planets. As we learn more about extrasolar planets, new search methods are sure to arise.

13.2 THE NATURE OF PLANETS AROUND OTHER STARS

The mere existence of planets around other stars has changed our perception of our place in the universe, because it shows that our planetary system is not unique. Scientifically, however, we want to know much more than just that these planets exist. In principle, the study of extrasolar planets and planetary systems should provide us with opportunities to vastly expand our general understanding of planets and how they form.

There are two basic scientific goals. First, we'd like to know the general characteristics of planets. For example, the planets of our own solar system come in only two basic types, terrestrial and jovian, and all orbit the Sun on nearly circular orbits in nearly the same plane. The study of extrasolar planets should tell us whether our solar system's planets are typical or whether there is a larger range of planetary types. Second, we'd like to know how extrasolar planets formed, and whether the nebular theory, which we use to explain the origin of our own solar system, can work equally well for explaining other planetary systems.

In this section, we'll discuss the types of data that we can acquire about extrasolar planets with present technology, and what these data tell us about the nature of planets around other stars. We'll then be prepared to turn our attention in the next section to the question of how well the nebular theory works for explaining other planetary systems.

What properties of extrasolar planets can we measure?

Despite the challenge of detecting extrasolar planets, we can actually learn a surprising amount about them. Depending on the method or methods used to study a planet, we can determine characteristics including orbital period and distance, orbital eccentricity, mass, size, density, and even a little bit about a planet's atmospheric composition and temperature. Let's briefly explore how we learn about each of these characteristics.

Orbital Period and Distance All three of the major indirect detection methods tell us a planet's orbital period. The astrometric method allows us to observe the star's motion as the star orbits the center of mass, which means we will

know the star's orbital period; the planet's orbital period must be the same. The orbital period for a planet detected by the Doppler method is simply the time between peaks in the velocity curve (see Figure 13.4a). For the transit method, the orbital period is the time between repeated transits.

Once we know the orbital period, we can determine average orbital distance (semimajor axis) with Newton's version of Kepler's third law [**Section 4.4**]. Recall that for a small object like a planet orbiting a much more massive object like a star, this law expresses a relationship between the star's mass, the planet's orbital period, and the planet's average distance. We generally know the masses of the stars with extrasolar planets (through methods we'll discuss in Chapter 15). Therefore, using the star's mass and the planet's orbital period, we can calculate the planet's average orbital distance (see Mathematical Insight 13.1).

Orbital Eccentricity and Inclination All planetary orbits are ellipses, but recall that ellipses vary in eccentricity, which is a measure of how "stretched out" they are [**Section 3.3**]. The planets in our solar system all have nearly circular orbits (low eccentricity), which means that their actual distances from the Sun are always relatively close to their average distances. Planets with higher eccentricity swing in close to their star on one side of their orbit and go much farther from their star on the other side.

MATHEMATICAL INSIGHT 13.1

Finding Orbital Distances for Extrasolar Planets

Recall that Newton's version of Kepler's third law reads

$$p^2 = \frac{4\pi^2}{G(M_1 + M_2)} a^3$$

In the case of a planet orbiting a star, p is the planet's orbital period, a is its average orbital distance (semimajor axis), and M_1 and M_2 are the masses of the star and planet, respectively. G is the gravitational constant; $G = 6.67 \times 10^{-11}$ m³/(kg × s²). Because a star is so much more massive than a planet, we can approximate $M_{star} + M_{planet} \approx M_{star}$ (see Mathematical Insight 4.3). If we make this approximation, then simple algebra allows us to solve for the average orbital distance a:

$$a \approx \sqrt[3]{\frac{GM_{star}}{4\pi^2} p_{planet}^2}$$

EXAMPLE: Doppler measurements show that the planet orbiting 51 Pegasi has an orbital period of 4.23 days; the star's mass is 1.06 times that of our Sun. What is the planet's average orbital distance?

SOLUTION:

Step 1 Understand: We are given both the planet's orbital period and the star's mass, so we can use Newton's version of Kepler's third law to find the average orbital distance. To make the units consistent, we convert the given stellar mass to kilograms (using the fact that the Sun's mass is 2×10^{30} kg) and the orbital period to seconds:

$$M_{star} = 1.06 \times M_{Sun} = 1.06 \times (2 \times 10^{30} \text{ kg}) = 2.12 \times 10^{30} \text{ kg}$$

$$p = 4.23 \text{ day} \times \frac{24 \text{ hr}}{1 \text{ day}} \times \frac{3600 \text{ s}}{1 \text{ hr}} = 3.65 \times 10^5 \text{ s}$$

Step 2 Solve: We use the above values to find the average distance a:

$$a \approx \sqrt[3]{\frac{GM_{star}}{4\pi^2} p_{planet}^2}$$

$$\approx \sqrt[3]{\frac{6.67 \times 10^{-11} \frac{\text{m}^3}{\text{kg} \times \text{s}^2} \times 2.12 \times 10^{30} \text{ kg}}{4 \times \pi^2} (3.65 \times 10^5 \text{ s})^2}$$

$$= 7.81 \times 10^9 \text{ m}$$

Step 3 Explain: The planet orbits its star at an average distance of 7.8 billion meters, or 7.8 million kilometers. It's easier to interpret this number if we convert it to astronomical units (1 AU $\approx 1.50 \times 10^{11}$ meters):

$$a = 7.81 \times 10^9 \text{ m} \times \frac{1 \text{ AU}}{1.50 \times 10^{11} \text{ m}} = 0.052 \text{ AU}$$

The planet's average orbital distance is 0.052 AU—small even compared to that of Mercury, which orbits the Sun at 0.39 AU. In fact, comparing the planet's 7.8-million-kilometer distance to the size of the star itself (presumably close to the 700,000-kilometer radius of our Sun), we estimate that the planet orbits its star at a distance only a little more than 10 times the star's radius.

The search for planets around other stars is one of the fastest growing and most exciting areas of astronomy. Although it has been less than two decades since the first discoveries, known extrasolar planets already number well into the thousands. This figure summarizes major techniques that astronomers use to search for and study extrasolar planets.

(1) **Gravitational Tugs:** We can detect a planet by observing the small orbital motion of its star as both the star and its planet orbit their mutual center of mass. The star's orbital period is the same as that of its planet, and the star's orbital speed depends on the planet's distance and mass. Any additional planets around the star will produce additional features in the star's orbital motion.

(1a) **The Doppler Method:** As a star moves alternately toward and away from us around the center of mass, we can detect its motion by observing alternating Doppler shifts in the star's spectrum: a blueshift as the star approaches and a redshift as it recedes.

(1b) **The Astrometric Method:** A star's orbit around the center of mass leads to tiny changes in the star's position in the sky. The *GAIA* mission is expected to discover many new planets with this method.

Current Doppler-shift measurements can detect an orbital velocity as small as 1 meter per second—walking speed.

The change in the Sun's apparent position, if seen from a distance of 10 light-years, would be similar to the angular width of a human hair at a distance of 5 kilometers.

Artist's conception of another planetary system, viewed near a ringed jovian planet.

(2) The Transit Method: If a planet's orbital plane happens to lie along our line of sight, the planet will transit in front of its star once each orbit, causing a dip in the star's visible-light brightness. An eclipse may occur half an orbit later, during which the system's infrared brightness will decline because the planet's contribution is blocked by the star.

*We observe a **transit** when the planet passes in front of the star.*

*When the planet passes behind the star, we say it is **eclipsed** by the star.*

planet

star

(3) Direct Detection: In principle, the best way to learn about an extrasolar planet is to observe directly either the visible starlight it reflects or the infrared light it emits. Current technology is capable of direct detection in some cases, but only with very low resolution.

The Keck Telescope imaged the region around the star HR 8799 in infrared light, discovering four planets labeled b through e (a refers to the star itself).

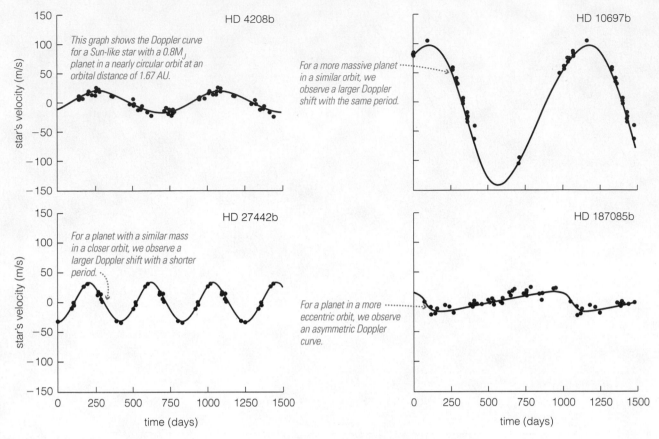

FIGURE 13.7 Sample data showing velocity curves obtained with the Doppler method for four extrasolar planet systems; dots are data and the curves are best-fits to the data. Shapes of the curves tell us about orbital eccentricity, and magnitudes of the velocity changes tell us about mass.

Currently, most of what we know about orbital eccentricities comes from detailed study of Doppler data. (In the future, astrometric data may provide similar information.) A planet with a perfectly circular orbit travels at a constant speed around its star, so its data curve is perfectly symmetric. Any asymmetry in the Doppler curve tells us that the planet is moving with varying speed and therefore must have a more eccentric elliptical orbit. **FIGURE 13.7** shows Doppler curves for four extrasolar planets; notice that the first three are fairly symmetric, implying low eccentricity, but the fourth planet must have a highly eccentric orbit.

Another important question about orbits in other planetary systems is whether the planets all share approximately the same orbital plane, as is the case in our solar system. Unfortunately, the only cases for which we can make strong statements about a shared orbital plane are those of systems with multiplanet transits, since planets observed transiting have very similar orbital planes. In the future, combining transit, astrometric, and Doppler data should in principle allow us to determine precise orbital inclinations for many more planets.

Planetary Mass Both the astrometric and Doppler methods measure motions caused by the gravitational tug of a planet, so both can in principle allow us to estimate planetary masses. However, to date nearly all extrasolar planet masses have come from the Doppler method. The Doppler method tells us about planetary masses because, for a given orbital distance, a more massive planet will cause its star to move at higher velocity around the center of mass; the Doppler velocity curves in Figure 13.7 show the basic idea.

THINK ABOUT IT

Study the four velocity curves in Figure 13.7. How would each be different if the planet were (a) closer to its star? (b) more massive? Explain.

However, there is an important caveat. Recall that Doppler shifts reveal only the part of a star's motion directed toward or away from us (see Figure 5.23). As a result, a planet whose orbit we view face-on (perpendicular to the plane of the orbit) does not cause a Doppler shift in the spectrum of its star, making it impossible to detect such a planet with the Doppler method (**FIGURE 13.8a**). We can observe Doppler shifts in a star's spectrum only if it has a planet orbiting at some angle other than face-on (**FIGURE 13.8b**), and the Doppler shift tells us a star's full orbital velocity only when we view an orbit precisely edge-on.

As a result, planetary masses that we infer from the Doppler method will be accurate only for planets in edge-on orbits, which are the same types of orbits that lead to transits. In other words, the Doppler method generally tells us a planet's precise mass only if we also observe transits of the same planet. In all other cases, the Doppler shift underestimates the true orbital speed of the star and therefore also leads to

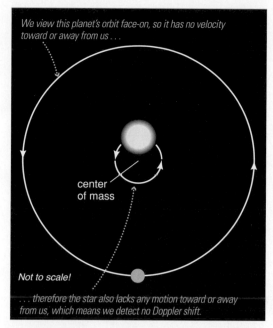

We view this planet's orbit face-on, so it has no velocity toward or away from us . . .

center of mass

Not to scale!

. . . therefore the star also lacks any motion toward or away from us, which means we detect no Doppler shift.

a If we view a planetary orbit face-on, we will not detect any Doppler shift at all.

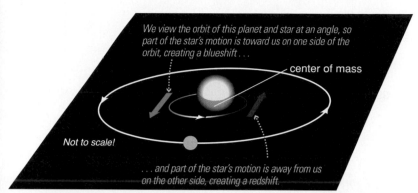

We view the orbit of this planet and star at an angle, so part of the star's motion is toward us on one side of the orbit, creating a blueshift . . .

center of mass

Not to scale!

. . . and part of the star's motion is away from us on the other side, creating a redshift.

b We can detect a Doppler shift only if some part of the orbital velocity is directed toward or away from us. The more an orbit is tilted toward edge-on, the greater the shift we observe.

FIGURE 13.8 The amount of Doppler shift we observe in a star's spectrum depends on the orientation of the planetary orbit that causes the star's motion.

an underestimate of the planet's true mass. Therefore, unless we have some way to determine the planet's precise orbital inclination (such as by combining Doppler data with astrometric data), the mass inferred from the Doppler method alone is the planet's *minimum* possible mass (or a "lower limit" mass). Statistically, however, we expect that the actual planetary mass should be no more than double the minimum mass in at least about 85% of all cases. The minimum masses obtained by the Doppler method are therefore relatively good estimates for most planets.

The transit method cannot by itself tell us masses for single planets, which is one reason scientists often try to follow up transit discoveries with the Doppler method, since the two methods together yield a precise planetary mass. However, the transit method can in some cases reveal planetary masses in multiplanet systems, because the gravitational tug of one planet on another can affect the timing of transits. For example, when another planet is tugging on a transiting planet in the direction in which it is moving, the tug will make the planet move a little faster so that the transit may begin up to tens of minutes earlier than it would otherwise; it may begin tens of minutes late when the planet is tugging from the opposite direction. Precise timing measurements can therefore allow us to measure slight variations in orbital speed and thereby determine the masses of the planets causing these variations. This capability is particularly important for planets with masses too small to be detectable by the Doppler method with current technology.

Planetary Size While knowing its mass gives us some insight into a planet's nature, we can learn much more if we also know a planet's size or radius. Transit observations are presently the only means by which we can measure sizes. The basic idea is easy to understand: The more of a star's light that a planet blocks during a transit, the larger the planet must be. (See Mathematical Insight 13.3 for details of the calculations.)

The Kepler 11 planetary system offers a great example. **FIGURE 13.9** shows the brightness of the star Kepler 11 over a 110-day period. Each downward dip represents a transit in which a planet blocks a small fraction of the starlight. Careful study shows that there are six different planets (represented by the colored dots) that transit the star with periods ranging from 10 to 120 days; sometimes more than one transits at the same time, causing a larger dip. The panels show a transit of each planet in more detail, along with size of the planet calculated from the depth of the dip; note that more distant planets have wider dips because they orbit more slowly, which means their transits last longer. When we combine these sizes with the orbital distances calculated from the orbital periods (which are the times between transits for each planet), we find that this system's planets have sizes ranging from 2 to 5 times Earth's size and all orbit between 0.1 and 0.5 AU from their star. Transported to our solar system, that would place six planets inside Venus's orbit.

THINK ABOUT IT

New data are coming in so quickly that the records for lowest-mass and smallest-size planets change rapidly. Do a web search to find the current record-holder for extrasolar planet with the lowest confirmed *mass* and for the planet with the smallest measured *size*.

time relative to transit (hours)

FIGURE 13.9 The Kepler 11 system. The black line shows the brightness of the star Kepler 11 over a 110-day period. Colored dots indicate transits by six different planets. The panels at right show the dips due to each planet separately, and indicate the planet's size in Earth radii.

Planetary Density No single method measures a planet's average density, but we can calculate it if we know a planet's size from the transit method and its mass from the Doppler method. Note that we can get a precise density value in this case, because transiting planets must have edge-on orbits, so the Doppler method gives an exact (rather than minimum) mass in these cases. The first planet for which scientists obtained both Doppler and transit data was HD 209458b (the planet that we discussed earlier as the first transit success). The planet was first discovered with the Doppler method, which revealed it to have a minimum mass of 0.69 times Jupiter's mass. Once its transits were discovered, we knew that this was an exact mass. The amount of starlight blocked during transits allowed us to calculate its radius, from which we could also find its volume (because $V = 4/3\pi r^3$). We then found its density simply by dividing the mass by the volume; the result is 0.37 g/cm^3. This density is even lower than that of Saturn, which is the lowest-density planet in our solar system.

At the opposite extreme, **FIGURE 13.10** shows how the density of Kepler 10b was found to be 8.8 g/cm^3, making it significantly more dense than Earth. With even just these two examples, we see that the variety of densities among extrasolar planets is much greater than that of the planets in our solar system, suggesting that extrasolar planets may come in a wider range of types than just the familiar terrestrial and jovian categories.

Atmospheric Composition and Temperature

For nearly all astronomical objects except extrasolar planets, we learn about their atmospheres and temperatures by studying spectra obtained through direct observations. Because these types of direct observations are not yet technologically feasible for most extrasolar planets, you might guess that we'd be unable to learn anything at all about their atmospheres

and temperatures. However, careful analysis of data collected during transits and eclipses allows us to obtain at least some information.

FIGURE 13.11 shows the basic ideas, using the example of HD 189733b, a hot Jupiter with an orbital period of only 2.2 days, which means it orbits only about 0.04AU from its star. The left graph shows that at most wavelengths, the transits create a 2.5% dip in the star's brightness, which tells us how the planet's radius out to its visible edge compares to its star's radius. However, we know from our own solar system

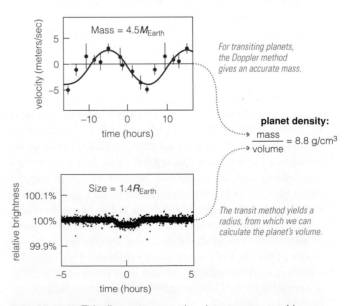

For transiting planets, the Doppler method gives an accurate mass.

planet density:
$$\frac{\text{mass}}{\text{volume}} = 8.8 \text{ g/cm}^3$$

The transit method yields a radius, from which we can calculate the planet's volume.

FIGURE 13.10 This diagram summarizes how we can combine data from the Doppler and transit methods to calculate a planet's average density. (The final density calculation requires converting the mass to grams and the volume to units of cubic centimeters.)

that planets can have atmospheric gases that extend above the visible disk, and these gases can absorb light at specific wavelengths. Infrared observations are especially powerful for planetary atmospheres, because many important molecules (such as those of greenhouse gases) have spectral lines in the infrared. By comparing the amount of starlight at these specific wavelengths blocked by the planet's upper atmosphere to the amount blocked by the main disk of the planet, we can identify gases in the planet's atmosphere. For HD 189733b, these types of observations have revealed the presence of water vapor in its atmosphere; in another star (HD 209458b), astronomers used the same technique to identify methane, hydrogen, sodium, and a few other gases.

We can learn more from the eclipses that occur as the planet goes behind the star. Before the eclipse, the light from the star system is the combined light of the star and the planet. During the eclipse, we see the light of the star without the added light of the planet. The difference between these two is the light from the planet by itself. Recall that, because planets emit at infrared wavelengths at which stars are less bright [Section 5.2], the dips that occur during eclipses are most easily measured with infrared telescopes. For example, the Spitzer Space Telescope measured the infrared brightness of HD 189733 dropping by about 0.3% during eclipses, telling us that the planet emits 0.3% as much infrared radiation

as the star. By combining this fact with the planet's radius measured during the transits, astronomers calculated the planet's surface (or cloudtop) temperature to be more than 1100 K. Astronomers then used the principles of planetary temperatures (see Mathematical Insight 10.1) to deduce that this planet has a very low reflectivity.

In some cases, changes in the total brightness of a system can even be measured as a planet travels along its orbit. Because planets reflect visible light, we can in principle notice changes in a system's visible-light brightness as the planet presents different phases to Earth: Much as we see phases of Venus from Earth (see Figure 3.22b), we see a planet as "new" during a transit and as nearly "full" just before and after an eclipse. *Kepler* has successfully measured changes in at least one planet's visible reflected light as the planet has gone from new to crescent to gibbous to full and back again. With infrared light, we can observe differences in brightness over a planet's day and night sides as it orbits. Moreover, all planets on close-in orbits are expected to have synchronous rotation so that they show the same face to their star at all times, just as the Moon always shows the same face to Earth [Section 4.5]. Therefore, we can in principle use infrared observations to create a crude "weather map" of an extrasolar planet. The first success of this technique came with Spitzer Space Telescope observations of HD 189733 and its planet, which showed that the planet

---

MATHEMATICAL INSIGHT 13.2

Finding Masses of Extrasolar Planets

The Doppler method allows us to find the mass of an extrasolar planet from the law of conservation of momentum [Section 4.3]; recall that momentum = mass × velocity. Consider a star with a single planet, both orbiting their common center of mass with the same orbital period. The system as a whole has no momentum relative to this center of mass (which stays in a fixed place between the star and planet), so the planet's momentum must equal the star's momentum (but in the opposite direction). Using M for mass and v for velocity relative to the center of mass, we write

$$M_{star} v_{star} = M_{planet} v_{planet}$$

Solving this equation for M_{planet}, we find

$$M_{planet} = \frac{M_{star} v_{star}}{v_{planet}}$$

The Doppler method tells us the star's velocity toward or away from us (v_{star}), and as discussed earlier, we generally know the star's mass (M_{star}). We can calculate the planet's orbital velocity (v_{planet}) from its orbital period p (the time between peaks in the Doppler curve) and its average orbital distance a (calculated with the method in Mathematical Insight 13.1). If we assume a circular orbit, the planet travels a distance $2\pi a$ during each orbit that takes time p, so its orbital velocity is $2\pi a_{planet}/p_{planet}$. Substituting this expression for the planet's velocity into the above equation for mass gives

$$M_{planet} = \frac{M_{star} v_{star} p_{planet}}{2\pi a_{planet}}$$

Remember that with velocity data from the Doppler method, this formula gives us the *minimum* mass of the planet.

EXAMPLE: Estimate the mass of the planet orbiting 51 Pegasi.

SOLUTION:

Step 1 Understand: From Mathematical Insight 13.1, we know the planet's orbital period ($p = 3.65 \times 10^5$ s) and orbital distance ($a = 7.81 \times 10^9$ m) and the star's mass ($M_{star} = 2.12 \times 10^{30}$ kg). The graph in Figure 13.4a shows that the star's velocity is about $v_{star} = 57$ m/s. We can therefore use the formula above to calculate the planet's mass.

Step 2 Solve: We enter the values into the mass formula:

$$M_{planet} = \frac{M_{star} v_{star} p_{planet}}{2\pi a_{planet}}$$

$$= \frac{(2.12 \times 10^{30} \text{ kg}) \times \left(57 \frac{\text{m}}{\text{s}}\right) \times (3.65 \times 10^5 \text{ s})}{2\pi \times (7.81 \times 10^9 \text{ m})}$$

$$\approx 9 \times 10^{26} \text{ kg}$$

Step 3 Explain: The minimum mass of the planet is about 9×10^{26} kilograms. This answer will be more meaningful if we convert it to Jupiter masses. From Appendix E, Jupiter's mass is 1.9×10^{27} kilograms, so the planet's minimum mass is

$$M_{planet} = 9 \times 10^{26} \text{ kg} \times \frac{1 M_{Jupiter}}{1.9 \times 10^{27} \text{ kg}} = 0.47 M_{Jupiter}$$

The planet orbiting 51 Pegasi has a minimum mass of just under half Jupiter's mass. Statistically, most planets will have an actual mass less than double the minimum mass, so the planet probably has a mass similar to or a little less than that of Jupiter.

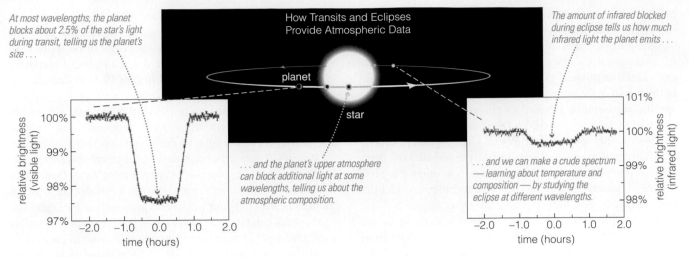

At most wavelengths, the planet blocks about 2.5% of the star's light during transit, telling us the planet's size . . .

The amount of infrared blocked during eclipse tells us how much infrared light the planet emits . . .

How Transits and Eclipses Provide Atmospheric Data

planet

star

. . . and the planet's upper atmosphere can block additional light at some wavelengths, telling us about the atmospheric composition.

. . . and we can make a crude spectrum — learning about temperature and composition — by studying the eclipse at different wavelengths.

FIGURE 13.11 This diagram summarizes how transit and eclipse observations can give us information about the composition and temperature of an extrasolar planet's atmosphere.

reaches a temperature of nearly 1200 K on its day side and 900 K on its night side (**FIGURE 13.12**). Note that the hottest point does not lie exactly at the center of the Sun-facing side, probably because of winds, which models suggest may blow at an incredible 10,000 km/hr. **TABLE 13.2** summarizes how we measure planetary properties.

How do extrasolar planets compare with planets in our solar system?

The number of known extrasolar planets for which we have measured many key properties is now large enough that we are beginning to gain insight into how these planets compare to the planets of our own solar system. Let's explore the general features of extrasolar planets as we know them today, focusing on orbital properties, masses and sizes, and what we can say about their general nature.

FIGURE 13.12 The first map of an extrasolar planet's temperatures, derived from infrared observations of HD 189733b made by the Spitzer Space Telescope.

Orbital Properties Much as Johannes Kepler first appreciated the true layout of our own solar system [**Section 3.3**], we now have enough data to allow us to step back and see the layout of many other solar systems. The two key orbital properties of extrasolar planets that help us understand planetary system layout are (1) average orbital distance (semimajor axis), which allows us to compare the locations of planets in other solar systems to the locations of planets in our own, and (2) orbital eccentricity, which allows us to determine whether or how much the orbits in other solar systems differ from the nearly circular orbits in our own solar system.

FIGURE 13.13 shows a graph of these two orbital properties for more than 500 extrasolar planets, with average orbital distance on the horizontal axis and orbital eccentricity on the vertical axis. Recall that eccentricity for an elliptical orbit is always between 0 and 1 (see Mathematical Insight 3.1); an eccentricity of 0 is a perfect circle, and higher eccentricity means a more stretched-out ellipse. The sizes of the dots on the graph represent approximate masses of the planets, as shown in the key at the upper left. These planets were detected by the Doppler method, so the indicated masses are actually minimum masses. Be sure to note that the distance scale increases by powers of 10, which makes it easier to plot the large number of planets with relatively close-in orbits.

Look first at the green dots representing the planets in our own solar system. To be sure you are reading the graph correctly, make sure you see how it shows that the four inner planets orbit within about 1.5 AU of the Sun, while the four outer planets are spread out to about 30 AU. Also notice that all but Mercury have very small eccentricities, meaning nearly circular orbits. Now look at the dots representing extrasolar planets. Two differences between many of the extrasolar planets and our solar system's planets should jump out at you:

1. Many of the known extrasolar planets orbit their stars more closely than Mercury orbits the Sun, and none are located as far from their stars as the jovian planets of our solar system.

TABLE 13.2 A Summary of How We Measure Properties of Extrasolar Planets

	Planetary Property	Method(s) Used	Explanation
Orbital Properties	**period**	Doppler, astrometric, or transit	We directly measure orbital period.
	distance	Doppler, astrometric, or transit	We calculate orbital distance from orbital period using Newton's version of Kepler's third law (Mathematical Insight 13.1).
	eccentricity	Doppler or astrometric	Velocity curves and astrometric star positions reveal eccentricity (Figure 13.7).
	inclination	transit or astrometric	Transits identify edge-on orbits; astrometric data measure any inclination angle.
Physical Properties	**mass**	astrometric or Doppler*	For single planets, we can calculate mass based on the amount of stellar motion caused by the planet's gravitational tugs (Mathematical Insight 13.2). The Doppler method gives *minimum* mass unless orbital inclination is known.
	size (radius)	transit	We calculate size based on the amount of dip in a star's brightness during a transit (Mathematical Insight 13.3).
	density	transit and Doppler	We can calculate density by dividing the mass (from Doppler method) by volume (using size from transit method). See Figure 13.10.
	atmospheric composition and temperature	transit or direct detection	Transits and the accompanying eclipses provide data on atmospheric composition and temperature (Figure 13.11); direct spectroscopic observations can in principle give more detailed information.

*Transits in some multiplanet systems allow us to determine masses from the mutual gravitational tugs of the planets.

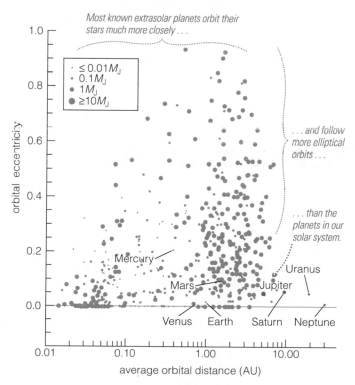

Orbital Properties of Extrasolar Planets

Most known extrasolar planets orbit their stars much more closely . . .

≤ 0.01M_J
0.1M_J
1M_J
≥10M_J

. . . and follow more elliptical orbits . . .

. . . than the planets in our solar system.

Mercury

Mars   Jupiter   Uranus

Venus   Earth   Saturn   Neptune

orbital eccentricity

average orbital distance (AU)

FIGURE 13.13 Orbital properties of the more than 500 extrasolar planets with known masses, distances, and eccentricities as of mid-2012. Dots closer to the left represent planets that orbit closer to their stars, and dots lower down represent planets with more circular orbits. Green dots are planets of our own solar system.

2. Many of these planets also have large orbital eccentricities, telling us that their elliptical orbits have very stretched-out shapes.

Both facts came as a surprise to astronomers when they were first discovered, and in the next section we'll see how they forced scientists to reconsider and ultimately modify the nebular theory.

THINK ABOUT IT

Should we be surprised that we haven't found many planets orbiting as far from their stars as Saturn, Uranus, and Neptune orbit the Sun? Why or why not?

We can learn even more from multiplanet systems, since those allow us to see a more complete layout rather than just a single planet. Several hundred stars have so far been found to have two or more planets (or *Kepler* planet candidates), including the six-planet Kepler 11 system (see Figure 13.9). This fact lends support to the general outline of the nebular theory, because its explanation of planet formation suggests that any star with planets is likely to have multiple planets. We therefore expect to find many more multiple-planet systems as observations improve.

Already, however, we've learned that other planetary systems can sometimes differ from ours in at least two ways. First, many have planets packed much closer to each other than our solar system's planets, often so close that they tug on each other gravitationally in measurable ways (such as the way in which transit timing allows us to measure masses). Second, the *Kepler* mission has identified candidate planets in binary star systems. Scientists had not been sure whether planetary orbits could remain stable in binary star systems, but the data now show that they can, at least in some cases. This has opened the possibility that some worlds might really have two (or more) "suns" in the sky, much like that the fictional planet Tatooine depicted in *Star Wars*.

Masses and Sizes Recall that most of our planetary mass measurements come from the Doppler method, while transits provide our only direct way of measuring sizes. We can therefore look independently at current data on masses and sizes, to see what they tell us about how extrasolar planets compare to the planets of our solar system.

FIGURE 13.14 shows mass data for more than 500 extrasolar planets; these are the same planets shown in Figure 13.13, but the mass data are much easier to study when displayed with a bar graph. The trend shown on the main graph is very clear: Low-mass planets outnumber massive planets, at least down to planets of Jupiter's or possibly Saturn's mass. The inset shows data for even lower-mass planets, and it appears to show a reversal of the trend. However, this reversal is almost certainly a consequence of the fact that it's much more difficult to detect low-mass planets than high-mass planets with the Doppler method. Statistical studies that take this fact into account suggest that Neptune-mass planets are indeed more abundant than Saturn- or Jupiter-mass planets, and Earth-mass planets may be even more common. In other words, the mass evidence is at least consistent with the idea that smaller worlds should outnumber larger worlds, which is what our own solar system should lead us to expect because it has vast numbers of small worlds (including terrestrial planets, moons, asteroids, and comets) and only four large jovian planets.

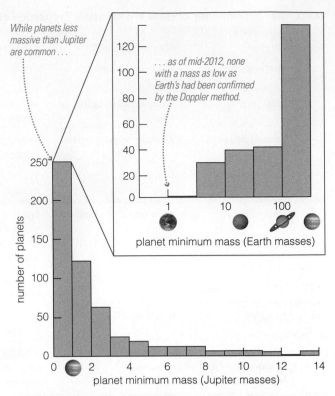

FIGURE 13.14 This bar chart shows the numbers of planets in different mass categories for extrasolar planets with masses measured by the Doppler method as of 2012; it does *not* include *Kepler* planet candidates, some of which are smaller in size and therefore likely to be lower in mass than Earth. Notice that the inset axis uses an exponential scale so that a wide range of masses can fit on the graph, and uses units of Earth masses rather than Jupiter masses.

FIGURE 13.15 shows planetary size data from the *Kepler* mission, with radius compared to Earth radius plotted along

MATHEMATICAL INSIGHT 13.3

Finding Sizes of Extrasolar Planets

We determine planet radii from the fraction of a star's light blocked during a transit. Viewed against the sky, both the star and the planet appear as tiny circular disks. These disks are far too small for our telescopes to resolve, but the fraction of the star's light that is blocked must be equal to the area of the planet's disk (πr^2_{planet}) divided by the area of the star's disk (πr^2_{star}). We generally know the approximate radius of the star (from methods we'll discuss in Chapter 15), so the fractional drop in the star's light during a transit is

$$\frac{\text{fraction of}}{\text{light blocked}} = \frac{\text{area of planet's disk}}{\text{area of star's disk}} = \frac{\pi r^2_{planet}}{\pi r^2_{star}} = \frac{r^2_{planet}}{r^2_{star}}$$

Solving for the planet's radius, we find

$$r_{planet} = r_{star} \times \sqrt{\text{fraction of light blocked}}$$

EXAMPLE: Figure 13.5 shows a transit of the star HD 189733. The star's radius is about 800,000 kilometers ($1.15 R_{Sun}$), and the planet

blocks 1.7% of the star's light during a transit. What is the planet's radius?

SOLUTION:

Step 1 Understand: The star's radius (800,000 km) and the fraction of its light blocked during a transit (1.7% = 0.017) are all we need to calculate the planet's radius.

Step 2 Solve: Plugging the numbers into the equation, we find

$$r_{planet} = r_{star} \times \sqrt{\text{fraction of light blocked}}$$
$$= 800,000 \text{ km} \times \sqrt{0.017}$$
$$\approx 100,000 \text{ km}$$

Step 3 Explain: The planet's radius is about 100,000 kilometers, which is about 1.4 times Jupiter's radius of 71,500 kilometers. That is, the planet is about 40% larger than Jupiter in radius.

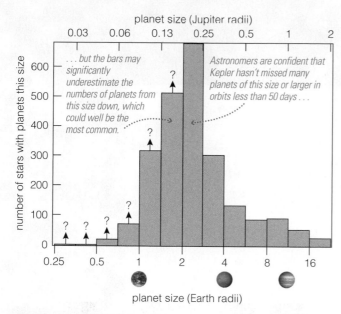

FIGURE 13.15 This bar chart shows the numbers of planets in different size categories for extrasolar planets detected by *Kepler* with orbital periods of less than 50 days. Notice that this plot uses a horizontal axis of Earth radii, so the scale is very different from the one based on Jupiter masses in Figure 13.14.

the horizontal axis. To understand the graph, recall that the *Kepler* mission is monitoring about 150,000 stars, and needs to observe at least three transits of an object before it becomes a planet candidate; at the time the graph was made, full data were available only for transiting planets with periods of less than 50 days. Note that, as we found with the mass data, there is a general trend toward smaller planets being more abundant than larger planets, at least down to a size of two Earth radii. The trend probably continues to smaller worlds, but we cannot be sure from the current data because such worlds are at the limit of *Kepler's* measurement sensitivity; that is, the actual numbers of small worlds are probably significantly higher than shown. Note that *Kepler* has already detected nearly a hundred planet candidates smaller than Earth, though all lie too close to their stars to be habitable. This is about the same number of planets it has found that are larger than Jupiter, even though Jupiter-size planets are more than a hundred times easier to detect than Earth-size planets (for any given orbital period).

Kepler scientists have used the number of planets discovered to make a rough estimate of the fraction of stars harboring planets, at least for those planets with periods short enough to have been detected so far. Current evidence is consistent with as many as a quarter to a third of all stars harboring at least one planet with an orbital period shorter than 50 days. Note that our own solar system would not appear on a list like this, because our closest-in planet (Mercury) has an orbital period of 88 days. Therefore, it seems likely that an estimate based on periods shorter than 50 days significantly underestimates the total fraction of stars with planets. If so, we are led to the remarkable conclusion that most stars harbor planets. Planet searches based on other methods are reaching the same conclusion.

The Nature of Extrasolar Planets We now come to the key question about extrasolar planets: Do they fall into the same terrestrial and jovian categories as the planets in our solar system, or do we find additional types of planets?

We cannot yet know for certain, because we do not yet have direct ways of obtaining spectra with sufficient detail to determine the bulk compositions of extrasolar planets. Nevertheless, because we have measured the abundances of different chemical elements among other stars, we know that all planetary systems must start out from gas clouds generally similar in composition to the solar nebula. That is, all star systems are born from gas clouds containing at least about 98% hydrogen and helium, sprinkled with much smaller amounts of ice, rock, and metal (see Table 8.1). Given this fact, knowing a planet's average density gives us great insight into its likely composition, even without being able to observe the planet directly.

More specifically, we can use our understanding of the behavior of different materials to create *models* that will tell us the expected composition of a planet based on its mass and radius, from which we can also calculate its average density. The results are shown in **FIGURE 13.16** for all planets for which both mass (usually from the Doppler method) and radius (from transits) are known as of mid-2012. Be sure you understand the following key features of the figure:

- The horizontal axis shows planetary mass, in units of Earth masses. (The top of the graph shows the equivalent values in Jupiter masses.) Notice that it uses an exponential scale because the masses vary over such a wide range.

- The vertical axis shows planetary radius in units of Earth radii. (The right side of the graph shows the equivalent values in Jupiter radii.)

- Each dot represents one planet for which both mass and radius have been measured. Planets discussed in this chapter are called out by name. Planets of our solar system are marked in green.

- The paintings around the graph show artist conceptions of what representative worlds might look like.

- Average density is easy to calculate from mass and radius, but the different scales used on the two axes make it difficult to read average density directly from the graph. To help with that, the three curves extending from the lower left to the top show three representative average densities.

- The colored regions indicate models representing the expected compositions of planets with the indicated combinations of mass and radius.

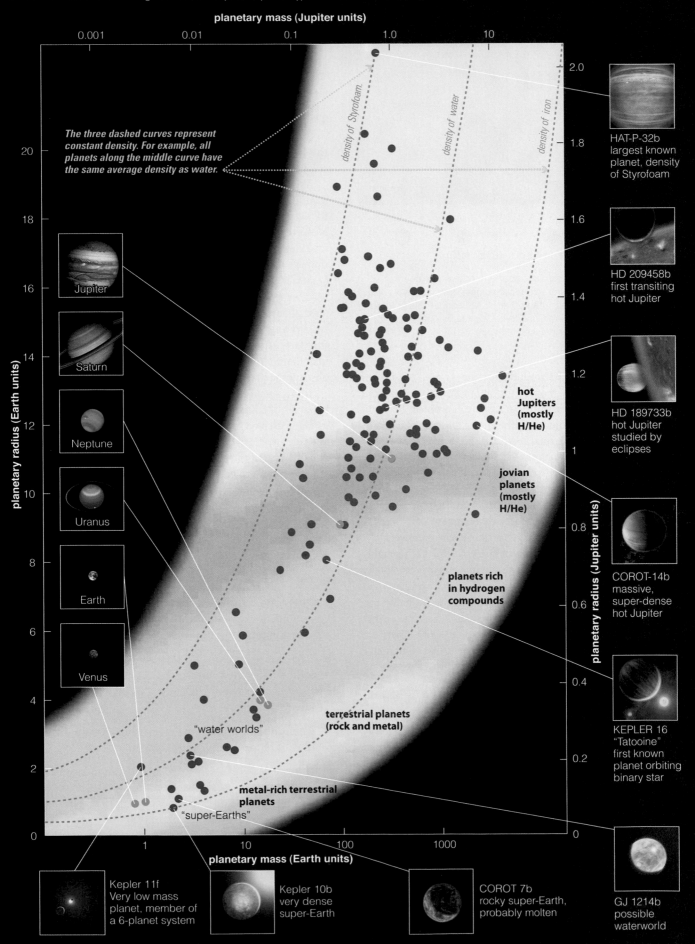

FIGURE 13.16 Masses and sizes of extrasolar planets for which both have been measured, compared to those of planets in our solar system. Each dot represents one planet. Dashed lines are lines of constant density for planets of different masses. Colored regions indicate expected planet types based on models of their compositions.

planetary mass (Jupiter units)

The three dashed curves represent constant density. For example, all planets along the middle curve have the same average density as water.

density of Styrofoam

density of water

density of iron

planetary radius (Earth units)

Jupiter

Saturn

Neptune

Uranus

Earth

Venus

hot Jupiters (mostly H/He)

jovian planets (mostly H/He)

planets rich in hydrogen compounds

terrestrial planets (rock and metal)

"water worlds"

metal-rich terrestrial planets

"super-Earths"

planetary radius (Jupiter units)

planetary mass (Earth units)

HAT-P-32b largest known planet, density of Styrofoam

HD 209458b first transiting hot Jupiter

HD 189733b hot Jupiter studied by eclipses

COROT-14b massive, super-dense hot Jupiter

KEPLER 16 "Tatooine" first known planet orbiting binary star

GJ 1214b possible waterworld

Kepler 11f Very low mass planet, member of a 6-planet system

Kepler 10b very dense super-Earth

COROT 7b rocky super-Earth, probably molten

As you study Figure 13.16, you'll notice that extrasolar planets show much more variety than the planets of our own solar system. For example, HAT-P-32b has more than twice Jupiter's radius despite having the same mass. Therefore it has one of the lowest densities: about 0.14 g/cm^3—similar to the density of Styrofoam. This low average density is probably a result of the fact that this planet orbits only 0.035 AU from its star, putting it more than 10 times closer to its star than Mercury is to the Sun. This close-in orbit gives the planet a very high temperature, which should puff up the planet's atmosphere and may explain why it has such a large size relative to its mass. Near the other extreme, the planet COROT-14b is only slightly larger than Jupiter but several times as massive, giving it an average density near 8 g/cm^3, about the same as the density of iron. Although such a high average density might seem surprising, it is not totally unexpected. Recall that jovian planets more massive than Jupiter are expected to have such strong gravity that they can be compressed to smaller sizes and much higher densities (see Figure 11.2b). Despite the wide spread in their densities, both HAT-P-32b and COROT-14b seem clearly to fall into the jovian planet category, being made largely of hydrogen and helium.

We similarly see many planets that appear to be terrestrial in nature, with compositions of rock and metal. For example, COROT-7b has an average density near 5 g/cm^3, comparable to Earth's density. Because it has a mass about 5 times that of Earth, COROT-7b is an example of what is sometimes called a "super-Earth"; it was the first super-Earth discovered (in 2009). COROT-7b orbits very close to its star, so its surface is probably molten. Kepler 10b is another super-Earth, but with only 4 times Earth's mass. Numerous other super-Earths also are now known; all of them are likely to have a rock/metal composition similar to that of the terrestrial worlds in our solar system.

Perhaps the most surprising planets shown in Figure 13.16 are the ones that are not clearly in either the terrestrial or the jovian category. Several planets cluster in the region of Uranus and Neptune, and perhaps share their composition of hydrogen compounds shrouded in an envelope of hydrogen and helium gas. Others (perhaps GJ 1214b) appear to fit the model for "water worlds," which may be made predominantly of either frozen or liquid water, or perhaps of other hydrogen compounds. Alternatively, some of these worlds might be composed of a dense rocky/metallic core and a thick envelope of low-density hydrogen-helium gas.

Finally, it's worth noting that while we've focused here on composition, orbital distance must also play a role in a planet's nature. In particular, because most of the planets known to date orbit fairly close to their stars, they will be much hotter than similar planets in our own solar system. We've already discussed how this could make hot Jupiters have puffed-up atmospheres and terrestrial worlds have molten surfaces. Hot Jupiters might also have very different clouds than we see on the actual Jupiter, such as clouds of mineral flakes instead of clouds of ammonia snow or water droplets. Worlds that might resemble larger versions of Ganymede or Titan in our solar system would be water worlds at closer orbital distances. Even warmer water worlds might become "steam planets" with

vast amounts of water vapor in their atmospheres. Also recall that all close-in planets are expected to be tidally locked in synchronous rotation, forever keeping one face toward their star.

The bottom line is that extrasolar planets are abundant and diverse. Unlike our solar system, with only two clear types of "typical planets," these new solar systems have additional types of planets that defy easy categorization.

Moons, Rings, and Other Remaining Questions

Although we now have a fair amount of information about planets around other stars, there are many questions that we are not yet able to address. For example, based on what we see in our own solar system, we might expect extrasolar jovian planets to be orbited by rings and many moons, but we do not yet know whether this is the case. Answering the question is important not only out of pure curiosity, but also because it might have implications for the search for life. For example, if you look back at Figure 13.14, you'll see that some extrasolar planets are jovian in mass but have Earth-like orbits. As we'll discuss in Chapter 24, such orbits may be conducive to life, though probably not on jovian worlds. However, if the jovian worlds have large moons, these moons might be hospitable to life.

13.3 THE FORMATION OF OTHER SOLAR SYSTEMS

The discovery of extrasolar planets presents us with an opportunity to test our theory of solar system formation. Can our existing theory explain other planetary systems, or do we have to go back to the drawing board?

As we discussed in Chapter 8, the nebular theory holds that our solar system's planets formed as a natural consequence of processes that accompanied the formation of our Sun. If the theory is correct, then the same processes should accompany the births of other stars, so the nebular theory clearly predicts the existence of other planetary systems. In that sense, the discovery of extrasolar planets means the theory has passed a major test, because its most basic prediction has been verified. Some details of the theory also seem supported. For example, the nebular theory says that planet formation begins with condensation of solid particles of rock and ice (see Figure 8.13), which then accrete to larger sizes. We might therefore expect that planets should form more easily in a nebula with a higher proportion of rock and ice, and some evidence supports this idea, at least for jovian planets.

Nevertheless, extrasolar planets have already presented at least two significant challenges to our theory. One concerns the categories of planets; as we saw in Figure 13.16, many extrasolar planets do not fall neatly into either the terrestrial or the jovian category. An even more significant challenge is posed by the orbits of extrasolar planets. According to the nebular theory, jovian planets form as gravity pulls in gas around large, icy planetesimals that accrete in a spinning disk of material around a young star. The theory therefore predicts that jovian planets should form only in the cold outer regions of star systems (because it must be cold for ice to condense), and that these planets should be born with nearly circular orbits (matching the orderly, circular motion of the spinning disk). The many known extrasolar planets that appear jovian in nature but have close-in or highly elliptical orbits present a direct challenge to these ideas.

Can we explain the surprising orbits of many extrasolar planets?

The nature of science demands that we question the validity of a theory whenever it is challenged by any observation or experiment [**Section 3.4**]. If the theory cannot explain the new observations, then we must revise or discard it. The surprising orbits of many known extrasolar planets have indeed caused scientists to reexamine the nebular theory of solar system formation.

Questioning began almost immediately upon the discovery of the first extrasolar planets. The close-in orbits of these massive planets made scientists wonder whether something might be fundamentally wrong with the nebular theory. For example, is it possible for jovian planets to form very close to a star? Astronomers addressed this question by studying many possible models of planet formation and reexamining the entire basis of the nebular theory. Several years of such reexamination did not turn up any good reasons to discard the basic theory. While we can't completely rule out the possibility that a major flaw has gone undetected, it seems much more likely that the basic outline of the nebular theory is correct. Scientists therefore suspect that extrasolar jovian planets were indeed born with circular orbits far from their stars, and that those that now have close-in or eccentric orbits underwent some sort of "planetary migration" or suffered gravitational interactions with other massive objects.

Planetary Migration If hot Jupiters formed in the outer regions of their star systems and then migrated inward, how did these planetary migrations occur? You might think that drag within the solar nebula could cause planets to migrate, much as atmospheric drag can cause satellites in low-Earth orbit to lose orbital energy and eventually spiral into the atmosphere. However, calculations show this drag effect to be negligible. A more likely scenario is that waves propagating through a gaseous disk lead to migration (**FIGURE 13.17**). The gravity of a planet orbiting in a disk can create waves that propagate through the disk, causing material to bunch up as the waves pass by. This "bunched up" matter (in the wave peaks) then exerts a gravitational pull on the planet that reduces its orbital energy, causing the planet to migrate inward toward its star.

Computer models confirm that waves in a nebula can cause young planets to spiral slowly toward their star. In our own solar system, this migration is not thought to have played a significant role because the nebular gas was cleared out before it could have much effect. But planets may form earlier in some other solar systems, allowing time for jovian planets to migrate substantially inward. In a few cases, the planets may form so early that they end up spiraling all the way into their stars. Indeed, astronomers have noted that

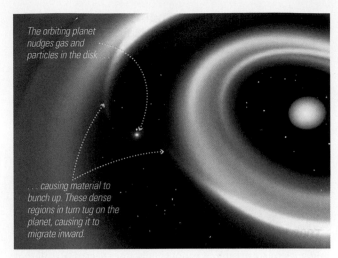

The orbiting planet nudges gas and particles in the disk . . .

. . . causing material to bunch up. These dense regions in turn tug on the planet, causing it to migrate inward.

FIGURE 13.17 This figure shows a simulation of waves created by a planet embedded in a disk of material surrounding its star.

some stars have an unusual assortment of elements in their outer layers, suggesting that they may have swallowed planets (including migrating jovian planets and possibly terrestrial planets shepherded inward along with the jovian planets). These ideas are not just hypothetical: One recently discovered planet appears to be on a million-year death spiral into its star.

Migration can also occur after the nebula has cleared, as long as small planetesimals are still abundant. Astronomers suspect that this type of migration affected the jovian planets in our own solar system. Recall that the Oort cloud is thought to consist of comets that were ejected outward by gravitational encounters with the jovian planets, especially Jupiter [**Section 12.2**]. In that case, the law of conservation of energy demands that Jupiter must have migrated inward, losing the same amount of orbital energy that the comets gained. It's not known if this kind of migration is common outside our solar system.

Encounters and Resonances Migration may explain close-in orbits, but why do so many extrasolar planets have highly eccentric orbits? Close encounters may be the answer. A close gravitational encounter between two massive planets can send one planet out of the star system entirely while the other is flung inward into a highly elliptical orbit. Repeated encounters can result in significant orbital migration.

Gravitational interactions can also affect orbits through resonances. Recall that Jupiter's moons Io, Europa, and Ganymede share orbital resonances that cause their orbits to be more elliptical than they would be otherwise (see Figure 11.19b). This type of resonance can also occur between planets, and several dozen cases of planetary resonances have been found among known multiplanet systems. Planetary migration may play a role in creating such resonances, forcing two planets into a resonance that they did not have originally. Simulations show that gravitational influences between planets can drive them into orbits that are very eccentric, highly tilted, or even backward. The question is whether resonances are common enough to explain the many planets now known with eccentric or other unusual orbits. On one hand, we have so far observed resonances among planets in only a relatively small number of multiplanet systems. On the other hand, it could be that most of the planets with unusual orbits are in resonance with other planets, but we have not yet detected the other planets.

Do we need to modify our theory of solar system formation?

We began this section by asking whether the nebular theory of solar system formation can hold up in light of our discoveries of planets around other stars. Assuming we are correct about the role of planetary migration in explaining the surprising orbits of hot Jupiters and other extrasolar planets, the basic tenets of the nebular theory still seem to hold. That is, we expect rocky terrestrial worlds to form in the inner regions of solar systems and hydrogen-rich jovian planets to form in the outer regions. Migration may later cause some of the jovian planets to spiral inward, and along the way the jovian planets are likely to alter the orbits of other, smaller planets. The remaining mystery, then, is why other solar systems seem to have planetary types that don't fall neatly into the terrestrial and jovian categories that we identify in our solar system.

Scientists still cannot fully explain the wide range of extrasolar planet properties, but we can envision possible explanations that seem to make sense and that would not fundamentally alter the nebular theory. For example, hydrogen-rich extrasolar planets vary in density by a factor of 100 (see Figure 13.16)—a range far greater than the density range we observe in our own solar system—but it seems reasonable to think that much of this range is attributable to differences in temperature caused by some jovian planets being very close to their stars. As we've discussed, this may puff up their atmospheres to large sizes and low densities, though models cannot yet account for the full density range.

Similarly, the lack of "water world" planets in our solar system may not be as mysterious as it seems. Water worlds may be similar to Uranus and Neptune in our solar system, though in some cases much smaller. These worlds may be much like the ice-rich planetesimals that seeded the formation of jovian planets in our solar system. In that case, perhaps whether water world planets exist depends on when a star clears its nebular gas, halting the epoch of planet formation. In our solar system, this did not occur until the ice-rich planetesimals pulled in vast quantities of hydrogen and helium gas from the solar nebula. Perhaps in other systems, an early solar wind blasted out the hydrogen and helium gas before it could be captured.

Super-Earths pose a different mystery: How did these planets gather so much rocky material, especially so close to their stars, given that rocky material represents such a small proportion of the material from which solar systems are born? The answer is not yet known, though perhaps we should not be too surprised. After all, even though rocky material comprised less than 2% of the solar nebula, this still in principle was enough to build planets much larger than Earth. Perhaps we only need a better understanding of the factors that determine the efficiency with which rocky

material can be incorporated into planets. We may similarly need a better understanding of the conditions under which super-Earths or water worlds may capture hydrogen and helium gas, since that could lead to planets whose true nature is hiding under enormous envelopes of gas. With hundreds of different planets discovered every year, other surprising planetary types are likely to be found in the future.

The bottom line is that discoveries of extrasolar planets have shown us that the nebular theory was incomplete. It explained the formation of planets and the simple layout of a solar system such as ours. However, it needs new features—such as planetary migration and variations in the basic planetary types—to explain the differing layouts of other solar systems. A much wider variety of solar system arrangements now seems possible than we had guessed before the discovery of extrasolar planets.

This leads us to what is perhaps the most profound question still to be addressed in our study of extrasolar planets: Given that we have not yet discovered other solar systems that are quite like ours, does this mean that our solar system is of a rare type, or does it simply mean that we have not yet acquired enough data to see how common solar systems like ours really are? This is a profound question because of its implications for the way we view our place in the universe. If solar systems like ours are common, then it seems reasonable to imagine that Earth-like planets—and perhaps life and civilizations—might also be common. But if our solar system is a rarity or even unique, then Earth might be the lone inhabited planet in our galaxy or even the universe. Only by continuing our search for planets around other stars can we hope to answer this question. As of 2012, we are on the brink of discovering Earth-size planets in Earth-like orbits and Jupiter-size planets in Jupiter-like orbits, so the answer may not be so far off.

13.4 THE FUTURE OF EXTRASOLAR PLANETARY SCIENCE

We have entered a new era in planetary science, one in which our understanding of planetary processes can be based on far more planets than just those of our own solar system. Although our current knowledge of extrasolar planets and their planetary systems is still quite limited, ingenious new observing techniques, dedicated observatories, and ambitious space telescope programs should broaden our understanding dramatically in the coming years and decades. In this section, we'll focus on the more dramatic improvements expected in coming years.

How will future observations improve our understanding?

As noted earlier, the deepest outstanding questions concern how common solar systems like ours may prove to be and, more specifically, whether other Earth-like planets exist. Keep in mind that there is an important distinction between Earth-*size* and Earth-*like*. The *Kepler* mission has already found many planet candidates that are Earth-size, but this does not necessarily mean that they are Earth-like in the sense of having features like continents, oceans, plate tectonics, or life.

Learning whether other planets are Earth-like will probably occur in two steps. First, we'll need to find planets that are Earth-size and have orbits that would in principle allow them to have liquid water on their surfaces. (We'll discuss the range of orbits that fall into this so-called *habitable zone* in Chapter 24.) Second, we'll need some type of direct imaging or spectroscopy to learn whether these planets actually have oceans and atmospheres that might be conducive to life.

THINK ABOUT IT

How do you think the discovery of other Earth-like planets would change our view of our place in the universe? Defend your opinions.

Scientists are developing a wide range of techniques for trying to accomplish these observations, including many that rely on ground-based observatories. However, for the remaining years of this decade, it's likely that the greatest advances will come primarily in three ways: with ongoing observations by the *Kepler* mission, with the *GAIA* mission, and with efforts to obtain direct imaging and spectroscopy of extrasolar planets. Let's briefly look at each.

More from *Kepler* The early results from the *Kepler* mission have already exceeded expectations, making scientists optimistic that even greater discoveries are still to come. **FIGURE 13.18** shows a graph of planet size versus orbital period, with dots indicating *Kepler's* candidate planets as of early 2012. Notice that any Earth-*like* planets would have to fall

FIGURE 13.18 This figure shows the orbital periods and sizes of all the *Kepler* planet candidates as of early 2012. We do not yet have enough data to learn about the region in the lower right, which would include possible Earth-like planets.

into the lower right portion of the graph, for which no data are yet available. The *Kepler* mission is currently scheduled to continue through 2016, which should give it enough time to fill in these regions of the graph (which require observations over a longer time period because of the longer orbital periods between transits).

GAIA We briefly discussed the European Space Agency's *GAIA* mission in Section 13.1. Slated for launch in 2013, *GAIA* is a remarkable space observatory designed primarily to make precise measurements of the positions of more than 1 billion stars in the Milky Way Galaxy. These astrometric measurements should enable scientists to calculate precise distances to these stars, thereby allowing the creation of a three-dimensional map of the galaxy. To make these measurements, *GAIA* actually has two telescopes, each with a collecting area of nearly one square meter, that will work together to give extremely precise measurements of stellar positions (**FIGURE 13.19**).

Because precise measurements of stellar position also allow scientists to detect changes in those positions, *GAIA* will be capable of discovering planets via the astrometric method, as well as some via the transit method. In fact, mission scientists estimate that they may be able to detect as many as 15,000 extrasolar planets during the life of the mission. Although sorting out the gravitational tug of Earth-size planets from the tugs of larger planets may be difficult, *GAIA* should provide enough data to give us a sense of the layout of thousands of planetary systems. Therefore, even if it does not find Earth-like planets itself, it should give us a much better sense of whether solar systems with layouts like our own are common or rare.

Direct Detection The indirect planet-hunting techniques we have discussed so far have started a revolution in planetary science by demonstrating that our solar system is just one of many planetary systems. But it can be very difficult to learn more than planets' most basic properties from indirect measurements. To learn more about their nature, it would be ideal to observe the planets themselves, obtaining images of their surfaces or spectra of their atmospheres.

FIGURE 13.19 Artist's conception of the *GAIA* spacecraft.

FIGURE 13.20 This infrared image from the Keck Telescope shows direct detection of a four-planet system (planets marked b, c, d, e) orbiting the star HR 8799. We know they are planets because they have moved slightly since their discovery. Light from the star itself (center) was mostly blocked out during the exposure, with its remaining light subtracted as much as possible. These planets are much larger, brighter, and farther from their star than jovian planets in our solar system.

The glare from stars makes direct detection of planets extremely difficult, especially given how close together the stars and planets lie when observed from Earth. To date astronomers have found only a few planets by direct detection. One confirmed direct detection made with infrared light is of a jovian planet orbiting the star Beta Pictoris (see Figure 8.4). This planet's existence was actually first suspected by study of ripples in the dust disk surrounding the star, and then later confirmed through imaging, which has also revealed the planet's orbital motion around its star. **FIGURE 13.20** shows another confirmed detection with infrared light, revealing at least four planets orbiting the star HR 8799. Astronomers are confident the planets are real because subsequent images have shown changes in their positions due to their orbits around the star.

Scientists are rapidly developing new observing capabilities that should allow for more direct observations in the future from large, ground-based observatories. Scientists are also optimistic that the James Webb Space Telescope, a large infrared space telescope with launch currently scheduled for 2018, will be able to make some direct observations.

Nevertheless, the quest to find Earth-like worlds is now in a bittersweet period. Thanks to recent discoveries and rapidly advancing technology, astronomers have begun to design advanced space observatories that should be able to obtain images and spectra of Earth-size planets around other stars, with enough resolution to be able to determine whether they are Earth-like, and perhaps even to detect spectral signatures that would indicate the presence of life. However, while some of these missions could in principle have flown in this decade, budgetary constraints have pushed them at least 10 to 20 years into the future. Answers to age-old questions lie tantalizingly close, but will not be obtained until we find the budget to pay for them.

THINK ABOUT IT

Suppose you were a member of the United States Congress. How much would you be willing to spend to build a space observatory capable of obtaining images and spectra of Earth-like planets around other stars? Defend your opinion.

The Big Picture

Putting Chapter 13 into Perspective

In this chapter, we have explored one of the newest areas of astronomy—the study of solar systems beyond our own. As you continue your studies, please keep in mind the following important ideas:

- In a period of barely two decades, we have gone from knowing of no other planets around other stars to knowing that many or most stars have one or more planets. As a result, there is no longer any question that planets are common in the universe.

- The discovery of other planetary systems represents a striking confirmation of a key prediction of the nebular theory of solar system formation. Nevertheless, the precise characteristics of other planets and planetary systems pose challenges to details of the theory that scientists are still investigating.

- While we have already identified thousands of extrasolar planets or candidate planets, nearly all of these have been found with indirect methods. These methods allow us to determine many properties of the planets, but we will need direct images or spectra to learn about them in much more detail.

- It is too soon to know if solar systems with layouts like ours are rare or common, but technology now exists that could allow us to answer this question and to learn if Earth-like planets are also common. It is only a matter of time until we know the answer to these fundamental questions.

13.1 DETECTING PLANETS AROUND OTHER STARS

■ **Why is it so challenging to learn about extrasolar planets?** The great challenge stems from the great distances to other stars, the small sizes of planets in comparison, and the vast difference in brightness between stars and planets. We can in principle look for planets directly or indirectly. The two major indirect approaches are (1) looking for subtle gravitational effects on stars due to orbiting planets and (2) looking for changes in a star's brightness as one of its planets passes in front of it.

■ **How can a star's motion reveal the presence of planets?**

 We can look for a planet's gravitational effect on its star through the **astrometric method**, which looks for small shifts in stellar position, or the **Doppler method**, which looks for the back-and-forth motion of stars revealed by Doppler shifts.

■ **How can changes in a star's brightness reveal the presence of planets?** A small fraction of all planetary

 systems should by chance be aligned in such a way that their planets can pass in front of their star as seen from Earth. Such a passage, called a **transit**, would occur once each orbit and cause a slight dimming in the measured brightness of the star. The planet may also pass behind the star in an **eclipse** on the other side of the orbit, potentially revealing even more information about the planet. The *Kepler* mission has discovered thousands of candidate planets with the transit method.

13.2 THE NATURE OF PLANETS AROUND OTHER STARS

■ **What properties of extrasolar planets can we measure?** All detection methods allow us to determine a planet's orbital period and distance from its star. The astrometric and Doppler methods can provide masses (or minimum masses), while the transit method can provide sizes. In cases where transit and Doppler methods are used together, we can determine average density. In some cases, transits (and eclipses) can provide other data, including limited data about atmospheric composition and temperature.

■ **How do extrasolar planets compare with planets in our solar system?** The known extrasolar planets have a

 much wider range of properties than the planets in our solar system. Many orbit much closer to their stars and with more eccentric orbital paths; some jovian planets, called *hot Jupiters*, are also found close to their stars. We also find properties indicating planetary types such as water worlds that do not fall neatly into the traditional terrestrial and jovian categories.

13.3 THE FORMATION OF OTHER SOLAR SYSTEMS

■ **Can we explain the surprising orbits of many extrasolar planets?** Jovian planets with close-in and eccentric

 orbits probably were born on orbits similar to those of the jovian planets in our solar system. Several different effects could later have changed their orbits: planetary migration induced by waves in the gaseous disk from which they formed, gravitational encounters with other objects, or orbital resonances with other massive planets.

■ **Do we need to modify our theory of solar system formation?** Our basic theory of solar system formation seems to be sound, but we have had to modify it to allow for planetary migration and a wider range of planetary types than we find in our solar system. Many mysteries remain, but they are unlikely to require major change to the nebular theory of solar system formation.

13.4 THE FUTURE OF EXTRASOLAR PLANETARY SCIENCE

■ **How will future observations improve our understanding?** In the future, scientists hope to learn whether solar systems with layouts like ours are rare or common, and whether

 Earth-like planets are common. Future observations by the *Kepler* and *GAIA* missions should help answer these questions, but ultimate answers will probably require direct detection with space observatories of the future.

Use the following questions to check your understanding of some of the many types of visual information used in astronomy. Answers are provided in Appendix J. For additional practice, try the Chapter 13 Visual Quiz at MasteringAstronomy®.

This plot, based on Figure 13.4a, shows the periodic variations in the Doppler shift of a star caused by a planet orbiting around it. Positive velocities mean the star is moving away from Earth, and negative velocities mean the star is moving toward Earth. (You can assume that the orbit appears edge-on from Earth.) Answer the following questions based on the information in the graph.

1. How long does it take the star and planet to complete one orbit around their center of mass?
2. What maximum velocity does the star attain?
3. Match the *star's* position at points 1, 2, 3, and 4 in the plot with the descriptions below.
   a. headed straight toward Earth
   b. headed straight away from Earth
   c. closest to Earth
   d. farthest from Earth

4. Match the *planet's* position at points 1, 2, 3, and 4 in the plot with the descriptions in question 3.
5. How would the plot change if the planet were more massive?
   a. It would not change, because it describes the motion of the star, not the planet.
   b. The peaks and valleys would get larger (greater positive and negative velocities) because of larger gravitational tugs.
   c. The peaks and valleys would get closer together (shorter period) because of larger gravitational tugs.

For instructor-assigned homework go to MasteringAstronomy®.

REVIEW QUESTIONS

Short-Answer Questions Based on the Reading

1. Why are *extrasolar planets* hard to detect directly?
2. What are the two major approaches to detecting extrasolar planets indirectly?
3. How can gravitational tugs from orbiting planets affect the motion of a star? Explain how alien astronomers could deduce the existence of planets in our solar system by observing the Sun's motion.
4. Briefly describe the *astrometric method* and its strengths and limitations.
5. Briefly describe the *Doppler method* and its strengths and limitations.
6. How does the *transit method* work? Could we use this method to find planets around all stars that have them? Why or why not?
7. Briefly describe the *Kepler* mission and how it meets three key challenges posed by the transit method.
8. Briefly summarize the planetary properties we can in principle measure with current detection methods, and state which methods allow us to measure each of these properties.
9. Why does the Doppler method generally allow us to determine only *minimum* planetary masses? In what cases can we be confident that we know precise masses? Explain.

10. Briefly describe what we can learn from careful study of a planet that undergoes transits and eclipses.
11. How do the orbits of known extrasolar planets differ from those of planets in our solar system? Why are these orbits surprising?
12. Summarize the current state of knowledge about extrasolar planet masses and sizes. Based on the evidence, is it likely that smaller planets or larger planets are more common?
13. Summarize the key features shown in Figure 13.16, and briefly describe the nature of planets that would fit each of the model curves shown on the graph.
14. Why do scientists suspect that planetary migration is behind the close-in and eccentric orbits of many extrasolar planets? How might this migration have occurred?
15. How can scientists account for the fact that extrasolar planets seem to come in a wider range of types than the planets of our own solar system?
16. Overall, does the nebular theory seem adequate for describing the origins of other planetary systems? Explain.
17. Briefly describe how *Kepler, GAIA,* and direct observations should improve our understanding of extrasolar planets in coming years.

TEST YOUR UNDERSTANDING

Does It Make Sense?

Decide whether the statement makes sense (or is clearly true) or does not make sense (or is clearly false). Explain clearly; not all these have definitive answers, so your explanation is more important than your chosen answer.

18. An extraterrestrial astronomer surveying our solar system with the Doppler method could discover the existence of Jupiter with just a few days of observation.

19. The fact that we have not yet discovered an Earth-size extrasolar planet in an Earth-like orbit tells us that such planets must be very rare.

20. Within the next few years, astronomers expect to confirm all the planet detections made with the astrometric and Doppler methods by observing transits of these same planets.

21. The infrared brightness of a star system decreases when a planet goes into eclipse.

22. Some extrasolar planets are likely to be made mostly of water.

23. Some extrasolar planets are likely to be made mostly of gold.

24. Current evidence suggests that there could be 100 billion or more planets in the Milky Way Galaxy.

25. It's the year 2018: The *Kepler* mission has announced the discovery of numerous planets with Neptune-like orbits around their stars.

26. It's the year 2020: Astronomers have successfully photographed an Earth-size planet, showing that it has oceans and continents.

27. It's the year 2040: Scientists announce that our first spacecraft to reach an extrasolar planet is now orbiting a planet around a star located near the center of the Milky Way Galaxy.

Quick Quiz

Choose the best answer to each of the following. Explain your reasoning with one or more complete sentences.

28. What method has detected the most extrasolar planets (or candidates) so far? (a) the transit method (b) Hubble images (c) the Doppler method

29. What fraction of extrasolar planets could in principle be detected by the transit method? (a) less than about 1% (b) about 20% (c) 100%

30. The astrometric method is best for finding massive planets that orbit (a) very close to their stars. (b) farther from their stars. (c) around extremely distant stars.

31. Which one of the following can the transit method tell us about a planet? (a) its mass (b) its size (c) the eccentricity of its orbit

32. Which method could detect a planet in an orbit that is face-on to the Earth? (a) Doppler method (b) transit method (c) astrometric method

33. Which detection method can be used with a backyard telescope? (a) Doppler method (b) transit method (c) astrometric method

34. To determine a planet's average density, we can use (a) the transit method alone. (b) the astrometric and Doppler methods together. (c) the transit and Doppler methods together.

35. Based on the model types shown in Figure 13.16, a planet made almost entirely of hydrogen compounds would be considered a (a) terrestrial planet. (b) jovian planet. (c) "water world."

36. What's the best explanation for the location of hot Jupiters? (a) They formed closer to their stars than Jupiter did. (b) They formed farther out like Jupiter but then migrated inward. (c) The strong gravity of their stars pulled them in close.

37. The major obstacle to NASA's building an observatory capable of obtaining moderately high-resolution images and spectra of extrasolar planets is that (a) no one has conceived of a technology that could do it. (b) it would require putting telescopes in space. (c) NASA lacks the necessary budget.

PROCESS OF SCIENCE

Examining How Science Works

38. *When Is a Theory Wrong?* As discussed in this chapter, in its original form the nebular theory of solar system formation does not explain the orbits of many known extrasolar planets, but it can explain them with modifications such as allowing for planetary migration. Does this mean the theory was "wrong" or only "incomplete" before the modifications were made? Explain. Be sure to look back at the discussion in Chapter 3 of the nature of science and scientific theories.

39. *Refuting the Theory.* Consider the following three hypothetical observations: (1) the discovery of a lone planet that is small and dense like a terrestrial planet but has a Jupiter-like orbit; (2) the discovery of a planetary system in which three terrestrial planets orbit the star beyond the orbital distance of two jovian planets; (3) the discovery that a majority of planetary systems have their jovian planets located nearer to their star than 1 AU and their terrestrial planets located beyond 5 AU. Each of these observations would challenge our current theory of solar system formation, but would any of them shake the very foundations of the theory? Explain clearly for each of the three hypothetical observations.

40. *Unanswered Questions.* As discussed in this chapter, we are only just beginning to learn about extrasolar planets. Briefly describe one important but unanswered question related to the study of planets around other stars. Then write 2–3 paragraphs in which you discuss how we might answer this question in the future. Be as specific as possible, focusing on the type of evidence necessary to answer the question and how the evidence could be gathered. What are the benefits of finding answers to this question?

GROUP WORK EXERCISE

41. *Time to Move On.* A common theme in science fiction is "leaving home" to find a new planet for humans to live on. Now that we know about thousands of planets, we can start imagining how to choose. Before you begin, assign the following roles to the people in your group: *Scribe* (takes notes on the group's activities), *Proposer* (proposes explanations to the group), *Skeptic* (points out weaknesses in proposed explanations), and *Moderator* (leads group discussion and makes sure everyone contributes). Then discuss the following questions.
   a. Examine the planets in Figure 13.13. Which kinds of planets might make good homes, or poor ones? Are planets missing from the graph that might be better still? Does this graph give enough information? What's missing? **b.** Now examine the planets in Figure 13.16, and again comment on the habitability of planets on different parts of the graph. **c.** Finally, evaluate the planets in Figure 13.18 by the same criteria. **d.** In the end, which graph or combination of graphs would help you make the best decision?

INVESTIGATE FURTHER

In-Depth Questions to Increase Your Understanding

Short-Answer/Essay Questions

42. *Why So Soon?* The detection of extrasolar planets came much sooner than astronomers expected. Was this a result of planets being different than expected or of technology improving faster than expected? Explain.

43. *Explaining the Doppler Technique.* Explain how the Doppler technique works in terms an elementary school child would understand. It may help to use an analogy to explain the difficulty of direct detection and the general phenomenon of the Doppler shift.

44. *Comparing Methods.* What are the strengths and limitations of the Doppler and transit techniques? What kinds of planets are easiest to

detect with each method? Are there certain planets that each method cannot detect, even if the planets are very large? Explain. What advantages are gained if a planet can be detected by both methods?

45. *No Hot Jupiters Here.* How do we think hot Jupiters formed? Why didn't one form in our solar system?

46. *Low-Density Planets.* Only one planet in our solar system has a density less than 1 g/cm^3, but many extrasolar planets do. Explain why in a few sentences. (*Hint:* Consider the densities of the jovian planets in our solar system, given in Figure 11.1.)

47. *A Year on HD 209458b.* Imagine you were visiting the planet that orbits the star HD 209458, hovering in the upper atmosphere in a suitable spacecraft. What would it be like? What would you see, and how would the view look different from the view you would have while floating in Jupiter's atmosphere? Consider factors like local conditions, clouds, how the Sun would appear, and orbital motion.

48. *Detect an Extrasolar Planet for Yourself.* Most colleges and many amateur astronomers have the equipment necessary to detect known extrasolar planets using the transit method. All that's required is a telescope 10 or more inches in diameter, a CCD camera system, and a computer system for data analysis. The basic method is to take exposures of a few minutes' duration over a period of several hours around the times of predicted transit, and to compare the brightness of the star being transited to that of other stars in the same CCD frame. For complete instructions, see the study area of Mastering Astronomy.

Quantitative Problems

Be sure to show all calculations clearly and state your final answers in complete sentences.

49. *Lost in the Glare.* How hard would it be for an alien astronomer to detect the light from planets in our solar system compared to the light from the Sun itself?
a. Calculate the fraction of the total emitted sunlight that is reflected by Earth. (*Hint:* Imagine a sphere around the Sun the size of the planet's orbit (area $= 4\pi a^2$). What fraction of that area does the disk of a planet (area $= \pi r_{planet}^2$) take up? Earth's reflectivity is 29%.)
b. Would detecting Jupiter be easier or harder than detecting Earth? Comment on whether you think Jupiter's larger size or greater distance has a stronger effect on its detectability. You may neglect any difference in reflectivity between Earth and Jupiter.

50. *Transit of TrES-1.* The planet TrES-1, orbiting a distant star, has been detected by both the transit and the Doppler technique, so we can calculate its density and get an idea of what kind of planet it is.
a. Using the method of Mathematical Insight 13.3, calculate the radius of the transiting planet. The planetary transits block 2% of the star's light. The star TrES 1 has a radius of about 85% of our Sun's radius. **b.** The mass of the planet is approximately 0.75 times the mass of Jupiter, and Jupiter's mass is about 1.9×10^{27} kilograms. Calculate the average density of the planet. Give your answer in grams per cubic centimeter. Compare this density to the average densities of Saturn (0.7 g/cm^3) and Earth (5.5 g/cm^3). Is the planet terrestrial or jovian in nature? (*Hint:* To find the volume of the planet, use the formula for the volume of a sphere: $V = \frac{4}{3}\pi r^3$. Be careful with unit conversions.)

51. *Planet Around 51 Pegasi.* The star 51 Pegasi has about the same mass as our Sun. A planet discovered orbiting it has an orbital period of 4.23 days. The mass of the planet is estimated to be 0.6 times the mass of Jupiter. Use Kepler's third law to find the planet's average distance (semimajor axis) from its star. (*Hint:* Because the mass of 51 Pegasi is about the same as the mass of our Sun, you can use Kepler's third law in its original form, $p^2 = a^3$ [**Section 3.3**]. Be sure to convert the period into years before using this equation.)

52. *Identical Planets?* Imagine two planets orbiting a star with orbits edge-on to the Earth. The peak Doppler shift for each is 50 m/s, but one has a period of 3 days and the other has a period of 300 days. Calculate the two minimum masses and say which, if either, is larger. (*Hint:* See Mathematical Insight 13.2.)

53. *Finding Orbit Sizes.* The Doppler method allows us to find a planet's semimajor axis using just the orbital period and the star's mass (Mathematical Insight 13.1).
a. Imagine that a new planet is discovered orbiting a $2M_{Sun}$ star with a period of 5 days. What is its semimajor axis? **b.** Another planet is discovered orbiting a $0.5M_{Sun}$ star with a period of 100 days. What is its semimajor axis?

54. *One Born Every Minute?* It's possible to make a rough estimate of how often planetary systems form by making some basic assumptions. For example, if you assume that the stars we see have been born at random times over the last 10 billion years, then the rate of star formation is simply the number of stars we see divided by 10 billion years. The fraction of stars with detected extrasolar planets is at least 5%, so this factor can be multiplied in to find the approximate rate of formation of planetary systems.
a. Using these assumptions, estimate how often a planetary system forms in our galaxy. (Our galaxy contains at least 100 billion stars.)
b. How often does a planetary system form somewhere in the observable universe, which contains at least 100 billion galaxies?
c. Write a few sentences describing your reaction to your results. Do you think the calculations are realistic? Are the rates larger or smaller than you expected?

55. *Habitable Planet Around 51 Pegasi?* The star 51 Pegasi is approximately as bright as our Sun and has a planet that orbits at a distance of only 0.052 AU.
a. Suppose the planet reflects 15% of the incoming sunlight. Using Mathematical Insight 10.1, calculate its "no greenhouse" average temperature. How does this temperature compare to that of Earth? **b.** Repeat part a, but assume that the planet is covered in bright clouds that reflect 80% of the incoming sunlight. **c.** Based on your answers to parts a and b, do you think it is likely that the conditions on this planet are conducive to life? Explain.

Discussion Questions

56. *So What?* What is the significance of the discovery of extrasolar planets, if any? Justify your answer in the context of this book's discussion of the history of astronomy.

57. *Is It Worth It?* The cost of the *Kepler* mission is several hundred million dollars. The cost of a mission that could obtain direct images or spectra of extrasolar planets would likely be several billion dollars. Are these expenses worth it, compared to the results expected? Defend your opinion.

Web Projects

58. *New Planets.* Research the latest extrasolar planet discoveries. Create a "planet journal," complete with illustrations as needed, with a page for each of at least three recently discovered planets. On each page, note the technique that was used to find the planet, give any information we have about the nature of the planet, and discuss how the planet does or does not fit in with our current understanding of planetary systems.

59. *Direct Detections.* In this chapter, we saw only a few examples of direct detection of possible extrasolar planets. Search for new information on these and any other direct detections now known. Have the detections discussed in this chapter been confirmed as planets? Have we made any other direct detections, and if so, how? Summarize your findings in a short written report, including images of the directly detected planets.

60. *Extrasolar Planet Mission.* Learn about a proposed future mission to study extrasolar planets, including its proposed design, capabilities, and goals. Write a short report on your findings.

Comparing the worlds in the solar system has taught us important lessons about Earth and why it is so suitable for life. This illustration summarizes some of the major lessons we've learned by studying other worlds both in our own solar system and beyond it.

(1) Comparing the terrestrial worlds shows that a planet's size and distance from the Sun are the primary factors that determine how it evolves through time [Chapters 9, 10].

Venus demonstrates the importance of distance from the Sun: If Earth were moved to the orbit of Venus, it would suffer a runaway greenhouse effect and become too hot for life.

Mars shows why size is important: A planet smaller than Earth loses interior heat faster, which can lead to a decline in geological activity and loss of atmospheric gas.

The smallest terrestrial worlds, Mercury and the Moon, became geologically dead long ago. They therefore retain ancient impact craters, which provide a record of how impacts must have affected Earth and other worlds.

2 Jovian planets are gas-rich and far more massive than Earth. They and their ice-rich moons have opened our eyes to the diversity of processes that shape worlds [Chapter 11].

Earth and the Moon

Our Moon led us to expect all small objects to be geologically dead . . .

The strong gravity of the jovian planets has shaped the asteroid and Kuiper belts, and flung comets into the distant Oort cloud, ultimately determining how frequently asteroids and comets strike Earth.

Jupiter **and** Europa

. . . but Europa—along with Io, Titan, and other moons—proved that tidal heating or icy composition can lead to geological activity, in some cases with subsurface oceans and perhaps even life.

3 Asteroids and comets may be small bodies in the solar system, but they have played major roles in the development of life on Earth [Chapter 12].

Comets or water-rich asteroids from the outer asteroid belt brought Earth the ingredients of its oceans and atmosphere.

Impacts of comets and asteroids have altered the course of life on Earth and may do so again.

4 The discovery of planets around other stars has shown that our solar system is not unique. Studies of other solar systems are teaching us new lessons about how planets form and about the likelihood of finding other Earth-like worlds [Chapter 13].

Rapid advances in extrasolar planet detection have allowed us to find some planets as small as Earth, and others at the right distances from their stars to be habitable. The discovery of a planet that meets both criteria may have happened by the time you read this book.

14 OUR STAR

464

Give me the splendid silent sun with all his beams full-dazzling.

—Walt Whitman (1819–1892), *from* Leaves of Grass

Astronomy today encompasses the study of the entire universe, but the root of the word *astronomy* comes from the Greek word for "star." Although we have learned a lot about the universe up to this point in the book, only now do we turn our attention to the study of the stars, the namesakes of astronomy.

When we think of stars, we usually think of the beautiful points of light visible on a clear night. But the nearest and most easily studied star is visible only in the daytime—our Sun. In this chapter, we will study the Sun in some detail. We will see how the Sun generates the energy that supports life on Earth. Equally important, we will study our Sun as a star so that it can serve as an introduction to subsequent chapters in which we will study stars throughout the universe.

 The Sun Tutorial, Lesson 1

14.1 A CLOSER LOOK AT THE SUN

Most people know that the Sun is a star—a giant ball of hot gas that generates light and shines it brightly in all directions. However, scientists realized this fact only quite recently in human history. In this section, we'll consider the age-old question of what makes the Sun shine, and then take an imaginary plunge into the Sun that will get us better acquainted with its general features.

Why does the Sun shine?

The Sun's energy is vital to human existence. Ancient peoples certainly recognized that fact. Some worshipped the Sun as a god. Others created mythologies to explain its daily rise and set. But no one who lived before the 20th century knew the energy source for the Sun's light and heat.

Ancient Ideas Most ancient thinkers imagined the Sun to be some type of fire, perhaps a lump of burning coal or wood. It was a reasonable suggestion for the times, since the idea could not yet be tested. Ancient people did not know the size or distance of the Sun, so they could not imagine how incredible its energy output really is. Nor did they know how long Earth had existed, so they had no way to realize that the Sun has provided light and heat for a very long time.

Modern scientists began to address the question of how the Sun shines around the middle of the 19th century, by which time the Sun's size and distance had been measured with reasonable accuracy. The ancient idea of the Sun being fueled by burning coal or wood was quickly ruled out: Calculations showed that such burning could not possibly account for the Sun's huge output of energy. Other ideas based on chemical processes were likewise ruled out.

Gravitational Contraction In the late 19th century, astronomers came up with an idea that seemed more plausible, at least at first. They suggested that the Sun generates energy by slowly contracting in size, a process called **gravitational contraction** (or *Kelvin-Helmholtz contraction*, after the scientists who proposed the mechanism). Recall that a shrinking gas cloud heats up because the gravitational potential energy of gas particles far from the cloud center is converted into thermal energy as the gas moves inward (see Figure 4.14b). A gradually shrinking Sun would always have some gas moving inward, converting gravitational potential energy into thermal energy. This thermal energy would keep the inside of the Sun hot.

Because of its large mass, the Sun would need to contract only very slightly each year to maintain its temperature—so slightly that the contraction would have been unnoticeable to 19th-century astronomers. Calculations showed that gravitational contraction could have kept the Sun shining steadily for up to about 25 million years. For a while, some astronomers thought that this idea had solved the ancient mystery of how the Sun shines. However, geologists pointed out a fatal flaw: Studies of rocks and fossils had already shown Earth to be far older than 25 million years, which meant that gravitational contraction could not be the mechanism by which the Sun generates its energy.

Einstein's Breakthrough With both chemical processes and gravitational contraction ruled out as possible explanations for why the Sun shines, scientists were at a loss. There was no known way that an object the size of the Sun could generate so much energy for billions of years. A completely new type of explanation was needed, and it came with Einstein's publication of his special theory of relativity in 1905.

Einstein's theory included his famous discovery of $E = mc^2$ [**Sections 4.3, S2.3**]. This equation shows that mass itself contains an enormous amount of potential energy. Calculations immediately showed that the Sun's mass contained more than enough energy to account for billions of years of sunshine, if the Sun could convert the energy of mass into thermal energy. It took a few decades for scientists to work out the details, but by the end of the 1930s we had learned that the Sun converts mass into energy through the process of *nuclear fusion* [**Section 1.2**].

How Fusion Started Nuclear fusion requires extremely high temperatures and densities (for reasons we will discuss in the next section). In the Sun, these conditions are found deep in the core. But how did the Sun become hot enough for fusion to begin in the first place?

The answer invokes the mechanism of gravitational contraction, which astronomers of the late 19th century mistakenly thought might be responsible for the Sun's heat today. Recall that our Sun was born about $4\frac{1}{2}$ billion years ago from a collapsing cloud of interstellar gas [**Section 8.2**]. The contraction of the cloud released gravitational potential energy, raising the interior temperature and pressure. This process continued until the core finally became hot enough to sustain nuclear fusion, because only then did the Sun produce enough energy to give it the stability that it has today.

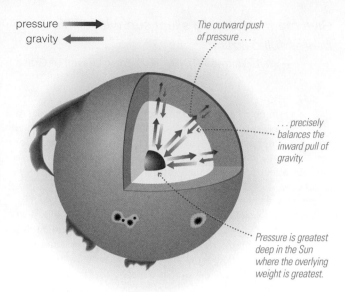

The outward push
of pressure . . .

. . . precisely
balances the
inward pull of
gravity.

Pressure is greatest
deep in the Sun
where the overlying
weight is greatest.

FIGURE 14.2 Gravitational equilibrium in the Sun: At each point inside, the pressure pushing outward balances the weight of the overlying layers.

THINK ABOUT IT

Earth's atmosphere is also in gravitational equilibrium, with the weight of upper layers supported by the pressure in lower layers. Use this idea to explain why the air gets thinner at higher altitudes.

The second kind of balance is **energy balance** between the rate at which fusion releases energy in the Sun's core and the rate at which the Sun's surface radiates this energy into space (**FIGURE 14.3**). Energy balance is important because without it, the balance between pressure and gravity would

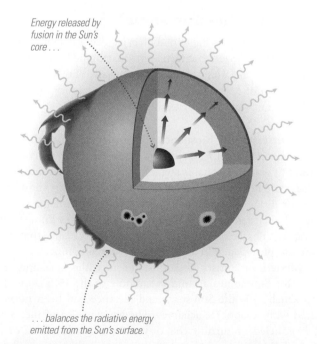

Energy released by
fusion in the Sun's
core . . .

. . . balances the radiative energy
emitted from the Sun's surface.

FIGURE 14.3 Energy balance in the Sun: Fusion must supply energy in the core at the same rate the Sun radiates energy from its surface.

FIGURE 14.1 An acrobat stack is in gravitational equilibrium: The lowest person supports the most weight and feels the greatest pressure, and the overlying weight and underlying pressure decrease for those higher up.

The Stable Sun The Sun continues to shine steadily today because it has achieved two kinds of balance that keep its size and energy output stable. The first kind of balance, called **gravitational equilibrium** (or *hydrostatic equilibrium*), is between the outward push of internal gas pressure and the inward pull of gravity. A stack of acrobats provides a simple example of gravitational equilibrium (**FIGURE 14.1**). The bottom person supports the weight of everybody above him, so his arms must push upward with enough pressure to support all this weight. At each higher level, the overlying weight is less, so it's a little easier for each additional person to hold up the rest of the stack.

Gravitational equilibrium works much the same way in the Sun, except the outward push against gravity comes from internal gas pressure rather than an acrobat's arms. The Sun's internal pressure precisely balances gravity at every point within it, thereby keeping the Sun stable in size (**FIGURE 14.2**). Because the weight of overlying layers is greater as we look deeper into the Sun, the pressure must increase with depth. Deep in the Sun's core, the pressure makes the gas hot and dense enough to sustain nuclear fusion. The energy released by fusion, in turn, heats the gas and maintains the pressure that keeps the Sun in balance against the inward pull of gravity.

not remain steady. If fusion in the core did not replace the energy radiated from the surface, thereby keeping the total thermal energy content constant, then gravitational contraction would cause the Sun to shrink and force its core temperature to rise.

In summary, the answer to the question "Why does the Sun shine?" is that about $4\frac{1}{2}$ billion years ago *gravitational contraction* made the Sun hot enough to sustain nuclear fusion in its core. Ever since, energy liberated by fusion has maintained *gravitational equilibrium* and *energy balance* within the Sun, keeping it shining steadily and supplying the light and heat that sustain life on Earth. Calculations show that the Sun was born with enough hydrogen in its core to shine steadily and maintain its gravitational equilibrium for about 10 billion years. The Sun is therefore only about halfway through its 10-billion-year "lifetime." About 5 billion years from now, when the Sun finally exhausts its nuclear fuel, gravitational contraction of the core will begin once again. As we will see in later chapters, some of the most important and spectacular processes in astronomy arise from the changes that occur as the crush of gravity begins to overcome a star's internal sources of pressure.

What is the Sun's structure?

We've already stated that the Sun is a giant ball of hot gas. To be more precise, the Sun is a ball of *plasma*—a gas in which many of the atoms are ionized because of the high temperature [Section 5.3]. The differing temperatures and densities of the plasma at different depths give the Sun the layered structure shown in FIGURE 14.4. To make sense of what you see in the figure, let's imagine that you have a spaceship that can somehow withstand the immense heat and pressure of the Sun, and take an imaginary journey from Earth to the center of the Sun. This journey will acquaint you with the basic properties of the Sun, which we'll discuss in greater detail later in this chapter.

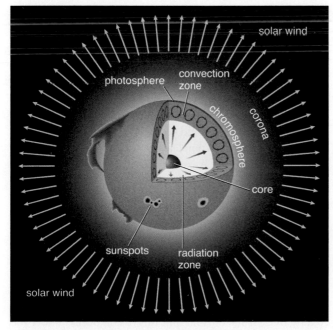

FIGURE 14.4 The basic structure of the Sun.

Basic Properties of the Sun As you begin your journey from Earth, the Sun appears as a whitish ball of glowing gas. Just as astronomers have done in real life, you can use simple observations to determine basic properties of the Sun. Spectroscopy [Section 5.4] tells you that the Sun is made almost entirely of hydrogen and helium. From the Sun's angular size and distance, you can determine that its radius is just under 700,000 kilometers, or more than 100 times the radius of Earth. Even **sunspots**, which appear as dark splotches on the Sun's surface, can be larger in diameter than Earth (FIGURE 14.5).

You can measure the Sun's mass using Newton's version of Kepler's third law [Section 4.4]. It is about 2×10^{30} kilograms, which is about 300,000 times the mass of Earth and nearly 1000 times the mass of all the planets in our solar system put together. You can observe the Sun's rotation rate by tracking the motion of sunspots or by measuring Doppler shifts on opposite sides of the Sun. Unlike a spinning ball, the entire Sun does *not* rotate at the same rate. Instead, the solar equator completes one rotation in about 25 days, and the rotation period increases with latitude to about 30 days near the solar poles.

The Sun releases an enormous amount of radiative energy into space, which you can measure through the window of your spacecraft. Recall that in science we measure energy in units of joules [Section 4.3]. We define **power** as the *rate* at which energy is used or released [Section 5.1]. The standard unit of power is the *watt*, defined as 1 joule of energy per second; that is, 1 watt = 1 joule/s. For example, a 100-watt light bulb requires 100 joules of energy for every second it is left turned on. The Sun's total power output, or **luminosity**, is an incredible 3.8×10^{26} watts. If we could somehow capture

FIGURE 14.5 This photo of the visible surface of the Sun shows several dark sunspots, each large enough to swallow our entire planet.

TABLE 14.1 Basic Properties of the Sun

Radius (R_{Sun})	696,000 km (about 109 times the radius of Earth)
Mass (M_{Sun})	2×10^{30} kg (about 300,000 times the mass of Earth)
Luminosity (L_{Sun})	3.8×10^{26} watts
Composition (by percentage of mass)	70% hydrogen, 28% helium, 2% heavier elements
Rotation rate	25 days (equator) to 30 days (poles)
Surface temperature	5800 K (average); 4000 K (sunspots)
Core temperature	15 million K

and store just 1 second's worth of the Sun's luminosity, it would be enough to meet current human energy demands for roughly the next 500,000 years. **TABLE 14.1** summarizes the basic properties of the Sun.

Of course, only a tiny fraction of the Sun's total energy output reaches Earth, with the rest dispersing in all directions into space. Most of this energy is radiated in the form of visible light, but after you've left the protection of Earth's atmosphere, you encounter significant amounts of other types of solar radiation, including dangerous ultraviolet and X rays.

The Sun's Atmosphere Even at great distances from the Sun, you and your spacecraft can feel slight effects from the **solar wind**—the stream of charged particles continually blown outward in all directions from the Sun. The solar wind helps shape the magnetospheres of planets [**Sections 10.1 and 11.1**] and blows back the material that forms the plasma tails of comets [**Section 12.2**].

As you approach the Sun more closely, you begin to encounter the low-density gas that represents what we usually think of as the Sun's atmosphere. The outermost layer of this atmosphere, called the **corona**, extends several million kilometers above the visible surface of the Sun. The temperature of the

corona is astonishingly high—about 1 million K—explaining why this region emits most of the Sun's X rays. However, the corona's density is so low that your spaceship absorbs relatively little heat despite the million-degree temperature [**Section 4.3**].

Nearer the surface, the temperature suddenly drops to about 10,000 K in the **chromosphere**, the middle layer of the solar atmosphere and the region that radiates most of the Sun's ultraviolet light. Then you plunge through the lowest layer of the atmosphere, or **photosphere**, which is the visible surface of the Sun. Although the photosphere looks like a well-defined surface from Earth, it consists of gas far less dense than Earth's atmosphere. The temperature of the photosphere averages just under 6000 K, and its surface seethes and churns like a pot of boiling water. The photosphere is also where you'll find sunspots, regions of intense magnetic fields that would cause your compass needle to swing about wildly.

The Sun's Interior Up to this point in your journey, you may have seen Earth and the stars when you looked back. But blazing light engulfs you as you slip beneath the photosphere. You are inside the Sun, and incredible turbulence tosses your spacecraft about. If you can hold steady long enough to see what is going on around you, you'll notice spouts of hot gas rising upward, surrounded by cooler gas cascading down from above. You are in the **convection zone**, where energy generated in the solar core travels upward, transported by the rising of hot gas and falling of cool gas called *convection* [**Section 9.1**]. The photosphere above you is the top of the convection zone, and convection is the cause of the Sun's seething, churning appearance.

About a third of the way down to the center, the turbulence of the convection zone gives way to the calmer plasma of the **radiation zone**, where energy moves outward primarily in the form of photons of light. The temperature rises to almost 10 million K, and your spacecraft is bathed in X rays trillions of times more intense than the visible light at the solar surface.

No real spacecraft could survive, but your imaginary one keeps plunging straight down to the solar **core**. There you finally find the source of the Sun's energy: nuclear fusion transforming hydrogen into helium. At the Sun's center, the temperature is about 15 million K, the density is more than 100 times that of water, and the pressure is 200 billion times that on the surface of Earth. The energy produced in the core today will take a few hundred thousand years to reach the surface.

With your journey complete, it's time to turn around and head back home. We'll continue this chapter by studying fusion in the solar core and then tracing the flow of the energy generated by fusion as it moves outward through the Sun.

 The Sun Tutorial, Lessons 2–3

14.2 NUCLEAR FUSION IN THE SUN

We've seen that the Sun shines because of energy generated by nuclear fusion, and that this fusion occurs under the extreme temperatures and densities found deep in the Sun's

FIGURE 14.6 Nuclear fission splits a nucleus into smaller nuclei, while nuclear fusion combines smaller nuclei into a larger nucleus.

core. But exactly how does fusion occur and release energy? And how can we claim to know about something taking place out of sight in the Sun's interior?

Before we begin to answer these questions, it's important to realize that the nuclear reactions that generate energy in the Sun are very different from those used to generate energy in human-built nuclear reactors on Earth. Our nuclear power plants generate energy by splitting large nuclei—such as those of uranium or plutonium—into smaller ones. The process of splitting an atomic nucleus is called **nuclear fission.** In contrast, the Sun makes energy by combining, or fusing, two or more small nuclei into a larger one. That is why we call the process **nuclear fusion. FIGURE 14.6** summarizes the difference between fission and fusion.

How does nuclear fusion occur in the Sun?

Fusion occurs within the Sun because the 15 million K plasma in the solar core is like a "soup" of hot gas, with bare, positively charged atomic nuclei (and negatively charged electrons) whizzing about at extremely high speeds. At any one time, some of these nuclei are on high-speed collision courses with each other. In most cases, electromagnetic forces deflect the nuclei, preventing actual collisions, because positive charges repel one another. If nuclei collide with sufficient energy, however, they can stick together to form a heavier nucleus (**FIGURE 14.7**).

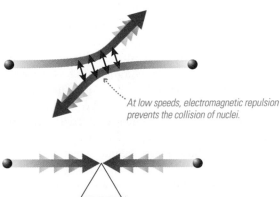

At low speeds, electromagnetic repulsion prevents the collision of nuclei.

At high speeds, nuclei come close enough for the strong force to bind them together.

FIGURE 14.7 Positively charged nuclei can fuse only if a high-speed collision brings them close enough for the strong force to come into play.

Sticking positively charged nuclei together is not easy. The **strong force**, which binds protons and neutrons together in atomic nuclei, is the only force in nature that can overcome the electromagnetic repulsion between two positively charged nuclei [**Section S4.2**]. In contrast to gravitational and electromagnetic forces, which drop off gradually as the distances between particles increase (by an inverse square law [**Section 4.4**]), the strong force is more like glue or Velcro: It overpowers the electromagnetic force over very small distances but is insignificant when the distances between particles exceed the typical sizes of atomic nuclei. The key to nuclear fusion, therefore, is to push the positively charged nuclei close enough together for the strong force to outmuscle electromagnetic repulsion.

The high pressures and temperatures in the solar core are just right for fusion of hydrogen nuclei into helium nuclei. The high temperature is important because the nuclei must collide at very high speeds if they are to come close enough together to fuse. (Quantum tunneling is also important to this process [**Section S4.4**].) The higher the temperature, the harder the collisions, making fusion reactions more likely. The high pressure of the overlying layers is necessary because without it, the hot plasma of the solar core would simply explode into space, shutting off the nuclear reactions.

The Proton-Proton Chain Let's investigate the fusion process in the Sun in a little more detail. Recall that hydrogen nuclei are simply individual protons, while the most common form of helium consists of two protons and two neutrons. The overall hydrogen fusion reaction therefore transforms four individual protons into a helium nucleus containing two protons and two neutrons:

$4\ ^1\text{H}$ $1\ ^4\text{He}$

This overall reaction actually proceeds through several steps involving just two nuclei at a time. The sequence of steps that occurs in the Sun is called the **proton-proton chain**, because it begins with collisions between individual protons (hydrogen nuclei). **FIGURE 14.8** illustrates the steps in the proton-proton chain:

Step 1: Two protons fuse to form a nucleus consisting of one proton and one neutron, which is the isotope of hydrogen known as *deuterium*. Note that this step

Hydrogen Fusion by the Proton-Proton Chain

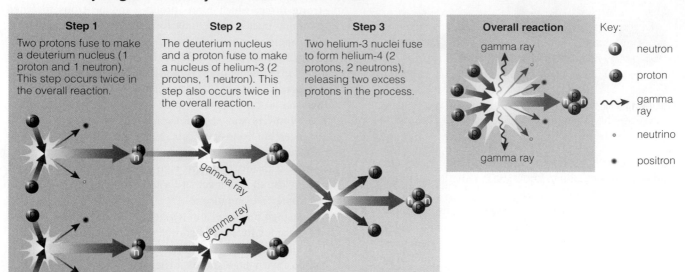

Step 1
Two protons fuse to make a deuterium nucleus (1 proton and 1 neutron). This step occurs twice in the overall reaction.

Step 2
The deuterium nucleus and a proton fuse to make a nucleus of helium-3 (2 protons, 1 neutron). This step also occurs twice in the overall reaction.

Step 3
Two helium-3 nuclei fuse to form helium-4 (2 protons, 2 neutrons), releasing two excess protons in the process.

Overall reaction
gamma ray

gamma ray

Key:
- n neutron
- p proton
- ⌇ gamma ray
- ○ neutrino
- • positron

FIGURE 14.8 interactive figure In the Sun, four hydrogen nuclei (protons) fuse into one helium-4 nucleus by way of the proton-proton chain. Gamma rays and subatomic particles known as neutrinos and positrons carry off the energy released in the reaction.

converts a proton into a neutron, reducing the total nuclear charge from +2 for the two fusing protons to +1 for the resulting deuterium nucleus. The lost positive charge is carried off by a *positron* (antielectron), the antimatter version of an electron with a positive rather than a negative charge [**Section S4.2**]. A **neutrino**—a subatomic particle with a very tiny mass—is also produced in this step.* (The positron won't last long, because it soon meets up with an ordinary electron, resulting in the creation of two gamma-ray photons through matter-antimatter annihilation.) This step must occur twice in the overall reaction, since it requires a total of four protons.

Step 2: A fair number of deuterium nuclei are always present along with the protons and other nuclei in the solar core, since Step 1 occurs so frequently in the Sun (about 10^{38} times per second). Step 2 occurs when one of these deuterium nuclei collides and fuses with a proton. The result is a nucleus of helium-3, a rare form of helium with two protons and one neutron, along with a gamma-ray photon. This step also occurs twice in the overall reaction.

Step 3: The third and final step of the proton-proton chain requires the addition of another neutron to the helium-3, thereby making normal helium-4. This final step can proceed in several different ways, but the most

common is through a collision of two helium-3 nuclei. Each of these helium-3 nuclei resulted from a prior, separate occurrence of Step 2 somewhere in the solar core. The final result is a normal helium-4 nucleus and two protons.

Notice that a total of six protons enter the reaction during Steps 1 and 2, with two coming back out in Step 3. The overall reaction therefore combines four protons to make one helium nucleus. The gamma rays and subatomic particles (neutrinos and positrons) carry off the energy released in the reaction.

Fusion of hydrogen into helium generates energy because a helium nucleus has a mass slightly less (by about 0.7%) than the combined mass of four hydrogen nuclei (see Mathematical Insight 14.1). That is, when four hydrogen nuclei fuse into a helium nucleus, a little bit of mass disappears. The disappearing mass becomes energy in accord with Einstein's formula $E = mc^2$. About 98% of the energy emerges as kinetic energy of the resulting helium nuclei and radiative energy of the gamma rays. As we will see, this energy slowly percolates to the solar surface, eventually emerging as the sunlight that bathes Earth. Neutrinos carry off the other 2% of the energy. Overall, fusion in the Sun converts about 600 million tons of hydrogen into 596 million tons of helium every second, which means that 4 million tons of matter is turned into energy each second. Although this sounds like a lot, it is a minuscule fraction of the Sun's total mass and does not affect the overall mass of the Sun in any measurable way.

The Solar Thermostat Nuclear fusion is the source of all the energy the Sun releases into space. If the fusion rate varied, so would the Sun's energy output, and large variations

---

*Producing a neutrino is necessary because of a law called *conservation of lepton number*: The number of leptons (e.g., electrons or neutrinos [**Chapter S4**]) must be the same before and after the reaction. The lepton number is zero before the reaction because there are no leptons. Among the reaction products, the positron (antielectron) has lepton number −1 because it is antimatter, and the neutrino has lepton number +1. Thus, the total lepton number remains zero.

in the Sun's luminosity would almost surely be lethal to life on Earth. Fortunately, the Sun fuses hydrogen at a steady rate, thanks to a natural feedback process that acts as a thermostat for the Sun's interior. To see how it works, let's examine what would happen if a small change were to occur in the core temperature (**FIGURE 14.9**).

Suppose the Sun's core temperature rose very slightly. The rate of nuclear fusion is extremely sensitive to temperature, so a slight temperature increase would cause the fusion rate to soar as protons in the core collided more frequently and with more energy. Because energy moves slowly through the Sun's interior, this extra energy would be bottled up in the core, temporarily forcing the Sun out of energy balance and raising the core pressure. The push of this pressure would temporarily exceed the pull of gravity, causing the core to expand and cool. This cooling, in turn, would cause the fusion rate to drop back down until the core returned to its original size and temperature, restoring both gravitational equilibrium and energy balance.

A slight drop in the Sun's core temperature would trigger an opposite chain of events. The reduced core temperature would lead to a decrease in the rate of nuclear fusion, causing a drop in pressure and contraction of the core. As the core shrank, its temperature would rise until the fusion rate returned to normal and restored the core to its original size and temperature.

The Gradually Brightening Sun While the processes involved in gravitational equilibrium prevent erratic changes in the fusion rate, they also ensure that the fusion

---

MATHEMATICAL INSIGHT 14.1

Mass-Energy Conversion in Hydrogen Fusion

Fusion of hydrogen into helium releases energy because the four hydrogen nuclei (protons) that go into the overall reaction have a slightly greater mass than the helium nucleus that comes out. When four protons (each with a mass of 1.6726×10^{-27} kg, for a total mass of 6.690×10^{-27} kg) fuse to make one helium-4 nucleus (mass of 6.643×10^{-27} kg), the amount of mass that "disappears" and becomes energy is

$$(6.690 \times 10^{-27} \text{ kg}) - (6.643 \times 10^{-27} \text{ kg}) = 0.047 \times 10^{-27} \text{ kg}$$

If we divide this lost mass by the original mass of the four protons, we find the fractional loss of mass:

$$\frac{\text{fractional mass loss}}{\text{in hydrogen fusion}} = \frac{0.047 \times 10^{-27} \text{ kg}}{6.69 \times 10^{-27} \text{ kg}} = 0.007$$

That is, a fraction 0.007, or 0.7%, of the original hydrogen mass is converted into energy according to Einstein's equation $E = mc^2$.

EXAMPLE 1: How much hydrogen is converted to helium each second in the Sun?

SOLUTION:

Step 1 Understand: The first step is to realize that we need to know the Sun's total energy output, or luminosity; Table 14.1 shows this to be 3.8×10^{26} watts, which means the Sun produces 3.8×10^{26} joules of energy each second. We can then use Einstein's equation $E = mc^2$ to calculate the total amount of mass converted into energy each second. We then use the fact that 0.7% of the hydrogen mass becomes energy to calculate how much hydrogen fuses each second.

Step 2 Solve: We start by solving Einstein's equation for the mass, m:

$$E = mc^2 \quad \Rightarrow \quad m = \frac{E}{c^2}$$

We plug in 3.8×10^{26} joules for the energy (E) that the Sun produces each second and recall that the speed of light is $c = 3 \times 10^8$ m/s; to make sure we can work with the units properly, we use the fact that 1 joule is equivalent to $1 \text{ kg} \times \text{m}^2/\text{s}^2$:

$$m = \frac{E}{c^2} = \frac{3.8 \times 10^{26} \dfrac{\text{kg} \times \text{m}^2}{\text{s}^2}}{\left(3 \times 10^8 \dfrac{\text{m}}{\text{s}}\right)^2} = 4.2 \times 10^9 \text{ kg}$$

We've found the amount of mass converted to energy in the Sun each second. Because this mass is 0.7% (= 0.007) of the mass of the hydrogen that fuses, the total mass of the hydrogen that undergoes fusion each second must be

$$\text{mass of hydrogen fused} = \frac{4.2 \times 10^9 \text{ kg}}{0.007} = 6.0 \times 10^{11} \text{ kg}$$

Step 3 Explain: The Sun fuses 600 billion kilograms of hydrogen each second, converting about 4 billion kilograms of this mass into energy; the rest, about 596 billion kilograms, becomes helium.

EXAMPLE 2: How many helium nuclei are created by fusion each second in the Sun?

SOLUTION:

Step 1 Understand: One way to approach this problem is to divide the total amount of mass lost each second in the Sun by the mass loss that occurs with each fusion reaction of four hydrogen nuclei into one helium nucleus. The result will tell us how many times this reaction occurs each second.

Step 2 Solve: From Example 1, the Sun converts 4.2×10^9 kilograms of mass into energy each second, and we learned earlier that each individual fusion reaction converts 0.047×10^{-27} kilogram of mass into energy. We divide to find the number of fusion reactions that occur each second in the Sun:

$$\begin{array}{l} \text{number of} \\ \text{fusion reactions} \\ \text{(per second)} \end{array} = \frac{\begin{array}{c} \text{total mass lost through fusion} \\ \text{(per second)} \end{array}}{\text{mass lost in each fusion reaction}}$$

$$= \frac{4.2 \times 10^9 \text{ kg}}{0.047 \times 10^{-27} \text{ kg}}$$

$$= 8.9 \times 10^{37}$$

Step 3 Explain: Notice that our result, 8.9×10^{37}, is just a little less than 10^{38}. In other words, the overall hydrogen fusion reaction occurs in the Sun nearly 10^{38} times each second.

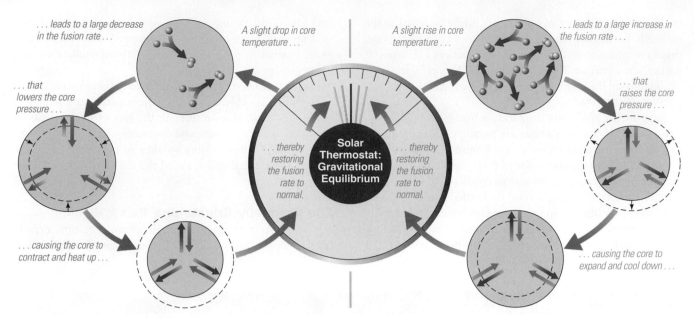

FIGURE 14.9 interactive figure The solar thermostat. Gravitational equilibrium regulates the Sun's core temperature. Everything is in balance if the amount of energy leaving the core equals the amount of energy produced by fusion. A rise in core temperature triggers a chain of events that causes the core to expand, lowering its temperature to the original value. A decrease in core temperature triggers the opposite chain of events, also restoring the original core temperature.

*Labels in figure (left to right):*
... leads to a large decrease in the fusion rate ...
A slight drop in core temperature ...
A slight rise in core temperature ...
... leads to a large increase in the fusion rate ...
... that lowers the core pressure ...
... thereby restoring the fusion rate to normal.
Solar Thermostat: Gravitational Equilibrium
... thereby restoring the fusion rate to normal.
... that raises the core pressure ...
... causing the core to contract and heat up ...
... causing the core to expand and cool down ...

rate gradually rises over billions of years. This rise in the fusion rate explains a fact we first encountered when we discussed climate change factors in Chapter 10: the gradual brightening of the Sun with time.

Remember that each fusion reaction converts *four* hydrogen nuclei into *one* helium nucleus. The total number of *independent particles* in the solar core therefore gradually decreases with time. This gradual reduction in the number of particles causes the solar core to shrink. The slow shrinkage, in turn, gradually increases the core temperature and fusion rate, keeping the core pressure high enough to counteract the stronger compression of gravity. Theoretical models indicate that the Sun's core temperature should have increased enough to raise its fusion rate and luminosity by about 30% since the Sun was born $4\frac{1}{2}$ billion years ago.

How does the energy from fusion get out of the Sun?

The solar thermostat balances the Sun's fusion rate so that the amount of nuclear energy generated in the core equals the amount of energy radiated from the surface as sunlight. However, the journey of solar energy from the core to the photosphere takes hundreds of thousands of years.

Most of the energy released by fusion starts its journey out of the solar core in the form of photons. Although photons travel at the speed of light, the path they take through the Sun's interior zigzags so much that it takes them a very long time to make any outward progress. Deep in the solar interior, the plasma is so dense that a photon can travel only a fraction of a millimeter in any one direction before it interacts with an electron. Each time a photon "collides" with an electron, the photon gets deflected into a new and random direction. The photon therefore bounces around the dense interior in a haphazard way (sometimes called a *random walk*) and only very gradually works its way outward from the Sun's center (**FIGURE 14.10**). The technical term for this slow outward migration of photons is **radiative diffusion**; to *diffuse* means to "spread out" and *radiative* refers to the photons of light, or radiation.

Energy released by fusion moves outward through the Sun's radiation zone (see Figure 14.4) primarily by way of these randomly bouncing photons. At the top of the radiation zone,

FIGURE 14.10 A photon in the solar interior bounces randomly among electrons, slowly working its way outward.

Bright spots appear on the Sun's surface where hot gas is rising . . .

. . . then the gas sinks after it has cooled off.

a This diagram shows convection beneath the Sun's surface. Hot gas (light yellow arrows) rises while cooler gas (black arrows) descends around it.

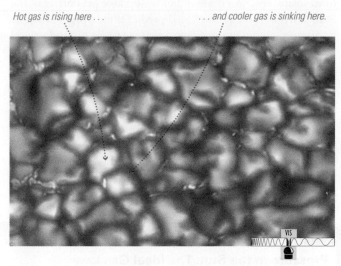

Hot gas is rising here . . .

. . . and cooler gas is sinking here.

b This photograph shows the mottled appearance of the Sun's photosphere. The bright spots, each about 1000 kilometers across, correspond to the rising plumes of hot gas in part a.

FIGURE 14.11 The Sun's photosphere churns with rising hot gas and falling cool gas as a result of underlying convection.

where the temperature has dropped to about 2 million K, the solar plasma absorbs photons more readily (rather than just bouncing them around). This absorption creates the conditions needed for convection [**Section 9.1**] and hence marks the bottom of the Sun's convection zone.

In the convection zone, convection occurs because hot gas is less dense than cool gas. Like a hot-air balloon, a hot bubble of solar plasma rises upward through the cooler plasma above it. Meanwhile, cooler plasma from above slides around the rising bubble and sinks to lower layers, where it is heated. The rising of hot plasma and sinking of cool plasma form a cycle that transports energy outward from the base of the convection zone to the solar surface, or photosphere (**FIGURE 14.11a**). There, the density of the gas becomes so low that photons can escape to space, which is why we see the photosphere as the "surface" of the Sun.

The convecting gas gives the photosphere the mottled appearance that we see in close-up photographs (**FIGURE 14.11b**): We see bright blobs where hot gas is welling up from inside the Sun and darker borders around those blobs where the cooler gas is sinking.* If we watched a movie of the photosphere, we'd see its surface bubbling rather like a pot of boiling water. Much like the way bubbles in a pot of boiling water burst on the surface and are replaced by new bubbles, each hot blob lasts only a few minutes before being replaced by others bubbling upward. The average temperature of the gas in the photosphere is about 5800 K, but convection causes the precise temperature to vary significantly from place to place.

To summarize, energy produced by fusion in the Sun's core works its way slowly through the radiation zone through random bounces of photons, then gets carried upward by convection in the convection zone. The photosphere lies at the top of the convection zone and marks the place where the density of gas has become low enough that photons can escape to space. The energy produced hundreds of thousands of years earlier in the solar core finally emerges from the Sun as thermal radiation [**Section 5.4**] produced by the 5800 K gas of the photosphere. Once in space, the photons travel straight away at the speed of light, bathing the planets in sunlight.

How do we know what is happening inside the Sun?

We cannot see inside the Sun, so you may wonder how we can claim to know so much about what goes on inside of it. In fact, we can study the Sun's interior in three different ways: through mathematical models of the Sun, observations of solar vibrations, and observations of solar neutrinos.

Mathematical Models Our primary way of learning about the interior of the Sun (and other stars) is by creating *mathematical models* that use the laws of physics to predict internal conditions. A basic model uses the Sun's observed composition and mass as inputs to equations that describe gravitational equilibrium, the solar thermostat, and the rate at which solar energy moves from the core to the photosphere. With the aid of a computer, we can use the model to calculate the Sun's temperature, pressure, and density at any depth. We can then predict the rate of nuclear fusion in the solar core by combining these calculations with knowledge about nuclear fusion gathered in laboratories on Earth.

If a model is a good description of the Sun's interior, it should correctly "predict" the radius, surface temperature,

---

*The blobs are formally called *granules,* and the photosphere's mottled appearance is sometimes referred to as *solar granulation.*

luminosity, age, and many other observable properties of the Sun. Current models do indeed predict these properties quite accurately, giving us confidence that we really do understand what is going on inside the Sun.

Solar Vibrations A second way to learn about the inside of the Sun is to observe vibrations of the Sun's surface that are somewhat similar to the vibrations that earthquakes cause on Earth. These vibrations result from movement of gas within the Sun, which generates waves of pressure that travel through the Sun like sound waves moving through air. We

can observe these vibrations on the Sun's surface by looking for Doppler shifts [**Section 5.4**]. Light from portions of the surface that are rising toward us is slightly blueshifted, while light from portions that are falling away from us is slightly redshifted. The vibrations are relatively small but measurable (**FIGURE 14.12**).

In principle, we can deduce a great deal about the solar interior by carefully analyzing these vibrations. (By analogy to seismology on Earth, this type of study of the Sun is called *helioseismology—helios* means "sun.") Results to date confirm that our mathematical models of the solar interior

---

MATHEMATICAL INSIGHT 14.2

Pressure in the Sun: The Ideal Gas Law

The pressure that resists gravity inside the Sun comes from the thermal motions of gas particles. Particles in a hot gas move more quickly and collide with more force than those in a cooler gas [**Section 5.3**], and therefore exert more pressure on any surface that tries to contain them. (Recall that pressure is defined as the force per unit area exerted on any surface.) However, pressure also depends on *how many* particles are colliding with each unit area of that surface each second, which means that pressure depends on the *number density* of gas particles—the number of gas particles contained in each cubic centimeter. Mathematically, we express this relationship between temperature, density, and pressure with the **ideal gas law,** which can be written as

$$P = nkT$$

where P represents gas pressure, n represents the number density of particles (in particles per cubic centimeter), T represents gas temperature (on the Kelvin scale), and $k = 1.38 \times 10^{-23}$ joule/K, which is known as *Boltzmann's constant*. This law applies to all gases consisting of simple, freely flying particles, like those in the Sun. In the units used here, the ideal gas law gives pressure in units of joules per cubic centimeter (J/cm^3); these units of energy per unit volume are equivalent to units of force per unit area.

EXAMPLE 1: The Sun's core contains about 10^{26} particles per cubic centimeter at a temperature of 15 million K. How does the gas pressure in the core of the Sun compare to the pressure of Earth's atmosphere at sea level, where there are about 2.4×10^{19} particles per cubic centimeter at a temperature of roughly 300 K?

SOLUTION:

Step 1 Understand: The ideal gas law allows us to calculate pressure from the temperature and the number density of particles in a gas, so we have all the information we need. We can compare the two pressures by dividing the larger (the Sun's core pressure) by the smaller (Earth's atmospheric pressure).

Step 2 Solve: Dividing the Sun's core pressure by Earth's atmospheric pressure, we find that their ratio is

$$\frac{P_{Sun(core)}}{P_{Earth(atmos)}} = \frac{n_{Sun}kT_{Sun}}{n_{Earth}kT_{Earth}} = \frac{10^{26} \text{ cm}^{-3} \times 1.5 \times 10^7 \text{ K}}{2.4 \times 10^{19} \text{ cm}^{-3} \times 300 \text{ K}}$$

$$= 2 \times 10^{11}$$

Step 3 Explain: The Sun's core pressure is about 200 billion (2×10^{11}) times as great as atmospheric pressure on Earth.

EXAMPLE 2: How would the Sun's core pressure change if the Sun fused all its core hydrogen into helium without shrinking and without its core temperature changing? Use your answer to explain what should happen as fusion occurs in the real Sun, in which the core *can* shrink and heat up. To simplify the problem, assume that the Sun's core begins with pure hydrogen that is fully ionized, so that there are two particles for every hydrogen nucleus (the proton plus one electron that must also be present for charge balance in the core), and ends as pure helium with three particles for each helium nucleus (the nucleus plus two electrons to balance the two protons in helium).

SOLUTION:

Step 1 Understand: The core pressure will change if either the temperature or the number density of particles changes. For this simplified example, we are told that the temperature does not change and that the core does not shrink, so pressure changes are due only to the declining number density that occurs as fusion reduces the total number of particles in the core. We can therefore find the pressure change simply by finding how fusion changes the number of particles in the core.

Step 2 Solve: In this simplified example, the core initially contained two particles for every hydrogen nucleus and finished with three particles for every helium nucleus. Therefore, for each fusion reaction that converts four hydrogen nuclei to one helium nucleus, there are eight particles before the reaction (four electrons and four protons) and three particles after it (one helium nucleus and two electrons); that is, after fusion there are $\frac{3}{8}$ as many particles as there were before it. Because the temperature and volume don't change in this example, the core pressure after all the hydrogen fuses into helium is also $\frac{3}{8}$ of its original value.

Step 3 Explain: If the core volume had stayed constant, the core pressure at the end of the fusion process would have decreased to $\frac{3}{8}$ of its initial value. This decrease in core pressure would tip the balance of gravitational equilibrium in favor of gravity, which is why the core shrinks as fusion progresses in the Sun. This shrinkage decreases the volume of the core, allowing the number density of particles to stay approximately constant. The gradual shrinkage also produces a slight but gradual rise in core temperature and therefore in the fusion rate, which is why the Sun's luminosity gradually increases with time.

-2500 -2000 -1500 -1000 500   0   500 1000 1500 2000
Velocity (m/s)

FIGURE 14.12 This image shows vibrations on the Sun's surface that have been measured from Doppler shifts. Shades of orange show how quickly each spot on the Sun's surface is moving toward or away from us at a particular moment. Dark shades (negative velocities) represent motion toward us; light shades (positive velocities) represent motion away from us. The large-scale change in color from left to right reflects the Sun's rotation, and the small-scale ripples reflect the surface vibrations.

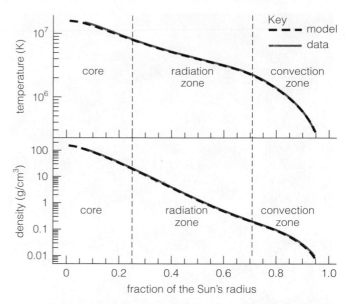

FIGURE 14.13 These graphs show the close agreement between predictions about the Sun's interior structure based on mathematical models and actual data obtained from observations of the Sun's surface vibrations. The agreement gives us confidence that the models are a good representation of the Sun's interior.

are on the right track (**FIGURE 14.13**). At the same time, analyses of the vibrations provide data that help to improve the models further.

Solar Neutrinos A third way to study the Sun's interior is to observe subatomic particles made by fusion reactions in the core. Remember that Step 1 of the proton-proton chain produces neutrinos. These tiny particles rarely interact with other forms of matter and therefore can pass through almost anything. For example, an inch of lead will stop an X ray, but stopping an average neutrino would require a slab of lead more than a light-year thick!

Neutrinos produced in the Sun's core pass outward through the solar interior almost as though it were empty space. Traveling at nearly the speed of light, they reach us in minutes, in principle giving us a way to monitor what is happening in the core of the Sun. In practice, their elusiveness makes neutrinos dauntingly difficult to monitor, because virtually all of them stream right through any detector built to capture and count them. About a thousand trillion solar neutrinos will zip through your body as you read this sentence, but don't panic—they will do no damage at all. In fact, nearly all of them will pass through the entire Earth as well.

Nevertheless, neutrinos *do* occasionally interact with matter, and it is possible to capture a few solar neutrinos with a large enough detector. To distinguish neutrino captures from reactions caused by other particles, scientists usually place neutrino detectors deep underground in mines. The overlying rock blocks most other particles, but the neutrinos have no difficulty passing through.

Early attempts to detect solar neutrinos were only partially successful, capturing only one-third of the number predicted by models of nuclear fusion in the Sun's core. This disagreement between model predictions and actual observations came to be called the *solar neutrino problem*. For more than three decades, the solar neutrino problem was one of the great mysteries in astronomy: Either something was wrong with our understanding of fusion in the Sun or some of the Sun's neutrinos were somehow escaping detection.

We now know that the missing solar neutrinos were going undetected. Neutrinos come in three distinct types, called *electron neutrinos, muon neutrinos,* and *tau neutrinos* [**Section S4.2**]. Fusion reactions in the Sun produce only electron neutrinos, and the early solar neutrino detectors could detect only electron neutrinos. More recent detectors, such as the Sudbury Neutrino Observatory (**FIGURE 14.14**), can detect all three neutrino types and their results confirm that the total number of solar neutrinos is equal to what we expect from our models of nuclear fusion in the Sun. In other words, some of the electron neutrinos produced by fusion change into neutrinos of the other two types during their trip from the Sun's core to its surface, a fact that has proved very important not only because it confirms our understanding of fusion in the Sun but also because of insights it provides into the fundamental physics of the subatomic world.

14.3 THE SUN-EARTH CONNECTION

Energy liberated by nuclear fusion in the Sun's core eventually reaches the solar surface, where it helps create a wide variety of phenomena that we can observe from Earth. Sunspots are only the most obvious of these phenomena. Because sunspots and other features of the Sun's surface change with time, they constitute what we call *solar weather*, or **solar activity**. The "storms" associated with solar weather are not just of academic interest. Sometimes they are so violent that they affect our day-to-day life on Earth. In this section, we'll explore solar activity and its far-reaching effects.

What causes solar activity?

Most of the Sun's surface churns constantly with rising and falling gas and looks like the close-up photo shown in Figure 14.11b. However, larger features sometimes appear, including sunspots, the huge explosions known as *solar flares*, and gigantic loops of hot gas that extend high into the Sun's corona. All these features are created by magnetic fields, which form and change easily in the convecting plasma in the outer layers of the Sun.

Sunspots and Magnetic Fields Sunspots are the most striking features of the solar surface (**FIGURE 14.15a**). If you could look directly at a sunspot without damaging your eyes, you would find it blindingly bright. Sunspots appear dark in photographs only because they are less bright than the surrounding photosphere. They are less bright because they are cooler: The temperature of the plasma in sunspots is about 4000 K, significantly cooler than the 5800 K of the plasma that surrounds them.

You may wonder how sunspots can be so much cooler than their surroundings. Gas can usually flow easily, so you might expect the hotter gas from outside a sunspot to mix with the cooler gas within it, quickly warming the sunspot. The fact

FIGURE 14.14 This photograph shows the main tank of the Sudbury Neutrino Observatory in Canada, located at the bottom of a mine shaft, more than 2 kilometers underground. The large sphere, 12 meters in diameter, contains 1000 tons of ultrapure *heavy water*. (Heavy water is water in which one or both hydrogen atoms are replaced by deuterium, making each molecule heavier than a molecule of ordinary water.) Neutrinos of all three types can cause reactions in the heavy water, and detectors surrounding the tank record these reactions when they occur.

Outside a sunspot we see a single spectral line . . .

. . . but the strong magnetic field inside a sunspot splits that line into three lines.

b Very strong magnetic fields split the absorption lines in spectra of sunspot regions. The dark vertical bands are absorption lines in a spectrum of the Sun. Notice that these lines split where they cross the dark horizontal bands corresponding to sunspots.

a This close-up view of the Sun's surface shows two large sunspots and several smaller ones. Each of the big sunspots is roughly as large as Earth.

FIGURE 14.15 Sunspots are regions of strong magnetic fields.

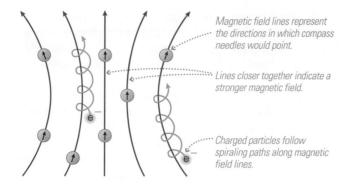

Magnetic field lines represent the directions in which compass needles would point.

Lines closer together indicate a stronger magnetic field.

Charged particles follow spiraling paths along magnetic field lines.

FIGURE 14.16 We draw magnetic field lines (red) to represent invisible magnetic fields. These lines also represent the directions along which charged particles tend to move.

that sunspots stay relatively cool means that something must prevent hot plasma from entering them, and that something turns out to be magnetic fields.

Detailed observations of the Sun's spectral lines reveal sunspots to be regions with strong magnetic fields. These magnetic fields can alter the energy levels in atoms and ions and therefore can alter the spectral lines they produce, causing some spectral lines to split into two or more closely spaced lines (**FIGURE 14.15b**). Wherever we see this effect (called the *Zeeman effect*), we know that magnetic fields must be present. Scientists can map magnetic fields on the Sun by looking for the splitting of spectral lines in light from different parts of the solar surface.

To understand how sunspots stay cooler than their surroundings, we must investigate the nature of magnetic fields in a little more depth. Magnetic fields are invisible, but we can represent them by drawing **magnetic field lines** (**FIGURE 14.16**). These lines represent the directions in which compass needles would point if we placed them within the magnetic field. The lines are closer together where the field is stronger and farther apart where the field is weaker. Because these imaginary field lines are so much easier to visualize than the magnetic field itself, we usually discuss magnetic fields by talking about how the field lines would appear. Charged particles, such as the ions and electrons in the solar plasma, cannot easily move perpendicular to the field lines, so they instead follow spiraling paths along them.

Solar magnetic field lines act somewhat like elastic bands, twisted into contortions and knots by turbulent motions in the solar atmosphere. Sunspots occur where tightly wound magnetic fields poke nearly straight out from the solar interior (**FIGURE 14.17a**). These tight magnetic field lines suppress convection within the sunspot and prevent surrounding plasma from entering the sunspot. With hot plasma unable to enter the region, the sunspot plasma becomes cooler than that of the rest of the photosphere. Individual sunspots typically last up to a few weeks, dissolving when their magnetic fields weaken and allow hotter plasma to flow in.

Sunspots tend to occur in pairs, connected by a loop of magnetic field lines that can arc high above the Sun's surface (**FIGURE 14.17b**). Gas in the Sun's chromosphere and corona becomes trapped in these loops, making giant **solar prominences**. Some prominences rise to heights of more than 100,000 kilometers above the Sun's surface (**FIGURE 14.18**).

Magnetic fields trap gas.

$T \approx 5800$ K   sunspots $T \approx 4500$ K $T \approx 5800$ K

convection cells

Magnetic fields of sunspots suppress convection and prevent surrounding plasma from sliding sideways into sunspot.

a Pairs of sunspots are connected by tightly wound magnetic field lines.

b This X-ray photo (from NASA's *TRACE* mission) shows hot gas trapped within looped magnetic field lines.

FIGURE 14.17 Strong magnetic fields keep sunspots cooler than the surrounding photosphere, and magnetic loops can arch from the sunspots to great heights above the Sun's surface.

Individual prominences can last for days or even weeks, disappearing only when the magnetic fields finally weaken and release the trapped gas.

Solar Storms The magnetic fields winding through sunspots and prominences sometimes undergo dramatic and sudden change, producing short-lived but intense storms on the Sun. The most dramatic of these storms are **solar flares**, which send bursts of X rays and fast-moving charged particles shooting into space (**FIGURE 14.19**).

Flares generally occur in the vicinity of sunspots, which is why we think they are created by changes in magnetic fields. The leading model for solar flares suggests that they occur when the magnetic field lines become so twisted and knotted

FIGURE 14.18 A gigantic solar prominence erupts from the solar surface in this ultraviolet-light image (from the Solar Dynamics Observatory), taken on December 8, 2011. The height of the prominence is more than 20 times the diameter of Earth.

that they can no longer bear the tension. They are thought to suddenly snap and reorganize themselves into a less twisted configuration. The energy released in the process heats the nearby plasma to 100 million K over the next few minutes or hours, generating X rays and accelerating some of the charged particles to nearly the speed of light.

Heating of the Chromosphere and Corona As we've seen, many of the most dramatic weather patterns and storms on the Sun involve the very hot gas of the Sun's chromosphere and corona. But why is that gas so hot in the first place?

Remember that temperatures gradually decline as we move outward from the Sun's core to the top of its photosphere. We might expect the decline to continue in the Sun's atmosphere, but instead it reverses, making the chromosphere and corona

much hotter than the Sun's surface. Some aspects of this atmospheric heating remain a mystery today, but we have at least a general explanation: The Sun's strong magnetic fields carry energy upward from the churning solar surface to the chromosphere and corona.

More specifically, the rising and falling of gas in the convection zone probably shakes tightly wound magnetic field lines beneath the solar surface. The magnetic field lines carry this energy upward to the solar atmosphere, where they deposit this energy as heat. The same magnetic fields that keep sunspots cool therefore make the overlying plasma of the chromosphere and corona hot.

Observations confirm the connection between magnetic fields and the structure of the chromosphere and corona. The density of gas in the chromosphere and corona is so low that we cannot see this gas with our eyes except during a total eclipse, when we observe the faint visible light scattered by electrons in the corona (see Figure 2.29). However, we can observe the chromosphere and corona at any time with ultraviolet and X-ray telescopes in space: The roughly 10,000 K plasma of the chromosphere emits strongly in the ultraviolet, and the 1 million K plasma of the corona is the source of virtually all X rays coming from the Sun. **FIGURE 14.20** shows an X-ray image of the Sun. The X-ray emission is brightest in regions where hot gas is being trapped and heated in magnetic field loops. Bright spots in the corona tend to be directly above sunspots in the photosphere, confirming that they are created by the same magnetic fields.

Notice that some regions of the corona barely show up in X-ray images; these regions, called **coronal holes**, are nearly devoid of hot coronal gas. More detailed analyses show that the magnetic field lines in coronal holes project out into space like broken rubber bands, allowing particles spiraling along them to escape the Sun altogether. These particles streaming outward from the corona make up the *solar wind,* which blows through the solar system at an average speed of

FIGURE 14.20 interactive photo An X-ray image of the Sun reveals the million-degree gas of the corona. Brighter regions of this image (yellow) correspond to areas of stronger X-ray emission. The darker regions (such as near the north pole at the top of this photo) are the coronal holes from which the solar wind escapes. (From the Yohkoh Space Observatory.)

FIGURE 14.19 This X-ray photo (from the *TRACE* spacecraft) shows a solar flare erupting from the Sun's surface.

about 500 kilometers per second and has important effects on planetary surfaces, atmospheres, and magnetospheres.

The solar wind also gives us something tangible to study. In the same way that meteorites provide us with samples of asteroids we've never visited, solar wind particles captured by satellites provide us with a sample of material from the Sun. Analysis of these solar particles has reassuringly verified that the Sun is made mostly of hydrogen, just as we conclude from studying the Sun's spectrum.

How does solar activity affect humans?

Flares and other solar storms sometimes eject large numbers of highly energetic charged particles from the Sun's corona. These particles travel outward from the Sun in huge bubbles called **coronal mass ejections** (**FIGURE 14.21**). The bubbles have strong magnetic fields and can reach Earth in a couple of days if they happen to be aimed in our direction. Once a coronal mass ejection reaches Earth, it can create a *geomagnetic storm* in Earth's magnetosphere. On the positive side, these storms can lead to unusually strong auroras (see Figure 10.11) that can be visible throughout much of the United States. On the negative side, they can hamper radio communications, disrupt electrical power delivery, and damage the electronic components in orbiting satellites.

During a particularly powerful magnetic storm on the Sun in March 1989, the U.S. Air Force temporarily lost track of more than 2000 satellites, and powerful currents induced in the ground circuits of the Quebec hydroelectric system caused it to collapse for more than 8 hours. The combined cost of the loss of power in the United States and Canada exceeded $100 million. In January 1997, AT&T lost contact with a $200 million communication satellite, probably because of damage

FIGURE 14.21 This X-ray image from the *SOHO* spacecraft shows a coronal mass ejection (the bright arc of gas headed almost straight upward) during the solar storms of 2003. The central red disk blocks the Sun itself, and the white circle represents the size of the Sun in this picture.

caused by particles coming from another powerful solar storm. More recently, a huge coronal mass ejection in early 2012 again threatened Earth's communication and electrical systems, but the systems escaped major damage, in part because of our improved preparedness for solar storms.

Satellites in low Earth orbit are particularly vulnerable during periods of strong solar activity, when the increase in solar X rays and energetic particles heats Earth's upper atmosphere, causing it to expand. The density of the gas surrounding low-flying satellites therefore rises, exerting drag that saps their energy and angular momentum. If this drag proceeds unchecked, the satellites ultimately plummet back to Earth.

THINK ABOUT IT

Solar activity is expected to peak between about 2013 and 2015, with a corresponding increase in the number of solar storms. Search for information on the most recent solar storm. Did it have any effects on satellites or electrical systems?

How does solar activity vary with time?

Solar weather is just as unpredictable as weather on Earth. Individual sunspots can appear or disappear at almost any time, and we have no way to know that a solar storm is coming until we observe it through our telescopes. However, long-term observations have revealed overall patterns in solar activity that make sunspots and solar storms more common at some times than at others.

The Sunspot Cycle The most notable pattern in solar activity is the **sunspot cycle**—a cycle in which the average number of sunspots on the Sun gradually rises and falls (**FIGURE 14.22**). At the time of *solar maximum*, when sunspots are most numerous, we may see dozens of sunspots on the Sun at one time. In contrast, we may see few if any sunspots at the time of *solar minimum*. The frequency of prominences, flares, and coronal mass ejections also follows the sunspot cycle, with these events being most common at solar maximum and least common at solar minimum.

As you can see in Figure 14.22a, the sunspot cycle varies from one period to the next. Some maximums have much greater numbers of sunspots than others. The length of time between maximums averages 11 years, but we have observed it to be as short as 7 years and as long as 15 years. The locations of sunspots on the Sun also vary with the sunspot cycle (Figure 14.22b). As a cycle begins at solar minimum, sunspots form primarily at mid-latitudes (30° to 40°) on the Sun. The sunspots tend to form at lower latitudes as the cycle progresses, appearing very close to the solar equator as the next solar minimum approaches. Then the sunspots of the next cycle begin to form near mid-latitudes again.

A less obvious feature of the sunspot cycle is that something peculiar happens to the Sun's magnetic field at each solar maximum: The Sun's entire magnetic field starts to flip, turning magnetic north into magnetic south and vice versa. We know this because the magnetic field lines connecting pairs of sunspots (see Figure 14.17) on the same side of the solar

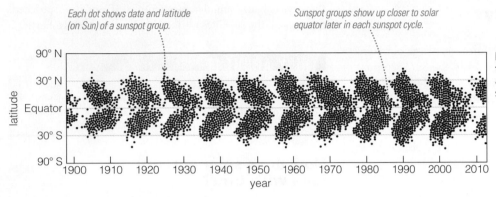

a This graph shows how the number of sunspots on the Sun changes with time. The vertical axis shows the percentage of the Sun's surface covered by sunspots. The cycle has a period of approximately 11 years.

b This graph shows how the latitudes at which sunspot groups appear tend to shift during a single sunspot cycle.

FIGURE 14.22 Sunspot cycle during the past century.

equator all tend to point in the same direction throughout an 11-year cycle. For example, all compass needles might point from the easternmost sunspot to the westernmost sunspot in each sunspot pair north of the solar equator. However, by the time the cycle ends at solar minimum, the magnetic field has reversed: In the subsequent solar cycle, the field lines connecting pairs of sunspots point in the opposite direction. The Sun's *complete* magnetic cycle (sometimes called the *solar cycle)* therefore averages 22 years, since it takes two 11-year sunspot cycles before the magnetic field is back the way it started.

Longer-Term Changes in the Sunspot Cycle

Figure 14.22 shows how the sunspot cycle has varied in length and intensity during the past century, the time during which we've had the most complete records of numbers of sunspots. However, astronomers have observed the Sun telescopically for nearly 400 years, and these longer-term observations suggest that the sunspot cycle can change even more dramatically (**FIGURE 14.23**). For example, astronomers observed virtually no sunspots between the years 1645 and 1715, a period sometimes called the *Maunder minimum* (after E. W. Maunder, who identified it in historical sunspot records).

Is it possible that the Maunder minimum is part of a longer-term cycle of solar activity, one lasting much longer

than 11 or 22 years? Some scientists have hypothesized that such cycles might exist, but little evidence has been found to back up these claims. Of course, searching for long-term variations is difficult, because we have very limited observational data predating the invention of the telescope.

The search for long-term cycles therefore relies on less direct evidence. For example, we can make educated guesses about past solar activity from historical descriptions of solar eclipses: When the Sun is more active, the corona tends to have longer and brighter "streamers" visible to the naked eye. We can also gauge solar activity further in the past (a few thousand years) by studying the amount of radioactive carbon-14 in tree rings. This amount varies because carbon-14 is produced in the atmosphere by interactions with high-energy *cosmic rays* [**Section 19.2**] coming from beyond our own solar system. During periods of high solar activity, the solar wind tends to grow stronger, shielding Earth from some of these cosmic rays and reducing the production of carbon-14. Because trees steadily incorporate atmospheric carbon into their rings (through their respiration of carbon dioxide), we can estimate the level of solar activity during each year of a tree's life by measuring the level of carbon-14 in the corresponding ring. These data have not yet turned up any clear evidence of longer-term cycles of solar activity, but the search goes on.

FIGURE 14.23 This graph reconstructs the sunspot cycle over the past 400 years, based on available data from telescopic observations.

The Cause of the Sunspot Cycle The precise reasons for the sunspot cycle are not fully understood, but the leading model ties the sunspot cycle to a combination of convection and the Sun's rotation. Convection is thought to dredge up weak magnetic fields generated in the solar interior, amplifying them as they rise. The Sun's rotation—faster at its equator than near its poles—then stretches and shapes these fields.

THINK ABOUT IT

Suppose you take a photograph of the Sun and notice two sunspots: one near the equator and one directly north of it at higher latitude. If you looked again at the Sun in a few days, would you still find one sunspot directly north of the other? Why or why not? Explain how you could use this type of observation to learn how the Sun rotates.

Imagine what happens to magnetic field lines that start out running along the Sun's surface from south to north (**FIGURE 14.24**). At the equator the lines circle the Sun every 25 days, but at higher latitudes they lag behind. As a result, the lines gradually get wound more and more tightly around the Sun. This process, operating at all times over the entire Sun, produces the contorted field lines that generate sunspots and other solar activity.

The detailed behavior of these magnetic fields is quite complex, so scientists attempt to study it with sophisticated computer models. Using these models, scientists have successfully replicated many features of the sunspot cycle, including changes in the number and latitude of sunspots and the magnetic field reversals that occur about every 11 years. However, much still remains mysterious, including why the period of the sunspot cycle varies and why solar activity is different from one cycle to the next.

Over extremely long time periods—hundreds of millions to billions of years—these theoretical models predict a gradual lessening of solar activity. Recall that, according to our theory of solar system formation, the Sun must have rotated much faster when it was young [**Section 8.3**], and a faster rotation rate should have meant much more activity. Observations of other stars that are similar to the Sun but rotate faster confirm that these stars are much more active. We find evidence for many more "starspots" on these stars than sunspots on the Sun, and their relatively bright ultraviolet and X-ray emissions suggest that they have brighter chromospheres and coronas—just as we would expect if they are more active than the Sun.

The Sunspot Cycle and Earth's Climate Despite the changes that occur during the sunspot cycle, the Sun's total output of energy barely changes at all—the largest measured changes have been less than 0.1% of the Sun's average luminosity. However, the ultraviolet and X-ray output of the Sun, which comes from the magnetically heated gas of the chromosphere and corona, can vary much more significantly. Could any of these changes affect the weather or climate on Earth?

Some data suggest connections between solar activity and Earth's climate. For example, the period from 1645 to 1715, when solar activity seems to have virtually ceased (see Figure 14.23), was a time of exceptionally low temperatures in Europe and North America known as the *Little Ice Age*. However, no one really knows whether the low solar activity caused these low temperatures or whether it was just a coincidence. Similarly, some researchers have claimed that certain weather phenomena, such as drought cycles or frequencies of storms, are correlated with the 11- or 22-year cycle of solar activity. A few scientists have even claimed that changes in the Sun may be responsible for Earth's recent global warming, though climate models indicate that the magnitude of the observed warming can be explained only by including human activity (through emissions of greenhouse gases) along with solar changes and other natural factors (see Figure 10.39). Nevertheless, the study of solar activity's possible effects on climate remains an active field of research.

Charged particles tend to push the field lines around with the Sun's rotation.

Because the Sun rotates faster near its equator than at its poles, the field lines bend ahead at the equator.

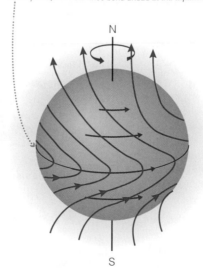

The field lines become more and more twisted with time, and sunspots form when the twisted lines loop above the Sun's surface.

FIGURE 14.24 The Sun rotates more quickly at its equator than it does near its poles. Because gas circles the Sun faster at the equator, it drags the Sun's north-south magnetic field lines into a more twisted configuration. The magnetic field lines linking pairs of sunspots, depicted here as dark blobs, trace out the directions of these stretched and distorted field lines.

Putting Chapter 14 into Context

In this chapter, we examined our Sun, the nearest star. When you look back at this chapter, make sure you understand these "big picture" ideas:

■ The ancient riddle of why the Sun shines has been solved. The Sun shines with energy generated by fusion of hydrogen into helium in the Sun's core. After a journey through the solar interior lasting several hundred thousand years and an 8-minute journey through space, a small fraction of this energy reaches Earth and supplies sunlight and heat.

■ The Sun shines steadily thanks to the balance between pressure and gravity (gravitational equilibrium) and the balance between energy production in the core and energy release at the surface

(energy balance). These two kinds of balance create a natural thermostat that regulates the Sun's fusion rate, keeping the Sun shining steadily and allowing life to flourish on Earth.

■ The Sun's atmosphere displays its own version of weather and climate, governed by solar magnetic fields. Some solar weather, such as coronal mass ejections, clearly affects Earth's magnetosphere. Other claimed connections between solar activity and Earth's climate may or may not be real.

■ The Sun is important not only because it is our source of light and heat, but also because it is the only star near enough for us to study in great detail. In the coming chapters, we will use what we've learned about the Sun to help us understand other stars.

SUMMARY OF KEY CONCEPTS

14.1 A CLOSER LOOK AT THE SUN

■ **Why does the Sun shine?** The Sun began to shine about

$4\frac{1}{2}$ billion years ago when **gravitational contraction** made its core hot enough to sustain nuclear fusion. It has shined steadily ever since because of two types of balance: (1) **gravitational equilibrium**, a balance between the outward push of pressure and the inward pull of gravity, and (2) **energy balance** between the energy released by fusion in the core and the energy radiated into space from the Sun's surface.

■ **What is the Sun's structure?** The Sun's interior layers, from the inside out, are the **core**, the **radiation zone**, and the

convection zone. Atop the convection zone lies the **photosphere**, the surface layer from which photons can freely escape into space. Above the photosphere are the warmer **chromosphere** and the very hot **corona**.

14.2 NUCLEAR FUSION IN THE SUN

■ **How does nuclear fusion occur in the Sun?** The core's extreme temperature and density are just right for fusion of

hydrogen into helium, which occurs via the **proton-proton chain**. Because the fusion rate is so sensitive to temperature, gravitational equilibrium acts as a thermostat that keeps the fusion rate steady.

■ **How does the energy from fusion get out of the Sun?** Energy moves through the deepest layers of the Sun—the core and the radiation zone—through **radiative diffusion**, in which photons bounce randomly among gas particles. After energy emerges from the radiation zone, convection carries it the rest of the way to the photosphere, where it is radiated into space as sunlight. Energy produced in the core takes hundreds of thousands of years to reach the photosphere.

■ **How do we know what is happening inside the Sun?** We can construct theoretical models of the solar interior using known laws of physics and then check the models against observations of the Sun's size, surface temperature, and energy output. We also use studies of solar vibrations and solar **neutrinos**.

14.3 THE SUN-EARTH CONNECTION

■ **What causes solar activity?** Convection combined with

the rotation pattern of the Sun—faster at the equator than at the poles—causes **solar activity** because these gas motions stretch and twist the Sun's magnetic field. These contortions of the magnetic field are responsible for phenomena such as **sunspots**, **flares**, **prominences**, and **coronal mass ejections**, and for heating the gas in the chromosphere and corona.

■ **How does solar activity affect humans?** Bursts of

charged particles ejected from the Sun during periods of high solar activity can hamper radio communications, disrupt electrical power generation, and damage orbiting satellites.

■ **How does solar activity vary with time?** The **sunspot**

cycle, or the variation in the number of sunspots on the Sun's surface, has an average period of 11 years. The magnetic field flip-flops every 11 years or so, resulting in a 22-year magnetic cycle. Sunspots first appear at mid-latitudes at solar minimum and then become increasingly more common near the Sun's equator as the next minimum approaches. The number of sunspots can vary dramatically from one cycle to the next, and sometimes sunspots seem to be absent altogether.

Use the following questions to check your understanding of some of the many types of visual information used in astronomy. Answers are provided in Appendix J. For additional practice, try the Chapter 14 Visual Quiz at MasteringAstronomy®.

Figure 14.22b, repeated above, shows the latitudes at which sunspots appeared on the surface of the Sun during the last century. Answer the following questions, using the information provided in the figure.

1. Which of the following years had the least sunspot activity?
   a. 1930
   b. 1949
   c. 1961
   d. 1987

2. What is the approximate range in latitude over which sunspots appear?

3. According the figure, how do the positions of sunspots appear to change during one sunspot cycle? Do they get closer to or farther from the equator with time?

EXERCISES AND PROBLEMS

MasteringAstronomy®

For instructor-assigned homework go to MasteringAstronomy®.

REVIEW QUESTIONS

Short-Answer Questions Based on the Reading

1. Briefly describe how *gravitational contraction* generates energy. When was it important in the Sun's history? Explain.

2. What two forces are balanced in *gravitational equilibrium*? What does it mean for the Sun to be in *energy balance*?

3. State the Sun's luminosity, mass, radius, and average surface temperature, and put the numbers into a perspective that makes them meaningful.

4. Briefly describe the distinguishing features of each of the layers of the Sun shown in Figure 14.4.

5. What is the difference between nuclear *fission* and nuclear *fusion*? Which one is used in nuclear power plants? Which one does the Sun use?

6. Why does nuclear fusion require high temperatures and pressures?

7. What is the overall nuclear fusion reaction in the Sun? Briefly describe the *proton-proton chain*.

8. Does the Sun's fusion rate remain steady or vary wildly? Describe the feedback process that regulates the fusion rate.

9. Why has the Sun gradually brightened with time?

10. Why does the energy produced by fusion in the solar core take so long to reach the solar surface? Describe the processes by which energy generated by fusion makes its way to the Sun's surface.

11. Explain how mathematical models allow us to predict conditions inside the Sun. How can we be confident that the models are on the right track?

12. What are *neutrinos*? What was the *solar neutrino problem*, and how was it solved?

13. What do we mean by *solar activity*? Describe some of the features of solar activity, including *sunspots*, *solar prominences*, *solar flares*, and *coronal mass ejections*.

14. Describe the appearance and temperature of the Sun's photosphere. Why does the surface look mottled? How are sunspots different from the surrounding photosphere?

15. How do magnetic fields keep sunspots cooler than the surrounding plasma? Explain.

16. Why are the chromosphere and corona best viewed with ultraviolet and X-ray telescopes, respectively? Briefly explain how we think the chromosphere and corona are heated.

17. What is the *sunspot cycle*? Why is it sometimes described as an 11-year cycle and sometimes as a 22-year cycle? Are there longerterm changes in solar activity?

18. Describe the leading model for explaining the sunspot cycle. Does the sunspot cycle influence Earth's climate? Explain.

TEST YOUR UNDERSTANDING

Does It Make Sense?

Decide whether the statement makes sense (or is clearly true) or does not make sense (or is clearly false). Explain clearly; not all these have definitive answers, so your explanation is more important than your chosen answer.

19. Before Einstein, gravitational contraction appeared to be a perfectly plausible mechanism for solar energy generation.

20. A sudden temperature rise in the Sun's core is nothing to worry about, because conditions in the core will soon return to normal.

21. If fusion in the solar core ceased today, worldwide panic would break out tomorrow as the Sun began to grow dimmer.

22. Astronomers have recently photographed magnetic fields churning deep beneath the solar photosphere.

23. I wear a lead vest to protect myself from solar neutrinos.

24. If there are few sunspots this year, we should expect many more in about 5 years.

25. News of a major solar flare today caused concern among professionals in the fields of communication and electrical power generation.

26. By observing solar neutrinos, we can learn about nuclear fusion deep in the Sun's core.

27. If the Sun's magnetic field somehow disappeared, there would be no more sunspots on the Sun.

28. Scientists are currently building an infrared telescope designed to observe fusion reactions in the Sun's core.

Quick Quiz

Choose the best answer to each of the following. Explain your reasoning with one or more complete sentences.

29. Which of these groups of particles has the greatest mass? (a) a helium nucleus with two protons and two neutrons (b) four electrons (c) four individual protons

30. Which of these layers of the Sun is coolest? (a) photosphere (b) chromosphere (c) corona

31. Which of these layers of the Sun is coolest? (a) core (b) radiation zone (c) photosphere

32. Scientists estimate the central temperature of the Sun using (a) probes that measure changes in Earth's atmosphere. (b) mathematical models of the Sun. (c) laboratories that create miniature versions of the Sun.

33. Why do sunspots appear darker than their surroundings? (a) They are cooler than their surroundings. (b) They block some of the sunlight from the photosphere. (c) They do not emit any light.

34. At the center of the Sun, fusion converts hydrogen into (a) plasma. (b) radiation and elements like carbon and nitrogen. (c) helium, energy, and neutrinos.

35. Solar energy leaves the core of the Sun in the form of (a) photons. (b) rising hot gas. (c) sound waves.

36. How does the number of neutrinos passing through your body at night compare with the number passing through your body during the day? (a) about the same (b) much smaller (c) much larger

37. What is the most common kind of element in the solar wind? (a) hydrogen (b) carbon (c) helium

38. Which of these things poses the greatest hazard to communication satellites? (a) photons from the Sun (b) solar magnetic fields (c) protons from the Sun

PROCESS OF SCIENCE

Examining How Science Works

39. *Inside the Sun.* Scientists claim to know what is going on inside the Sun, even though they cannot observe the solar interior directly. What is the basis for these claims, and how are they aligned with the hallmarks of science outlined in Section 3.4?

40. *The Solar Neutrino Problem.* Early solar neutrino experiments detected only about a third of the number of neutrinos predicted by the theory of fusion in the Sun. Why didn't scientists simply abandon their models at this point? What features of the Sun did the model get right? What alternatives were there for explaining the mismatch between the predictions and the observations?

GROUP WORK EXERCISE

41. *The Sun's Future.* In this chapter, you learned that gravitational contraction caused the interior of the solar nebula to heat up until hydrogen fusion began in the Sun. When the fusion rate rose to match the energy radiated from the Sun's surface, the Sun came into a long-lasting state of balance. In this exercise, you will discuss what will happen inside the Sun *after* it converts all the hydrogen in the core to helium. Before you begin, assign the following roles to the people in your group: *Scribe* (takes notes on the group's activities), *Proposer* (proposes explanations to the group), *Skeptic* (points out weaknesses in proposed explanations), and *Moderator* (leads group discussion and makes sure everyone contributes). Then discuss the following questions. **a.** What will happen to the core temperature of the Sun after the core runs out of hydrogen for fusion? Will the temperature go up or down? **b.** If you think the temperature will go up, will it rise forever? What could eventually stop the temperature from rising? If you think the temperature will go down, will it decrease forever? What could eventually stop it from falling? **c.** Propose and describe an Earth-based experiment or a set of stellar observations that could test your hypothesis from part b.

INVESTIGATE FURTHER

In-Depth Questions to Increase Your Understanding

Short-Answer/Essay Questions

42. *The End of Fusion I.* Describe what would happen in the Sun if fusion reactions abruptly shut off.

43. *The End of Fusion II.* If fusion reactions were to suddenly shut off in the Sun, how would we be able to tell?

44. *A Really Strong Force.* How would the interior temperature of the Sun be different if the strong force that binds nuclei together were 10 times as strong?

45. *Covered with Sunspots.* Describe what the Sun would look like from Earth if the entire photosphere were the same temperature as a sunspot.

46. *Inside the Sun.* Describe how scientists determine what the interior of the Sun is like. Why haven't we sent a probe into the Sun to measure what is happening there?

47. *Solar Energy Output.* Observations over the past century show that the Sun's visible-light output varies by less than 1%, but the Sun's maximum X-ray output can be as much as 10 times as great as its minimum X-ray output. Explain why changes in X-ray output can be so much more pronounced than those in the output of visible light.

48. *An Angry Sun.* A *Time* magazine cover once suggested that an "angry Sun" was becoming more active as human activity changed Earth's climate through global warming. It's certainly possible for the Sun to become more active at the same time that humans are affecting Earth, but is it possible that the Sun could be responding to human activity? Can humans affect the Sun in any significant way? Explain.

Quantitative Problems

Be sure to show all calculations clearly and state your final answers in complete sentences.

49. *Chemical Burning and the Sun.* Estimate how long the Sun would last if it were merely a huge fire that was releasing chemical energy. Assume that the Sun begins with roughly 10^8 joules per kilogram, a chemical energy content typical of atomic matter.

50. *The Lifetime of the Sun.* The total mass of the Sun is about 2×10^{30} kilograms, of which about 70% was hydrogen when the Sun formed. However, only about 13% of this hydrogen ever becomes available for fusion in the core. The rest remains in layers of the Sun where the temperature is too low for fusion. **a.** Use the given data to calculate the total mass of hydrogen available for fusion over the lifetime of the Sun. **b.** The Sun fuses about 600 billion kilograms of hydrogen each second. Based on your result from part a, calculate how long the Sun's initial supply of hydrogen can last. Give your answer in both seconds and years. **c.** Given that our solar system is now about 4.6 billion years old, when will we need to worry about the Sun running out of hydrogen for fusion?

51. *Solar Power Collectors.* This problem leads you through the calculation and discussion of how much solar power can be collected by solar cells on Earth. **a.** Imagine a giant sphere with a radius of 1 AU surrounding the Sun. What is the surface area of this sphere, in square meters? (*Hint:* The formula for the surface area of a sphere is $4\pi r^2$.) **b.** Because this imaginary giant sphere surrounds the Sun, the Sun's entire luminosity of 3.8×10^{26} watts must pass through it. Calculate the power passing through each square meter of this imaginary sphere in *watts per square meter.* Explain why this number represents the maximum power per square meter that a solar collector in Earth orbit can collect. **c.** List several reasons why the average power per square meter collected by a solar collector on the ground will always be less than what you found in part b. **d.** Suppose you want to put a solar collector on your roof. If you want to optimize the amount of power you can collect, how should you orient the collector? (*Hint:* The optimum orientation depends on both your latitude and the time of year and day.)

52. *Solar Power for the United States.* Total annual U.S. energy consumption is about 2×10^{20} joules. **a.** What is the average power requirement for the United States, in watts? (*Hint:* 1 watt = 1 joule/s.) **b.** With current technologies and solar collectors on the ground, the best we can hope for is that solar cells will generate an average (day and night) power of about 200 watts/m^2. (You might compare this to the maximum power per square meter you found in Problem 51.) What total area would we need to cover with solar cells to supply all the power needed for the United States? Give your answer in both square meters and square kilometers.

53. *The Color of the Sun.* The Sun's average surface temperature is about 5800 K. Use Wien's law (see Mathematical Insight 5.2) to calculate the wavelength of peak thermal emission from the Sun. What color does this wavelength correspond to in the visible-light spectrum? Why do you think the Sun appears white or yellow to our eyes?

54. *The Color of a Sunspot.* The typical temperature of a sunspot is about 4000 K. Use Wien's law (see Mathematical Insight 5.2) to calculate the wavelength of peak thermal emission from a sunspot. What color does this color correspond to in the visible-light spectrum? How does this color compare with that of the Sun?

55. *Solar Mass Loss.* Estimate how much mass the Sun loses through fusion reactions during its 10-billion-year life. You can simplify the problem by assuming the Sun's energy output remains constant. Compare the amount of mass lost with Earth's mass.

56. *Pressure of the Photosphere.* The gas pressure of the photosphere changes substantially from its upper levels to its lower levels. Near the top of the photosphere, the temperature is about 4500 K and there are about 1.6×10^{16} gas particles per cubic centimeter. In the middle, the temperature is about 5800 K and there are about 1.0×10^{17} gas particles per cubic centimeter. At the bottom of the photosphere, the temperature is about 7000 K and there are about 1.5×10^{17} gas particles per cubic centimeter. Compare the pressures of each of these layers and explain the reason for the trend in pressure that you find. How do these gas pressures compare with Earth's atmospheric pressure at sea level?

57. *Tire Pressure.* Air pressure at sea level is about 15 pounds per square inch. The recommended air pressure in your car tires is about 30 pounds per square inch. How does the density of gas particles inside your tires compare with the density of gas particles in the air outside your tires? What happens to gas pressure in the tire if it springs a leak and loses gas particles? How does your tire respond to this loss of gas particles? How is the tire's response like the response of the Sun's core to the slowly declining number of independent particles within it? How is the tire's response different?

58. *Your Energy Content.* The power needed to operate your body is about 100 watts. Suppose your body could run on fusion power and could convert 0.7% of its mass into energy. How much energy would be available through fusion? For how long could your body then operate on fusion power?

Discussion Questions

59. *The Role of the Sun.* Briefly discuss how the Sun affects us here on Earth. Be sure to consider not only factors such as its light and warmth but also how the study of the Sun has led us to new understandings in science and to technological developments. Overall, how important has solar research been to our lives?

60. *The Sun and Global Warming.* One of the most pressing environmental issues on Earth is the extent to which human emissions of greenhouse gases are warming our planet. Some people claim that part or all of the observed warming over the past century may be due to changes in the Sun, rather than to anything humans have done. Discuss how a better understanding of the Sun might help us comprehend the threat posed by greenhouse gas emissions. Why is it so difficult to develop a clear understanding of how the Sun affects Earth's climate?

Web Projects

61. *Current Solar Weather.* Daily information about solar activity is available at NASA's website spaceweather.com. Where are we in the sunspot cycle right now? When is the next solar maximum or minimum expected? Have there been any major solar storms in the past few months? If so, did they have any significant effects on Earth? Summarize your findings in a one- to two-page report.

62. *Solar Observatories in Space.* Visit NASA's website for the Sun-Earth connection and explore some of the current and planned space missions designed to observe the Sun. Choose one mission to study in greater depth, and write a one- to two-page report on the status and goals of the mission and what it has taught or will teach us about the Sun.

63. *Sudbury Neutrino Observatory.* Visit the website for the Sudbury Neutrino Observatory (SNO) and learn how it has helped to solve the solar neutrino problem. Write a one- to two-page report describing the observatory, any recent results, and what we can expect from it in the future.

64. *Nuclear Power.* There are two basic ways to generate energy from atomic nuclei: through nuclear fission (splitting nuclei) and through nuclear fusion (combining nuclei). All current nuclear reactors are based on fission, but fusion would have many advantages if we could develop the technology. Research some of the advantages of fusion and some of the obstacles to developing fusion power. Do you think fusion power will be a reality in your lifetime? Explain.

24 LIFE IN THE UNIVERSE

LEARNING GOALS

We, this people, on a small and lonely planet
Travelling through casual space
Past aloof stars, across the way of indifferent suns
To a destination where all signs tell us
It is possible and imperative that we learn
A brave and startling truth.

—Maya Angelou, from A Brave and Startling Truth

We have covered a lot of ground in this book, discussing fundamental questions about the nature and origin of our planet, our star, our galaxy, and our universe. But we have not yet discussed the most profound question of all: Are we alone? The universe contains worlds beyond imagination—more than 100 billion star systems in our galaxy alone, and some 100 billion galaxies in the observable universe—yet we do not know whether any world other than our own has ever been home to life.

In this chapter, we will discuss the possibility of life beyond Earth. We'll begin by considering the history of life on Earth, which will help us understand the prospects of finding life elsewhere. We'll then consider the possibility of finding microbial life elsewhere in our solar system or beyond, and examine efforts to search for extraterrestrial intelligence (SETI). Finally, we'll discuss the astonishing implications that the search for life may hold for our future.

24.1 LIFE ON EARTH

It may seem that aliens are everywhere. Aliens abound in television shows and movies, and it's not hard to find websites claiming alien atrocities or a government conspiracy to hide alien corpses in "Area 51."

Scientists are interested in aliens too, although most scientists remain deeply skeptical about reports of aliens on Earth (see Special Topic, page 712). Scientists are therefore searching for life elsewhere, looking for evidence of life on other worlds in our solar system, trying to learn whether we should expect to find life on planets orbiting other stars, and searching for signals broadcast by other civilizations.

Scientific interest in extraterrestrial life is not new. Belief in life on other worlds was common even among ancient Greek philosophers, and many famous scientists of the past few centuries took it as a given that intelligent beings exist on other planets. For example, Kepler suggested that the Moon was inhabited, and William Herschel (co-discoverer, with his sister Caroline, of Uranus) spoke of life on virtually all the planets in our solar system. Most famously, in the late 19th century, Percival Lowell claimed to see networks of canals on Mars, which he argued were the mark of an advanced civilization [**Section 9.4**] and which formed the basis for H. G. Wells's novel *The War of the Worlds*.

However, until fairly recently, the question of life beyond Earth could be addressed only through guesswork, because there was little actual evidence to go on. The situation has changed dramatically in recent decades, making the search for life beyond Earth a topic of serious scientific study. One major factor has been the discovery of planets around other stars [**Section 13.1**], which confirms that there are plenty of places to look for life beyond our own solar system. But new discoveries about life on Earth have played an even more important role. In particular, three recent developments in the study of life on Earth have made it seem much more likely that life might exist elsewhere:

1. We have learned that life arose quite early in Earth's history, suggesting that life might also form quickly on other worlds with the right conditions.

2. Laboratory experiments have shown that the chemical constituents thought to have been common on the young Earth combine readily into complex organic molecules. These experiments suggest that life might have arisen through naturally occurring chemistry—in which case the same chemistry could have given rise to life on many other worlds.

3. We have discovered microscopic living organisms that probably could survive in conditions similar to those on at least some other worlds in our solar system, suggesting that the necessities of life may be common in the universe.

Because these three ideas are so important to understanding the modern science of life in the universe, often called *astrobiology*, let's examine each of them in greater detail. Along the way, we'll also discuss the nature and history of life on Earth. We will then be prepared to consider how we might search for life on other worlds.

When did life arise on Earth?

The first of our three major ideas above is that life arose quite early on Earth. But how do we know? We learn about the history of life on Earth through the study of **fossils**, relics of organisms that lived and died long ago. Most fossils form when dead organisms fall to the bottom of a sea (or other body of water) and are gradually buried by layers of sediments. The sediments are produced by erosion on land and carried by rivers to the sea. Over millions of years, sediments pile up on the seafloor, and the weight of the upper layers compresses underlying layers into rock. Erosion or tectonic activity can later expose the fossils (**FIGURE 24.1**). In some places, such as the Grand Canyon, the sedimentary layers record hundreds of millions of years of Earth's history (**FIGURE 24.2**). Some fossils are remarkably well preserved (**FIGURE 24.3**), though the vast majority of dead organisms decay completely and leave no fossils behind.

The Geological Time Scale The key to reconstructing the history of life is to determine the dates at which fossil organisms lived. The *relative* ages of fossils found in different layers are easy to determine: Deeper layers formed earlier and contain more ancient fossils. Radiometric dating [**Section 8.3**] confirms these relative ages and gives us fairly precise absolute ages for fossils. Based on the layering of rocks and fossils, geologists divide Earth's history into a set of distinct intervals that make up what we call the **geological time scale**. **FIGURE 24.4** shows the names of the various intervals on a timeline, along with numerous important events in Earth's history.

Rivers carry sediment to the oceans.

Sedimentary rocks containing fossils form on the ocean floor.

Over time, more layers are added, containing fossils from each time period.

Tectonic stresses and sea level changes push the seafloor upward, exposing sedimentary rocks.

Erosion by rivers reveals layers; deeper layers contain older fossils.

FIGURE 24.1 Formation of sedimentary rock. Each layer represents a particular time and place in Earth's history and is characterized by fossils of organisms that lived in that time and place.

Fossil Evidence for the Early Origin of Life You might wonder why the geological time scale shows so much more detail for the last few hundred million years than it does for earlier times. The answer is that fossils become increasingly difficult to find as we look deeper into Earth's history, for three major reasons. First, older rocks are much rarer than younger rocks, because most of Earth's surface is geologically young. Second, even when we find very old rocks, they often have been subject to transformations (caused by heat and pressure) that would have destroyed any fossil evidence they may have contained. Third, all life prior to a few hundred million years ago was microscopic, and microscopic fossils are much more difficult to identify.

Despite these difficulties, geologists have found a few very old rocks that suggest life was already thriving on Earth 3.5 billion years ago, and possibly for several hundred million years before that. One strong line of evidence comes from rocks called *stromatolites* (**FIGURE 24.5**), which are strikingly similar in size, shape, and interior structure to sections of mats known as "living stromatolites" that are formed today by colonies of microbes. The clear implication is that the stromatolites are fossils of ancient microbes, although there is some controversy about this conclusion because geological processes of sedimentation can mimic the layering found in stromatolites. Moreover, if the microbes that made the stromatolites are like the microbes in the living mats today, then the implication is that at least some of these ancient microbes produced energy by photosynthesis. Because photosynthesis is a fairly sophisticated metabolic process, we presume that it must have taken at least a moderately long time for this process to evolve in living organisms. In other words, if we are correct in concluding that stromatolites tell us that photosynthetic life already existed some 3.5 billion years ago, then we can infer that more primitive life must have existed even earlier, and that the origin of life itself substantially predates this.

THINK ABOUT IT

How does the length of time during which animals and plants have lived on land compare to the length of time during which life has existed? How does the length of time during which humans have existed compare to the length of time since mammals and dinosaurs first arose?

Other evidence for early life comes from recognizable fossils of microbes, known as *microfossils*. Although it is often difficult to distinguish a true microfossil from a microscopic structure formed by mineral processes, scientists are confident that they have found microfossils dating to at least 3.0 billion years ago, and some rocks that contain structures that appear to be microfossils date to 3.5 billion years ago.

More intriguingly, careful study of some of the most ancient rocks on Earth suggests that life was already present

FIGURE 24.2 The rock layers of the Grand Canyon record more than 500 million years of Earth's history.

FIGURE 24.3 Dinosaur fossils preserved in sandstone in Dinosaur National Monument, which straddles Utah and Colorado.

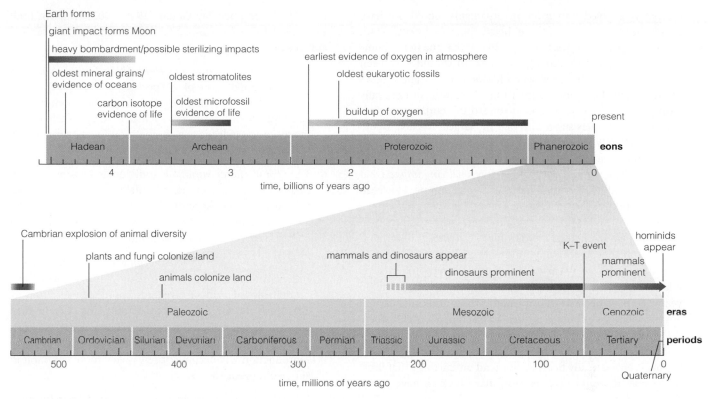

FIGURE 24.4 The geological time scale. Notice that the lower timeline is an expanded view of the last portion of the upper timeline. The eons, eras, and periods are defined by changes observed in the fossil record. The absolute ages come from radiometric dating. (The K-T event is the geological term for the impact linked to the mass extinction of the dinosaurs [**Section 12.4**].)

at least 3.85 billion years ago. These rocks have undergone too much change for them to hold intact fossils, but careful analysis of carbon isotopes within them suggests that they once held living organisms. Carbon has two stable isotopes: carbon-12, with six protons and six neutrons in its nucleus, and carbon-13, which has one extra neutron (see Figure 5.9). Living organisms incorporate carbon-12 slightly more easily than carbon-13. As a result, the fraction of carbon-13 is always a bit lower in fossils than in rock samples that lack fossils. All life and all fossils tested to date show the same characteristic ratio of the

two carbon isotopes, and this ratio has also been found in rocks that are more than 3.85 billion years old, suggesting that they contained life when they formed. (The carbon isotope evidence is found in sedimentary rocks that cannot be dated precisely; however, they must be at least 3.85 billion years old, because that is the age obtained from radiometric dating of volcanic [igneous] rock that cuts through them.)

Could Life Have Arisen Even Earlier? Until recently, many scientists thought that life (or, at least, the life that

a These knee-high mats at Shark Bay, Western Australia, are colonies of microbes known as "living stromatolites."

b The banded structure in this section from one of the Shark Bay stromatolites is formed by layers of sediment attaching to the microbial mats.

c This section of a 3.5-billion-year-old stromatolite (found in the Strelley Pool Formation in Western Australia) shows the same type of structure found in living stromatolites. The black layers are organic deposits that are the remains of ancient microbial mats. (The ruler is marked in centimeters.)

FIGURE 24.5 Rocks called *stromatolites* offer evidence of microbial life existing as early as 3.5 billion years ago.

ultimately evolved into modern organisms) could not have arisen before the end of the heavy bombardment, about 3.9 billion years ago [**Section 8.2**]. Recall that the heavy bombardment was a time during which Earth should have been repeatedly struck by large asteroids or comets, and studies of craters on the Moon suggest that there was an especially strong *late heavy bombardment* toward the end of that time period. Past models suggested that the largest impacts would have completely vaporized the oceans and raised the global surface temperature high enough to melt the upper crust. In that case, these events would have been *sterilizing impacts* that would have killed off any life on Earth at the time. However, more recent models suggest that these effects were overstated, and that while large impacts would have sterilized substantial portions of our planet, microscopic life living underground and in some deep ocean environments could have survived.

If life did arise very early, evidence will be hard to find. The relatively few known rocks that date to earlier than 3.5 billion years ago have been so transformed by heat and pressure that any evidence they contain is likely to be controversial. Indeed, if the carbon isotope evidence really does indicate life some 3.85 billion years ago, the rarity of such old rocks suggests that life must have already been widespread on Earth by that time; otherwise, it would take extraordinary luck to have found the evidence.

To summarize, fossil evidence points to life on Earth thriving 3.5 billion years ago, and carbon isotope evidence may point to widespread life more than 3.85 billion years ago. Life could have arisen even earlier, though evidence would be hard to find. Still, no matter how we look at it, it seems that life on Earth arose within no more than a few hundred million years after Earth's birth about $4\frac{1}{2}$ billion years ago. Geologically speaking, that is a remarkably short time.

How did life arise on Earth?

We have discussed *when* life arose on Earth, but *how* did it arise? To answer this question, we must first consider how life today differs from the earliest life on Earth. Then we can discuss how the first organisms may have originated.

The Theory of Evolution The fossil record clearly shows that life has gone through great changes over time. If we are going to understand how life arose, we must understand what causes these changes, so that we can trace life back to its origin. The unifying theory through which scientists understand the history of life on Earth is the **theory of evolution**, first published by Charles Darwin in 1859.

Evolution simply means "change with time." Even before Darwin, numerous scientists (some going back to ancient times) had recognized evidence for evolution in the fossil record, but no one before Darwin successfully explained how species might undergo change. In essence, the fossil record provides strong evidence that evolution *has* occurred, while Darwin's theory of evolution explains *how* it occurs.

Darwin provided extensive evidence for the occurrence of evolution and, based on this evidence, put forth a simple model to explain how evolution occurs. The underlying logic of Darwin's model is easy to understand. As described

by biologist Stephen Jay Gould (1941–2002), Darwin built his model from "two undeniable facts and an inescapable conclusion":

- *Fact 1: overproduction and competition for survival.* Any localized population of a species has the potential to produce far more offspring than the local environment can support with resources such as food and shelter. This overproduction leads to a competition for survival among the individuals of the population.

- *Fact 2: individual variation.* Individuals in a population of any species vary in many heritable traits (traits passed from parents to offspring). No two individuals are exactly alike, and some individuals possess traits that make them better able to compete for food and other vital resources.

- *The inescapable conclusion: unequal reproductive success.* In the struggle for survival, those individuals whose traits best enable them to survive and reproduce will, on average, leave the largest number of offspring that in turn survive to reproduce. Therefore, in any local environment, heritable traits that enhance survival and successful reproduction will become progressively more common in succeeding generations.

It is this unequal reproductive success that Darwin called **natural selection:** Over time, advantageous genetic traits will naturally win out (be "selected") over less advantageous traits because they are more likely to be passed down through many generations. This process explains how species can change in response to their environment—by favoring traits that improve adaptation—and it is the primary mechanism of evolution. By studying the variety of species in places like the Galápagos Islands, Darwin backed his model with so much evidence that it quickly gained the status of a scientific *theory* [**Section 3.4**].

The Mechanism of Evolution In the more than 150 years since Darwin published his theory, ongoing research has only given it further support. Perhaps the strongest support for the theory has come from study of **DNA** (short for *deoxyribonucleic acid*), the genetic material of all life on Earth, which has allowed scientists to understand the mechanism of evolution on a molecular level.

Living organisms reproduce by copying DNA and passing these copies on to their descendants. A molecule of DNA consists of two long strands—somewhat like the interlocking strands of a zipper—wound together in the spiral shape known as a *double helix* (**FIGURE 24.6**). The instructions for assembling a living organism are written in the precise order of four chemical bases that make up the interlocking portions of the DNA "zipper": adenine (A), thymine (T), guanine (G), and cytosine (C). These bases pair up in a way that ensures that both strands of a DNA molecule contain the same genetic information. By unwinding and allowing new strands (made from chemicals floating around inside a cell) to form alongside the original ones, a single DNA molecule can give rise to two identical copies of itself. That is how genetic material is copied and passed on to future generations.

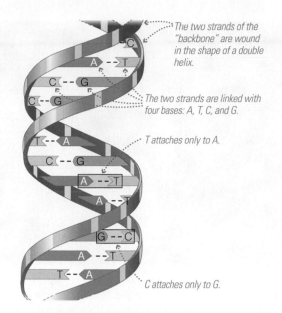

The two strands of the "backbone" are wound in the shape of a double helix.

The two strands are linked with four bases: A, T, C, and G.

T attaches only to A.

C attaches only to G.

FIGURE 24.6 This diagram represents a small piece of a DNA molecule, which looks like a zipper twisted into a spiral. Hereditary information is contained in the "teeth" linking the strands. These "teeth" are the DNA bases. Only four DNA bases are used, and they can link up between the two strands only in specific ways: T attaches only to A, and C attaches only to G. (The color coding is arbitrary and is used only to represent different types of chemical groups; in the backbone, blue and yellow represent sugar and phosphate groups, respectively.)

Evolution occurs because the transfer of genetic information from one generation to the next is not always perfect. An organism's DNA may occasionally be altered by copying errors or by external influences, such as ultraviolet light from the Sun or exposure to toxic or radioactive chemicals. Any change in an organism's DNA is called a **mutation**. Many mutations are lethal, killing the cell in which the mutation occurs. Some, however, may improve a cell's ability to survive and reproduce. The cell then passes on this improvement to its offspring.

Our understanding of the molecular mechanism of natural selection has put the theory of evolution on a stronger foundation than ever. While no theory can ever be proved true beyond all doubt, the theory of evolution is as solid as any theory in science, including the theory of gravity and the theory of atoms. Biologists routinely witness evolution occurring before their eyes among laboratory microorganisms, or over periods of just a few decades among plants and animals subjected to environmental stress. Moreover, the theory of evolution has become the underpinning of virtually all modern biology, medicine, and agriculture. For example, agricultural scientists apply the idea of natural selection to develop pest control strategies that reduce the populations of harmful insects without harming the populations of beneficial ones; medical researchers test new drugs on animals that are genetically similar to humans, because the theory of evolution tells us that genetically similar species should have somewhat similar physiological responses; and biologists study the genetic relationships between organisms by comparing their DNA.

The First Living Organisms The basic chemical nature of DNA is virtually identical among all living organisms. This fact, along with other biochemical similarities shared by all living organisms, tells us that all life on Earth today can trace its origins to a common ancestor that lived long ago.

We are unlikely to find fossils of the earliest organisms, but we can learn about early life through careful studies of the DNA of living organisms. Biologists can determine the evolutionary relationships among living species by comparing the sequences of bases in their DNA. For example, two organisms whose DNA sequences differ in five places for a particular gene are probably more distantly related than two organisms whose gene sequences differ in only one place. Many such DNA comparisons suggest that all living organisms are related in a way depicted schematically by the "tree of life" in **FIGURE 24.7**. Although details in the structure of this tree remain uncertain (particularly because we now know that different species can swap genes), it indicates that life on Earth is divided into three major groupings, or *domains*, called *bacteria*, *archaea*, and *eukarya*, and that all three domains share a common ancestor. Notice that plants and animals represent only two tiny branches of the domain *eukarya*.

Organisms on branches located closer to the root of the tree of life must contain DNA that is evolutionarily older, suggesting that they more closely resemble the organisms that lived early in Earth's history. Some of the modern-day organisms that appear to be evolutionarily oldest are microbes that live in very hot water around seafloor volcanic vents in the deep oceans (**FIGURE 24.8**). Unlike most life at Earth's surface, which depends on sunlight, these organisms get energy from chemical reactions in the hot, mineral-rich water around the volcanic vents.

The idea that early organisms might have lived in such "extreme" conditions may at first seem surprising, but it makes sense when we think about it. The deep ocean environment

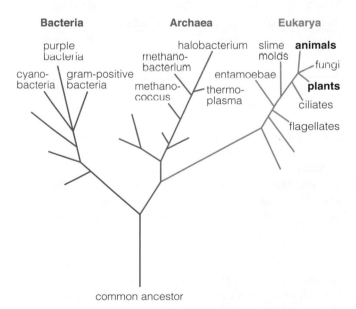

FIGURE 24.7 The tree of life, showing evolutionary relationships determined by comparison of DNA sequences in different organisms. Just two small branches represent *all* plant and animal species. (Only a few of the many relationships are shown and labeled.)

FIGURE 24.8 This photograph shows a volcanic vent on the ocean floor that spews out hot, mineral-rich water. DNA studies indicate that the microbes living near these vents are evolutionarily older than most other living organisms, hinting that early life may have arisen in similar environments.

would have been protected from the effects of impacts and from harmful ultraviolet radiation that bathed Earth's surface before our atmosphere had oxygen or an ozone layer. Moreover, the chemical pathways used to extract energy from mineral-rich hot water are simpler than other chemical pathways (such as photosynthesis) used by living organisms to obtain energy. For these reasons, many scientists suspect that life arose near deep-sea vents, or in similar environments that provided chemical energy.

The Transition from Chemistry to Biology The theory of evolution explains how the earliest organisms evolved into the great diversity of life on Earth today. But where did the first organisms come from? We may never know for sure, because we are unlikely to find fossils that would show the transition from nonlife to life. However, over the past several decades, scientists have conducted laboratory experiments designed to mimic the conditions that existed on the young Earth. These experiments suggest that life could have arisen through natural chemical reactions.

Such experiments were first performed in the 1950s (FIGURE 24.9), and they have been refined and improved since that time. In essence, the experiments mix chemicals thought to have been present on the early Earth and then "spark" the chemicals with electricity to simulate lightning or other energy sources. The chemical reactions that follow have produced nearly all the major molecules of life, including amino acids and DNA bases. Many of these molecules are also found in meteorites, suggesting that some organic molecules may have arrived from space.

Laboratory studies of the possible origins of life have advanced greatly in the past few years. Scientists have found that naturally occurring clays, which should have been abundant

FIGURE 24.9 Stanley Miller poses with a reproduction of the experimental setup he first used in the 1950s to study pathways to the origin of life. (He worked with Harold Urey, so the experiment is called the Miller-Urey experiment.)

in the early oceans, can catalyze the formation of short strands of RNA, a molecule much like a single strand of DNA. Some of these short RNA molecules can in turn catalyze other reactions. Because some RNA molecules (though not those so far produced in the laboratory) are capable of self-replication, many biologists now presume that RNA was the original genetic material of life on Earth, with DNA coming later.

Equally significantly, experiments have also shown that microscopic enclosed membranes can form on the surfaces of the same clay minerals that help assemble RNA molecules, sometimes with RNA inside them (FIGURE 24.10). Self-replicating RNA molecules enclosed in such membranes would have in essence made "pre-cells" in which the RNA might have undergone rapid changes. Pre-cells in which RNA replicated faster and more accurately were more likely to spread, leading to a type of positive feedback that would have encouraged even faster and more accurate replication. Given millions of years of chemical reactions occurring all over Earth, it seems plausible that RNA-based life would have arisen first and then eventually evolved into DNA-based life.

FIGURE 24.10 This microscopic photo (made with the aid of fluorescent dyes) shows short strands of RNA (red) contained within an enclosed membrane (green circle), both of which formed spontaneously with the aid of clay minerals beneath them.

FIGURE 24.11 summarizes the steps by which chemical processes might have led to life on Earth. We may never know whether life really arose in this way, but it seems possible.

Could Life Have Migrated to Earth? Our scenario suggests that life could have arisen naturally here on Earth. However, an alternative possibility is that life arose somewhere else first—perhaps on Venus or Mars—and then migrated to Earth on meteorites. Remember that we have collected meteorites that were blasted by impacts from the surfaces of the Moon and Mars [**Section 12.1**]. Calculations suggest that Venus, Earth, and Mars all should have exchanged many tons of rock, especially in the early days of the solar system when impacts were more common.

The idea that life could travel through space to Earth once seemed outlandish. After all, it's hard to imagine a more forbidding environment than that of space, with no air, no water, and constant bombardment by dangerous radiation from the Sun and stars. However, the presence of organic molecules in meteorites and comets tells us that the building blocks of life can survive in space, and tests have shown that some microbes can survive in space for years.

SPECIAL TOPIC

Evolution and the Schools

You're probably familiar with the public debate about teaching evolution in schools. Much of the controversy comes from misunderstandings on both sides. Scientists sometimes fail to show respect for people's individual religious beliefs, while those opposed to teaching evolution often misunderstand the overwhelming amount of scientific evidence that supports the theory.

The debate centers around the question of what counts as science and what does not, as you can see by considering a common question: Do you believe in creationism or evolution? This question may be philosophically and theologically interesting, but it is poorly phrased from a scientific standpoint: Science is about evidence, not beliefs. Scientific evidence clearly shows that the universe is much older than Earth, that Earth is some $4\frac{1}{2}$ billion years old, that life on Earth has undergone dramatic change since it first arose nearly 4 billion years ago, and that this change has occurred through the mechanism of evolution by natural selection. Nevertheless, all the evidence in the world can never prove a scientific theory to be true beyond all doubt. To take a somewhat extreme case, you could accept the overwhelming evidence for evolution and still believe that God created Earth and the universe a mere 6000 years ago by, for example, assuming that God put all the evidence for evolution in place. Because the evidence would look the same in this case as if evolution actually occurred, science cannot say anything about the validity of such a belief.

We can gain deeper insight by examining the theory of evolution using the hallmarks of science from Chapter 3. The first hallmark says that science seeks explanations for observed phenomena that rely on natural causes, and the theory of evolution does exactly that: It explains the fossil record and observed differences between species in terms of natural selection. Note that this hallmark does *not* preclude the possibility of divine intervention; it just says that if divine intervention has occurred, we cannot study it scientifically. The second and third hallmarks remind us that science progresses through the creation and testing of models designed to explain our observations, and that we must modify or discard models that fail our tests. The theory of evolution exhibits these hallmarks because it is continually subject to testing and verification, and it has indeed been modified as new evidence has come to light. For example, Darwin's original theory has been improved and strengthened by modern molecular understanding of DNA and by observations of evolution in microbes (such as bacteria gaining resistance to antibiotics) and in small, localized populations of plants and animals.

The fact that evolution is such a well-established and important scientific theory makes it a key part of the science curriculum for schools, and few people disagree. However, many people have suggested teaching alternative ideas such as *creationism* or *intelligent design* alongside evolution in school curricula. The scientific response is very simple: Scientists take no position on the general school curriculum, but the *science curriculum* should not include these alternative ideas because they are not science, which we can see by again considering the hallmarks of science.

Let's start with *creationism*, the idea that the world was divinely created. In fact, we can't even define this idea clearly enough to test it against the hallmarks of science. Many religions incorporate stories of creation, and these stories are often quite different from one another. For example, many Native American religious beliefs speak of creation in terms that bear little resemblance to the story in Genesis. Even among Christians who claim a literal belief in the Bible, there are differences of interpretation. Some biblical literalists argue that the creation must have occurred in just 6 days, as the first chapter of Genesis seems to say, but others suggest that the term "day" in Genesis does not necessarily mean 24 hours and that Genesis is therefore compatible with a much older Earth.

The related idea of *intelligent design* holds that the complexity of life is too great to have arisen naturally, and therefore there must have been a "designer." Intelligent design can sound much like science, especially because most of its proponents accept both the idea of an old Earth and the idea that some living things have changed through time as indicated by the fossil record. However, the assumption of an intelligent designer does not yield testable predictions, because it posits that the designer is beyond our scientific comprehension; in essence, the idea of intelligent design means that there is no point in continuing to look for natural causes, because a supernatural process made the universe and life. Whether or not this idea is true, it isn't science. As an analogy, consider the collapse of a bridge. If an engineer declares that the collapse was an act of God, he could well be right—but that belief won't help him learn how to design a better bridge. It is the scientific quest for a natural understanding of life—embodied in the theory of evolution—that has led to the discovery of genetics, DNA, and virtually all modern medicine.

So where does this discussion leave us with regard to the question of teaching evolution in schools? Let's start with what our discussion does *not* tell us. It does not tell us that everyone needs to accept the theory of evolution—whether you choose to do so is up to you. Nor does it tell us whether divine intervention has guided evolution—science does not say anything about the existence or role of God. What it *does* tell us is that the theory of evolution is a clear and crucial part of the study of science, and that without it we cannot fully understand modern science and its impact on our world.

1. Naturally forming organic molecules are the building blocks of life.

2. Clay minerals catalyze production of RNA and membranes that form pre-cells.

3. Molecular natural selection favors efficient, self-replicating RNA molecules.

4. True living cells with RNA genome give rise to "RNA World."

5. DNA evolves from RNA and biological evolution continues.

RNA

RNA genome

primitive cell

DNA genome

FIGURE 24.11 A summary of the steps by which chemistry on the early Earth may have led to the origin of life.

In a sense, Earth, Venus, and Mars have been "sneezing" on each other for billions of years. Life could conceivably have originated on any of these three planets and been transported to the others. It's an intriguing thought, but it does not change our basic scenario for the origin of life—it simply moves it from one planet to another.

A Brief History of Life on Earth We are now ready to take a quick look at the history of life on Earth, summarized in Figure 24.4. Earth formed about $4\frac{1}{2}$ billion years ago, and the giant impact thought to have formed the Moon [**Section 8.2**] probably happened soon after. Mineral evidence suggests that Earth had oceans by 4.3 to 4.4 billion years ago, and those early oceans would have been natural laboratories for chemical reactions that could have led to life. Once life started, evolution rapidly diversified it.

Despite the rapid pace of evolution, the most complex organisms remained single-celled for at least a billion years after life first arose. All of these organisms probably lived in the oceans, because the lack of a protective ozone layer made the land inhospitable. Things began to change only when oxygen started building up in Earth's atmosphere.

Nearly all the oxygen in our atmosphere was originally released through photosynthesis by single-celled organisms known as *cyanobacteria* (**FIGURE 24.12**). Fossil evidence indicates that cyanobacteria were producing oxygen via

photosynthesis by at least 2.7 billion years ago, and possibly for hundreds of millions of years before that. However, oxygen did not immediately begin to accumulate in the atmosphere. For hundreds of millions of years, chemical reactions with surface rocks pulled oxygen back out of the atmosphere nearly as fast as the cyanobacteria could produce it. But these tiny organisms were abundant and persistent, and eventually the surface rock was so saturated with oxygen that the rate of oxygen removal slowed down. Oxygen then began to accumulate in the atmosphere, though it may not have reached a level that we could have breathed until just a few hundred million years ago.

Today, we often think of oxygen as a necessity for life. However, oxygen was probably poisonous to most organisms living before about 2 billion years ago (and remains a poison to many microbes still living today). The rise of atmospheric oxygen therefore caused tremendous evolutionary pressure and may have been a major factor in the evolution of complex plants and animals.

There were undoubtedly many crucial changes as primitive microbes gradually evolved into multicellular organisms and early plants and animals, but the fossil record does not allow us to pinpoint the times at which all these changes occurred. However, we see a dramatic change in the fossil record during the *Cambrian period,* beginning about 542 million years ago. During this period, animal life evolved from tiny and primitive organisms into all the basic body types (phyla) that we find on Earth today. This remarkable diversification occurred in such a short time relative to Earth's history that it is often called the *Cambrian explosion.*

Early dinosaurs and mammals arose some 225 to 250 million years ago, but dinosaurs at first proved more successful and dominated for well over 100 million years. Their sudden demise 65 million years ago paved the way for the evolution of large mammals—including humans. The earliest humans appeared on the scene only a few million years ago, or after 99.9% of Earth's history to date had already gone by. Our few centuries of industry and technology have come after 99.99999% of Earth's history.

What are the necessities of life?

If we are going to look for other worlds that might harbor life, the first step is to understand the necessities of life. While it is possible that life on other worlds could be quite different

FIGURE 24.12 This photo shows microscopic chains of modern cyanobacteria. The ancestors of these living organisms produced essentially all the oxygen in Earth's atmosphere.

from life on Earth, it's easiest to begin the search for life by looking for conditions in which organisms from Earth could live. So what exactly does Earth life need to survive?

If we think about ourselves, the requirements for life seem fairly stringent: We need abundant oxygen in an atmosphere that is otherwise not poisonous, we need temperatures in a fairly narrow range of conditions, and we need abundant and varied food sources. However, the discovery of life in "extreme" environments—such as in the hot water near undersea volcanic vents—shows that many microbes (often called *extremophiles*) can survive in a much wider range of conditions.

Organisms living in hot water prove that at least some microbes can survive in much higher temperatures than we would have guessed. Other organisms live in other extremes. In the freezing cold but very dry valleys of Antarctica, scientists have found microbes that live *inside* rocks, surviving on tiny droplets of liquid water and energy from sunlight. Microscopic life has also been found deep underground in water that fills pores within subterranean rock. We have found life thriving in environments so acidic, alkaline, or salty that humans would be poisoned almost instantly. We have even found microbes that can survive high doses of radiation, making it possible for them to survive for many years in the radiation-filled environment of space.

If we compare all the different forms of life on Earth, we find that life as a whole has only three basic requirements:

1. A source of nutrients (atoms and molecules) from which to build living cells

2. Energy to fuel the activities of life, whether from sunlight, from chemical reactions, or from the heat of Earth itself

3. Liquid water

These requirements give us a basic road map for the search for life elsewhere. If we want to find life on other worlds, it makes sense to start by searching for worlds that offer these basic necessities.

Interestingly, only the third requirement (liquid water) seems to pose much of a constraint. Organic molecules are present almost everywhere—even on meteorites and comets. Many worlds are large enough to retain internal heat that could provide energy for life, and virtually all worlds have sunlight (or starlight) bathing their surfaces, although the inverse square law for light [**Section 15.1**] means that light provides less energy to worlds farther from their star. Nutrients and energy should therefore be available to some degree on almost every planet and moon. In contrast, liquid water is relatively rare. Therefore, the search for liquid water drives the current search for life in our universe.

SPECIAL TOPIC

What Is Life?

You may have noticed that while we've been talking about *life*, we haven't actually defined the term. It's surprisingly difficult to draw a clear boundary between life and nonlife. Life can be so difficult to define that we may be tempted to fall back on the famous words of Supreme Court Justice Potter Stewart, who, in avoiding the difficulty of defining pornography, wrote: "I shall not today attempt further to define [it]. . . . But I know it when I see it." If living organisms on other worlds turn out to be much like those found on Earth, it may prove true that we'll know them when we see them. But if the organisms are fairly different from those on Earth, we'll need clearer guidelines to decide whether they are truly "living."

One approach seeks to identify distinguishing features common to all known life. For example, most or all living organisms on Earth appear to share the following six key properties:

1. *Order:* Living organisms are not random collections of molecules but rather have molecules arranged in orderly patterns that form cell structures.

2. *Reproduction:* Living organisms are capable of reproducing.

3. *Growth and development:* Living organisms grow and develop in patterns determined in part by heredity.

4. *Energy utilization:* Living organisms use energy to fuel their many activities.

5. *Response to the environment:* Living organisms actively respond to changes in their surroundings. For example, organisms may alter their chemistry or movements in the presence of a food source.

6. *Evolutionary adaptation:* Life evolves through natural selection, in which organisms pass on traits that make them better adapted to survival in their local environments.

These six properties are all important, but biologists today regard evolution as the most fundamental and unifying of them. Evolutionary adaptation is the only property that can explain the great diversity of life on Earth. Moreover, understanding how evolution works allows us to understand how all the other properties came to be. A simple definition of life might therefore be "something that can reproduce and evolve through natural selection."

This definition of life can probably suffice for most practical purposes, but some cases may still challenge it. For example, computer scientists can now write programs (that is, lines of computer code) that can reproduce themselves (that is, create additional sets of identical lines of code). By adding instructions that allow random changes to the programs, they can even make "artificial life" that evolves on a computer. Should this "artificial life," which consists of nothing but electronic signals processed by computer chips, be considered alive?

The fact that we have such difficulty distinguishing the living from the nonliving on Earth suggests that we should be very cautious about constraining our search for life elsewhere. No matter what definition of life we choose, there's always the possibility that we'll someday encounter something that challenges it. Nevertheless, the properties of reproduction and evolution seem likely to be shared by most if not all life in the universe and therefore provide a useful starting point as we consider how to explore the possibility of extraterrestrial life.

24.2 LIFE IN THE SOLAR SYSTEM

If we assume that life elsewhere would be at least a little bit like life on Earth, then the search for life begins with a search for **habitable worlds**—worlds that contain the basic necessities for life as we know it, including liquid water. The liquid water requirement rules out most of the worlds in our solar system. Mercury and the Moon are barren and dry, Venus is far too hot to have liquid water on its surface (though its clouds contain droplets of water laced with acid), and most of the small bodies of the outer solar system are too cold. The jovian planets may have droplets of liquid water in some of their clouds, but the strong vertical winds on these planets make them unlikely homes for life. That leaves two major possibilities besides Earth for habitability in our solar system: (1) Mars and (2) a few of the large moons orbiting jovian planets, most notably Europa [**Section 11.2**]. Let's discuss the current evidence about the potential habitability of these worlds.

THINK ABOUT IT

Is it possible for a world to be habitable but not actually have life? Is it possible for a world to have life but not be habitable? Explain.

Could there be life on Mars?

We now have sufficiently detailed images of Mars to be quite confident that no civilizations have ever existed there. The "canals" seen by Percival Lowell were some type of mirage, presumably formed by a combination of real Martian features and his own imagination. Nevertheless, we have good reason to believe that liquid water flowed on Mars in the past [**Section 9.4**], making it seem possible that life could once have found a home there. Moreover, since Mars contains abundant ice today, it's conceivable that life could still survive near sources of volcanic heat, where pockets of liquid water might persist underground.

Missions to Mars Mars is not only the best candidate in our solar system for life beyond Earth but also the only place where we've begun an actual search for life. Our first attempt came with the two *Viking* landers that reached Mars in 1976. Each was equipped with a robotic arm for scooping up soil samples, which were fed into several on-board robotically controlled experiments. Although some of the experiments produced results that seemed suggestive of life, taken as a set the results seem more likely to have been due to nonbiological chemical reactions. Most scientists have therefore concluded that the *Viking* landers did not detect life, though a few scientists take issue with this conclusion.

More recent missions have taught us much more about the possibility of life on Mars, with three rovers providing some of the most important information. The *Spirit* and *Opportunity* rovers landed on Mars in 2004, with *Spirit* operating for about six years and *Opportunity* still collecting data as this book went to press in 2012. Both missions provided an abundance of data about Martian surface conditions and found clear evidence that water was once plentiful in their locations on Mars [**Section 9.4**], giving a boost to the hypothesis that Mars

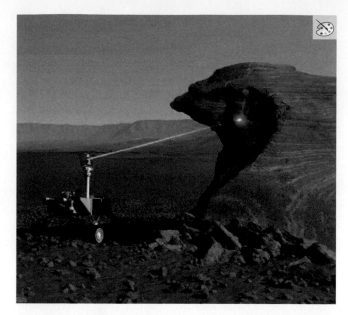

FIGURE 24.13 This is an artist's conception of *Curiosity*'s ChemCam instrument firing its laser at a rock on Mars, so that it can spectroscopically analyze the composition of the vaporized rock.

was once habitable. Scientists expect to learn far more from the *Curiosity* rover, which landed on Mars in August 2012. *Curiosity* carries sophisticated instruments designed to learn about the habitability of Mars. Its "ChemCam" instrument fires a laser at rocks, vaporizing some of the rocky material so that it can then analyze it spectroscopically (**FIGURE 24.13**). Other instruments can study soil samples scooped up by the rover.

THINK ABOUT IT

Find recent results from the *Curiosity* rover. Has it found anything of significance to the search for life on Mars? Explain.

Methane on Mars Atmospheric studies can also provide clues about potential life, and scientists are intrigued by the claimed detection of methane (CH_4) in the Martian atmosphere. Atmospheric methane is quickly destroyed by ultraviolet light from the Sun, so its presence would mean that it is being continually released from the Martian surface. Moreover, while some scientists think the detection is a measurement error, the amount of methane in the atmosphere appears to vary regionally across Mars, and also seems to vary with the Martian seasons.

If there really is methane in Mars's atmosphere, could it be produced by life, as is the case for most of the methane in Earth's atmosphere? It's possible, but methane can also be produced by nonbiological processes. For example, volcanoes release methane, so low levels of volcanic activity could be responsible for the methane on Mars, and the seasonal variations may be due to escape routes for gas from underground being plugged in the winter and clear in the summer. However, even if the source is volcanic rather than biological, the methane would be significant to the possibility of life. The amount of volcanic heat necessary for methane release would probably also be sufficient to maintain pockets of liquid water underground, which could potentially provide a habitat for life.

Scientists are hoping that the *Curiosity* rover will help to either confirm or disprove the detection of methane. If *Curiosity* does confirm the presence of methane in the Martian atmosphere, it may also be able to investigate the ground source of the methane and help us discover whether the source is volcanic or biological.

The Debate over Martian Meteorites An entirely different approach to the search for life on Mars relies on studies of meteorites whose chemical composition suggests they came from Mars [**Section 12.1**]. One particular meteorite, designated ALH84001, has been the subject of controversial claims that it contains evidence of past life on Mars.

ALH84001 was found on the Antarctic ice in 1984. Careful study of the meteorite shows that it landed in Antarctica about 13,000 years ago, following a 16-million-year journey through space after being blasted from Mars by an impact. The rock itself dates to about 4.1 billion years ago, which means that it resided on Mars during times when the climate may have been warmer and wetter. Analysis of the meteorite reveals several lines of evidence that could indicate the past presence of life on Mars. For example, the rock contains layered carbonate minerals and complex organic molecules (polycyclic aromatic hydrocarbons, or PAHs) that are associated with life when found in Earth rock, as well as microscopic chains of magnetite crystals quite similar to chains made in Earth rocks by living bacteria. However, each of the tantalizing hints of Martian life can also be explained in a nonbiological way. Perhaps more important, terrestrial bacteria have been found living inside the meteorite, indicating that it was contaminated by Earth life during the 13,000 years it resided in Antarctica.

Overall, most scientists now doubt that the Martian meteorite shows true evidence of Martian life. Nevertheless, studies of ALH84001 and other Martian meteorites are continuing, and they may yet turn up surprises. Scientists are particularly excited about study of a meteorite that landed in Morocco in 2011 and appears to be one of the rare meteorites from Mars. Because this meteorite was collected soon after its arrival, it is unlikely to have suffered the same type of contamination as meteorites that have resided longer on Earth's surface.

Could there be life on Europa or other jovian moons?

After Mars, the next most likely candidates for life in our solar system are some of the moons of the jovian planets—especially Jupiter's moons Europa, Ganymede, and Callisto, and Saturn's moons Titan and Enceladus [**Section 11.2**]. Europa is the strongest candidate, because we have good reason to think it contains a deep ocean beneath its icy crust (see Figure 11.21). The ice and rock from which Europa formed undoubtedly included the necessary chemical ingredients for life, and Europa's internal heating (primarily due to tidal heating) is strong enough to power volcanic vents on its seafloor. If life on Earth arose near such undersea volcanic vents, Europa would seem to have everything needed for life to originate.

The possibility of life on Europa is especially interesting because, unlike any potential life on Mars, it would not necessarily have to be microscopic. After all, the several kilometers of surface ice that hide Europa's suspected ocean could also hide large creatures swimming within it. However, the potential energy sources for life on Europa are far more limited than the energy sources for life on Earth, mainly because sunlight could not fuel photosynthesis in the subsurface ocean. As a result, most scientists suspect that any life that might exist on Europa would probably be quite small and primitive.

As we discussed in Chapter 11, some evidence suggests that Jupiter's moons Ganymede and Callisto may also have subsurface oceans. However, these moons would have even less energy for life than Europa. If they have life at all, it is almost certainly small and primitive. Nevertheless, Europa, Ganymede, and Callisto offer the astonishing possibility that Jupiter alone could be orbited by more worlds with life than we find in all the rest of the solar system.

Titan offers another enticing place to look for life. Its surface is far too cold for liquid water, but it has lakes and rivers of liquid methane and ethane [**Section 11.2**]. Although many biologists think it unlikely, it is possible that this liquid could support life as water does on Earth. Titan might also have liquid water or a colder ammonia–water mixture deep underground, so even water-based life is possible.

The discovery of ice fountains on Saturn's moon Enceladus suggests that it, too, could be habitable if these fountains are powered by subsurface liquids. If this proves to be the case, it would open up the possibility that similar liquids—and potentially life—could exist on other solar system bodies, including Neptune's moon Triton and some of the moons of Uranus.

 Detecting Extrasolar Planets Tutorial, Lessons 1–3

24.3 LIFE AROUND OTHER STARS

Studies of extrasolar planets suggest that our galaxy contains billions of planetary systems [**Section 13.2**], which might make prospects for life elsewhere seem quite good. But numbers alone don't tell the whole story. In this section, we'll consider the prospects for life on worlds orbiting other stars.

Before we begin, we must distinguish between *surface* life like that on Earth and *subsurface* life like that we envision as a possibility on Mars or Europa. While large telescopes could in principle allow us to discover surface life on extrasolar planets, no foreseeable technology will allow us to find life that is hidden deep underground in other star systems (unless the subsurface life has a noticeable effect on the planet's atmosphere). We therefore will focus on the search for life on planets with habitable surfaces—surfaces with temperatures and pressures that could allow liquid water to exist.

What kinds of stars might have habitable planets?

The billions of planetary systems likely to exist in our galaxy offer both an opportunity and a challenge. The opportunity is that we have a lot of places to look for planets that are habitable or that have life; the challenge is in narrowing down the search

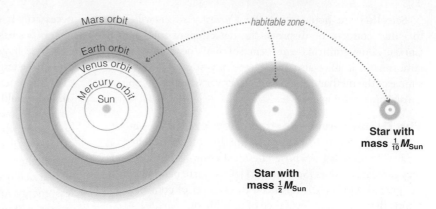

FIGURE 24.14 interactive figure The approximate habitable zones around our Sun, a star with one-half the mass of the Sun (spectral type K), and a star with one-tenth the mass of the Sun (spectral type M), shown to scale. The habitable zone becomes increasingly smaller and close-in for stars of lower mass and luminosity.

to the most promising candidates. Although we'll see that it is difficult to narrow the search down by much, the obvious place to start is in deciding which types of star systems are most likely to have habitable planets. In other words, which stars would make good "suns," providing heat and light to the surfaces of terrestrial planets that happen to orbit them?

Star System Constraints The first key constraint in searching for potential suns concerns stellar lifetimes. Although life on Earth started quite early relative to Earth's history, the actual time was still probably a few hundred million years. If the same idea holds true elsewhere, then we would expect life to be possible only on planets that orbit stars with lifetimes of hundreds of millions of years or more. This would rule out stars with more than a few times the mass of our Sun, because these stars have short lifetimes [**Section 17.3**]. However, because low-mass stars are far more common than high-mass stars, the lifetime constraint rules out only about 1% of all stars.

A second constraint is that the star system must allow planets to have orbits that remain stable for hundreds of millions to billions of years. About half of all stars are in binary or multiple star systems, around which long-term stable planetary orbits are less likely than around single stars. If life were not possible in such systems, then we would rule out about half the stars in our galaxy as potential homes to life. Of course, the other half—which still means 100 billion or more stars—would remain possible homes for life. Moreover, under some circumstances, stable planetary orbits are possible in multiple star systems, and planets in such systems have already been found.

All in all, it seems that the vast majority of stars are at least potentially capable of having life-bearing planets. Even very conservative assumptions suggest enormous numbers of possibilities. For example, limiting the search for habitable planets to stars similar to our Sun (that is, spectral type G) would still mean billions of potential suns in the Milky Way Galaxy.

The Habitable Zone Once we've narrowed the search to star systems with sufficiently long lifetimes and the possibility of stable planetary orbits, the next issue is the size of a star's **habitable zone**—the region in which a terrestrial planet of the right size could have a surface temperature that might allow for liquid water and life. In our own solar system, for

example, the habitable zone must be some donut-shaped region around Earth's orbit, because Earth has oceans and life.

Our Sun's habitable zone extends inward to the point where it would become too hot for liquid water to exist. This point must be somewhere between the orbits of Venus and Earth, because moving Earth to the orbit of Venus would cause a runaway greenhouse effect that would boil the oceans away [**Section 10.5**]. On the other end, the habitable zone extends outward to the point at which it would be too cold for liquid water to exist on a planet's surface. This outer bound to the habitable zone is probably somewhere near the orbit of Mars; it may even be beyond Mars, because if Mars had been larger (such as the size of Earth), it might have been able to retain liquid water to this day [**Section 10.4**].

Based on current models, **FIGURE 24.14** shows the approximate sizes, to scale, of the habitable zones around our Sun, around a star with about one-half the mass of our Sun (spectral type K), and around a star with about one-tenth the mass of our Sun (spectral type M). Notice that less massive and hence less luminous main-sequence stars have smaller habitable zones. This might make it seem that habitable planets are less likely around lower mass stars, though some models suggest that terrestrial planets might also form more easily around such stars, offsetting the effects of the smaller zones. Moreover, even if an individual low-mass star is less likely to have habitable planets, the fact that low-mass stars are the most common type of star may still make them "suns" to vast numbers of habitable planets.

The *Kepler* mission has already identified dozens of candidate planets that appear to orbit within their stars' habitable zones and to be similar enough in size to Earth that they are likely to be terrestrial planets. The first of these candidates to be confirmed through other methods, Kepler-22b, has a radius a little more than twice that of Earth (making it a "super-Earth") and orbits in the middle of the habitable zone around a star only slightly cooler and less massive than our Sun. As discussed in Chapter 13, it is seeming increasingly likely that terrestrial planets in habitable zones are very common in our galaxy, and may number in the tens of billions.

Confirming Habitability or Life While we are rapidly learning about the number of planets in habitable zones and the sizes of these planets, we can so far at best say that a planet is *potentially* habitable. In order to confirm that it is habitable, we would need to confirm that it actually has surface conditions that allow for liquid water. This will probably require direct images or spectra with at least moderate resolution, so that we might, for example, detect water on the surface. Longer term, images from future telescopes may tell us whether extrasolar planets have continents and oceans like Earth and perhaps even allow us to monitor seasonal changes.

Spectra should prove even more important to the search for life. Moderate-resolution infrared spectra can reveal the presence and abundance of many atmospheric gases, including carbon dioxide, ozone, methane, and water vapor (**FIGURE 24.15**). Careful analysis of atmospheric makeup might tell us whether a planet has life. On Earth, for example, the large abundance of oxygen (21% of our atmosphere) is a direct result of photosynthetic life. Abundant oxygen in the atmosphere of a distant world might similarly indicate the presence of life, since we know of no nonbiological way to produce an oxygen abundance as high as Earth's. Other evidence might come from the ratio of oxygen to the other detected gases. Scientists are working to improve our understanding of how life influences atmospheric chemistry, in the hope that we will be able to recognize particular gas combinations as signatures of life.

Are Earth-like planets rare or common?

We will not know for certain whether habitable planets exist or how common such planets may be until we are able to study the candidates with much more powerful telescopes. Nevertheless, the fact that we are already finding Earth-size planets within the habitable zones of other stars suggests that *potentially* habitable planets will prove to be very common. But should we expect these planets to have Earth-like conditions in which life is likely to arise and evolve?

Most scientists think so, but a few have raised some interesting questions. In essence, these scientists suggest that Earth's hospitable environment is the result of several rare kinds of planetary luck. According to this idea, sometimes called the "rare Earth hypothesis," the specific circumstances that have allowed life on Earth to survive and evolve into complex forms (such as trees and people) might be so rare that ours could be the only planet in the galaxy that harbors anything but the simplest life. Let's briefly examine some of the key issues in the rare Earth hypothesis.

Galactic Constraints Proponents of the rare Earth hypothesis suggest that Earth-like planets can form in only a relatively small region of the Milky Way Galaxy, making the number of potential homes for life far smaller than we might otherwise expect it to be. In essence, they argue that there is a fairly narrow ring at about our solar system's distance from the center of the Milky Way Galaxy that makes up a *galactic habitable zone* analogous to the habitable zone around an individual star (**FIGURE 24.16**).

According to the arguments for a galactic habitable zone, outer regions of our galaxy are unlikely to have terrestrial planets because of a low abundance of elements other than hydrogen and helium. Recall that the fraction of heavy elements varies among stars, from less than 0.1% among the old stars in globular clusters to more than 2% among young stars in the galactic disk [**Section 19.3**]. Even within the galactic disk, the abundance tends to decline with distance from the center of the galaxy. Because terrestrial planets are made almost entirely of heavy elements, a lower abundance of these elements might lessen the chance that terrestrial planets

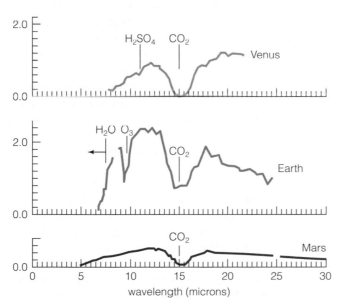

FIGURE 24.15 The infrared spectra of Venus, Earth, and Mars as they might be seen from afar, showing absorption features that point to the presence of carbon dioxide (CO_2), ozone (O_3), and sulfuric acid (H_2SO_4) in their atmospheres. While carbon dioxide is present in all three spectra, only our own planet has appreciable oxygen (and hence ozone)—a product of photosynthesis. If we could make similar spectral analyses of distant planets, we might detect atmospheric gases that would indicate life.

FIGURE 24.16 The highlighted green ring in this painting of the Milky Way Galaxy represents what some scientists suspect to be a galactic habitable zone—the only region of the galaxy in which Earth-like planets are likely to be found. However, other scientists think that Earth-like planets could be far more widespread.

could form. The argument against finding habitable planets in the inner regions of the galaxy concerns supernova rates. Supernovae are more common in the more crowded, inner regions of the galactic disk, making it more likely that a terrestrial planet would be exposed to the intense radiation from a nearby supernova. By assuming that this radiation would be detrimental to life, the proponents of a galactic habitable zone argue against finding habitable planets in the inner regions of the galaxy. Together, the constraints on finding Earth-like planets in the inner and outer regions of the galaxy leave the galactic habitable zone as a relatively narrow ring encompassing no more than about 10% of the stars in the galactic disk.

However, other scientists offer counterarguments to both sets of galactic constraints. Although discoveries of extrasolar planets to date suggest that *large* planets are indeed rarer around stars with very low fractions of heavy elements, we cannot yet say whether terrestrial planets are rare in such systems. Remember that Earth's mass is less than $\frac{1}{100,000}$ of the mass of the Sun, so even a very small heavy-element abundance could be enough to make one or more Earth-like planets. Regarding the radiation danger from supernovae, we do not really know whether such radiation would be detrimental to life. A planet's atmosphere might protect life against the effects of this radiation. It is even possible that the radiation could be beneficial to life by increasing the rate of mutations and thereby accelerating the pace of evolution. If these counterarguments are correct, then habitable planets might be found throughout much or all of the galaxy.

Impact Rates and Jupiter Another issue raised by rare Earth proponents is the impact rate on planets in other star systems. Earth was subjected to numerous large impacts during the heavy bombardment that went on during the first half-billion years or so after our planet's birth, and the evolution of complex life probably could not have occurred if these impacts had continued at such a high rate. In our solar system, the impact rate had lessened dramatically by about 3.9 billion years ago. But could the impact rate remain high much longer in other planetary systems?

The most numerous small objects in our solar system are the trillion or so comets of the distant Oort cloud [**Section 12.2**]. Fortunately for us, nearly all these myriad objects are essentially out of reach, posing no threat to our planet. However, the reason they are out of reach can be traced directly to Jupiter: Recall that the Oort cloud's comets are thought to have formed among the jovian planets, later being "kicked out" to their current orbits by close encounters with these planets, especially Jupiter (see Figure 12.29). If Jupiter did not exist, many of these comets might have remained in regions of the solar system where they could pose an ongoing danger to Earth. In that case, the heavy bombardment might never have ended, and huge impacts would continue to this day. From this viewpoint, our existence on Earth has been possible only because of the "luck" of having Jupiter as a planetary neighbor.

The primary question in this case is just how "lucky" this situation might be. Our discoveries of extrasolar planets suggest that Jupiter-size planets are quite common, although we've also found evidence that many of these planets migrate inward [**Section 13.3**], perhaps disrupting terrestrial planet orbits along the way. Whether most star systems have a Jupiter-like planet in the right place for ejecting comets remains an open question.

Climate Stability Another issue affecting the rarity of Earth-like planets is climate stability. Earth's climate has been stable enough for liquid water to exist throughout the past 4 billion years. This climate stability has almost certainly played a major role in allowing complex life to evolve. If our planet had frozen over like Mars or overheated like Venus, we would not be here today. Advocates of the rare Earth hypothesis point to at least two pieces of "luck" related to Earth's stable climate.

The first piece of "luck" is the existence of plate tectonics. Recall that plate tectonics plays a major role in regulating Earth's climate through the carbon dioxide cycle [**Section 10.6**]. Plate tectonics probably was not necessary to the origin of life, but it seems to have been important in keeping the climate stable enough for the subsequent evolution of plants and animals. But are we really "lucky" to have plate tectonics, or should this geological process be common on similar-size planets elsewhere? We do not yet know. The lack of plate tectonics on Venus [**Section 9.5**], which is quite similar in size to Earth, might seem to argue for plate tectonics being rare. On the other hand, if Venus's lack of plate tectonics is due to its runaway greenhouse effect, as many scientists suspect [**Section 10.5**], then Venus might have had plate tectonics if it had been born a little farther from the Sun. In that case, it's possible that any Earth-size planet within a star's habitable zone would have plate tectonics.

The second piece of "luck" in climate stability is the existence of Earth's relatively large Moon. Models of Earth's rotation and orbit show that if the Moon did not exist, gravitational tugs from other planets would cause large swings in Earth's axis tilt over periods of tens to hundreds of thousands of years. (Such swings in axis tilt are thought to occur on Mars.) Changes in axis tilt would affect the severity of the seasons, which in turn could cause deeper ice ages and more intense periods of warmth. Given that the Moon probably formed as a result of a random giant impact [**Section 8.2**], we might seem to be very lucky to have the Moon and the climate stability it brings.

Again, however, there are other ways to look at the issue. First, the expected changes in axis tilt also depend on rotation rate; if Earth rotated in less than about 12 hours, the axis would be fairly stable even without the Moon. Second, the Moon's presumed formation in a random giant impact does not necessarily mean that large moons are rare, because solar system formation models indicate that at least a few giant impacts should be expected in any planetary system. Finally, even if the axis tilt changes dramatically, this may not have a major impact on potential life. Changes in axis tilt might warm or cool different parts of the planet dramatically, but the changes would probably occur slowly enough for life to adapt or migrate as the climate changed.

The Bottom Line The bottom line is that while the rare Earth hypothesis offers some intriguing arguments, it is too early to say whether any of them will hold up over time. For each potential argument that Earth has been lucky, we've seen counterarguments suggesting otherwise. There's no doubt that our solar system and our world have "personality"—they exhibit properties that might be found only occasionally in other star systems—but we have no clear reason to think that these special properties are essential to the existence of complex or even intelligent life. Indeed, it may be that we have missed out on some helpful phenomena that could have sped evolution on Earth. Until we learn more about other planets in the universe, we cannot know whether Earth-like planets and complex life are common or rare.

24.4 THE SEARCH FOR EXTRATERRESTRIAL INTELLIGENCE

So far, we have focused on search strategies for microbial or other nonintelligent life. However, if intelligent beings and civilizations exist elsewhere, we might be able to find them with a completely different type of strategy. Instead of searching for hard-to-find spectroscopic signs of life, we might simply listen for signals that intelligent beings are sending into interstellar space, either in deliberate attempts to contact other civilizations or as a means of communicating among themselves. The search for signals from other civilizations is generally known as the **search for extraterrestrial intelligence**, or **SETI** for short.

How many civilizations are out there?

SETI efforts have a chance to succeed only if other advanced civilizations are broadcasting signals that we could receive. To judge the chances of SETI success, we'd need to know how many civilizations are broadcasting such signals right now.

Given that we do not even know whether microbial life exists anywhere beyond Earth, we certainly don't know whether other civilizations exist, let alone how many there might be. Nevertheless, for the purposes of planning a search for extraterrestrial intelligence, it is useful to have an organized way of thinking about the number of civilizations that might be out there. To keep our discussion simple, let's consider only the number of potential civilizations in our own galaxy. We can always extend our estimate to the rest of the universe by simply multiplying the result we find for our galaxy by 100 billion, the approximate number of galaxies in the observable universe.

The Drake Equation The first scientific conference on the search for extraterrestrial intelligence was held in 1961 in Green Bank, West Virginia, where a pioneering search for an alien signal had already been conducted. During the meeting, astronomer Frank Drake wrote a simple equation designed to summarize the factors that would determine the number of civilizations we might contact (**FIGURE 24.17**).

This equation is now known as the **Drake equation**, and in principle it gives us a simple way to calculate the number of civilizations capable of interstellar communication that are currently sharing the Milky Way Galaxy with us. In a form

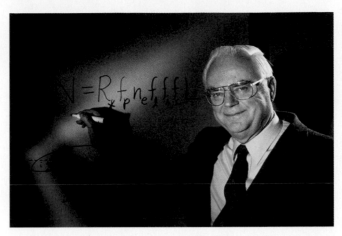

FIGURE 24.17 Astronomer Frank Drake, with the equation he first wrote in 1961. (With Dr. Drake's approval, we use a slightly modified form of his equation in this book.)

slightly modified from the original, the Drake equation looks like this:

$$\text{Number of civilizations} = N_{HP} \times f_{life} \times f_{civ} \times f_{now}$$

This equation will make sense once you understand the meaning of each factor:

- N_{HP} is the number of habitable planets in the galaxy—that is, the number of planets that could potentially have life.

- f_{life} is the fraction of habitable planets that actually *have* life. For example, $f_{life} = 1$ would mean that all habitable planets have life, and $f_{life} = \frac{1}{1,000,000}$ would mean that only 1 in a million habitable planets has life. Therefore, the product $N_{HP} \times f_{life}$ tells us the number of life-bearing planets in the galaxy.

- f_{civ} is the fraction of life-bearing planets on which a civilization capable of interstellar communication *has at some time* arisen. For example, $f_{civ} = \frac{1}{1000}$ would mean that such a civilization has existed on 1 out of 1000 planets with life, while the other 999 out of 1000 have not had a species that learned to build radio transmitters, high-powered lasers, or other devices for interstellar conversation. When we multiply this factor by the first two factors to form the product $N_{HP} \times f_{life} \times f_{civ}$, we get the total number of planets on which intelligent beings have evolved and developed a communicating civilization at some time in the galaxy's history.

- f_{now} is the fraction of these civilization-bearing planets that happen to have a civilization *now*, as opposed to, say, millions or billions of years in the past. This factor is important because we can hope to contact only civilizations that are broadcasting signals we could receive at present. (In estimating f_{now}, we assume that the light-travel time for signals from other stars has been taken into account.)

Because the product of the first three factors tells us the total number of civilizations that have *ever* arisen in the galaxy, multiplying by f_{now} tells us how many civilizations we could potentially make contact with today. In other words, the result of the Drake equation should in principle tell us the number of civilizations that we might hope to contact. Unfortunately, we don't yet know the value of any of the factors in the Drake equation, so we cannot actually calculate its result. Nevertheless, the equation is a useful way of organizing our thinking, as we can see by considering the potential values for each of its factors.

The Number of Life-Bearing Planets Let's begin with the first two factors in the Drake equation, whose product ($N_{HP} \times f_{life}$) tells us the number of life-bearing planets in our galaxy. We can make a reasonably educated guess only about the first factor, the number of habitable planets (N_{HP}). Current understanding of solar system formation (see Figure 8.13) suggests that terrestrial planets ought to form easily and, as discussed earlier, there ought to be billions of stars in our galaxy with habitable zones large enough to have Earth-like planets. Unless some of the "rare Earth" ideas prove to be correct, it seems reasonable to suppose our galaxy has billions of habitable planets.

The factor f_{life} presents more difficulty. At present, we have no reliable way to estimate the fraction of habitable planets on which life actually arose. The problem is that we cannot generalize when we have only one example to study—our own Earth. Still, the fact that life arose rapidly on Earth suggests that the origin of life was fairly "easy." In that case, we might expect most or all habitable planets to also have life, making the fraction f_{life} close to 1. Of course, until we have solid evidence that life arose anywhere else, such as on Mars, it is also possible that Earth has been very lucky and that f_{life} is so close to zero that life has never arisen on any other planet in our galaxy.

SPECIAL TOPIC

Are Aliens Already Here?

In this chapter, we discuss aliens as a possibility, not a reality. However, opinion polls suggest that up to half the American public believes that aliens are already visiting us. What can science say about this remarkable notion?

The bulk of the claimed evidence for alien visitation consists of sightings of UFOs—unidentified flying objects. Many thousands of UFOs are reported each year, and no one doubts that unidentified objects are being seen. The question is whether they are alien spacecraft.

Aliens have long been a staple of science fiction, but modern interest in UFOs began with a widely reported sighting in 1947. While flying a private plane near Mount Rainier in Washington State, businessman Kenneth Arnold saw nine mysterious objects streaking across the sky. He told a reporter that the objects "flew erratic, like a saucer if you skip it across the water." (One possible explanation is that he saw meteors skipping across the atmosphere.) He did *not* say that the objects were saucer-shaped, but the reporter wrote of "flying saucers." The story was front-page news throughout America, and "flying saucers" soon invaded popular culture, if not our planet.

The flying saucer reports also interested the U.S. Air Force, largely out of concern that the UFOs might represent new types of aircraft developed by the Soviet Union. For two decades, the Air Force hired teams of academics to study UFO reports. In most cases, these experts were able to specify a plausible identification of the UFO. The explanations included bright stars and planets, aircraft rockets, balloons, birds, meteors, atmospheric phenomena, and the occasional hoax. In a few cases, the investigators could not deduce what was seen, but their overall conclusion was that there was no reason to believe the UFOs were either highly advanced Soviet craft or visitors from other worlds.

Believers discounted the Air Force denials, claiming to have other evidence of alien visitation. So far, none of this evidence has withstood scientific scrutiny. Photographs and film clips are nearly always either fuzzy or obviously faked. Crop circles are easily made by pranksters, and claimed pieces of alien spacecraft turn out to have more mundane origins. Champions of alien visitation generally explain away the lack of clear evidence either as a government cover-up or as a failure of the scientific community to take the subject seriously, but neither explanation is compelling.

It's conceivable that a government might *try* to put the lid on evidence of alien visits, though the motivation for doing so is unclear. The usual explanations are that the public couldn't handle the news or that the government is taking secret advantage of the alien materials to design new military hardware (via "reverse engineering"). However, given that half the population already believes in alien visitors, the shock of discovery would seem unlikely to cause panic. As for reverse-engineering extraterrestrial spacecraft, any society that could routinely cross interstellar distances would be far beyond us technologically. Reverse engineering their spaceships is as unlikely as Neandertals constructing personal computers just because a laptop somehow landed in their cave. In addition, while a government might successfully hide evidence for a short time, over decades the lure of talk show fame and riches would surely cause someone to reveal the conspiracy. Moreover, unless the aliens landed only in the United States, the conspiracy would have to include all governments on Earth, which seems highly unlikely given the world's political conditions.

Claims of disinterest on the part of the scientific community also fall apart on scrutiny. Scientists are constantly competing with one another to be the first to make a great discovery, and clear evidence of alien visitors would rank high on the all-time list. The fact that few scientists are engaged in such study reflects not a lack of interest, but a lack of evidence worthy of study.

Of course, absence of evidence is not evidence of absence. Most scientists are open to the possibility that we might someday find evidence of alien visits, and given what we've learned about the prospects of life in the universe, it is at least plausible to imagine that such visits occur. So far, however, we have no hard evidence to support the belief that aliens are already here.

The Question of Intelligence Even if life-bearing planets are very common, civilizations capable of interstellar communication might not be. The fraction of life-bearing planets that at some time have such civilizations, f_{civ}, depends on at least two things. First, a planet would have to have a species evolve with sufficient intelligence to develop interstellar communication. In other words, the planet needs a species at least as smart as we are. Second, that species would have to develop a civilization with technology at least as advanced as ours.

Although we cannot really be sure, most scientists suspect that only the first requirement is difficult to meet. A fundamental assumption in nearly all science today is that we are not "special" in any particular way. We live on a fairly typical planet orbiting an ordinary star in a normal galaxy, and we assume that living creatures elsewhere—whether they prove to be rare or common—would be subjected to evolutionary pressures similar to those that have operated on Earth. Therefore, if species with intelligence similar to ours have arisen elsewhere, we assume that they would have similar sociological drives that would eventually lead them to develop the technology necessary for interstellar communication.

If this assumption is correct, then the fraction f_{civ} depends primarily on the question of whether sufficient intelligence is rare or common among life-bearing planets. As with the question of life of any kind, the short answer is that we just don't know, but we can get at least some insight by considering what happened on Earth.

Look again at Figure 24.4. While life arose quite quickly on Earth, nearly all life remained microbial until just a few hundred million years ago, and it took nearly 4 billion years for us to arrive on the scene. This slow progress toward intelligence might suggest that producing a civilization is very difficult even when life is present. On the other hand, roughly half the stars in the Milky Way are older than our Sun, so if Earth's case is typical, then plenty of planets have existed long enough for intelligence to arise.

Another way to address the question is by considering our level of intelligence in comparison to that of other animals on Earth. We can get a rough measure of intelligence by comparing brain mass to total body mass (a measure sometimes called the *encephalization quotient*, or EQ). FIGURE 24.18 shows the brain weights for a sampling of birds and mammals plotted against their body weight. There is a clear and expected trend in that heavier animals have heavier brains. By drawing a straight line that fits these data, we can define an average value of brain mass for each body mass. Animals whose brain mass falls above the line are smarter than average, while animals whose brain mass falls below the line are less mentally agile. Keep in mind that it is the *vertical* distance above the line that tells us how much smarter a species is than the average, and that the scale goes in powers of 10 on both axes. If you look closely, you'll see that the data point for humans lies significantly farther above the line than the data point for any other species. By this measure of intelligence, we are far smarter than any other species that has ever existed on Earth.

Some people use this fact to argue that even on a planet with complex life, a species as intelligent as we are would be

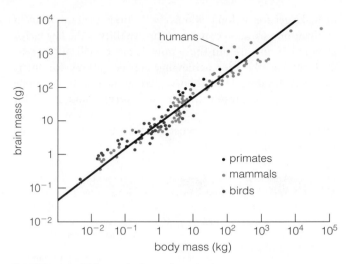

FIGURE 24.18 This graph shows how brain mass compares to body mass for some mammals (including primates) and birds. The straight line represents an average of the ratio of brain mass to body mass, so animals that fall above the line are smarter than average and animals that fall below the line are less smart. Note that the scale uses powers of 10 on both axes. (Data from Harry J. Jerison.)

very rare. They say that even if there is an evolutionary drive toward intelligence in general, it takes extreme luck to reach our level of intelligence. After all, while it's evolutionarily useful to have enough intelligence to capture prey and evade other predators, it's not clear why natural selection would lead to brains big enough to build spacecraft. However, the same data can be used to reach an opposite conclusion. The scatter in the levels of intelligence among different animals tells us that some variation should be expected, and statistical analysis shows that we are not unreasonably far above the average. It might therefore be inevitable that some species would develop our level of intelligence on any planet with complex life.

Technological Lifetimes For the sake of argument, let's assume that life and intelligence are reasonably likely, so thousands or millions of planets in our galaxy have at some time given birth to a civilization. In that case, the final factor in the Drake equation, f_{now}, determines the likelihood of there being someone whom we could contact now. The value of this factor depends on how long civilizations survive.

Consider our own example. In the roughly 12 billion years during which our galaxy has existed, we have been capable of interstellar communication via radio for only about 60 years. Therefore, if we were to destroy ourselves tomorrow, then other civilizations could have received signals from us during only 60 years out of the galaxy's 12-billion-year existence, equivalent to 1 part in 200 million of the galaxy's history. If such a short technological lifetime is typical of civilizations, then f_{now} would be only $\frac{1}{200,000,000}$, and some 200 million civilization-bearing planets would need to have existed at one time or another in the Milky Way in order for us to have a good chance of finding another civilization out there now.

However, we'd expect f_{now} to be so small only if we are on the brink of self-destruction—after all, the fraction will grow larger for as long as our civilization survives. Thus, if civilizations are at all common, survivability is the key factor in whether any are out there now. If most civilizations self-destruct shortly after achieving the technology for interstellar communication, then we are almost certainly alone in the galaxy at present. But if most survive and thrive for thousands or millions of years, the Milky Way may be brimming with civilizations—most of them far more advanced than our own.

THINK ABOUT IT

Describe a few reasons why a civilization capable of interstellar communication would also be capable of self-destruction. Overall, do you believe our civilization can survive for thousands or millions of years? Defend your opinion.

How does SETI work?

If there are indeed other civilizations out there, then in principle we ought to be able to make contact with them. Based on our current understanding of physics, it seems likely that even very advanced civilizations would communicate much as we do—by encoding signals in radio waves or other forms of light. Most SETI researchers use large radio telescopes to search for alien radio signals (**FIGURE 24.19**). A few researchers are studying other parts of the electromagnetic spectrum. For example, some scientists use visible-light telescopes to search for communications encoded as laser pulses. Of course, advanced civilizations may well have invented communication technologies that we cannot even imagine, let alone detect.

A good way to think about our chances of picking up an alien signal is to imagine what aliens would need to do to pick up signals from us. We have been sending relatively

FIGURE 24.20 In 1974, a short message was broadcast to the globular cluster M13 using the Arecibo radio telescope. The picture shown here was encoded by using two different radio frequencies, one for "on" and one for "off" (the colors used here are arbitrary). To decode the message, the aliens would need to realize that the bits are meant to be arranged in a rectangular grid as shown, but that should not be difficult: The grid has 73 rows and 23 columns, and aliens would presumably know that these are both prime numbers. The picture represents the Arecibo radio dish, our solar system, a human stick figure, and a schematic of DNA and the eight simple molecules used in its construction.

high-power transmissions into space since about the 1950s in the form of television broadcasts. In principle, anyone within about 60 light-years of Earth could watch our old television shows (perhaps a frightening thought). However, in order to detect our broadcasts, they would need far larger and more sensitive radio telescopes than we have today. If their technology were at the same level as ours, they could receive a signal from us only if we deliberately broadcast an unusually high-powered transmission.

To date, humans have made only a few attempts to broadcast our existence in this way. The most famous was a 3-minute transmission made in 1974 with a planetary radar transmitter on the Arecibo radio telescope, in which a simple pictorial message was beamed toward the globular cluster M13 (**FIGURE 24.20**). This target was chosen in part because it contains a few hundred thousand stars, seemingly offering a good chance that at least one has a civilization around it. However, M13 is about 25,000 light-years from Earth, so it will take some 25,000 years for our signal to get there and another 25,000 years for any response to make its way back to Earth.

Several SETI projects under way or in development would be capable of detecting signals like the one we broadcast from Arecibo if they came from civilizations within a few hundred light-years. These SETI efforts scan millions of radio frequency bands simultaneously. If anyone nearby is deliberately broadcasting on an ongoing basis, we have a good chance of detecting the signals.

FIGURE 24.19 The Allen Telescope Array in Hat Creek, California, is being used to search for radio signals from extraterrestrial civilizations.

THINK ABOUT IT

SETI efforts are often controversial because of their cost and uncertain chance of success, but supporters say the cost is justified because contact with an extraterrestrial intelligence would be such an important discovery. What do *you* think?

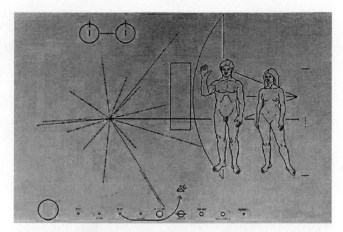

a The *Pioneer* plaque, about the size of an automobile license plate. The human figures are shown in front of a drawing of the spacecraft to give them a sense of scale. The "prickly" graph to their left shows the Sun's position relative to nearby pulsars, and Earth's location around the Sun is shown below. Binary code indicates the pulsar periods; because pulsars slow with time, the periods will allow someone reading the plaque to determine when the spacecraft was launched.

b *Voyagers 1* and *2* carry a phonograph record—a 12-inch gold-plated copper disk containing music, greetings, and images from Earth.

FIGURE 24.21 Messages aboard the *Pioneer* and *Voyager* spacecraft, which are bound for the stars.

24.5 INTERSTELLAR TRAVEL AND ITS IMPLICATIONS FOR CIVILIZATION

So far, we have discussed ways of detecting distant civilizations without ever leaving the comfort of our own planet. Could we ever actually visit other worlds in other star systems? A careful analysis of this question turns out to have profound implications for the future of our civilization. To see why, we first need to consider the prospects for achieving interstellar travel.

How difficult is interstellar travel?

In many science fiction movies, our descendants travel among the stars as routinely as we jet about our planet today in airplanes. They race around the galaxy in starships of all sizes and shapes, circumventing nature's prohibition on faster-than-light travel by entering hyperspace, wormholes, or warp drive. They witness firsthand incredible cosmic phenomena such as stars and planets in all stages of development, accretion disks around white dwarfs and neutron stars, and the distortion of spacetime near black holes. Along the way they encounter numerous alien species, most of which happen to look and act a lot like us.

Unfortunately, Einstein's theory of relativity tells us that real interstellar travel must be at speeds slower than the speed of light [**Section S2.2**], and even to approach that speed we will need to overcome huge technological hurdles. Nevertheless, we have already sent out our first emissaries to the stars, and there's no reason to believe that we won't develop better technologies in the future.

The Challenge of Interstellar Travel To date, we have launched five spacecraft that will leave our solar system and eventually travel among the stars: the planetary probes *Pioneer 10, Pioneer 11, Voyager 1, Voyager 2,* and *New Horizons*

(the spacecraft currently en route to Pluto). These spacecraft are traveling about as fast as anything ever built by humans, but their speeds are still less than $\frac{1}{10,000}$ of the speed of light. It would take each of them some 100,000 years just to reach the next nearest star system (Alpha Centauri), but their trajectories won't take them anywhere near it. Instead, they will simply continue their journey without passing close to any nearby stars, wandering the Milky Way for millions or even billions of years to come. The *Pioneer* and *Voyager* spacecraft carry greetings from Earth, just in case someone comes across one of them someday (**FIGURE 24.21**).

If we want to make interstellar journeys within human lifetimes, we will need starships that can travel at speeds close to the speed of light. We will need entirely new types of engines to reach such high speeds. The energy requirements of interstellar spacecraft may pose an even more daunting challenge. For example, the energy needed to accelerate a single ship the size of *Star Trek*'s *Enterprise* to just half the speed of light would be more than 2000 times the total annual energy use of the world today. Clearly, interstellar travel will require vast new sources of energy. In addition, fast-moving starships will require new types of shielding to protect crew members from instant death. As a starship travels through interstellar gas at near-light speed, ordinary atoms and ions will hit it like a deadly flood of high-energy cosmic rays.

If we succeed in building starships capable of traveling at speeds close to the speed of light, the crews will face significant social challenges. According to the well-tested principles of Einstein's theory of relativity, time will run much more slowly on a spaceship that travels at high speed to the stars than it does here on Earth [**Section S2.4**]. For example, in a ship traveling at an average speed of 99.9% of the speed of light, the 50-light-year round trip to the star Vega would take the travelers only about 2 years—but more than 50 years would pass on Earth while they were gone. The crew would therefore

FIGURE 24.22 Artist's conception of the *Project Orion* starship, showing one of the small hydrogen-bomb detonations that would propel it. Debris from the detonation strikes the "pusher plate" at the back of the spaceship. The central sections (enclosed in a lattice) hold the bombs, and the front sections house the crew.

FIGURE 24.23 Artist's conception of a spaceship powered by an interstellar ramjet. The giant scoop in the front (left) collects interstellar hydrogen for use as fusion fuel.

need only 2 years' worth of provisions and would age only 2 years during the voyage, but they would return to a world quite different from the one they left. Family and friends would be older or deceased, new technologies might have made their knowledge and skills obsolete, and many political and social changes might have occurred in their absence. The crew would face a difficult adjustment when they came home to Earth.

Starship Design Despite all the difficulties, some scientists and engineers have already proposed designs that could in principle take us to nearby stars. In the 1960s, a group of scientists proposed *Project Orion,* which envisioned accelerating a spaceship with repeated detonations of relatively small hydrogen bombs. Each explosion would take place a few tens of meters behind the spaceship and would propel the ship forward as the vaporized debris impacted a "pusher plate" on the back of the spacecraft (**FIGURE 24.22**). Calculations showed that a spaceship accelerated by the rapid-fire detonation of a million H-bombs could reach Alpha Centauri in just over a century. In principle, we could build an *Orion* spacecraft with existing technology, though it would be very expensive and would require an exception to the international treaty banning nuclear detonations in space.

A century to the nearest star isn't bad, but it still wouldn't make interstellar travel easy. Unfortunately, no available technology could go much faster. The problem is mass: Making a rocket faster requires more fuel, but adding fuel adds mass and makes it more difficult for the rocket to accelerate. Calculations show that even in the best case, rockets carrying nuclear fuel could achieve speeds no more than a few percent of the speed of light. Nevertheless, we can envision some possible future technologies that might get around this problem.

One idea suggests powering starships with engines that generate energy through matter–antimatter annihilation. While nuclear fusion converts less than 1% of the mass of atomic nuclei into energy, matter–antimatter annihilation [**Section 22.1**] converts *all* the annihilated mass into energy. Starships with matter–antimatter engines could probably reach speeds of 90% or more of the speed of light. At these speeds, the slowing of time predicted by relativity becomes noticeable, putting many nearby stars within a few years' journey for the crew members. However, because no natural reservoirs of antimatter exist, we would have to be able to manufacture many tons of antimatter and then store it safely for the trip—capabilities that are far beyond our present means.

An even more speculative and futuristic design, known as an *interstellar ramjet,* would collect interstellar hydrogen with a gigantic scoop, using the collected gas as fuel for its nuclear engines (**FIGURE 24.23**). By collecting fuel along the way, the ship could avoid carrying the weight of fuel on board. However, because the density of interstellar gas is so low, the scoop would need to be enormous. As astronomer Carl Sagan said, we are talking about "spaceships the size of worlds."

The bottom line is that while we face enormous obstacles to achieving interstellar travel, there's no reason to think it's impossible. If we can avoid self-destruction and if we continue to explore space, our descendants might well journey to the stars.

Where are the aliens?

Imagine that we survive long enough to become interstellar travelers and that we begin to colonize habitable planets around nearby stars. As each colony grows, it may send out explorers to other star systems. Even if our starships traveled at relatively low speeds—say, a few percent of the speed of light—we could have dozens of outposts around nearby stars within a few centuries. In 10,000 years, our descendants would be spread among stars within a few hundred light-years of Earth. In a few million years, we could have outposts throughout the Milky Way Galaxy. We would have become a true galactic civilization.

Now, if we take the idea that *we* could develop a galactic civilization within a few million years and combine it with the reasonable (though unproved) idea that civilizations ought to be common, we are led to an astonishing conclusion: Someone else should already have created a galactic civilization.

To see why, let's take some sample numbers. Suppose the overall odds of a civilization arising around a star are about the same as your odds of winning the lottery, or 1 in a million. Using a low estimate of 100 billion stars in the Milky Way Galaxy, this would mean some 100,000 civilizations in our galaxy alone. Moreover, current evidence suggests that stars and planetary systems like our own could have formed for at least 5 billion years before our solar system was even born, in which case the first of these 100,000 civilizations would have arisen at least 5 billion years ago. Others would have arisen, on average, about every 50,000 years. Under these assumptions, we would expect the youngest civilization besides ours to be some 50,000 years ahead of us technologically, and most would be millions or billions of years ahead of us.

We thereby encounter a strange paradox: Plausible arguments suggest that a galactic civilization should already exist, yet we have so far found no evidence of such a civilization. This paradox is often called *Fermi's paradox,* after the Nobel Prize–winning physicist Enrico Fermi. During a 1950 conversation with other scientists about the possibility of extraterrestrial intelligence, Fermi responded to speculations by asking, "So where is everybody?"

This paradox has many possible solutions, but broadly speaking we can group them into three categories:

1. *We are alone.* There is no galactic civilization because civilizations are extremely rare—so rare that we are the first to have arisen on the galactic scene, perhaps even the first in the universe.

2. *Civilizations are common, but no one has colonized the galaxy.* There are at least three possible reasons why this might be the case. Perhaps interstellar travel is much harder or more expensive than we have guessed, and civilizations are unable to venture far from their home worlds. Perhaps the desire to explore is unusual, and other societies either never leave their home star systems or stop exploring before they've colonized much of the galaxy. Most ominously, perhaps many civilizations have

arisen, but they have all destroyed themselves before achieving the ability to colonize the stars.

3. *There IS a galactic civilization,* but it has not yet revealed its existence to us.

We do not know which, if any, of these explanations is the correct solution to the question "Where are the aliens?" However, each category of solution has astonishing implications for our own species.

Consider the first solution—that we are alone. If this is true, then our civilization is a remarkable achievement. It implies that through all of cosmic evolution, among countless star systems, we are the first matter in our galaxy or the universe ever to know that the rest of the universe exists. Through us, the universe has attained self-awareness. Some philosophers and many religions argue that the ultimate purpose of life is to become truly self-aware. If so, and if we are alone, then the destruction of our civilization and the loss of our scientific knowledge would represent an inglorious end to something that took the universe some 14 billion years to achieve. From this point of view, humanity becomes all the more precious, and the collapse of our civilization would be all the more tragic.

The second category of solutions has much more terrifying implications. If thousands of civilizations before us have all failed to achieve interstellar travel on a large scale, what hope do we have? Unless we somehow think differently than all other civilizations, this solution says that we will never go far in space. Because we have always explored when the opportunity arose, this solution almost inevitably leads to the conclusion that failure will come about because we destroy ourselves. We hope that this answer is wrong.

The third solution is perhaps the most intriguing. It says that we are newcomers on the scene of a galactic civilization that has existed for millions or billions of years before us. Perhaps this civilization is deliberately leaving us alone for the time being and will someday decide the time is right to invite us to join it.

No matter what the answer turns out to be, learning it will surely mark a turning point in the brief history of our species. Moreover, this turning point is likely to be reached within the next few decades or centuries. We already have the ability to destroy our civilization. If we do so, then our fate is sealed. But if we survive long enough to develop technology that can take us to the stars, the possibilities seem almost limitless.

The Big Picture

Putting Chapter 24 into Context

Throughout our study of astronomy, we have taken a "big picture" view of understanding how we fit into the universe. Here, at last, we have returned to Earth and examined the role of our own generation in the big picture of human history. Tens of thousands of past human generations have walked this Earth. Ours is the first generation with the technology to study the far reaches of our universe, to search for life elsewhere, and to travel beyond our home planet. It is up to us to decide whether we will use this technology to advance our species or to destroy it.

Imagine for a moment the grand view, a gaze across the centuries and millennia from this moment forward. Picture our descendants living among the stars, having created or joined a great galactic civilization. They will have the privilege of experiencing ideas, worlds, and discoveries far beyond our wildest imagination. Perhaps, in their history lessons, they will learn of our generation—the generation that history placed at the turning point and that managed to steer its way past the dangers of self-destruction and onto the path to the stars.

24.1 LIFE ON EARTH

■ **When did life arise on Earth?** Fossil evidence puts the origin of life at least 3.5 billion years ago, and carbon isotope evidence pushes this date to more than 3.85 billion years ago. Life therefore arose within a few hundred million years after Earth's birth and possibly in a much shorter time.

■ **How did life arise on Earth?** Genetic evidence suggests that all life on Earth evolved from a common ancestor, which may have resembled microbes that live today in hot water near undersea volcanic vents. We do not know how this first organism arose, but laboratory experiments suggest that it may have been the result of natural chemical processes on the early Earth. Once life arose, it rapidly diversified and evolved through **natural selection**.

■ **What are the necessities of life?** Life on Earth thrives in a wide range of environments, and in general seems to require only three things: a source of nutrients, a source of energy, and liquid water.

24.2 LIFE IN THE SOLAR SYSTEM

■ **Could there be life on Mars?** Mars once had conditions that may have been conducive to the origin of life. If life arose, it might still survive in pockets of liquid water underground.

■ **Could there be life on Europa or other jovian moons?** Europa probably has a subsurface ocean of liquid water and may have seafloor volcanoes. If so, it has conditions much like those in which early life arose on Earth, making it a good candidate for life. Ganymede and Callisto might have oceans as well. Titan has other liquids on its surface, though it is too cold for liquid water. Perhaps life can survive in these other liquids, or perhaps Titan has liquid water deep underground. Enceladus also shows evidence of subsurface liquids, offering yet another possibility for life.

24.3 LIFE AROUND OTHER STARS

■ **What kinds of stars might have habitable planets?** Stars must have sufficiently long lifetimes and the possibility of stable planetary orbits in order to have habitable planets, but these constraints still leave billions of stars with **habitable zones** in which life-bearing planets might exist. Smaller (lower mass and luminosity) stars have smaller habitable zones.

■ **Are Earth-like planets rare or common?** We don't know. Arguments can be made on both sides of the question, and at present we lack the data to determine which answer is correct.

24.4 THE SEARCH FOR EXTRATERRESTRIAL INTELLIGENCE

■ **How many civilizations are out there?** We don't know, but the **Drake equation** gives us a way to organize our thinking about the question. The equation (in a modified form) says that the number of civilizations in the Milky Way Galaxy with whom we could potentially communicate is $N_{HP} \times f_{life} \times f_{civ} \times f_{now}$, where N_{HP} is the number of habitable planets in the galaxy, f_{life} is the fraction of habitable planets that actually have life on them, f_{civ} is the fraction of life-bearing planets on which a civilization capable of interstellar communication has at some time arisen, and f_{now} is the fraction of all these civilizations that exist now.

■ **How does SETI work?** **SETI**, the search for extraterrestrial intelligence, generally involves efforts to detect signals—such as radio or laser communications—coming from civilizations on other worlds.

24.5 INTERSTELLAR TRAVEL AND ITS IMPLICATIONS FOR CIVILIZATION

■ **How difficult is interstellar travel?** Convenient interstellar travel remains well beyond our technological capabilities because of the technological requirements for engines, the enormous energy needed to accelerate spacecraft to speeds near the speed of light, and the difficulties of shielding the crew from radiation. Nevertheless, people have proposed ways around all these difficulties, and it seems reasonable to think that we will someday achieve interstellar travel if we survive long enough.

■ **Where are the aliens?** A civilization capable of interstellar travel ought to be able to colonize the galaxy in a few million years or less, and the galaxy was around for billions of years before Earth was even born. It therefore seems that some civilization should have colonized the galaxy long ago—yet we have no evidence of other civilizations. Every possible explanation for this surprising fact has astonishing implications for our species and our place in the universe.

Use the following questions to check your understanding of some of the many types of visual information used in astronomy. Answers are provided in Appendix J. For additional practice, try the Chapter 24 Visual Quiz at MasteringAstronomy®.

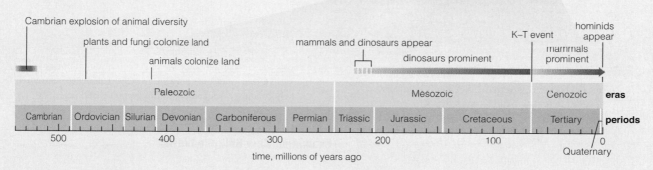

The figure above, which repeats Figure 24.4, shows the geological time scale. Use this figure to answer the following questions.

1. List the following events in the order in which they occurred, from first to last.
   a. earliest humans   b. earliest animals   c. impact causes extinction of dinosaurs   d. earliest mammals   e. earliest plants living on land   f. first time there is significant oxygen in Earth's atmosphere   g. first life on Earth

2. List the following time frames in order of how long they lasted, from longest to shortest.
   a. Hadean eon   b. Proterozoic eon   c. Paleozoic era   d. Cretaceous period

3. Which time frame(s) do we live in today? (More than one may apply.)

   a. Quaternary period   b. Tertiary period   c. Cenozoic era   d. Phanerozoic eon   e. Paleozoic era

4. How long did the Cambrian explosion last?
   a. less than 1 year   b. about a decade   c. about 10,000 years   d. about 40 million years   e. about 500 million years

5. When did the heavy bombardment end?
   a. about 4.5 billion years ago   b. between about 4.3 and 4.5 billion years ago   c. between about 3.8 and 4.0 billion years ago   d. exactly 3.85 billion years ago

6. How long have mammals been present on Earth?
   a. about 1 million years   b. about 65 million years   c. about 225 million years   d. about 510 million years

For instructor-assigned homework go to MasteringAstronomy®.

MasteringAstronomy®

REVIEW QUESTIONS

Short-Answer Questions Based on the Reading

1. Describe three recent developments in the study of life on Earth that make life elsewhere seem more plausible.

2. How do we study the history of life on Earth? Describe the *geological time scale* and a few of the major events along it.

3. Summarize the evidence pointing to an early origin of life on Earth. How far back in Earth's history did life exist?

4. Why is the *theory of evolution* so critical to our understanding of the history of life on Earth? Explain how evolution proceeds by *natural selection*, and what happens to DNA that allows species to evolve.

5. Give a brief overview of the history of life on Earth. What evidence points to a common ancestor for all life? How and when did oxygen accumulate in Earth's atmosphere? When did larger animals diversify on Earth?

6. How are laboratory experiments helping us study the origin of life on Earth? Explain.

7. Is it possible that life migrated to Earth from elsewhere? Explain.

8. Describe the range of environments in which life thrives on Earth. What three basic requirements apply to life in all these environments?

9. What is a *habitable world*? Which worlds in our solar system seem potentially habitable, and why?

10. Briefly summarize the current status of the search for life on Mars.

11. What do we mean by a star's *habitable zone*? Do we expect many stars to be capable of having habitable planets? Explain.

12. What is the "rare Earth hypothesis"? Summarize the arguments on both sides regarding the validity of this hypothesis.

13. What is *SETI*? Describe the capabilities of current SETI efforts.

14. What is the *Drake equation*? Define each of its factors, and describe the current state of understanding about the potential values of each factor.

15. Why is interstellar travel so difficult? Describe a few technologies that might someday make it possible.

16. What is *Fermi's paradox*? Describe several potential solutions to the paradox and the implications of each to our civilization.

TEST YOUR UNDERSTANDING

Fantasy or Science Fiction?

For each of the following futuristic scenarios, decide whether it is plausible according to our present understanding of science or whether it is unlikely to be possible. Explain clearly; not all these have definitive answers, so your explanation is more important than your chosen answer.

17. The first human explorers on Mars discover the ruins of an ancient civilization, including remnants of tall buildings and temples.

18. The first human explorers on Mars drill a hole into a Martian volcano to collect a sample of soil from several meters underground. Upon analysis of the soil, they discover that it holds living microbes resembling terrestrial bacteria but with a different biochemistry.

19. In 2040, a spacecraft lands on Europa and melts its way through the ice into the Europan ocean. It finds numerous strange, living microbes, along with a few larger organisms that feed on the microbes.

20. It's the year 2075. A giant telescope on the Moon, consisting of hundreds of small telescopes linked together across a distance of 500 kilometers, has just captured a series of images of a planet around a distant star that clearly show seasonal changes in vegetation.

21. A century from now, after completing a careful study of planets around stars within 100 light-years of Earth, astronomers discover that the most diverse life exists on a planet orbiting a young star that formed just 100 million years ago.

22. In 2040, a brilliant teenager working in her garage builds a coal-powered rocket that can travel at half the speed of light.

23. In the year 2750, we receive a signal from a civilization around a nearby star telling us that the *Voyager 2* spacecraft recently crash-landed on its planet.

24. Crew members of the matter–antimatter spacecraft *Star Apollo*, which left Earth in the year 2165, return to Earth in the year 2450, looking only a few years older than when they left.

25. Aliens from a distant star system invade Earth with the intent to destroy us and occupy our planet, but we successfully fight them off when their technology proves no match for ours.

26. A single great galactic civilization exists. It originated on a single planet long ago but is now made up of beings from many different planets, all assimilated into the galactic culture.

Quick Quiz

Choose the best answer to each of the following. Explain your reasoning with one or more complete sentences.

27. Fossil evidence suggests that life on Earth arose (a) almost immediately after Earth formed. (b) very soon after the end of the heavy bombardment. (c) about a billion years before the rise of the dinosaurs.

28. The theory of evolution is (a) a scientific theory backed by extensive evidence. (b) one of several competing scientific models that all seem equally successful in explaining the nature of life on Earth. (c) essentially just a guess about how life changes through time.

29. Plants and animals are (a) the two major forms of life on Earth. (b) the only organisms that have DNA. (c) just two small branches of the diverse "tree of life" on Earth.

30. Which of the following is a reason why early living organisms on Earth could not have survived on the surface? (a) the lack of an ozone layer (b) the lack of oxygen in the atmosphere (c) the fact that these organisms were single-celled

31. According to current understanding, the key requirement for life is (a) photosynthesis. (b) liquid water. (c) an ozone layer.

32. Which of the following worlds is *not* considered a candidate for harboring life? (a) Europa (b) Mars (c) the Moon

33. How does the habitable zone around a star of spectral type G compare to that around a star of spectral type M? (a) It is larger. (b) It is hotter. (c) It is closer to its star.

34. In the Drake equation, suppose that the term f_{life} was equal to $\frac{1}{2}$. What would this mean? (a) Half the stars in the Milky Way Galaxy have a planet with life. (b) Half of all life-forms in the universe are intelligent. (c) Half of the habitable worlds in the galaxy actually have life, while the other half don't.

35. The amount of energy that would be needed to accelerate a large spaceship to half the speed of light is (a) about 100 times the energy needed to launch the Space Shuttle. (b) more than 2000 times the current annual world energy consumption. (c) more than the amount of energy released by a supernova.

36. According to current scientific understanding, the idea that the Milky Way Galaxy might be home to a civilization millions of years more advanced than ours is (a) a virtual certainty. (b) extremely unlikely. (c) one reasonable answer to Fermi's paradox.

PROCESS OF SCIENCE

Examining How Science Works

37. *Extraordinary Claims.* As discussed in the chapter, both the *Viking* results and the study of a Martian meteorite led some scientists to think we had found evidence of life on Mars, even while most scientists disagreed. Those who disagree often point to Carl Sagan's dictum that "extraordinary claims require extraordinary evidence." Briefly discuss one of these cases, and decide whether you think extraordinary evidence should be required before it is accepted. What follow-up evidence might lead scientists to reevaluate their positions?

38. *Unanswered Questions.* In a sense, this entire chapter was about one big unanswered question: Are we alone in the universe? But as we attempt to answer that "big" question, there are many smaller questions that we might wish to answer along the way. Describe one currently unanswered question about life in the universe that we might be able to answer with new missions or experiments over the next couple of decades. What kinds of evidence will we need to answer the question? How will we know when it is answered?

GROUP WORK EXERCISE

39. *Habitable Planets?* In this exercise, you will rank several extrasolar planetary systems in order of the probability that they support life. Before you begin, assign the following roles to the people in your group: *Scribe* (takes notes on the group's activities), *Proposer* (proposes explanations to the group), *Skeptic* (points out weaknesses in proposed explanations), and *Moderator* (leads group discussion and makes sure everyone contributes). List the systems in order of most probable to least probable to support life, and explain the reasons for your ranking in each case.
a. a planet orbiting a star of spectral type B (approximately 10 solar masses) in a circular orbit with an average temperature expected to be 300 K **b.** a planet orbiting a Sun-like star in a circular orbit at a distance twice Earth's orbital distance from the Sun **c.** a planet orbiting a star with a luminosity one-quarter the Sun's luminosity in a circular orbit at a distance one-half Earth's orbital distance from the Sun **d.** a planet orbiting a Sun-like star in an elliptical orbit ranging from Earth's orbital distance to ten times Earth's orbital distance from the Sun

INVESTIGATE FURTHER

In-Depth Questions to Increase Your Understanding

Short-Answer/Essay Questions

40. *Artificial Selection.* Suppose you lived hundreds of years ago (before we knew about genetic engineering) and wanted to breed a herd of cows that would provide more milk than the cows in your current herd. How would you have gone about it? How would this process of "artificial selection" be similar to natural selection? How would it be different?

41. *Statistics of One.* Much of the search for life in the universe is based on what we know about life on Earth. Write a short essay discussing the pros and cons of basing our general understanding of biology on the example of Earth.

42. *Most Likely to Have Life.* Suppose you were asked to vote in a contest to name the world in our solar system (besides Earth) "most likely to have life." Which world would you cast your vote for? Explain and defend your choice in a one-page essay.

43. *Likely Suns.* Study the stellar data for nearby stars given in Appendix F, Table F.1. Which star on the list would you expect to have the largest habitable zone? Which would have the second-largest habitable zone? If we rule out multiple-star systems, which star would you expect to have the highest probability of having a habitable planet? Explain your answers.

44. *Are Earth-Like Planets Common?* Based on what you have learned in this book, do you think Earth-like planets will ultimately prove to be rare, common, or something in between? Write a one- to two-page essay explaining and defending your opinion.

45. *Solution to the Fermi Paradox.* Which of the various possible solutions to the question "Where are the aliens?" do you think is most likely? (If you have no opinion on their likelihood, which do you like best?) Write a one- to two-page essay in which you explain why you favor this solution.

46. *What's Wrong with This Picture?* Many science fiction stories have imagined the galaxy divided into a series of empires, each having arisen from a different civilization on a different world, that hold one another at bay because they all have about the same level of military technology. Is this a realistic scenario? Explain.

47. *Aliens in the Movies.* Choose a science fiction movie (or television show) that involves an alien species. Do you think aliens like this could really exist? Write a one- to two-page critical review of the movie, focusing primarily on the question of whether the movie portrays the aliens in a scientifically reasonable way.

Quantitative Problems

Be sure to show all calculations clearly and state your final answers in complete sentences.

48. *Nearest Civilization.*
    a. Suppose there are 10,000 civilizations in the Milky Way Galaxy. If the civilizations were randomly distributed throughout the disk of the galaxy, about how far (on average) would it be to the nearest civilization? (*Hint:* Start by finding the area of the Milky Way's disk, assuming that it is circular and 100,000 light-years in diameter. Then find the average area per civilization, and use the distance across this area to estimate the distance between civilizations.) **b.** Repeat part a, but this time assume that there are only 100 civilizations in the galaxy.

49. *SETI Search.* Suppose there are 10,000 civilizations broadcasting radio signals in the Milky Way Galaxy right now. On average, how many stars would we have to search before we would expect to hear a signal? Assume there are 500 billion stars in the galaxy. How does the answer change if there are only 100 civilizations instead of 10,000?

50. *SETI Signal.* Consider a civilization broadcasting a signal with a power of 10,000 watts. The Arecibo radio telescope, which is about 300 meters in diameter, could detect this signal if it was coming from as far away as 100 light-years. Suppose instead that the signal is being broadcast from the other side of the Milky Way Galaxy, about 70,000 light-years away. How large a radio telescope would we need to detect this signal? (*Hint:* Use the inverse square law for light.)

51. *Cruise Ship Energy.* Suppose we have a spaceship about the size of a typical ocean cruise ship today, which means it has a mass of about 100 million kilograms, and we want to accelerate the ship to a speed of 10% of the speed of light.
    a. How much energy would be required? (*Hint:* You can find the answer simply by calculating the kinetic energy of the ship when it reaches its cruising speed; because 10% of the speed of light is still small compared to the speed of light, you can use the formula that tells us that kinetic energy $= \frac{1}{2} \times m \times v^2$.) **b.** How does your answer compare to total world energy use at present, which is about 5×10^{22} joules per year? **c.** The typical cost of energy today is roughly 5¢ per 1 million joules. At this price, how much would it cost to generate the energy needed by this spaceship?

52. *Matter–Antimatter Engine.* Consider the spaceship from Problem 51. Suppose you wanted to generate the energy to get it to cruising speed using matter–antimatter annihilation. How much antimatter would you need to produce and take on the ship? (*Hint:* When matter and antimatter meet, they turn all their mass into energy equivalent to mc^2.)

Discussion Questions

53. *Funding the Search for Life.* Imagine that you are a member of Congress and your job includes deciding how much government funding goes to research in different areas of science. How much would you allot to the search for life in the universe compared to the amount allotted to research in other areas of astronomy and planetary science? Why?

54. *Distant Dream or Near Reality?* Considering all the issues surrounding interstellar flight, when (if ever) do you think we are likely to begin traveling among the stars? Why?

55. *The Turning Point.* Discuss the idea that our generation has acquired a greater responsibility for the future than any previous generation. Do you agree with this assessment? If so, how should we deal with this responsibility? Defend your opinions.

Web Projects

56. *Astrobiology News.* Go to NASA's astrobiology site and read some of the recent news about the search for life in the universe. Choose one article and write a one- to two-page summary of the research.

57. *Martian Meteorites.* Research the latest discoveries concerning Martian meteorites. Choose one discovery and write a short summary of how you think it alters the debate about the habitability of Mars.

58. *The Search for Extraterrestrial Intelligence.* Learn more about SETI at the SETI Institute home page, and summarize your findings in one page or less.

59. *Starship Design.* Read about a proposal for starship propulsion or design. How would the starship work? What new technologies would be needed, and what existing technologies could be applied? Write a one- to two-page report on your research.

60. *Advanced Spacecraft Technologies.* NASA supports many efforts to incorporate new technologies into spaceships. Although few of them are suitable for interstellar colonization, most are innovative and fascinating. Learn about one such project, and write a short overview of your findings.

Throughout this book, we have seen that the history of the universe has proceeded in a way that has made our existence on Earth possible. This figure summarizes some of the key ideas, and leads us to ask: If life arose here, shouldn't it also have arisen on many other worlds? We do not yet know the answer, but scientists are actively seeking to learn whether life is rare or common in the universe.

(1) The protons, neutrons, and electrons in the atoms that make up Earth and life were created out of pure energy during the first few moments after the Big Bang, leaving the universe filled with hydrogen and helium gas [Section 22.1].

electron

gamma-ray photons

antielectron

Matter can be created from energy: $E = mc^2$.

(2) Ripples in the density of the early universe were necessary for life to form later on. Without those ripples, matter would never have collected into galaxies, stars, and planets [Section 22.3].

RADIO

We observe the seeds of structure formation in the cosmic microwave background.

(3) The attractive force of gravity pulls together the matter that makes galaxies, stars, and planets [Section 4.4].

M_1 $F_g = G \dfrac{M_1 M_2}{d^2}$ M_2

d

Every piece of matter in the universe pulls on every other piece.

(4) Our planet and all the life on it is made primarily of elements formed by nuclear fusion in high-mass stars and dispersed into space by supernovae [Section 17.3].

^{16}O ^{28}Si

^{31}S
(16p, 15n)

^{56}Fe
(26p, 30n)

^{16}O ^{28}Si

High-mass stars have cores hot enough to make elements heavier than carbon.

5 Our galaxy is large enough to retain the elements ejected by supernovae, and it recycles them into new stars and planetary systems [Section 19.2].

New elements mix with the interstellar medium, which then forms new stars and planets.

6 Planets can form in gaseous disks of material around newly formed stars. Earth was built from heavy elements that condensed from the gas as particles of metal and rock, which then gradually accreted to become our planet [Section 8.2].

Terrestrial planets formed in warm, inner regions of the solar nebula; jovian planets formed in cooler, outer regions.

7 Life as we know it requires liquid water, so we define the habitable zone around a star to be the zone in which a suitably large planet can have liquid water on its surface [Section 24.3].

The Sun's habitable zone (green) occupies a region from beyond the orbit of Venus to near the orbit of Mars.

8 Early life has had the time needed to evolve into complex forms—including us—because the solar thermostat has kept the Sun shining steadily for billions of years [Section 14.2].

The solar thermostat keeps the Sun's fusion rate stable.

APPENDIXES

A   USEFUL NUMBERS

Astronomical Distances

1 AU $\approx 1.496 \times 10^8$ km $= 1.496 \times 10^{11}$ m

1 light-year $\approx 9.46 \times 10^{12}$ km $= 9.46 \times 10^{15}$ m

1 parsec (pc) $\approx 3.09 \times 10^{13}$ km ≈ 3.26 light-years

1 kiloparsec (kpc) $= 1000$ pc $\approx 3.26 \times 10^3$ light-years

1 megaparsec (Mpc) $= 10^6$ pc $\approx 3.26 \times 10^6$ light-years

Universal Constants

Speed of light: $\quad\quad\quad\quad c = 3.00 \times 10^5$ km/s $= 3 \times 10^8$ m/s

Gravitational constant: $\quad G = 6.67 \times 10^{-11} \dfrac{m^3}{kg \times s^2}$

Planck's constant: $\quad\quad\quad h = 6.63 \times 10^{-34}$ joule $\times$ s

Stefan-Boltzmann constant: $\quad \sigma = 5.67 \times 10^{-8} \dfrac{watt}{m^2 \times K^4}$

Mass of a proton: $\quad\quad\quad m_p = 1.67 \times 10^{-27}$ kg

Mass of an electron: $\quad\quad m_e = 9.11 \times 10^{-31}$ kg

Astronomical Times

1 solar day (average) $= 24^h$

1 sidereal day $\approx 23^h 56^m 4.09^s$

1 synodic month (average) ≈ 29.53 solar days

1 sidereal month (average) ≈ 27.32 solar days

1 tropical year ≈ 365.242 solar days

1 sidereal year ≈ 365.256 solar days

Energy and Power Units

Basic unit of energy: 1 joule $= 1 \dfrac{kg \times m^2}{s^2}$

Basic unit of power: 1 watt $= 1$ joule/s

Electron-volt: 1 eV $= 1.60 \times 10^{-19}$ joule

Useful Sun and Earth Reference Values

Mass of the Sun: $1 M_{Sun} \approx 2 \times 10^{30}$ kg

Radius of the Sun: $1 R_{Sun} \approx 696,000$ km

Luminosity of the Sun: $1 L_{Sun} \approx 3.8 \times 10^{26}$ watts

Mass of Earth: $1 M_{Earth} \approx 5.97 \times 10^{24}$ kg

Radius (equatorial) of Earth: $1 R_{Earth} \approx 6378$ km

Acceleration of gravity on Earth: $g = 9.8$ m/s^2

Escape velocity from surface of Earth: $v_{escape} = 11.2$ km/s $= 11,200$ m/s

B USEFUL FORMULAS

- Universal law of gravitation for the force between objects of mass M_1 and M_2, with distance d between their centers:

$$F = G\frac{M_1 M_2}{d^2}$$

- Newton's version of Kepler's third law, which applies to any pair of orbiting objects, such as a star and planet, a planet and moon, or two stars in a binary system; p is the orbital period, a is the distance between the centers of the orbiting objects, and M_1 and M_2 are the object masses:

$$p^2 = \frac{4\pi^2}{G(M_1 + M_2)}a^3$$

- Escape velocity at distance R from center of object of mass M:

$$v_{escape} = \sqrt{\frac{2GM}{R}}$$

- Relationship between a photon's wavelength (λ), frequency (f), and the speed of light (c):

$$\lambda \times f = c$$

- Energy of a photon of wavelength λ or frequency f:

$$E = hf = \frac{hc}{\lambda}$$

- Stefan-Boltzmann law for thermal radiation at temperature T (in Kelvin):

$$\text{emitted power per unit area} = \sigma T^4$$

- Wien's law for the peak wavelength (λ_{max}) thermal radiation at temperature T (on the Kelvin scale):

$$\lambda_{max} = \frac{2{,}900{,}000}{T}\text{ nm}$$

- Doppler shift (radial velocity is positive if the object is moving away from us and negative if it is moving toward us):

$$\frac{\text{radial velocity}}{\text{speed of light}} = \frac{\text{shifted wavelength} - \text{rest wavelength}}{\text{rest wavelength}}$$

- Angular separation (α) of two points with an actual separation s, viewed from a distance d (assuming d is much larger than s):

$$\alpha = \frac{s}{2\pi d} \times 360°$$

- Inverse square law for light (d is the distance to the object):

$$\text{apparent brightness} = \frac{\text{luminosity}}{4\pi d^2}$$

- Parallax formula (distance d to a star with parallax angle p in arcseconds):

$$d \text{ (in parsecs)} = \frac{1}{p \text{ (in arcseconds)}}$$

$$\text{or } d \text{ (in light-years)} = 3.26 \times \frac{1}{p \text{ (in arcseconds)}}$$

- The orbital velocity law, to find the mass M_r contained within the circular orbit of radius r for an object moving at speed v:

$$M_r = \frac{r \times v^2}{G}$$

C A FEW MATHEMATICAL SKILLS

This appendix reviews the following mathematical skills: powers of 10, scientific notation, working with units, the metric system, and finding a ratio. You should refer to this appendix as needed while studying the textbook.

C.1 Powers of 10

Powers of 10 indicate how many times to multiply 10 by itself. For example:

$$10^2 = 10 \times 10 = 100$$

$$10^6 = 10 \times 10 \times 10 \times 10 \times 10 \times 10 = 1{,}000{,}000$$

Negative powers are the reciprocals of the corresponding positive powers. For example:

$$10^{-2} = \frac{1}{10^2} = \frac{1}{100} = 0.01$$

$$10^{-6} = \frac{1}{10^6} = \frac{1}{1{,}000{,}000} = 0.000001$$

TABLE C.1 lists powers of 10 from 10^{-12} to 10^{12}. Note that powers of 10 follow two basic rules:

1. A positive exponent tells how many zeros follow the 1. For example, 10^0 is a 1 followed by no zeros, and 10^8 is a 1 followed by eight zeros.

2. A negative exponent tells how many places are to the right of the decimal point, including the 1. For example, $10^{-1} = 0.1$ has one place to the right of the decimal point; $10^{-6} = 0.000001$ has six places to the right of the decimal point.

Multiplying and Dividing Powers of 10

Multiplying powers of 10 simply requires adding exponents, as the following examples show:

$$10^4 \times 10^7 = \underbrace{10{,}000}_{10^4} \times \underbrace{10{,}000{,}000}_{10^7} = \underbrace{100{,}000{,}000{,}000}_{10^{4+7} = 10^{11}} = 10^{11}$$

$$10^5 \times 10^{-3} = \underbrace{100{,}000}_{10^5} \times \underbrace{0.001}_{10^{-3}} = \underbrace{100}_{10^{5+(-3)} = 10^2} = 10^2$$

$$10^{-8} \times 10^{-5} = \underbrace{0.00000001}_{10^{-8}} \times \underbrace{0.00001}_{10^{-5}} = \underbrace{0.0000000000001}_{10^{-8+(-5)} = 10^{-13}} = 10^{-13}$$

TABLE C.1   Powers of 10

Zero and Positive Powers			Negative Powers		
Power	Value	Name	Power	Value	Name
10^0	1	One			
10^1	10	Ten	10^{-1}	0.1	Tenth
10^2	100	Hundred	10^{-2}	0.01	Hundredth
10^3	1000	Thousand	10^{-3}	0.001	Thousandth
10^4	10,000	Ten thousand	10^{-4}	0.0001	Ten-thousandth
10^5	100,000	Hundred thousand	10^{-5}	0.00001	Hundred-thousandth
10^6	1,000,000	Million	10^{-6}	0.000001	Millionth
10^7	10,000,000	Ten million	10^{-7}	0.0000001	Ten-millionth
10^8	100,000,000	Hundred million	10^{-8}	0.00000001	Hundred-millionth
10^9	1,000,000,000	Billion	10^{-9}	0.000000001	Billionth
10^{10}	10,000,000,000	Ten billion	10^{-10}	0.0000000001	Ten-billionth
10^{11}	100,000,000,000	Hundred billion	10^{-11}	0.00000000001	Hundred-billionth
10^{12}	1,000,000,000,000	Trillion	10^{-12}	0.000000000001	Trillionth

Dividing powers of 10 requires subtracting exponents, as in the following examples:

$$\frac{10^5}{10^3} = \underbrace{100,000}_{10^5} \div \underbrace{1000}_{10^3} = \underbrace{100}_{10^{5-3} = 10^2} = 10^2$$

$$\frac{10^3}{10^7} = \underbrace{1000}_{10^3} \div \underbrace{10,000,000}_{10^7} = \underbrace{0.0001}_{10^{3-7} = 10^{-4}} = 10^{-4}$$

$$\frac{10^{-4}}{10^{-6}} = \underbrace{0.0001}_{10^{-4}} \div \underbrace{0.000001}_{10^{-6}} = \underbrace{100}_{10^{-4-(-6)} = 10^2} = 10^2$$

Powers of Powers of 10

We can use the multiplication and division rules to raise powers of 10 to other powers or to take roots. For example:

$$(10^4)^3 = 10^4 \times 10^4 \times 10^4 = 10^{4+4+4} = 10^{12}$$

Note that we can get the same end result by simply multiplying the two powers:

$$(10^4)^3 = 10^{4 \times 3} = 10^{12}$$

Because taking a root is the same as raising to a fractional power (e.g., the square root is the same as the $\frac{1}{2}$ power; the cube root is the same as the $\frac{1}{3}$ power, etc.), we can use the same procedure for roots, as in the following example:

$$\sqrt{10^4} = (10^4)^{1/2} = 10^{4 \times (1/2)} = 10^2$$

Adding and Subtracting Powers of 10

Unlike multiplying and dividing powers of 10, there is no shortcut for adding or subtracting powers of 10. The values must be written in longhand notation. For example:

$$10^6 + 10^2 = 1{,}000{,}000 + 100 = 1{,}000{,}100$$

$$10^8 + 10^{-3} = 100{,}000{,}000 + 0.001 = 100{,}000{,}000.001$$

$$10^7 - 10^3 = 10{,}000{,}000 - 1000 = 9{,}999{,}000$$

Summary

We can summarize our findings using n and m to represent any numbers:

- To *multiply* powers of 10, *add* exponents: $10^n \times 10^m = 10^{n+m}$

- To *divide* powers of 10, *subtract* exponents: $\dfrac{10^n}{10^m} = 10^{n-m}$

- To *raise* powers of 10 to other powers, multiply exponents: $(10^n)^m = 10^{n \times m}$

- To add or subtract powers of 10, first write them out longhand.

C.2 Scientific Notation

When we are dealing with large or small numbers, it's generally easier to write them with powers of 10. For example, it's much easier to write the number 6,000,000,000,000 as 6×10^{12}. This format, in which a number *between* 1 and 10 is multiplied by a power of 10, is called **scientific notation**.

Converting a Number to Scientific Notation

We can convert numbers written in ordinary notation to scientific notation with a simple two-step process:

1. Move the decimal point to come after the *first* nonzero digit.

2. The number of places the decimal point moves tells you the power of 10; the power is *positive* if the decimal point moves to the left and *negative* if it moves to the right.

Examples:

$$3042 \xrightarrow[\text{3 places to left}]{\text{decimal needs to move}} 3.042 \times 10^3$$

$$0.00012 \xrightarrow[\text{4 places to right}]{\text{decimal needs to move}} 1.2 \times 10^{-4}$$

$$226 \times 10^2 \xrightarrow[\text{2 places to left}]{\text{decimal needs to move}} (2.26 \times 10^2) \times 10^2 = 2.26 \times 10^4$$

Converting a Number from Scientific Notation

We can convert numbers written in scientific notation to ordinary notation by the reverse process:

1. The power of 10 indicates how many places to move the decimal point; move it to the *right* if the power of 10 is positive and to the *left* if it is negative.

2. If moving the decimal point creates any open places, fill them with zeros.

Examples:

$$4.01 \times 10^2 \xrightarrow[\text{2 places to right}]{\text{move decimal}} 401$$

$$3.6 \times 10^6 \xrightarrow[\text{6 places to right}]{\text{move decimal}} 3{,}600{,}000$$

$$5.7 \times 10^{-3} \xrightarrow[\text{3 places to left}]{\text{move decimal}} 0.0057$$

Multiplying or Dividing Numbers in Scientific Notation

Multiplying or dividing numbers in scientific notation simply requires operating on the powers of 10 and the other parts of the number separately.

Examples:

$$(6 \times 10^2) \times (4 \times 10^5) = (6 \times 4) \times (10^2 \times 10^5) = 24 \times 10^7 = (2.4 \times 10^1) \times 10^7 = 2.4 \times 10^8$$

$$\frac{4.2 \times 10^{-2}}{8.4 \times 10^{-5}} = \frac{4.2}{8.4} \times \frac{10^{-2}}{10^{-5}} = 0.5 \times 10^{-2-(-5)} = 0.5 \times 10^3 = (5 \times 10^{-1}) \times 10^3 = 5 \times 10^2$$

Note that, in both these examples, we first found an answer in which the number multiplied by a power of 10 was *not* between 1 and 10. We therefore followed the procedure for converting the final answer to scientific notation.

Addition and Subtraction with Scientific Notation

In general, we must write numbers in ordinary notation before adding or subtracting.

Examples:

$$(3 \times 10^6) + (5 \times 10^2) = 3{,}000{,}000 + 500 = 3{,}000{,}500 = 3.0005 \times 10^6$$

$$(4.6 \times 10^9) - (5 \times 10^8) = 4{,}600{,}000{,}000 - 500{,}000{,}000 = 4{,}100{,}000{,}000 = 4.1 \times 10^9$$

When both numbers have the *same* power of 10, we can factor out the power of 10 first.

Examples:

$$(7 \times 10^{10}) + (4 \times 10^{10}) = (7 + 4) \times 10^{10} = 11 \times 10^{10} = 1.1 \times 10^{11}$$

$$(2.3 \times 10^{-22}) - (1.6 \times 10^{-22}) = (2.3 - 1.6) \times 10^{-22} = 0.7 \times 10^{-22} = 7.0 \times 10^{-23}$$

C.3 Working with Units

Showing the units of a problem as you solve it usually makes the work much easier and also provides a useful way of checking your work. If an answer does not come out with the units you expect, you probably did something wrong. In general, working with units is very similar to working with numbers, as the following guidelines and examples show.

Five Guidelines for Working with Units

Before you begin any problem, think ahead and identify the units you expect for the final answer. Then operate on the units along with the numbers as you solve the problem. The following five guidelines may be helpful when you are working with units:

1. Mathematically, it doesn't matter whether a unit is singular (e.g., meter) or plural (e.g., meters); we can use the same abbreviation (e.g., m) for both.

2. You cannot add or subtract numbers unless they have the *same* units. For example, 5 apples + 3 apples = 8 apples, but the expression 5 apples + 3 oranges cannot be simplified further.

3. You *can* multiply units, divide units, or raise units to powers. Look for key words that tell you what to do.

   - *Per* suggests division. For example, we write a speed of 100 kilometers per hour as

   $$100 \, \frac{km}{hr} \quad \text{or} \quad 100 \, \frac{km}{1 \, hr}$$

   - *Of* suggests multiplication. For example, if you launch a 50-kg space probe at a launch cost *of* $10,000 per kilogram, the total cost is

   $$50 \, kg \times \frac{\$10{,}000}{kg} = \$500{,}000$$

- *Square* suggests raising to the second power. For example, we write an area of 75 square meters as 75 m^2.

- *Cube* suggests raising to the third power. For example, we write a volume of 12 cubic centimeters as 12 cm^3.

4. Often the number you are given is not in the units you wish to work with. For example, you may be given that the speed of light is 300,000 km/s but need it in units of m/s for a particular problem. To convert the units, simply multiply the given number by a *conversion factor:* a fraction in which the numerator (top of the fraction) and denominator (bottom of the fraction) are equal, so that the value of the fraction is 1; the number in the denominator must have the units that you wish to change. In the case of changing the speed of light from units of km/s to m/s, you need a conversion factor for kilometers to meters. Thus, the conversion factor is

$$\frac{1000 \text{ m}}{1 \text{ km}}$$

Note that this conversion factor is equal to 1, since 1000 meters and 1 kilometer are equal, and that the units to be changed (km) appear in the denominator. We can now convert the speed of light from units of km/s to m/s simply by multiplying by this conversion factor:

$$\underbrace{300{,}000 \ \frac{\text{km}}{\text{s}}}_{\substack{\text{speed of light} \\ \text{in km/s}}} \times \underbrace{\frac{1000 \text{ m}}{1 \text{ km}}}_{\substack{\text{conversion from} \\ \text{km to m}}} = \underbrace{3 \times 10^8 \ \frac{\text{m}}{\text{s}}}_{\substack{\text{speed of light} \\ \text{in m/s}}}$$

Note that the units of km cancel, leaving the answer in units of m/s.

5. It's easier to work with units if you replace division with multiplication by the reciprocal. For example, suppose you want to know how many minutes are represented by 300 seconds. We can find the answer by dividing 300 seconds by 60 seconds per minute:

$$300 \text{ s} \div 60 \ \frac{\text{s}}{\text{min}}$$

However, it is easier to see the unit cancellations if we rewrite this expression by replacing the division with multiplication by the reciprocal (this process is easy to remember as "invert and multiply"):

$$300 \text{ s} \div 60 \ \frac{\text{s}}{\text{min}} = 300 \text{ s} \times \underbrace{\frac{1 \text{ min}}{60 \text{ s}}}_{\substack{\text{invert} \\ \text{and multiply}}} = 5 \text{ min}$$

We now see that the units of seconds (s) cancel in the numerator of the first term and the denominator of the second term, leaving the answer in units of minutes.

More Examples of Working with Units

Example 1. How many seconds are there in 1 day?

Solution: We can answer the question by setting up a *chain* of unit conversions in which we start with 1 *day* and end up with *seconds*. We use the facts that there are 24 hours per day (24 hr/day), 60 minutes per hour (60 min/hr), and 60 seconds per minute (60 s/min):

$$1 \text{ day} \times \underbrace{\frac{24 \text{ hr}}{\text{day}}}_{\substack{\text{conversion} \\ \text{from} \\ \text{day to hr}}} \times \underbrace{\frac{60 \text{ min}}{\text{hr}}}_{\substack{\text{conversion} \\ \text{from} \\ \text{hr to min}}} \times \underbrace{\frac{60 \text{ s}}{\text{min}}}_{\substack{\text{conversion} \\ \text{from} \\ \text{min to s}}} = 86{,}400 \text{ s}$$

where 1 day is the starting value.

Note that all the units cancel except *seconds*, which is what we want for the answer. There are 86,400 seconds in 1 day.

Example 2. Convert a distance of 10^8 cm to km.

Solution: The easiest way to make this conversion is in two steps, since we know that there are 100 centimeters per meter (100 cm/m) and 1000 meters per kilometer (1000 m/km):

$$10^8 \text{ cm} \times \underbrace{\frac{1 \text{ m}}{100 \text{ cm}}}_{\substack{\text{conversion} \\ \text{from} \\ \text{cm to m}}} \times \underbrace{\frac{1 \text{ km}}{1000 \text{ m}}}_{\substack{\text{conversion} \\ \text{from} \\ \text{m to km}}} = 10^8 \text{ cm} \times \frac{1 \text{ m}}{10^2 \text{ cm}} \times \frac{1 \text{ km}}{10^3 \text{ m}} = 10^3 \text{ km}$$

where 10^8 cm is the starting value.

Alternatively, if we recognize that the number of kilometers should be smaller than the number of centimeters (because kilometers are larger), we might decide to do this conversion by dividing as follows:

$$10^8 \text{ cm} \div \frac{100 \text{ cm}}{\text{m}} \div \frac{1000 \text{ m}}{\text{km}}$$

In this case, before carrying out the calculation, we replace each division with multiplication by the reciprocal:

$$10^8 \text{ cm} \div \frac{100 \text{ cm}}{\text{m}} \div \frac{1000 \text{ m}}{\text{km}} = 10^8 \text{ cm} \times \frac{1 \text{ m}}{100 \text{ cm}} \times \frac{1 \text{ km}}{1000 \text{ m}}$$

$$= 10^8 \text{ cm} \times \frac{1 \text{ m}}{10^2 \text{ cm}} \times \frac{1 \text{ km}}{10^3 \text{ m}}$$

$$= 10^3 \text{ km}$$

Note that we again get the answer that 10^8 cm is the same as 10^3 km, or 1000 km.

Example 3. Suppose you accelerate at 9.8 m/s^2 for 4 seconds, starting from rest. How fast will you be going?

Solution: The question asked "how fast?" so we expect to end up with a speed. Therefore, we multiply the acceleration by the amount of time you accelerated:

$$9.8 \, \frac{\text{m}}{\text{s}^2} \times 4 \text{ s} = (9.8 \times 4) \, \frac{\text{m} \times \text{s}}{\text{s}^2} = 39.2 \, \frac{\text{m}}{\text{s}}$$

Note that the units end up as a speed, showing that you will be traveling 39.2 m/s after 4 seconds of acceleration at 9.8 m/s^2.

Example 4. A reservoir is 2 km long and 3 km wide. Calculate its area, in both square kilometers and square meters.

Solution: We find its area by multiplying its length and width:

$$2 \text{ km} \times 3 \text{ km} = 6 \text{ km}^2$$

Next we need to convert this area of 6 km² to square meters, using the fact that there are 1000 meters per kilometer (1000 m/km). Note that we must square the term 1000 m/km when converting from km² to m²:

$$6 \text{ km}^2 \times \left(1000 \, \frac{\text{m}}{\text{km}} \right)^2 = 6 \text{ km}^2 \times 1000^2 \, \frac{\text{m}^2}{\text{km}^2} = 6 \text{ km}^2 \times 1{,}000{,}000 \, \frac{\text{m}^2}{\text{km}^2}$$

$$= 6{,}000{,}000 \text{ m}^2$$

The reservoir area is 6 km², which is the same as 6 million m².

C.4 The Metric System (SI)

The modern version of the metric system, known as *Système Internationale d'Unites* (French for "International System of Units") or **SI**, was formally established in 1960. Today, it is the primary measurement system in nearly every country in the world with the exception of the United States. Even in the United States, it is the system of choice for science and international commerce. The basic units of length, mass, and time in the SI are

- The **meter** for length, abbreviated m

- The **kilogram** for mass, abbreviated kg

- The **second** for time, abbreviated s

Multiples of metric units are formed by powers of 10, using a prefix to indicate the power. For example, *kilo* means 10^3 (1000), so a kilometer is 1000 meters; a microgram is 0.000001 gram, because *micro* means 10^{-6}, or one millionth. Some of the more common prefixes are listed in **TABLE C.2**.

TABLE C.2 SI (Metric) Prefixes

	Small Values			Large Values	
Prefix	Abbreviation	Value	Prefix	Abbreviation	Value
Deci	d	10^{-1}	Deca	da	10^{1}
Centi	c	10^{-2}	Hecto	h	10^{2}
Milli	m	10^{-3}	Kilo	k	10^{3}
Micro	μ	10^{-6}	Mega	M	10^{6}
Nano	n	10^{-9}	Giga	G	10^{9}
Pico	p	10^{-12}	Tera	T	10^{12}

Metric Conversions

TABLE C.3 lists conversions between metric units and units used commonly in the United States. Note that the conversions between kilograms and pounds are valid only on Earth, because they depend on the strength of gravity.

Example 1. International athletic competitions generally use metric distances. Compare the length of a 100-meter race to that of a 100-yard race.

Solution: Table C.3 shows that 1 m = 1.094 yd, so 100 m is 109.4 yd. Note that 100 meters is almost 110 yards; a good "rule of thumb" to remember is that distances in meters are about 10% longer than the corresponding number of yards.

TABLE C.3 Metric Conversions

To Metric	From Metric
1 inch = 2.540 cm	1 cm = 0.3937 inch
1 foot = 0.3048 m	1 m = 3.28 feet
1 yard = 0.9144 m	1 m = 1.094 yards
1 mile = 1.6093 km	1 km = 0.6214 mile
1 pound = 0.4536 kg	1 kg = 2.205 pounds

Example 2. How many square kilometers are in 1 square mile?

Solution: We use the square of the miles-to-kilometers conversion factor:

$$(1 \text{ mi}^2) \times \left(\frac{1.6093 \text{ km}}{1 \text{ mi}}\right)^2 = (1 \text{ mi}^2) \times \left(1.6093^2 \frac{\text{km}^2}{\text{mi}^2}\right) = 2.5898 \text{ km}^2$$

Therefore, 1 square mile is 2.5898 square kilometers.

C.5 Finding a Ratio

Suppose you want to compare two quantities, such as the average density of Earth and the average density of Jupiter. The way we do such a comparison is by dividing, which tells us the *ratio* of the two quantities. In this case, Earth's average density is 5.52 g/cm^3 and Jupiter's average density is 1.33 g/cm^3 (see Figure 11.1), so the ratio is

$$\frac{\text{average density of Earth}}{\text{average density of Jupiter}} = \frac{5.52 \text{ g/cm}^3}{1.33 \text{ g/cm}^3} = 4.15$$

Notice how the units cancel on both the top and bottom of the fraction. We can state our result in two equivalent ways:

- The ratio of Earth's average density to Jupiter's average density is 4.15.

- Earth's average density is 4.15 times Jupiter's average density.

Sometimes, the quantities that you want to compare may each involve an equation. In such cases, you could, of course, find the ratio by first calculating each of the two quantities individually and then dividing. However, it is much easier if you first express the ratio as a fraction, putting the equation for one quantity on top and the other on the bottom. Some of the terms in the equation may then cancel out, making any calculations much easier.

Example 1. Compare the kinetic energy of a car traveling at 100 km/hr to that of the same car traveling at 50 km/hr.

Solution: We do the comparison by finding the ratio of the two kinetic energies, recalling that the formula for kinetic energy is $\frac{1}{2}mv^2$. Since we are not told the mass of the car, you might at first think that we don't have enough information to find the ratio. However, notice what happens when we put the equations for each kinetic energy into the ratio, calling the two speeds v_1 and v_2:

$$\frac{\text{K.E. car at } v_1}{\text{K.E. car at } v_2} = \frac{\frac{1}{2}m_{\text{car}} v_1^2}{\frac{1}{2}m_{\text{car}} v_2^2} = \frac{v_1^2}{v_2^2} = \left(\frac{v_1}{v_2}\right)^2$$

All the terms cancel except those with the two speeds, leaving us with a very simple formula for the ratio. Now we put in 100 km/hr for v_1 and 50 km/hr for v_2:

$$\frac{\text{K.E. car at 100 km/hr}}{\text{K.E. car at 50 km/hr}} = \left(\frac{100 \text{ km/hr}}{50 \text{ km/hr}}\right)^2 = 2^2 = 4$$

The ratio of the car's kinetic energies at 100 km/hr and 50 km/hr is 4. That is, the car has four times as much kinetic energy at 100 km/hr as it has at 50 km/hr.

Example 2. Compare the strength of gravity between Earth and the Sun to the strength of gravity between Earth and the Moon.

Solution: We do the comparison by taking the ratio of the Earth–Sun gravity to the Earth–Moon gravity. In this case, each quantity is found from the equation of Newton's law of gravity. (See Section 4.4.) Thus, the ratio is

$$\frac{\text{Earth–Sun gravity}}{\text{Earth–Moon gravity}} = \frac{G\dfrac{M_{\text{Earth}}M_{\text{Sun}}}{(d_{\text{Earth–Sun}})^2}}{G\dfrac{M_{\text{Earth}}M_{\text{Moon}}}{(d_{\text{Earth–Moon}})^2}} = \frac{M_{\text{Sun}}}{(d_{\text{Earth–Sun}})^2} \times \frac{(d_{\text{Earth–Moon}})^2}{M_{\text{Moon}}}$$

Note how all but four of the terms cancel; the last step comes from replacing the division with multiplication by the reciprocal (the "invert and multiply" rule for division). We can simplify the work further by rearranging the terms so that we have the masses and distances together:

$$\frac{\text{Earth–Sun gravity}}{\text{Earth–Moon gravity}} = \frac{M_{\text{Sun}}}{M_{\text{Moon}}} \times \frac{(d_{\text{Earth–Moon}})^2}{(d_{\text{Earth–Sun}})^2}$$

Now it is just a matter of looking up the numbers (see Appendix E) and calculating:

$$\frac{\text{Earth–Sun gravity}}{\text{Earth–Moon gravity}} = \frac{1.99 \times 10^{30} \text{ kg}}{7.35 \times 10^{22} \text{ kg}} \times \frac{(384.4 \times 10^3 \text{ km})^2}{(149.6 \times 10^6 \text{ km})^2} = 179$$

In other words, the Earth–Sun gravity is 179 times stronger than the Earth–Moon gravity.

D THE PERIODIC TABLE OF THE ELEMENTS

Key

12	—— Atomic number
Mg	—— Element's symbol
Magnesium	—— Element's name
24.305	—— Atomic mass*

*Atomic masses are fractions because they represent a weighted average of atomic masses of different isotopes—in proportion to the abundance of each isotope on Earth.

1 **H** Hydrogen 1.00794																	2 **He** Helium 4.003
3 **Li** Lithium 6.941	4 **Be** Beryllium 9.01218											5 **B** Boron 10.81	6 **C** Carbon 12.011	7 **N** Nitrogen 14.007	8 **O** Oxygen 15.999	9 **F** Fluorine 18.988	10 **Ne** Neon 20.179
11 **Na** Sodium 22.990	12 **Mg** Magnesium 24.305											13 **Al** Aluminum 26.98	14 **Si** Silicon 28.086	15 **P** Phosphorus 30.974	16 **S** Sulfur 32.06	17 **Cl** Chlorine 35.453	18 **Ar** Argon 39.948
19 **K** Potassium 39.098	20 **Ca** Calcium 40.08	21 **Sc** Scandium 44.956	22 **Ti** Titanium 47.88	23 **V** Vanadium 50.94	24 **Cr** Chromium 51.996	25 **Mn** Manganese 54.938	26 **Fe** Iron 55.847	27 **Co** Cobalt 58.9332	28 **Ni** Nickel 58.69	29 **Cu** Copper 63.546	30 **Zn** Zinc 65.39	31 **Ga** Gallium 69.72	32 **Ge** Germanium 72.59	33 **As** Arsenic 74.922	34 **Se** Selenium 78.96	35 **Br** Bromine 79.904	36 **Kr** Krypton 83.80
37 **Rb** Rubidium 85.468	38 **Sr** Strontium 87.62	39 **Y** Yttrium 88.9059	40 **Zr** Zirconium 91.224	41 **Nb** Niobium 92.91	42 **Mo** Molyhdenum 95.94	43 **Tc** Technetium (98)	44 **Ru** Ruthenium 101.07	45 **Rh** Rhodium 102.906	46 **Pd** Palladium 106.42	47 **Ag** Silver 107.868	48 **Cd** Cadmium 112.41	49 **In** Indium 114.82	50 **Sn** Tin 118.71	51 **Sb** Antimony 121.75	52 **Te** Tellurium 127.60	53 **I** Iodine 126.905	54 **Xe** Xenon 131.29
55 **Cs** Cesium 132.91	56 **Ba** Barium 137.34		72 **Hf** Hafnium 178.49	73 **Ta** Tantalum 180.95	74 **W** Tungsten 183.85	75 **Re** Rhenium 186.207	76 **Os** Osmium 190.2	77 **Ir** Iridium 192.22	78 **Pt** Platinum 195.08	79 **Au** Gold 196.967	80 **Hg** Mercury 200.59	81 **Tl** Thallium 204.383	82 **Pb** Lead 207.2	83 **Bi** Bismuth 208.98	84 **Po** Polonium (209)	85 **At** Astatine (210)	86 **Rn** Radon (222)
87 **Fr** Francium (223)	88 **Ra** Radium 226.0254		104 **Rf** Rutherfordium (263)	105 **Db** Dubnium (262)	106 **Sg** Seaborgium (266)	107 **Bh** Bohrium (267)	108 **Hs** Hassium (277)	109 **Mt** Meitnerium (268)	110 **Ds** Darmstadtium (281)	111 **Rg** Roentgenium (272)	112 **Cn** Copernicium (285)	113 **Uut** Ununtrium (284)	114 **Uuq** Ununquadium (289)	115 **Uup** Ununpentium (288)	116 **Uuh** Ununhexium (292)	117 **Uus** Ununseptium (294)	118 **Uuo** Ununoctium (294)

Lanthanide Series

57. **La** Lanthanum 138.906	58 **Ce** Cerium 140.12	59 **Pr** Praseodymium 140.908	60 **Nd** Neodymium 144.24	61 **Pm** Promethium (145)	62 **Sm** Samarium 150.36	63 **Eu** Europium 151.96	64 **Gd** Gadolinium 157.25	65 **Tb** Terbium 158.925	66 **Dy** Dysprosium 162.50	67 **Ho** Holmium 164.93	68 **Er** Erbium 167.26	69 **Tm** Thulium 168.934	70 **Yb** Ytterbium 173.04	71 **Lu** Lutetium 174.967

Actinide Series

89 **Ac** Actinium 227.028	90 **Th** Thorium 232.038	91 **Pa** Protactinium 231.036	92 **U** Uranium 238.029	93 **Np** Neptunium 237.048	94 **Pu** Plutonium (244)	95 **Am** Americium (243)	96 **Cm** Curium (247)	97 **Bk** Berkelium (247)	98 **Cf** Californium (251)	99 **Es** Einsteinium (252)	100 **Fm** Fermium (257)	101 **Md** Mendelevium (258)	102 **No** Nobelium (259)	103 **Lr** Lawrencium (260)

E PLANETARY DATA

TABLE E.1 **Physical Properties of the Sun and Planets**

Name	Radius (Eq[a]) (km)	Radius (Eq) (Earth units)	Mass (kg)	Mass (Earth units)	Average Density (g/cm³)	Surface Gravity (Earth = 1)	Escape Velocity (km/s)
Sun	695,000	109	1.99×10^{30}	333,000	1.41	27.5	—
Mercury	2440	0.382	3.30×10^{23}	0.055	5.43	0.38	4.43
Venus	6051	0.949	4.87×10^{24}	0.815	5.25	0.91	10.4
Earth	6378	1.00	5.97×10^{24}	1.00	5.52	1.00	11.2
Mars	3397	0.533	6.42×10^{23}	0.107	3.93	0.38	5.03
Jupiter	71,492	11.19	1.90×10^{27}	317.9	1.33	2.36	59.5
Saturn	60,268	9.46	5.69×10^{26}	95.18	0.70	0.92	35.5
Uranus	25,559	3.98	8.66×10^{25}	14.54	1.32	0.91	21.3
Neptune	24,764	3.81	1.03×10^{26}	17.13	1.64	1.14	23.6
Pluto[b]	1160	0.181	1.31×10^{22}	0.0022	2.05	0.07	1.25
Eris[b]	1430	0.22	1.66×10^{22}	0.0028	2.30	0.08	1.4

[a]Eq = equatorial.
[b]Under the IAU definitions of August 2006, Pluto and Eris are officially designated "dwarf planets."

TABLE E.2   Orbital Properties of the Sun and Planets

Name	Distance from Sun[a] (AU)	Distance from Sun[a] (10^6 km)	Orbital Period (years)	Orbital Inclination[b] (degrees)	Orbital Eccentricity	Sidereal Rotation Period (Earth days)[c]	Axis Tilt (degrees)
Sun	—	—	—	—	—	25.4	7.25
Mercury	0.387	57.9	0.2409	7.00	0.206	58.6	0.0
Venus	0.723	108.2	0.6152	3.39	0.007	−243.0	177.3
Earth	1.00	149.6	1.0	0.00	0.017	0.9973	23.45
Mars	1.524	227.9	1.881	1.85	0.093	1.026	25.2
Jupiter	5.203	778.3	11.86	1.31	0.048	0.41	3.08
Saturn	9.54	1427	29.5	2.48	0.056	0.44	26.73
Uranus	19.19	2870	84.01	0.77	0.046	−0.72	97.92
Neptune	30.06	4497	164.8	1.77	0.010	0.67	29.6
Pluto	39.48	5906	248.0	17.14	0.248	6.39	112.5
Eris	67.67	10,120	557	44.19	0.442	15.8	78

[a]Semimajor axis of the orbit.

[b]With respect to the ecliptic.

[c]A negative sign indicates rotation is backward relative to other planets.

TABLE E.3 Satellites of the Solar System (as of 2012)[a]

Planet Satellite	Radius or Dimensions[b] (km)	Distance from Planet (10³ km)	Orbital Period[c] (Earth days)	Mass[d] (kg)	Density[d] (g/cm³)	Notes About the Satellite
Earth						
Moon	1738	384.4	27.322	7.349×10^{22}	3.34	*Moon:* Probably formed in giant impact.
Mars						
Phobos	13 × 11 × 9	9.38	0.319	1.3×10^{16}	1.9	*Phobos, Deimos:* Probable captured asteroids.
Deimos	8 × 6 × 5	23.5	1.263	1.8×10^{15}	2.2	
Jupiter						
Small inner moons (4 moons)	8–83	128–222	0.295–0.674	—	—	*Metis, Adrastea, Amalthea, Thebe:* Small moonlets within and near Jupiter's ring system.
Io	1821	421.6	1.769	8.933×10^{22}	3.57	*Io:* Most volcanically active object in the solar system.
Europa	1565	670.9	3.551	4.797×10^{22}	2.97	*Europa:* Possible oceans under icy crust.
Ganymede	2634	1070.0	7.155	1.482×10^{23}	1.94	*Ganymede:* Largest satellite in solar system; unusual ice geology.
Callisto	2403	1883.0	16.689	1.076×10^{23}	1.86	*Callisto:* Cratered iceball.
Irregular group 1 (7 moons)	4–85	7500–17,000	130–457	—	—	*Themisto, Leda, Himalia, Lysithea, Elara, and others:* Probable captured moons with inclined orbits.
Irregular group 2 (52 moons)	1–30	17,000–29,000	490–980	—	—	*Ananke, Carme, Pasiphae, Sinope, and others:* Probable captured moons in inclined backward orbits.
Saturn						
Small inner moons (12)	3–89	117–212	0.5–1.2	—	—	*Pan, Atlas, Prometheus, Pandora, Epimetheus, Janus, and others:* Small moonlets within and near Saturn's ring system.
Mimas	199	185.52	0.942	3.70×10^{19}	1.17	*Mimas, Enceladus, Tethys:* Small and medium-size iceballs, many with interesting geology.
Enceladus	249	238.02	1.370	1.2×10^{20}	1.24	
Tethys	530	294.66	1.888	6.17×10^{20}	1.26	
Calypso and Telesto	8–12	294.66	1.888	—	—	*Calypso and Telesto:* Small moonlets sharing Tethys's orbit.
Dione	559	377.4	2.737	1.08×10^{21}	1.44	*Dione:* Medium-size iceball, with interesting geology.
Helene and Polydeuces	2–16	377.4	2.737	1.6×10^{16}	—	*Helene and Polydeuces:* Small moonlets sharing Dione's orbit.
Rhea	764	527.04	4.518	2.31×10^{21}	1.33	*Rhea:* Medium-size iceball, with interesting geology.
Titan	2575	1221.85	15.945	1.35×10^{23}	1.88	*Titan:* Dense atmosphere shrouds surface; ongoing geological activity.
Hyperion	180 × 140 × 112	1481.1	21.277	2.8×10^{19}	—	*Hyperion:* Only satellite known not to rotate synchronously.
Iapetus	718	3561.3	79.331	1.59×10^{21}	1.21	*Iapetus:* Bright and dark hemispheres show greatest contrast in the solar system.

Name	Radius (km)	Distance (10³ km)	Period (days)	Mass (kg)	Density (g/cm³)	Description
Phoebe	110	12,952	−550.4	1×10^{19}	—	*Phoebe:* Very dark; material ejected from Phoebe may coat one side of Iapetus.
Irregular groups (37 moons)	2–16	11,300–25,200	450–930 / −550 to −150	—	—	Probable captured moons with highly inclined and/or backward orbits.
Uranus						
Small inner moons (13 moons)	5–81	49–98	0.3–0.9	—	—	*Cordelia, Ophelia, Bianca, Cressida, Desdemona, Juliet, Portia, Rosalind, Cupid, Belinda, Perdita, Puck, Mab:* Small moonlets within and near Uranus's ring system.
Miranda	236	129.8	1.413	6.6×10^{19}	1.26	
Ariel	579	191.2	2.520	1.35×10^{21}	1.65	*Miranda, Ariel, Umbriel, Titania, Oberon:* Small and medium-size
Umbriel	584.7	266.0	4.144	1.17×10^{21}	1.44	iceballs, with some interesting geology.
Titania	788.9	435.8	8.706	3.52×10^{21}	1.59	
Oberon	761.4	582.6	13.463	3.01×10^{21}	1.50	
Irregular group (9 moons)	5–95	4280–21,000	260–2800	—	—	*Francisco, Caliban, Stephano, Trinculo, Sycorax, Margaret, Prospero, Setebos, Ferdinand:* Probable captured moons; several in backward orbits.
Neptune						
Small inner moons (5 moons)	29–96	48–74	0.30–0.55	—	—	*Naiad, Thalassa, Despina, Galatea, Larissa:* Small moonlets within and near Neptune's ring system.
Proteus	218 × 208 × 201	117.6	1.121	6×10^{19}	—	
Triton	1352.6	354.59	−5.875	2.14×10^{22}	2.0	*Triton:* Probable captured Kuiper belt object—largest captured object in solar system.
Nereid	170	5588.6	360.125	3.1×10^{19}	—	*Nereid:* Small, icy moon; very little known.
Irregulars (5 moons)	15–27	16,600–49,300	1880–9750	—	—	*2002 N1, N2, N3, N4, 2003 N1:* Possible captured moons in inclined or backward orbit.
Pluto						
Charon	593	17.5	6.4	1.56×10^{21}	1.6	*Charon:* Unusually large compared to Pluto; may have formed in giant impact.
P5	5	42.0	20.2	—	—	
Nix	50	48.7	24.9	—	—	*P5, Nix, P4, Hydra:* Newly discovered moons outside Charon's orbit.
P4	10	59.0	32.1	—	—	
Hydra	75	64.7	38.2	—	—	
Eris						
Dysnomia	50	37.4	15.8	—	—	*Dysnomia:* Approximate properties determined in June 2007.

[a] *Note:* Authorities differ substantially on many of the values in this table.

[b] $a \times b \times c$ values for the dimensions are the approximate lengths of the axes (center to edge) for irregular moons.

[c] Negative sign indicates backward orbit.

[d] Masses and densities are most accurate for those satellites visited by a spacecraft on a flyby. Masses for the smallest moons have not been measured but can be estimated from the radius and an assumed density.

F STELLAR DATA

TABLE F.1   Stars Within 12 Light-Years

Star	Distance (ly)	Spectral Type		RA h	RA m	Dec °	Dec '	Luminosity (L/L_{Sun})
Sun	0.000016	G2	V	—	—	—	—	1.0
Proxima Centauri	4.2	M5.0	V	14	30	−62	41	0.0006
α Centauri A	4.4	G2	V	14	40	−60	50	1.6
α Centauri B	4.4	K0	V	14	40	−60	50	0.53
Barnard's Star	6.0	M4	V	17	58	+04	42	0.005
Wolf 359	7.8	M5.5	V	10	56	+07	01	0.0008
Lalande 21185	8.3	M2	V	11	03	+35	58	0.03
Sirius A	8.6	A1	V	06	45	−16	42	26.0
Sirius B	8.6	DA2	White dwarf	06	45	−16	42	0.002
BL Ceti	8.7	M5.5	V	01	39	−17	57	0.0009
UV Ceti	8.7	M6	V	01	39	−17	57	0.0006
Ross 154	9.7	M3.5	V	18	50	−23	50	0.004
Ross 248	10.3	M5.5	V	23	42	+44	11	0.001
ϵ Eridani	10.5	K2	V	03	33	−09	28	0.37
Lacaille 9352	10.7	M1.0	V	23	06	−35	51	0.05
Ross 128	10.9	M4	V	11	48	+00	49	0.003
EZ Aquarii A	11.3	M5	V	22	39	−15	18	0.0006
EZ Aquarii B	11.3	—	—	22	39	−15	18	0.0004
EZ Aquarii C	11.3	—	—	22	39	−15	18	0.0003
61 Cygni A	11.4	K5	V	21	07	+38	42	0.17
61 Cygni B	11.4	K7	V	21	07	+38	42	0.10
Procyon A	11.4	F5	IV–V	07	39	+05	14	8.6
Procyon B	11.4	DA	White dwarf	07	39	+05	14	0.0005
Gliese 725 A	11.5	M3	V	18	43	+59	38	0.02
Gliese 725 B	11.5	M3.5	V	18	43	+59	38	0.01
GX Andromedae	11.6	M1.5	V	00	18	+44	01	0.03
GQ Andromedae	11.6	M3.5	V	00	18	+44	01	0.003
ϵ Indi A	11.8	K5	V	22	03	−56	45	0.30
ϵ Indi B	11.8	T1.0	Brown dwarf	22	04	−56	46	—
ϵ Indi C	11.8	T6.0	Brown dwarf	22	04	−56	46	—
DX Cancri	11.8	M6.0	V	08	30	+26	47	0.0003
τ Ceti	11.9	G8.5	V	01	44	−15	57	0.67
GJ 1061	12.0	M5.0	V	03	36	−44	31	0.001

Note: These data were provided by the RECONS project, courtesy of Dr. Todd Henry (January, 2010). The luminosities are all total (bolometric) luminosities. The DA stellar types are white dwarfs. The coordinates are for the year 2000. The bolometric luminosity of the brown dwarfs is primarily in the infrared and has not been measured accurately yet.

TABLE F.2 Twenty Brightest Stars

Star	Constellation	RA h	RA m	Dec °	Dec ′	Distance (ly)	Spectral Type		Apparent Magnitude	Luminosity (L/L_{Sun})
Sirius	Canis Major	6	45	−16	42	8.6	A1	V	−1.46	26
Canopus	Carina	6	24	−52	41	313	F0	Ib–II	−0.72	13,000
α Centauri	Centaurus	14	40	−60	50	4.4	G2	V	−0.01	1.6
							K0	V	1.3	0.53
Arcturus	Boötes	14	16	+19	11	37	K2	III	−0.06	170
Vega	Lyra	18	37	+38	47	25	A0	V	0.04	60
Capella	Auriga	5	17	+46	00	42	G0	III	0.75	70
							G8	III	0.85	77
Rigel	Orion	5	15	−08	12	772	B8	Ia	0.14	70,000
Procyon	Canis Minor	7	39	+05	14	11.4	F5	IV–V	0.37	7.4
Betelgeuse	Orion	5	55	+07	24	427	M2	Iab	0.41	38,000
Achernar	Eridanus	1	38	−57	15	144	B5	V	0.51	3600
Hadar	Centaurus	14	04	−60	22	525	B1	III	0.63	100,000
Altair	Aquila	19	51	+08	52	17	A7	IV–V	0.77	10.5
Acrux	Crux	12	27	−63	06	321	B1	IV	1.39	22,000
							B3	V	1.9	7500
Aldebaran	Taurus	4	36	+16	30	65	K5	III	0.86	350
Spica	Virgo	13	25	−11	09	260	B1	V	0.91	23,000
Antares	Scorpio	16	29	−26	26	604	M1	Ib	0.92	38,000
Pollux	Gemini	7	45	+28	01	34	K0	III	1.16	45
Fomalhaut	Piscis Austrinus	22	58	29	37	25	A3	V	1.19	18
Deneb	Cygnus	20	41	+45	16	2500	A2	Ia	1.26	170,000
β Crucis	Crux	12	48	−59	40	352	B0.5	IV	1.28	37,000

Note: Three of the stars on this list, Capella, α Centauri, and Acrux, are binary systems with members of comparable brightness. They are counted as single stars because that is how they appear to the naked eye. All the luminosities given are total (bolometric) luminosities. The coordinates are for the year 2000.

G GALAXY DATA

TABLE G.1 Galaxies of the Local Group

Galaxy Name	Distance (millions of ly)	Type[a]	RA h	RA m	Dec °	Dec ′	Luminosity (millions of L_{Sun})
Milky Way	—	Sbc	—	—	—	—	15,000
WLM	3.0	Irr	00	02	−15	30	50
IC 10	2.7	dIrr	00	20	+59	18	160
Cetus	2.5	dE	00	26	−11	02	0.72
NGC 147	2.4	dE	00	33	+48	30	131
And III	2.5	dE	00	35	+36	30	1.1
NGC 185	2.0	dE	00	39	+48	20	120
NGC 205	2.7	E	00	40	+41	41	370
And VIII	2.7	dE	00	42	+40	37	240
M32	2.6	E	00	43	+40	52	380
M31	2.5	Sb	00	43	+41	16	21,000
And I	2.6	dE	00	46	+38	00	4.7
SMC	0.19	Irr	00	53	−72	50	230
And IX	2.9	dE	00	52	+43	12	—
Sculptor	0.26	dE	01	00	−33	42	2.2
LGS 3	2.6	dIrr	01	04	+21	53	1.3
IC 1613	2.3	Irr	01	05	+02	08	64
And V	2.9	dE	01	10	+47	38	—
And II	1.7	dE	01	16	+33	26	2.4
M33	2.7	Sc	01	34	+30	40	2800
Phoenix	1.5	dIrr	01	51	−44	27	0.9
Fornax	0.45	dE	02	40	−34	27	15.5
EGB0427 + 63	4.3	dIrr	04	32	+63	36	9.1
LMC	0.16	Irr	05	24	−69	45	1300
Carina	0.33	dE	06	42	−50	58	0.4
Canis Major	0.025	dIrr	07	15	−28	00	—
Leo A	2.2	dIrr	09	59	+30	45	3.0
Sextans B	4.4	dIrr	10	00	+05	20	41
NGC 3109	4.1	Irr	10	03	−26	09	160
Antlia	4.0	dIrr	10	04	−27	19	1.7
Leo I	0.82	dE	10	08	+12	18	4.8
Sextans A	4.7	dIrr	10	11	−04	42	56
Sextans	0.28	dE	10	13	−01	37	0.5
Leo II	0.67	dE	11	13	+22	09	0.6
GR 8	5.2	dIrr	12	59	+14	13	3.4
Ursa Minor	0.22	dE	15	09	+67	13	0.3
Draco	2.7	dE	17	20	+57	55	0.3
Sagittarius	0.08	dE	18	55	−30	29	18
SagDIG	3.5	dIrr	19	30	−17	41	6.8
NGC 6822	1.6	Irr	19	45	−14	48	94
DDO 210	2.6	dIrr	20	47	−12	51	0.8
IC 5152	5.2	dIrr	22	03	−51	18	70
Tucana	2.9	dE	22	42	−64	25	0.5
UKS2323-326	4.3	dE	23	26	−32	23	5.2
And VII	2.6	dE	23	38	+50	35	—
Pegasus	3.1	dIrr	23	29	+14	45	12
And VI	2.8	dE	23	52	+24	36	—

[a]Types beginning with S are spiral galaxies classified according to Hubble's system (see Chapter 20). Type E galaxies are elliptical or spheroidal. Type Irr galaxies are irregular. The prefix d denotes a dwarf galaxy. This list is based on a list originally published by M. Mateo in 1998 and augmented by discoveries of Local Group galaxies made between 1998 and 2005.

TABLE G.2 Nearby Galaxies in the Messier Catalog[a, b]

Galaxy Name (M / NGC)[c]	RA h	RA m	Dec °	Dec ′	RV_{hel}[d]	RV_{gal}[e]	Type[f]	Nickname
M31 / NGC 224	00	43	+41	16	−300 ± 4	−122	Spiral	Andromeda
M32 / NGC 221	00	43	+40	52	−145 ± 2	32	Elliptical	
M33 / NGC 598	01	34	+30	40	−179 ± 3	−44	Spiral	Triangulum
M49 / NGC 4472	12	30	+08	00	997 ± 7	929	Elliptical/Lenticular/Seyfert	
M51 / NGC 5194	13	30	+47	12	463 ± 3	550	Spiral/Interacting	Whirlpool
M58 / NGC 4579	12	38	+11	49	1519 ± 6	1468	Spiral/Seyfert	
M59 / NGC 4621	12	42	+11	39	410 ± 6	361	Elliptical	
M60 / NGC 4649	12	44	+11	33	1117 ± 6	1068	Elliptical	
M61 / NGC 4303	12	22	+04	28	1566 ± 2	1483	Spiral/Seyfert	
M63 / NGC 5055	13	16	+42	02	504 ± 4	570	Spiral	Sunflower
M64 / NGC 4826	12	57	+21	41	408 ± 4	400	Spiral/Seyfert	Black Eye
M65 / NGC 3623	11	19	+13	06	807 ± 3	723	Spiral	
M66 / NGC 3627	11	20	+12	59	727 ± 3	643	Spiral/Seyfert	
M74 / NGC 628	01	37	+15	47	657 ± 1	754	Spiral	
M77 / NGC 1068	02	43	−00	01	1137 ± 3	1146	Spiral/Seyfert	
M81 / NGC 3031	09	56	+69	04	−34 ± 4	73	Spiral/Seyfert	
M82 / NGC 3034	09	56	+69	41	203 ± 4	312	Irregular/Starburst	
M83 / NGC 5236	13	37	−29	52	516 ± 4	385	Spiral/Starburst	
M84 / NGC 4374	12	25	+12	53	1060 ± 6	1005	Elliptical	
M85 / NGC 4382	12	25	+18	11	729 ± 2	692	Spiral	
M86 / NGC 4406	12	26	+12	57	−244 ± 5	−298	Elliptical/Lenticular	
M87 / NGC 4486	12	30	+12	23	1307 ± 7	1254	Elliptical/Central Dominant/Seyfert	Virgo A
M88 / NGC 4501	12	32	+14	25	2281 ± 3	2235	Spiral/Seyfert	
M89 / NGC 4552	12	36	+12	33	340 ± 4	290	Elliptical	
M90 / NGC 4569	12	37	+13	10	−235 ± 4	−282	Spiral/Seyfert	
M91 / NGC 4548	12	35	+14	30	486 ± 4	442	Spiral/Seyfert	
M94 / NGC 4736	12	51	+41	07	308 ± 1	360	Spiral	
M95 / NGC 3351	10	44	+11	42	778 ± 4	677	Spiral/Starburst	
M96 / NGC 3368	10	47	+11	49	897 ± 4	797	Spiral/Seyfert	
M98 / NGC 4192	12	14	+14	54	−142 ± 4	−195	Spiral/Seyfert	
M99 / NGC 4254	12	19	+14	25	2407 ± 3	2354	Spiral	
M100 / NGC 4321	12	23	+15	49	1571 ± 1	1525	Spiral	
M101 / NGC 5457	14	03	+54	21	241 ± 2	360	Spiral	
M104 / NGC 4594	12	40	−11	37	1024 ± 5	904	Spiral/Seyfert	Sombrero
M105 / NGC 3379	10	48	+12	35	911 ± 2	814	Elliptical	
M106 / NGC 4258	12	19	+47	18	448 ± 3	507	Spiral/Seyfert	
M108 / NGC 3556	11	09	+55	57	695 ± 3	765	Spiral	
M109 / NGC 3992	11	55	+53	39	1048 ± 4	1121	Spiral	
M110 / NGC 205	00	40	+41	41	−241 ± 3	−61	Elliptical	

[a]Galaxies identified in the catalog published by Charles Messier in 1781; these galaxies are relatively easy to observe with small telescopes.

[b]Data obtained from NED: NASA/IPAC Extragalactic Database (http://ned.ipac.caltech.edu). The original Messier list of galaxies was obtained from SED, and the list data were updated to 2001 and M102 was dropped.

[c]The galaxies are identified by their Messier number (M followed by a number) and NGC number, which comes from the *New General Catalog* published in 1888.

[d]Radial velocity in kilometers per second, with respect to the Sun (heliocentric). Positive values mean motion away from the Sun; negative values are toward the Sun.

[e]Radial velocity in kilometers per second, with respect to the Milky Way Galaxy, calculated from the RV_{hel} values with a correction for the Sun's motion around the galactic center.

[f]Galaxies are first listed by their primary type (spiral, elliptical, or irregular) and then by any other special categories that apply (see Chapter 20).

TABLE G.3   Nearby, X-Ray Bright Clusters of Galaxies

Cluster Name	Redshift	Distance[a] (billions of ly)	Temperature of Intracluster Medium (millions of K)	Average Orbital Velocity of Galaxies[b] (km/s)	Cluster Mass[c] ($10^{15} M_{Sun}$)
Abell 2142	0.0907	1.26	101. ± 2	1132 ± 110	1.6
Abell 2029	0.0766	1.07	100. ± 3	1164 ± 98	1.5
Abell 401	0.0737	1.03	95.2 ± 5	1152 ± 86	1.4
Coma	0.0233	0.32	95.1 ± 1	821 ± 49	1.4
Abell 754	0.0539	0.75	93.3 ± 3	662 ± 77	1.4
Abell 2256	0.0589	0.82	87.0 ± 2	1348 ± 86	1.4
Abell 399	0.0718	1.00	81.7 ± 7	1116 ± 89	1.1
Abell 3571	0.0395	0.55	81.1 ± 3	1045 ± 109	1.1
Abell 478	0.0882	1.23	78.9 ± 2	904 ± 281	1.1
Abell 3667	0.0566	0.79	78.5 ± 6	971 ± 62	1.1
Abell 3266	0.0599	0.84	78.2 ± 5	1107 ± 82	1.1
Abell 1651a	0.0846	1.18	73.1 ± 6	685 ± 129	0.96
Abell 85	0.0560	0.78	70.9 ± 2	969 ± 95	0.92
Abell 119	0.0438	0.61	65.6 ± 5	679 ± 106	0.81
Abell 3558	0.0480	0.67	65.3 ± 2	977 ± 39	0.81
Abell 1795	0.0632	0.88	62.9 ± 2	834 ± 85	0.77
Abell 2199	0.0314	0.44	52.7 ± 1	801 ± 92	0.59
Abell 2147	0.0353	0.49	51.1 ± 4	821 ± 68	0.56
Abell 3562	0.0478	0.67	45.7 ± 8	736 ± 49	0.48
Abell 496	0.0325	0.45	45.3 ± 1	687 ± 89	0.47
Centaurus	0.0103	0.14	42.2 ± 1	863 ± 34	0.42
Abell 1367	0.0213	0.30	41.3 ± 2	822 ± 69	0.41
Hydra	0.0126	0.18	38.0 ± 1	610 ± 52	0.36
C0336	0.0349	0.49	37.4 ± 1	650 ± 170	0.35
Virgo	0.0038	0.05	25.7 ± 0.5	632 ± 41	0.20

Note: This table lists the 25 brightest clusters of galaxies in the X-ray sky from a catalog by J. P. Henry (2000).

[a]Cluster distances were computed using a value for Hubble's constant of 21.5 km/s/million light-years.

[b]The average orbital velocities given in this column are the velocity component along our line of sight. This velocity should be multiplied by the square root of 2 to get the average orbital velocity used in Mathematical Insights 23.2 and 23.3.

[c]This column gives each cluster's mass within the largest radius at which the intracluster medium can be in gravitational equilibrium. Because our estimates of that radius depend on Hubble's constant, these masses are inversely proportional to Hubble's constant, which we have assumed to be 21.5 km/s/million light-years.

Constellation Locations

These two charts each show half of the celestial sphere in projection, so that you can use them to learn the approximate locations of the constellations. The grid lines are marked by the right ascension and declination.

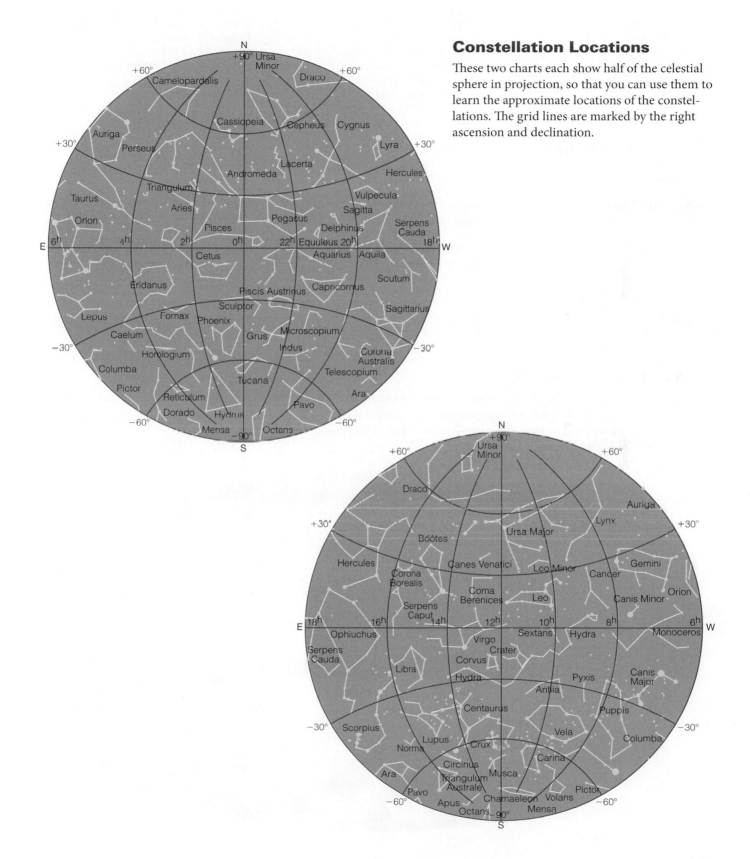

Constellation Names (English Equivalent in Parentheses)

Andromeda (The Chained Princess)
Antlia (The Air Pump)
Apus (The Bird of Paradise)
Aquarius (The Water Bearer)
Aquila (The Eagle)
Ara (The Altar)
Aries (The Ram)
Auriga (The Charioteer)
Boötes (The Herdsman)
Caelum (The Chisel)
Camelopardalis (The Giraffe)
Cancer (The Crab)
Canes Venatici (The Hunting Dogs)
Canis Major (The Great Dog)
Canis Minor (The Little Dog)
Capricornus (The Sea Goat)
Carina (The Keel)
Cassiopeia (The Queen)
Centaurus (The Centaur)
Cepheus (The King)
Cetus (The Whale)
Chamaeleon (The Chameleon)
Circinus (The Drawing Compass)
Columba (The Dove)
Coma Berenices (Berenice's Hair)
Corona Australis (The Southern Crown)
Corona Borealis (The Northern Crown)
Corvus (The Crow)
Crater (The Cup)
Crux (The Southern Cross)

Cygnus (The Swan)
Delphinus (The Dolphin)
Dorado (The Goldfish)
Draco (The Dragon)
Equuleus (The Little Horse)
Eridanus (The River)
Fornax (The Furnace)
Gemini (The Twins)
Grus (The Crane)
Hercules
Horologium (The Clock)
Hydra (The Sea Serpent)
Hydrus (The Water Snake)
Indus (The Indian)
Lacerta (The Lizard)
Leo (The Lion)
Leo Minor (The Little Lion)
Lepus (The Hare)
Libra (The Scales)
Lupus (The Wolf)
Lynx (The Lynx)
Lyra (The Lyre)
Mensa (The Table)
Microscopium (The Microscope)
Monoceros (The Unicorn)
Musca (The Fly)
Norma (The Level)
Octans (The Octant)
Ophiuchus (The Serpent Bearer)
Orion (The Hunter)

Pavo (The Peacock)
Pegasus (The Winged Horse)
Perseus (The Hero)
Phoenix (The Phoenix)
Pictor (The Painter's Easel)
Pisces (The Fish)
Piscis Austrinus (The Southern Fish)
Puppis (The Stern)
Pyxis (The Compass)
Reticulum (The Reticle)
Sagitta (The Arrow)
Sagittarius (The Archer)
Scorpius (The Scorpion)
Sculptor (The Sculptor)
Scutum (The Shield)
Serpens (The Serpent)
Sextans (The Sextant)
Taurus (The Bull)
Telescopium (The Telescope)
Triangulum (The Triangle)
Triangulum Australe (The Southern Triangle)
Tucana (The Toucan)
Ursa Major (The Great Bear)
Ursa Minor (The Little Bear)
Vela (The Sail)
Virgo (The Virgin)
Volans (The Flying Fish)
Vulpecula (The Fox)

All-Sky Constellation Map

This map of the entire sky shows the locations of all the constellations, in much the same way that a world map shows all of the countries on Earth. It does not use the usual celestial coordinate system of right ascension and declination, but instead is oriented so that the Milky Way Galaxy's center is at the center of the map and the Milky Way's disk (shown in shades of lighter blue) stretches from left to right across the map.

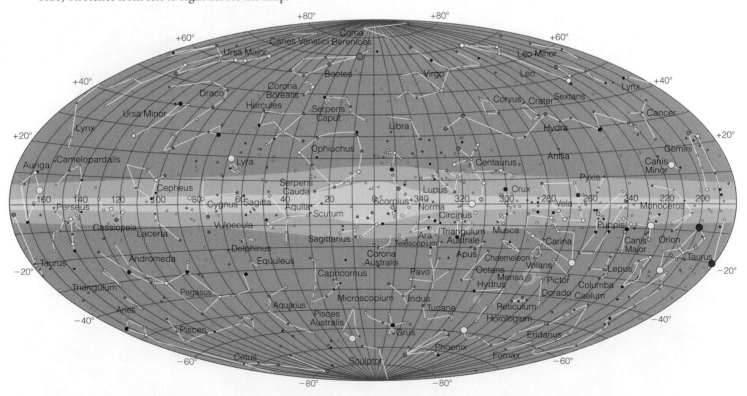

STAR CHARTS

How to use the star charts:

Check the times and dates under each chart to find the best one for you. Take it outdoors within an hour or so of the time listed for your date. Bring a dim flashlight to help you read it.

On each chart, the round outside edge represents the horizon all around you. Compass directions around the horizon are marked in yellow. Turn the chart around so that the edge marked with the direction you're facing (for example, north, southeast) is down. The stars above this horizon now match the stars you are facing. Ignore the rest until you turn to look in a different direction.

The center of the chart represents the sky overhead, so a star plotted on the chart halfway from the edge to the center can be found in the sky halfway from the horizon to straight up.

The charts are drawn for 40°N latitude (for example, Denver, New York, Madrid). If you live far south of there, stars in the southern part of your sky will appear higher than on the chart and stars in the north will be lower. If you live far north of there, the reverse is true.

Jan.–March
© Sky Publishing Corp.

Use this chart January, February, and March.

Early January—1 A.M.   Early February—11 P.M.   Early March—9 P.M.
Late January—Midnight   Late February—10 P.M.   Late March—Dusk

Apr.–June

© Sky Publishing Corp.

© 1999 Sky & Telescope

Use this chart April, May, and June.

Early April—3 A.M.*   Early May—1 A.M.*   Early June—11 P.M.*
Late April—2 A.M.*   Late May—Midnight*   Late June—Dusk

*Daylight Saving Time

July–Sept.

Use this chart July, August, and September.

Early July—1 A.M.*   Early August—11 P.M.*   Early September—9 P.M.*
Late July—Midnight*   Late August—10 P.M.*   Late September—Dusk

*Daylight Saving Time

© 1999 Sky & Telescope

Use this chart October, November, and December.

Early October—1 A.M.* Early November—10 P.M. Early December—8 P.M.
Late October—Midnight* Late November—9 P.M. Late December—7 P.M.

*Daylight Saving Time

J ANSWERS TO VISUAL SKILLS CHECK QUESTIONS

Chapter 1
1. b
2. c
3. c
4. No; the nearest stars would not fit on Earth on this scale.

Chapter 2
1. B
2. D
3. C
4. d
5. b
6. d
7. c
8. c

Chapter 3
1. d
2. a
3. d
4. a
5. a
6. a
7. a

Chapter S1
1. a
2. b
3. b, c, f
4. 66.5°S (the Antarctic Circle)
5. d
6. a, b, c
7. Declination is +30°; right ascension cannot be characterized without further information.

Chapter 4
1. b
2. c
3. d
4. d
5. c

Chapter 5
1. 1
2. 5
3. b
4. b
5. c

Chapter 6
1. b
2. a, c, e
3. all
4. a
5. a

Chapter 7
1. From A to H, the planets are Mercury, Mars, Venus, Earth, Neptune, Uranus, Saturn, and Jupiter.
2. The pairs are Mercury and Mars, Venus and Earth, Neptune and Uranus, Saturn and Jupiter.
3. d
4. a. The exponential plot shows information on low-mass planets that can't be seen on the linear plot.
   b. The linear plot
   c. The exponential plot

Chapter 8
1. a
2. d
3. c

Chapter 9
1. a
2. c
3. b
4. b, c, a

Chapter 10
1. d
2. b
3. d
4. c
5. b

Chapter 11
1. c
2. e
3. b
4. b

Chapter 12
1. about 5 km, though its unusual shape could lead to answers between 4 and 8 km
2. b
3. a
4. c
5. b

Chapter 13
1. about 4 days
2. about 50 meters/sec
3. a. 2 b. 4 c. 3 d. 1
4. a. 4 b. 2 c. 1 d. 3
5. b

Chapter 14
1. d
2. Sunspots appear over a range of 40–50°N latitude to 40–50°S latitude.
3. Sunspots get closer to the equator during a sunspot cycle.

Chapter 15
1. b
2. d
3. c
4. luminosity: about $10,000L_{Sun}$; lifetime: slightly longer than 10 million years
5. luminosity: about $100L_{Sun}$; lifetime: slightly shorter than 1 billion years
6. luminosity: about $10L_{Sun}$; lifetime: approximately 1 billion years

Chapter 16
1. 200–300 light-years
2. About 1000 light-years
3. Cool gas is mostly in the centers of the streams.
4. The cold gas heats up when it enters the central cloud.

Chapter 17

1. approximately $10L_{Sun}$
2. approximately 3500 K
3. approximately $10^4 L_{Sun}$
4. approximately $10^{-4} L_{Sun}$

Chapter 18

1. b
2. d
3. b
4. c
5. d

Chapter 19

1. brightest: white; lowest levels of brightness: black/dark blue
2. white
3. Regions with strong radio emission are dark in the visible-light image.
4. Regions with strong radio emission are brighter in the infrared image than they are in the visible-light image.
5. c
6. yes

Chapter 20

1. Cepheids
2. parallax
3. approximately 10 million to 100 million light-years
4. from about 30 million light-years to more than 10 billion light-years
5. Cepheids
6. white dwarf supernovae (distant standards)

Chapter 21

1. white
2. purple
3. approximately 400,000 light-years
4. approximately 20,000 light-years

Chapter 22

1. b
2. c
3. d

Chapter 23

1. accelerating
2. accelerating
3. coasting
4. decelerating

Chapter 24

1. g, f, b, e, d, c, a
2. b, a, c, d
3. a, c, d
4. d
5. c
6. c

GLOSSARY

absolute magnitude A measure of an object's luminosity; defined to be the apparent magnitude the object would have if it were located exactly 10 parsecs away.

absolute zero The coldest possible temperature, which is 0 K.

absorption (of light) The process by which matter absorbs radiative energy.

absorption line spectrum A spectrum that contains absorption lines.

accelerating universe A universe in which a repulsive force (*see* cosmological constant) causes the expansion of the universe to accelerate with time. Its galaxies will recede from one another increasingly faster, and it will become cold and dark more quickly than a coasting universe.

acceleration The rate at which an object's velocity changes. Its standard units are m/s².

acceleration of gravity The acceleration of a falling object. On Earth, the acceleration of gravity, designated by *g*, is 9.8 m/s².

accretion The process by which small objects gather together to make larger objects.

accretion disk A rapidly rotating disk of material that gradually falls inward as it orbits a starlike object (e.g., white dwarf, neutron star, or black hole).

active galactic nuclei The unusually luminous centers of some galaxies, thought to be powered by accretion onto supermassive black holes. Quasars are the brightest type of active galactic nuclei; radio galaxies also contain active galactic nuclei.

active galaxy A term sometimes used to describe a galaxy that contains an *active galactic nucleus.*

adaptive optics A technique in which telescope mirrors flex rapidly to compensate for the bending of starlight caused by atmospheric turbulence.

Algol paradox A paradox concerning the binary star Algol, which contains a subgiant star that is less massive than its main-sequence companion.

altitude (above horizon) The angular distance between the horizon and an object in the sky.

amino acids The building blocks of proteins.

analemma The figure-8 path traced by the Sun over the course of a year when viewed at the same place and the same time each day; it represents the discrepancies between apparent and mean solar time.

Andromeda Galaxy (M31; the Great Galaxy in Andromeda) The nearest large spiral galaxy to the Milky Way.

angular momentum Momentum attributable to rotation or revolution. The angular momentum of an object moving in a circle of radius *r* is the product $m \times v \times r$.

angular resolution (of a telescope) The smallest angular separation that two pointlike objects can have and still be seen as distinct points of light (rather than as a single point of light).

angular size (or **angular distance**) A measure of the angle formed by extending imaginary lines outward from our eyes to span an object (or the space between two objects).

annihilation *See* matter–antimatter annihilation.

annular solar eclipse A solar eclipse during which the Moon is directly in front of the Sun but its angular size is not large enough to fully block the Sun; thus, a ring (or *annulus*) of sunlight is still visible around the Moon's disk.

Antarctic Circle The circle on Earth with latitude 66.5°S.

antielectron The antimatter equivalent of an electron. It is identical to an electron in virtually all respects, except it has a positive rather than a negative electrical charge.

antimatter Any particle with the same mass as a particle of ordinary matter but whose other basic properties, such as electrical charge, are precisely opposite.

aphelion The point at which an object orbiting the Sun is farthest from the Sun.

apogee The point at which an object orbiting Earth is farthest from Earth.

apparent brightness The amount of light reaching us *per unit area* from a luminous object; often measured in units of watts/m².

apparent magnitude A measure of the apparent brightness of an object in the sky, based on the ancient system developed by Hipparchus.

apparent retrograde motion The apparent motion of a planet, as viewed from Earth, during the period of a few weeks or months when it moves westward relative to the stars in our sky.

apparent solar time Time measured by the actual position of the Sun in your local sky, defined so that noon is when the Sun is *on* the meridian.

arcminute (or **minute of arc**) 1/60 of 1°.

arcsecond (or **second of arc**) 1/60 of an arcminute, or 1/3600 of 1°.

Arctic Circle The circle on Earth with latitude 66.5°N.

asteroid A relatively small and rocky object that orbits a star; asteroids are officially considered part of a category known as "small solar system bodies."

asteroid belt The region of our solar system between the orbits of Mars and Jupiter in which asteroids are heavily concentrated.

astrobiology The study of life on Earth and beyond; it emphasizes research into questions of the origin of life, the conditions under which life can survive, and the search for life beyond Earth.

astrometric technique The detection of extrasolar planets through the side-to-side motion of a star caused by gravitational tugs from the planet.

astronomical unit (AU) The average distance (semimajor axis) of Earth from the Sun, which is about 150 million km.

atmosphere A layer of gas that surrounds a planet or moon, usually very thin compared to the size of the object.

atmospheric pressure The surface pressure resulting from the overlying weight of an atmosphere.

atmospheric structure The layering of a planetary atmosphere due to variations in temperature with altitude. For example, Earth's atmospheric structure from the ground up consists of the troposphere, stratosphere, thermosphere, and exosphere.

atomic hydrogen gas Gas composed mostly of hydrogen atoms, though in space it is generally mixed with helium and small amounts of other elements as well; it is the most common form of interstellar gas.

atomic mass number The combined number of protons and neutrons in an atom.

atomic number The number of protons in an atom.

atoms Consist of a nucleus made from protons and neutrons, surrounded by a cloud of electrons.

aurora Dancing lights in the sky caused by charged particles entering our atmosphere; called the *aurora borealis* in the Northern Hemisphere and the *aurora australis* in the Southern Hemisphere.

axis tilt (of a planet in our solar system) The amount by which a planet's axis is tilted with respect to a line perpendicular to the ecliptic plane.

azimuth (usually called **direction** in this book) Direction around the horizon from due north, measured clockwise in degrees. For example, the azimuth of due north is 0°, due east is 90°, due south is 180°, and due west is 270°.

bar The standard unit of pressure, approximately equal to Earth's atmospheric pressure at sea level.

barred spiral galaxies Spiral galaxies that have a straight bar of stars cutting across their centers.

baryonic matter Ordinary matter made from atoms (so called because the nuclei of atoms contain protons and neutrons, which are both baryons).

baryons Particles, including protons and neutrons, that are made from three quarks.

basalt A type of dark, high-density volcanic rock that is rich in iron and magnesium-based silicate minerals; it forms a runny (easy flowing) lava when molten.

belts (on a jovian planet) Dark bands of sinking air that encircle a jovian planet at a particular set of latitudes.

Big Bang The name given to the event thought to mark the birth of the universe.

Big Bang theory The scientific theory of the universe's earliest moments, stating that all the matter in our observable universe came into being at a single moment in time as an extremely hot, dense mixture of subatomic particles and radiation.

Big Crunch The name given to the event that would presumably end the universe if gravity ever reverses the universal expansion and the universe someday begins to collapse.

binary star system A star system that contains two stars.

biosphere The "layer" of life on Earth.

blackbody radiation *See* thermal radiation.

black hole A bottomless pit in spacetime. Nothing can escape from within a black hole, and we can never again detect or observe an object that falls into a black hole.

black smokers Structures around seafloor volcanic vents that support a wide variety of life.

BL Lac objects A class of active galactic nuclei that probably represent the centers of radio galaxies whose jets happen to be pointed directly at us.

blowout Ejection of the hot, gaseous contents of a superbubble when it grows so large that it bursts out of the cooler layer of gas filling the galaxy's disk.

blueshift A Doppler shift in which spectral features are shifted to shorter wavelengths, observed when an object is moving toward the observer.

bosons Particles, such as photons, to which the exclusion principle does not apply.

bound orbits Orbits on which an object travels repeatedly around another object; bound orbits are elliptical in shape.

brown dwarf An object too small to become an ordinary star because electron degeneracy pressure halts its gravitational collapse before fusion becomes self-sustaining; brown dwarfs have mass less than $0.08M_{Sun}$.

bubble (interstellar) An expanding shell of hot, ionized gas driven by stellar winds or supernovae, with very hot and very low density gas inside.

bulge (of a spiral galaxy) The central portion of a spiral galaxy that is roughly spherical (or football shaped) and bulges above and below the plane of the galactic disk.

Cambrian explosion The dramatic diversification of life on Earth that occurred between about 540 and 500 million years ago.

carbonate rock A carbon-rich rock, such as limestone, that forms underwater from chemical reactions between sediments and carbon dioxide. On Earth, most of the outgassed carbon dioxide currently resides in carbonate rocks.

carbon dioxide cycle (CO_2 cycle) The process that cycles carbon dioxide between Earth's atmosphere and surface rocks.

carbon stars Stars whose atmospheres are especially carbon-rich, thought to be near the ends of their lives; carbon stars are the primary sources of carbon in the universe.

Cassini division A large, dark gap in Saturn's rings, visible through small telescopes on Earth.

CCD (charge coupled device) A type of electronic light detector that has largely replaced photographic film in astronomical research.

celestial coordinates The coordinates of right ascension and declination that fix an object's position on the celestial sphere.

celestial equator (CE) The extension of Earth's equator onto the celestial sphere.

celestial navigation Navigation on the surface of the Earth accomplished by observations of the Sun and stars.

celestial sphere The imaginary sphere on which objects in the sky appear to reside when observed from Earth.

Celsius (temperature scale) The temperature scale commonly used in daily activity internationally, defined so that, on Earth's surface, water freezes at 0°C and boils at 100°C.

center of mass (of orbiting objects) The point at which two or more orbiting objects would balance if they were somehow connected; it is the point around which the orbiting objects actually orbit.

central dominant galaxy A giant elliptical galaxy found at the center of a dense cluster of galaxies, apparently formed by the merger of several individual galaxies.

Cepheid *See* Cepheid variable stars.

Cepheid variable stars A particularly luminous type of pulsating variable star that follows a period–luminosity relation and hence is very useful for measuring cosmic distances.

Chandrasekhar limit *See* white dwarf limit.

charged particle belts Zones in which ions and electrons accumulate and encircle a planet.

chemical enrichment The process by which the abundance of heavy elements (heavier than helium) in the interstellar medium gradually increases over time as these elements are produced by stars and released into space.

chemical potential energy Potential energy that can be released through chemical reactions; for example, food contains chemical potential energy that your body can convert to other forms of energy.

chondrites Another name for primitive meteorites. The name comes from the round chodrules within them. *Achondrites,* meaning "without chondrules," is another name for processed meteorites.

chromosphere The layer of the Sun's atmosphere below the corona; most of the Sun's ultraviolet light is emitted from this region, in which the temperature is about 10,000 K.

circulation cells (or **Hadley cells**) Large-scale cells (similar to convection cells) in a planet's atmosphere that transport heat between the equator and the poles.

circumpolar star A star that always remains above the horizon for a particular latitude.

climate The long-term average of weather.

close binary A binary star system in which the two stars are very close together.

closed universe A universe in which spacetime curves back on itself to the point where its overall shape is analogous to that of the surface of a sphere.

cluster of galaxies A collection of a few dozen or more galaxies bound together by gravity; smaller collections of galaxies are simply called *groups.*

cluster of stars A group of anywhere from several hundred to a million or so stars; star clusters come in two types—open clusters and globular clusters.

CNO cycle The cycle of reactions by which intermediate- and high-mass stars fuse hydrogen into helium.

coasting universe A universe that will keep expanding forever with little change in its rate of expansion; in the absence of a repulsive force (*see* cosmological constant), a coasting universe is one in which the actual mass density is *smaller* than the critical density.

coma (of a comet) The dusty atmosphere of a comet, created by sublimation of ices in the nucleus when the comet is near the Sun.

comet A relatively small, icy object that orbits a star. Like asteroids, comets are officially considered part of a category known as "small solar system bodies."

comparative planetology The study of the solar system by examining and understanding the similarities and differences among worlds.

compound (chemical) A substance made from molecules consisting of two or more atoms with different atomic numbers.

condensates Solid or liquid particles that condense from a cloud of gas.

condensation The formation of solid or liquid particles from a cloud of gas.

conduction (of energy) The process by which thermal energy is transferred by direct contact from warm material to cooler material.

conjunction (of a planet with the Sun) An event in which a planet and the Sun line up in our sky.

conservation of angular momentum (law of) The principle that, in the absence of net torque (twisting force), the total angular momentum of a system remains constant.

conservation of energy (law of) The principle that energy (including mass-energy) can be neither created nor destroyed, but can only change from one form to another.

conservation of momentum (law of) The principle that, in the absence of net force, the total momentum of a system remains constant.

constellation A region of the sky; 88 official constellations cover the celestial sphere.

continental crust The thicker lower-density crust that makes up Earth's continents. It is made when remelting of seafloor crust allows lower-density rock to separate and erupt to the surface. Continental crust ranges in age from very young to as old as about 4 billion years (or more).

continuous spectrum A spectrum (of light) that spans a broad range of wavelengths without interruption by emission or absorption lines.

convection The energy transport process in which warm material expands and rises while cooler material contracts and falls.

convection cell An individual small region of convecting material.

convection zone (of a star) A region in which energy is transported outward by convection.

Copernican revolution The dramatic change, initiated by Copernicus, that occurred when we learned that Earth is a planet orbiting the Sun rather than the center of the universe.

core (of a planet) The dense central region of a planet that has undergone differentiation.

core (of a star) The central region of a star, in which nuclear fusion can occur.

Coriolis effect The effect due to rotation that causes air or objects on a rotating surface or planet to deviate from straight-line trajectories.

corona (solar) The tenuous uppermost layer of the Sun's atmosphere; most of the Sun's X rays are emitted from this region, in which the temperature is about 1 million K.

coronal holes Regions of the corona that barely show up in X-ray images because they are nearly devoid of hot coronal gas.

coronal mass ejections Bursts of charged particles from the Sun's corona that travel outward into space.

cosmic microwave background The remnant radiation from the Big Bang, which we detect using radio telescopes sensitive to microwaves (which are short-wavelength radio waves).

cosmic rays Particles such as electrons, protons, and atomic nuclei that zip through interstellar space at close to the speed of light.

cosmological constant The name given to a term in Einstein's equations of general relativity. If it is not zero, then it represents a repulsive force or a type of energy (sometimes called *dark energy* or *quintessence*) that might cause the expansion of the universe to accelerate with time.

cosmological horizon The boundary of our observable universe, which is where the lookback time is equal to the age of the universe. Beyond this boundary in spacetime, we cannot see anything at all.

Cosmological Principle The idea that matter is distributed uniformly throughout the universe on very large scales, meaning that the universe has neither a center nor an edge.

cosmological redshift The redshift we see from distant galaxies, caused by the fact that expansion of the universe stretches all the photons within it to longer, redder wavelengths.

cosmology The study of the overall structure and evolution of the universe.

cosmos An alternative name for the universe.

crescent (phase) The phase of the Moon (or of a planet) in which just a small portion (less than half) of the visible face is illuminated by sunlight.

critical density The precise average density for the entire universe that marks the dividing line between a recollapsing universe and one that will expand forever.

critical universe A universe that will never collapse, but that expands more and more slowly as time progresses; in the absence of a repulsive force (*see* cosmological constant), a critical universe is one in which the average mass density *equals* the critical density.

crust (of a planet) The low-density surface layer of a planet that has undergone differentiation.

curvature of spacetime A change in the geometry of space that is produced in the vicinity of a massive object and is responsible for the force we call gravity. The overall geometry of the universe may also be curved, depending on its overall mass-energy content.

cycles per second Units of frequency for a wave; describes the number of peaks (or troughs) of a wave that pass by a given point each second. Equivalent to *hertz*.

dark energy Name sometimes given to energy that could be causing the expansion of the universe to accelerate. *See* cosmological constant.

dark matter Matter that we infer to exist from its gravitational effects but from which we have not detected any light; dark matter apparently dominates the total mass of the universe.

daylight saving time Standard time plus 1 hour, so that the Sun appears on the meridian around 1 p.m. rather than around noon.

decay (radioactive) *See* radioactive decay.

declination (dec) Analogous to latitude, but on the celestial sphere; it is the angular north-south distance between the celestial equator and a location on the celestial sphere.

deferent The large circle upon which a planet follows its circle-upon-circle path around Earth in the (Earth-centered) Ptolemaic model of the universe. *See also* epicycle.

degeneracy pressure A type of pressure unrelated to an object's temperature, which arises when electrons (electron degeneracy pressure) or neutrons (neutron degeneracy pressure) are packed so tightly that the exclusion and uncertainty principles come into play.

degenerate object An object, such as a brown dwarf, white dwarf, or neutron star, in which degeneracy pressure is the primary pressure pushing back against gravity.

density (mass) The amount of mass per unit volume of an object. The average density of any object can be found by dividing its mass by its volume. Standard metric units are kilograms per cubic meter, but in astronomy density is more commonly stated in units of grams per cubic centimeter.

deuterium A form of hydrogen in which the nucleus contains a proton and a neutron, rather than only a proton (as is the case for most hydrogen nuclei).

differential rotation Rotation in which the equator of an object rotates at a different rate than the poles.

differentiation The process by which gravity separates materials according to density, with high-density materials sinking and low-density materials rising.

diffraction grating A finely etched surface that can split light into a spectrum.

diffraction limit The angular resolution that a telescope could achieve if it were limited only by the interference of light waves; it is smaller (i.e., better angular resolution) for larger telescopes.

dimension (mathematical) Describes the number of independent directions in which movement is possible; for example, the surface of Earth is two-dimensional because only two independent directions of motion are possible (north-south and east-west).

direction (in local sky) One of the two coordinates (the other is altitude) needed to pinpoint an object in the local sky. It is the direction, such as north, south, east, or west, in which you must face to see the object. *See also* azimuth.

disk (of a galaxy) The portion of a spiral galaxy that looks like a disk and contains an interstellar medium with cool gas and dust; stars of many ages are found in the disk.

disk population The stars that orbit within the disk of a spiral galaxy; sometimes called *Population I*.

DNA (deoxyribonucleic acid) The molecule that represents the genetic material of life on Earth.

Doppler effect (shift) The effect that shifts the wavelengths of spectral features in objects that are moving toward or away from the observer.

Doppler technique The detection of extrasolar planets through the motion of a star toward and away from the observer caused by gravitational tugs from the planet.

double shell–fusing star A star that is fusing helium into carbon in a shell around an inert carbon core and is fusing hydrogen into helium in a shell at the top of the helium layer.

down quark One of the two quark types (the other is the up quark) found in ordinary protons and neutrons. It has a charge of $-\frac{1}{3}$.

Drake equation An equation that lays out the factors that play a role in determining the number of communicating civilizations in our galaxy.

dust (or **dust grains**) Tiny solid flecks of material; in astronomy, we often discuss interplanetary dust (found within a star system) or interstellar dust (found between the stars in a galaxy).

dust tail (of a comet) One of two tails seen when a comet passes near the Sun (the other is the plasma tail). It is composed of small solid particles pushed away from the Sun by the radiation pressure of sunlight.

dwarf elliptical galaxy A small elliptical galaxy with less than about a billion stars.

dwarf galaxies Relatively small galaxies, consisting of less than about 10 billion stars.

dwarf planet An object that orbits the Sun and is massive enough for its gravity to have made it nearly round in shape, but that does not qualify as an official planet because it has not cleared its orbital neighborhood. The dwarf planets of our solar system include the asteroid Ceres and the Kuiper belt objects Pluto, Eris, Haumea, and Makemake.

Earth-orbiters (spacecraft) Spacecraft designed to study Earth or the universe from Earth orbit.

eccentricity A measure of how much an ellipse deviates from a perfect circle; defined as the center-to-focus distance divided by the length of the semimajor axis.

eclipse An event in which one astronomical object casts a shadow on another or crosses our line of sight to the other object.

eclipse seasons Periods during which lunar and solar eclipses can occur because the nodes of the Moon's orbit are aligned with Earth and the Sun.

eclipsing binary A binary star system in which the two stars happen to be orbiting in the plane of our line of sight, so that each star will periodically eclipse the other.

ecliptic The Sun's apparent annual path among the constellations.

ecliptic plane The plane of Earth's orbit around the Sun.

ejecta (from an impact) Debris ejected by the blast of an impact.

electrical charge A fundamental property of matter that is described by its amount and as either positive or negative; more technically, a measure of how a particle responds to the electromagnetic force.

electromagnetic field An abstract concept used to describe how a charged particle would affect other charged particles at a distance.

electromagnetic radiation Another name for light of all types, from radio waves through gamma rays.

electromagnetic spectrum The complete spectrum of light, including radio waves, infrared light, visible light, ultraviolet light, X rays, and gamma rays.

electromagnetic wave A synonym for *light*, which consists of waves of electric and magnetic fields.

electromagnetism (or **electromagnetic force**) One of the four fundamental forces; it is the force that dominates atomic and molecular interactions.

electron degeneracy pressure Degeneracy pressure exerted by electrons, as in brown dwarfs and white dwarfs.

electrons Fundamental particles with negative electric charge; the distribution of electrons in an atom gives the atom its size.

electron-volt (eV) A unit of energy equivalent to 1.60×10^{-19} joule.

electroweak era The era of the universe during which only three forces operated (gravity, strong force, and electroweak force), lasting from 10^{-38} second to 10^{-10} second after the Big Bang.

electroweak force The force that exists at high energies when the electromagnetic force and the weak force exist as a single force.

element (chemical) A substance made from individual atoms of a particular atomic number.

ellipse A type of oval that happens to be the shape of bound orbits. An ellipse can be drawn by moving a pencil along a string whose ends are tied to two tacks; the locations of the tacks are the *foci* (singular: *focus*) of the ellipse.

elliptical galaxies Galaxies that appear rounded in shape, often longer in one direction, like a football. They have no disks and contain very little cool gas and dust compared to spiral galaxies, though they often contain very hot, ionized gas.

elongation (greatest) For Mercury or Venus, the point at which it appears farthest from the Sun in our sky.

emission (of light) The process by which matter emits energy in the form of light.

emission line spectrum A spectrum that contains emission lines.

emission nebula Another name for an ionization nebula. *See* ionization nebula.

energy Broadly speaking, what can make matter move. The three basic types of energy are kinetic, potential, and radiative.

energy balance (in a star) The balance between the rate at which fusion releases energy in the star's core and the rate at which the star's surface radiates this energy into space.

epicycle The small circle upon which a planet moves while simultaneously going around a larger circle (the *deferent*) around Earth in the (Earth-centered) Ptolemaic model of the universe.

equation of time An equation describing the discrepancies between apparent and mean solar time.

equinox *See* fall equinox *and* spring equinox.

equivalence principle The fundamental starting point for general relativity, which states that the effects of gravity are exactly equivalent to the effects of acceleration.

era of atoms The era of the universe lasting from about 500,000 years to about 1 billion years after the Big Bang, during which it was cool enough for neutral atoms to form.

era of galaxies The present era of the universe, which began with the formation of galaxies when the universe was about 1 billion years old.

era of nuclei The era of the universe lasting from about 3 minutes to about 380,000 years after the Big Bang, during which matter in the universe was fully ionized and opaque to light. The cosmic background radiation was released at the end of this era.

era of nucleosynthesis The era of the universe lasting from about 0.001 second to about 3 minutes after the Big Bang, by the end of which virtually all of the neutrons and about one-seventh of the protons in the universe had fused into helium.

erosion The wearing down or building up of geological features by wind, water, ice, and other phenomena of planetary weather.

eruption The process of releasing hot lava on a planet's surface.

escape velocity The speed necessary for an object to completely escape the gravity of a large body such as a moon, planet, or star.

evaporation The process by which atoms or molecules escape into the gas phase from a liquid.

event Any particular point along a worldline; all observers will agree on the reality of an event but may disagree about its time and location.

event horizon The boundary that marks the "point of no return" between a black hole and the outside universe; events that occur within the event horizon can have no influence on our observable universe.

evolution (biological) The gradual change in populations of living organisms responsible for transforming life on Earth from its primitive origins to the great diversity of life today.

exchange particle A type of subatomic particle that transmits one of the four fundamental forces; according to the standard model of physics, these particles are always exchanged whenever two objects interact through a force.

excited state (of an atom) Any arrangement of electrons in an atom that has more energy than the ground state.

exclusion principle The law of quantum mechanics that states that two fermions cannot occupy the same quantum state at the same time.

exosphere The hot, outer layer of an atmosphere, where the atmosphere "fades away" to space.

expansion (of the universe) The idea that the space between galaxies or clusters of galaxies is growing with time.

exposure time The amount of time during which light is collected to make a single image.

extrasolar planet A planet orbiting a star other than our Sun.

extremophiles Living organisms that are adapted to conditions that are "extreme" by human standards, such as very high or low temperature or a high level of salinity or radiation.

Fahrenheit (temperature scale) The temperature scale commonly used in daily activity in the United States; defined so that, on Earth's surface, water freezes at 32°F and boils at 212°F.

fall (September) equinox Refers both to the point in Virgo on the celestial sphere where the ecliptic crosses the celestial equator and to the moment in time when the Sun appears at that point each year (around September 21).

false-color image An image displayed in colors that are *not* the true, visible-light colors of an object.

fault (geological) A place where rocks slip sideways relative to one another.

feedback processes Processes in which a small change in some property (such as temperature) leads to changes in other properties that either amplify or diminish the original small change.

fermions Particles, such as electrons, neutrons, and protons, that obey the exclusion principle.

Fermi's paradox The question posed by Enrico Fermi about extraterrestrial intelligence—"So where is everybody?"—which asks why we have not observed other civilizations even though simple arguments would suggest that some ought to have spread throughout the galaxy by now.

field An abstract concept used to describe how a particle would interact with a force. For example, the idea of a gravitational field describes how a particle would react to the local strength of gravity, and the idea of an electromagnetic field describes how a charged particle would respond to forces from other charged particles.

filter (for light) A material that transmits only particular wavelengths of light.

fireball A particularly bright meteor.

first-quarter (phase) The phase of the Moon that occurs one-quarter of the way through each cycle of phases, in which precisely half of the visible face is illuminated by sunlight.

fission The process by which one atomic nucleus breaks into two smaller nuclei. It releases energy if the two smaller nuclei together are less massive than the original nucleus.

flare star A small, spectral type M star that displays particularly strong flares on its surface.

flat (or Euclidean) geometry The type of geometry in which the rules of geometry for a flat plane hold, such as that the shortest distance between two points is a straight line and that the sum of the angles in a triangle is 180°.

flat universe A universe in which the overall geometry of spacetime is flat (Euclidean), as would be the case if the density of the universe was equal to the critical density.

flybys (spacecraft) Spacecraft that fly past a target object (such as a planet), usually just once, as opposed to entering a bound orbit of the object.

focal plane The place where an image created by a lens or mirror is in focus.

foci Plural of *focus*.

focus (of a lens or mirror) The point at which rays of light that were initially parallel (such as those from a distant star) converge.

focus (of an ellipse) One of two special points within an ellipse that lie along the major axis; these are the points around which we could stretch a pencil and string to draw an ellipse. When one object orbits a second object, the second object lies at one focus of the orbit.

force Anything that can cause a change in momentum.

formation properties (of planets) In this book, for the purpose of understanding geological processes, planets are defined to be born with four formation properties: size (mass and radius), distance from the Sun, composition, and rotation rate.

fossil Any relic of an organism that lived and died long ago.

frame of reference *See* reference frame.

free-fall The condition in which an object is falling without resistance; objects are weightless when in free-fall.

free-float frame A frame of reference in which all objects are weightless and hence float freely.

frequency The rate at which peaks of a wave pass by a point, measured in units of 1/s, often called *cycles per second* or *hertz*.

frost line The boundary in the solar nebula beyond which ices could condense; only metals and rocks could condense within the frost line.

fundamental forces There are four known fundamental forces in nature: gravity, the electromagnetic force, the strong force, and the weak force.

fundamental particles Subatomic particles that cannot be divided into anything smaller.

fusion The process by which two atomic nuclei fuse together to make a single, more massive nucleus. It releases energy if the final nucleus is less massive than the two nuclei that went into the reaction.

galactic cannibalism The term sometimes used to describe the process by which large galaxies merge with other galaxies in collisions. *Central dominant galaxies* are products of galactic cannibalism.

galactic fountain A model for the cycling of gas in the Milky Way Galaxy in which fountains of hot, ionized gas rise from the disk into the halo and then cool and form clouds as they sink back into the disk.

galactic wind A wind of low-density but extremely hot gas flowing out from a starburst galaxy, created by the combined energy of many supernovae.

galaxy A huge collection of anywhere from a few hundred million to more than a trillion stars, all bound together by gravity.

galaxy cluster *See* cluster of galaxies.

galaxy evolution The formation and development of galaxies.

Galilean moons The four moons of Jupiter that were discovered by Galileo: Io, Europa, Ganymede, and Callisto.

gamma-ray burst A sudden burst of gamma rays from deep space; such bursts apparently come from distant galaxies, but their precise mechanism is unknown.

gamma rays Light with very short wavelengths (and hence high frequencies)—shorter than those of X rays.

gap moons Tiny moons located within a gap in a planet's ring system. The gravity of a gap moon helps clear the gap.

gas phase The phase of matter in which atoms or molecules can move essentially independently of one another.

gas pressure The force (per unit area) pushing on any object due to surrounding gas. *See also* pressure.

general theory of relativity Einstein's generalization of his special theory of relativity so that the theory also applies when we consider effects of gravity or acceleration.

genetic code The "language" that living cells use to read the instructions chemically encoded in DNA.

geocentric model Any of the ancient Greek models that were used to predict planetary positions under the assumption that Earth lay in the center of the universe.

geocentric universe The ancient belief that Earth is the center of the entire universe.

geological activity Processes that change a planet's surface long after formation, such as volcanism, tectonics, and erosion.

geological processes The four basic geological processes are impact cratering, volcanism, tectonics, and erosion.

geological time scale The time scale used by scientists to describe major eras in Earth's past.

geology The study of surface features (on a moon, planet, or asteroid) and the processes that create them.

geostationary satellite A satellite that appears to stay stationary in the sky as viewed from Earth's surface, because it orbits in the same time it takes Earth to rotate and orbits in Earth's equatorial plane.

geosynchronous satellite A satellite that orbits Earth in the same time it takes Earth to rotate (one sidereal day).

giant galaxies Galaxies that are unusually large, typically containing a trillion or more stars. Most giant galaxies are elliptical, and many contain multiple nuclei near their centers.

giant impact A collision between a forming planet and a very large planetesimal, such as is thought to have formed our Moon.

giant molecular cloud A very large cloud of cold, dense interstellar gas, typically containing up to a million solar masses worth of material. *See also* molecular clouds.

giants (luminosity class III) Stars that appear just below the supergiants on the H-R diagram because they are somewhat smaller in radius and lower in luminosity.

gibbous (phase) The phase of the Moon (or of a planet) in which more than half but less than all of the visible face is illuminated by sunlight.

global positioning system (GPS) A system of navigation by satellites orbiting Earth.

global warming An expected increase in Earth's global average temperature caused by human input of carbon dioxide and other greenhouse gases into the atmosphere.

global wind patterns (or **global circulation**) Wind patterns that remain fixed on a global scale, determined by the combination of surface heating and the planet's rotation.

globular cluster A spherically shaped cluster of up to a million or more stars; globular clusters are found primarily in the halos of galaxies and contain only very old stars.

gluons The exchange particles for the strong force.

grand unified theory (GUT) A theory that unifies three of the four fundamental forces—the strong force, the weak force, and the electromagnetic force (but not gravity)—in a single model.

granulation (on the Sun) The bubbling pattern visible in the photosphere, produced by the underlying convection.

gravitation (law of) *See* universal law of gravitation.

gravitational constant The experimentally measured constant G that appears in the law of universal gravitation:

$$G = 6.67 \times 10^{-11} \frac{m^3}{kg \times s^2}$$

gravitational contraction The process in which gravity causes an object to contract, thereby converting gravitational potential energy into thermal energy.

gravitational encounter An encounter in which two (or more) objects pass near enough so that each can feel the effects of the other's gravity and they can therefore exchange energy.

gravitational equilibrium A state of balance in which the force of gravity pulling inward is precisely counteracted by pressure pushing outward.

gravitational lensing The magnification or distortion (into arcs, rings, or multiple images) of an image caused by light bending through a gravitational field, as predicted by Einstein's general theory of relativity.

gravitationally bound system Any system of objects, such as a star system or a galaxy, that is held together by gravity.

gravitational potential energy Energy that an object has by virtue of its position in a gravitational field; an object has more gravitational potential energy when it has a greater distance that it can potentially fall.

gravitational redshift A redshift caused by the fact that time runs slowly in gravitational fields.

gravitational time dilation The slowing of time that occurs in a gravitational field, as predicted by Einstein's general theory of relativity.

gravitational waves Waves, predicted by Einstein's general theory of relativity, that travel at the speed of light and transmit distortions of space through the universe. Although they have not yet been observed directly, we have strong indirect evidence that they exist.

gravitons The exchange particles for the force of gravity.

gravity One of the four fundamental forces; it is the force that dominates on large scales.

grazing incidence (in telescopes) Reflections in which light grazes a mirror surface and is deflected at a small angle; commonly used to focus high-energy ultraviolet light and X rays.

great circle A circle on the surface of a sphere whose center is at the center of the sphere.

greatest elongation *See* elongation (greatest).

Great Red Spot A large, high-pressure storm on Jupiter.

greenhouse effect The process by which greenhouse gases in an atmosphere make a planet's surface temperature warmer than it would be in the absence of an atmosphere.

greenhouse gases Gases, such as carbon dioxide, water vapor, and methane, that are particularly good absorbers of infrared light but are transparent to visible light.

Gregorian calendar Our modern calendar, introduced by Pope Gregory in 1582.

ground state (of an atom) The lowest possible energy state of the electrons in an atom.

group (of galaxies) A few to a few dozen galaxies bound together by gravity. *See also* cluster of galaxies.

GUT era The era of the universe during which only two forces operated (gravity and the grand-unified-theory, or GUT, force), lasting from 10^{-43} second to 10^{-38} second after the Big Bang.

GUT force The proposed force that exists at very high energies when the strong force, the weak force, and the electromagnetic force (but not gravity) all act as one.

H II region Another name for an ionization nebula. *See* ionization nebula.

habitable world A world with environmental conditions under which life could *potentially* arise or survive.

habitable zone The region around a star in which planets could potentially have surface temperatures at which liquid water could exist.

Hadley cells *See* circulation cells.

half-life The time it takes for half of the nuclei in a given quantity of a radioactive substance to decay.

halo (of a galaxy) The spherical region surrounding the disk of a spiral galaxy.

Hawking radiation Radiation predicted to arise from the evaporation of black holes.

heavy bombardment The period in the first few hundred million years after the solar system formed during which the tail end of planetary accretion created most of the craters found on ancient planetary surfaces.

heavy elements In astronomy, generally all elements *except* hydrogen and helium.

helium-capture reactions Fusion reactions that fuse a helium nucleus into some other nucleus; such reactions can fuse carbon into oxygen, oxygen into neon, neon into magnesium, and so on.

helium flash The event that marks the sudden onset of helium fusion in the previously inert helium core of a low-mass star.

helium-fusing star A star that is currently fusing helium into carbon in its core.

helium fusion The fusion of three helium nuclei into one carbon nucleus; also called the *triple-alpha reaction*.

hertz (Hz) The standard unit of frequency for light waves; equivalent to units of 1/s.

Hertzsprung-Russell (H-R) diagram A graph plotting individual stars as points, with stellar luminosity on the vertical axis and spectral type (or surface temperature) on the horizontal axis.

high-mass stars Stars born with masses above about $8M_{Sun}$; these stars will end their lives by exploding as supernovae.

horizon A boundary that divides what we can see from what we cannot see.

horizontal branch The horizontal line of stars that represents helium-fusing stars on an H-R diagram for a cluster of stars.

horoscope A predictive chart made by an astrologer; in scientific studies, horoscopes have never been found to have any validity as predictive tools.

hot Jupiter A class of planet that is Jupiter-like in size but orbits very close to its star, causing it to have a very high surface temperature.

hot spot (geological) A place within a plate of the lithosphere where a localized plume of hot mantle material rises.

hour angle (HA) The angle or time (measured in hours) since an object was last on the meridian in the local sky; defined to be 0 hours for objects that are on the meridian.

Hubble's constant A number that expresses the current rate of expansion of the universe; designated H_0, it is usually stated in units of km/s/Mpc. The reciprocal of Hubble's constant is the age the universe would have *if* the expansion rate had never changed.

Hubble's law Mathematical expression of the idea that more distant galaxies move away from us faster: $v = H_0 \times d$, where v is a galaxy's speed away from us, d is its distance, and H_0 is Hubble's constant.

hydrogen compounds Compounds that contain hydrogen and were common in the solar nebula, such as water (H_2O), ammonia (NH_3), and methane (CH_4).

hydrogen shell fusion Hydrogen fusion that occurs in a shell surrounding a stellar core.

hydrosphere The "layer" of water on Earth consisting of oceans, lakes, rivers, ice caps, and other liquid water and ice.

hydrostatic equilibrium *See* gravitational equilibrium.

hyperbola The precise mathematical shape of one type of unbound orbit (the other is a parabola) allowed under the force of gravity; at great distances from the attracting object, a hyperbolic path looks like a straight line.

hypernova A term sometimes used to describe a supernova (explosion) of a star so massive that it leaves a black hole behind.

hyperspace Any space with more than three dimensions.

hypothesis A tentative model proposed to explain some set of observed facts, but which has not yet been rigorously tested and confirmed.

ice ages Periods of global cooling during which the polar caps, glaciers, and snow cover extend closer to the equator.

ices (in solar system theory) Materials that are solid only at low temperatures, such as the hydrogen compounds water, ammonia, and methane.

ideal gas law The law relating the pressure, temperature, and number density of particles in an ideal gas.

image A picture of an object made by focusing light.

imaging (in astronomical research) The process of obtaining pictures of astronomical objects.

impact The collision of a small body (such as an asteroid or comet) with a larger object (such as a planet or moon).

impact basin A very large impact crater, often filled by a lava flow.

impact crater A bowl-shaped depression left by the impact of an object that strikes a planetary surface (as opposed to burning up in the atmosphere).

impact cratering The excavation of bowl-shaped depressions (*impact craters*) by asteroids or comets striking a planet's surface.

impactor The object responsible for an impact.

inflation (of the universe) A sudden and dramatic expansion of the universe thought to have occurred at the end of the GUT era.

infrared light Light with wavelengths that fall in the portion of the electromagnetic spectrum between radio waves and visible light.

inner solar system Generally considered to encompass the region of our solar system out to about the orbit of Mars.

intensity (of light) A measure of the amount of energy coming from light of specific wavelength in the spectrum of an object.

interferometry A telescopic technique in which two or more telescopes are used in tandem to produce much better angular resolution than the telescopes could achieve individually.

intermediate-mass stars Stars born with masses between about $2M_{Sun}$ and $8M_{Sun}$; these stars end their lives by ejecting a planetary nebula and becoming a white dwarf.

interstellar cloud A cloud of gas and dust between the stars.

interstellar dust grains Tiny solid flecks of carbon and silicon minerals found in cool interstellar clouds; they resemble particles of smoke and form in the winds of red giant stars.

interstellar medium The gas and dust that fills the space between stars in a galaxy.

interstellar ramjet A hypothesized type of spaceship that uses a giant scoop to sweep up interstellar gas for use in a nuclear fusion engine.

interstellar reddening The change in the color of starlight as it passes through dusty gas. The light appears redder because dust grains absorb and scatter blue light more effectively than red light.

intracluster medium Hot, X-ray-emitting gas found between the galaxies within a cluster of galaxies.

inverse square law A law followed by any quantity that decreases with the square of the distance between two objects.

inverse square law for light The law stating that an object's apparent brightness depends on its actual luminosity and the inverse square of its distance from the observer:

$$\text{apparent brightness} = \frac{\text{luminosity}}{4\pi \times (\text{distance})^2}$$

inversion (atmospheric) A local weather condition in which air is colder near the surface than higher up in the troposphere—the opposite of the usual condition, in which the troposphere is warmer at the bottom.

ionization The process of stripping an electron from an atom.

ionization nebula A colorful, wispy cloud of gas that glows because neighboring hot stars irradiate it with ultraviolet photons that can ionize hydrogen atoms.

ionosphere A portion of the thermosphere in which ions are particularly common (because of ionization by X rays from the Sun).

ions Atoms with a positive or negative electrical charge.

Io torus A donut-shaped charged-particle belt around Jupiter that approximately traces Io's orbit.

irregular galaxies Galaxies that look neither spiral nor elliptical.

isotopes Forms of an element that have the same number of protons but different numbers of neutrons.

jets High-speed streams of gas ejected from an object into space.

joule The international unit of energy, equivalent to about 1/4000 of a Calorie.

jovian nebulae The clouds of gas that swirled around the jovian planets, from which the moons formed.

jovian planets Giant gaseous planets similar in overall composition to Jupiter.

Julian calendar The calendar introduced in 46 B.C. by Julius Caesar and used until the Gregorian calendar replaced it.

Kelvin (temperature scale) The most commonly used temperature scale in science, defined such that absolute zero is 0 K and water freezes at 273.15 K.

Kepler's first law Law stating that the orbit of each planet about the Sun is an ellipse with the Sun at one focus.

Kepler's laws of planetary motion Three laws discovered by Kepler that describe the motion of the planets around the Sun.

Kepler's second law The principle that, as a planet moves around its orbit, it sweeps out equal areas in equal times. This tells us that a planet moves faster when it is closer to the Sun (near perihelion) than when it is farther from the Sun (near aphelion) in its orbit.

Kepler's third law The principle that the square of a planet's orbital period is proportional to the cube of its average distance from the Sun (semimajor axis), which tells us that more distant planets move more slowly in their orbits; in its original form, written $p^2 = a^3$. *See also* Newton's version of Kepler's third law.

kinetic energy Energy of motion, given by the formula $\frac{1}{2}mv^2$.

Kirchhoff's laws A set of rules that summarizes the conditions under which objects produce thermal, absorption line, or emission line spectra. In brief: (1) An opaque object produces thermal radiation. (2) An absorption line spectrum occurs when thermal radiation passes through a thin gas that is cooler than the object emitting the thermal radiation. (3) An emission line spectrum occurs when we view a cloud of gas that is warmer than any background source of light.

Kirkwood gaps On a plot of asteroid semimajor axes, regions with few asteroids as a result of orbital resonances with Jupiter.

K–T event (or impact) The collision of an asteroid or comet 65 million years ago that caused the mass extinction best known for wiping out the dinosaurs. *K* and *T* stand for the geological layers above and below the event.

Kuiper belt The comet-rich region of our solar system that spans distances of about 30–100 AU from the Sun. Kuiper belt comets have orbits that lie fairly close to the plane of planetary orbits and travel around the Sun in the same direction as the planets.

Kuiper belt object Any object orbiting the Sun within the region of the Kuiper belt, although the term is most often used for relatively large objects. For example, Pluto and Eris are considered large Kuiper belt objects.

Large Magellanic Cloud One of two small, irregular galaxies (the other is the Small Magellanic Cloud) located about 150,000 light-years away; it probably orbits the Milky Way Galaxy.

large-scale structure (of the universe) Generally refers to the structure of the universe on size scales larger than that of clusters of galaxies.

latitude The angular north-south distance between Earth's equator and a location on Earth's surface.

leap year A calendar year with 366 rather than 365 days. Our current calendar (the Gregorian calendar) incorporates a leap year every 4 years (by adding February 29) except in century years that are not divisible by 400.

length contraction The effect in which you observe lengths to be shortened in reference frames moving relative to you.

lens (gravitational) *See* gravitational lensing.

lenticular galaxies Galaxies that look lens-shaped when seen edge-on, resembling spiral galaxies without arms. They tend to have less cool gas than normal spiral galaxies but more gas than elliptical galaxies.

leptons Fermions *not* made from quarks, such as electrons and neutrinos.

life track A track drawn on an H-R diagram to represent the changes in a star's surface temperature and luminosity during its life; also called an *evolutionary track*.

light-collecting area (of a telescope) The area of the primary mirror or lens that collects light in a telescope.

light curve A graph of an object's intensity against time.

light gases (in solar system theory) Hydrogen and helium, which never condense under solar nebula conditions.

light pollution Human-made light that hinders astronomical observations.

light-year (ly) The distance that light can travel in 1 year, which is 9.46 trillion km.

liquid phase The phase of matter in which atoms or molecules are held together but move relatively freely.

lithosphere The relatively rigid outer layer of a planet; generally encompasses the crust and the uppermost portion of the mantle.

Local Bubble (interstellar) The bubble of hot gas in which our Sun and other nearby stars apparently reside. *See also* bubble (interstellar).

Local Group The group of about 40 galaxies to which the Milky Way Galaxy belongs.

local sidereal time (LST) Sidereal time for a particular location, defined according to the position of the spring equinox in the local sky. More formally, the local sidereal time at any moment is defined to be the hour angle of the spring equinox.

local sky The sky as viewed from a particular location on Earth (or another solid object). Objects in the local sky are pinpointed by the coordinates of *altitude* and *direction* (or azimuth).

local solar neighborhood The portion of the Milky Way Galaxy that is located relatively close (within a few hundred to a couple thousand light-years) to our Sun.

Local Supercluster The supercluster of galaxies to which the Local Group belongs.

longitude The angular east-west distance between the prime meridian (which passes through Greenwich, England) and a location on Earth's surface.

lookback time The amount of time since the light we see from a distant object was emitted. If an object has a lookback time of 400 million years, we are seeing it as it looked 400 million years ago.

low-mass stars Stars born with masses less than about $2M_{Sun}$; these stars end their lives by ejecting a planetary nebula and becoming a white dwarf.

luminosity The total power output of an object, usually measured in watts or in units of solar luminosities ($L_{Sun} = 3.8 \times 10^{26}$ watts).

luminosity class A category describing the region of the H-R diagram in which a star falls. Luminosity class I represents supergiants, III represents giants, and V represents main-sequence stars; luminosity classes II and IV are intermediate to the others.

lunar eclipse An event that occurs when the Moon passes through Earth's shadow, which can occur only at full moon. A lunar eclipse may be total, partial, or penumbral.

lunar maria The regions of the Moon that look smooth from Earth and actually are impact basins.

lunar month *See* synodic month.

lunar phase *See* phase (of the Moon or a planet).

MACHOs One possible form of dark matter in which the dark objects are relatively large, like planets or brown dwarfs; stands for *massive compact halo objects*.

magma Underground molten rock.

magnetic braking The process by which a star's rotation slows as its magnetic field transfers its angular momentum to the surrounding nebula.

magnetic field The region surrounding a magnet in which it can affect other magnets or charged particles.

magnetic field lines Lines that represent how the needles on a series of compasses would point if they were laid out in a magnetic field.

magnetosphere The region surrounding a planet in which charged particles are trapped by the planet's magnetic field.

magnitude system A system for describing stellar brightness by using numbers, called *magnitudes*, based on an ancient Greek way of describing the brightnesses of stars in the sky. This system uses *apparent magnitude* to describe a star's apparent brightness and *absolute magnitude* to describe a star's luminosity.

main sequence The prominent line of points (representing *main-sequence stars*) running from the upper left to the lower right on an H-R diagram.

main-sequence fitting A method for measuring the distance to a cluster of stars by comparing the apparent brightness of the cluster's main sequence with that of the standard main sequence.

main-sequence lifetime The length of time for which a star of a particular mass can shine by fusing hydrogen into helium in its core.

main-sequence stars (luminosity class V) Stars whose temperature and luminosity place them on the main sequence of the H-R diagram. Main-sequence stars are all releasing energy by fusing hydrogen into helium in their cores.

main-sequence turnoff point The point on a cluster's H-R diagram where its stars turn off from the main sequence; the age of the cluster is equal to the main-sequence lifetime of stars at the main-sequence turnoff point.

mantle (of a planet) The rocky layer that lies between a planet's core and crust.

Martian meteorites Meteorites found on Earth that are thought to have originated on Mars.

mass A measure of the amount of matter in an object.

mass-energy The potential energy of mass, which has an amount $E = mc^2$.

mass exchange (in close binary star systems) The process in which tidal forces cause matter to spill from one star to a companion star in a close binary system.

mass extinction An event in which a large fraction of the species living on Earth go extinct, such as the event in which the dinosaurs died out about 65 million years ago.

mass increase (in relativity) The effect in which an object moving past you seems to have a mass greater than its rest mass.

massive star supernova A supernova that occurs when a massive star dies, initiated by the catastrophic collapse of its iron core; often called a *Type II supernova*.

mass-to-light ratio The mass of an object divided by its luminosity, usually stated in units of solar masses per solar luminosity. Objects with high mass-to-light ratios must contain substantial quantities of dark matter.

matter–antimatter annihilation An event that occurs when a particle of matter and a particle of antimatter meet and convert all of their mass-energy to photons.

mean solar time Time measured by the average position of the Sun in your local sky over the course of the year.

meridian A half-circle extending from your horizon (altitude 0°) due south, through your zenith, to your horizon due north.

metallic hydrogen Hydrogen that is so compressed that the hydrogen atoms all share electrons and thereby take on properties of metals, such as conducting electricity. It occurs only under very high-pressure conditions, such as those found deep within Jupiter.

metals (in solar system theory) Elements, such as nickel, iron, and aluminum, that condense at fairly high temperatures.

meteor A flash of light caused when a particle from space burns up in our atmosphere.

meteorite A rock from space that lands on Earth.

meteor shower A period during which many more meteors than usual can be seen.

Metonic cycle The 19-year period, discovered by the Babylonian astronomer Meton, over which the lunar phases occur on the same dates.

microwaves Light with wavelengths in the range of micrometers to millimeters. Microwaves are generally considered to be a subset of the radio wave portion of the electromagnetic spectrum.

mid-ocean ridges Long ridges of undersea volcanoes on Earth, along which mantle material erupts onto the ocean floor and pushes apart the existing seafloor on either side. These ridges are essentially the source of new seafloor crust, which then makes its way along the ocean bottom for millions of years before returning to the mantle at a subduction zone.

Milankovitch cycles The cyclical changes in Earth's axis tilt and orbit that can change the climate and cause ice ages.

Milky Way Used both as the name of our galaxy and to refer to the band of light we see in the sky when we look into the plane of the Milky Way Galaxy.

millisecond pulsars Pulsars with rotation periods of a few thousandths of a second.

minor planets An alternative name for *asteroids*.

model (scientific) A representation of some aspect of nature that can be used to explain and predict real phenomena without invoking myth, magic, or the supernatural.

molecular bands The tightly bunched lines in an object's spectrum that are produced by molecules.

molecular cloud fragments (or **molecular cloud cores**) The densest regions of molecular clouds, which usually go on to form stars.

molecular clouds Cool, dense interstellar clouds in which the low temperatures allow hydrogen atoms to pair up into hydrogen molecules (H_2).

molecular dissociation The process by which a molecule splits into its component atoms.

molecule Technically, the smallest unit of a chemical element or compound; in this text, the term refers only to combinations of two or more atoms held together by chemical bonds.

momentum The product of an object's mass and velocity.

moon An object that orbits a planet.

moonlets Very small moons that orbit within the ring systems of jovian planets.

mutations Errors in the copying process when a living cell replicates itself.

natural selection The process by which mutations that make an organism better able to survive get passed on to future generations.

neap tides The lower-than-average tides on Earth that occur at first- and third-quarter moon, when the tidal forces from the Sun and Moon oppose each other.

nebula A cloud of gas in space, usually one that is glowing.

nebular capture The process by which icy planetesimals capture hydrogen and helium gas to form jovian planets.

nebular theory The detailed theory that describes how our solar system formed from a cloud of interstellar gas and dust.

net force The overall force to which an object responds; the net force is equal to the rate of change in the object's momentum, or equivalently to the object's mass × acceleration.

neutrino A type of fundamental particle that has extremely low mass and responds only to the weak force; neutrinos are leptons and come in three types—electron neutrinos, mu neutrinos, and tau neutrinos.

neutron degeneracy pressure Degeneracy pressure exerted by neutrons, as in neutron stars.

neutrons Particles with no electrical charge found in atomic nuclei, built from three quarks.

neutron star The compact corpse of a high-mass star left over after a supernova; it typically contains a mass comparable to the mass of the Sun in a volume just a few kilometers in radius.

newton The standard unit of force in the metric system:

$$1 \text{ newton} = 1 \frac{\text{kg} \times \text{m}}{\text{s}^2}$$

Newton's first law of motion Principle that, in the absence of a net force, an object moves with constant velocity.

Newton's laws of motion Three basic laws that describe how objects respond to forces.

Newton's second law of motion Law stating how a net force affects an object's motion. Specifically, force = rate of change in momentum, or force = mass × acceleration.

Newton's third law of motion Principle that, for any force, there is always an equal and opposite reaction force.

Newton's universal law of gravitation See universal law of gravitation.

Newton's version of Kepler's third law A generalization of Kepler's third law used to calculate the masses of orbiting objects from measurements of orbital period and distance; usually written as

$$p^2 = \frac{4\pi^2}{G(M_1 + M_2)}a^3$$

nodes (of Moon's orbit) The two points in the Moon's orbit where it crosses the ecliptic plane.

nonbaryonic matter Matter that is not part of the normal composition of atoms, such as neutrinos or the hypothetical WIMPs. (More technically, particles that are not made from three quarks.)

nonscience As defined in this book, any way of searching for knowledge that makes no claim to follow the scientific method, such as seeking knowledge through intuition, tradition, or faith.

north celestial pole (NCP) The point on the celestial sphere directly above Earth's North Pole.

nova The dramatic brightening of a star that lasts for a few weeks and then subsides; it occurs when a burst of hydrogen fusion ignites in a shell on the surface of an accreting white dwarf in a binary star system.

nuclear fission The process in which a larger nucleus splits into two (or more) smaller particles.

nuclear fusion The process in which two (or more) smaller nuclei slam together and make one larger nucleus.

nucleus (of a comet) The solid portion of a comet—the only portion that exists when the comet is far from the Sun.

nucleus (of an atom) The compact center of an atom made from protons and neutrons.

observable universe The portion of the entire universe that, at least in principle, can be seen from Earth.

Occam's razor A principle often used in science, holding that scientists should prefer the simpler of two models that agree equally well with observations; named after the medieval scholar William of Occam (1285–1349).

Olbers' paradox A paradox pointing out that if the universe were infinite in both age and size (with stars found throughout the universe), then the sky would not be dark at night.

Oort cloud A huge, spherical region centered on the Sun, extending perhaps halfway to the nearest stars, in which trillions of comets orbit the Sun with random inclinations, orbital directions, and eccentricities.

opacity A measure of how much light a material absorbs compared to how much it transmits; materials with higher opacity absorb more light.

opaque Describes a material that absorbs light.

open cluster A cluster of up to several thousand stars; open clusters are found only in the disks of galaxies and often contain young stars.

open universe A universe in which spacetime has an overall shape analogous to the surface of a saddle.

opposition The point at which a planet appears opposite the Sun in our sky.

optical quality The ability of a lens, mirror, or telescope to obtain clear and properly focused images.

orbit The path followed by a celestial body because of gravity; an orbit may be *bound* (elliptical) or *unbound* (parabolic or hyperbolic).

orbital energy The sum of an orbiting object's kinetic and gravitational potential energies.

orbital resonance A situation in which one object's orbital period is a simple ratio of another object's period, such as 1/2, 1/4, or 5/3. In such cases, the two objects periodically line up with each other, and the extra gravitational attractions at these times can affect the objects' orbits.

orbital velocity law A variation on Newton's version of Kepler's third law that allows us to use a star's orbital speed and distance from the galactic center to determine the total mass of the galaxy contained *within* the star's orbit; mathematically,

$$M_r = \frac{r \times v^2}{G}$$

where M_r is the mass contained within the star's orbit, r is the star's distance from the galactic center, v is the star's orbital velocity, and G is the gravitational constant.

orbiters (of other worlds) Spacecraft that go into orbit of another world for long-term study.

outer solar system Generally considered to encompass the region of our solar system beginning at about the orbit of Jupiter.

outgassing The process of releasing gases from a planetary interior, usually through volcanic eruptions.

oxidation Chemical reactions, often with rocks on the surface of a planet, that remove oxygen from the atmosphere.

ozone The molecule O_3, which is a particularly good absorber of ultraviolet light.

ozone depletion The decline in levels of atmospheric ozone found worldwide on Earth, especially in Antarctica, in recent years.

ozone hole A place where the concentration of ozone in the stratosphere is dramatically lower than is the norm.

pair production The process in which a concentration of energy spontaneously turns into a particle and its antiparticle.

parabola The precise mathematical shape of a special type of unbound orbit allowed under the force of gravity. If an object in a parabolic orbit loses only a tiny amount of energy, it will become bound.

paradigm (in science) A general pattern of thought that tends to shape scientific study during a particular time period.

paradox A situation that, at least at first, seems to violate common sense or contradict itself. Resolving paradoxes often leads to deeper understanding.

parallax The apparent shifting of an object against the background, due to viewing it from different positions. *See also* stellar parallax.

parallax angle Half of a star's annual back-and-forth shift due to stellar parallax; related to the star's distance according to the formula

$$\text{distance in parsecs} = \frac{1}{p}$$

where p is the parallax angle in arcseconds.

parsec (pc) The distance to an object with a parallax angle of 1 arcsecond; approximately equal to 3.26 light-years.

partial lunar eclipse A lunar eclipse during which the Moon becomes only partially covered by Earth's umbral shadow.

partial solar eclipse A solar eclipse during which the Sun becomes only partially blocked by the disk of the Moon.

particle accelerator A machine designed to accelerate subatomic particles to high speeds in order to create new particles or to test fundamental theories of physics.

particle era The era of the universe lasting from 10^{-10} second to 0.001 second after the Big Bang, during which subatomic particles were continually created and destroyed, and ending when matter annihilated antimatter.

peculiar velocity (of a galaxy) The component of a galaxy's velocity relative to the Milky Way that deviates from the velocity expected by Hubble's law.

penumbra The lighter, outlying regions of a shadow.

penumbral lunar eclipse A lunar eclipse during which the Moon passes only within Earth's penumbral shadow and does not fall within the umbra.

perigee The point at which an object orbiting Earth is nearest to Earth.

perihelion The point at which an object orbiting the Sun is closest to the Sun.

period–luminosity relation The relation that describes how the luminosity of a Cepheid variable star is related to the period between peaks in its brightness; the longer the period, the more luminous the star.

phase (of matter) The state determined by the way in which atoms or molecules are held together; the common phases are solid, liquid, and gas.

phase (of the Moon or a planet) The state determined by the portion of the visible face of the Moon (or of a planet) that is illuminated by sunlight. For the Moon, the phases cycle through new, waxing crescent, first-quarter, waxing gibbous, full, waning gibbous, third-quarter, waning crescent, and back to new.

photon An individual particle of light, characterized by a wavelength and a frequency.

photosphere The visible surface of the Sun, where the temperature averages just under 6000 K.

pixel An individual "picture element" on a CCD.

Planck era The era of the universe prior to the Planck time.

Planck's constant A universal constant, abbreviated h, with a value of $h = 6.626 \times 10^{-34}$ joule $\times$ s.

Planck time The time when the universe was 10^{-43} second old, before which random energy fluctuations were so large that our current theories are powerless to describe what might have been happening.

planet A moderately large object that orbits a star and shines primarily by reflecting light from its star. More precisely, according to a definition approved in 2006, a planet is an object that (1) orbits a star (but is itself neither a star nor a moon); (2) is massive enough for its own gravity to give it a nearly round shape; and (3) has cleared the neighborhood around its orbit. Objects that meet the first two criteria but not the third, including Ceres, Pluto, and Eris, are designated *dwarf planets*.

planetary geology The extension of the study of Earth's surface and interior to apply to other solid bodies in the solar system, such as terrestrial planets and jovian planet moons.

planetary migration A process through which a planet can move from the orbit on which it is born to a different orbit that is closer to or farther from its star.

planetary nebula The glowing cloud of gas ejected from a low-mass star at the end of its life.

planetesimals The building blocks of planets, formed by accretion in the solar nebula.

plasma A gas consisting of ions and electrons.

plasma tail (of a comet) One of two tails seen when a comet passes near the Sun (the other is the dust tail). It is composed of ionized gas blown away from the Sun by the solar wind.

plates (on a planet) Pieces of a lithosphere that apparently float upon the denser mantle below.

plate tectonics The geological process in which plates are moved around by stresses in a planet's mantle.

polarization (of light) The property of light describing how the electric and magnetic fields of light waves are aligned; light is said to be *polarized* when all of the photons have their electric and magnetic fields aligned in some particular way.

Population I *See* disk population.

Population II *See* spheroidal population.

positron *See* antielectron.

potential energy Energy stored for later conversion into kinetic energy; includes gravitational potential energy, electrical potential energy, and chemical potential energy.

power The rate of energy usage, usually measured in watts (1 watt = 1 joule/s).

precession The gradual wobble of the axis of a rotating object around a vertical line.

precipitation Condensed atmospheric gases that fall to the surface in the form of rain, snow, or hail.

pressure The force (per unit area) pushing on an object. In astronomy, we are generally interested in pressure applied by surrounding gas (or plasma). Ordinarily, such pressure is related to the temperature of the gas (*see* thermal pressure). In objects such as white dwarfs and neutron stars, pressure may arise from a quantum effect (*see* degeneracy pressure). Light can also exert pressure (*see* radiation pressure).

primary mirror The large, light-collecting mirror of a reflecting telescope.

prime focus (of a reflecting telescope) The first point at which light focuses after bouncing off the primary mirror; located in front of the primary mirror.

prime meridian The meridian of longitude that passes through Greenwich, England; defined to be longitude 0°.

primitive meteorites Meteorites that formed at the same time as the solar system itself, about 4.6 billion years ago. Primitive meteorites from the inner asteroid belt are usually stony, and those from the outer belt are usually carbon-rich.

processed meteorites Meteorites that apparently once were part of a larger object that "processed" the original material of the solar nebula into another form. Processed meteorites can be rocky if chipped from the surface or mantle, or metallic if blasted from the core.

proper motion The motion of an object in the plane of the sky, perpendicular to our line of sight.

protogalactic cloud A huge, collapsing cloud of intergalactic gas from which an individual galaxy formed.

proton–proton chain The chain of reactions by which low-mass stars (including the Sun) fuse hydrogen into helium.

protons Particles found in atomic nuclei with positive electrical charge, built from three quarks.

protoplanetary disk A disk of material surrounding a young star (or protostar) that may eventually form planets.

protostar A forming star that has not yet reached the point where sustained fusion can occur in its core.

protostellar disk A disk of material surrounding a protostar; essentially the same as a protoplanetary disk, but may not necessarily lead to planet formation.

protostellar wind The relatively strong wind from a protostar.

protosun The central object in the forming solar system that eventually became the Sun.

pseudoscience Something that purports to be science or may appear to be scientific but that does not adhere to the testing and verification requirements of the scientific method.

Ptolemaic model The geocentric model of the universe developed by Ptolemy in about 150 A.D.

pulsar A neutron star from which we see rapid pulses of radiation as it rotates.

pulsating variable stars Stars that grow alternately brighter and dimmer as their outer layers expand and contract in size.

quantum laws The laws that describe the behavior of particles on a very small scale; *see also* quantum mechanics.

quantum mechanics The branch of physics that deals with the very small, including molecules, atoms, and fundamental particles.

quantum state The complete description of the state of a subatomic particle, including its location, momentum, orbital angular momentum, and spin, to the extent allowed by the uncertainty principle.

quantum tunneling The process in which, thanks to the uncertainty principle, an electron or other subatomic particle appears on the other side of a barrier that it does not have the energy to overcome in a normal way.

quarks The building blocks of protons and neutrons; quarks are one of the two basic types of fermions (leptons are the other).

quasar The brightest type of active galactic nucleus.

radar mapping Imaging of a planet by bouncing radar waves off its surface, especially important for Venus and Titan, where thick clouds mask the surface.

radar ranging A method of measuring distances within the solar system by bouncing radio waves off planets.

radial motion The component of an object's motion directed toward or away from us.

radial velocity The portion of any object's total velocity that is directed toward or away from us. This part of the velocity is the only part that we can measure with the Doppler effect.

radiation pressure Pressure exerted by photons of light.

radiation zone (of a star) A region of the interior in which energy is transported primarily by radiative diffusion.

radiative diffusion The process by which photons gradually migrate from a hot region (such as the solar core) to a cooler region (such as the solar surface).

radiative energy Energy carried by light; the energy of a photon is Planck's constant times its frequency, or $h \times f$.

radioactive decay The spontaneous change of an atom into a different element, in which its nucleus breaks apart or a proton turns into an electron. It releases heat in a planet's interior.

radioactive element (or **radioactive isotope**) A substance whose nucleus tends to fall apart spontaneously.

radio galaxy A galaxy that emits unusually large quantities of radio waves; thought to contain an active galactic nucleus powered by a supermassive black hole.

radio lobes The huge regions of radio emission found on either side of radio galaxies. The lobes apparently contain plasma ejected by powerful jets from the galactic center.

radiometric dating The process of determining the age of a rock (i.e., the time since it solidified) by comparing the present amount of a radioactive substance to the amount of its decay product.

radio waves Light with very long wavelengths (and hence low frequencies)—longer than those of infrared light.

random walk A type of haphazard movement in which a particle or photon moves through a series of bounces, with each bounce sending it in a random direction.

recession velocity (of a galaxy) The speed at which a distant galaxy is moving away from us because of the expansion of the universe.

recollapsing universe A universe in which the collective gravity of all its matter eventually halts and reverses the expansion, causing the galaxies to come crashing back together and the universe to end in a fiery Big Crunch.

red giant A giant star that is red in color.

red-giant winds The relatively dense but slow winds from red giant stars.

redshift (Doppler) A Doppler shift in which spectral features are shifted to longer wavelengths, observed when an object is moving away from the observer.

reference frame (or **frame of reference**) What two people (or objects) share if they are *not* moving relative to one another.

reflecting telescope A telescope that uses mirrors to focus light.

reflection (of light) The process by which matter changes the direction of light.

reflection nebula A nebula that we see as a result of starlight reflected from interstellar dust grains. Reflection nebulae tend to have blue and black tints.

refracting telescope A telescope that uses lenses to focus light.

resonance *See* orbital resonance.

rest wavelength The wavelength of a spectral feature in the absence of any Doppler shift or gravitational redshift.

retrograde motion Motion that is backward compared to the norm. For example, we see Mars in apparent retrograde motion during the periods of time when it moves westward, rather than the more common eastward, relative to the stars.

revolution The orbital motion of one object around another.

right ascension (RA) Analogous to longitude, but on the celestial sphere; the angular east-west distance between the spring equinox and a location on the celestial sphere.

rings (planetary) The collections of numerous small particles orbiting a planet within its Roche tidal zone.

Roche tidal zone The region within two to three planetary radii (of any planet) in which the tidal forces tugging an object apart become comparable to the gravitational forces holding it together; planetary rings are always found within the Roche tidal zone.

rocks (in solar system theory) Materials common on the surface of Earth, such as silicon-based minerals, that are solid at temperatures and pressures found on Earth but typically melt or vaporize at temperatures of 500–1300 K.

rotation The spinning of an object around its axis.

rotation curve A graph that plots rotational (or orbital) velocity against distance from the center for any object or set of objects.

runaway greenhouse effect A positive feedback cycle in which heating caused by the greenhouse effect causes more greenhouse gases to enter the atmosphere, which further enhances the greenhouse effect.

saddle-shaped (or **hyperbolic**) **geometry** The type of geometry in which the rules—such as that two lines that begin parallel eventually diverge—are most easily visualized on a saddle-shaped surface.

Sagittarius Dwarf A small dwarf elliptical galaxy that is currently passing through the disk of the Milky Way Galaxy.

saros cycle The period over which the basic pattern of eclipses repeats, which is about 18 years $11\frac{1}{3}$ days.

satellite Any object orbiting another object.

scattered light Light that is reflected into random directions.

Schwarzschild radius A measure of the size of the event horizon of a black hole.

science The search for knowledge that can be used to explain or predict natural phenomena in a way that can be confirmed by rigorous observations or experiments.

scientific method An organized approach to explaining observed facts through science.

scientific theory A model of some aspect of nature that has been rigorously tested and has passed all tests to date.

seafloor crust On Earth, the thin, dense crust of basalt created by seafloor spreading.

seafloor spreading On Earth, the creation of new seafloor crust at mid-ocean ridges.

search for extraterrestrial intelligence (SETI) The name given to observing projects designed to search for signs of intelligent life beyond Earth.

secondary mirror A small mirror in a reflecting telescope, used to reflect light gathered by the primary mirror toward an eyepiece or instrument.

sedimentary rock A rock that formed from sediments created and deposited by erosional processes.

seismic waves Earthquake-induced vibrations that propagate through a planet.

selection effect (or **selection bias**) A type of bias that arises from the way in which objects of study are selected and that can lead to incorrect conclusions. For example, when you are counting animals in a jungle it is easiest to see brightly colored animals, which could mislead you into thinking that these animals are the most common.

semimajor axis Half the distance across the long axis of an ellipse; in this text, it is usually referred to as the *average* distance of an orbiting object, abbreviated *a* in the formula for Kepler's third law.

Seyfert galaxies The name given to a class of galaxies that are found relatively nearby and that have nuclei much like those of quasars, except that they are less luminous.

shepherd moons Tiny moons within a planet's ring system that help force particles into a narrow ring; a variation on *gap moons*.

shield volcano A shallow-sloped volcano made from the flow of low-viscosity basaltic lava.

shock wave A wave of pressure generated by gas moving faster than the speed of sound.

sidereal day The time of 23 hours 56 minutes 4.09 seconds between successive appearances of any particular star on the meridian; essentially, the true rotation period of Earth.

sidereal month The time required for the Moon to orbit Earth once (as measured against the stars); about $27\frac{1}{4}$ days.

sidereal period (of a planet) A planet's actual orbital period around the Sun.

sidereal time Time measured according to the position of stars in the sky rather than the position of the Sun in the sky. *See also* local sidereal time.

sidereal year The time required for Earth to complete exactly one orbit as measured against the stars; about 20 minutes longer than the tropical year on which our calendar is based.

silicate rock A silicon-rich rock.

singularity The place at the center of a black hole where, in principle, gravity crushes all matter to an infinitely tiny and dense point.

Small Magellanic Cloud One of two small, irregular galaxies (the other is the Large Magellanic Cloud) located about 150,000 light-years away; it probably orbits the Milky Way Galaxy.

small solar system body An asteroid, comet, or other object that orbits a star but is too small to qualify as a planet or dwarf planet.

snowball Earth Name given to a hypothesis suggesting that, some 600–700 million years ago, Earth experienced a period in which it became cold enough for glaciers to exist worldwide, even in equatorial regions.

solar activity Short-lived phenomena on the Sun, including the emergence and disappearance of individual sunspots, prominences, and flares; sometimes called *solar weather*.

solar circle The Sun's orbital path around the galaxy, which has a radius of about 28,000 light-years.

solar day 24 hours, which is the average time between appearances of the Sun on the meridian.

solar eclipse An event that occurs when the Moon's shadow falls on Earth, which can occur only at new moon. A solar eclipse may be total, partial, or annular.

solar flares Huge and sudden releases of energy on the solar surface, probably caused when energy stored in magnetic fields is suddenly released.

solar luminosity The luminosity of the Sun, which is approximately 4×10^{26} watts.

solar maximum The time during each sunspot cycle at which the number of sunspots is the greatest.

solar minimum The time during each sunspot cycle at which the number of sunspots is the smallest.

solar nebula The piece of interstellar cloud from which our own solar system formed.

solar neutrino problem The disagreement between the predicted and observed number of neutrinos coming from the Sun.

solar prominences Vaulted loops of hot gas that rise above the Sun's surface and follow magnetic field lines.

solar sail A large, highly reflective (and thin, to minimize mass) piece of material that can "sail" through space using pressure exerted by sunlight.

solar system (or **star system**) A star (sometimes more than one star) and all the objects that orbit it.

solar thermostat See stellar thermostat; the solar thermostat is the same idea applied to the Sun.

solar wind A stream of charged particles ejected from the Sun.

solid phase The phase of matter in which atoms or molecules are held rigidly in place.

solstice See summer solstice and winter solstice.

sound wave A wave of alternately rising and falling pressure.

south celestial pole (SCP) The point on the celestial sphere directly above Earth's South Pole.

spacetime The inseparable, four-dimensional combination of space and time.

spacetime diagram A graph that plots a spatial dimension on one axis and time on another axis.

special theory of relativity Einstein's theory that describes the effects of the fact that all motion is relative and that everyone always measures the same speed of light.

spectral lines Bright or dark lines that appear in an object's spectrum, which we can see when we pass the object's light through a prismlike device that spreads out the light like a rainbow.

spectral resolution The degree of detail that can be seen in a spectrum; the higher the spectral resolution, the more detail we can see.

spectral type A way of classifying a star by the lines that appear in its spectrum; it is related to surface temperature. The basic spectral types are designated by a letter (OBAFGKM, with O for the hottest stars and M for the coolest) and are subdivided with numbers from 0 through 9.

spectrograph An instrument used to record spectra.

spectroscopic binary A binary star system whose binary nature is revealed because we detect the spectral lines of one or both stars alternately becoming blueshifted and redshifted as the stars orbit each other.

spectroscopy (in astronomical research) The process of obtaining spectra from astronomical objects.

spectrum (of light) See electromagnetic spectrum.

speed The rate at which an object moves. Its units are distance divided by time, such as m/s or km/hr.

speed of light The speed at which light travels, which is about 300,000 km/s.

spherical geometry The type of geometry in which the rules—such as that lines that begin parallel eventually meet—are those that hold on the surface of a sphere.

spheroidal component (of a galaxy) The portion of any galaxy that is spherical (or football-like) in shape and contains very little cool gas; it generally contains only very old stars. Elliptical galaxies have only a spheroidal component, while spiral galaxies also have a disk component.

spheroidal galaxy Another name for an *elliptical galaxy.*

spheroidal population Stars that orbit within the spheroidal component of a galaxy; sometimes called *Population II.* Elliptical galaxies have only a spheroidal population (they lack a disk population), while spiral galaxies have spheroidal population stars in their bulges and halos.

spin (quantum) See spin angular momentum.

spin angular momentum The inherent angular momentum of a fundamental particle; often simply called *spin.*

spiral arms The bright, prominent arms, usually in a spiral pattern, found in most spiral galaxies.

spiral density waves Gravitationally driven waves of enhanced density that move through a spiral galaxy and are responsible for maintaining its spiral arms.

spiral galaxies Galaxies that look like flat white disks with yellowish bulges at their centers. The disks are filled with cool gas and dust, interspersed with hotter ionized gas, and usually display beautiful spiral arms.

spreading centers (geological) Places where hot mantle material rises upward between plates and then spreads sideways, creating new seafloor crust.

spring (March) equinox Refers both to the point in Pisces on the celestial sphere where the ecliptic crosses the celestial equator and to the moment in time when the Sun appears at that point each year (around March 21).

spring tides The higher-than-average tides on Earth that occur at new and full moon, when the tidal forces from the Sun and Moon both act along the same line.

standard candle An object for which we have some means of knowing its true luminosity, so that we can use its apparent brightness to determine its distance with the luminosity–distance formula.

standard model (of physics) The current theoretical model that describes the fundamental particles and forces in nature.

standard time Time measured according to the internationally recognized time zones.

star A large, glowing ball of gas that generates energy through nuclear fusion in its core. The term *star* is sometimes applied to objects that are in the process of becoming true stars (e.g., protostars) and to the remains of stars that have died (e.g., neutron stars).

starburst galaxy A galaxy in which stars are forming at an unusually high rate.

star cluster See cluster of stars.

star–gas–star cycle The process of galactic recycling in which stars expel gas into space, where it mixes with the interstellar medium and eventually forms new stars.

star system See solar system.

state (quantum) See quantum state.

steady state theory A now-discredited theory that held that the universe had no beginning and looks about the same at all times.

Stefan–Boltzmann constant A constant that appears in the laws of thermal radiation, with value

$$\sigma = 5.7 \times 10^{-8} \frac{\text{watt}}{\text{m}^2 \times \text{Kelvin}^4}$$

stellar evolution The formation and development of stars.

stellar parallax The apparent shift in the position of a nearby star (relative to distant objects) that occurs as we view the star from different positions in Earth's orbit of the Sun each year.

stellar thermostat The regulation of a star's core temperature that comes about when a star is in both energy balance (the rate at which fusion releases energy in the star's core is balanced with the rate at which the star's surface radiates energy into space) and gravitational equilibrium.

stellar wind A stream of charged particles ejected from the surface of a star.

stratosphere An intermediate-altitude layer of Earth's atmosphere that is warmed by the absorption of ultraviolet light from the Sun.

stratovolcano A steep-sided volcano made from viscous lavas that can't flow very far before solidifying.

string theory New ideas, not yet well-tested, that attempt to explain all of physics in a much simpler way than current theories.

stromatolites Large bacterial "colonies."

strong force One of the four fundamental forces; it is the force that holds atomic nuclei together.

subduction (of tectonic plates) The process in which one plate slides under another.

subduction zones Places where one plate slides under another.

subgiant A star that is between being a main-sequence star and being a giant; subgiants have inert helium cores and hydrogen-fusing shells.

sublimation The process by which atoms or molecules escape into the gas phase from a solid.

summer (June) solstice Refers both to the point on the celestial sphere where the ecliptic is farthest north of the celestial equator and to the moment in time when the Sun appears at that point each year (around June 21).

sunspot cycle The period of about 11 years over which the number of sunspots on the Sun rises and falls.

sunspots Blotches on the surface of the Sun that appear darker than surrounding regions.

superbubble Essentially a giant interstellar bubble, formed when the shock waves of many individual bubbles merge to form a single giant shock wave.

superclusters The largest known structures in the universe, consisting of many clusters of galaxies, groups of galaxies, and individual galaxies.

supergiants The very large and very bright stars (luminosity class I) that appear at the top of an H-R diagram.

supermassive black holes Giant black holes, with masses millions to billions of times that of our Sun, thought to reside in the centers of many galaxies and to power active galactic nuclei.

supernova The explosion of a star.

Supernova 1987A A supernova witnessed on Earth in 1987; it was the nearest supernova seen in nearly 400 years and helped astronomers refine theories of supernovae.

supernova remnant A glowing, expanding cloud of debris from a supernova explosion.

surface area–to–volume ratio The ratio defined by an object's surface area divided by its volume; this ratio is larger for smaller objects (and vice versa).

synchronous rotation The rotation of an object that always shows the same face to an object that it is orbiting because its rotation period and orbital period are equal.

synchrotron radiation A type of radio emission that occurs when electrons moving at nearly the speed of light spiral around magnetic field lines.

synodic month (or lunar month) The time required for a complete cycle of lunar phases, which averages about $29\frac{1}{2}$ days.

synodic period (of a planet) The time between successive alignments of a planet and the Sun in our sky; measured from opposition to opposition for a planet beyond Earth's orbit, or from superior conjunction to superior conjunction for Mercury and Venus.

tangential motion The component of an object's motion directed across our line of sight.

tangential velocity The portion of any object's total velocity that is directed across (perpendicular to) our line of sight. This part of the velocity cannot be measured with the Doppler effect. It can be measured only by observing the object's gradual motion across our sky.

tectonics The disruption of a planet's surface by internal stresses.

temperature A measure of the average kinetic energy of particles in a substance.

terrestrial planets Rocky planets similar in overall composition to Earth.

theories of relativity (special and general) Einstein's theories that describe the nature of space, time, and gravity.

theory (in science) See scientific theory.

theory of evolution The theory, first advanced by Charles Darwin, that explains how evolution occurs through the process of natural selection.

thermal emitter An object that produces a thermal radiation spectrum; sometimes called a *blackbody*.

thermal energy The collective kinetic energy, as measured by temperature, of the many individual particles moving within a substance.

thermal escape The process in which atoms or molecules in a planet's exosphere move fast enough to escape into space.

thermal pressure The ordinary pressure in a gas arising from motions of particles that can be attributed to the object's temperature.

thermal pulses The predicted upward spikes in the rate of helium fusion, occurring every few thousand years, that occur near the end of a low-mass star's life.

thermal radiation The spectrum of radiation produced by an opaque object that depends only on the object's temperature; sometimes called *blackbody radiation*.

thermosphere A high, hot, X-ray-absorbing layer of an atmosphere, just below the exosphere.

third-quarter (phase) The phase of the Moon that occurs three-quarters of the way through each cycle of phases, in which precisely half of the visible face is illuminated by sunlight.

tidal force A force that occurs when the gravity pulling on one side of an object is larger than that on the other side, causing the object to stretch.

tidal friction Friction within an object that is caused by a tidal force.

tidal heating A source of internal heating created by tidal friction. It is particularly important for satellites with eccentric orbits such as Io and Europa.

time dilation The effect in which you observe time running more slowly in reference frames moving relative to you.

timing (in astronomical research) The process of tracking how the light intensity from an astronomical object varies with time.

torque A twisting force that can cause a change in an object's angular momentum.

total apparent brightness See apparent brightness. The word "total" is sometimes added to make clear that we are talking about light across all wavelengths, not just visible light.

totality (eclipse) The portion of a total lunar eclipse during which the Moon is fully within Earth's umbral shadow or a total solar eclipse during which the Sun's disk is fully blocked by the Moon.

total luminosity See luminosity. The word "total" is sometimes added to make clear that we are talking about light across all wavelengths, not just visible light.

total lunar eclipse A lunar eclipse in which the Moon becomes fully covered by Earth's umbral shadow.

total solar eclipse A solar eclipse during which the Sun becomes fully blocked by the disk of the Moon.

transit An event in which a planet passes in front of a star (or the Sun) as seen from Earth. Only Mercury and Venus can be seen in transit of our Sun. The search for transits of extrasolar planets is an important planet detection strategy.

transmission (of light) The process in which light passes through matter without being absorbed.

transparent Describes a material that transmits light.

tree of life (evolutionary) A diagram that shows relationships between different species as inferred from genetic comparisons.

triple-alpha reaction See helium fusion.

Trojan asteroids Asteroids found within two stable zones that share Jupiter's orbit but lie 60° ahead of and behind Jupiter.

tropical year The time from one spring equinox to the next, on which our calendar is based.

Tropic of Cancer The circle on Earth with latitude 23.5°N, which marks the northernmost latitude at which the Sun ever passes directly overhead (which it does at noon on the summer solstice).

Tropic of Capricorn The circle on Earth with latitude 23.5°S, which marks the southernmost latitude at which the Sun ever passes directly overhead (which it does at noon on the winter solstice).

tropics The region on Earth surrounding the equator and extending from the Tropic of Capricorn (latitude 23.5°S) to the Tropic of Cancer (latitude 23.5°N).

troposphere The lowest atmospheric layer, in which convection and weather occur.

Tully–Fisher relation A relationship among spiral galaxies showing that the faster a spiral galaxy's rotation speed, the more luminous it is. It is important because it allows us to determine the distance to a spiral galaxy once we measure its rotation rate and apply the luminosity–distance formula.

turbulence Rapid and random motion.

21-cm line A spectral line from atomic hydrogen with wavelength 21 cm (in the radio portion of the spectrum).

ultraviolet light Light with wavelengths that fall in the portion of the electromagnetic spectrum between visible light and X rays.

umbra The dark central region of a shadow.

unbound orbits Orbits on which an object comes in toward a large body only once, never to return; unbound orbits may be parabolic or hyperbolic in shape.

uncertainty principle The law of quantum mechanics that states that we can never know both a particle's position and its momentum, or both its energy and the time it has the energy, with absolute precision.

universal law of gravitation The law expressing the force of gravity (F_g) between two objects, given by the formula

$$F_g = G\,\frac{M_1 M_2}{d^2}$$

$$\left(\text{where } G = 6.67 \times 10^{-11}\ \frac{\text{m}^3}{\text{kg} \times \text{s}^2}\right)$$

universal time (UT) Standard time in Greenwich, England (or anywhere on the prime meridian).

universe The sum total of all matter and energy.

up quark One of the two quark types (the other is the down quark) found in ordinary protons and neutrons; has a charge of $+\frac{2}{3}$.

velocity The combination of speed and direction of motion; it can be stated as a speed in a particular direction, such as 100 km/hr due north.

virtual particles Particles that "pop" in and out of existence so rapidly that, according to the uncertainty principle, they cannot be directly detected.

viscosity The thickness of a liquid described in terms of how rapidly it flows; low-viscosity liquids flow quickly (e.g., water), while high-viscosity liquids flow slowly (e.g., molasses).

visible light The light our eyes can see, ranging in wavelength from about 400 to 700 nm.

visual binary A binary star system in which both stars can be resolved through a telescope.

voids Huge volumes of space between superclusters that appear to contain very little matter.

volatiles Substances, such as water, carbon dioxide, and methane, that are usually found as gases, liquids, or surface ices on the terrestrial worlds.

volcanic plains Vast, relatively smooth areas created by the eruption of very runny lava.

volcanism The eruption of molten rock, or lava, from a planet's interior onto its surface.

waning (phases) The set of phases in which less and less of the visible face of the Moon is illuminated; the phases that come after full moon but before new moon.

watt The standard unit of power in science; defined as 1 watt = 1 joule/s.

wavelength The distance between adjacent peaks (or troughs) of a wave.

waxing (phases) The set of phases in which more and more of the visible face of the Moon is becoming illuminated; the phases that come after new moon but before full moon.

weak bosons The exchange particles for the weak force.

weak force One of the four fundamental forces; it is the force that mediates nuclear reactions, and it is the only force besides gravity felt by weakly interacting particles.

weakly interacting particles Particles, such as neutrinos and WIMPs, that respond only to the weak force and gravity; that is, they do not feel the strong force or the electromagnetic force.

weather The ever-varying combination of winds, clouds, temperature, and pressure in a planet's troposphere.

weight The net force that an object applies to its surroundings; in the case of a stationary body on the surface of Earth, it equals mass × acceleration of gravity.

weightlessness A weight of zero, as occurs during free-fall.

white dwarf limit (or **Chandrasekhar limit**) The maximum possible mass for a white dwarf, which is about $1.4 M_{Sun}$.

white dwarfs The hot, compact corpses of low-mass stars, typically with a mass similar to that of the Sun compressed to a volume the size of Earth.

white dwarf supernova A supernova that occurs when an accreting white dwarf reaches the white-dwarf limit, ignites runaway carbon fusion, and explodes like a bomb; often called a *Type Ia supernova*.

WIMPs A possible form of dark matter consisting of subatomic particles that are dark because they do not respond to the electromagnetic force; stands for *weakly interacting massive particles*.

winter (December) solstice Refers both to the point on the celestial sphere where the ecliptic is farthest south of the celestial equator and to the moment in time when the Sun appears at that point each year (around December 21).

worldline A line that represents an object on a spacetime diagram.

wormholes The name given to hypothetical tunnels through hyperspace that might connect two distant places in our universe.

X-ray binary A binary star system that emits substantial amounts of X rays, thought to be from an accretion disk around a neutron star or black hole.

X-ray burster An object that emits a burst of X rays every few hours to every few days; each burst lasts a few seconds and is thought to be caused by helium fusion on the surface of an accreting neutron star in a binary system.

X-ray bursts Bursts of X rays coming from sudden ignition of fusion on the surface of an accreting neutron star in an X-ray binary system.

X rays Light with wavelengths that fall in the portion of the electromagnetic spectrum between ultraviolet light and gamma rays.

Zeeman effect The splitting of spectral lines by a magnetic field.

zenith The point directly overhead, which has an altitude of 90°.

zodiac The constellations on the celestial sphere through which the ecliptic passes.

zones (on a jovian planet) Bright bands of rising air that encircle a jovian planet at a particular set of latitudes.

CREDITS

PHOTO CREDITS

Chapter 1
Opener NASA Earth Observing System 1.2 David Malin, Australian Astronomical Observatory 1.3 Jerry Lodriguss/Photo Researchers, Inc. 1.5 and p. 20 Jeffrey Bennett 1.7 Goddard Institute for Space Studies p. 8 Michael Williamson, The Washington Post/Getty Images 1.8 Akira Fujii 1.9 Jeffrey Bennett 1.11 Terraced hills: Blakeley Kim/Pearson Science; pyramid: Corel Corporation; Earth: NASA Earth Observing System; telescope: Jon Arnold Images Ltd/Alamy 1.14 NASA Earth Observing System 1.15 NASA Earth Observing System 1.17 NASA Earth Observing System p. 21 NASA Earth Observing System

Chapter 2
Opener and p. 49 David Nunuk/Photo Researchers, Inc. 2.1 Wally Pacholka 2.11b geophoto/Alamy 2.17 Frank Zullo/Photo Researchers, Inc. 2.18 Arnulf Husmo/Getty Images Inc. 2.19 The GeoSphere Project 2.21 NASA/JPL-Caltech/University of Arizona 2.26 (top and bottom) Akira Fujii; (center) epa european pressphoto agency b.v./Alamy 2.27 Itahisa González Álvarez 2.28b NASA 2.29 Akira Fujii 2.31 Akira Fujii 2.32 Tunc Tezel

Chapter 3
Opener NASA Earth Observing System 3.2 Karl Kost/Alamy 3.3 and p. 79 Timothy Kohlbacher/Shutterstock.com 3.4 SuperStock/SuperStock 3.5 National Park Service/U.S. Department of the Interior 3.6 William E. Woolam 3.7 Jeffrey Bennett 3.8 Kristian Peetz/Alamy 3.9 Amy Nichole Harris/Shutterstock.com 3.10 WALTER MEAYERS EDWARDS/National Geographic Stock Image Collection 3.11a and b Courtesy of Carl Sagan Production, Inc.; c Bibliotheca Alexandrina p. 64 (left) Image Asset Management Ltd./SuperStock; (right) SuperStock/SuperStock 3.15 Gianni Tortoli/Photo Researchers, Inc. p. 65 Science Source/Photo Researchers, Inc. p. 68 Fine Art Images/SuperStock 3.20 Stuart J. Robbins 3.21 Galileo Galilei 3.25 (top left) Tunc Tezel; (bottom left) University of Oklahoma History of Science Collections; (bottom right) The University of Sydney Library p. 74 The Science Museum of Virginia 3.26 INTERFOTO/Alamy

Chapter S1
Opener Arnulf Husmo/Stone Allstock/Getty Images Inc. S1.5 Rex Features via AP Images S1.6 Dennis Mcdonald/Alamy S1.24a Science and Society/SuperStock; b Bettmann/Corbis; c Photo Researchers, Inc.; d Deepak Budhraja/iStockphoto.com p. 108 (top left) Timothy Kohlbacher/Shutterstock.com; (top right) SuperStock/SuperStock p. 109 (top left) Yerkes Observatory; (top center) Pictorial Press Ltd/Alamy; (top right) NASA/Johnson Space Center

Chapter 4
Opener NASA Earth Observing System p. 115 Sir Godfrey Kneller 4.5 (left and right) NASA/Goddard Institute for Space Studies; (center) Bruce Kluckhohn/Getty Images 4.10 and p. 132 Bikes: EyeWire/gettyimages.com; sunflower: Eric Gevaert/Alamy; gas pump: Don Hammond/Design Pics/Corbis 4.15 U.S. Department of Energy 4.19 NASA/Goddard Institute for Space Studies 4.24 (left) Travel Pix Collection/Jon Arnold Images/SuperStock; (right) Photononstop/SuperStock

Chapter 5
Opener NASA Earth Observing System 5.1 Richard Megna/Fundamental Photographs, NYC

Chapter 6
Opener and p. 184 Jon Arnold Images Ltd/Alamy 6.3 Solvin Zankl/Nature Picture Library 6.8 Jeffrey Bennett 6.9 Space Telescope Science Institute 6.10 Yerkes Observatory 6.11b National Optical Astronomy Observatories 6.12 (left) Richard Wainscoat/Alamy; (right) C.A.R.A./W. M. Keck Observatory 6.13 DMI David Malin Images 6.14 NASA Earth Observing System 6.18 NASA/Jet Propulsion Laboratory 6.19 CFHT Corporation 6.20 Richard Wainscoat/Alamy 6.21 NASA/Johnson Space Center 6.23 Spencer Grant/Photoedit 6.24 David Parker/Science Photo Library/National Astronomy and Ionosphere Center 6.25 Jim Ross/NASA 6.26 NASA Earth Observing System 6.27a NASA Earth Observing System 6.28 NASA Headquarters 6.30 Jon Arnold Images Ltd/Alamy 6.31 Jon Arnold Images Ltd/Alamy p. 188 EyeWire/gettyimages.com; Eric Gevaert/Alamy; Don Hammond/Design Pics/Corbis

Chapter 7
Opener The Planetary Society 7.2a and b NASA/Marshall Space Flight Center 7.3 NASA/Johns Hopkins University Applied Physics Laboratory/Carnegie Institution of Washington 7.4 (left) NASA/Marshall Space Flight Center; (right) Southern Methodist University 7.5a ARC Science Simulations; b NASA Earth Observing System 7.6 NASA Earth Observing System 7.7 ARC Science Simulations 7.8 and p. 210 NASA/Jet Propulsion Laboratory 7.9 ARC Science Simulations 7.10 ARC Science Simulations 7.11 NASA/STScI 7.12 NASA/Goddard Institute for Space Studies 7.13 Niescja Turner and Carter Emmart 7.15 and p. 211 NASA/JPL-Caltech/Univ. of Arizona

Chapter 8
8.2 NASA Earth Observing System 8.4a NASA Earth Observing System; b NASA Headquarters 8.5 and p. 229 NASA/Jet Propulsion Laboratory 8.7 Robert A. Haag 8.10 NASA Earth Observing System 8.11a and b NASA Earth Observing System

Chapter 9
Opener Tom Pfeiffer/VolcanoDiscovery/Getty Images 9.1 Mecury, Venus, Earth, Mars: NASA/Jet Propulsion Laboratory; Earth's Moon: Akira Fujii 9.3 Richard Megna/Fundamental Photographs, NYC 9.7 Don Davis 9.8a Brad Snowder; b Lunar & Planetary Institute and NASA; 9.9a NASA/Jet Propulsion

Laboratory; b NASA/Earth Observing System; c NASA/ Jet Propulsion Laboratory 9.10 Paul Chesley/Getty Images 9.11a NASA Earth Observing System; b NASA/Jet Propulsion Laboratory; c zschnepf/shutterstock.com 9.12a U.S. Geological Survey, Denver; b Cornforth Images/Alamy 9.13 (left) NASA Earth Observing System; (right) NASA/Jet Propulsion Laboratory 9.14a Gene Ahrens/Bruce Coleman Inc./Photoshot Holdings, Ltd.; b Joachim Messerschmidt/Bruce Coleman Inc./Photoshot Holdings, Ltd.; c Mytho/Fotolia; d Florian Werner/age fotostock 9.15a and p. 265 NASA; b NASA Earth Observing System 9.17 (left) NASA/Johns Hopkins University Applied Physics Laboratory/Carnegie Institution of Washington; (right) Akira Fujii 9.18 (left) Frank Barrett; (right) malcolm park/Oxford Scientific/gettyimages.com 9.19 Pearson Education 9.20 NASA Earth Observing System 9.21a NASA/Goddard Institute for Space Studies; b NASA/Goddard Space Flight Center 9.22a–c NASA/Johns Hopkins University Applied Physics Laboratory/ Carnegie Institution of Washington 9.23b NASA/Jet Propulsion Laboratory 9.24 (left) NASA Earth Observing System; (right) Lowell Observatory 9.25 NASA/Goddard Institute for Space Studies 9.26 and p. 265 (bottom) NASA/Goddard Institute for Space Studies; (top) NASA/Jet Propulsion Laboratory 9.27 (bottom) NASA/Goddard Institute for Space Studies; (top) NASA 9.28 NASA/Jet Propulsion Laboratory 9.29a EROS Data Center, U.S. Geological Survey; b National Air and Space Museum Smithsonian Institution; c NASA EOS Earth Observing System 9.30 and p. 265 (left) NASA/Jet Propulsion Laboratory; (right) Dr. Marjorie A Chan 9.31 NASA/JPL-Caltech/Cornell/ASU 9.32 NASA/JPL-Caltech/MSSS 9.33 NASA Earth Observing System 9.34 and p. 265 NASA/Jet Propulsion Laboratory 9.35 European Space Agency 9.36 Ted Stryk/NASA 9.37 National Geophysical Data Center 9.41 NASA/Jet Propulsion Laboratory 9.42 NASA/Jet Propulsion Laboratory 9.43 NASA/Jet Propulsion Laboratory 9.44 NASA/Jet Propulsion Laboratory 9.45 (top left) Ronen Zilberman/AP Images; (top center) Nina B/Shutterstock. com; (top right) University of Hawaii/SOEST; (bottom right) Alexander Malahoff p. 266 NASA Earth Observing System

Chapter 10
Opener NASA Earth Observing System 10.1 Mercury: NASA; surface of Venus: from the Voyage scale model solar system, developed by Challenger Center for Space Science Education, the Smithsonian Institution, and NASA, ©2001 David P. Anderson, Southern Methodist University; Earth valley: Dmitry/ Shutterstock; Moon, Mars, Earth: NASA Earth Observing System 10.2 and p. 303 NASA Earth Observing System 10.5 NASA Earth Observing System 10.11b Doug Duncan 10.25 NASA Earth Observing System 10.26 NASA/Jet Propulsion Laboratory 10.27 NASA/Jet Propulsion Laboratory 10.28 NASA Earth Observing System 10.29 NASA/JPL/University of Arizona 10.31 European Space Agency 10.32 NASA 10.33 NASA

Chapter 11
Opener NASA/Jet Propulsion Laboratory 11.1 NASA/Jet Propulsion Laboratory 11.10 NASA/Jet Propulsion Laboratory 11.11a NASA Earth Observing System; b NASA/Jet Propulsion Laboratory; c Lawrence Sromovsky; d NASA/Jet Propulsion Laboratory 11.12 NASA/Jet Propulsion Laboratory 11.14 NASA/Jet Propulsion Laboratory 11.15 Calypso, Janus, Telesto: NASA Earth Observing System; Pandora, Hyperion, Phoebe: NASA/Jet Propulsion Laboratory 11.16 NASA/Jet Propulsion Laboratory 11.17 NASA/Jet Propulsion Laboratory 11.18

NASA/Jet Propulsion Laboratory 11.20 NASA/Jet Propulsion Laboratory 11.21 NASA/Jet Propulsion Laboratory 11.22 NASA/ Jet Propulsion Laboratory 11.23 NASA/Jet Propulsion Laboratory 11.24 (left) NASA/Jet Propulsion Laboratory; (right) NASA Earth Observing System 11.25 NASA/Jet Propulsion Laboratory 11.26 NASA Earth Observing System 11.27 NASA/Jet Propulsion Laboratory 11.28 NASA/Jet Propulsion Laboratory 11.29 NASA/ Jet Propulsion Laboratory 11.30 NASA/Jet Propulsion Laboratory 11.31 NASA/Jet Propulsion Laboratory 11.32 NASA/Jet Propulsion Laboratory 11.33a Lunar and Planetary Laboratory; b NASA/Jet Propulsion Laboratory 11.34 NASA/Jet Propulsion Laboratory 11.35 NASA/Jet Propulsion Laboratory 11.36 Jupiter: Imke de Pater; Saturn, Uranus, Neptune: NASA/Jet Propulsion Laboratory p. 335 NASA/Jet Propulsion Laboratory p. 336 NASA/Jet Propulsion Laboratory

Chapter 12
Opener Photography and Image Processing by Milsolav Druckmuller 12.1 NASA/Jet Propulsion Laboratory 12.2 NASA, ESA, JAXA and Emily Lakdawalla 12.5 NASA/JPL-Caltech/ UCLA/MBS/DLR/IDA 12.7 Ivan and Colby Navarro, NASA 12.8 Jonathan Blair/Corbis 12.9 Robert A. Haag 12.10 (top) ISAS/ JAXA; (bottom) NASA/Jet Propulsion Laboratory 12.11a Peter Ceravolo; b Frank Zullo/Photo Researchers, Inc. 12.12 and p. 362 NASA/Jet Propulsion Laboratory 12.13a NASA/Jet Propulsion Laboratory; b European Space Agency 12.14a BiStar Astronomical Observatory; b European Space Agency 12.15 NASA 12.16 ICSTARS Inc. 12.17 NASA 12.19 ESO 12.21b NASA/Goddard Space Flight Center 12.22a NASA/Jet Propulsion Laboratory; b Joe Bergeron; c MSSO, ANU/Science Library/Photo Researchers, Inc.; d NASA/Goddard Space Flight Center 12.23 Kirk R. Johnson 12.24 Virgil L. Sharpton 12.25 and p. 362 Pearson Education 12.26 Sovfoto/Eastfoto 12.27 La Republica Newspaper/AP Images p. 363 NASA Earth Observing System

Chapter 13
Opener National Research Council Canada, C. Marois & Keck Observatory p. 368 The Print Collector/Alamy 13.6 National Research Council Canada, C. Marois & Keck Observatory 13.12 NASA Earth Observing System 13.20 European Southern Observatory p. 396 Earth's Moon: NASA Earth Observing System; Europa: NASA/Jet Propulsion Laboratory; impact: Pearson Education, Pearson Science

Chapter S2
Opener Bettmann/Corbis

Chapter S3
Opener Andrew Fruchter/NASA S3.20 and p. 439 ESA/ Hubble & NASA

Chapter S4
S4.1 CERN/European Organization for Nuclear Research S4.6 Terry Oakley/Alamy p. 463 NASA/Jet Propulsion Laboratory

Chapter 14
Opener NASA 14.1 Corel Corporation 14.5 NASA/Marshall Space Flight Center 14.11b Institute for Solar Physics 14.12 National Optical Astronomy Observatories (NOAA) 14.14 Lawrence Berkeley National Laboratory 14.15a Institute for Solar Physics; b National Solar Observatory 14.17b NASA/Jet

Propulsion Laboratory 14.18 SDO/GSFC/NASA 14.19 NASA Transition Region and Coronal Explorer 14.20 Lockheed Martin Solar & Astrophysics Laboratory

Chapter 15
Opener Till Credner/AllTheSky.com 15.4 NASA Jet Propulsion Laboratory 15.5 Harvard College Observatory p. 494 (top right) Harvard College Observatory 15.16 and p. 506 Robert Gendler/Photo Researchers 15.17 and p. 506 NASA Jet Propulsion Laboratory

Chapter 16
Opener Far-infrared: ESA/Herschel/PACS/SPIRE/Hill, Motte, HOBYS Key Programme Consortium; X ray: ESA/XMM-Newton/EPIC/XMM-Newton-SOC/Boulanger 16.1 and p. 527 DMI David Malin Images 16.3 NASA Jet Propulsion Laboratory 16.4 Annual Reviews, Inc. 16.5 European Southern Observatory 16.6 (left) ESA; (right) Paul A. Scowen, Research Professional 16.7 NASA Jet Propulsion Laboratory 16.8 Far-infrared: ESA/Herschel/PACS/SPIRE/Hill, Motte, HOBYS Key Programme Consortium; X ray: ESA/XMM-Newton/EPIC/XMM-Newton-SOC/Boulanger 16.10 and p. 527 Matthew Bate, Ph.D. 16.11 NASA/Jet Propulsion Laboratory 16.12 Tom Abel 16.13 NASA Earth Observing System 16.15 NASA Jet Propulsion Laboratory 16.19 European Southern Observatory 16.20 STScI/NASA p. 528 Tom Abel

Chapter 17
Opener NASA Earth Observing System 17.1 NASA/Jet Propulsion Laboratory 17.7 and p. 551 NASA Earth Observing System 17.17 and p. 551 European Southern Observatory 17.18 DMI David Malin Images 17.19 NASA/Jet Propulsion Laboratory

Chapter 18
Opener NASA Earth Observing System 18.1a NASA; b NASA/Jet Propulsion Laboratory 18.3 and p. 571 Pearson Education, Pearson Science 18.4b NASA/Jet Propulsion Laboratory 18.6 NASA/Jet Propulsion Laboratory 18.8 European Southern Observatory p. 567 Chandrasekhar: Bettmann/Corbis; Eddington and Landau: Argelander Institut Fur Astronomie; Oppenheimer: Corbis; Bell: The Regents of the University of California 18.17 NASA/Goddard Space Flight Center 18.18 National Radio Astronomy Observatory

Chapter 19
Opener NASA/Jet Propulsion Laboratory 19.3 NASA/Jet Propulsion Laboratory; NASA; Eagle Nebula: Paul A. Scowen, Research Professional; gas bubble: Robert J. Vanderbei 19.4 NASA, ESA, HEIC, and The Hubble Heritage Team (STScI/AURA) 19.5 Robert J. Vanderbei 19.6 NASA/Jet Propulsion Laboratory 19.7a Stocktrek Images/SuperStock; b NASA 19.8a National Radio Astronomy Observatory; b NASA/CXC/MIT/UMass Amherst/M. D. Stage et al. 19.9b ESO/Manu Mejias 19.10 NASA 19.11 NASA Earth Observing System 19.12a NASA/Jet Propulsion Laboratory; b NASA; c NASA/Jet Propulsion Laboratory; d NASA/Jet Propulsion Laboratory; e Axel Mellinger, A Color All-Sky Panorama Image of the Milky Way, *Publ. Astron. Soc. Pacific 121*, 1180–1187 (2009); f NASA/Jet Propulsion Laboratory; g Fermi-LAT Collaboration/NASA 19.13 NASA 19.14 DMI David Malin Images 19.15 DMI David Malin Images 19.16 and p. 597 NASA/Jet Propulsion Laboratory 19.17 NASA/Jet Propulsion Laboratory 19.20a NASA Headquarters;

b–c National Radio Astronomy Observatory; d European Southern Observatory 19.22 NASA/Jet Propulsion Laboratory p. 598 NASA/Jet Propulsion Laboratory; Axel Mellinger, A Color All-Sky Panorama Image of the Milky Way, *Publ. Astron. Soc. Pacific 121*, 1180–1187 (2009)

Chapter 20
Opener NASA Earth Observing System 20.1 NASA/Jet Propulsion Laboratory 20.2a DMI David Malin Images; b NASA/Jet Propulsion Laboratory; c DMI David Malin Images 20.3 NASA/Jet Propulsion Laboratory 20.4 NASA/Jet Propulsion Laboratory 20.5 NASA/Jet Propulsion Laboratory 20.6a DMI David Malin Images; b NASA/Jet Propulsion Laboratory 20.7b NASA/Jet Propulsion Laboratory; c DMI David Malin Images 20.8 National Optical Astronomy Observatories 20.9 NOAO Gemini Science Center 20.10 NASA/Jet Propulsion Laboratory 20.14 NASA, ESA, A. Riess (STScI and JHU), and S. Rodney (JHU) 20.15 NASA Jet Propulsion Laboratory 20.16 Carnegie Institution of Washington 20.17 Huntington Library/SuperStock p. 620 NASA/Jet Propulsion Laboratory; DMI David Malin Images

Chapter 21
Opener NASA, ESA, and the Hubble Heritage Team (STScI/AURA) 21.2 and p. 641 Volker Springel/Max Planck Institute for Astrophysics 21.4 NASA 21.5 NASA, ESA, and the Hubble Heritage Team (STScI/AURA)-ESA/Hubble Collaboration 21.6 NASA/Jet Propulsion Laboratory 21.7 Space Telescope Science Institute 21.8a DMI David Malin Images; b E. R. Carrasco et al. (2010, ApJ, 816, L160), Gemini Observatory 21.10 NASA/Jet Propulsion Laboratory 21.11a and p. 641 NASA; b Subaru Telescope/National Astronomical Observatory of Japan NAOJ 21.12 NASA, ESA, the Hubble Heritage Team (STScI/AURA), and A. Aloisi (STScI/ESA) 21.13 NASA/Jet Propulsion Laboratory 21.16 National Radio Astronomy Observatory 21.17 National Radio Astronomy Observatory 21.19 NASA/Jet Propulsion Laboratory 21.22 NASA/Jet Propulsion Laboratory

Chapter 22
Opener E. Bunn 22.5 Gas: E. Bunn; galaxy: DMI David Malin Images; Hubble spacecraft: NASA 22.9 E. Bunn 22.18a Photo by Greg Thow/Getty Images; b WDG Photo/Shutterstock.com

Chapter 23
Opener NASA 23.6 California Institute of Technology Archives 23.7 and p. 690 NASA Earth Observing System 23.8 and p. 690 NASA EOS Earth Observing System 23.10 NASA, ESA, J. Richard (Center for Astronomical Research/Observatory of Lyon, France), and J.-P. Kneib (Astrophysical Laboratory of Marseille, France) 23.11 NASA 23.13 Michael Strauss 23.15 and p. 690 Andrey Kravtsov 23.16 NASA Jet Propulsion Laboratory

Chapter 24
Opener Seth Shostak 24.2 Darlene Cutshall/Shutterstock 24.3 James L. Amos/Photo Researchers, Inc. 24.5a Jane Gould/Alamy; b B Christopher/Alamy; c Abigail Allwood/NASA 24.8 and p. 718 Woods Hole Oceanographic Institution 24.9 REUTERS/David McNew 24.10 Martin M. Hanczyc 24.12 M I (Spike) Walker/Alamy 24.13 and p. 718 M. Di Lorenzo, NASA Earth Observing System 24.16 Yeshe Fenner 24.17 Seth Shostak 24.19 and p. 718 Seth Shostak 24.20 National Astronomy

TEXT AND ILLUSTRATION CREDITS

INDEX

nuclei
    atomic, 145
    of comets, 349–350, 356
    era of, 649, 651, 653
nucleosynthesis, era of, 649, 651, 653
NuStar, t178, 181

OBAFGKM sequence, 493–494
obelisks, and time measurement, 55
Oberon (Uranus), 329
observable universe, 4, 5
    galaxies in, 10
    limit of, 5, 6
    number of stars in, 10, 11
observations, astronomical
    archaeoastronomy and, 58–59
    direct, 367, 377
    indirect, 367
    length of day from, 55
    length of month and year from, 56
    lunar calendars from, 57–58
    marking seasons through, 56–57
    modern science and, 59–61
    naked-eye, 65
    scientific theory and, 76
    time of day from, 55–56
    verifiable, 74–76
    weather predictions from, 54
observatories
    ground-based, 176
    naked-eye, 65
    in space, t178, 180
Occam's razor, 74
occultation, stellar, 332
ocean trench, 260
Olbers, Heinrich, 661
Olbers' paradox, 661–662
Old Royal Greenwich Observatory, 30
Olympus Mons (Mars), 243, 252, 253
Oort cloud, 193, 206, 215, 223, 351–352, 353, 710
opaque material, 139
open clusters, of stars, 504–505, 580
Ophiuchus, 32
Oppenheimer, Robert, 567
Opportunity rover, 198, 209, t210, 254–255, 706
opposition vs. conjunction, 86
orbit(s), 4, 14, 205
    of asteroids, 340, 342–343
    atmospheric drag and, 127
    backward of Neptune's moon, 202, 356
    bound vs. unbound, 124
    of Charon, 130–131
    of disk, bulge, and halo stars, 580–581
    eccentricity of, 66, 68
    of extrasolar planets, 375, 378
    free-fall in, 114
    of galaxy clusters, 672–673
    of Galilean moons, 322–323
    gravity and, 124–125
    of Io, 136
    of Mercury, 195
    of Moon, 136
    planetary, 66–67, 124–125, 191–192, A-15
    of Pluto, 353
    speeds of, 17, 41n, 91
    of Sun, 369, 581
orbital angular momentum, 113, 118
orbital distance, 70
    of extrasolar planets, 375, 378
    of geosynchronous satellite, 126
    Newton's version of Kepler's third law and, 126, 375
    seasons and, 36–37
orbital eccentricity, 66, 68, A-15
orbital energy, 125–127
orbital inclination, 496n, A-15
orbital period, 67, 70, A-15
    distance calculations from, 126
    of extrasolar planets, 375
    synodic period and, 88–89
orbital resonance, 323
    asteroid belt and, 344
    of extrasolar planets, 388–389
    of Galilean moons, 323, 331
    of Saturn's moons, 331

orbital speed, 17, 67n. See also orbital velocity law
    of gas, 670
    length of solar day on Earth and, 91
    of Moon, 41n
    in solar system, 669–670
    of stars, 495–496
orbital velocity law, 581, 673, A-3
orbiter, 208
Orcus, 354
order of magnitude estimates, 10
ordinary dark matter, 677
Orion Nebula, 4, 5, 590, 695
    declination/RA of, 106
    emission line spectrum of, 153
    formation of stars in, 216, 217
Osiris-REX mission, 346
outer core region, of Earth, 235
outgassing, 243
    atmosphere formation and, 284–285
    in carbon dioxide cycle, 295–296
    on Earth, 244
    on Venus, 246, 286
    from volcanism, 243
overall force, 112
oxidation reactions, 294
oxygen, 270
    in CNO cycle, 541–542
    early organisms and, 704
    in Earth's atmosphere, 270, 271, 294–295, 704
    ionization of, 148
    molecular, 270, 294–295
ozone, 295, 296
ozone layer, 302

P waves, 237
PAHs, 707
pair production, 447
Pallas (asteroid), 341, 343
Palomar Observatory (California), 672
Pangaea, 263
parabolic orbital paths, 124
paradigm, 76
paradox, 401
parallax, 47, 207, 489–490, 607, 614. See also stellar parallax
parallax angle, 489
parallax formula, 490, A-3
parent isotope, 225
parsec (pc), 489, A-2
partial lunar eclipse, 43
partial solar eclipse, 36
particle(s), 140. See also particle accelerators; subatomic particle
    anti-, 447
    creation and annihilation of, 648
    era, 649, 650–651, 653
    exchange, 447
    fundamental, 445–446
    of light. See photon
    quantum state of, 452
    virtual, 444, 456
    waves and, 140–141
    weakly interacting massive, 678, 685
particle accelerators, 123
    discovery of new particles with, 444
    pair production in, 447
    particles with quarks and, 446
    testing time dilation with, 411–412
particle radiation, 142
Pauli, Wolfgang, 452
Pawnee lodges, 58
Payne-Gaposchkin, Cecilia, 494
pc, 489, A-2
peak thermal velocity, 285
peculiar velocity, 679
pendulum, Foucault, 74
penumbra, 43
penumbral lunar eclipse, 43
Penzias, Arno, 654
perihelion, 66, 67
period-luminosity relation, 609, 611–612
periodic table of elements, A-13
Perseids meteor shower, 351
Philosophiae Naturalis Principia Mathematica (Newton), 115
Phobos (Mars), 198, 289

Phoebe (Saturn), 328
Phoenix lander, t210, 288, 289
photoelectric effect, 399
photometry, 172n
photon, 141–142
    in absorption and emission line spectra, 152–153
    as boson, 445
    in cosmic microwave background, 654
    cosmological redshifts and, 618
    energy of, 142, A-3
    as exchange particle, 449
    gamma-ray, 470
    migration of, from solar core, 472–473
    radiation pressure and, 526
    striking pixels with, 167
    Wien's law and, 155
photosphere, 152, 468, 473
photosynthesis, 294, 704
physics, universality of, 188–189
Pickering, Edward, 493–494
picture elements (pixels), 167
Pioneer missions, 196, 309, 715
pixels, 167
Planck, Max, 649
Planck era, 649–650, 652
Planck observatory, t178
Planck's constant, 144, 445n, 451, A-2
plane, two-dimensional, 424
planet(s), 4, 7. See also extrasolar planet(s); jovian planets; terrestrial planets
    axis tilt of, A-15
    density of, t204, 220, 235–236, A-14
    destiny of, 225
    distance from Sun of, 246, 247, A-15
    dwarf, 4, 8, 78, 203, t204, 312
    escape velocity of, A-14
    features of, t204
    formation of, 220–222
    fundamental properties of, 247
    infrared light from, 178
    masses of, 211, A-14
    minor. See asteroids
    orbits of, 66–67, 124–125, 191–192, A-15
    physical properties of, A-14
    radii of, 211, A-14
    reflectivity of, 274–275, 283, 284
    rings of, 330–334
    size of, and geological history, 239, 246, 247
    surface gravity of, A-14
    time periods of, 85–86
    types of, 205, 218–219
    visible to naked eye, 46
planetary geology, 235
planetary motion. See also Newton's version of Kepler's third law
    ancient Greeks and, 46
    apparent retrograde, 46
    gravity and, 124–125
    Kepler's laws of, 66–67, 70, 73, 119
    Ptolemaic model of, 63, 69
planetary nebula, 538–539, 549
planetary periods, 85–86
planetesimals, 221, 340. See also asteroids; comets
planetology, 191, 205–206
planispheres, 101
plasma, 148
    compressing, 454
    in Sun, 467
plasma tail (of comet), 350
plate tectonics, 243, 260–261, 295, 710
plates, 243
Plato, 60, 62, 84
playas, 326
Pleiades, 504, 505
Plutinos, 356
Pluto, 8, 203, t204, 353
    atmosphere of, 355
    axis tilt of, A-15
    composition of, 203
    density of, 203, A-14
    discovery of, 312
    distance from Sun of, 203, A-15
    escape velocity of, A-14
    mass of, 203, A-14
    moons of, 203, 355
    orbit of, 353, A-15